THIRTEENTH EDITION

Human
GEOGRAPHY

LANDSCAPES OF HUMAN ACTIVITIES

MARK D. BJELLAND

Calvin University

DANIEL R. MONTELLO

University of California-Santa Barbara

ARTHUR GETIS

San Diego State University

Mc Graw Hill Education

HUMAN GEOGRAPHY: LANDSCAPES OF HUMAN ACTIVITIES

Some ancillaries, including electronic and print components, may not be available to customers outside the United States.

This book is printed on acid-free paper.

4 5 6 7 8 9 GPC 21

ISBN 978-1-260-56605-5
MHID 1-260-56605-6

Cover Image: ©Ken Welsh/Pixtal/AGEfotostock (Map); ©View Stock/Getty Images (Shanghai, China); ©Feng Wei Photography/Getty Images (Bhaktapur, Nepal); ©NASA/NOAA/GOES Project (Earth); ©Shutterstock/Miki Studio (Woman at Floating Market); ©Space Images/Blend Images (Rural Landscape)

All credits appearing on page or at the end of the book are considered to be an extension of the copyright page.

The Internet addresses listed in the text were accurate at the time of publication. The inclusion of a website does not indicate an endorsement by the authors or McGraw-Hill Education, and McGraw-Hill Education does not guarantee the accuracy of the information presented at these sites.

mheducation.com/highered

Brief Contents

Contents

Part One

THEMES AND FUNDAMENTALS OF HUMAN GEOGRAPHY

©Goodshoot/Getty Images

Part Two

PATTERNS OF DIVERSITY AND UNITY

©Spondylolithesis/iStock/Getty Images

Part Three

DYNAMIC PATTERNS OF THE GLOBAL ECONOMY

©*Jeremy Woodhouse/Getty Images*

Nine ECONOMIC GEOGRAPHY: MANUFACTURING AND SERVICES 287

Ten ECONOMIC DEVELOPMENT AND CHANGE 316

Part Four

LANDSCAPES OF FUNCTIONAL ORGANIZATION

©Ingram Publishing/SuperStock

Part Five

HUMAN ACTIONS AND ENVIRONMENTAL IMPACTS

©RAJESH NIRGUDE/AP Images

List of Boxes

Preface

This thirteenth edition of *Human Geography* retains the organization and structure of its earlier versions. Like them, it seeks to introduce its users to the scope and excitement of geography and its relevance to their daily lives and roles as informed citizens. We recognize that for many students, human geography may be their first or only work in geography and this, their first or only textbook in the discipline. For these students particularly, we seek to convey the richness and breadth of human geography and to give insight into the nature and intellectual challenges of the field of geography itself. Our goals are to be inclusive in content, current in data, and relevant in interpretations. These goals are elusive. Because of the time lapse between world events and the publication of a book, events inevitably outpace analysis. We therefore depend on a continuing partnership with classroom instructors to incorporate and interpret current events and emerging geographic patterns.

Organization

The text can easily be read in a one-semester course. The emphasis is on key concepts and theories in human geography, which can then be applied to understanding patterns of human activities and current events. Chapter 1 sets the stage by briefly introducing students to the scope, methods, and background basics of geography as a discipline and to the tools—especially maps—that all geographers employ. It is supplemented by Appendix A, which gives a more detailed treatment of map projections than is appropriate in a general introductory chapter. Both are designed to be helpful, with content supportive of, not essential to, the later chapters of the text.

The arrangement of those chapters reflects our own sense of logic and teaching experiences. Chapters 2 through 4 introduce major themes and fundamental concepts in human geography. Chapter 2 examines the basis of culture, culture change, and cultural regionalism. Chapter 3 offers a comprehensive review of concepts of spatial interaction and spatial behavior. Chapter 4 considers population structures, patterns, and change. Chapters 5 through 7 discuss the foundations for different social and cultural identities: language and religion (Chapter 5), ethnicity (Chapter 6), and folk and popular culture (Chapter 7). Chapter 7 also examines the landscape expressions of different cultures. Chapters 8, 9, and 10 focus on economic geography, beginning with activities connected to the Earth and natural resources (Chapter 8), then exploring the changing geographies of manufacturing and service industries (Chapter 9), and concluding with an examination of inequality and issues in economic development (Chapter 10).

Chapter 11 examines the organization of urban systems and urban space, while Chapter 12 explores the political organization of territory. Chapter 13 draws together in sharper focus selected aspects of the human impact on the environment, demonstrating the relevance of human geographic concepts and patterns to matters of current national and global environmental concern.

Among those concepts is the centrality of gender issues that underlie all facets of human geographic inquiry. Because they are so pervasive and significant, we felt it unwise to relegate their consideration to a single separate chapter, thus artificially isolating women and women's concerns from all the topics of human geography for which gender distinctions and interests are relevant. Instead, we have incorporated significant gender/female issues within the several chapters where those issues apply—either within the running text of the chapter or, very often, highlighted in boxed discussions.

We hope by means of this structure to convey to students the logical integration that we recognize in the broad field of human geography. We realize that our sense of organization and continuity is not necessarily that of instructors using this text and have designed each chapter to be reasonably self-contained, able to be assigned in any sequence that satisfies the arrangement preferred by the instructor.

New to This Edition

We are pleased to introduce this newly updated and revised edition of *Human Geography: Landscapes of Human Activity*. Although the text's established framework has been retained in this thirteenth edition, each chapter has been revised to improve readability, and every chapter contains at least brief text additions or modifications to reflect current data. All chapters contain new or revised illustrations, maps, and photos.

The thirteenth edition contains many new and updated topics, including the following:

New Maps

Many existing maps have been updated for the thirteenth edition of *Human Geography*. In addition, new maps introduced in this edition include:

- A new choropleth map of the religiously unaffiliated in the United States
- A new map of metropolitan regions specializing in the information economy
- Satellite maps of urban growth in Las Vegas, Nevada
- Renewable freshwater resources per capita by country

New Boxes

Many of the boxed elements in the text have been updated, and the following new boxes have been introduced:

- "The Burning Man Festival of Art and Music: Subcultural Landscape in the Great Basin Desert" in Chapter 2
- "Health Geography," in Chapter 2
- "Geography and Citizenship: Changing Toponyms," in Chapter 5
- "Religious Attire in Secular Spaces," in Chapter 5
- "Geography and Citizenship: Monuments, Memorials, and Civic Spaces," in Chapter 7
- "Sustainable Development Goals," in Chapter 10
- "Geography and Citizenship: Gerrymandering in the United States," in Chapter 12

New/Revised Topics

New and revised topics in this book include the following:

- Updated population data and forecasts throughout
- New comparison of the components of projected population change in the developed and developing regions of the world
- Updated information on legal and undocumented immigration into the United States and proposals for construction of a wall along the Mexico border
- New material on China's and India's population policies and prospects
- An updated discussion of successful efforts to slow the spread of HIV in sub-Saharan Africa
- New population pyramid graphs for Nigeria, New Zealand, Japan, and East St. Louis, Ilinois
- New material on the 2014–2015 Ebola outbreak in West Africa

- Updated data for ethnic groups in the United States from the American Community Survey
- Updated material on refugee movements, immigration policy debates, and independence movements
- A new opening vignette for Chapter 7 on China's exuberant urban landscapes
- New material on the New Urbanism movement in urban planning
- New discussion of quaternary economic activities
- New material on the use of genetically modified crops in agriculture
- Updated information on the success of the United Nations, Millennium Development Goals (MDGs) and new Sustainable Development Goals (SDGs) for 2030
- A new presentation of Borchert's transportation epochs
- New material on Mackinder's heartland theory of geopolitics
- New discussion of the U.K. vote to withdraw from the European Union
- New set of four trend graphs for key indicators of global climate change
- The latest information on global climate change drawn from the 5th Assessment Report issued by the Intergovernmental Panel on Climate Change

The Art of Human Geography

Most of the world maps use a Robinson projection, which permits some exaggeration of size in the high latitudes in order to improve the shapes of landmasses. Size and shape are most accurate in the temperature and tropical zones. The color palette for the maps was specifically chosen to accommodate most colorblind readers.

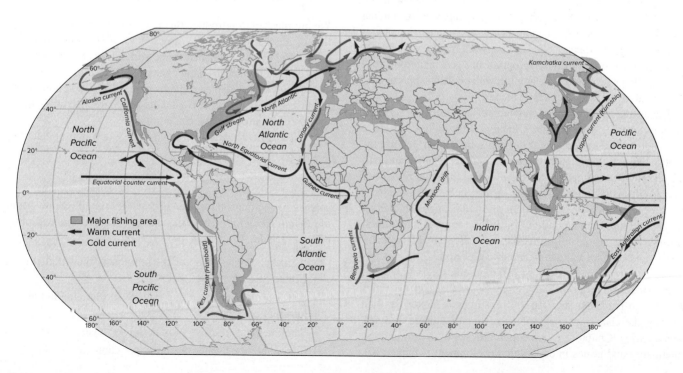

Acknowledgments

It is with great pleasure that we again acknowledge our debts of gratitude to both departmental colleagues—at Calvin University; at University of California–Santa Barbara; and at San Diego State University—and all others who have given generously of their time and knowledge in response to our requests. Special thanks go to undergraduate students Seth Haase and Hannah Fertich who assisted with map production. Other colleagues have been identified in earlier editions, and although their names are not repeated here, they know of our continuing appreciation.

We specifically, however, wish to recognize with gratitude the advice, suggestions, corrections, and general assistance in matters of content and emphasis provided by Johnathan Bascom, *Calvin University* and Dr. Susan Cassels, *University of California–Santa Barbara,* for this edition.

We appreciate their invaluable help, as well as that of the many other previous reviewers recognized in earlier editions of this book. No one, except the authors, of course, is responsible for final decisions on content, or for errors of fact or interpretation that the reader may detect.

A final note of thanks is reserved for the publisher's "book team" members separately named on the copyright page. It is a privilege to emphasize here their professional competence, unflagging interest, and always courteous helpfulness:

Mark D. Bjelland
Daniel R. Montello
Arthur Getis

Meet the Authors

Mark D. Bjelland

Mark Bjelland received his B.S. and M.S. degrees from the University of Minnesota and University of Washington, respectively. He worked for six years as a consultant on transportation systems in Minneapolis–Saint Paul, water management in Washington State, environmental cleanups on First Nations reservations in British Columbia, Canada, and major urban redevelopments in Vancouver, British Columbia. Intrigued by the geographical questions he encountered in his work, he earned a Ph.D. in geography from the University of Minnesota. He wrote his dissertation comparing deindustrialization, environmental justice, and brownfields redevelopment in the United States and Canada. He is professor of geography at Calvin University, where he has taught for 6 years, after 15 years at Gustavus Adolphus College. He has been a visiting scholar in the geography department at the University of British Columbia, received a Fulbright Foundation German Studies award to study urban planning in post-reunification Germany, and was awarded a U.S.-U.K. Fulbright Scholar Award to spend a year at Cardiff University in Wales researching the redevelopment of derelict industrial land and the creation of eco-communities. He loves to take geography students into the field, and in addition to local field trips to farms, small towns, and cities, he has led field courses to the Pacific Northwest, Hawai'i, and Europe. His research interests include urban and economic geography, environmental justice, urban sustainability, and religious diversity. His interests have been reflected in book chapters and articles published in *The Professional Geographer, The Geographical Review, The Encyclopedia of Geography, Research Journal of the Water Pollution Control Federation, Urban Geography,* and elsewhere. He is a co-author of McGraw-Hill's *Introduction to Geography.*

Daniel R. Montello

Daniel R. Montello received his B.A. degree from the Johns Hopkins University and his M.A. and Ph.D. degrees in environmental psychology from Arizona State University. He was also a postdoctoral fellow at the University of Minnesota and a visiting professor at North Dakota State University. He is currently professor of geography and affiliated professor of psychology at the University of California–Santa Barbara (UCSB), where he has been on the faculty since 1992. Dan teaches graduate and undergraduate courses in human geography, behavioral geography, cognitive science, statistics, research methods, cognitive issues in cartography and GIS, and environmental perception and cognition. His research is in the areas of spatial, environmental, and geographic perception, cognition, affect, and behavior. Specific research topics he and his colleagues have investigated include how people navigate in built and natural environments, how people find and lose their way, how children and adults develop an understanding of space and place (including how they acquire and use distance and direction knowledge), how people perceive and reason with maps and other visualizations, how people express their experience of place and space in language, how individuals and groups of people are similar and different in spatial thinking, how spatial relations interrelate with social relations, how people and information systems conceptualize geographic reality, and how human psychology relates to aspects of Earth science (including climate science and geology). Dan has co-authored or edited six books, including 2006's *An Introduction to Scientific Research Methods in Geography* by SAGE Publications, with Paul C. Sutton, and 2018's *Handbook of Behavioral and Cognitive Geography* by Edward Elgar Publishing. He and his co-authors have also published around 10 articles and book chapters. Dan is currently co-editor of the academic journal *Spatial Cognition and Computation* and sits on the editorial boards of *Environment and Behavior* and the *Journal of Environmental Psychology.* He has served as a reviewer for several funding agencies around the world and more than 50 academic journals in the fields of geography, cartography, geographic information science, psychology, education, cognitive science, computer science, anthropology, communication, economics, and planning. He is a member of the Association of American Geographers, the Psychonomic Society, and the Sigma Xi Scientific Honor Society.

Arthur Getis

Arthur Getis received his B.S. and M.S. degrees from Pennsylvania State University and his Ph.D. from the University of Washington. He is the co-author of several geography textbooks, as well as two books dealing with map pattern analysis. He has also published widely in the areas of urban geography, spatial analysis, and geographical information systems. He is coeditor of *Journal of Geographical Systems* and for many years served on the editorial boards of *Geographical Analysis* and *Papers in Regional Science*. He has held administrative appointments at Rutgers University, the University of Illinois, and San Diego State University (SDSU) and currently holds the Birch Chair of Geographical Studies at SDSU. In 2002 he received the Association of American Geographers Distinguished Scholarship Award. Professor Getis is a member of many professional organizations and has served as an officer in, among others, the Western Regional Science Association and the University Consortium for Geographic Information Science.

In Memory Of

Jerome D. Fellmann (1926–2010) was the lead author for the first ten editions of this book. He earned his B.S., M.S., and Ph.D. degrees at the University of Chicago and spent over 50 years teaching geography at the University of Illinois at Urbana–Champaign. His research specializations were in urban geography and economic geography, Russia, and geographic education. His contributions to undergraduate education are honored by the annual Fellmann prize, given to top graduating geography and GIS students at the University of Illinois.

Judith M. Getis (1938–2010) contributed to the early editions of this book. She did her undergraduate studies at Radcliffe College (Harvard University) and the University of Michigan and completed her M.A. in geography at Michigan State University. She taught cartography at Rutgers University and developed educational materials for Educational Testing Services. She was a co-investigator on the National Science Foundation's original High School Geography Project. In addition to this book, she co-authored *Introduction to Geography; Environments, Peoples, and Inequalities; The United States and Canada;* and *You Can Make a Difference*.

Students—study more efficiently, retain more and achieve better outcomes. Instructors—focus on what you love—teaching.

SUCCESSFUL SEMESTERS INCLUDE CONNECT

FOR INSTRUCTORS

You're in the driver's seat.

Want to build your own course? No problem. Prefer to use our turnkey, prebuilt course? Easy. Want to make changes throughout the semester? Sure. And you'll save time with Connect's auto-grading too.

65%
Less Time Grading

They'll thank you for it.

Adaptive study resources like SmartBook® help your students be better prepared in less time. You can transform your class time from dull definitions to dynamic debates. Hear from your peers about the benefits of Connect at **www.mheducation.com/highered/connect**

Make it simple, make it affordable.

Connect makes it easy with seamless integration using any of the major Learning Management Systems—Blackboard®, Canvas, and D2L, among others—to let you organize your course in one convenient location. Give your students access to digital materials at a discount with our inclusive access program. Ask your McGraw-Hill representative for more information.

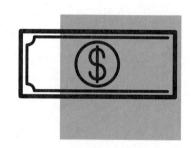

Solutions for your challenges.

A product isn't a solution. Real solutions are affordable, reliable, and come with training and ongoing support when you need it and how you want it. Our Customer Experience Group can also help you troubleshoot tech problems—although Connect's 99% uptime means you might not need to call them. See for yourself at **status.mheducation.com**

FOR STUDENTS

Effective, efficient studying.

Connect helps you be more productive with your study time and get better grades using tools like SmartBook, which highlights key concepts and creates a personalized study plan. Connect sets you up for success, so you walk into class with confidence and walk out with better grades.

©Shutterstock/wavebreakmedia

" I really liked this app—it made it easy to study when you don't have your textbook in front of you. **"**

—Jordan Cunningham,
Eastern Washington University

Study anytime, anywhere.

Download the free ReadAnywhere app and access your online eBook when it's convenient, even if you're offline. And since the app automatically syncs with your eBook in Connect, all of your notes are available every time you open it. Find out more at **www.mheducation.com/readanywhere**

No surprises.

The Connect Calendar and Reports tools keep you on track with the work you need to get done and your assignment scores. Life gets busy; Connect tools help you keep learning through it all.

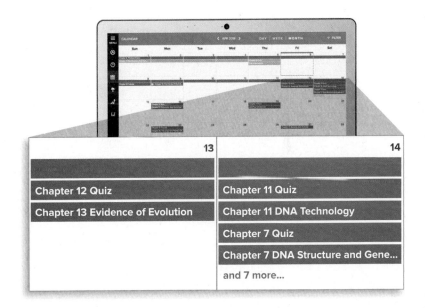

	13		14
Chapter 12 Quiz		Chapter 11 Quiz	
Chapter 13 Evidence of Evolution		Chapter 11 DNA Technology	
		Chapter 7 Quiz	
		Chapter 7 DNA Structure and Gene...	
		and 7 more...	

Learning for everyone.

McGraw-Hill works directly with Accessibility Services Departments and faculty to meet the learning needs of all students. Please contact your Accessibility Services office and ask them to email accessibility@mheducation.com, or visit **www.mheducation.com/about/accessibility.html** for more information.

INTRODUCTION:
Some Background Basics

The imprint of human activity created by this windfarm dominates this California landscape.

©Thinkstock Images/Stockbyte/Getty Images

Key Concepts

1.1 The nature of geography and the role of human geography

1.2 Seven fundamental geographic observations and the basic concepts that underlie them

1.3 The regional concept and the characteristics of regions

1.4 Why geographers use maps and how maps show spatial information

1.5 Other means of visualizing and analyzing spatial data: mental maps, systems, and models

1.1 Getting Started

The fundamental question asked by geographers is, "What difference does it make where things are located?" For example, it matters a great deal that languages of a certain kind are spoken in certain places. But knowledge of the location of a specific language group is not of itself particularly significant. Geographic study of a language requires that we try to answer questions about why and how the language shows different characteristics in different locations and how the present distribution of its speakers came about. In the course of our study, we would logically discuss such concepts as migration, acculturation, the diffusion of innovation, the effect of physical barriers on communication, and the relationship of language to other aspects of culture. As geographers, we are interested in how things are interrelated in different regions and give evidence of the existence of "spatial systems."

What Is Geography?

Many people associate the word *geography* simply with describing *where* things are and the characteristics of things at particular locations; where are countries such as Myanmar and Uruguay, what are the populations of cities such as Timbuktu or Almaty, or where are large deposits of natural resources such as petroleum or iron ore? Some people pride themselves on knowing which are the longest rivers, the tallest mountains, and the largest deserts. Such factual knowledge about the world has value, permitting us to place current events in their proper spatial setting. When we hear of an earthquake in Turkey or an assault in Timor-Leste, we at least can visualize where they occurred. Knowing *why* they occurred in those places, however, is considerably more important.

Geography is much more than place names and locations. It is the study of spatial variation, of how and why things differ from place to place on the surface of the Earth. It is, further, the study of how observable spatial patterns evolved through time. Just as knowing the names and locations of organs in the human body does not equip one to perform open-heart surgery, and just as memorizing the periodic table does not enable one to formulate new medications, so knowing where things are located geographically is only the first step toward understanding why things are where they are, and what events and processes determine or change their distribution. Why is Chechnya but not Tasmania wracked by insurgency, and why do you find a concentration of French speakers in Quebec but not in other parts of Canada? Why are famines so frequent and severe in East Africa and why, among all the continents, has African food production and distribution failed to keep pace with population growth over the past half century?

In answering questions such as these, geographers focus on the interaction of people and social groups with their environment—planet Earth—and with one another; they seek to understand how and why physical and cultural spatial patterns evolved through time and continue to change. Because geographers study both the physical environment and the human use of that environment, they are sensitive to the variety of forces affecting a place and to the interactions among them. To explain why Brazilians burn a significant portion of the tropical rain forest each year, for example, geographers draw on their knowledge of the climate and soils of the Amazon Basin; population pressures, landlessness, and the need for more agricultural area in rural Brazil; the country's foreign debt status; midlatitude markets for lumber, beef, and soy beans; and economic development objectives. Understanding the environmental consequences of the burning requires knowledge of, among other things, the oxygen and carbon balance of the Earth; the contribution of the fires to the greenhouse effect, acid rain, and depletion of the ozone layer; and the relationship among deforestation, soil erosion, and floods. Thus, one might say that geography is the "study of the Earth as the home of humanity."

Geography, therefore, is about geographic space and its content. We think of and respond to places from the standpoint not only of where they are but, rather more importantly, of what they contain or what we think they contain. Reference to a place or an area usually calls up images about its physical nature or what people do there and often suggests, without conscious thought, how those physical objects and human activities are related."Colorado," "mountains," and "skiing" might be a simple example. The content of area, that is, has both physical and cultural aspects, and geography is always concerned with understanding both (**Figure 1.1**).

Although space is central to geography, time is important, too. How do places change over time, how do structures and processes change location over time, and how do patterns of interaction change over time? Buffalo, New York, was one of the ten largest cities in the United States around 1900. Its location at the western terminus of the Erie Canal, and then along rail lines for transporting the manufacturing and agricultural products of the Midwest, attracted job seekers and investors. Now it is around the 80th largest city in the United States and continuing to shrink in population (in both absolute terms and relative to fast-growing cities such as Houston and Phoenix). Manufacturing in the United States decreased dramatically at the end of the 20th century, and agricultural products have found other routes to move. In other words, geography is about both static and dynamic aspects of space and place.

Evolution of the Discipline

The fundamental inspiration for geographical thought probably originated with the recognition of *areal differentiation*—that one place is different than another. Climate varies, plants vary, people vary. This insight surely occurred in prehistoric times. Early developments in the study of geography took place in ancient Egypt, China, Mesopotamia, the Arab world, Greece, and Rome. This early work was motivated by practical problems in astronomy, land surveying and agriculture, trade, and military activity. From the beginning, geographic thought was characterized by three scholarly traditions: a literary tradition, including travel logs written about foreign places; a cartographic tradition, in which places were mapped; and a mathematical tradition, which involved measuring and calculating spatial and nonspatial information about places. Although their relative importance to

Figure 1.1 The ski development at Whistler Mountain, British Columbia, Canada, site of 2010 Winter Olympic events, clearly shows the interaction of physical environment and human activity. Climate and terrain have made specialized human use attractive and possible. Human exploitation has placed a cultural landscape on the natural environment, thereby altering it.

©*Karl Weatherly/Corbis Documentary/Getty Images*

geographic scholarship has varied over time, all three traditions are still active parts of the study of geography.

Geography, the "mother of sciences," initiated in antiquity lines of inquiry that led to the development of separate disciplines such as anthropology, demography, geology, ecology, and economics. Geography's combination of interests was apparent even in the work of the early Greek geographers who first gave structure to the discipline. Geography's name was coined by the Greek scientist Eratosthenes more than 2,200 years ago from the words *geo,* "the Earth," and *graphein,* "to write." From the beginning, that writing focused both on the physical structure of the Earth and on the nature and activities of the people who inhabited the different lands of the known world. To Strabo (*ca.* 64 BCE–CE 20), the task of geography was to, "describe the several parts of the inhabited world . . . to write the assessment of the countries of the world [and] to treat the differences between countries." Even earlier, Herodotus (*ca.* 484–425 BCE) had found it necessary to devote much of his book to the lands, peoples, economies, and customs of the various parts of the Persian Empire as necessary background to an understanding of the causes and course of the Persian wars.

Greek (and, later, Roman) geographers measured the Earth, devised the global grid of parallels and meridians (marking latitudes and longitudes—see Section 1.4), and drew upon that grid surprisingly sophisticated maps of their known world (**Figure 1.2**). They explored the apparent latitudinal variations in climate and described in numerous works the familiar Mediterranean basin and the more remote, partly rumored lands of northern Europe, Asia, and equatorial Africa. Employing nearly modern concepts, they described river systems, explored causes of erosion and patterns of deposition, cited the dangers of deforestation, described areal variations in the natural landscape, and noted the consequences of environmental abuse. Against that physical backdrop, they focused their attention on what humans did in home and distant areas—how they lived; what their distinctive similarities and differences were in language, religion, and custom; and how they used, altered, and perhaps destroyed the lands they inhabited. Strabo, indeed, cautioned against the assumption that the nature and actions of humans were determined solely by the physical environment they inhabited. He observed that humans were active elements in a human–environmental partnership.

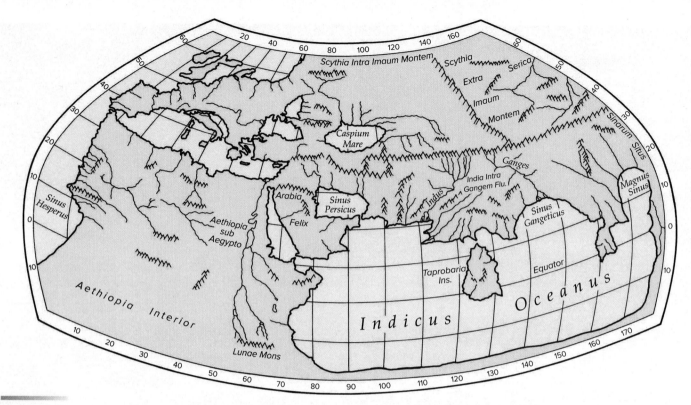

Figure 1.2 World map of the 2nd-century CE by Greco–Egyptian geographer–astronomer Ptolemy. Ptolemy (Claudius Ptolemaeus) adopted a previously developed map grid of latitude and longitude based on the division of the circle into 360°, permitting a precise mathematical location for every recorded place. Unfortunately, errors of assumption and measurement rendered both the map and its accompanying six-volume gazetteer inaccurate. Ptolemy's map, accepted in Europe as authoritative for nearly 1,500 years, was published in many variants in the 15th and 16th centuries. The version shown here summarizes the extent and content of the original. Its underestimation of the Earth's size convinced Columbus that a short westward voyage would carry him to Asia.

These are enduring and universal interests. The ancient Chinese, for example, were as involved in geography as an explanatory viewpoint as were Westerners, though there was no exchange between them. Further, as Christian Europe entered its Middle Ages between CE 500 and 1400 and lost its knowledge of Greek and Roman geographical work, Muslim scholars—who retained that knowledge—undertook to describe and analyze their known world in its physical, cultural, and regional variation (see the feature "Roger's Book").

Modern geography had its origins in the surge of scholarly inquiry that, beginning in the 17th century, gave rise to many of the traditional academic disciplines we know today. In its European rebirth, geography from the outset was recognized—as it always had been—as a broadly based integrative study. Patterns and processes of the physical landscape were early interests, as was concern with humans as part of the Earth's variation from place to place. The rapid development of geology, botany, zoology, and other natural sciences by the end of the 18th century strengthened regional geographic investigation and increased scholarly and popular awareness of the intricate interconnections of items in space and between places. By that same time, accurate determination of latitude and longitude and scientific mapping of the Earth made assignment of place information more reliable and comprehensive.

During the 19th century, national censuses, trade statistics, and ethnographic studies gave firmer foundation to human geographic investigation. By the end of the 19th century, geography had become a distinctive and respected discipline in universities throughout Europe and in other regions of the world where European academic examples were followed. The proliferation of professional geographers and geography programs resulted in the development of a whole series of increasingly specialized disciplinary subdivisions.

Geography and Human Geography

Geography's specialized subfields are not entirely distinct but are interrelated. Geography in all its subdivisions is characterized by three dominating interests. The first is in the areal variation of physical and human phenomena on the surface of the Earth. Geography examines relationships between human societies and the natural environments that they occupy and modify. The second is a focus on the spatial systems[1] that link physical phenomena and human activities in one area of the Earth with other areas. Together, these interests lead to a third enduring theme, that of regional analysis: geography studies human-environment—*ecological*—relationships and spatial systems in specific locational settings. This areal orientation pursued by some geographers is called **regional geography.** Similar to many of the articles in *National Geographic Magazine*, regional geography typically focuses on a comprehensive understanding of physical and human characteristics of particular regions. For some, the regions of interest may be large: Southeast Asia or Latin America, for

---

[1]A *system* is simply a group of elements organized in a way that every element is to some degree directly or indirectly interdependent with every other element. For geographers, the systems of interest are those that distinguish or characterize different regions or areas of the Earth.

Roger's Book

The Arab geographer Idrisi, or Edrisi (*ca.* CE 1099–1154), a descendant of the Prophet Mohammed, was directed by Roger II, the Christian king of Sicily in whose court he served, to collect all known geographical information and assemble it in a truly accurate representation of the world. An academy of geographers and other scholars was gathered to assist Idrisi in the project. Books and maps of classical and Islamic origins were consulted, mariners and travelers interviewed, and scientific expeditions dispatched to foreign lands to observe and record. Data collection took 15 years before the final world map was fabricated on a silver disc some 200 centimeters (80 inches) in diameter and weighing more than 135 kilograms (300 pounds). Lost to looters in 1160, the map is survived by "Roger's Book," containing the information amassed by Idrisi's academy and including a world map, 71 part maps, and 70 sectional itinerary maps.

Idrisi's "inhabited earth" is divided into the seven "climates" of Greek geographers, beginning at the equator and stretching northward to the limit at which, it was supposed, the Earth was too cold to be inhabited. Each climate was then subdivided by perpendicular lines into 11 equal parts beginning with the west coast of Africa and ending with the east coast of Asia. Each of the resulting 77 square compartments was then discussed in sequence in "Roger's Book."

Though Idrisi worked in one of the most prestigious courts of Europe, there is little evidence that his work had any impact on European geographic thought. He was strongly influenced by Ptolemy's work and misconceptions and shared the then common Muslim fear of the unknown western ocean. Yet Idrisi's clear understanding of such scientific truths as the roundness of the Earth, his grasp of the scholarly writings of his Greek and Muslim predecessors, and the faithful recording of information on little-known portions of Europe, the Near East, and North Africa set his work far above the mediocre standards of contemporary Christian geography.

---

example; others may focus on smaller areas differently defined, such as Alpine France or the Corn Belt in the United States.

Other geographers choose to identify particular classes of things, rather than segments of the Earth's surface, for specialized study. These **systematic geographers** may focus their attention on one or a few related aspects of the physical environment or of human populations and societies. In each case, the topic selected for study is examined in its interrelationships with other spatial systems and areal patterns. **Physical geographers** practice systematic geography by directing their attention to the natural environmental side of the human-environment structure. Their concerns are with landforms and their distribution, with atmospheric conditions and climatic patterns, with soils or vegetation associations, and the like. The other systematic branch of geography—and the subject of this book—is **human geography**.

Human Geography

Human geography deals with the world as it is and with the world as it might be made to be. Its emphasis is on people: where they are, what they are like, how they interact over space, and what kinds of landscapes of human use they erect on the natural landscapes they occupy. It encompasses all those interests and topics of geography that are not directly concerned with the physical environment or, like cartography, are concerned with geographic techniques that apply to all domains of geography. Its content provides integration for all of the social sciences, for it gives to those sciences the necessary spatial and systems viewpoint that they might otherwise lack. For example, economists are often concerned with trends and patterns over time but do not fully appreciate that many of their interests concern patterns over space, too. Similarly, psychologists have long been interested in mind

and behavior but have often failed to recognize the spatial context of this mind and behavior. At the same time, human geography draws on other social sciences in the analyses identified with its subfields, such as *behavioral, political, economic,* and *social geography* (**Figure 1.3**). As Figure 1.3 suggests, human

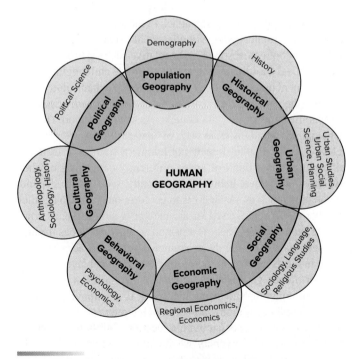

Figure 1.3 Some of the subdivisions of human geography and the allied fields to which they are related. Geography retains its ties to them and shares their insights and data, reinforcing its role as an essential synthesizer of data, concepts, and models that have integrative regional and spatial implications.

geographers also carry out work in areas traditionally recognized as part of the humanities, including history, philosophy, languages, and literature.

Human geography admirably serves the objectives of a liberal education. It helps us to understand the world we occupy and to appreciate the circumstances affecting peoples and countries other than our own. It clarifies the contrasts and similarities in societies and cultures and in the human landscapes they have created in different regions of the Earth. Its models and explanations of how things are interrelated in particular places and regions give us a clearer understanding of the economic, social, and political systems within which we live and operate. Its analyses of those spatial systems make us more aware of the realities and prospects of our own society in an increasingly connected and competitive world. Our study of human geography, therefore, can help make us better-informed citizens, more able to understand the important issues facing our communities and our countries and better prepared to contribute to their solutions. Importantly, it can also help open the way to wonderfully rewarding and diverse careers as professional geographers (see the feature "Careers in Geography").

background basics

1.2 Core Geographic Concepts

The topics included in human geography are diverse, but that very diversity emphasizes the reality that all geographers—whatever their particular topical or regional specialties—are united by the similar questions they ask and the common set of basic concepts they employ to consider their answers. Of either a physical or cultural phenomenon they will inquire: What is it? Where is it? How did it come to be what and where it is? Where is it in relation to other things that affect it or are affected by it? How is it changing? How is it part of a functioning whole? How do people affect it? How does its location affect people's lives and the content of the area in which it is found? These and similar questions are rooted in geography's central concern with **space** and **place** and in the special meanings geographers attach to those terms.

For geographers, *space* implies areal extent and may be understood in both an absolute and a relative sense. *Absolute space* is about fixed coordinate systems, like latitude and longitude, and measurement units, like miles or kilometers. Such absolute space remains the same in all contexts. In contrast, *relative space* is comparative and varies with context. As such, it is more flexible in recognizing that different ways of "measuring" space are more relevant for particular domains of human activity. For instance, different cultures think of space differently depending on their livelihood and travel habits. Economically, spaces vary depending on how much money it costs to get from one place to another. In other cases, relative spaces are mental or subjective, as when a route seems longer because a person thinks that it goes through a dangerous area. In all these examples, relative space measures space in terms other than those of a fixed physical layout.

For human geographers, *place* is the companion concept to *space*. In common understanding, *place* is a synonym for *location*. In human geography, however, *place* refers to the attributes and meanings we associate with a location. Our home town and neighborhood, the university we attend or the high school from which we graduated, a favorite downtown shopping area, and the like are all examples. Clearly, our *sense of place*—the impressions, feelings, and attitudes we have regarding specific locations and their complex of attributes—is unique to each of us, though we often share some aspects of our sense of place with other members of our culture or subculture. And clearly, too, we can even have a well-developed sense of place about locations we may never have personally experienced: Rome or Mecca or Jerusalem, for example, or—closer to home—Mount Rushmore or the Washington Mall. Of course, our sense of place may largely reflect a **place stereotype** rather than reality.

Our individual or group sense of place and attachments can, of course, set us off from others. Our home neighborhood that we find familiar and view favorably may equally be seen as alien and, perhaps, dangerous by others. The attributes and culture of places shape the lives and outlooks of those who inhabit them in ways basic to the socioeconomic patterning of the world. The viewpoints, normative behavior, religious and cultural beliefs, and ways of life absorbed and expressed by a middle-class, suburban American are undoubtedly vastly different from the understandings, cultural convictions, and life expectations of, for example, a young, unemployed male resident of Baghdad or the slums of Cairo. The implicit, ingrained, place-induced differences between the two help us understand one reason for the resistance to the globalization of Western social and economic values by those of different cultural backgrounds and place identification.

The sense of place is reinforced by recognized local and regional distinctiveness. It may be diminished or lost and replaced by a feeling of **placelessness** as the uniformity of brand-name fast-food outlets, national retail store chains, uniform shopping malls, repetitive highway billboards, and the like spread nationally and even internationally, reducing or eliminating the uniqueness of formerly separated locales and cultures. We'll examine some aspects of the sense of place and placelessness as we look at folk and popular cultures in Chapter 7.

Geographers use the word *spatial* as an essential modifier in framing their questions and forming their concepts. Geography, they say, is a *spatial* science. It is concerned with *spatial behavior* of people, with the *spatial relationships* that are observed between places on the Earth's surface, and with the *spatial processes* that create or maintain those behaviors and relationships. The word *spatial* comes, of course, from *space,* and to geographers, it always carries the idea of the way items are distributed, the way movements occur, and the way processes operate over the whole or a part of the surface of the Earth. The geographer's space, then, is Earth space, the surface area occupied or available to be occupied by humans. Spatial phenomena have locations on that surface, and spatial interactions occur among places, things, and people within the Earth area available to them. The need to understand those relationships, interactions, and processes helps frame the questions that geographers ask.

Additionally, those questions have their starting point in basic observations about the location and nature of places and about how places are similar to or different from one another. Such observations, though simply stated, are profoundly important to our comprehension of the world we occupy.

- Places have location, direction, and distance with respect to other places.
- A place has size; it may be large or small. Scale is important.
- A place offers both a physical setting and a social setting.
- The attributes of places develop and change over time.
- Places are connected to other places.
- The content of places is structured and explainable.
- Places may be generalized into regions of similarities and differences.

These are basic notions understandable to everyone. They also are the means by which geographers express fundamental observations about the Earth spaces they examine and put those observations into a common framework of reference. Each of the concepts is worth further discussion, for they are not quite as simple as they at first seem.

Geographic Features

Of course, space and (especially) place are not empty. **Geographic features** include natural features such as mountains, rivers, forests, oceans, and atmospheric fronts. They also include cultural features such as buildings, roads, cornfields, cities, and countries. Although all geographic features, like all material entities of any kind, are in reality three-dimensional, we often think of them or depict them on a map as if their dimensionality were less. So zero-dimensional features are thought of as points; an example might be a water well or a mountain peak. One-dimensional features are like lines, whether curved or straight; an example might be a river or a highway. Two-dimensional features are like areas or polygons; an example might be a forest or a neighborhood. Finally, some features are best thought of as being fully three-dimensional or volumetric. An oil deposit and a cloud are examples of this. It is important to recognize that the most appropriate way to think about a feature's dimensionality can depend greatly on the scale with which you examine it. A city may be a point when looking at a map of an entire country, but it becomes much more like an area when you zoom in to it.

Our discussion of feature dimensionality suggests something else about the way we conceptualize geographic features. They are typically thought of as being like discrete objects or like continuous fields. **Objects** are discrete entities that we think of as having sharp boundaries and being separated by space that may be conceived of as empty. Features like mountain peaks or roads are objects. **Fields** are continuously varying surfaces on the Earth that we think of as completely covering the space of the landscape they occupy without overlapping other fields. Features like average precipitation and landform elevations are fields. The distinction between objects and fields is admittedly abstract, and

there are features like water bodies that can readily be thought of in either way, or as a combination of the two. Human population is another intriguing example. At one scale, people are discrete objects; but at another scale, we can treat populations as a density field that may be said to have a nonzero value anywhere that is inhabited. However, it usually seems to make more sense to treat features and properties as more like objects or more like fields, even if we accept that this is sometimes imperfect. And if the distinction seems esoteric, we discuss below how it is quite important for the practical issue of how best to represent and model the world in computerized geographic information systems.

Location, Direction, and Distance

Location, direction, and *distance* are everyday ways of assessing the space around us and identifying our position in relation to other items and places of interest. They are also essential in understanding the processes of spatial interaction that figure so importantly in the study of human geography.

Location

The location of places and objects is the starting point of all geographic study, as well as all our personal movements and spatial actions in everyday life. We think of and refer to location in at least two different senses, *absolute* and *relative*.

Absolute location is the identification of place by some precise and accepted system of coordinates; it therefore is sometimes called *mathematical location*. We have several such accepted systems of pinpointing positions. One of them is the global grid of parallels and meridians (discussed later, beginning in Section 1.4). With it, the absolute location of any point on the Earth can be accurately described by reference to its degrees, minutes, and seconds of **latitude** and **longitude** (**Figure 1.4**).

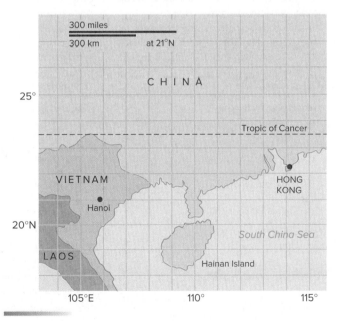

Figure 1.4 The latitude and longitude of Hong Kong is 22° 15′ N, 114° 10′ E (read as 22 degrees, 15 minutes north; 114 degrees, 10 minutes east). The circumference of the Earth measures 360 degrees; each degree contains 60 minutes, and each minute has 60 seconds of latitude or longitude. What are the coordinates of Hanoi?

Careers in Geography

The study of geography is an essential part of a liberal arts education and helps students become better citizens as they come to understand local, national, and global issues.

Can it, as well, be a pathway to employment for those who wish to specialize in the discipline? The answer is "Yes," in a number of different types of jobs. One broad cluster is concerned with supporting the field itself through teaching and research. Teaching opportunities exist at all levels, from elementary to university postgraduate. Teachers with some training in geography are increasingly in demand in elementary and high schools throughout the United States, reflecting geography's inclusion as a core subject in the federally adopted *Educate America Act* (Public Law 103-227) and the national determination to create a geographically literate society. At the college level, specialized teaching and research in all branches of geography have long been established, and geographically trained scholars are prominently associated with urban, global, community, and environmental studies, regional science, locational economics, and other interdisciplinary programs.

Because of the breadth and diversity of the field, training in geography involves the acquisition of techniques and approaches applicable to a wide variety of jobs outside the academic world. Modern geography is both a physical and social science, as well as part of the humanities, and fosters a wealth of technical skills. The employment possibilities it presents are as many and varied as are the agencies and enterprises dealing with the natural environment and human activities, and with the acquisition and analysis of spatial data.

Many professional geographers work in government, either at the state or local level, or in a variety of federal agencies and international organizations. Although many positions do not carry a geography title, physical geographers serve as water, mineral, and other natural resource analysts; weather and climate experts; soil scientists; and the like. An area of recent high demand is for environmental managers and technicians. Geographers who have specialized in environmental studies find jobs in both public and private agencies. Their work may include assessing the environmental impact of proposed development projects on such things as air and water quality and endangered species, as well as preparing the environmental impact statements required before construction can begin.

Human geographers work in many different roles in the public sector. Jobs include data acquisition and analysis in health care, transportation, population studies, economic development, and international economics. Many geography graduates find positions as planners in local and state governmental agencies concerned with housing and community development, park and recreation planning, and urban and regional planning. They map and analyze land-use plans and transportation systems, monitor urban land development, make informed recommendations about the location of public facilities, and engage in basic research.

Most of these same specializations are also found in the private sector. Geographic training is ideal for such tasks as business planning and market analysis; factory, store, and shopping-center site selection; community and economic development programs for banks, public utilities, and railroads; and similar applications. Publishers of maps, atlases, news and travel magazines, and the like employ geographers as writers, editors, and mapmakers.

The combination of a traditional, broadly based liberal arts perspective with the technical skills required in geographic research and analysis gives geography graduates a competitive edge in the labor market. These field-based skills include familiarity with geographic information systems (GISs), cartography and computer mapping, remote sensing and photogrammetry, and competence in data analysis and problem solving. In particular, students with expertise in GIS, who are knowledgeable about data sources, hardware, and software, are finding that they have ready access to employment opportunities. The following table, based on the booklet "Careers in Geography,"* summarizes some of the professional opportunities open to students who have specialized in one (or more) of the various subfields of geography. Also, be sure to read the informative discussions under the "Careers in Geography" option on the home page of the Association of American Geographers at *www.aag.org/*.

Geographic Field of Concentration	Employment Opportunities
Geographic technology	Cartographer for federal government (agencies such as Defense Mapping Agency, U.S. Geological Survey, or Environmental Protection Agency) or private sector (e.g., Environmental Systems Research Institute, ERDAS, Intergraph, or Bentley Systems); map librarian; GIS specialist for planners, land developers, real estate agencies, utility companies, local government; remote-sensing analyst; surveyor
Physical geography	Weather forecaster; outdoor guide; coastal zone manager; hydrologist; soil conservation/agricultural extension agent
Environmental geography	Environmental manager; forestry technician; park ranger; hazardous waste planner
Cultural geography	Community developer; Peace Corps volunteer; map librarian
Economic geography	Site selection analyst for business and industry; market researcher; traffic/route delivery manager; real estate agent/broker/appraiser; economic development researcher
Urban and regional planning	Urban and community planner; transportation planner; housing, park, and recreation planner; health services planner
Regional geography	Area specialist for federal government; international business representative; travel agent; travel writer
Geographic education or general geography	Elementary/secondary school teacher; college professor; overseas teacher

*"Careers in Geography," by Richard G. Boehm. Washington, DC: National Geographic Society, 1996. Previously published by Peterson's Guides, Inc.

Figure 1.5 The reality of *relative location* on the globe may be strikingly different from the impressions we form from maps with conventional projections like the Mercator. The position of Russia with respect to North America when viewed from a polar perspective emphasizes a closer relative location than many people realize.

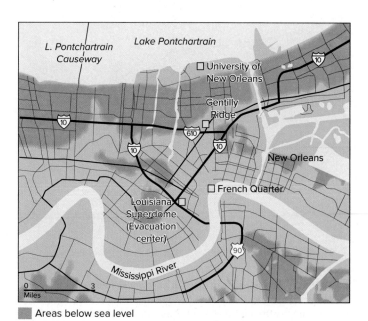

Areas below sea level

Figure 1.6 The *site* of New Orleans is hardly ideal for building a city. The city was built by the French on the most suitable high ground they could find near the mouth of the Mississippi River. The site extends north from the "high ground" along the Mississippi River to former swamp and marshland near Lake Pontchartrain. Much of the city and its suburbs are located below sea level on sinking soils composed of soft sediments deposited by past river floods.

Other coordinate systems are also in use. Survey systems such as the township, range, and section description of property in much of the United States give mathematical locations on a regional level, while street address precisely defines a building according to the reference system of an individual town. For convenience or special purposes, locational grid references may be superimposed on the basic global grid. The Universal Transverse Mercator (UTM) system, for example, based on a set of 60 longitude zones, is widely used in geographic information system (GIS) applications and, with different notations, as a military grid reference system. Absolute location is unique to each described place, is independent of any other characteristic or observation about that place, and has obvious value in the legal or scientific description of places, in measuring the distance separating places, or in finding directions between places on the Earth's surface.

When geographers—or real estate agents—remark that "location matters," their reference is usually not to absolute but to **relative location**—the position of a place in relation to that of other places or activities (**Figure 1.5**). Relative location expresses spatial interconnection and interdependence and may carry social (neighborhood character) and economic (assessed valuations of vacant land) implications. On an immediate and personal level, we think of the location of the school library not in terms of its street address or room number but where it is relative to our classrooms, or the cafeteria, or some other reference point. On the larger scene, relative location tells us that people, things, and places exist not in a spatial vacuum but in a world of physical and cultural characteristics that differ from place to place.

New York City, for example, may in absolute terms be described as located at (approximately) latitude 40° 439 N and longitude 73° 589 W. We have a better understanding of the *meaning* of its location, however, when reference is made to its spatial relationships: to the continental interior through the Hudson–Mohawk lowland corridor or to its position on the eastern seaboard of the United States. Within the city, we gain understanding of the locational significance of Central Park or the Lower East Side not solely by reference to the street addresses or city blocks they occupy, but by their spatial and functional relationships to the total land use, activity, and population patterns of New York City.

In view of these different ways of looking at location, geographers make a distinction between the *site* and the *situation* of a place. **Site** refers to the physical and cultural characteristics and attributes of the place itself. It is more than mathematical location, for it tells us something about the internal features of that place. The site of New Orleans, for example, extends from the natural levee on the Mississippi River to Lake Pontchartrain, much of which lies below sea level (**Figure 1.6**). **Situation,** on the other hand, refers to the external relations of a locale. It is an expression of relative location with particular reference to items of significance to the place in question. The situation of New Orleans might be described as being as close as possible to the mouth of the Mississippi River, which drains 41 percent of the land area of the continental United States, taking in much of the area from the Appalachian Mountains to the Rocky Mountains. Waterways on the Upper Mississippi, Missouri, Arkansas-Red-White, Ohio, and Tennessee River systems drain through the Lower Mississippi, connecting New Orleans to many of the country's important agricultural and manufacturing regions

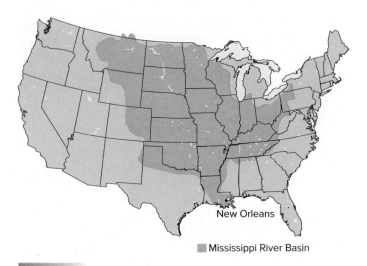

Figure 1.7 The *situation* of New Orleans is ideal for building a city. As the port at the mouth of the Mississippi River, New Orleans receives imports from Europe, Latin America, Asia, and Africa and exports grain, other food products, and petroleum from the United States. New Orleans is connected to 14,500 miles (23,335 km) of waterways as the Mississippi River drains a basin that stretches from the Rocky Mountains to the Appalachian Mountains.

(**Figure 1.7**). Although the flood-prone site makes it a challenging place to build a city, the incredible advantages offered by its situation have inspired generations of residents to make it their home.

Direction

Direction is a second universal spatial concept. Like location, it has more than one meaning and can be expressed in absolute or relative terms. **Absolute direction** is based on global or macroscopic features such as the cardinal points of north, south, east, and west, or on the directions to prominent stars. These appear uniformly and independently in all cultures, derived from the obvious "givens" of nature: the rising and setting of the sun for east and west, the sky location of the noontime sun and of certain fixed stars for north and south, or the direction toward or away from the center of an island.

We also commonly use **relative** or *relational* **directions.** In the United States, we worry about conflict in the "Near East" or economic competition from the "Far Eastern countries." These directional references are culturally based and locationally variable, despite their reference to cardinal compass points. The Near and the Far East locate parts of Asia from the European perspective; they are retained in the Americas by custom and usage, even though one would normally travel westward across the Pacific, for example, to reach the "Far East" from California, British Columbia, or Chile. Another important example of relative directional terms include body-centered terms like *left, right, in front of,* and *behind.*

Distance

Distance joins location and direction as a commonly understood term that has dual meanings for geographers. Like its two companion spatial concepts, distance may be viewed in both an absolute and a relative sense.

Absolute distance refers to the physical separation between two points on the Earth's surface measured by some accepted standard unit such as miles or kilometers for widely separated locales, feet or meters for more closely spaced points. **Relative distance** transforms those linear measurements into other units that could be more meaningful for the spatial relationship in question.

To know that two competing malls are about equidistant in miles from your residence is perhaps less important in planning your shopping trip than is knowing that because of street conditions or traffic congestion, one is 5 minutes and the other 15 minutes away (**Figure 1.8**). Many people, in fact, think of time distance rather than physical distance in their daily activities; downtown is 20 minutes by bus, the library is a 5-minute walk. In some instances, money rather than time may be the distance transformation. An urban destination might be estimated to be a $10 cab ride away, information that may affect either the decision to make the trip at all or the choice of travel mode to get there. As a college student, you already know that rooms and apartments are less expensive at a greater distance from campus. And a walk uphill may well seem longer than one that slopes gently downhill; in some situations, effort is an expression of relative distance between places.

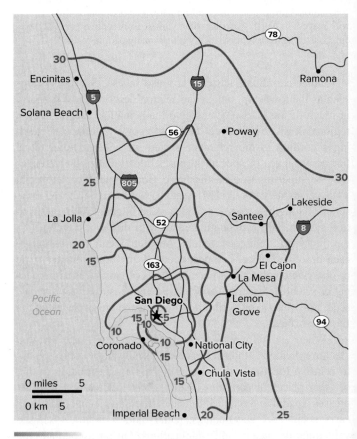

Figure 1.8 Lines of equal travel time (*isochrones*: from Greek, *isos*, "equal" , and *chronos*, "time") mark off the different linear distances accessible within given spans of time from a starting point. The fingerlike outlines of isochrone boundaries reflect variations in road conditions, terrain, traffic congestion, and other aids or impediments to movement. On this map, the areas within 5–30 minutes' travel time from downtown San Diego are recorded for the year 2002. Note the effect of freeways on travel time.

A *psychological* transformation of distance is also frequent. The solitary late-night walk back to the car through an unfamiliar or dangerous neighborhood may seem far longer than a daytime stroll of the same distance through familiar and friendly territory. A first-time trip to a new destination frequently seems much longer than the return trip over the same path. Distance relationships, their measurement, and their meaning for human spatial interaction are fundamental to our understanding of human geography. They are a subject of Chapter 3, and reference to them recurs throughout this book.

Size and Scale

When we say that a place may be large or small, we speak both of the nature of the place itself and of the generalizations that can be made about it. In either instance, geographers are concerned with **scale**. Although scale is always about relative size (whether spatial or temporal), we use the term in different ways. We can, for example, study a problem—say, population or agriculture—at the local scale, the regional scale, or on a global scale. Here the reference is purely to the size of unit studied. In this sense, *large-scale* means large units or areas studied, and *small-scale* means small units or areas studied. In a technical, cartographic sense, scale tells us the ratio between the length of physical distance on a map and the actual length of the mapped distance on the surface of the Earth (see Appendix A). Whatever the scale of a map, it is a feature of every map and important to recognizing the areal meaning of what is shown on that map.

In both senses of the word, *scale* implies something about the degree of generalization involved (**Figure 1.9**). Generalization is averaging over details, so that a large-scale unit of study (and a small-scale map) generalizes more than a small-scale unit of study (and large-scale map). Geographic inquiry may be broad or narrow; it occurs at many different size scales. Climate may be an object of study, but research and generalization focused on climates of the world will differ in degree and kind from study of the microclimates of a city. Awareness of scale is very important. In geographic work, concepts, relationships, and understandings that have meaning at one scale may not apply at another.

As another example, the study of world agricultural patterns may refer to global climatic regimes, cultural food preferences, levels of economic development, and patterns of world trade. These large-scale relationships are of little concern to the study of crop patterns within single counties of the United States, where topography, soil and drainage conditions, farm size, ownership, and capitalization, or even personal management preferences, may be of greater explanatory significance.

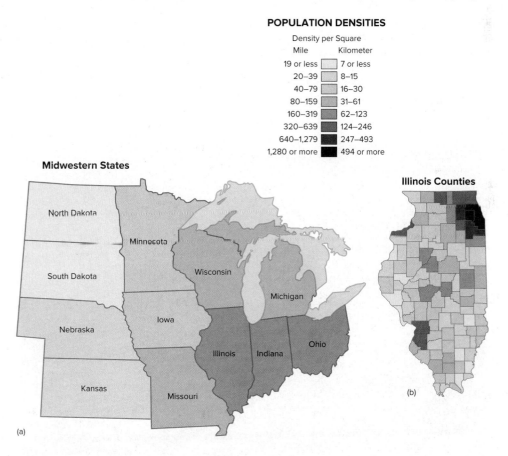

POPULATION DENSITIES

Density per Square	
Mile	Kilometer
19 or less	7 or less
20–39	8–15
40–79	16–30
80–159	31–61
160–319	62–123
320–639	124–246
640–1,279	247–493
1,280 or more	494 or more

Midwestern States

Illinois Counties

(a)

(b)

Figure 1.9 Population density patterns depend upon map scale. "Truth" depends on one's scale of inquiry. Map (*a*) reveals that the maximum year 2010 population density of Midwestern states was no more than 123 people per square kilometer (319 per sq mi). From map (*b*), however, we see that population densities in three Illinois counties exceeded 494 people per square kilometer (1,280 per sq mi) in 2010. If we were to reduce our scale of inquiry even further, examining individual city blocks in Chicago, we would find densities reaching 2,500 or more people per square kilometer (10,000 per sq mi). Scale matters!

Physical and Cultural Attributes

All places have physical and cultural attributes that distinguish them from other places and give them character, potential, and meaning. Geographers are concerned with identifying and analyzing the details of those attributes and, particularly, with recognizing the interrelationship between the physical and cultural components of area: the human-environmental interface.

Physical characteristics refer to such natural aspects of a locale as its climate and soil, the presence or absence of water supplies and mineral resources, its terrain features, and the like. These **natural landscape** attributes provide the setting within which human action occurs. They help shape—but do not dictate—how people live. The resource base, for example, is physically determined, though how resources are perceived and utilized is, to some extent, culturally conditioned.

Environmental circumstances directly affect agricultural potential and reliability; indirectly they may influence such matters as employment patterns, trade flows, population distributions, and national diets. The physical environment simultaneously presents advantages and drawbacks with which humans must deal. Thus, the danger of typhoons in central China or monsoonal floods in Bangladesh must be balanced against the agricultural bounty derived from the regions' favorable terrain, soil, and moisture conditions.

At the same time, by occupying a given place, people modify its environmental conditions. The existence of the U.S. Environmental Protection Agency (EPA; and its counterparts elsewhere) is a reminder that humans are the active and frequently harmful agents in the continuing interplay between the cultural and physical worlds (**Figure 1.10**). Virtually every human activity leaves its imprint on an area's soils, water, vegetation, animal life, and other resources and on the atmosphere common to all Earth space. The impact of humans has been so universal and so long exerted that essentially no purely "natural landscape" any longer exists. One can even find human-made debris in the middle of the ocean and air pollution in the Arctic.

The visible expression of that human activity is the **cultural landscape**. It, too, exists at different scales and different levels of visibility. Differences in agricultural practices and land use between Mexico and southern California are evident in **Figure 1.11,** while the signs, structures, and people of, for instance, Los Angeles's Chinatown leave a smaller, more confined imprint within the larger cultural landscape of the metropolitan area itself.

Although the focus of this book is on the human characteristics of places, geographers are ever aware that the physical content of an area is also important in understanding the activity patterns of people and the interconnections between people and the environments they occupy and modify. Those interconnections and modifications are not static or permanent, however, but are subject to continual change. For example, marshes and wetlands, when drained, may be transformed into productive, densely settled farmland, while the threat or occurrence of eruption of a long-dormant volcano may quickly and drastically alter established patterns of farming, housing, and transportation on or near its flanks.

The Changing Attributes of Place

The physical environment surrounding us seems eternal and unchanging but, of course, it is not. In the framework of geologic time, change is both continuous and pronounced. Islands

Figure 1.10 Sites such as this landfill are all too-frequent reminders of the adverse environmental impacts of humans and their waste products. Here, bulldozers compact solid waste and spread a daily cover at a "sanitary" landfill.

©Doug Sherman/Geofile

Figure 1.11 This NASA image reveals contrasting cultural landscapes along the Mexico-California border. Move your eyes from the Salton Sea (the dark patch at the top of the image) southward to the agricultural land extending to the edge of the image. Notice how the regularity of the fields and the bright colors (representing growing vegetation) give way to a marked break, where irregularly shaped fields and less prosperous agriculture are evident. Above the break is the Imperial Valley of California; below the border is Mexico.

Source: NASA

form and disappear; mountains rise and are worn low to swampy plains; vast continental glaciers form, move, and melt away, and sea levels fall and rise in response. Geologic time is long, but the forces that give shape to the land are timeless and relentless.

Even within the short period of time since the most recent retreat of continental glaciers—some 11,000 or 12,000 years ago—the environments occupied by humans have been subject to change. Glacial retreat itself marked a period of climatic alteration, extending the area habitable by humans to include vast reaches of northern Eurasia and North America formerly covered by thousands of feet of ice. With moderating climatic conditions came associated changes in vegetation and fauna. On the global scale, these were natural environmental changes; humans were as yet too few in numbers and too limited in technology to alter materially the course of physical events. On the regional scale, however, even early human societies exerted an impact on the environments they occupied. Fire was used to clear forest undergrowth, to maintain or extend grassland for grazing animals and to drive them in the hunt, and later to clear openings for rudimentary agriculture.

With the dawn of civilizations and the invention and spread of agricultural technologies, humans accelerated their management and alteration of the now no longer "natural" environment. Even the classical Greeks noted how the landscape they occupied differed—for the worse—from its former condition. With growing numbers of people and particularly with industrialization

and the spread of European exploitative technologies throughout the world, the pace of change in the content of area accelerated. The built landscape—the product of human effort—increasingly replaced the natural landscape. Each new settlement or city, each agricultural assault on forests, each new mine, dam, or factory changed the content of regions and altered the temporarily established spatial interconnections between humans and the environment.

Characteristics of places today are the result of constantly changing past conditions. They are, as well, the forerunners of differing human-environmental balances yet to be struck. Geographers are concerned with places at given moments of time. But to understand fully the nature and development of places, to appreciate the significance of their relative locations, and to comprehend the interplay of their physical and cultural characteristics, geographers must view places as the present result of the past operation of distinctive physical and cultural processes (**Figure 1.12**).

You will recall that one of the questions geographers ask about a place or thing is, "How did it come to be what and where it is?" This is an inquiry about process and about becoming. The forces and events shaping the physical and cultural environment of places today are an important focus of geography. They are, particularly in their human context, the subjects of most of the separate chapters of this book. To understand them is to appreciate more fully the changing human spatial order of our world.

(a)

(b)

Figure 1.12 The process of change in a cultural landscape can be dramatic. (*a*) In 1913, Miami, Florida, was just a small settlement on the banks of the Miami River amidst woodlands and wetlands. (*b*) By the end of the 20th century, it had grown from a few thousand inhabitants to some 350,000, with buildings, streets, and highways completely transforming its natural landscape.

Sources: (a) Library of Congress, Prints & Photographs Division, Reproduction number LC-DIG-det-4a24101 (digital file from original); (b) South Florida Water Management District

Interrelations Between Places

The concepts of relative location and distance that we earlier introduced lead directly to a fundamental spatial reality: places interact with other places in structured and comprehensible ways. In describing the processes and patterns of that **spatial interaction,** geographers add *accessibility* and *connectivity* to the ideas of location and distance.

Tobler's First Law of Geography tells us that in a spatial sense, everything is related to everything else, but that relationships are stronger when items are near one another. Our observation, therefore, is that interaction between places tends to diminish in intensity and frequency as distance between them increases—a statement of the idea of *distance decay,* which we explore in Chapter 3. Think about it—are you more likely to go to a fast-food outlet next door or to a nearly identical restaurant across town? Human decision making is unpredictable in many ways and decisions are frequently made for obscure reasons, but in this case you can see how you will probably frequent the nearer place more often.

Consideration of distance implies assessment of **accessibility.** How easy or difficult is it to overcome the *friction of distance*? That is, how easy or difficult is it to surmount the barrier of the time and space separating places? Distance isolated North America from Europe until the development of ships (and aircraft) that reduced the effective distance between the continents. All parts of the ancient and medieval city were accessible by walking; they were *pedestrian cities,* a status lost as cities expanded in area and population with industrialization. Accessibility between city districts could be maintained only by the development of public transit systems whose fixed lines of travel increased ease of movement between connected points and reduced it between areas not on the transit lines themselves. Later, the invention and widespread adoption of the automobile had its own profound effects on urban form and activity patterns.

Accessibility, therefore, suggests the idea of **connectivity,** a broader concept implying all the tangible and intangible ways in which places are linked: by physical telephone lines, street and road systems, pipelines and sewers; by unrestrained walking across open countryside; by radio and TV broadcasts beamed outward from a central source. Where routes are fixed and flow is channelized, *networks*—the patterns of routes connecting sets of places—determine the efficiency of movement and the connectedness of points. Very rapid and uniform accessibility and connectivity are expected in today's advanced societies. Technologies and devices to achieve it proliferate, as our own lifestyles show. Cell phones, e-mail, broadband wireless Internet access, instant messaging, and more have considerably reduced time and distance barriers to communication that formerly separated and isolated individuals and groups, especially in the developed world, and have reduced our dependence on physical movement and on networks fixed in the landscape. The realities of accessibility and connectivity, that is, clearly change over time (**Figure 1.13**).

There is, inevitably, interchange between connected places. **Spatial diffusion** is the process of dispersion of an idea or an item from a center of origin to more distant points with which it is directly or indirectly connected. The rate and extent of that diffusion are affected by the distance separating the originating center of, say, a new idea or technology and other places where it is eventually adopted. Diffusion rates are also affected by population densities, means of communication, obvious advantages of the innovation, and importance or prestige of the originating *node*. These ideas of diffusion are further explored in Chapter 2.

Geographers study the dynamics of spatial relationships. Movement, connection, and interaction are part of the social and economic processes that give character to places and regions. Geography's study of those relationships recognizes that spatial interaction is not just an awkward necessity but a fundamental organizing principle of human life on Earth. That recognition has become universal, repeatedly expressed in the term *globalization*. **Globalization** implies the increasing interconnection of peoples and societies in all parts of the world as the full range

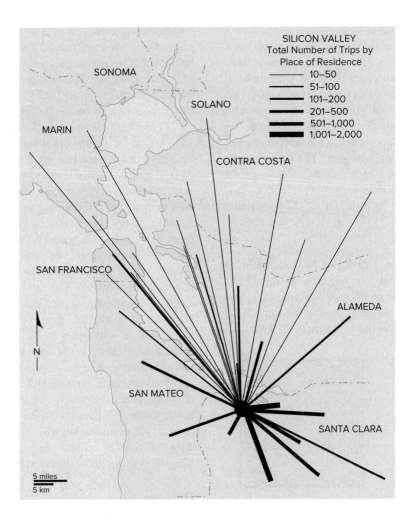

Figure 1.13 An indication of one form of spatial interaction and *connectivity* is suggested by this map recording the volume of daily work trips within the San Francisco Bay area to the Silicon Valley employment node (geographers sometimes refer to these displays as *spider diagrams,* for obvious reasons). The ends of the lines define the outer reaches of a physical interaction region defined by the network of connecting roads and routes. The region changed in size and shape over time as the network was enlarged and improved, the Valley employment base expanded, and the commuting range of workers increased. The map, of course, gives no indication of the global reach of the Valley's *accessibility* and interaction through other means of communication and interchange.

Source: Robert Cervero, Suburban Gridlock. Published by the Center for Urban Policy Research, Rutgers, the State University of New Jersey, 1986.

of social, cultural, political, economic, and environmental processes become international in scale and effect. Promoted by continuing advances in worldwide accessibility and connectivity related in part to developments in the technologies of transportation and communication, globalization encompasses other core geographic concepts of spatial interaction, accessibility, connectivity, and diffusion. More detailed implications of globalization will be touched on in later chapters of this text.

The Structured Content of Place

A starting point for geographic inquiry is how objects are distributed in area—for example, the placement of churches or supermarkets within a town. That interest distinguishes geography from other sciences, physical or social, and underlies many of the questions geographers ask: Where is a thing located? How is that location related to other items? How did the location we observe come to exist? Such questions carry the conviction that the contents of an area are comprehensibly arranged or structured.

The arrangement of items on the Earth's surface is called **spatial distribution** and may be analyzed by the elements common to all spatial distributions: *density, dispersion,* and *pattern.* In addition, pairs or larger sets of distributions often show *spatial association.*

Density

Density is usually thought of as a measure of the number or quantity of a specific feature within a defined unit of area. In fact, density does not apply only to areas. Thus, one can speak of the density of point (zero-dimensional) features, like gas stations, within a unit of a linear (one-dimensional) feature, like a highway, or within a unit of an areal (two-dimensional) feature, like a county. Similarly, one can speak of the density of linear features, like highways, within a unit of an areal feature, like a county. It is therefore not simply a count of items but of items in relation to the space in which they are found. When the relationship is absolute, as in population per square kilometer, for example, or dwelling units per acre, we are defining *arithmetic density* (see Figure 1.9).

Sometimes it is more meaningful to relate item numbers to a specific kind of space. *Physiological density,* for example, is a measure of the number of persons per unit area of arable land. Density defined in population terms is discussed in Chapter 4.

A density figure is a statement of fact, but not necessarily one useful in itself. Densities are normally employed comparatively, relative to one another. High or low density implies a comparison with a known standard, with an average, or with a different area. Ohio, with 107 persons per square kilometer (277 per sq mi) in 2000, might be thought to have a high density compared with neighboring Michigan at 68 per square kilometer (175 per sq mi), and a low one in relation to New Jersey at 438 (1,134 per sq mi).

Dispersion

Dispersion (or its opposite, **concentration**) is a statement of the extent to which features within a distribution are spread out (dispersed or scattered) from one another, or clustered (agglomerated) together (**Figure 1.14**). Dispersion is an entirely separate distributional property from density. **Figure 1.15** shows the

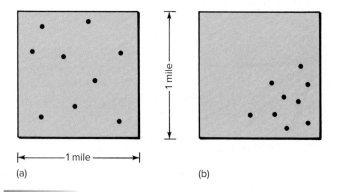

(a)   1 mile   (b)

Figure 1.14 Density and dispersion each tell us something different about how items are distributed in an area. *Density* is simply the number of items or observations within a defined area; it remains the same no matter how the items are distributed. The density of houses per square mile, for example, is the same in both (*a*) and (*b*). *Dispersion* is a statement about nearness or separation. The houses in (*a*) are more *dispersed* than those shown *clustered* in (*b*).

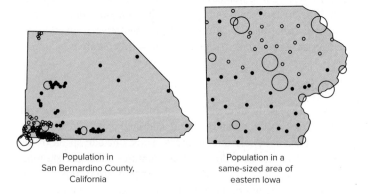

Population in San Bernardino County, California   Population in a same-sized area of eastern Iowa

Figure 1.15 Density and dispersion are separate properties of spatial distributions. The population distribution of San Bernardino County, California, on the left is highly concentrated, whereas that of eastern Iowa on the right is highly dispersed, even though both regions have about equal area and population, and thus equal arithmetic density.

population distributions of San Bernardino County in California and an eastern portion of the state of Iowa. Both have similar spatial areas and populations, so that they both have similar arithmetic densities. But the population in eastern Iowa is rather evenly spread out, while that of San Bernardino is quite concentrated (near the city of Los Angeles to the west).

The phenomenon of dispersion or concentration must always be judged against the standard of how a distribution would be arranged if it were generated randomly—if it were not based on a systematic cause. For example, a police department may want to know whether crimes are clustered or randomly located. Although it is sometimes easy to see with the naked eye whether a distribution is dispersed or clustered, often it is not. Systematic processes often act against a background of "noise," so they are not so evident to simple inspection. At the same time, people are good at seeing meaningful arrangements, but they are not good at seeing randomness. That is, people tend to see clusters, for instance, even when there is no more concentration than would be expected by chance alone. Geographers have developed formal statistical tests to try to verify the existence of nonrandom dispersion or clustering that is stronger than would be expected by chance alone. A comparison to chance also makes it clear that distributions can be nonrandomly dispersed either by being more clustered than expected by chance, or by being *less clustered* (more dispersed) than would be expected by chance. In a later chapter, for example, we will see that central place theory predicts that cities of a given size will be spread more widely across a region than would be expected by chance alone.

Pattern

The geometric arrangement of feature in space is called **pattern.** Like dispersion, pattern refers to distribution, but in a way, that emphasizes the design or shape of feature locations rather than just their spacing (**Figure 1.16**). The distribution of towns along a railroad or houses along a street may be seen as *linear*. A *centralized* pattern may involve items concentrated around a single node. A *random* pattern may be the best description of an unstructured irregular distribution. Like dispersion, patterns created by systematic processes are sometimes easy to confuse with unpatterned distributions created by random processes. That is, people tend to see patterns even when no systematic process has operated to create a pattern (for instance, have you ever seen a

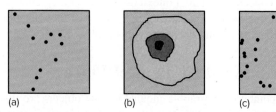

(a)   (b)   (c)

Figure 1.16 *Pattern* describes spatial arrangement and design. The *linear* pattern of towns in (*a*) perhaps traces the route of a road or railroad or the course of a river. The urban land use in (*b*) with its categories of residential and commercial activities represents a circular pattern, while the dots in (*c*) are *randomly* distributed.

face in the clouds?). Again, geographers have statistical tests to try to verify the existence of nonrandom patterns that are stronger than would be expected by chance alone.

An example of a patterned distribution includes that resulting from the rectangular system of land survey adopted in much of the United States under the Ordinance of 1785, which created a checkerboard rural pattern of "sections" and "quarter-sections" of farmland (see Figure 7.21 in Chapter 7). As a result, in most American cities, streets display a *grid* or *rectilinear* pattern. The same is true of cities in Canada, Australia, New Zealand, and South Africa, which adopted similar geometric survey systems. The *hexagonal* pattern of service areas of farm towns is a mainstay of central place theory discussed in Chapter 11. These references to the geometry of distribution patterns help us visualize and describe the structured arrangement of items in space. They help us make informed comparisons between areas and use the patterns we discern to ask further questions about the interrelationship of things.

Spatial Association

Two distributions of features often spatially correspond with each other. That is, places where one feature is found are more likely (or less likely) than chance to be the places where a different type of feature is found. This is **spatial association** or covariation. For example, counties in Texas where consuming alcoholic beverages is allowed by law tend to be the same counties that have a majority of Catholic residents, while so-called dry counties are more likely to have a majority of Protestant residents (**Figure 1.17**). Spatial association is very similar to looking for spatial pattern, except that it is a spatial pattern involving two or more distributions simultaneously. Like pattern, geographers attempt to identify associations that are clearly stronger than would be expected by chance alone. It is critical to remember, however, that finding an association does not in itself tell you *why* two distributions covary—it does not explain what caused it.

Place Similarity and Regions

The distinctive characteristics of places in content and structure immediately suggest two geographically important ideas. The first is that no two places on the surface of the Earth can be *exactly* the same. Not only do they have different absolute locations, but—as in the features of the human face—the precise mix of physical and cultural characteristics of a place is never exactly duplicated.

Because geography is a spatial science, the inevitable uniqueness of place would seem to impose impossible problems of generalizing spatial information. That this is not the case results from the second important idea: the physical and cultural content of an area and the dynamic interconnections of people and places show patterns of spatial similarity. For example, a geographer doing fieldwork in France might find that all farmers in one area use a similar specialized technique to build fences around their fields. Often, such similarities are striking enough for us to conclude that

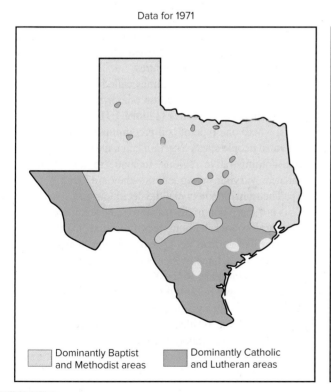

Data for 1971

Dominantly Baptist and Methodist areas

Dominantly Catholic and Lutheran areas

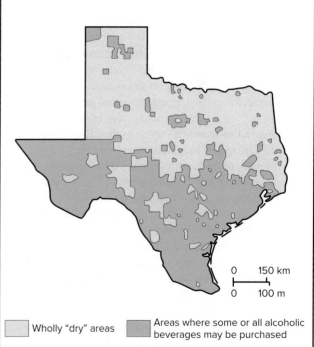

Status in 1972

0   150 km

0   100 m

Wholly "dry" areas

Areas where some or all alcoholic beverages may be purchased

Figure 1.17 The distributions of religion and alcohol sales in Texas show a spatial correlation. Catholic and Lutheran areas generally choose to be "wet," and Baptist-Methodist areas retain prohibition. Both the Baptist and Methodist churches have traditionally taken a stand against alcoholic beverages. (From 38th Annual Report of the Texas Alcoholic Beverage Commission, Austin, 1972, p. 49; and Churches and Church Membership in the United States: 1971, National Council of the Churches of Christ in the U.S.A., 1974.)

spatial regularities exist. They permit us to recognize and define **regions**—Earth areas that display significant elements of internal uniformity and external difference from surrounding territories. Places are, therefore, both unlike and like other places, creating patterns of areal differences and of coherent spatial similarity.

The problem of the historian and the geographer is similar. Each must generalize about items of study that are essentially unique. The historian creates arbitrary but meaningful and useful historical periods for reference and study. The "Roaring Twenties" and the "Victorian Era" are shorthand summary names for specific time spans, internally quite complex and varied but significantly distinct from what went before or followed after. The region is the geographer's equivalent of the historian's era. It is a device of areal generalization that segregates into component parts the complex reality of the Earth's surface. In both the time and the space needed for generalization, attention is focused on key unifying elements or similarities of the era or area selected for study. In both the historical and geographical cases, the names assigned to those times and places serve to identify the time span or region and to convey between speaker and listener a complex set of interrelated attributes.

All of us have a general idea of the meaning of region, and all of us refer to regions in everyday speech and action. We visit "the old neighborhood" or "go downtown"; we plan to vacation or retire in the "Sunbelt"; or we speculate about the effects of weather conditions in the "Corn Belt" on next year's food prices. In each instance, we have mental images of the areas mentioned, and in each, we have engaged in an informal place classification to pass along quite complex spatial, organizational, or content ideas. We have applied the **regional concept** to bring order to the immense diversity of the Earth's surface. In the end, we can see that there is nothing particularly exotic or peculiar about using similarities to group unique entities into similarity classes. All people in all cultures and time periods (including all of us!) do it constantly when they recognize that two unique objects they can sit on are both "chairs" or two unique woody plants are both "trees." In other words, regionalizing is spatial *categorization,* and categorization appears to be culturally and historically universal to all people.

Regions are not just "given" in nature, any more than "eras" are entirely given in the course of human events. Regions are, in part, devised; they are spatial summaries designed to bring order to the immense diversity of the Earth's surface. At their root, they are based on the recognition and mapping of *spatial distributions*— the territorial occurrence of environmental, human, or organizational features selected for study. For example, the location of Welsh speakers in Britain is a distribution that can be identified and mapped. As many spatial distributions exist as there are imaginable physical, cultural, or connectivity elements of area to examine. Because regions are partially mental constructs, different observers employing different criteria may bestow the same regional identity on differently bounded areal units. In each case, however, the key characteristics that are selected for study are those that contribute to the understanding of a specific topic or problem.

Types of Regions

All regions share certain properties. They are all two-dimensional geographic features. They all have location and size (area). All regions have boundaries, which divide places inside the region from places outside the region. These boundaries vary in their degree of sharpness or *vagueness,* however, as we discuss below. Regions also vary in the permeability of their boundaries. That is, boundaries allow material, energy, people, and information to move across them more or less easily (with more or less effort, time, and cost), and permeability in one direction may be different than in the other. We also observe that regions are often *hierarchically organized.* Regions at one scale of size or importance are often contained by larger or more important regions in one direction, and contain smaller or less important regions in the other direction. The simplest example of region hierarchies is probably the relationship of larger and smaller political regions: states or provinces are within countries, while entities like counties and cities are within states or provinces.

Regions may be *administrative, thematic, functional,* or *perceptual.* An **administrative region** is created by law, treaty, or regulation. It includes political regions such as countries and states, bureaucratic regions such as school and voting districts, and cadastral (real estate) regions. Even the end zone on a football field is an administrative region. The boundaries of administrative regions are different than the boundaries of the other three types of regions, in that they are as sharp as measurement precision allows, or at least potentially (as soon as someone cares about the location of administrative boundaries, they can be made very precise by diplomats, lawyers, and surveyors). Given these precise boundaries, administrative regions have *uniform membership functions*—every place within the region is fully and equally representative of the region. Interestingly, this is an exception to the rule stated above that "no two places are identical," insofar as all places within the boundaries of an administrative region are generally treated identically with respect to administrative rules and procedures.

Thematic regions (sometimes called formal or uniform regions in other texts) are based on one or more objectively measurable themes or properties (**Figure 1.18a**). Examples are soil regions, where one type of soil predominates, or dialect regions, where most people speak a certain language using a given dialect. Unlike administrative regions, thematic regions typically have boundaries varying in vagueness—they are "fuzzy" rather than sharp. The transition between the deciduous forest and the grasslands is not sharp but gradual, as fewer and smaller trees give way to larger grassy areas without trees. Of course, maps usually show boundaries like these as being sharp, but that sharpness is largely a handy fiction that makes it easier to display and think about the regions. As a corollary to these fuzzy or vague boundaries, thematic regions have *non-uniform membership functions*. At particular places within a given dialect region, for example, virtually 100 percent of the residents speak the dialect, while in other places, only a slight majority do. In such a case, which is quite common, one can say that certain places within the region are more strongly or clearly representative of the region.

Functional regions emerge from patterns of interaction over space and time that connect places (**Figure 1.18b**). Examples include the region in which most people shop at a particular shopping center or listen to a particular radio station. In physical geography, the movement of air and water currents defines

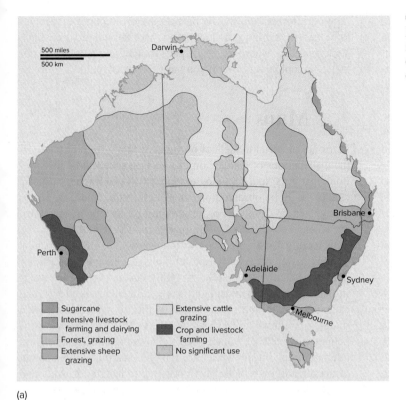

Figure 1.18 (*a*) This generalized land use map of Australia is composed of thematic *regions* whose internal physical and cultural characteristics show essential similarities, setting them off from adjacent territories of different conditions of use.

500 miles
500 km

Darwin

Brisbane

Perth

Adelaide

Sydney

Melbourne

- Sugarcane
- Intensive livestock farming and dairying
- Forest, grazing
- Extensive sheep grazing
- Extensive cattle grazing
- Crop and livestock farming
- No significant use

(a)

(b)

Figure 1.18 (*b*) The functional (or nodal) regions shown on this map were based on linkages among large banks of major central cities and the "correspondent" banks they served in smaller towns in the 1970s, before the advent of electronic banking and bank consolidation.

Source: (b) Annals of the Association of American Geographers, *John R. Borchert, vol. 62, p. 358, Association of American Geographers, 1972.*

functional regions; a watershed is an important example. Often, functional regions have a pointlike core from which interaction originates, and thus they are sometimes called *nodal regions,* but they need not originate from a point. Certain famous wind patterns in various parts of the world form functional regions when they originate from a linear feature rather than a point. Of course, like thematic regions, functional regions generally have vague boundaries and non-uniform membership functions; you can literally hear the "fuzziness" of functional regions defined by radio stations as you drive further from the transmission tower.

Finally, **perceptual regions** (also called *cognitive regions*) are the informal subjective regions defined by people's beliefs, feelings, and images. They reflect the universal tendency for humans to regionalize parts of the Earth's surface, even though the particular regions identified certainly vary across cultures and historical times, even across individual people. Again, like thematic and functional regions, perceptual regions typically have vague boundaries and non-uniform membership functions. Two places may both be thought of as "downtown," for example, but one is seen to represent downtown more clearly than the other. In addition, perceptual regions like downtown are often culturally shared. In this case, perceptual regions may be called *vernacular regions.* Vernacular regions are real in the minds of cultural group members and are often reflected in regionally based names employed by businesses, by sports teams, or in advertising slogans. The frequency of references to "Dixie" in the southeastern United States represents that kind of regional consensus and awareness. Geographer Wilbur Zelinsky created his map of the perceptual regions of North America by counting the frequency that regional terms were used in the names of businesses (**Figure 1.19**). At a different scale, such urban ethnic enclaves (see Chapter 6) as "Little Italy" or "Chinatown" have comparable regional identities in the minds of their inhabitants. However, perceptual regions are not always culturally shared; they

may be idiosyncratic to just one person or a couple of people. For example, a child growing up in a rural community may have a special place where he or she goes to hang out and daydream. What perceptual regions do you have clearly in mind?

1.3 Maps

Maps are pictorial models of portions of the Earth's surface and the distributions of features on that surface. The spatial distributions, patterns, and relations of interest to geographers usually cannot easily be observed or interpreted in the landscape itself. Many, such as landform or agricultural regions or major cities, are so extensive spatially that they cannot be seen or studied in their totality from one or a few vantage points. Others, such as regions of language usage or religious belief, are spatial phenomena, but are not tangible or readily visible to someone walking around in the environment. Various interactions, flows, and exchanges imparting the dynamic quality to spatial interaction may not be directly observable at all. And even if all matters of geographic interest could be seen and measured through field examination, the infinite variety of tangible and intangible content of area would make it nearly impossible to isolate for study and interpretation the few items of interest selected for special investigation in any particular situation.

Therefore, the map has become one of the essential and distinctive tools of geographers. Maps allow spatial distributions and interactions of whatever nature to be reduced to an observable scale, isolated for individual study, and combined or recombined to reveal relationships not directly measurable in the landscape itself. Maps highlight and clarify relevant properties, but at the same time, they omit or downplay irrelevant properties. For instance, subway maps in most major cities focus on showing connections between stops but intentionally leave out

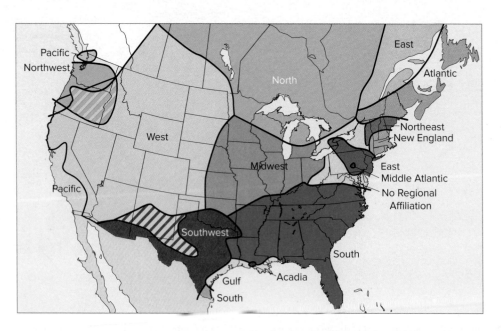

Figure 1.19 This map of the perceptual (cognitive) regions of North America was created based on the names of local businesses.

Source: Annals of the Association of American Geographers, *Wilbur Zelinsky, "North America's Vernacular Regions," Vol. 70, No. 1, p. 14, 1980.*

accurate information about distances and directions because most riders do not need this information (**Figure 1.20**). But maps can serve their purpose only if their users have a clear idea of their strengths and limitations, the diversity of map styles, and the conventions observed in their preparation and interpretation.

Map Scale

We have already seen that scale is a vital element of every map. Because it is a much reduced version of the reality it summarizes, a map generalizes the data it displays. *Scale*—the relationship between size or length of a feature on the map and the same item on the Earth's surface—determines the amount of that generalization. The smaller the scale of the map, the larger is the area it covers and the more generalized are the data it portrays. The larger the scale, the smaller is the depicted area and the more precisely can its content be represented (**Figure 1.21**). It may seem backward, but large-scale maps show small areas, and small-scale maps show large areas.

Map scale is selected according to the amount of generalization of data that is acceptable and the size of area that must be depicted. The user must consider map scale in evaluating the reliability of the spatial data that are presented. Regional boundary lines drawn on the world maps in this and other books or atlases

Figure 1.21 Map scale affects both the areal extent that can be shown and the level of detail. The larger the scale, the greater the number and kinds of features that can be included. Notice how individual buildings are visible in the large-scale map (upper left), while the city of Boston is a mere point symbol in the smallest scale map (lower right). Scale may be reported to the map user in one (or more) of three ways. A *verbal* scale is given in words ("1 centimeter to 1 kilometer" or "1 inch to 1 mile"). A **representative fraction** (such as that placed at the left, below each of the four maps shown here) is a numerical expression of how many linear units on the Earth's surface are represented by one unit on the map. In the upper-left map, for example, one map inch represents 25,000 inches on the ground. A **graphic scale** (such as that placed at the right and below each of these maps) is a line or bar marked off in map units but labeled in ground units.

would cover many kilometers or miles on the Earth's surface. They obviously distort the reality they are meant to define, and on small-scale maps major distortion is inevitable. In fact, a general rule of thumb is that the larger the Earth area depicted on a map, the greater is the distortion built into the map.

This is so not only because geographic features must be generalized more on smaller-scale maps, but because a map has to depict the curved surface of the three-dimensional Earth on a two-dimensional sheet of paper. The term **projection** designates the method chosen to represent the Earth's curved surface as a flat map—to **develop** the Earth's surface. Because absolutely accurate representation is impossible, all projections inevitably distort. To start with, all maps break a continuous Earth surface at some arbitrarily chosen juncture (the places shown on the right and left sides of most maps are next to each other in reality). If that weren't enough, projections also distort one or more of the four main spatial properties of maps—area, shape, distance, or direction.[2] Specific projections may be selected, however, to minimize distortion of a particular spatial property or compromise the amount of distortion in one property by increasing distortion in another.

The Globe Grid

As we have seen, geography is about the planet Earth and the natural and human structures and processes found there. The Earth is the third planet from the sun in our solar system. It revolves around the sun about once every 365 days, and it rotates once approximately every 24 hours around an axis that stretches from one pole to the other. The **equator** is the imaginary circle around the middle that separates the Earth into Northern and Southern hemispheres. The Earth is sometimes called the *water planet* because its surface is about 71 percent water and only 29 percent land. The shape of the Earth is close to a ball, but it is not perfectly spherical; instead, it is a *bumpy oblate spheroid.* We say "bumpy" because of the topographic features like mountains and canyons. We say "oblate" because the physics of spinning objects causes the Earth to bulge slightly around its equator. That is, while the Earth's circumference is about 25,000 miles around, and its diameter is about 8,000 miles, the Earth is approximately 27 miles wider at the equator than it is from pole to pole. So the Earth is not a perfect sphere but a flattened "spheroid." That said, the Earth is very nearly a perfectly smooth ball; if it were shrunk to the size of a billiard ball, it would be as round and about as smooth!

In order to represent and communicate information about the Earth, we construct models of it. Because the Earth is three-dimensional and nearly spherical, it makes sense to construct a three-dimensional spherical model to understand it. This is called a **globe.** In order to identify locations on the Earth surface in a precise and standardized way, a grid of lines is laid over the globe, called the **graticule.** The graticule identifies two dimensions of Earth-surface location with lines running horizontally, in the direction of the equator, and lines running vertically, from pole to pole (**Figure 1.22**). The horizontal lines are called **parallels,** and they encode **latitude,** which refers to locations north or south of

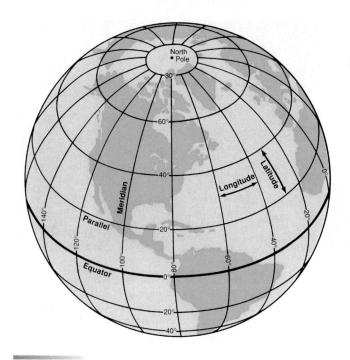

Figure 1.22 The grid system of parallels of latitude and meridians of longitude. Because the meridians converge at the poles, parallels become increasingly shorter away from the equator. On the globe, the 60th parallel (60°) is only one-half as long as the equator, and a degree of longitude along it measures only about 55 1/2 kilometers (about 34 1/2 miles) compared with about 111 kilometers (about 69 miles) at the equator (0°).

the equator. Because the Earth is a spherical shape, the locations of latitude lines are expressed in angular terms as degrees north or south (°N or °S). Intermediate latitude lines are expressed in terms of angular minutes and seconds, or tenths and hundredths of degrees. The equator is 0° (neither north nor south), the North Pole is 90°N, and the South Pole is 90°S. The vertical lines are called **meridians** and they encode **longitude.** Meridians express location east or west, but there is no natural vertical line equivalent to the equator to use as an origin line. Instead, an international treaty from more than a century ago established an origin line passing through the Greenwich Observatory outside London; this longitude line is called the **prime meridian.** Like latitude, the locations of longitude lines are expressed in angular terms as degrees, either east or west of the prime meridian (°E or °W). The prime meridian is 0° (neither east nor west), the city of Beijing in China is about 116°E, and the city of Vancouver in Canada is about 123°W. On the opposite side of the Earth from the prime meridian, running vertically through the Pacific Ocean, is the 180° meridian line (also neither east nor west). Incidentally, many people confuse the 180° meridian line with the International Date Line (where the date of the day of the year changes forward or back, depending on your direction of travel), but the two are different and coincide only in some places.

Maps are geographers' primary tools of spatial analysis. All spatial analysis starts with locations, and all locations are related to the global grid of latitude and longitude. Because these lines of reference are drawn on the spherical Earth, their projection onto a map distorts their grid relationships. The extent of variance between the globe grid and a map grid helps tell us the kind and degree of distortion that the map will contain.

---

[2] A more detailed discussion of map projections, including examples of their different types and purposes, may be found in Appendix A.

Only the globe grid itself retains all of these characteristics. To project it onto a surface that can be laid flat is to distort some or all of these properties and consequently to distort the reality the map attempts to portray.

How Maps Show Data

The properties of the globe grid and of various projections are the concern of the cartographer; **cartography** is the art and science of maps and map-making. Geographers are more interested in the depiction of spatial data and in the analysis of the patterns and interrelationships those data present. Out of the myriad items comprising the content of an area, the geographer must, first, select those that are of concern to the problem at hand and, second, decide on how best to display them for study or demonstration. In that effort, geographers can choose among different types of maps and different systems of symbolization. These symbols use properties like shape, color, and size to represent geographic meaning, in somewhat the same way that words represent meaning in language.

Maps can first be classified as either reference maps or thematic maps. **Reference maps** are *general-purpose maps*. Their purpose is simply to show without analysis or interpretation a variety of natural or human-made features of an area or of the world as a whole, including showing the locations of features accurately. Reference maps answer the question, "What is there?" Familiar examples are highway maps, city street maps, topographic maps (**Figure 1.23**), atlas maps, and the like. Until about the middle of the 18th century, the general-purpose or reference map was the only type of map, for the function of the mapmaker (and the explorer who supplied the new data) was to "fill in" the world's unknown areas with reliable locational information. With the passage of time, scholars saw the possibility of using the accumulation of locational information to display and study the spatial patterns of social and physical data. The maps they made of climate, vegetation, soil, population, and other distributions introduced the thematic map, the second major class of maps.

Thematic maps are *specific-purpose maps*—they present a specific spatial distribution or a single category of data—that is, a graphic theme. Thematic maps could be called *statistical* or *graph maps;* they answer the question, "What is the spatial pattern of this variable?" The way the information is shown on such a map may vary according to the type of information to be conveyed, the level of generalization that is desired, and the symbolization selected. Thematic maps may be either *qualitative* or *quantitative.* The principal purpose of the qualitative map is to show the distribution of a particular class of information. The world location of producing oil fields, the distribution of U.S. national parks, or the pattern of areas of agricultural specialization within a state or country are examples. The interest is in where things are in an approximate way, and nothing is reported about—in the examples cited—barrels of oil extracted or in reserve, number of park visitors, or value or volume of crops or livestock produced.

In contrast, quantitative thematic maps show the spatial characteristic of numerical data. Usually, a single variable such

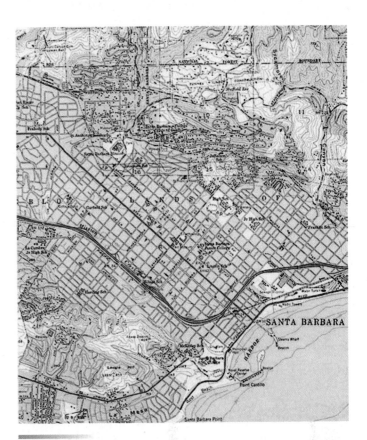

Figure 1.23 A portion of the Santa Barbara, California, topographic quadrangle of the U.S. Geological Survey 1:24,000 series. Topographic maps portray landscape features of relatively small areas. Elevations and shapes of landforms, streams and other water bodies, vegetation, and coastal features are recorded, often with great accuracy. Because cultural items that people have added to the physical landscape, such as roads, railroads, buildings, political boundaries, and the like, are also frequently depicted on them, topographic maps are classed as general-purpose or reference maps by the International Cartographic Association. Incidentally, the scale of the original map no longer applies to this photographic reduction.

Source: U.S. Geological Survey.

as population, average rainfall, median income, annual wheat production, or average land value is chosen, and the map displays differences from place to place in that variable. Important types of quantitative thematic maps include graduated circle, dot, isoline, and choropleth maps (**Figure 1.24**).

Graduated circle maps use circles of different size to show the magnitude of a variable of interest in different places; the larger the circle, the greater the magnitude of the variable. They are examples of the more general class of thematic symbols called **proportional area symbols** that include shapes other than circles. On **dot maps,** a single or specified number of occurrences of the item studied is recorded by a single dot. An **isoline map** features lines that connect points registering equal values of the item mapped (*iso* means "equal"). For example, *isotherms* shown on daily weather maps connect points recording the same temperature at the same moment of time or the same average temperature during the day. Identical elevations above sea level may be shown by a form of isoline called a *contour line.*

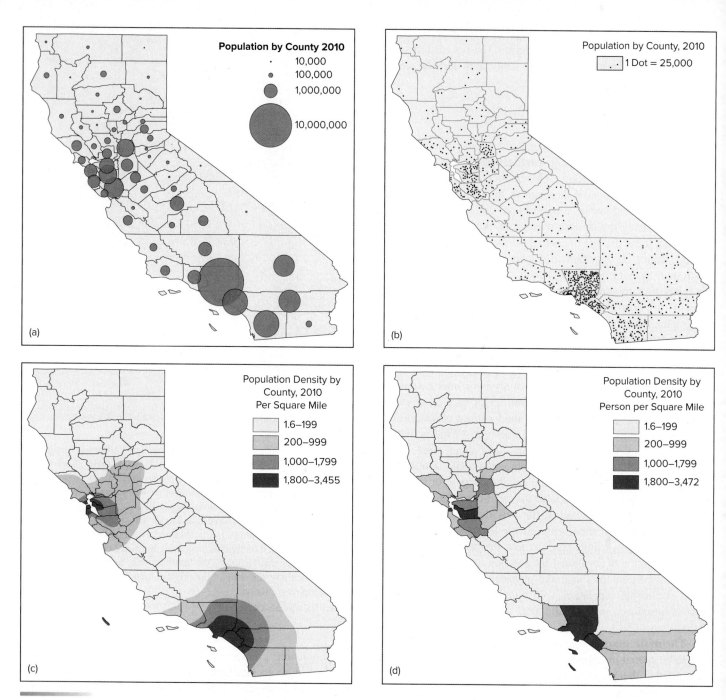

Figure 1.24 Four thematic map types display the same data. Although population is the theme of each, these different California maps present their information in strikingly different ways. (*a*) In the graduated circle map, the area of the circle is approximately proportional to the absolute number of people within each county. (*b*) In a dot-distribution map where large numbers of items are involved, the value of each dot is identical and stated in the map legend. The placement of dots on this map does not indicate precise locations of people within the county, but simply their total number. (*c*) Population density is recorded by the isoline map, while the choropleth map (*d*) may show absolute values, as here, or, more usually, ratio values such as population per square kilometer.

A **choropleth map** presents average value of the data studied per preexisting areal unit—dwelling unit rents or assessed values by city block, for example, or (in the United States) population densities by individual townships within counties. Each unit area on the map is then shaded or colored to suggest the magnitude of the event or item found within its borders. Where the choropleth map is based on the absolute number of items within the unit area, as it is in Figure 1.24*d*, rather than on areal averaging (total numbers, for example, instead of numbers per square kilometer), a misleading

statement about density may be conveyed. That is, if the same magnitude is shown with the same shading or color, large areal units will dominate the visual field, even though they actually represent a much less dense concentration of the feature in question.

A **statistical map** records the actual numbers or occurrences of the mapped item per established unit area or location. The actual count of each state's colleges and universities shown on an outline map of the United States or the number of traffic accidents at each street intersection within a city are examples of

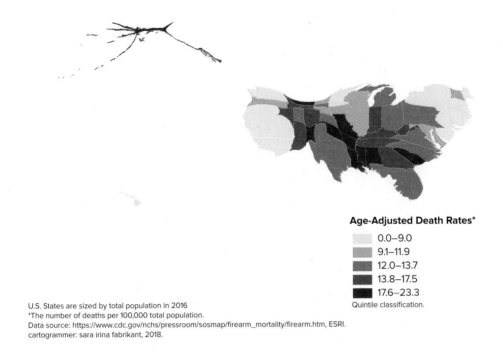

Firearm Mortality by U.S. State in 2016

Age-Adjusted Death Rates*

0.0–9.0
9.1–11.9
12.0–13.7
13.8–17.5
17.6–23.3

Quintile classification.

U.S. States are sized by total population in 2016
*The number of deaths per 100,000 total population.
Data source: https://www.cdc.gov/nchs/pressroom/sosmap/firearm_mortality/firearm.htm, ESRI.
cartogrammer: sara irina fabrikant, 2018.

Figure 1.25 A cartogram in which each state is sized according to its number of residents in the year 2016. The cartogram also shows firearm mortality for 2016.

statistical maps. A **cartogram** uses such statistical data to transform space so that the largest areal unit on the map is the one showing the greatest statistical value (**Figure 1.25**).

Maps communicate information but, as in all forms of communication, the message conveyed by a map reflects the intent and, perhaps, the biases of its author. Maps are persuasive because of the implied precision of their lines, scales, color and symbol placement, and information content. But maps, as communication devices, can subtly or blatantly manipulate the message they impart or contain intentionally false information (**Figure 1.26**). In fact, maps cannot help but be selective in the information they present, and they simplify this information by using cartographic symbols. Maps always show features at a different scale than they exist in reality, and the necessity to use projections to show the Earth's surface as flat always introduces some spatial distortion, as we discussed above. Maps, then, can distort and lie as readily as they can convey verifiable spatial data or scientifically valid analyses. The more that map users are aware of those possibilities, and the more understanding of map projections, symbolization, and common forms of thematic and reference mapping standards they possess, the more likely they are to reasonably question and clearly understand the messages maps communicate.

1.4 Contemporary Geospatial Technologies

The growth and advancement of three interrelated geospatial technologies—global positioning systems, remote sensing, and geographic information systems—has revolutionized geography and increased the geographer's ability to collect, analyze, and visually represent some forms of geographic data. Global positioning systems (GPSs) rely upon a system of 24 orbiting satellites, Earth-bound tracking stations that control the satellites, and portable receivers that determine exact geographic locations based on the time delay in signals received from three or more satellites (technically, they determine location from an inference based on the time required for several signals to travel from the satellite to the Earth and back). Remote sensing allows the collection of vast amounts of geographic data, while geographic information systems (GISs) can integrate GPS, remote sensing, and other forms of spatial data. Google Earth and interactive mapping and navigation web sites such as MapQuest are everyday uses of contemporary geographic research technologies.

Remote Sensing

Remote sensing is a relatively new term, but the process that it describes—detecting the nature of an object and the content of an area from a distance—is more than 150 years old. Soon after the development of the camera, photographs were being taken from balloons and kites. In the early 20th century, fixed-wing aircraft provided a platform for the camera and photographer, and by the 1930s, aerial photography from planned positions and routes permitted reliable data gathering for large and small area mapping purposes. Even today, high- and low-altitude aerial photography with returned film remains a widely used remote sensing technique. Standard photographic film detects reflected energy within the visible portion of the electromagnetic spectrum. It

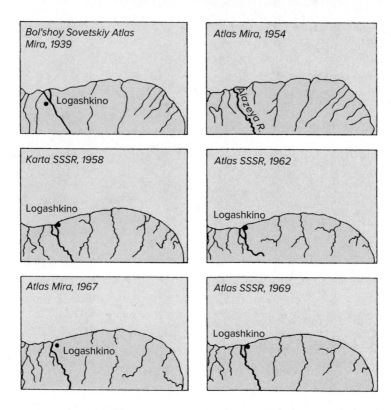

Figure 1.26 The wandering town of Logashkino, as traced in various Soviet atlases by Mark Monmonier. Deliberate, extensive cartographic "disinformation" and locational falsification, he reports, became a Cold War tactic of the Soviet Union. We usually use—and trust—maps to tell us exactly where things are located. On the maps shown, however, Logashkino migrates from west of the river away from the coast to east of the river on the coast, while the river itself gains and loses a distributary and, in 1954, the town itself disappears. The changing misinformation, Monmonier suggests, was intended to obscure from potential enemies the precise location of possible military targets.

Source: Mark Monmonier, How to Lie with Maps, 2nd Ed. University of Chicago Press, 1996.

can be supplemented by special sensitized infrared film that has proved particularly useful for the recording of vegetation and hydrographic features and by nonphotographic imaging techniques, including thermal scanning (widely used for studying various aspects of water features such as ocean currents and water pollution and, because it can be employed during nighttime hours, for military surveillance and energy budget observations) and radar mapping (also operative night and day and useful for penetrating clouds and haze).

For more than 30 years, spacecraft (with or without crews) have supplemented the airplane as the vehicle for imaging Earth features. Among the advantages of satellites are the speed of coverage and the fact that views of large regions can be obtained. In addition, they are equipped to record and report back to Earth digitized information from multiple parts of the electromagnetic spectrum, including some that are outside the range of human eyesight. Satellites enable us to map the invisible, including atmospheric and weather conditions, in addition to providing images with applications in agriculture and forest inventory, land-use classification, identification of geologic structures and mineral deposits, and more. The different sensors of the American Landsat satellites, first launched in 1972 (Landsat 8 was put aloft in 2013), are capable of resolving objects between 15 and 60 meters (50 and 200 ft) in size. Even sharper images are yielded by the French SPOT satellite (SPOT 7 was launched in

2014); its sensors can show objects that are as small as 2 meters (6 1/2 ft). Satellite imagery is relayed by electronic signals to receiving stations, where computers convert them into photolike images for use in scientific research and in mapping programs.

Geographic Information Systems (GISs)

Increasingly, digital computers, mapping software, and computer-based display units and printers are employed in the design and production of maps and in the development of databases used in map production. In computer-assisted cartography, the content of standard maps—reference and thematic—is digitized and stored in computers. The use of computers and printers in map production permits increases in the speed, flexibility, and accuracy of many steps in the mapmaking process but in no way reduces the obligation of the mapmaker to employ sound judgment in the design of the map or the communication of its content.

Geographic information systems (GISs) extend the use of digitized data and computer manipulation to investigate and display spatial information of all types. A GIS is both an integrated software package for handling, processing, and analyzing geographical data and a computer database in which every item of information is tied to a precise geographic location. In the section above that introduced geographic features, we discussed the fact that some features are better conceived of as objects and others as fields. As we mentioned there, this has implications for how we represent geographic information in the GIS. In the **vector approach,** reminiscent of object conceptualization, the precise location of each object—point, line, or area—in a distribution is described. In the **raster approach,** reminiscent of the field conceptualization, the study area is divided into a set of small (usually) square cells, with the content of each cell described or quantified (the rasters are analogous to pixels on a computer screen). The vector approach is more often suitable for human or cultural data, such as roads or cities, whereas the raster approach is more often suitable for natural geographic data, like elevations or rainfall. In either approach, a vast amount of different spatial information can be stored, accessed, compared, processed, analyzed, and displayed.

A GIS database, then, can be envisioned as a set of discrete informational overlays linked by reference to a basic locational grid of latitude and longitude (**Figure 1.27**) or some other coordinate system. The system then permits the separate display of the spatial information contained in the database. It allows the user to overlay maps of different themes, analyze the relations revealed, and compute spatial relationships. It shows aspects of spatial associations otherwise difficult to display on conventional maps, such as flows, interactions, and three-dimensional characteristics. In short, a GIS database, as a structured set of spatial information, has become a powerful tool for performing geographical analysis and synthesis.

A GIS data set may contain a great amount of place-specific information collected and published by the U.S. Census Bureau,

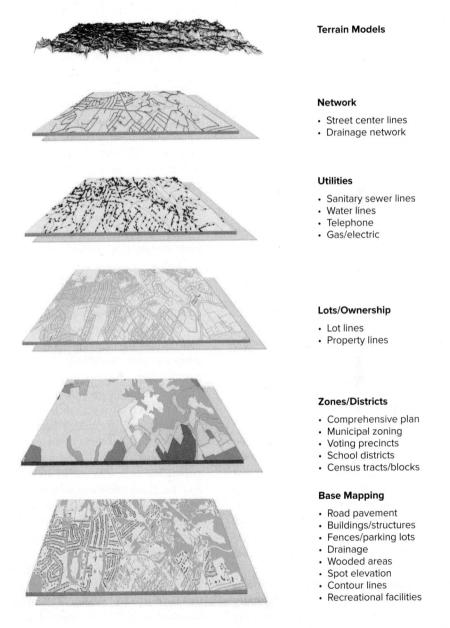

Terrain Models

Network
- Street center lines
- Drainage network

Utilities
- Sanitary sewer lines
- Water lines
- Telephone
- Gas/electric

Lots/Ownership
- Lot lines
- Property lines

Zones/Districts
- Comprehensive plan
- Municipal zoning
- Voting precincts
- School districts
- Census tracts/blocks

Base Mapping
- Road pavement
- Buildings/structures
- Fences/parking lots
- Drainage
- Wooded areas
- Spot elevation
- Contour lines
- Recreational facilities

Figure 1.27 A model of a GIS. A GIS incorporates three primary components: data storage capability, computer graphics programs, and statistical packages. In this example, the different layers of information are to be used in different combinations for city planning purposes. Different data sets, all selected for applicability to the questions asked, may be developed and used in urban geography, economic geography, transportation planning, industrial location work, and similar applications.

Source: Shaoli Huang.

including population distribution, race, ethnicity, income, housing, employment, industry, farming, and so on. It may also hold environmental information downloaded from satellite imagery or taken from Geological Survey maps and other governmental and private sources. In human geography, the vast and growing array of spatial data has encouraged the use of GIS to explore models of regional economic and social structure; to examine transportation systems and urban growth patterns; and to study patterns of voting behavior, disease incidence, accessibility of public services, and a vast array of other topics. GISs are essential tools in the process of voter redistricting. For physical geographers, the analytic and modeling capabilities of GIS are fundamental to the understanding of processes and interrelations in the natural environment.

Because of the growing importance of GIS in all manner of public and private spatial inquiries, demand in the job market is high and growing for those skilled in its techniques. Most university courses in GIS are taught in Geography departments, and "GIS/remote sensing" is a primary occupational specialty for which many geography undergraduate and graduate majors seek preparation.

Mental Maps

Maps that shape our understanding of distributions and locations or influence our perception of the world around us are not always drawn on paper. We carry with us *mental maps* that in some ways are more accurate in reflecting our view of spatial reality than the formal maps created by geographers or cartographers. **Mental maps (cognitive maps)** are internal models or representations of an area or an environment developed by an individual on the basis of information or impressions received, interpreted, and stored. Sometimes, they leave out what are believed to be unnecessary details, and only important elements are incorporated. We use this information—this mental map—in organizing our daily activities: selecting our destinations and the sequence in which they will be visited, deciding on our routes of travel, recognizing where we are in relation to where we wish to be. A mental route map may also include reference points to be encountered on the chosen path of connection or on alternate lines of travel.

Such mental maps are every bit as real to their creators (and we all have them) as are the street or highway maps commercially available, and they are a great deal more immediate in their impact on our spatial decisions. We may choose routes or avoid neighborhoods not on objective grounds but on emotional or perceptual ones. In those choices, characteristics of people such as age or gender play an important role. For instance, the mental maps of women may contain danger zones where fear of sexual assault, harassment, or threatening persons?

To parallel the other two items is a determinant in routes chosen or times of journey. Whole sections of a community may be voids on our mental maps, as unknown as the interiors of Africa and South America were to Western Europeans two centuries ago. Our areas of awareness generally increase with the increasing mobility that comes with age, affluence, and education and may be enlarged or restricted for different social groups within the city (**Figure 1.28**).

1.5 Systems, Maps, and Models

The contents of areas are interrelated and constitute a **spatial system** that, in common with all systems, functions as a unit because its component parts are interdependent. Only rarely do individual elements of area operate in isolation, and to treat them as if they do is to lose touch with spatial reality. The systems of geographic concern are those in which the functionally important variables are spatial: location, distance, direction, density, connectivity, and the other basic concepts that we have reviewed. The systems that they define are not the same as regions, though spatial systems may be the basis for regional identification.

Systems have components, and the analysis of the role of components helps reveal the operation of the system as a whole. To conduct that analysis, individual system elements must be isolated for separate identification and, perhaps, manipulated to see their function within the structure of the system or subsystem. Maps and models are devices that geographers use to achieve that isolation and separate study.

Maps, as we have seen, are effective to the degree that they can segregate at an appropriate level of generalization those system elements selected for examination. By compressing, simplifying, and abstracting reality, maps record in manageable dimension the real-world conditions of interest. A model is a simplified abstraction of reality, designed to clarify relationships among its elements. Maps are a type of **model,** representing reality in an idealized form to make certain aspects more clear.

The complexities of spatial systems analysis—and the opportunities for quantitative analysis of systems made possible by computers and sophisticated statistical techniques—have led geographers to use other kinds of models in their work. An important example is the computational model that represents reality as a set of mathematical or computer programming statements. Model building is the technique scientists use to simplify complex situations, to eliminate (as does the map) unimportant details, and to isolate for special study and analysis the role of one or more interacting elements in a total system. Models also allow geographers to conduct experiments on a simulation of a portion of reality instead of the reality itself, which is often very difficult, unethical, or impossible to do.

An interaction model discussed in Chapter 3, for instance, suggests that the amount of exchange expected between two places depends on the distance separating them and on their population size. The model indicates that the larger the places and the closer their distance, the greater is the amount of interaction. Such a model helps us to isolate the important components of the spatial system, to manipulate them separately, and to reach conclusions concerning their relative importance. When a model satisfactorily predicts the volume of intercity interaction in the majority of cases, the lack of agreement in a particular case leads to an examination of the circumstances contributing to the disparity. The quality of connecting roads, political barriers, or other variables may affect the specific places examined, and these causative elements may be isolated for further study.

Indeed, the steady pursuit of more refined and definitive analysis of human geographic questions—the "further study" that continues to add to our understanding of how people occupy and utilize the Earth, interact with one another, and organize and alter Earth

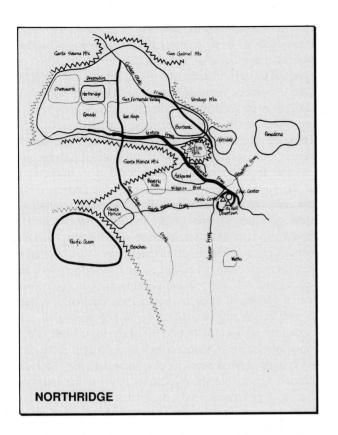

NORTHRIDGE

BOYLE HEIGHTS

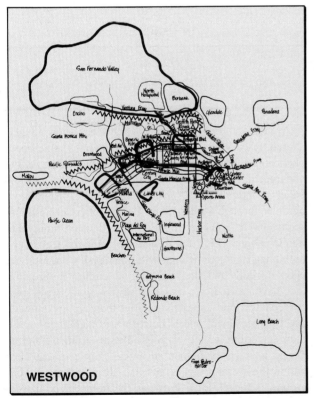

WESTWOOD

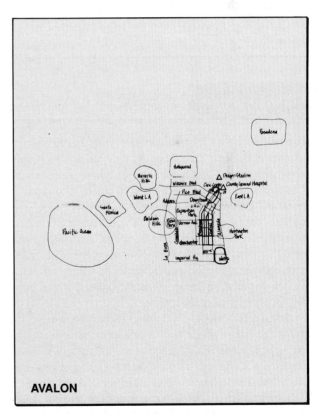

AVALON

Figure 1.28 Four mental maps of Los Angeles. The upper-middle-income residents of the Northridge and Westwood communities have expansive views of the metropolis, reflecting their mobility and area of travel. Inner-city residents of the Avalon community display a more restricted view of the city because of their lower incomes, social isolation, and reliance on public transportation. Even more limited is the mental map of Hispanic residents of the Boyle Heights community, which reflects their spatial and linguistic isolation within the metropolitan area.

Source: From Department of City Planning, City of Los Angeles, The Visual Environment of Los Angeles, *1971.*

space—has led to the remarkably diversified, fascinating, and interesting yet coherent field of modern human geography. With the content of this introductory chapter as background to the nature, traditions, and tools of geography, we are ready to begin its exploration.

1.6 The Structure of This Book

By way of getting started, it is useful for you to know how the organization and topics of this text have been structured to help you reach the kinds of understandings we seek.

We begin in Chapters 2–4 with introductory material on cultural processes and spatial interactions among an unevenly distributed and expanding world population. Chapter 2 introduces the components and structure of culture, culture change, diffusion, and divergence. Chapter 3 presents characteristics of spatial interaction and spatial behavior that are common to all peoples and cultures. Chapter 4 examines population geography and the factors driving patterns of population movement, growth, and distribution.

While the book's first four chapters focus on geographic themes common to all peoples and cultures, Chapters 5–7 turn to the features that distinguish societies and culture realms and create patterns of unity and diversity in the cultural landscape. Although there are innumerable ways in which human populations differ, we focus on spatial patterns of three major points of contrast: language and religion (Chapter 5), and ethnicity (Chapter 6).

Further, we examine the diversity of cultural identities and the expressions of those cultures in the landscape (Chapter 7). We pay particular attention to folk cultures—the material and nonmaterial aspects of daily life among groups insulated from outside influences through spatial isolation or cultural barriers. Patterns of cultural diversity are in constant tension with unifying forces as the world experiences greater spatial interaction. Thus, we also examine ways in which folk cultures are undergoing erosion under the influence of globalized popular cultures.

Our focus shifts in Chapters 8–10 to the dynamic patterns of the global economy, examining spatial patterns of food and raw material production (Chapter 8), manufacturing and services (Chapter 9), and finally measures, spatial patterns, and models of economic development (Chapter 10). Economic development is generally accompanied by more formal systems of organizing society, resources, and territory. Thus, in Chapters 11 and 12, we examine systems of functional organization within systems of cities and inside individual cities (Chapter 11), as well as systems of political control of geographic space that range from the local to the international scale (Chapter 12). Human impact on the environment is an integral part of each chapter and is the topic of the concluding chapter of the book (Chapter 13). In concluding with human impacts, we return to the underlying concern of all geographic study: the relationship between human geographic patterns and processes and both the present conditions and future prospects of the physical and cultural landscapes we inhabit.

SUMMARY

Geography is the study of the Earth's surface and its physical and cultural content. Throughout its long history, geography has remained consistent in its focus on human-environmental interactions, the interrelatedness of places, and the likenesses and differences in physical and cultural content of area that exist from place to place. The collective interests of geographers are summarized by the spatial and systems analytical questions they ask. The responses to those questions are interpreted through basic concepts of space and place, location, distance, direction, content evolution, spatial interaction, and regional organization.

Geographers employ maps and models to abstract the complex reality of space and to isolate its components for separate study. Maps are imperfect renderings of the three-dimensional Earth and its parts on a two-dimensional surface. In that rendering, some or all of the characteristics of the globe grid are distorted, but convenience and data manageability are gained. Spatial information may be depicted visually in a number of ways, each designed to simplify and to clarify the infinite complexity of spatial content. Geographers also use verbal and mathematical models for the same purpose, to abstract and analyze.

In their study of the Earth's surface as the occupied and altered space within which humans operate, some geographers concentrate on the integration of physical and cultural phenomena in a specific Earth area (regional geography). Other geographers may, instead, emphasize systematic geography through study of the Earth's physical systems of spatial and human concern or, as here, devote primary attention to people. This is a text in *human geography*. Its focus is on human interactions both with the physical environments people occupy and alter and with the cultural environments they have created. We are concerned with the ways people perceive the landscapes and regions they occupy, act within and between them, make choices about them, and organize them according to the varying cultural, political, and economic interests of human societies. This is a text clearly within the social sciences, but like all geography, its background is the physical Earth as the home of humans. As human geographers, our concern is with how that physical Earth has nurtured particular cultural choices and patterns, that physical Earth has been altered by societies and cultures. In the next chapter, we begin with an inquiry about the roots and nature of culture.

KEY WORDS

absolute direction
absolute distance
absolute location
accessibility
administrative region
cartogram
cartography
choropleth map
concentration
connectivity
cultural landscape
density
dispersion
develop
dot map
equator
field
functional region
geographic feature
geographic information system (GIS)
globalization
globe
graduated circle map

graphic scale
graticule
human geography
isoline map
latitude
longitude
mental (cognitive) map
meridian
model
natural landscape
object
parallel
pattern
perceptual region
physical geography
place
place stereotype
placelessness
prime meridian
projection
proportional area symbol
raster approach
reference map

region
regional concept
regional geography
relative direction
relative distance
relative location
remote sensing
representative fraction
scale
site
situation
space
spatial association
spatial diffusion
spatial distribution
spatial interaction
spatial system
statistical map
systematic geography
thematic map
thematic region
vector approach

FOR REVIEW

1. In what two meanings and for what different purposes do we refer to *location?*

2. Describe the *site* and the *situation* of the town where you live, work, or go to school.

3. What kinds of distance transformations are suggested by the term *relative distance?* How is the concept of *psychological distance* related to relative distance?

4. What are the common elements of *spatial distribution?* What different aspects of the spatial arrangement of things do they address?

5. What are the common characteristics of *regions?* How are *administrative, thematic,* and *functional* regions different in concept and definition? What is a *perceptual region?*

6. List at least four properties of the globe grid (graticule). Why are globe grid properties distorted on maps?

7. What does *prime meridian* mean? What happens to the length of a degree of longitude as one approaches the poles?

8. What different ways of displaying statistical data on maps can you name and describe?

KEY CONCEPTS REVIEW

1. **What is the nature of geography and the role of human geography?** Section 1.1
Geography is a spatial science concerned with how the content of Earth areas differs from place to place. It is the study of spatial variation in the world's physical and cultural (human) features. The emphasis of human geography is on the spatial variations in characteristics of peoples and cultures, on the way humans interact over space, and the ways they utilize and alter the natural landscapes they occupy, even as those landscapes influence culture.

2. **What are the fundamental geographic observations and their underlying concepts?** Section 1.2
Basic geographic observations all concern the characteristics, content, and interactions of places. Their underlying concepts involve such place specifics as location, direction, distance, size, scale, physical and cultural attributes, interrelationships, and regional similarities and differences.

3. **What are the regional concept and the generalized characteristics of regions?** Section 1.2

The regional concept tells us that physical and cultural features of the Earth's surface are rationally arranged by understandable processes. All recognized regions are characterized by location, spatial extent, defined boundaries, and position within a hierarchy of regions. Regions may be *administrative, thematic, functional, or perceptual* in nature.

4. **Why do geographers use maps, and how do maps show spatial information?** Section 1.3

Maps are tools that geographers use to perform spatial analyses such as identifying regions and analyzing their content. They permit the study of areas and areal features too extensive to be completely viewed or understood on the Earth's surface itself. Thematic (statistical) maps may be either qualitative or quantitative. Their data may be shown in graduated circle, dot distribution, isoline, choropleth, or cartogram form.

5. **In what ways in addition to maps may spatial data be visualized or analyzed?** Section 1.4

Informally, we all create *mental maps* reflecting highly personalized but culturally overlapping impressions and information about the spatial arrangement of things (for example, buildings, streets, landscape features). More formally, geographers recognize the content of areas as forming a spatial system to which techniques of spatial systems analysis and model building are applicable.

CHAPTER 2

ROOTS AND MEANING OF CULTURE:

Introduction

South African San hunter-gatherers are modern-day followers of the world's oldest, most enduring livelihood system.

©*franco lucato/Shutterstock*

Key Concepts

2.1–2.2 Culture components and the nature of human-environmental relations

2.3–2.5 Culture origins and culture hearths

2.6–2.8 The structure of culture and forms of culture change

*T*hey buried him there in the cave where they were working, less than 6 kilometers (4 miles) from the edge of the ice sheet. Outside stretched the tundra, summer feeding grounds for the mammoths whose ivory they had come so far to collect. Inside, near where they dug his grave, were stacked the tusks that they had gathered and were cutting and shaping. They prepared the body carefully and dusted it with red ocher, then buried it in an elaborate grave with tundra flowers and offerings of food, a bracelet on its arm, a pendant about its throat, and 40 to 50 polished rods of ivory by its side. It rested there, in modern Wales, undisturbed for some 18,000 years until discovered early in the 19th century. The 25-year-old hunter had died far from the group's home, some 650 kilometers (400 miles) away, near present-day Paris, France. He had been part of a routine annual summer expedition overland from the forested south across the as-yet-unflooded English Channel to the mammoths' grazing grounds at the edge of the glacier.

As always, they were well prepared for the trip. Their boots were carefully made. Their sewn-skin leggings and tunics served well for travel and work; heavier fur parkas warded off the evening chill. They carried emergency food, fire-making equipment, and braided cord that they could fashion into nets, fishing lines, ropes, or thread. They traveled by reference to sun and stars, recognizing landmarks from past journeys and occasionally consulting a crude map etched on bone.

Although the hunters returned bearing the sad news of their companion's death, they also brought the ivory to be carved and traded among the scattered peoples of Europe, from the Atlantic Ocean to the Ural Mountains.

As shown by their tools and equipment, their behaviors and beliefs, these Stone Age travelers displayed highly developed and distinctive characteristics, primitive only from the vantage point of our own different technologies and customs. They represented the culmination of a long history of development of skills, of invention of tools, and of creation of lifestyles that set them apart from peoples elsewhere in Europe, Asia, and Africa, who possessed still different cultural heritages.

To writers in newspapers and the popular press, the word *culture* means the arts (literature, painting, music, and the like). To a social scientist, however, **culture** is the specialized behavioral patterns, understandings, adaptations, and social systems that summarize a group of people's learned way of life. In this broader sense, culture is an ever-present part of the regional differences that are the essence of human geography. The visible and invisible evidences of culture—buildings and farming patterns, language, political organization, and ways of earning a living, for example—are all parts of the spatial diversity that human geographers study. Cultural differences over time may present contrasts as great as those between the Stone Age ivory hunters and modern urban Americans. Cultural differences in space result in human landscapes with variations as subtle as the differing "feel" of urban Paris, Moscow, or New York, or as obvious as the sharp contrasts of rural and the U.S. Midwest (**Figure 2.1**).

Because such tangible and intangible cultural differences exist and have existed in various forms for thousands of years, human geography attempts first to describe the pattern

(a)

(b)

Figure 2.1 Culture is reflected in agricultural practices and in the look of the landscape. Compare (*a*) farmers using oxen to plow fields for planting in Bhutan and (*b*) the extensive fields and mechanized farming of the U.S. Midwest.

(a) ©Lissa Harrison; (b) Source: Tim McCabe, USDA Natural Resource Conservation Service/U.S. Department of Agriculture (USDA)

of cultural practices (traits) across the Earth's surface. Then it addresses the question of why? Why, because humankind constitutes a single species, are cultures so varied? What and where were the origins of the different culture traits we now observe? How, from whatever limited areas individual culture traits developed, were they diffused over a wider portion of the globe? How did people who had roughly similar origins come to display significant areal differences in technology, social structure, ideology, and the innumerable other expressions of human cultural diversity? In what ways and why are there distinctive cultural variations even in presumed "melting pot" societies, such as the United States and Canada, or in the historically homogeneous, long-established countries of Europe? How is knowledge about cultural differences important to us today? Some of the answers to these questions are to be found in the way that separate human groups developed techniques to solve regionally varied problems such as securing food, clothing, and shelter, and in the process, created distinctive customs and ways of life.

2.1 Components of Culture

Culture is transmitted within a society to succeeding generations by imitation, instruction, suggestion, and example. In short, while the capacity for culture is biological, culture itself is learned. As members of a social group, individuals acquire integrated sets of behavioral patterns, environmental and social perceptions, and knowledge of existing technologies. Of necessity, each of us learns the culture in which we are born and reared. But we need not—indeed, cannot—learn its totality. Age, sex, status, or occupation may dictate aspects of the cultural whole to which an individual is more or less likely to be exposed, and which it is more or less appropriate for an individual to internalize.

A culture, that is, despite overall generalized and identifying characteristics and even an outward appearance of uniformity, displays a social structure—a framework of roles and interrelationships of individuals and established groups. Each individual learns and is expected to adhere to the rules and conventions not only of the culture as a whole, but also of those specific to the subculture to which he or she belongs. And that subgroup may have its own recognized social structure (**Figure 2.2**). Think back to the different subgroups and aspects of your own national culture that you became part of (and left) as you progressed from childhood through high school and on to college-age adulthood and, perhaps, to first employment.

Many different cultures, then, can coexist within a given area, each with its own influence on the thoughts and behaviors of their separate members. Subcultures are groups that can be distinguished from the wider society by their cultural patterns. Within the United States, for example, we can readily recognize a variety of subcultures within the larger "American" culture: white, black, Hispanic, Asian American, or other ethnic groups; gay and straight; urban and rural; and many others (see the feature "The Burning Man Festival of Art and Music: Subcultural Landscape in the Great Basin Desert"). Human geography increasingly recognizes the pluralism of cultures within regions. In addition to examining the separate content and influence of those subcultures, it attempts to record and analyze the varieties of contested cultural interactions among them, including those of a political and economic nature.

Culture is a complexly interlocked web of behaviors, attitudes, and material artifacts. Realistically, its full and diverse content cannot be appreciated, and in fact may be wholly misunderstood, if we concentrate our attention only on limited, obvious traits. Distinctive eating utensils, the use of gestures, or the ritual of religious ceremony may summarize and characterize a culture for the casual observer. These are, however, individually insignificant parts of a much more complex structure that can be appreciated only when the whole is experienced.

(a)

(b)

Figure 2.2 Both the traditional rice farmer of rural Japan and the Tokyo commuter are part of a common Japanese culture. They occupy, however, vastly different positions in its social structure.
(a) ©KIRAYONAK YULIYA/Shutterstock; (b) ©Jane Rix/123RF

Out of the richness and intricacy of human life, we seek to isolate for special study those more fundamental cultural variables that give structure and spatial order to societies. We begin with *culture traits,* the smallest distinctive units of culture. **Culture traits** range from the language spoken to the tools used or the games played. A trait may be an object (a fishhook, for example), a technique (weaving and knotting of a fishnet), or a belief (in the spirits resident in water bodies). Traits are the most elementary expression of culture, the building blocks of the complex behavioral patterns of distinctive groups of peoples. Of course, the same trait—the Christian religion, perhaps, or the Spanish language—may be part of more than one culture. Similarly, traits are sometimes clearly distinct, as in two completely different rules about the appropriateness of marrying first cousins, or somewhat overlapping, as in preferences for one or the other of two styles of rap music that share many common aspects.

The Burning Man Festival of Art and Music: Subcultural Landscape in the Great Basin Desert

Is it an art festival? A music festival? Just a gathering of like-minded people, or a large yet ephemeral city that springs up anew every late summer in the Black Rock desert playa of northwest Nevada? Maybe it's the world's largest rave. In terms of movie references, it may well remind you of a mashup of *Fear and Loathing in Las Vegas; The Adventures of Priscilla, Queen of the Desert; Mad Max;* and almost any film by Federico Fellini or Terry Gilliam. A little surrealism, creativity, and positive vibes set to an electronica soundtrack, with some hedonism and fire thrown in for good measure.

This is the Burning Man Festival of Art and Music ("Burning Man" for short). As we have pointed out, people are not only members of a main national culture but of various subcultures. These subcultures are based on a dizzying potential array of characteristics, interests, and facets of identity. They are distinguishable from the wider culture by their particular membership and their distinctive cultural patterns of behaviors, attitudes, and material artifacts. Burning Man is a fascinating example of such a subculture.

Burning Man started in 1986, when the first "Man" effigy was created by Larry Harvey and Jerry James and set on fire as an artistic statement on a San Francisco beach. This spontaneous (and, at the time, illegal) performance was witnessed by just a few handfuls of onlookers who happened to be there. But attendance grew, and in 1990, what had now become an annual event moved to a broad playa in the Black Rock desert of northwest Nevada (within the broader Great Basin). Attendance hit 1,000 "Burners" in 1993; 10,000 in 1997; and 50,000 in 2010. As of 2017, attendance reached nearly 70,000! This makes it temporarily the second largest population agglomeration in all of Nevada, after Las Vegas (**Figure 2A***a*). Such a large gathering motivates the Burning Man Web site (*https://burningman.org/*) to refer to itself as "one of the great cities of the world". (Technically, our analysis of cities in Chapter 11 suggests that Burning Man would more accurately be considered something like a bedroom suburb with many different central cities located around the world—who says life in the burb's is no fun?)

Burning Man now lasts a little over a week, giving its "residents" ample opportunity to sample its physical and cultural environment. Physically, the playa at Black Rock is a very flat plain of highly alkaline soil. Most days of the festival see high temperatures and copious sun. Temperatures at night typically drop 30°F or more. Wind is also common, with robust dust storms often occurring at least once or twice during the event. But rain can happen too, and when it does, the playa becomes astoundingly mucky.

Clearly, the Burning Man subculture integrates all three components of ideological, technological, and sociological culture. Architectural and sculptural art installations abound; some of them have engines installed that turn the art into transportation—cars and shuttles that might spew flames or emit ominous tones (**Figure 2A***b*). Burners wear costumes and headdresses and sometimes very little at all. They hang out in theme camps organized around statements of creativity, nonconformity, humor, or simply libertinism. Activities include fireworks and

Figure 2A(a) The sun rises over the 2013 Festival.
(a) ©Lukas Bischoff/123RF

Figure 2A(b) Example of a typical art car and some "Burners" from the 2013 Festival.
(b) ©Lukas Bischoff/123RF

artistic burning events, culminating with the ceremonial burning of the "Man" on the final Saturday evening. Various forms of merriment occur, all accompanied by a nearly constant soundtrack of various types of rock and electronica music, which some might describe as clamor when they must listen to it at 4 A.M. All of this describes the surface appearance of Burning Man, but the ideas underneath it are perhaps what is most intriguing. The Festival claims to carry the torch for guerrilla resistance to the social norms of the wider main national culture, a modern echo of the social movements of the Beats, the hippies, and other radical free thinkers and doers of times past.

At the end of the Festival, Burners trudge off to their regular lives in the various corners of the globe from which they've come. For the great majority of them, these lives must be considerably more sober and conventional than their week or so in the Black Rock desert. One assumes that the last traces of the Festival are hardly cleaned up before the organizers start planning for the next year's event, the heavily planned reality for what started as a spontaneous act of performance art.

But our discussion should not imply that culture traits exist in isolation—they are always interrelated. Individual cultural traits that are functionally interrelated comprise a **culture complex.** The existence of such complexes is universal. Keeping cattle was a *culture trait* of the Masai of Kenya and Tanzania. Related traits included the measurement of personal wealth by the number of cattle owned, a diet containing the milk and blood of cattle, and disdain for labor unrelated to herding. The assemblage of these and other related traits yielded a culture complex descriptive of one aspect of Masai society (**Figure 2.3**). In exactly the same way, religious complexes, business behavior complexes, sports complexes, and others can easily be recognized in any society.

In the United States, for example, some environmentalists would like to wean Americans from their automobiles. However, a study of American culture reveals the difficulty in making such a change because automobiles are part of an interrelated cultural complex. Automobile brands and models speak of a person's status in U.S. society. Movies, video games, and sports often give automobiles a central role; movies such as *The Fast and the Furious* series, video games such as *Grand Theft Auto,* and NASCAR are familiar examples. Entire suburban landscapes have been built around the needs of the automobile. Even rites of passage often focus on the automobile: driver's education, passing the driver's exam, getting one's first automobile, picking up a date in one's automobile, and the practice of decorating the newlywed couple's automobile at the end of the wedding ceremony.

A **cultural system** is a broader generalization than a culture complex and refers to the collection of interacting culture traits and complexes that are shared by a group within a particular territory. Multiethnic societies, perhaps further subdivided by linguistic differences, varied food preferences, and a host of other internal differentiations, may nonetheless share enough joint characteristics to be recognizably distinctive cultural systems to themselves and others. Certainly, citizens of the "melting pot" United States would identify themselves as *Americans,* together constituting a unique culture system on the world scene.

Culture traits, complexes, and systems have a real extent. When they are plotted on maps, the regional character of the components of culture is revealed. Although human geographers are interested in the spatial distribution of these individual elements of culture, their usual concern is with a type of thematic region known as the **culture region,** a portion of the Earth's surface occupied by populations sharing recognizable and distinctive cultural characteristics. Examples include the political organizations societies devise, the religions they espouse, the form of economy they pursue, and even the type of clothing they wear, eating utensils they use, or kind of housing they occupy. There are as many such conceptual culture regions as there are culture traits and complexes recognized for population groups. Their recognition will be particularly important in the discussions of ethnic, folk, and popular culture that will occur in later chapters of this book. In those later reviews, as within the present chapter, we must keep in mind that within any single recognized culture region, groups united by the specific mapped characteristics may be competing and distinctive in other important culture traits.

Figure 2.3 The formerly migratory Masai of Kenya are now largely sedentary, partially urbanized, and frequently owners of fenced farms. Cattle formed the traditional basis of Masai culture and were evidence of wealth and social status.

©*The McGraw-Hill Education/Barry Barker, photographer*

Finally, a set of culture regions showing related culture complexes and landscapes may be grouped to form a **culture realm.** The term recognizes a large segment of the Earth's surface having an assumed fundamental uniformity in its cultural characteristics and showing a significant difference in them from adjacent realms. Culture realms are, in a sense, culture regions at the broadest scale of recognition. In fact, the scale is so broad and the diversity within the recognized realms so great that the very concept of realm may mislead more than it informs.

Indeed, the current validity of distinctive culture realms has been questioned in light of an assumed globalization of all aspects of human society and economy. The result of that globalization, it has been suggested, is a homogenization of cultures as economies are integrated and uniform consumer demands are satisfied by standardized commodities produced by international corporations. Certainly, the increasing mobility of people, goods, and information has reduced the rigidly compartmentalized ethnicities, languages, and religions of earlier periods. Cultural flows and exchanges have increased over the recent decades, and with them has come a growing worldwide intermixture of peoples and customs. Despite that growing globalism in all facets of life and economy, however, the world is far from homogenized. Although an increased sameness of commodities and experiences is encountered in distant places, even common and standardized items of everyday life—branded soft drinks, for example, or American fast-food franchises—take on unique regional meanings and roles, conditioned by the total cultural mix they enter. Those multiple regional cultural mixes are often defiantly distinctive and separatist as recurring incidents of ethnic conflict, civil war, and strident regionalism attest. Rather than successfully leveling and removing all regional contrasts, as frequently predicted, globalization continues to be countered by powerful forces of regionalism, place identity, and ethnicity.

If a global culture can be discerned, it may best be seen as a combination of multiple territorial cultures rather than a standardized uniformity. It is those territorially different cultural mixtures that are recognized by the culture realms suggested on **Figure 2.4,** which is only one of many such possible divisions. The spatial pattern and characteristics of these generalized realms will help us place the discussions and examples of human geography of later chapters in their regional contexts. They are commonly employed in courses on world regional geography.

2.2 Interaction of People and Environment

Culture develops in a physical environment that, in its way, contributes to differences among people. In premodern subsistence societies, the acquisition of food, shelter, and clothing, all parts of culture, depends on the utilization of the natural resources at hand. The interrelations of people to the environment of a given area, their perceptions and utilization of it, and their impact on it are interwoven themes of **cultural ecology**—the study of the relationship between a culture group and the natural environment that it occupies.

Cultural ecologists see evidence that subsistence pastoralists, hunter-gatherers, and gardeners adapted their productive activities—and, by extension, their social organizations and relationships—to the specific physical limitations of their different local habitats. Presumably, similar natural environmental conditions influenced the development of similar adaptive responses and cultural outcomes in separate, unconnected locales. That initial influence, of course, does not predetermine the details of the subsequent culture.

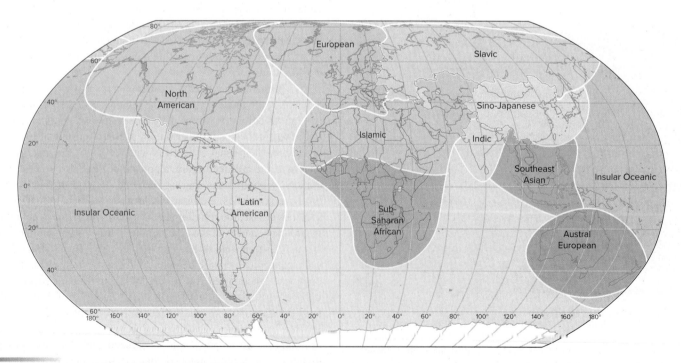

Figure 2.4 Culture realms of the modern world. This is just one of many possible subdivisions of the world into multifactor cultural regions.

Environments as Controls

Geographers have long focused on the role of the physical environment in shaping human culture. However, most geographers today dismiss as intellectually limiting and demonstrably invalid extreme versions of **environmental determinism,** the belief that the physical environment exclusively shapes humans and their cultures. Environmental factors alone cannot account for the cultural variations that occur around the world. Levels of technology, systems of organization, and ideas about what is true and right are not dictated by environmental circumstances.

The environment does place certain limitations on the human use of territory. Such limitations, however, must be seen not as absolute, enduring restrictions but as relative to technologies, cost considerations, national aspirations, and linkages with the larger world. Human choices in the use of landscapes are affected by group perception of the feasibility and desirability of their settlement and exploitation. These are not circumstances inherent in the land. Mines, factories, and cities have been created in the formerly nearly unpopulated tundra and forests of Siberia as a reflection of Russian developmental programs, not in response to recent environmental improvement.

Possibilism is the viewpoint better the original way, as it is now Vidal de la Blache. He argued that the natural environment constrains or limits culture, making some cultural variants more or less possible than others, but it does not strictly determine culture. The needs, traditions, and level of technology of a culture affect how that culture assesses the possibilities of an area and shape what choices the culture makes regarding them. Each society uses natural resources in accordance with its circumstances. Changes in a group's technical abilities or objectives bring about changes in its perceptions of the usefulness of the land. Possibilism is thus a form of environment-culture interactionism.

Unlike some social scientists, however, geographers generally do not accept the extreme opposite of environmental determinism—**cultural autonomy**—that cultures are equally likely to develop any particular set of cultural traits no matter what the environmental circumstances. For example, evidence suggests the nature of some environmental limitations on the use of land area. The vast majority of the world's population is differentially concentrated on less than one-half of the Earth's land surface, as Figure 4.21 indicates. Areas with relatively mild climates and flat topographies that offer a supply of fresh water, fertile soil, and abundant mineral resources are densely settled, reflecting in part the different potentials of the land under earlier technologies to support population. Even today, the polar regions, high and rugged mountains, deserts, and some hot and humid lowland areas contain very few people. If resources for feeding, clothing, or housing ourselves within an area are lacking, or if we do not recognize them there, there is no inducement for people to occupy a territory.

Environments that do contain such recognized resources provide the framework within which a culture operates. Coal, oil, and natural gas have been in their present locations throughout human history, but they were rarely of use to preindustrial cultures and did not impart any understood advantage to their sites of occurrence. Not until the Industrial Revolution did coal deposits gain importance and come to influence the location of such great industrial complexes as the Midlands in England, the Ruhr in Germany, and the steel-making districts formerly so important in parts of northeastern United States. Native Americans made one use of the environment around Pittsburgh, while 19th-century industrialists made quite another. The environment influences the chances that a thriving steel industry will develop in one place rather than another, but only when the technological and economic conditions in the culture support steel-making as an activity.

Human Impacts

People are also able to modify their environment, and this is the other half of the human-environment relationship of geographic concern. Geography, including cultural geography, examines both the reactions of people to the physical environment and their impact on that environment. By using our environment, we modify it—in part, through the material objects we place on the landscape: cities, farms, roads, and so on (**Figure 2.5**). The form these take is the product of the kind of culture group in which we live. The **cultural landscape,** the Earth's surface as modified by human action, is the tangible physical record of a given culture. House types, transportation networks, parks and cemeteries, and the size and distribution of settlements are among the indicators of the use that humans have made of the land.

Human actions, both deliberate and inadvertent, have modified or even destroyed the environment for perhaps as long as humankind has existed. People have used, altered, and replaced the vegetation in wide areas of the tropics and midlatitudes. They have hunted to extinction vast herds and whole species of animals. They have, through overuse and abuse of the Earth and its resources, rendered sterile and unpopulated formerly productive and attractive regions.

Fire has been called the first great tool of humans, and the impact of its early and continuing use is found on nearly every continent. Poleward of the great rain forests of equatorial South America, Africa, and South Asia lies the *tropical savanna* of extensive grassy vegetation separating scattered trees and forest

Figure 2.5 The physical and cultural landscapes of Cape Town, South Africa, in juxtaposition. Advanced societies are capable of so altering the circumstances of nature that the cultural landscapes that they create become the main controlling environment.

©Dereje/Shutterstock

Chaco Canyon Desolation

It is not certain when they first came, but by 1000 CE, the Anasazi people were building a flourishing civilization in present-day Arizona and New Mexico. They were corn farmers, thriving during the 300 years or so of the medieval warm period, beginning about 900 CE in the American Southwest. In Chaco Canyon alone, they erected as many as 75 towns, all centered around pueblos, huge stone-and-adobe apartment buildings as tall as five stories and with as many as 800 rooms. These were the largest and tallest buildings of North America prior to the construction of iron-framed "cloudscrapers" in major cities at the end of the 19th century. An elaborate network of roads and irrigation canals connected and supported the pueblos. About 1200 CE, the settlements were abruptly abandoned. The Anasazi, advanced in their skills of agriculture and communal dwelling, were—according to some scholars—forced to move on by the ecological disaster their pressures had brought to a fragile environment.

They needed forests for fuel and for the hundreds of thousands of logs used as beams and bulwarks in their dwellings. The pinyon-juniper woodland of the canyon was quickly depleted. For larger timbers needed for construction, the Anasazi first harvested stands of ponderosa pine found some 40 kilometers (25 miles) away. As early as 1030 CE these, too, were exhausted, and the community switched to spruce and Douglas fir from mountaintops surrounding the canyon. When they were gone by 1200 CE, the Anasazi fate was sealed—not only by the loss of forest but by the irreversible ecological changes deforestation and agriculture had occasioned. With forest loss came erosion that destroyed the topsoil. The surface water channels that had been built for irrigation were deepened by accelerated erosion, converting them into enlarging arroyos useless for agriculture.

The material roots of their culture destroyed, the Anasazi turned upon themselves; warfare convulsed the region and, compelling evidence suggests, cannibalism was practiced. Smaller groups sought refuge elsewhere, re-creating on reduced scale their pueblo way of life but now in nearly inaccessible, highly defensible mesa and cliff locations. The destruction they had wrought destroyed the Anasazi in turn.

Figure 2B Chaco Canyon
©*Jon Malinowski/Human Landscape Studio*

---

groves (**Figure 2.6**). The trees appear to be the remnants of naturally occurring tropical dry forests, thorn forests, and scrub now largely obliterated by the use, over many millennia, of fire to remove the unwanted and unproductive trees and to clear off old grasses for more nutritious new growth. The grasses supported the immense herds of grazing animals that were the basis of hunting societies. After independence, the government of Kenya in East Africa sought to protect its national game preserves by prohibiting the periodic use of fire. It quickly found that the immense herds of gazelles, zebras, antelope, and other grazers (and the lions and other predators that fed on them) that tourists came to see were being replaced by less appealing browsing species—rhinos, hippos, and elephants. With fire prohibited, the forests began to reclaim their natural habitat and the grassland fauna was replaced.

The same form of vegetation replacement occurred in midlatitudes. The grasslands of North America were greatly extended by Native Americans who burned the forest margin to extend grazing areas and to drive animals in the hunt. The control of fire in modern times has resulted in the advance of the forest once again in formerly grassy areas of Colorado, northern Arizona, and other parts of the U.S. West.

Examples of adverse human impact abound. The *Pleistocene overkill*—the Stone Age loss of whole species of large animals on all inhabited continents—is often ascribed to the unrestricted hunting to extinction carried on by societies familiar with fire who drove animals and used hafted weapons (with handles) to slaughter them. With these methods, according to one estimate, about 40 percent of African large-animal genera passed to extinction. The majority of large mammal, reptile, and flightless bird species had disappeared from Australia around 46,000 years ago; in North America, some two-thirds of original large mammals had succumbed by 11,000 years ago under pressure from the hunters migrating to and

Figure 2.6 The parklike landscape of grasses and trees characteristic of the tropical savanna is seen in this view from Kenya, in Africa.
©Fuse/Getty Images

Figure 2.7 Now treeless, Easter Island once was lushly forested. The statues (some weighing up to 85 tons) dotting the island were rolled to their locations and lifted into place with logs.
©Steve Allen/Stockbyte/Getty Images

spreading across the continent. Although some have suggested that climatic changes or pathogens carried by dogs, rats, and other camp followers were at least partially responsible, human action is the more generally accepted explanation for the abrupt faunal changes. No uncertainty exists in the record of faunal destruction by the Maori of New Zealand or of Polynesians who had exterminated some 80 to 90 percent of South Pacific bird species—as many as 2,000 in all—by the time Captain James Cook of Britain arrived in the 18th century. Similar destruction of key marine species—Caribbean sea turtles, sea cows off the coast of Australia, sea otters near Alaska, and others elsewhere—as early as 10,000 years ago resulted in environmental damage whose effects continue to the present.

Not only the destruction of animals but of the life-supporting environment itself has been a frequent consequence of human misuse of areas (see the feature "Chaco Canyon Desolation"). North Africa, the "granary of Rome" during the empire, became wasted and sterile in part because of mismanagement. Roman roads standing high above the surrounding barren wastes give testimony to the erosive power of wind and water when natural vegetation is unwisely removed and farming techniques are inappropriate. Easter Island in the South Pacific was covered lushly with palms and other trees when Polynesians settled there about 400 CE. By the beginning of the 18th century, Easter Island had become the barren wasteland that it remains today. Deforestation increased soil erosion, removed the supply of timbers needed for the vital dugout fishing canoes, and made it impossible to move the massive stone statues that were significant in the islanders' religion (**Figure 2.7**). With the loss of livelihood resources and the collapse of religion, warfare broke out and the population was decimated. A similar tragic sequence is occurring on Madagascar in the Indian Ocean today. Despite current romantic notions, early societies did not necessarily live in harmony with their environment.

However, economic and technological developments in culture allow much greater impacts on the natural environment, and in general, the more technologically advanced and complex the culture, the more apparent is its impact on the natural landscape. In sprawling urban-industrial societies, the cultural landscape has come to outweigh the natural physical environment in its impact on most people's daily lives. It interposes itself between "nature" and humans, and residents of the cities of such societies—living

and working in climate-controlled buildings, driving to enclosed shopping malls—can go through life with much less contact with or concern about the physical environment.

2.3 Roots of Culture

Earlier humans found the physical environment more immediate and controlling than we do today. Some 11,000 years ago, the massive glaciers—moving ice sheets of great depth—that had covered much of the land and water of the Northern Hemisphere (**Figure 2.8**) began to retreat. Animal, plant, and human

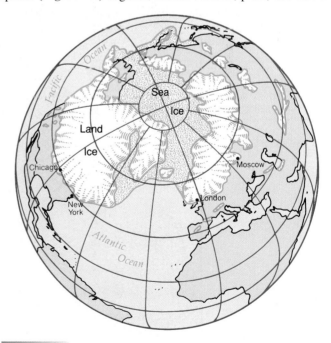

Figure 2.8 Maximum extent of glaciation. In their fullest development, glaciers of the most recent Ice Age covered large parts of Eurasia and North America. Even areas not covered by ice were affected as ocean levels dropped and rose, and climate and vegetation regions changed with glacial advance and retreat.

populations that had been spatially confined by both the ice margin and the harsh climates of middle-latitude regions began to spread, colonizing newly opened territories. The end of the *Paleolithic* (Old Stone Age) is the period near the end of glaciation during which small and scattered groups like the ivory hunters at this chapter's start began to develop regional variations in their ways of life and livelihood.

All were **hunter-gatherers,** preagricultural people dependent on the year-round availability of plant and animal foodstuffs that they could secure with the rudimentary stone tools and weapons at their disposal. Even during the height of the Ice Age, the unglaciated sections of western, central, and northeastern Europe, which today are home to productive farms, forests, and cities, were then covered with tundra vegetation, the mosses, lichens, and low shrubs typical of areas too cold to support forests. Southeastern Europe and southern Russia had forest, tundra, and steppe (grasslands), and the Mediterranean areas,

which today have shrub and scrub-oak vegetation, had forest cover (**Figure 2.9**). Gigantic herds of herbivores—reindeer, bison, mammoth, and horses—browsed, bred, and migrated throughout the tundra and the grasslands. Abundant animal life filled the forests.

Human migration northward into present-day Sweden, Finland, and Russia demanded a much more elaborate set of tools and provision for shelter and clothing than had previously been required. It necessitated the crossing of a number of ecological barriers and the occupation of previously avoided difficult environments. By the end of the Paleolithic period, humans had spread to all the continents but Antarctica, carrying with them their adaptive hunting-gathering cultures and social organizations. The settlement of the lands bordering the Pacific Ocean is suggested in **Figure 2.10.** As they occupied different regions, hunter-gatherers focused on a diversity of foodstuffs. Some specialized in marine or river resources, while others were wholly

Figure 2.9 Late Paleolithic environments of Europe. During the late Paleolithic period, new food-gathering, shelter, and clothing strategies were developed to cope with harsh and changing environments, so different from those in Europe today.

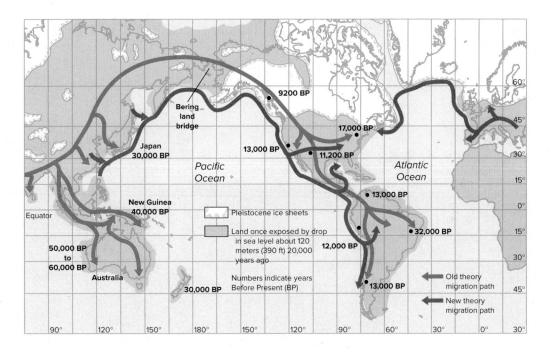

Figure 2.10 Settlement of the Americas and the Pacific basin. Genetic studies suggest humans spread around the globe from their Old World origins beginning some 100,000 years ago. Their time of arrival in the Western Hemisphere, however, is uncertain. The older view claimed that earliest migrants to the Americas, the ancestors of modern Amerindian groups, crossed the Bering land bridge in three different waves beginning 11,500 years ago. Recent evidence suggests that those North Asian land migrants encountered (and conquered or absorbed) earlier occupants who had arrived from Europe, Polynesia, and coastal East Asia by boat traveling along frozen or open shorelines. Although genetic and linguistic research yields mixed conclusions, physical evidence considered solid by some investigators indicates that the first Asian arrivals came at least 22,000 years and more likely 30,000 or more years ago. Eastern United States artifacts that have been assigned dates of 17,000 to 30,000 years ago hint at European arrivals as early as those of coastal Asians; a South Carolina site found in 2004 has been dated at 50,000 years ago. Many researchers, however, caution that any New World population dates earlier than 18,000 years ago are questionable, and that the first migrants from that period probably were most closely related to prehistoric Jomon and later Ainu groups of Japan who crossed over the Bering land bridge.

dependent on land plants and animals. In all cases, their material culture reflected the different climate and vegetation regions they occupied, the tools they developed to exploit the resources on which they depended, and the housing and clothing solutions they differently adopted. Even today, African Bushmen have few cultural similarities with Inuit (Eskimo) hunting-fishing societies, though both culture groups are spoken of as "hunter-gatherers."

While spreading, the total population also increased. But hunting and foraging bands require considerable territory to support a relatively small number of individuals. There were contacts between groups and, apparently, even planned gatherings for trade, socializing, and selecting mates from outside the home group. Nevertheless, the bands tended to live in isolation. Estimates place the Paleolithic population of the entire island of Great Britain, which was on the northern margin of habitation, at only some 400–500 persons living in widely separated families of 20–40 people. Total world population at about 9000 BCE probably ranged from 5 to 10 million. Variations in the types of tools characteristic of different population groups steadily increased as people migrated and encountered new environmental problems.

Improved tool technology greatly extended the range of possibilities in the use of locally available materials. The result was more efficient and extensive exploitation of the physical environment than had been possible earlier. At the same time, regional contrasts in plant and animal life and in environmental conditions accelerated the differentiation of culture among isolated groups who under earlier, less varied conditions had shared common characteristics.

Within many environments, even harsh ones, the hunting and foraging process was not particularly demanding of either time or energy. Recent studies of South African San people (Bushmen), for example, indicate that such bands survive well on the equivalent of a 2½-day workweek. Time was available for developing skills in working flint and bone for tools, in developing regionally distinctive art and sculpture, and in making decorative beads and shells for personal adornment and trade. By the end of the Ice Age (about 11,000 to 12,000 years ago), language, religion, long-distance trade, permanent settlements, and social stratification within groups appear to have been well developed in many culture areas.

What was learned and created was transmitted within the cultural group. The increasing variety of adaptive strategies and technologies and the diversity of noneconomic creations in art, religion, language, and custom meant an inevitable cultural variation of humankind. That diversification began to replace the rough social uniformity among hunting and gathering people that had been based on their similar livelihood challenges, informal leadership structures, small-band kinship groups, and the like (**Figure 2.11**).

Figure 2.11 Hunter-gatherers practiced the most enduring lifestyle in human history, trading it for the sometimes more arduous life of farmers due to the necessity to provide larger quantities of less diversified foodstuffs for a growing population. For hunter-gatherers (unlike their settled farmer rivals and successors), age and sex differences, not caste or economic status, were and are the primary basis for the division of labor and of interpersonal relations. Here, a San (Bushman) hunter of Botswana, Africa, stalks his prey. Men also help collect the gathered food that constitutes 80 percent of the San diet.
©Ben McRae/123RF

2.4 Seeds of Change

The retreat of the last glaciers marked the end of the Paleolithic era and the beginning of successive periods of cultural evolution, leading from basic hunting and gathering economies at the outset through the development of agriculture and animal husbandry to, ultimately, the urbanization and industrialization of modern societies and economies. Because not all cultures passed through all stages at the same time, or even at all, **cultural divergence** between human groups became evident.

Glacial recession brought new ecological conditions to which people had to adapt. The weather became warmer and forests began to appear on the open plains and tundras of Europe and northern China. In the Middle East, where much plant and animal domestication would later occur, savanna (grassland) vegetation replaced more arid landscapes. Populations grew and, through hunting, depleted the large herds of grazing animals already retiring northward with the retreating glacial front.

Further population growth demanded new food bases and production techniques, for the **carrying capacity**—the number of persons supportable within a given area by the technologies at their disposal—of the Earth for hunter-gatherers is low. The *Mesolithic* (Middle Stone Age) period, from about 11,000 to 5000 BCE in Europe, marked the transition from the collection of food to its production. These stages of the Stone Age—occurring during different time spans in different world areas—mark distinctive changes in tools, tasks, and social complexities of the cultures that experienced the transition from "Old" to "Middle" to "New" Stone Age.

Agricultural Origins and Spread

The population of hunter-gatherers rose slowly at the end of the glacial period. As rapid climatic fluctuation adversely affected their established plant and animal food sources, people independently, in more than one world area, experimented with the

domestication of plants and animals. There is no agreement on whether the domestication of animals preceded or followed that of plants. The sequence may well have been different in different areas. What appears certain is that animal domestication—the successful breeding of species to serve human needs—began during the Mesolithic, not as a conscious economic effort by humans but as outgrowths of the keeping of small or young wild animals as pets and the attraction of scavenger animals to the refuse of human settlements. The assignment of religious significance to certain animals and the docility of others to herding by hunters all strengthened the human-animal connections that ultimately led to full domestication. Eventually, nearly everyone in the world would come to obtain their food via agriculture rather than hunting and gathering (see discussion of Innovation below). This had such dramatic effects on the way humans spent their daily lives, as well as the way they interrelated with the natural environment, that it is often referred to as the *Agricultural Revolution.*

Radiocarbon dates suggest the occurrence of the domestication of pigs in southeastern Turkey and of goats in the Near East as early as 8000–8400 BCE, of sheep in Turkey by about 7500 BCE, and of cattle and pigs in both Greece and the Near East about 7000 BCE North Africa, India, and southeastern Asia were other Old World domestication sources, as were—less successfully—Mesoamerica and the Andean Uplands. Although there is evidence that the concept of animal domestication diffused from limited source regions, once its advantages were learned, numerous additional domestications were accomplished elsewhere. The widespread natural occurrence of species able to be domesticated made that certain. Cattle of different varieties, for example, were domesticated in India, north-central Eurasia, Southeast Asia, and Africa. Pigs and various domestic fowl are other examples.

The domestication of plants, like that of animals, appears to have occurred independently in more than one world region over a time span of between 10,000 and perhaps as long as 20,000 years ago. A strong case can be made that most widespread Eurasian food crops were first cultivated in the Near East beginning some 12,000 years ago, and dispersed rapidly from there across the mid-latitudes of the Old World. However, clear evidence also exists that African peoples were raising crops of wheat, barley, dates, lentils, and chickpeas on the floodplains of the Nile River as early as 18,500 years ago. In other world regions, farming began more recently; the first true farmers in the Americas appeared in Mexico no more than 5,000 years ago.

Familiarity with plants of desirable characteristics is universal among hunter-gatherers. In those societies, females were assigned the primary food-gathering role and thus developed the greatest familiarity with nutritive plants. Their fundamental role in initiating crop production to replace less reliable food gathering seems certain. Indeed, women's major contributions as innovators of technology—in food preparation and clothing production, for example—or as inventors of such useful and important items as baskets and other containers, baby slings, yokes for carrying burdens, and the like are unquestioned.

Agriculture itself, however, may not have been an "invention" as such, but the logical extension to food species of plant selection and nurturing habits developed for nonfood varieties. Plant poisons applied to hunting arrows or spread on lakes or streams to stun fish made food gathering easier and more certain. Plant dyes and pigments were universally collected or prepared for personal adornment or article decoration. Medicinal and mood-altering plants and derivatives were known, gathered, protected, and cultivated by all early cultures. Indeed, persuasive evidence exists to suggest that early gathering and cultivation of grains was not for grinding and baking as bread but for brewing as beer, a beverage that became so important in some cultures (for religious and nutritional reasons) that it may well have been a first and continuing reason for sedentary agricultural activities.

Nevertheless, full-scale domestication of food plants, like that of animals, can be traced to a limited number of origin areas identified by geographer Carl Sauer and other scientists (**Figure 2.12**). Although there were several source regions, certain uniformities united them. In each, domestication focused on plant species selected apparently for their capability of providing large quantities of storable calories or protein. In each, there was a population that was already well fed and able to devote time to the selection, propagation, and improvement of plants available from a diversified vegetation. Agricultural innovation would likely have occurred in relatively fertile and productive areas like river valleys and coastal plains. Some speculate, however, that grain domestication in the Near East may have been a forced inventive response, starting some 13,000 years ago, to food shortages reflecting abrupt increases in summertime temperatures and aridity in the Jordan Valley. That environmental stress—reducing summer food supplies and destroying habitats of wild game—favored the selection and cultivation of short-season annual grains and legumes whose seeds could be stored and planted during cooler, wetter winter growing seasons.

In the tropics and humid subtropics, selected plants were apt to be those that reproduced vegetatively—from roots, tubers, or cuttings. Outside of those regions, wild plants reproducing from seeds were more common and the objects of domestication. Although there was some duplication, each of the origin areas developed crop complexes characteristic of itself alone, as Figure 2.12 summarizes. From each, there was dispersion of crop plants to other areas, slowly at first under primitive systems of population movement and communication (**Figure 2.13**), more rapidly and extensively with the onset of more modern transportation and communication technologies.

While adapting wild plant stock to agricultural purposes, the human cultivators, too, adapted. They assumed sedentary residence to protect the planted areas from animal, insect, and human predators. They developed labor specializations and created more formalized and expansive religious structures in which fertility and harvest rites became important elements. The regional contrasts between hunter-gatherer and sedentary agricultural societies increased. Where the two groups came in contact, farmers were the victors and hunter-gatherers the losers in the competition for territorial control. The contest continued into modern times. During the past 500 years, European expansion totally dominated the hunting and gathering cultures it encountered in large parts of the world such as North America and Australia (see the feature "Is Geography Destiny?"). Even today, in the rain forests of central Africa, Bantu farmers put continuing pressure on hunting and gathering Pygmies,

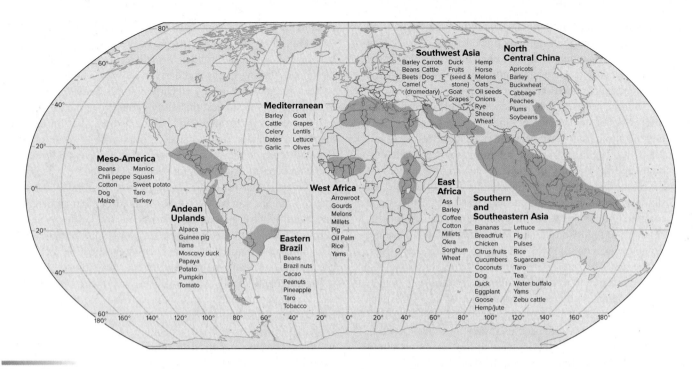

Figure 2.12 Chief centers of plant and animal domestication. The Southern and Southeastern Asia center was characterized by the domestication of plants such as taro, which are propagated by the division and replanting of existing plants (vegetative reproduction). Reproduction by the planting of seeds (e.g., maize and wheat) was more characteristic of Mesoamerica and Southwest Asia. The African and Andean areas developed crops reproduced by both methods. The lists of crops and livestock are selective, not exhaustive.

and in southern Africa, Hottentot herders and Bantu farmers constantly advance on the territories of the San (Bushmen) hunter-gatherer bands. The contrast and conflict between the hunter-gatherers and agriculturalists provide dramatic evidence of cultural divergence.

Neolithic Innovations

The domestication of plants and animals began during the Meso-lithic period, but in its refined form it marked the onset of the *Neolithic* (New Stone Age). Like other Stone Age levels, the Neolithic

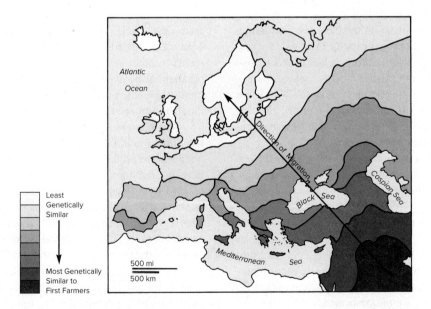

Figure 2.13 The migration of the first farmers out of the Middle East into Europe, starting about 10,000 years ago is presumably traced by blood and gene markers. If the gene evidence interpretation is valid, the migrants spread at a rate of about 1 kilometer (five-eighths of a mile) per year, gradually interbreeding with and replacing the indigenous European hunter-gatherers throughout that continent.

Source: Adapted from L. Luca Cavalli-Sforza, Paolo Menozzi, and Alberto Piazza. The History and Geography of Human Genes. *Princeton, N.J.: Princeton University Press (1994).*

Is Geography Destiny?

In his 1997 Pulitzer Prize–winning book *Guns, Germs, and Steel: The Fates of Human Societies,* Jared Diamond argues that "history followed different courses for different peoples because of differences among peoples' environments, not because of biological differences among peoples themselves." The environmental differences that counted—and that led to world dominance by Eurasians—were the availability in Eurasia of an abundance of plants and animals suitable for domestication on a landmass whose east-west orientation made easy the long-distance transfer of animals, food crops, and technologies. No other continent had either of those advantages, according to Diamond.

Food production was the key. Although agriculture was independently developed in several world areas after the end of the Ice Age, the inhabitants of the Middle East were fortunate in having an abundance of plants suitable for domestication. These included six of the eight most important food grasses, among them ancestral wheat, plants that adapted easily to cultivation, grew rapidly, and had high nutritive value and high population-supporting potential. Eurasia also

had an abundance of large animals that could be domesticated, including the cow, goat, pig, sheep, and horse, giving a further spur to population growth. In addition, by living in close proximity to animals, Eurasians contracted and developed immunities to the epidemic diseases that would later devastate the inhabitants of other continents when the diseases were brought to their shores by Eurasian explorers and colonizers.

The food-producing technologies developed in such hearth regions as the Middle East were easily diffused along the immense east-west axis of Eurasia, where roughly similar climates suited to the same crop and livestock mix were encountered from China to Spain. In addition, Eurasia's great size meant it had a great number of different people, each capable of developing new technologies that in turn could be diffused over long distances. Population growth, agricultural productivity, and inventive minds led to civilizations—central governments, cities, labor specializations, textiles, pottery, writing, mathematics, long-distance trade, metal working, and eventually, the guns that conquering Eurasians carried to other continents.

No other world region enjoyed Eurasia's environmental and subsequent population and technological advantages. The few food crops developed in Africa or the Americas could not effectively diffuse across the climatic and ecological barriers in those north-south aligned continents. Because of accidents of nature or massive predation of large animals by early inhabitants, sub-Saharan Africa and Australia yielded no domesticated animals, and the Americas had only the localized llama. Without the food bases and easy latitudinal movement of Eurasia, populations elsewhere remained smaller, more isolated, and collectively less inventive. When the voyages of discovery and colonization began in the 15th century, Eurasian advantages proved overwhelming. Decimated by diseases against which they had no resistance, without the horses, armor, firearms, or organization of their conquerors, inhabitants of other continents found themselves quickly subdued and dominated—not, in Diamond's opinion, because of innate inferiority, but because of geographical disadvantages that limited or delayed their developmental prospects.

---

was more a stage of cultural development than a specific span of time. The term implies the creation of an advanced set of tools and technologies to deal with the conditions and needs encountered by an expanding, sedentary population whose economy was based on the agricultural management of the environment (**Figure 2.14**).

Not all peoples in all areas of the Earth made the same cultural transition at the same time. In the Near East, from which most of our knowledge of this late prehistoric period comes, the Neolithic lasted from approximately 8000 to 3500 BCE There, as elsewhere, it brought complex and revolutionary changes in human life. Culture began to alter at an accelerating pace, and change itself became a way of life. In an interconnected adaptive web, technological and social innovations came with a speed and genius surpassing all previous periods.

Humans learned the arts of spinning and weaving plant and animal fibers. They learned to use the potter's wheel and to fire clay and make utensils. They developed techniques of brick making, mortaring, and construction, and they discovered the skills of mining, smelting, and casting metals. On the foundation of such technical advancements, a more complex exploitative culture

appeared and a more formal economy emerged. A stratified society based on labor and role specialization replaced the rough equality of adults in hunting and gathering economies (various forms of sex-role division of labor largely continued, however). Special local advantages in resources or products promoted the development of long-distance trading connections, which the invention of the sailboat helped to maintain.

By the end of the Neolithic period, certain spatially restricted groups, having created a food-producing rather than a foraging society, undertook the purposeful restructuring of their environment. They began to modify plant and animal species; to manage soil, terrain, water, and mineral resources; and to use animal energy to supplement that of humans. They used metal to make refined tools and superior weapons—first pure copper, and later the alloy of tin and copper that produced the harder, more durable bronze. Humans had moved from adopting and shaping to the art of creating.

As people gathered together in larger communities, new and more formalized rules of conduct and control emerged, especially important where the use of land was involved. We see the beginnings of governments to enforce laws and specify punishments

(a)

(b)

Figure 2.14 (*a*) The Mediterranean scratch plow, the earliest form of this basic agricultural tool, was essentially an enlarged digging stick dragged by an ass, an ox, or—as here in ancient Palestine—by a ox and a donkey. The scratch plow represented a significant technological breakthrough in human use of tools and animal power in food production. (*b*) Its earliest evidence is found in Egyptian tomb drawings and in art preserved from the ancient Middle East, but it was elsewhere either independently invented or introduced by those familiar with its use. (See also Figure 2.17a.)

(a) Source: Library of Congress Prints and Photographs Division [LC-DIG-ppmsca-02754]; (b) ©Bojan Brecelj/Corbis Historical/Getty Images

for wrongdoers. The protection of private property, so much greater in amount and variety than that carried by the nomad, demanded more complex legal codes, as did the enforcement of the rules of societies increasingly stratified by social privileges and economic status.

Religions became more formalized. For the hunter, religion could be individualistic, and his worship was concerned with personal health and safety. The collective concerns of farmers were based on the calendar: the cycle of rainfall, the seasons of planting and harvesting, the rise and fall of waters to irrigate the crops, and so on. Religions responsive to those concerns developed rituals appropriate to seasons of planting, irrigation, harvesting, and thanksgiving. An established priesthood was required, one that stood not only as an intermediary between people and the forces of nature, but also as an authenticator of the timing and structure of the needed rituals. In daily life, occupations became increasingly specialized. Metal-workers, potters, sailors, priests, merchants, scribes, and in some areas, warriors complemented the work of farmers and hunters.

2.5 Culture Hearths

The social and technical revolutions that began in and characterized the Neolithic period were initially spatially confined. The new technologies, the new ways of life, and the new social structures diffused from those points of origin and were selectively adopted by people who were not a party to their creation. The term **culture hearth** is used to describe such centers of innovation and invention, from which clusters of key culture traits and elements moved to exert an influence on surrounding regions.

The hearth may be viewed as the "cradle" of any culture group whose developed systems of livelihood and life created a distinctive cultural landscape. Most of the hearths that evolved across the world remained at low levels of social and technical development. Only a few developed the trappings of *civilizations*. The definition of that term is not precise, but indicators of its achievement are commonly assumed to be writing, metallurgy, long-distance trade connections, astronomy and mathematics, social stratification and labor specialization, formalized governmental systems, and a structured urban culture.

Several major culture hearths emerged in the Neolithic period. Prominent centers of early creativity were found in Egypt, Crete, Mesopotamia, the Indus Valley of the Indian subcontinent, northern China, southeastern Asia, several locations in sub-Saharan Africa, in the Americas, and elsewhere (**Figure 2.15**). They arose in widely separated areas of the world, at different times, and under differing ecological circumstances. Each displayed its own unique mix of culture traits.

All major culture hearths were centered around relatively urbanized landscapes, the indisputable mark of civilization first encountered in the Near East 5,500–6,000 years ago, but the urbanization of each was somewhat differently arrived at and expressed (**Figure 2.16**). In some hearth areas, such as Mesopotamia and Egypt, the transition from settled agricultural village to urban form was gradual and prolonged. In Minoan Crete, urban life was less explicitly developed than in the Indus Valley, where early trade contacts with the Near East suggest the importance of exchange in fostering urban growth (see the feature "Social Collapse"). Trade seems particularly important in the development of West African culture hearths, such as Ghana and Kanem. Coming later (from the 8th to the 10th centuries) than the Nile or Mesopotamian centers, their numerous stone-built towns seem to have been supported both by an extensive agriculture whose origins were probably as early as those of the Middle East and, particularly, by long-distance trade across the

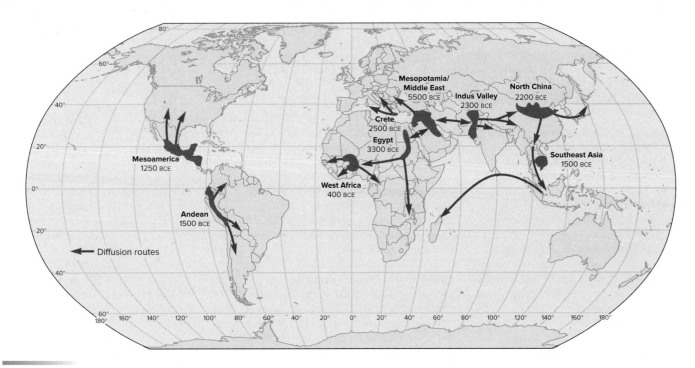

Figure 2.15 Early culture hearths of the Old World and the Americas. The BCE (Before the Common Era, equivalent to BC) dates approximate times when the hearths developed complex social, intellectual, and technological bases and served as cultural diffusion centers.

Figure 2.16 Urbanization was invariably a characteristic of culture hearths of both the Old and the New Worlds. Pictured is the Pyramid of the Sun and Avenue of the Dead at Teotihuacán, a city that at its height between 300 CE and 700 CE spread over nearly 18 square kilometers (7 square miles). Located some 50 kilometers (30 miles) northeast of Mexico City in the Valley of Mexico, the planned city of Teotihuacán featured broad, straight avenues and an enormous pyramid complex. The Avenue of the Dead, bordered with low stone-faced buildings, was some 3 kilometers (nearly 2 miles) in length.

©Glow Images

Social Collapse

Sustainable development requires a long-term balance between human actions and environmental conditions. When either poor management of resources by an exploiting culture or natural environmental alteration unrelated to human actions destroys that balance, a society's use of a region is no longer "sustainable" in the form previously established. Recent research shows that more than 4,000 years ago an unmanageable natural disaster spelled the death of half a dozen ancient civilizations from the Mediterranean Sea to the Indus Valley.

That disaster took the form of an intense, 300-year drought that destroyed the rain-based agriculture on which many of the early civilizations were dependent. Although they prospered through trade, urban societies were sustained by the efforts of farmers. When, about 2200 BCE, fields dried and crops failed through lack of rain, urban and rural inhabitants alike were forced to flee the dust storms and famine of intolerable environmental deterioration.

Evidence of the killer drought that destroyed so many Bronze Age cultures—for example, those of Mesopotamia, early Minoan Crete, and the Old Kingdom in Egypt—includes cities abandoned in 2200 BCE and not reoccupied for more than 300 years; deep accumulations (20–25 cm, or 8–10 in.) of windblown sand over farmlands during the same three centuries; abrupt declines in lake water levels; and thick lake-bed and sea-bed deposits of windblown debris.

Similar, but differently timed drought periods—such as the catastrophic aridity between 800 CE and 1000 CE that destroyed Mayan culture in Mesoamerica—have been blamed for the collapse of advanced societies in the New World as well. Not even the most thriving of early urban cultures were immune to restrictions imposed by nature.

Sahara. The Shang kingdom on the middle course of the Huang He (Yellow River) on the North China Plain had walled cities containing wattle-and-daub buildings but no monumental architecture.

Each culture hearth showed a rigorous organization of agriculture resulting in local productivity sufficient to enable a significant number of people to engage in nonfarm activities. Therefore, each hearth region saw the creation of a stratified society that included artisans, warriors, merchants, scholars, priests, and administrators. Each also developed or adopted astronomy, mathematics, and the all-essential calendar. Each, while advancing in cultural diversity and complexity, exported technologies, skills, and learned behaviors far beyond its own boundaries.

Writing appeared first in Mesopotamia and Egypt at least 5,000 years ago, as cuneiform in the former and as hieroglyphics in the latter. The separate forms of writing have suggested to some that they arose independently, in separate hearths. Others maintain that the idea of writing originated in Mesopotamia and spread outward to Egypt, to the Indus Valley, to Crete, and perhaps even to China, though independent development of Chinese ideographic writing is usually assumed. The systems of record keeping developed in New World hearths were not related to those of the Old, but once created, they spread widely in areas under the influence of Andean and Mesoamerican hearths. In Mesoamerica, distinctive Zapotec, Olmec, and Maya writing systems apparently emerged between 2,600 and 2,300 years ago. Skill in working iron, so important in Near Eastern kingdoms, was an export of sub-Saharan African hearths.

The anthropologist Julian Steward proposed the concept of **multilinear evolution** to explain the common characteristics of widely separated cultures developed under similar ecological circumstances. He suggested that each major environmental zone—arid, high-altitude, midlatitude steppe, tropical forest, and so on—tends to induce common adaptive traits in the cultures of those who exploit it. Those traits were, at base, founded on the development of agriculture and the emergence of similar cultural and administrative structures in the several culture hearths. But *similar* does not imply *identical*. Steward simply suggested that because comparable sequences of developmental events cannot always or even often be explained on the basis of borrowing or exporting of ideas and techniques (because of time and space differences in the cultures sharing them), they must be regarded as evidence of parallel creations based on similar ecologies. From similar origins, but through separate adaptations and independent innovations, distinctive cultures emerged.

In contrast, spatially separated cultures also show similarities because of the spatial spread (diffusion) of cultural traits from common origin sites. In some cases, cultural innovations are passed on along trade routes and through group contact rather than being the result of separate and independent creation. Although the extreme form of this idea, that cultures show similarities primarily—perhaps even solely—because of diffusion from one or only a very few common origin sites (a view known as *diffusionism*) is long out of favor, recent archaeological discoveries apparently document some very long-distance transfer of ideas, technologies, and language by migrating peoples.

In any event, the common characteristics deriving from multilinear evolution and the spread of specific culture traits and complexes contained the roots of **cultural convergence.** That term describes the sharing of technologies, organizational structures, and even cultural traits and artifacts that is so evident among widely separated societies in a modern world united by rapid communication and efficient transportation. Convergence in those worldwide traits is, for many observers, proof of the pervasive globalization of culture.

2.6 The Structure of Culture

Understanding a culture fully is, perhaps, impossible for one who is not part of it (and it is difficult for those inside it, as well). For analytical purposes, however, the traits and complexes of culture—its building blocks and expressions—may be grouped and examined as subsets of the whole. The anthropologist Leslie White suggested that for analytical purposes, a culture could be viewed as a three-part structure composed of subsystems that he termed *ideological, technological,* and *sociological.* Specific traits within each subsystem of culture can be labeled *mentifacts, artifacts,* or *sociofacts.* Together, according to these interpretations, the subsystems—identified by their separate components—comprise the system of culture as a whole. But we emphasize that this subdivision is for analytic purposes only; in reality, they are integrated, with each acting on the others and, in turn, affected by them. The **ideological subsystem** consists of ideas, beliefs, and knowledge of a culture and of the ways in which these things are expressed in speech or other forms of communication. Mythologies and theologies, legend, language, literature, philosophy, ethical systems, and folk wisdom make up this category. Passed on from generation to generation, these abstract belief systems, or **mentifacts,** tell us what we ought to believe, what we should value, and how we ought to act. Two basic strands of the ideological subsystem—language and religion—are the subject of Chapter 5.

The **technological subsystem** is composed of the material objects, together with the techniques of their use, by means of which people live. The objects are the tools and other instruments that enable us to feed, clothe, house, defend, transport, and amuse ourselves. We must have food, we must be protected from the elements, and we must be able to defend ourselves. The material objects that we use to fill these needs are **artifacts** (**Figure 2.17**). In Chapter 10, we will examine the relationship between technological subsystems and regional patterns of economic development.

The **sociological subsystem** of a culture is the sum of those expected and accepted patterns of interpersonal relations and social rituals that find their outlet in economic, political, military, religious, kinship and mating, and other associations. These **sociofacts** define the social organization of a culture. They

(a)

(b)

Figure 2.17 Artifacts are an important component of culture. (*a*) This Chinese farmer plowing with an ox uses artifacts (tools) typical of the lower technological levels of subsistence agriculture. (*b*) Cultures with advanced technological subsystems use complex machinery to harness inanimate energy for productive use.

(a) ©The McGraw-Hill Education/Barry Barker, photographer; (b) ©Vevchic/Shutterstock

regulate how the individual functions relative to the group—whether it be family, church, or state. There are no "givens" as far as the patterns of interaction in any of these associations are concerned, except that most cultures possess a variety of formal and informal ways of structuring behavior (**Figure 2.18**).

Classifications are necessarily arbitrary to some degree, and these classifications of the subsystems and components of culture are no exception. The three-part categorization of subsystems of culture, while helping us to appreciate its structure and complexity, can simultaneously obscure the many-sided nature of individual elements of culture. A dwelling, for example, is an artifact providing shelter for its occupants. It is, simultaneously, a sociofact reflecting the nature of the family or kinship group it is designed to house, and a mentifact summarizing a culture group's convictions about appropriate design and orientation of dwelling units. In the same vein, clothing serves as an artifact of bodily protection appropriate to climatic conditions, available materials and techniques, or the activity in which the wearer is engaged. But garments also may be sociofacts, identifying an individual's role in the social structure of the community or culture, and mentifacts, evoking larger community value systems (**Figure 2.19**).

Nothing in a culture stands totally alone. Changes in the ideas that a society holds may affect the sociological and technological systems just as changes in technology force adjustments in the social system. The abrupt alteration of the ideological structure of Russia following the 1917 communist revolution from a monarchical, agrarian system to an industrialized, communistic society involved sudden, interrelated alteration of all facets of that country's culture system. The equally abrupt disintegration of Russian communism in the early 1990s was similarly disruptive of all its established economic, social, and administrative structures. The interlocking nature of all aspects of a culture is termed **cultural integration.**

Figure 2.18 All societies prepare their children for membership in the culture group. In each of these settings, certain values, beliefs, skills, and proper ways of acting are being transmitted to the youngsters.

(a) ©Majority World/Universal Images Group/Getty Images; (b) ©Somos Images LLC/Alamy Stock Photo; (c) ©Steve AllenUK/123RF; (d) ©Ryan McVay/Photodisc/Getty Images

(a)

(c)

(b)

Figure 2.19 (*a*) When clothing serves primarily to cover, protect, or assist in activities, it is an *artifact*. (*b*) Some garments are *sociofacts*, identifying a role or position within the social structure: the distinctive "uniforms" of the soldier, the cleric, or the beribboned ambassador immediately proclaim their respective roles in a culture's social organizations. (*c*) The hijabs worn by many Muslim women are *mentifacts*, indicative not specifically of the fashion preferences of the wearer but of the values of the wearer's culture.

(a) ©Sanchai Rattakunchorn/123RF; (b) ©Rob Melnychuk/Stockbyte/Getty Images; (c) ©Grigvovan/Shutterstock

2.7 Culture Change

The recurring theme of cultural geography is change. No culture is, or has been, characterized by a permanently fixed set of material objects, systems of organization, or even ideologies. Admittedly, all of these may be long enduring within a stable, isolated society at equilibrium with its resource base. Such isolation and stability have always been rare. On the whole, while cultures are essentially conservative, they are always in a state of flux. Some changes are major and pervasive. The transition from hunter-gatherer to sedentary farmer, as we have seen, affected markedly every facet of the cultures experiencing that change and the experiences of the people living in those cultures. Profound, too, was the impact of the Industrial Revolution and its associated urbanization on all the societies that it has touched.

Not all change is so extensive as that following the introduction of agriculture or mechanized manufacturing. Many changes are so slight individually as to go almost unnoticed at their inception, though cumulatively they may substantially alter the affected culture. Think of how the culture of the United States differs today from what you know it to have been in 1940—not in essentials, perhaps, but in the innumerable electrical, digital, and transportational devices that have been introduced and in the social, behavioral, and recreational changes they and other technological changes have wrought. Among these latter have been shifts in employment patterns to include greater participation by women in the waged workforce and associated adjustments in attitudes toward the role of women in the society at large. Such cumulative changes occur because the cultural traits of any group are not independent; they are clustered in a coherent and integrated pattern. Change on a small scale can have wide repercussions as associated traits accommodate to the adopted adjustment. Change, both major and minor, within cultures is induced by *innovation, diffusion,* and *acculturation.*

Innovation

Innovation implies changes to a culture that result from ideas created within the social group itself and adopted by the culture. The novelty may be an invented improvement in material technology, like the bow and arrow or the jet engine. It may involve the development of nonmaterial forms of social structure and interaction: feudalism, for example, or Christianity. It may be a new form of popular music or a new way of styling hair.

Many innovations are of little consequence by themselves, but sometimes the widespread adoption of seemingly inconsequential innovations may bring about large changes when viewed over a period of time. A new musical style such as hip-hop "adopted" by a few people may spread to others and bring with it changes to vernacular speech, clothing styles, dance styles, graffiti art, and other forms of entertainment, which, in turn, may affect retailers' advertising campaigns and consumers' spending. Eventually, a new cultural form will be identified that may have an important impact on the thinking processes of the adopters and on those who come into contact with the adopters. Notice that a broad definition of innovation is used, but notice also that what is important is whether or not innovations are accepted and adopted.

Premodern and traditional societies characteristically are not very innovative. In societies at equilibrium with their environment and with no unmet needs, change has no obvious or immediate adaptive value and little reason to occur. Indeed, all societies have an innate resistance to change because innovation inevitably creates tensions between the new reality and other established socioeconomic conditions. Those tensions can be resolved only by adaptive changes elsewhere in the total system. The gap that may develop between, for example, a newly adopted technology and other, slower-paced social traits has been called *cultural lag.* Complaints about youthful fads or the glorification of times past are familiar examples of reluctance to accept or adjust to change.

Innovation and invention—frequently under stressful conditions—have marked the history of humankind. As we have seen, growing populations at the end of the Ice Age necessitated an expanded food base. In response, domestication of plants and animals appears to have occurred independently in more than one world area. Indeed, a most striking fact about early agriculture is the near universality of its development or adoption within a very short span of human history. In 10,000 BCE, the world population of no more than 10 million was nearly exclusively hunter-gatherers. By 1500 CE, only 1 percent of the world's 350 million people still followed that way of life. Today, much less than 1 percent of the world's 7 billion people do. The revolution in food production represented by the advent of agriculture affected every facet of the threefold subsystems of culture of every society accepting it. All innovation has a radiating impact on the web of culture; the more basic the innovation, the more pervasive its consequences.

In most modern societies, innovative change has become common, expected, and inevitable. The rate of invention, at least as measured by the number of patents granted, has steadily increased, and the period between idea conception and product availability has been decreasing. A general axiom is that the more

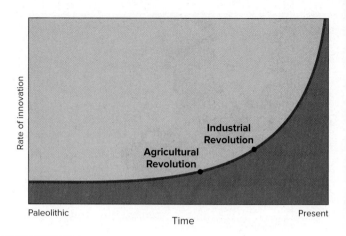

Figure 2.20 The trend of innovation through human history. Hunter-gatherers, living in equilibrium with their environment and their resource base during the Paleolithic period, had little need for innovation and no necessity for cultural change. The Agricultural Revolution accelerated the diffusion of the ideas and techniques of domestication, urbanization, and trade. With the Industrial Revolution, dramatic increases in all aspects of socioeconomic innovation began to alter cultures throughout the world.

ideas available and the more minds able to exploit and combine them, the greater the rate of innovation. The spatial implication is that larger urban centers of advanced technologies tend to be centers of innovation. This is not just because of their size but because of the opportunities they provide for the exchange of ideas. Indeed, ideas not only stimulate new thoughts and viewpoints but also create circumstances in which societies come to develop new solutions increasingly as a cultural practice in itself (**Figure 2.20**).

Diffusion

Diffusion is the process by which an idea or innovation is transmitted from one individual or group to another across space. Diffusion may assume a variety of forms, each different in its impact on social groups. Basically, however, two processes are involved: (1) People move, for any of a number of reasons, to a new area and take their culture with them. For example, immigrants to the American colonies brought with them crops and farming techniques, building styles, and concepts of government that were alien to their new home. (2) Information about an innovation (e.g., hybrid corn or compact discs) may spread throughout a society, perhaps aided by local or mass media advertising; or new adopters of an ideology or way of life—for example, a new religious creed—may be inspired or recruited by immigrant or native converts. The former is known as *relocation diffusion,* the latter as *expansion diffusion* (**Figure 2.21**).

Expansion diffusion involves the spread of a culture trait from one place to others when people who did not formerly practice the trait adopt it after direct or indirect contact with those who do practice the trait. In the process, the thing diffused also remains—and is frequently intensified—in the origin area. Islam, for example, expanded from its Arabian Peninsula origin locale across much of Asia and North Africa.

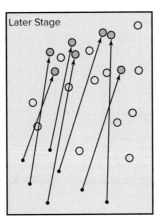

(a) RELOCATION DIFFUSION

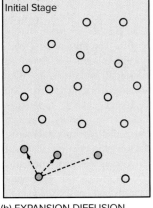

(b) EXPANSION DIFFUSION

O Non-Adopter
◑ Adopter of innovation

Figure 2.21 Processes of diffusion. (*a*) In *relocation diffusion,* innovations or ideas are transported to new areas by carriers who permanently leave the home locale. (*b*) In *expansion diffusion,* a phenomenon spreads from one place to another, but in the process remains and is often intensified in the place of origin.

Source: Redrawn from Spatial Diffusion, by Peter R. Gould, Resource Paper no. 4, page 4, Association of American Geographers, 1969.

At the same time, it strengthened its hold over its Near Eastern birthplace by displacing practitioners of tribal, Christian, and Jewish religions. The term *expansion* refers to the fact that this process of diffusion necessarily involves an increase in the number of people or societies practicing the trait. Furthermore, as potential adopters become adopters and serve to pass on the innovation to others, the number of contacts of adopters with potential adopters will compound. Consequently, the innovation will spread slowly at first and then more and more rapidly until saturation occurs or a barrier is reached. The incidence of adoption under expansion diffusion is represented by the S-shaped curve in **Figure 2.22.**

When adoption of an innovation spreads from an area where it is practiced to a neighboring area, reflecting *distance decay* (Chapter 3), it is termed **contagious diffusion.** This pattern of

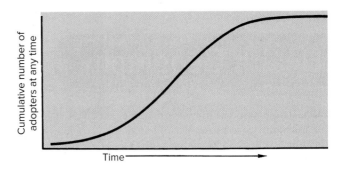

Figure 2.22 The diffusion of innovations over time. The number of adopters of an innovation rises at an increasing rate until the point at which about one-half of the total who ultimately decide to adopt the innovation have made the decision. At that point, the number of adopters increases at a decreasing rate.

diffusion reflects the importance of direct contact between those who developed or have adopted the innovation and those who newly encounter it, and is reminiscent of the course of infectious diseases (**Figure 2.23**); in fact, some geographers apply the concept of diffusion to the study of spatial patterns of disease (**see the feature "Health Geography"**). Contagious diffusion results in the continuous spread of innovations, like "waves." In some instances, however, geographic distance is less important in the transfer of ideas than is communication between major centers or important people. News of new clothing styles, for

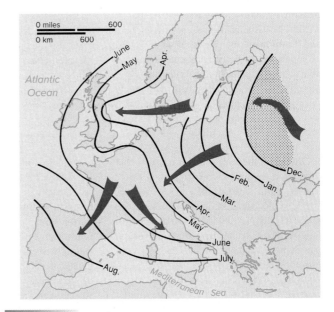

Figure 2.23 The pattern of contagious *diffusion* is sensitive to both time and distance, as suggested by the diffusion pathways of the European influenza pandemic of 1781. The pattern there was a wavelike radiation from a Russian origin area.

Source: Based on Gerald F. Pyle and K. David Patterson, Ecology of Disease 2, no. 3 (1984): 179.

Health Geography

Although our focus on diffusion in this chapter has mostly concerned the spread of cultural traits, diseases also spread across the landscape. Scholars in *health geography* are interested in the spatial patterns of diseases. Health geographers incorporate theories and methods from a number of different disciplines in addition to geography, including epidemiology and other social sciences, in order to examine health-related topics. They look beyond individual-level behaviors to study how social and physical environments (i.e., contextual factors), and individual's interactions with one another and the environment (i.e., social epidemiology), influence the health of populations and the spread of disease.

Consider an infectious disease, or a disease that can be transmitted from one person to another by some type of contact. Often these diseases spread over space and time following the rules of spatial diffusion, such as distance decay and hierarchies of nodes. Health geographers additionally consider how individuals and groups have different patterns of movement and types of contacts, which may result in unique but predictable patterns of disease diffusion. For example, wealth is associated with activity spaces and travel patterns. Wealthier individuals, or populations from richer countries, may be able to travel farther in a shorter amount of time (such as on airplanes) before spreading infection. On the other hand, diffusion of infectious disease may present differently in an impoverished neighborhood, with cases of infection clustering closely together.

No matter the scale—within a city, within a country, or across countries at a global scale—health inequalities exist. Wealth, education, occupation, and social status all predict better health. In the United States, long-standing racial disparities in health persist in a number of infectious and non-infectious diseases. For example, African Americans suffer from HIV (the virus that causes AIDS) at a rate nearly 10 times higher than their non-Hispanic white counterparts. The root of this disparity is still unknown but is at least partially due to unequal access to health care, stigma, and racism. But the persistence of the disparity is also partly explained by predictable qualities of disease diffusion. Within sexually transmitted infections, sexual contacts between people are not randomly determined. Rather, contacts are structured based on race/ethnicity and age—people have sexual contact with others that are similar to them. According to the National Longitudinal Study of Adolescent to Adult Health, around 90 percent of sexual partnerships among young adults are within the same racial category (e.g., white/white or black/black), and sexually transmitted infections circulate within these segregated networks. This results in a contact network that dictates how infections spread through populations and partly sustains racial disparities in HIV.

Social processes also influence who gets infected after being exposed to an infection. Patterns of vaccination can predict where clusters of infection may or may not appear. In many resource-poor settings, higher levels of vaccine-preventable diseases, such as measles, are due to inadequate coverage of vaccines. However, even in the United States, pockets of infection appear in places where levels of vaccination were once sufficiently high to prevent an epidemic. Recently, one of the largest outbreaks of measles occurred in a tight-knit community in Minnesota where immunization rates plummeted from 92 percent to 42 percent over 5 years based on discredited fears that vaccines trigger autism. Diffusion of fear and mistrust in the community led to the decline in vaccination rates, and thus the measles outbreak.

Now consider a noninfectious disease or health condition, such as obesity. A medical doctor would likely claim that obesity was a result of a poor diet and lack of exercise, and thus would prescribe a healthy diet and plenty of exercise to treat the problem. A health geographer would approach the obesity epidemic with a broader lens. She would consider how and why obesity could be associated with a persons' natural and social environment, culture, and lived experiences throughout his life, in addition to individual behavior (such as diet and exercise); interventions for reducing obesity would have a similar broad, inclusive approach.

Descriptive health geography can reveal interesting spatial patterns in the health of populations over space and time. More than one-third (36.5 percent) of adults in the United States have obesity, whereas in 1990, obese adults made up less than 15 percent of the population. Further, maps of obesity by state clearly show that high rates of obesity cluster in the Deep South—in Arkansas, Louisiana, Mississippi, and Alabama. Why is obesity unevenly distributed across the United States? Spatial analytic techniques reveal that patterns of obesity correlate with race, income, and smoking behaviors, among other factors. In some international settings, being overweight is a sign of status and prestige. Descriptive health geography can describe patterns and lay the foundation for further scientific study.

Physical space and the built environment clearly affect the health of populations, from exposure to air pollution to the ability to walk safely in a neighborhood. Health geographers examine nutritional environments to assess whether access to affordable, healthy food influences diet and rates of obesity. Interestingly, the results of research on this have been mixed. Some research has found a link between the density of healthy food options and a healthier diet. On the other hand, a study in Glasgow showed that when a large supermarket opened in a poor neighborhood that had previously lacked one, local residents did not begin to shop there—they perceived that the market was not designed for them. Links between the physical environment and health can be straightforward, or can be more complex and depend on other concepts of culture, tradition, diffusion of knowledge, and history.

By Susan Cassels, University of California, Santa Barbara. All rights reserved. Used with permission.

example, quickly spreads internationally among major cities and only later filters down irregularly to smaller towns and rural areas. The pattern of transferring ideas first between larger places or prominent people and only later to smaller or less important points or people is known as **hierarchical diffusion.** The Christian faith in Europe, for example, spread from Rome as the principal center to provincial capitals and thence to smaller Roman settlements in largely pagan-occupied territories (see Figure 5.21). Today, new discoveries are shared among scientists at leading universities before they appear in textbooks or become general knowledge through the public press. The hierarchical pattern takes place because, in many cases, distance is largely overcome by communication networks; this implies contact through indirect means, such as the Internet or mass media. Big cities or leading scientists, connected by strong information flows, are "closer" than their simple distance separation suggests. Hierarchical diffusion results in patterns of innovation "jumping" over space, and often results in the formation of new clusters of the trait.

While the diffusion of ideas may be slowed by the need to overcome distance, their speed of spread may be increased to the point of becoming almost instantaneous through the *space-time compression* made possible by modern communication. Given access to radios; telephones; worldwide transmission of television news, sports, and entertainment programs; and—perhaps most importantly—to computers and the Internet, people and areas distantly separated can rapidly share in a common fund of thought and innovation. Modern communication technology, that is, has encouraged and facilitated the globalization of culture.

In contrast to expansion diffusion, **relocation diffusion** occurs when the innovation is carried to new areas by migrating individuals or populations that possess it (Figure 2.21a); in Chapter 3, we learn that migration is also known by geographers as *residential relocation*. Mentifacts, sociofacts, or artifacts are therefore introduced into new locales by new settlers who become part of populations not themselves associated or in contact with the origin area of the innovation. The spread of religions by settlers or conquerors is a clear example of relocation diffusion, as was the diffusion of agriculture to Europe from the Middle East (Figure 2.13). Christian Europeans brought their faiths to areas of colonization or economic penetration throughout the world. At the world scale, massive relocation diffusion resulted from the European colonization and economic penetration that began in the 16th century. More localized relocation diffusion continues today as Asian refugees or foreign "guest workers" bring their cultural traits to their new areas of settlement in Europe or North America. Like expansion diffusion, relocation diffusion spreads cultural traits across the landscape, but unlike it, relocation diffusion does not entail an increase in the number of people or societies practicing the trait. Also, because migration often reflects patterns of distance decay, wherein people move over short distances, or it reflects patterns of long-distance relocation over transportation networks like highways or airline routes, relocation diffusion can also be characterized as reflecting patterns of contagious or hierarchical diffusion.

For either expansion or relocation diffusion, innovations may be relatively readily diffused to, and accepted by, cultures that have basic similarities and compatibilities. Continental Europe and North America, for example, could easily and quickly adopt the innovations of the Industrial Revolution diffused from England with which they shared a common ethnic, economic, and technological background. Industrialization was not quickly accepted in Asian and African societies of totally different cultural conditioning. On the ideological level, too, successful diffusion depends on acceptability of the innovations. The attempt by Mohammed Reza Pahlavi, the shah of Iran, at rapid westernization of traditional Iranian Islamic culture after World War II provoked a traditionalist backlash and revolution that deposed him and reestablished clerical control of the state in 1979.

The conclusion must be, therefore, that diffusion cannot be viewed solely as the outcome of contact and knowledge dispersal. The acceptance of new traits, artifacts, or ways of doing or thinking by a potential receiving population depends not just on information flow to that population but also upon its entire cultural and economic structure. Innovation may be rejected not because of lack of knowledge but because the new trait violates the established cultural norms of the culture to which it is introduced. For example, cash crop specialization recommended to a peasant agricultural society may be rejected not because it is not understood, but because it unacceptably disrupts the knowledge base and culture complex devoted to assured food security in a subsistence farming economy. Similarly, less disruptive new production ideas—chemical fertilizers, deep-well irrigation, hybrid seeds, and the like—may be rejected simply because, though understood, they are not affordable. Culture is a complex organized system and culture change involves alteration of the system's established structure in ways that may be rejected even after knowledge of an innovation is received and understood.

It is not always possible, of course, to determine the precise point of origin or the routes of diffusion of innovations now widely adopted (see the feature "Documenting Diffusion"). Nor is it always certain whether the existence of a cultural trait in two different areas is the result of diffusion or of **independent** (or *parallel*) **invention.** Cultural similarities do not necessarily prove that diffusion has occurred. The pyramids of Egypt and of the Central American Maya civilization most likely were separately conceived and are not necessarily evidence, as some have proposed, of pre-Columbian voyages from the Mediterranean to the Americas (or of visits by aliens from outer space, as has also been suggested). A Neolithic monument-building culture, after all, has only a limited number of shapes from which to choose.

Historical examples of independent, parallel invention are numerous: logarithms by John Napier (1614) and Jost Bürgi (1620), the calculus by Isaac Newton (1672) and Gottfried Leibniz (1675), and the telephone by Elisha Gray and Alexander Graham Bell (1876) are commonly cited. It appears beyond doubt that agriculture was independently developed not only in both the New World and the Old, but also in more than one culture hearth in each of the hemispheres.

Documenting Diffusion

The places of origin of many ideas, items, and technologies important in contemporary cultures are only dimly known or supposed, and their routes of diffusion are speculative at best. Gunpowder and printing are presumed to be the products of Chinese inventiveness; the lateen sail has been traced to the Near Eastern culture world. The moldboard plow is ascribed to 6th-century Slavs of northeastern Europe. The sequence and routes of the diffusion of these innovations has not been documented.

In other cases, such documentation exists, and the process of diffusion is open to analysis. Clearly marked is the diffusion path of the custom of smoking tobacco, a practice that originated among Amerindians. Sir Walter Raleigh's Virginia colonists, returning home in 1586, introduced smoking to English court circles, and the habit very quickly spread among the general populace. England became the source region of the new custom for northern Europe; smoking was introduced to Holland by English medical students in 1590. The Dutch and English together spread the habit by sea to the Baltic and Scandinavian areas and overland through Germany to Russia. The innovation continued its eastward diffusion, and within a hundred years, tobacco had spread across Siberia and was, in the 1740s, reintroduced to the American continent at Alaska by Russian fur traders. A second route of diffusion for tobacco

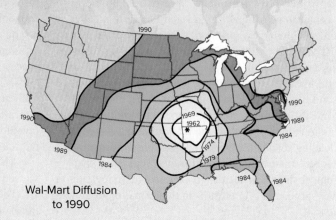

Wal-Mart Diffusion to 1990

Figure 2C

Source: *Map based on data from Thomas O. Graff and Dub Ashton, "Spatial Diffusion of Wal-Mart: Contagious and Reverse Hierarchical Elements."* Professional Geographer *46, 1 (1994): 19–29.*

smoking can be traced from Spain, where the custom was introduced in 1558, and from which it spread more slowly through the Mediterranean area into Africa, the Near East, and Southeast Asia.

In more recent times, hybrid corn was first adopted by imaginative farmers of northern Illinois and eastern Iowa in the mid-1930s. By the late 1930s and early 1940s, the new seeds were being planted as far east as Ohio and north to Minnesota, Wisconsin, and northern Michigan. By the late 1940s, all commercial corn-growing districts of the

United States and southern Canada were cultivating hybrid corn varieties.

A similar pattern of diffusion marked the expansion of the Wal-Mart store chain. From its origin in northwest Arkansas in 1962, the discount chain had spread throughout the United States by the 1990s to become the country's largest retailer in sales volume (Figure 2C). In its expansion, Wal-Mart displayed a "reverse hierarchical" diffusion, initially spreading by being price-competitive with small-town merchants before opening its first stores in larger cities and metropolitan areas.

Acculturation and Cultural Modification

A culture group may undergo major modifications in its own identifying traits by adopting some or all of the characteristics of another, dominant culture group. Such is the case in **acculturation**—discussed at greater length in Chapter 6 (Section 6.3)—as immigrant populations take on the values, attitudes, customs, and speech of the receiving society, which itself undergoes change from absorption of the arriving group. A different form of contact and subsequent cultural alteration may occur in a conquered or colonized region where the subordinate or subject population is either forced to adopt the culture of the new ruling group, introduced through relocation diffusion, or does so voluntarily, overwhelmed by the superiority in numbers or

the technical level of the conqueror. Tribal Europeans in areas of Roman conquest, native populations in the wake of Slavic occupation of Siberia, and Native Americans stripped of their lands following European settlement of North America experienced this kind of cultural modification or adoption.

In extreme cases, of course, small and, particularly, primitive indigenous groups brought into contact with conquering or absorbing societies may simply cease to exist as separate cultural entities. Although presumably such cultural loss has been part of all of human history, its increasing occurrence has been noted over the past 500 years. By one informed estimate, at least one-third of the world's inventory of human cultures has totally disappeared since 1500 CE, along with their languages, traditions, ways of life, and, indeed, their very identity and memory of their existence.

In many instances, close contact between two different groups may involve adjustments of the original cultural patterns of both rather than disappearance of either. For example, changes in Japanese political organization and philosophy were imposed by occupying Americans after World War II, and the Japanese voluntarily adopted some of the more frivolous aspects of American life (**Figure 2.24**). In turn, American society was enriched by the selective importation of Japanese cuisine, architecture, and philosophy, demonstrating the two-way nature of cultural diffusion. Where that two-way flow reflects a more equal exchange of cultural outlooks and ways of life, a process of *transculturation* has occurred. That process is observable within the United States as massive South and Central American immigration begins to intertwine formerly contrasting cultures, altering both.

2.8 Contact Between Regions

Virtually all cultures are amalgams of innumerable innovations spread spatially from their points of origin and integrated into the structure of the receiving societies. It has been estimated that no more than 10 percent of the cultural items of any society are traceable to innovations created by its members and that the other 90 percent come to the society through diffusion (see the feature "A Homemade Culture"). Because, as we have seen, the pace of innovation is affected strongly by the mixing of ideas among alert, responsive people and is increased by exposure to a variety of cultures, the most active and innovative historical hearths of culture were those at crossroad locations and those deeply involved in distant trade and colonization. Ancient Mesopotamia and classical Greece and Rome had such locations and involvements, as did the West African culture hearth after the 5th century and, much later, England during the Industrial Revolution and the spread of the British Empire.

Recent changes in technology permit us to travel farther than ever before, with greater safety and speed, and to communicate without physical contact more easily and completely than was previously possible. This intensification of contact has resulted in an acceleration of innovation and in the rapid spread of goods and ideas. Several millennia ago, innovations such as smelting of metals took hundreds of years to diffuse. Today, worldwide

Figure 2.24 Baseball, an import from America, is one of the most popular sports in Japan, attracting millions of spectators annually.

©Masterpress - Samurai Japan/Getty Images

diffusion—through Internet interest groups, for example—may be almost instantaneous.

Obstacles do exist, of course. *Diffusion barriers* are any conditions that hinder either the flow of information or the movement of people and thus retard or prevent the acceptance of an innovation. Because of the *friction of distance,* generally the farther two areas are from each other, the less likely is interaction to occur, an observation earlier (Section 2.7) summarized by the term *distance decay.* Distance and barriers as factors in spatial interaction are further explored in Chapter 3. For now, it is sufficient to note that distance may be an *absorbing barrier,* halting the spread of an innovation.

Political restrictions, religious taboos, and other social customs are cultural barriers to diffusion. The French Canadians, although close geographically to many Anglo-Canadian centers of diffusion such as Ottawa and Toronto, strive to be only minimally influenced by them. Both their language and culture complex govern their selective acceptance of Anglo influences, and restrictive French-only language regulations are enforced to preserve the integrity of their distinctive French culture. In a more extreme fashion, the Afghan Taliban and other Mideast fundamentalist groups adamantly or violently reject Western sociocultural values, seeking to preserve their religious and cultural purity through isolation from secular, non-Islamic influences. Traditional groups, perhaps controlled by firm religious conviction, may largely reject culture traits and technologies of the larger society in whose midst they live (see Figure 7.2 in Chapter 7).

Adopting cultures do not usually accept items intact that originate outside the receiving society. Diffused ideas and artifacts commonly undergo some alteration of meaning or form that makes them acceptable to a borrowing group. The process of the fusion of the old and new is called **syncretism** and is a major feature of culture change. It can be seen in alterations to religious ritual and dogma made by convert societies seeking acceptable conformity between old and new beliefs. For example, enslaved people brought voodoo from West Africa to the Americas where it thrived in Haiti and Louisiana. Over the years, it absorbed influences from French and Spanish Catholicism, American Indian spiritual practices, and even Masonic tradition. Despite those adaptive mixings, many believers consider themselves to be Catholics and see no contradiction between Christianity and their faith in protective spirits and other tenets of voodoo. On a more familiar level, syncretism is reflected in subtle or blatant alterations of imported cuisines to make them conform to the demands of the American palate and fast-food franchises (**Figure 2.25**).

Figure 2.25 Foreign foods modified for American tastes represent syncretism in action.
©Mark Bjelland

SUMMARY

The web of culture is composed of many strands. Together, culture traits and complexes in their spatial patterns create human landscapes, define culture regions, and distinguish culture groups. Those landscapes, regions, and group characteristics change through time as human societies interact with their environments, develop for themselves new solutions to collective needs, or are altered through innovations adopted from outside the group itself. The cultural similarity of a preagricultural world composed solely of hunter-gatherers was lost as domestication of plants and animals in many world areas led to the emergence of culture hearths of wide-ranging innovation and to a cultural divergence between farmers and gatherers. Innovations spread outward from their origin points, carried by migrants through relocation diffusion or adopted by others through a variety of expansion diffusion and acculturation processes. Although diffusion barriers exist, most successful or advantageous innovations find adopters, and both cultural modification and cultural convergence of different societies result. The details of the technological, sociological, and ideological subsystems of culture define the differences that still exist between world areas.

The ivory hunters who opened our chapter showed how varied and complex the culture of even a primitive group can be. Their artifacts of clothing, fire making, hunting, and fishing displayed diversity and ingenuity. They were part of a structured kinship system and engaged in organized production and trade. Their artistic efforts and ritual burial customs speak of a sophisticated set of abstract beliefs and philosophies. Their culture complex did not develop in isolation; it reflected at least in part their contacts with other groups, even those far distant from their Paris Basin homeland. As have culture groups always and everywhere, the hunters continued their own pursuits and interacted with others in spatial settings. They exhibited and benefited from structured *spatial behavior,* the topic to which we next turn our attention.

KEY WORDS

acculturation
artifact
carrying capacity
contagious diffusion
cultural autonomy
cultural convergence
cultural divergence
cultural ecology
cultural integration
cultural landscape
cultural system
culture

culture complex
culture hearth
culture realm
culture region
culture trait
diffusion
domestication
environmental determinism
expansion diffusion
hierarchical diffusion
hunter-gatherer
ideological subsystem

independent invention
innovation
mentifact
multilinear evolution
possibilism
relocation diffusion
sociofact
sociological subsystem
syncretism
technological subsystem

FOR REVIEW

1. What is included in the concept of *culture?* How is culture transmitted? What personal characteristics affect the aspects of culture that any single individual acquires or fully masters?

2. What do we mean by *domestication?* When and where did the domestication of plants and animals occur? What impact on culture and population numbers did plant domestication have?

3. What is a *culture hearth?* What new traits of culture characterized the early hearths? Identify and locate some of the major culture hearths that emerged at the close of the Neolithic period.

4. What do we mean by *innovation?* By *diffusion?* What different processes and patterns of diffusion can you describe? Discuss the role played by innovation and diffusion in altering the cultural structure in which you are a participant from that experienced by your great-grandparents.

5. Differentiate between *culture traits* and *culture complexes,* and between *environmental determinism* and *possibilism.*

6. What are the components or subsystems of the three-part system of culture? What characteristics are included in each of the subsystems?

KEY CONCEPTS REVIEW

1. **What are the components of culture and nature of culture–environment interactions?** Section 2.1–2.2.

   Culture traits and complexes may be grouped into culture regions and realms. Differing developmental levels color human perceptions of environmental opportunities. In general, as the active agents in the relationship, humans exert adverse impacts on the natural environment.

2. **How did cultures develop and diverge and where did cultural advances originate?** Section 2.3–2.5.

   From Paleolithic hunting and gathering to Neolithic farming and then to city civilizations, different groups made differently timed cultural transitions. All early cultural advances had their origins in a few areally distinct *hearths.*

3. **What are the structures of culture and forms of culture change?** Section 2.6–2.8.

   All cultures contain ideological, technological, and sociological components that work together to create cultural integration. Cultures change through innovations that they themselves invent or that diffuse from other areas and are accepted or adapted.

SPATIAL INTERACTION AND SPATIAL BEHAVIOR

The blurred lights of traffic on the Avenue des Champs Elysees in Paris, France, typify spatial interaction in contemporary society.
©Goodshoot/Getty Images

Key Concepts

3.1 The three bases for spatial interaction and measuring the likelihod of spatial interaction

3.2–3.4 The forms and nature of human spatial behavior

3.5 Information and cognition in human spatial behavior, and migration patters, types and controls

*E*arly in January of 1849 we first thought of migrating to California. It was a period of National hard times . . . and we longed to go to the new El Dorado and "pick up" gold enough with which to return and pay off our debts.

Our discontent and restlessness were enhanced by the fact that my health was not good. . . . The physician advised an entire change of climate thus to avoid the intense cold of Iowa, and recommended a sea voyage, but finally approved of our contemplated trip across the plains in a "prairie schooner."

Full of the energy and enthusiasm of youth, the prospects of so hazardous an undertaking had no terror for us, indeed, as we had been married but a few months, it appealed to us as a romantic wedding tour.[1]

So begins Catherine Haun's account of their nine-month journey from Iowa to California, just two of the quarter-million people who traveled across the continent on the Overland Trail in one of the world's great migrations. The migrants faced months of grueling struggle over badly marked routes that crossed swollen rivers, deserts, and mountains. The weather was often foul, with hailstorms, drenching rains, and burning summer temperatures. Graves along the route were a silent testimony to the lives claimed by buffalo stampedes, Indian skirmishes, cholera epidemics, and other disasters.

What inducements were so great as to make emigrants leave behind all that was familiar and risk their lives on an uncertain venture? Catherine Haun alludes to economic hard times gripping the country and to their hope for riches to be found in California. Like other migrants, the Hauns were attracted by the climate in the West, which was said to be always sunny and free of disease. Finally, like most who undertook the perilous journey west, the Hauns were young, moved by restlessness, a sense of adventure, and a belief that greater opportunities awaited in a new land. They, like their predecessors back to the beginnings of humankind, were acting in space and across space on the basis of acquired information and anticipation of opportunity—prepared to pay the price in time, money, and hardship costs of overcoming distance.

A fundamental question in human geography is: What considerations influence how individual human beings use space and act within it? Related queries include: Are there discernible controls on human spatial behavior? How does distance affect human interaction? How do our beliefs about places influence our spatial activities? How do we overcome the consequences of distance in the exchange of commodities and information? How are movement and migration decisions (like that of the Hauns) reached? How have new technologies enabled increased spatial interaction across great distances and contributed to globalization? These questions address geography's concern with understanding spatial interaction.

Spatial interaction is contact between places, and in human geography, means the movement of people, ideas, and commodities (goods bought and sold) from place to place. The Hauns were engaging in spatial interaction (**Figure 3.1**). International trade, the movement of semitrailers on the expressways, radio broadcasts, and business or personal telephone calls are other familiar examples. Such movements and exchanges are designed to achieve

Figure 3.1 A public bus negotiates a washed-out section of highway on one of the major routes connecting the capital city of Kathmandu with southern Nepal and India. Movement in Nepal is more difficult than in developed countries because of the limited road network, narrow, winding mountain roads, and frequent landslides. A ride on a public bus in Nepal can be an adventure in sharing space with people, agricultural produce, and livestock.

©Mark Bjelland

effective integration between different points of human activity. Movement and contact of whatever nature between places serve to smooth out the spatially differing availability of required resources, commodities, information, or opportunities. Whatever the particular purpose of a movement, there is inevitably some manner of trade-off balancing the benefit of the interaction with the costs that are incurred in overcoming spatial separation. Because commodity movements represent simple demonstrations of the principles underlying all spatial interactions, let us turn to them first.

3.1 Bases for Interaction

Neither the world's resources nor products are uniformly distributed. Commodity flows are responses to these differences; they are links between points of supply and locales of demand. Such response may not be immediate, or even direct. Factors such as the awareness of supplies or markets, the presence or absence of transportation connections, costs of movement, and the ability to pay for things wanted and needed—all that and more affect the structure of commodity exchange. Underlying these, however, is a basic set of controlling principles governing spatial interaction.

A Summarizing Model

Geographer Edward Ullman (1912–1976) speculated on the essential conditions affecting such interactions and proposed an explanatory model. He observed that spatial interaction is effectively controlled by three flow-determining factors that he called *complementarity, transferability,* and *intervening opportunity.* Although Ullman's model deals with commodity flows, it has—as we shall see—applicability to information transfers and human movements as well.

Complementarity

For two places to interact, one place must have what another place wants and can secure. That is, one place must have a supply

[1]From Catherine Haun, "A Woman's Trip Across the Plains in 1849," in Lillian Schlissel, *Women's Diaries of the Westward Journey* (New York: Schocken Books, 1982).

of an item for which there is an effective demand in the other, as evidenced by desire for the item, purchasing power to acquire it, and means to transport it. The word describing this circumstance is **complementarity.** Effective supply and demand are important considerations; mere differences from place to place in commodity surplus or deficit are not enough to initiate exchange. For example, Greenland and the Amazon basin are notably unlike in their natural resources and economies, but their amount of interaction is minimal. Supply and market must come together, as they do in the flow of seasonal fruits and vegetables from California's Imperial Valley to the urban markets of the American Midwest and East or in the movement of manganese from Ukraine to the steel mills of Western Europe. The massive movement of crude and refined petroleum clearly demonstrates complementarity in international trade (**Figure 3.2**). More generalized patterns of complementarity underlie the exchanges of the raw materials and agricultural goods of less developed countries for the money or industrial commodities of the developed states.

Transferability

Even when complementarity exists, spatial interaction occurs only when conditions of **transferability**—acceptable costs of an exchange—are met. Spatial movement responds not just to availability and demand but to considerations of time and cost. Transferability is an expression of the mobility of a commodity and is a function of three interrelated conditions: (1) the characteristics and value of the product; (2) the distance, measured in time and money penalties, over which it must be moved; and (3) the ability of the commodity to bear the costs of movement. If the time

and money costs of traversing a distance are too great, exchange does not occur. That is, mobility is not just a physical matter but an economic one as well. If a given commodity is not affordable upon delivery to an otherwise willing buyer, it will not move in trade, and the potential buyer must seek a substitute or go without.

Transferability is not a constant condition. It differs between places over time, depending upon what is being transferred and how it is to be moved. In the 1820s, the newly opened Erie Canal cut shipping costs from Buffalo to New York City by 90 percent. However, the growth of railroads and highways across New York State provided quicker alternatives and caused great declines in commercial canal traffic during the early 20th century. More recently, containerized shipping has had a similar effect on the global shipments of goods. An increasing scarcity of high-quality ores will enhance the transferability of lower-quality mine outputs by increasing their value. Low-cost bulk commodities not economically moved by air may be fully transferable by rail or water. Poorly developed and costly transportation may inhibit exchanges even at short distance among otherwise willing traders. In short, transferability expresses the changing relationships between the costs of transportation and the value of the product to be shipped.

Intervening Opportunity

Complementarity can be effective only in the absence of more attractive alternative sources of supply or demand closer at hand or cheaper. **Intervening opportunities** serve to reduce supply/demand interactions that otherwise might develop between distant complementary areas. A supply of Saharan sand is not enough to assure its flow to sand-deficient Manhattan because

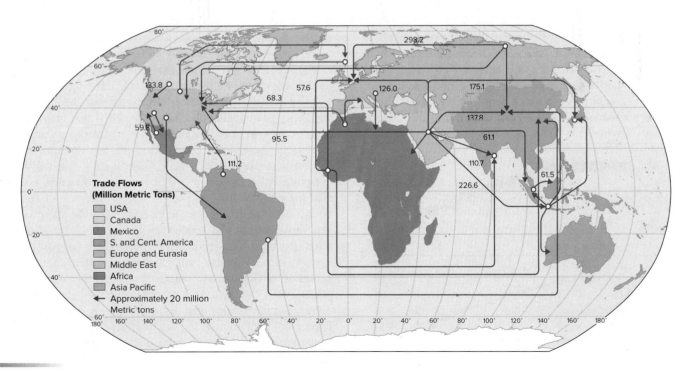

Figure 3.2 Major international crude oil and other product exports flow, 2007. Complementarity is so basic in initiating interaction that even relatively low-value bulk commodities such as coal, fertilizer, and grain move in trade over long distances. For many years, despite fluctuating prices, petroleum has been the most important commodity in international trade, moving long distances in response to effective supply and demand considerations.

Source: Adapted from The BP Amoco Statistical Review of World Energy, 2008.

supplies of sand are more easily and cheaply available within the New York metropolitan region. For reasons of cost and convenience, a purchaser is unlikely to buy identical commodities at a distance when a suitable nearby supply is available. When it is, the intervening opportunity demonstrates complementarity at a shorter distance.

Similarly, markets and destinations are sought, if possible, close at hand. Growing metropolitan demand in California reduces the importance of Midwestern markets for western fruit growers. The intervening opportunities offered by Chicago or Philadelphia reduce the number of job seekers from Iowa searching for employment in New York City. People from New England are more likely to take winter vacations in Florida, which is relatively near and accessible, than in Southern California, which is not. That is, opportunities that are discerned closer at hand reduce the pull of opportunities offered by a distant destination (**Figure 3.3**). Patterns of spatial interaction are dynamic, reflecting the changeable structure of apparent opportunity.

Measuring Interaction

Complementarity, transferability, and intervening opportunity— the controlling conditions of commodity movement—help us understand all forms of spatial interaction, including choosing a restaurant, where to go to college, or where to buy a house—even the once-in-a-lifetime transcontinental adventure of the Hauns. The study of unique spatial interactions, such as the discovery of an Inuit carving in a Miami gift shop, is interesting but does not establish general patterns. In this chapter, we focus on general principles that govern the frequency and intensity of interaction both to validate the three preconditions of spatial exchange and to establish the probability that any given potential interaction will actually occur. Our interest is similar to that of the physical scientist investigating, for example, the response of a gas to variations in temperature and pressure. The concern there is with all of the gas molecules and the probability of their collective reactions; the action of any particular molecule is of little interest. Similarly, we are mostly concerned here with the probability of aggregate, not individual (disaggregate), behavior.

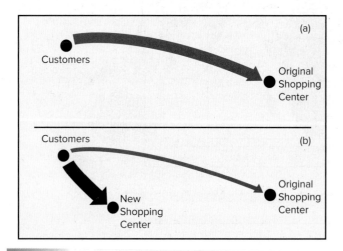

Figure 3.3 (*a*) The volume of expected customers for a shopping mall based solely on their complementarity and distance apart, may be (*b*) reduced if a new mall opens as an intervening opportunity nearer to the customers.

That concern with aggregate behavior conceals or ignores a great deal of spatial interaction of vital importance both in the real world and in human geography. Most theoretical and observational studies of spatial interaction have focused on the standard normative spatial behavior of fully physically and economically capable Western-culture adults. That standard does not address the individual or collective spatial problems and actions of such others as children, the poor, the elderly, the disabled, or socially disadvantaged individuals or groups, nor does it recognize the very real, though often subtle, differences between male and female spatial action responses and decisions. Our orientation to the North American culture realm means also that the aggregate spatial behavioral norms that we discern there fail to recognize the many and varied sociocultural, economic, legal, and similar constraints on spatial behavior operative in other culture areas of the world. Nevertheless, observational evidence suggests that the same basic influences on personal spatial behavior that we recognize here have universal applicability, despite their inevitable modification in different contexts.

Distance Decay

In all manner of ways, the lives and activities of people everywhere are influenced by the **friction of distance.** That phrase reminds us that distance has a retarding effect on human interaction because there are increasing penalties in time and cost associated with longer-distance, more expensive interchanges. We visit nearby friends more often than distant friends; we go more frequently to the neighborhood convenience store cluster than to the farther regional shopping center. Telephone calls or mail deliveries between nearby towns are greater in volume than those to more distant locations. An informal study showed that college students living in dormitories near the cafeteria are more likely to use the cafeteria; students living farther away do not visit the cafeteria as often (we assume that students who dislike cafeteria food do not choose to live in more distant dormitories!).

Our common experience, clearly supported by maps and statistical data tracking all kinds of flows, is that most interactions occur over short distances. That is, interaction decreases as distance increases, a reflection of the fact that transferability costs increase with distance. More generally stated, **distance decay** describes the decline of an activity or function with increasing distance from its point of origin. Because of distance decay, closer places interact more and tend to be more similar in a variety of ways. This is a very general trend that is seen so often in geographic phenomena (including those of physical geography such as weather patterns) that it is widely known as the **First Law of Geography,** so named by geographer Waldo Tobler (1930–2018): "Everything is related to everything else, but near things are more related than distant things." As the examples in **Figure 3.4** demonstrate, for instance, near destinations have a disproportionate pull over more distant points in commodity movements. However, it is also evident that the rate of distance decay varies with the type of activity. It usually declines nonlinearly at a decelerating rate so that the reduction is fastest at close distances, reducing more and more slowly as one gets farther and farther away. Thus, the amount of interaction between two points

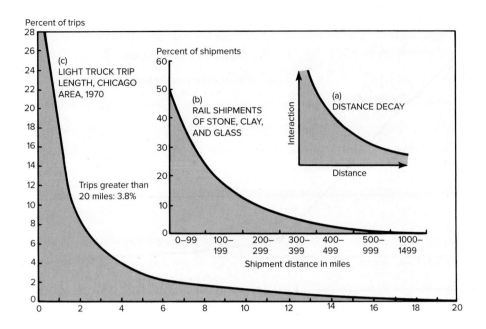

Figure 3.4 The shape of distance decay. The geographer Waldo Tobler summarized the concept of distance decay in proposing his First Law of Geography: "Everything is related to everything else, but near things are more related than distant things." Distance decay curves vary with the type of flow. Graph *(a)* is a generalized statement of distance decay, *(b)* summarizes U.S. data for a single year, and *(c)* suggests the primary use of light trucks as short-haul pickup and delivery vehicles.

Source: (c) Data from Chicago Area Transportation Study, A Summary of Travel Characteristics, *1977.*

80 kilometers (50 miles) apart would usually be less than half that between points 40 kilometers (25 miles) apart, rather than exactly one half. It is also interesting to note that while distance decay in human geography occurs primarily because of the extra "cost" caused by the friction of distance, just as in other areas of geography, other factors can contribute to declining interaction in economic or social contexts, such as declining familiarity and interest with increasing distance.

When the friction of distance is reduced by lowered costs or increased ease of flow, the slope of the distance decay curve is flattened and more total area is effectively united than when those costs are high. This, of course, is the phenomenon of **space-time compression** (convergence) that we introduced in Chapter 2 (**Figure 3.5**). When automobiles and expressways became widely available in the second half of the 20th century, U.S. cities underwent massive geographic expansion as the friction of distance was sharply reduced and large areas of rural land were brought within a reasonable commute time from the city. Figure 3.4 shows that distance decay is evident for both truck and rail shipments, but that the more expensive mode (trucking) is typically used for shorter distances.

The Gravity Concept

Interaction decisions are not based on distance or distance/cost considerations alone. The large regional shopping center attracts customers from a wide radius because of the variety of shops and goods its very size promises. We go to distant big cities "to seek our fortune," rather than to the nearer small town. That is, we are attracted by the expectation of opportunity that we associate with larger rather than smaller places. That expectation is summarized by another model of spatial interaction, the **gravity model,** also drawn from the physical sciences.

In the 1850s, Henry C. Carey (1793–1879), in his *Principles of Social Science,* observed that the physical laws of gravity and motion developed by Sir Isaac Newton (1642–1727) were applicable to the aggregate actions of humans. Newton's law of universal gravitation states that the gravitational pull between any two objects is proportional to the product of their masses and inversely proportional to the square of the distance between them. More simply put, Newton's law tells us that big things have a stronger attraction force (greater gravitational pull) than do small objects, and that things close to each other have stronger mutual attraction than do objects at greater distance—and that the attraction decreases very rapidly with even small increases in separation. Here is Newton's law of physical gravity:

$$F_{ij} = g \frac{M_i M_j}{d_{ij}^2}.$$

This equation states that the amount of gravitational force between two bodies *i* and *j* equals the product of their masses, divided by the square of the distance between them, and multiplied by the gravitational constant. By analogy, Carey's law of social gravity is

$$I_{ij} = k \frac{P_i P_j}{D_{ij}^B}.$$

This equation states that the amount of interaction (frequency or rate) between two places *i* and *j* equals the product of

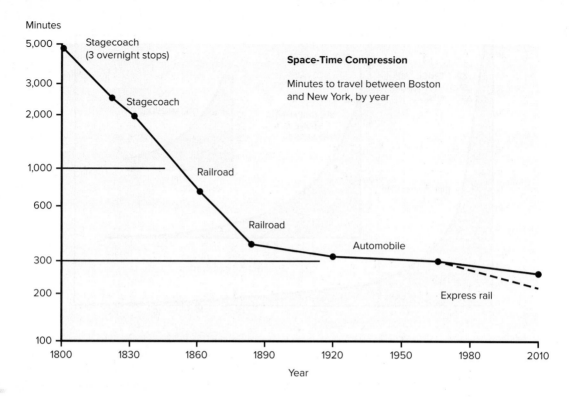

Minutes

5,000 — Stagecoach (3 overnight stops)

Space-Time Compression

Minutes to travel between Boston and New York, by year

3,000 —

2,000 — Stagecoach

1,000 — Railroad

600 —

Railroad

300 — Automobile

200 —

Express rail

100 —

1800   1830   1860   1890   1920   1950   1980   2010

Year

Figure 3.5 Space-time compression is clearly shown in this plot of travel time between the U.S. cities of Boston and New York, from the year 1800 to the year 2010, resulting from improvements to transportation technologies and infrastructure.

Source: Donald G. Janelle, UCSB.

their "interaction masses," divided by the distance between them raised to some exponent, and multiplied by the interaction constant. Notice some differences between the physical and social laws. Unlike the physical law, the mass of a place in the social law is expressed in terms of some measure of how attractive the place is to interaction; population is often used, as suggested by the letter *P*, but many other measures are possible, such as size, social status, the diversity of commodities offered, and so on (for instance, one might prefer to go to a shopping center with more different stores). Importantly, distance may not be straight-line physical distance but some other measure of physical separation, such as distance along the roads connecting places (*route distance*). Alternatively, it is often more relevant to express distance in nonspatial terms, such as travel time, cost, or effort (for most of us, walking downhill is more alluring than walking uphill). Unlike in physical gravity, where the force of attraction declines as the square of distance (the *inverse-square law*), in social gravity, the force of interaction attraction declines at somewhat different rates depending on the domain of interaction (visiting a friend versus visiting a relative, for instance, or shopping for bread versus shopping for caviar). In many domains, the distance exponent is a fractional number (but almost always greater than 1). Finally, social gravity requires multiplication by a constant at the end just like physical gravity does. However, this social "constant" is actually variable across interaction domains, and unlike the *g* in physical gravity that represents something profound about the nature of the physical world, the *k* in social gravity is probably best thought of as just a mathematical way to get the interaction numbers to work out properly as the output of the equation.

Carey's second observation—that large cities have greater drawing power for individuals than do small ones—was subsequently addressed by the **law of retail gravitation,** proposed by William J. Reilly (1899–1970) in 1931. Using the population and distance inputs of the gravity model, Reilly determined the breaking-point (BP) location between two towns where one would expect to find the boundary separating the market areas of the two towns. The market areas are functional regions for each town that enclose the area where each town exerts controlling influence over retail trade; beyond that boundary, the other town dominates. Residents within a town's market area will likely travel to that town to shop. **Reilly's Breaking-Point Law** is:

$$BP_{(from\ i)} = \frac{D_{ij}}{1 + \sqrt{\dfrac{P_j}{P_i}}}.$$

This equation states that the BP location (the market area boundary) between two towns *i* and *j*—expressed as distance from town *i*, equals the distance between the two towns divided by 1, plus the square root of the population of town *j* divided by the population of town *i*. As with all social gravity models, distance may be measured as various forms of physical or nonphysical separation. Because the breaking point between cities of unequal size will lie farther from the larger of the two, its spatially greater drawing power is captured (**Figure 3.6**).

Later studies in location theory, city systems, trade area analysis, and other social topics all suggest that the gravity

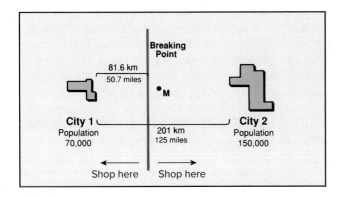

Figure 3.6 The *law of retail gravitation* provides a quick determination of the trade boundary (or breaking point) between two cities. In the diagram, Cities 1 and 2 are 201 kilometers (125 miles) apart. Reilly's law tells us that the breaking point between them lies 81.6 kilometers (50.7 miles) distant from City 1. A potential customer located at *M*, midway (100.5 km or 62.5 mi) between the cities, would lie well within the trade zone of City 2. A series of such calculations would define the "trade area" of any single city.

model can be used to account for a wide variety of interaction flow patterns in human geography, including population migration, commodity flows, journeys to work or to shop, telephone call volumes, and the like. Each flow pattern suggests that size as well as distance influences spatial interaction. Carey's observation, made some 150 years ago, initiated a type of analysis that in modified form is used today for a variety of practical studies that help us better understand the *friction of distance*.

Interaction Potential

The spatial interaction models of distance decay and gravitational pull we have considered thus far deal with only two places at a time. The world of reality is rather more complex. All cities, not just city pairs, within a regional system of cities have the possibility of interacting with one another. Indeed, the more specialized the goods produced in each separate center—that is, the greater their collective complementarity—the more likely it is that such multiple interactions will occur. Similarly, shoppers often have more than two stores or shopping centers to choose from on any given shopping trip.

A **potential model,** also based on the principles of social gravity, provides an estimate of the interaction opportunities available to a center in such a multicentered network. It tells us the relative interaction pull of each point in relation to all other places within a region. It does so by summing the interaction attractiveness and distance relationships between all points of potential interaction within an area. The concept of potential is applicable whenever the measurement of the intensity of spatial interaction is of concern—as it is in studies of retail behavior, marketing, land values, broadcasting, commuting patterns, and the like.

Movement Biases

Distance decay and the gravity models help us understand the bases for interaction in an idealized area without natural or cultural barriers to movement or restrictions on routes followed, and in which only rational interaction decisions are made. Even under those model conditions, the pattern of spatial interaction that develops for whatever reason inevitably affects the conditions under which future interactions will occur. An initial structure of centers and connecting flows will tend to freeze into the landscape a mutually reinforcing continuation of that same pattern. The predictable flows of shoppers to existing shopping centers make those centers attractive to other merchants. New store openings increase customer flow; increased flow strengthens the developed pattern of spatial interaction. And increased road traffic calls for the highway improvement that encourages additional traffic volume.

Such an aggregate regularity of flow is called a **movement bias.** We have already noted a distance bias that favors short movements over long ones. There is also direction bias, in which of all possible directions of movement, actual flows are restricted to only one or a few. Direction bias is simply a statement that from a given origin, flows are not random (**Figure 3.7**); rather, certain places have a greater attraction than do others. The movement patterns from an isolated farmstead are likely oriented to a favored shopping town. On a larger scale, in North America or

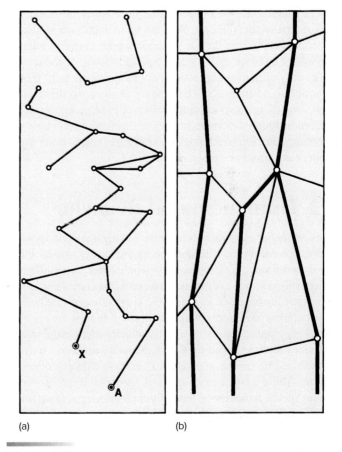

(a)   (b)

Figure 3.7 Direction bias. (*a*) When direction bias is absent, movements tend to be almost random, occurring in all possible directions, but less likely between points, such as *A* and *X,* that are not directly connected. (*b*) Direction bias indicating predominantly north-south movements, likely as a result of transportation routes and/or major destinations being aligned north to south.

Siberia, long-distance freight movements are directionally biased in favor of east-west flows. Direction bias is reflected in not just the orientation but also the intensity of flow. Movements from a single point—from Novosibirsk in Siberia, for example, or from Winnipeg, Canada, or Kansas City in the United States—may occur in all directions; they are in reality more intense along the east-west axis, in part because of the primary landmass orientations of these large countries.

Such directional biases are in part a reflection of **network bias,** a shorthand way of saying that the presence or absence of connecting channels strongly affects the likelihood that spatial interaction, including transportation and communication, will occur. A set of routes and the places that they connect are collectively called a **network;** in abstract terms, the places are called **nodes** and the connecting routes are called **links.** Networks facilitate spatial interaction along their links and between nodes they directly connect. Flows cannot occur between all nodes if not all nodes are linked, at least indirectly. In Figure 3.6a, the interchange between *A* and *X,* though not necessarily impossible, is unlikely because the route between them is indirect and circuitous. In information flows, a worker on the assembly line is less likely to know of company production plans than is a secretary in the executive offices; these two workers are tied into quite different information networks.

In Chapter 2, where we focused on cultural diffusion, we were reminded that there are obstacles to interaction in particular directions besides just distance. **Barriers** impede spatial interaction, either by blocking it totally, slowing it down, or redirecting it. Barriers to human interaction may be physical, as in mountains, oceans, or freeways (when you try to go *across* them rather than *along* them); sociocultural, as in groups of people speaking different languages or zoning laws that disallow certain land-use activities; and even psychological, as in areas people don't know about or are afraid of.

3.2 Human Spatial Behavior

Humans are not commodities and individually do not necessarily respond predictably to the impersonal dictates of spatial interaction constraints. Yet, to survive, people must be mobile and collectively do react to distance, time, and cost considerations of movement in space and to the implications of complementarity, transferability, and intervening opportunity. Indeed, an exciting line of geographic inquiry involves how individuals make spatial behavioral decisions and how those separate decisions may be summarized by models and generalizations to explain collective actions. These geographers approach geographic problems of human spatial **behavior** at the individual or *disaggregate* level. Geographers taking this **behavioral approach** believe they can improve models of spatial behavior such as those based on social gravity or economic rationality (see Chapter 10) by learning more about the way people actually make decisions about where and how to travel and perform other geographic acts. They also recognize that problems involving the psychology of place, space, and environment are geographic problems in their own right because geographers study all aspects of space, place, and

environment. As such, behavioral geographers appreciate that the spatial behaviors of individual people depend, in part, on what people believe to be true about themselves and the world, and how people actually remember and reason about properties of themselves and the world—their **cognition.** They further appreciate that humans do not make decisions only rationally or consciously; many decisions result from nonconscious mental activities and are influenced by emotions and **attitudes.** Finally, an individual analysis of human geography reminds us that no two people are exactly the same; whether based on age, culture, gender, genetic heritage, social class, training, or some other aspect of personal background or experience, people may differ somewhat in their geographic behavior. Models that treat all people as identical are oversimplifications, according to this approach.

Mobility is the general term applied to all types of human movement through space and time. Two aspects of that mobility behavior concern us. The first is the daily or temporary use of space—the journeys to stores, to work, or to school, or for longer periods on vacation or visiting friends and relatives. These types of mobility are often designated as **temporary travel** and have no suggestion of relocation of residence (**Figure 3.8**); people intend to return home after such trips, if not the same day, then relatively soon thereafter. The second type of mobility is the longer-term commitment related to decisions to leave the home territory permanently and find residence in a new location (residential relocation). This second form of spatial behavior is termed **migration.** The distinction between temporary travel and migration is important to a geographic analysis of spatial mobility and is usually fairly clear, but there are ambiguous cases. College students "migrate" to their campus housing from home but may return to that home every summer and after they graduate. Seasonal farm workers may be away from their home for much of the year, following the ripening fruits and vegetables, only to return home at the end of the harvest season. Refugees forced out of their home may intend to return when political, economic, or environmental conditions (see below) permit, but

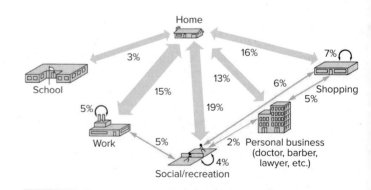

Figure 3.8 Seven County Minneapolis–St. Paul Metropolitan Area travel patterns. The numbers are the percentage of all urban trips taken on a typical weekday. In recent decades, the relative importance of work and school trips has decreased, while other types of trips have risen in importance.

Source: Data from Metropolitan Council: The 2000 Travel Behavior Inventory.

for many of these forced migrants, that time never comes. And of course, nomadic people permanently live in temporary housing that moves with them in a never-ending cycle of residential mobility.

Both aspects of temporary and permanent mobility imply a time dimension. Humans' spatial actions are not instantaneous, even with improvements to transportation technologies. They operate over time, frequently imparting a rhythm to individual and group activity patterns and imposing choices among time-consuming behaviors. Elements of both aspects of human spatial behavior are also embodied in how individuals conceive of space and act within it, and how they respond to information affecting their space-behavioral decisions. The nature of those cognitions and responses affect us all in our daily movements. The more permanent movement embodied in migration involves additional and less common decisions and behaviors, as we shall see later in this chapter.

3.3 Individual Activity Space

One of the realities of life is that groups and countries draw boundaries around themselves and divide space into territories that are, if necessary, defended. Some see the concept of **territoriality**—the emotional attachment to and the defense of home ground—as a root explanation of much of human action and response. It is true that some individual and collective activity appears to be governed by territorial defense responses: the conflict between street groups in claiming and protecting their "turf" (and their fear for their lives when venturing beyond it) and the sometimes violent rejection by an ethnic urban neighborhood of any different advancing population group that it considers threatening. On a more individualized basis, each of us claims as **personal space** the zone of privacy and separation from others that our culture or our physical circumstances require or permit. Anglo Americans strive for greater face-to-face separation in conversations than do Latin Americans. Personal space on a crowded beach or in a department store is acceptably more limited than it is in our homes or when we are studying in a library (**Figure 3.9**).

For most of us, our personal sense of territoriality is a tempered one. We regard our homes and property as defensible private domains but open them to innocent visitors, known and unknown, or to those on private or official business. Nor do we confine our activities so exclusively within controlled home territories as street-gang members do within theirs. Rather, we have a more or less extended home range, an **activity space** or area within which we typically move freely on our rounds of regular activity, sharing that space with others who are also about their daily affairs. **Figure 3.10** depicts activity spaces for a suburban family of five for an average weekday. Note that the activity space is different for each individual.

The types of trips that individuals make, and thus the extent of their activity space, depend on at least three interrelated variables: their stage in life course; the means of mobility at their command; and the demands or opportunities implicit in their daily activities. The first variable, stage in life, refers to membership in specific age groups. School-age children usually travel short distances to lower schools and longer distances to upper-level schools. After-school activities tend to be limited to walking or to bicycle trips to nearby locations. Greater mobility is characteristic of high-school students. Adults responsible for household duties make shopping trips and trips related to child care, as well as journeys away from home for social, cultural,

(a)

(b)

Figure 3.9 Our demanded *personal space* is not necessarily uniform in shape or constant in size. We tolerate strangers closer to our sides than directly in front of us; we accept more crowding in an elevator than in a store. We accept the press of the crowd on a popular beach *(a)*—as do these students on spring break in the Florida Keys *(b)*, but tend to distance ourselves from others in a public square.

(a) ©McGraw-Hill Higher Education; (b) ©Daniel Montello

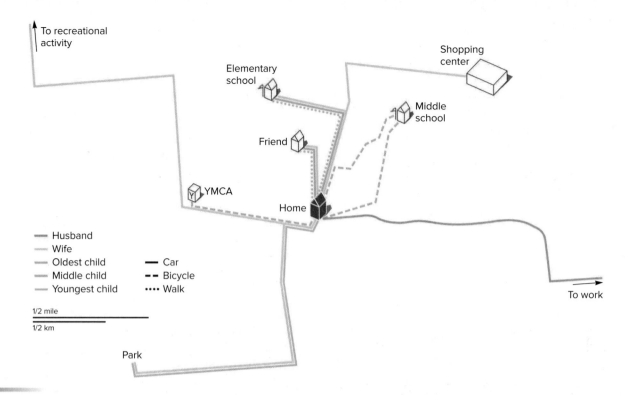

Figure 3.10 Activity space for each member of a family of five for a typical weekday. Routes of regular movement and areas recurrently visited help to foster a sense of territoriality and to affect one's cognitions of space.

or recreational purposes. Wage-earning adults usually travel farther from home than other family members. Elderly people may, through infirmity or interests, have less extensive activity spaces.

The second variable that affects the extent of activity space is mobility, or the ability to travel. An informal consideration of the cost and effort required to overcome the friction of distance is implicit. Where incomes are high, automobiles are available, and the cost of fuel is reckoned as minor in the family budget, mobility may be great and individual activity space large. In societies or neighborhoods where cars are not a standard means of conveyance, the daily non-emergency activity space may be limited to walking, bicycling, or taking infrequent trips on public transportation. Wealthy suburbanites are far more mobile than are residents of inner-city slums, a circumstance that affects ability to learn about, seek, or retain work and to have access to medical care, educational facilities, and social services.

A third factor limiting activity space is the individual assessment of the existence of possible activities or opportunities. In subsistence economies where the needs of daily life are satisfied at home, the impetus for journeys away from home is minimal. If there are no stores, schools, factories, or even roads, expectations and opportunities are limited. Not only are activities spatially restricted, but **awareness space**—knowledge of opportunity locations beyond normal activity space—is minimal, distorted, or absent. In low-income neighborhoods of modern cities in any country, poverty and isolation limit the

inducements, opportunities, destinations, and necessity of travel (Figure 1.28). Opportunities plus mobility conditioned by life stage bear heavily on the amount of spatial interaction in which individuals engage.

3.4 The Tyranny of Time

Whatever their activity, people are always located somewhere. Furthermore, they are always located somewhere at some time of the day or night, and they spend various periods of time at each location before moving through space and time to new locations. The study of the temporal characteristics of activities in conjunction with their spatial characteristics is known as **time geography.** We can depict this by drawing a graph that shows activity locations and movements around the landscape plotted against time on the vertical axis. This graph of a person's activity locations at certain times is a **space-time path. Figure 3.11** shows a daily space-time path, but weekly, yearly, or even lifetime paths are possible. Whatever the temporal scale, space-time paths are notable for their tendency to show cyclic, nonrandom patterns of recurring activity locations at particular times, such as daily or seasonal patterns.

The activities of humans—eating, sleeping, traveling between home and destination, working, or attending classes—all consume time and involve space. An individual's spatial reach is restricted because one cannot be in two different places at the

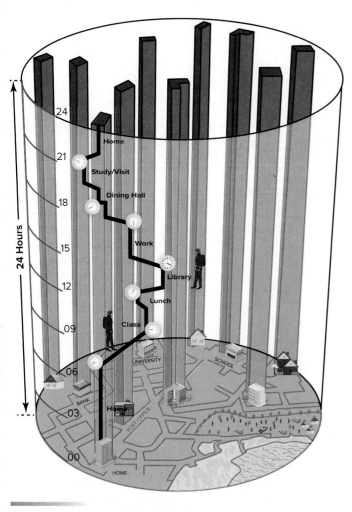

Figure 3.11 The school-day space-time path for a hypothetical college student. Vertical segments are times when the student stays in the same place for some time. Sloped segments indicate movement—changes of location over time. Shallower lines show faster travel.

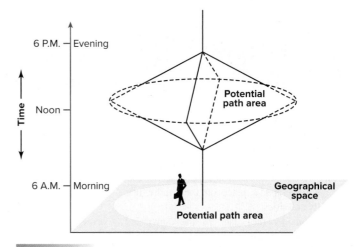

Figure 3.12 The space-time prism. An individual's daily prism has both geographical limits and totally surrounding space-time walls. The *time* (vertical axis) involved in movement affects the space that is accessible, along with the time and space available for other purposes than travel. If we can travel only slowly, such as by walking, we will not be able to move as far from home each day or have as much time available for other activities besides travel. This is suggested by steeper sides on our prism. If we can travel more quickly, such as by driving, we will be able to move farther from home each day and have more time available for other activities. This is suggested by prisms with shallower sides.

same moment or engage simultaneously in activities that are spatially separate. Further, because there is a finite amount of time within a day and each of us is biologically bound to a daily rhythm of day and night, sleeping and eating, time tyrannically limits the spatial choices we can make and the activity space that we can command.

The daily space-time constraints of our time geography may be represented by a **space-time prism,** the volume of space and length of time within which our activities must be confined. Its size and shape are determined by our mobility and our locational responsibilities; its boundaries define what we can or cannot accomplish spatially and temporally (**Figure 3.12**). If our circumstances demand that we walk to work or school, the sides of our prism are steep and the space available for our activities is narrow; steep diagonals mean slow travel and indicate that we cannot go as far in a given time period. We cannot use time spent in transit for most other activities, and the area reasonably accessible to

the pedestrian is limited. The space-time prism for the driver has more shallow-angled sides and the individual's spatial range is wide; shallow diagonals mean rapid travel and indicate that we can go farther in a given time period. The area of the prism determines what spatially defined activities are possible, for no activity can exceed the bounds of the prism (see the feature "Space, Time, and Women"). That is, space-time paths are constrained to fit within space-time prisms. Because most activities have their own time constraints, the choices of things you can do and the places in which you can do them are strictly limited. Scheduled class hours, travel time from residence to campus, and dining hall location and opening and closing hours, for example, may be the constraints on your space-time path. If you also need part-time work, your choice of jobs is restricted by their respective locations and work hours, for the job, too, must fit within your daily space-time prism. Parenting responsibilities, particularly for single parents, place major constraints on the spatial range of individuals. In households where one partner (typically the woman) bears greater responsibilities for childcare and household chores, their job choices may be limited by their narrow time-geographic constraints, and they may be forced to accept lower pay and/or a less prestigious job. In other words, all people have *space-time budgets* that dictate how far they can travel in a given time period. Geographers apply the study of these space-time budgets and other aspects of time geography to problems such as traffic control, mass transit, and highway and parking structure design.

Space, Time, and Women

From a time-geography perspective, it is apparent that many of the limitations that women face in their choices of employment or other activities outside the home reflect the restrictions that women's time budgets and travel paths place on their individual daily activities.

Consider the case* of the unmarried working woman with one or more pre-school-age children. The location and operating hours of available child-care facilities may have more of an influence on her choice of job than do her labor skills or the relative merits of alternative employment opportunities. For example, the woman may not be able to leave her home before a given hour because the only full-day child-care service available to her is not open earlier. She must return at a specified time to pick up her child and arrive home to prepare food at a reasonable (for the child) dinner time. Her travel mode and speed determine the outer limits of her daily space-time prism.

Suppose both of two solid job offers have the same working hours and fall within her possible activity space. She cannot accept the preferred, better-paying job because drop-off time at the child-care center would make her late for work, and work hours would make her miss the center's closing time. On the other hand, although the other job is acceptable from a child-care standpoint, it leaves no time (or store options) for shopping or errands except during the lunch break. Job choice and shopping opportunities are thus determined not by the woman's labor skills or awareness of store price comparisons, but by her time-geography constraints. Other women in other job skill, parenthood, locational, or mobility circumstances experience different but comparable space-path restrictions.

Mobility is a key to activity mix, time-budget, and activity space configurations. Again, research indicates that women are frequently disadvantaged. Because of their multiple work, child-care, and home maintenance tasks, women on average make more—though shorter—trips than men, leaving less time for alternate activities.

The lower income level of many single women with or without children limits their ability to own cars and leads them to use public transit disproportionately to their numbers—to the detriment of both their money and time-space budgets. They are, it has been observed, "transportation deprived and transit dependent."

Geographer Mei-Po Kwan used geographic information system (GIS) and travel diaries to create three-dimensional diagrams of the time-geography patterns of a sample of men and women who all had driver's licenses and access to automobiles. Despite their relative affluence, Kwan found that women experience more time-geography constraints than men because of their child-care, school drop-off, or other responsibilities. Women with other adults in the household to share domestic responsibilities experienced fewer constraints, and women with the most time-geography constraints were more likely to have to accept part-time work.

*Suggested by Risa Palm and Allan Pred, "A Time-Geographic Perspective on Problems of Inequality for Women," Institute of Urban and Regional Development, Working Paper No. 236, University of California, Berkeley, 1974.

Source: *Mei-Po Kwan, 1999. "Gender, the Home-Work Link, and Space-Time Patterns of Non-Employment Activities,"* Economic Geography, *75(4): 370–394 (1999).*

3.5 Distance and Human Interaction

People make many more short-distance trips than long ones, a statement in human behavioral terms of the concept of distance decay. If we drew a boundary line around our activity space, it would be evident that trips to the boundary are taken much less often than short-distance trips around the home. The tendency is for the frequency of trips to fall off very rapidly beyond an individual's **critical distance**—the distance beyond which cost, effort, and means strongly influence our willingness to travel. **Figure 3.13** illustrates the point with regard to journeys from the homesite.

Regular movements defining our individual activity space are undertaken for different purposes and are differently influenced by time and distance considerations. The kinds of activities that individuals engage in can be classified according to type of trip: journeys to work, to school, to shop, for recreation, and so on. People in nearly all parts of the world make these same types of journeys, though the spatially variable requirements of culture, economy, and personal circumstance dictate their frequency, duration, and

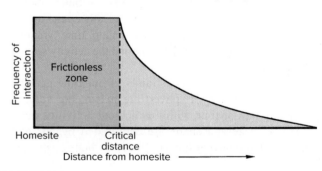

Figure 3.13 Critical distance. This general diagram indicates how most people observe distance. For each activity, there is a distance beyond which the intensity of contact declines. This is called the *critical distance*. The distance up to the critical distance is identified as a *frictionless zone*, in which time or distance considerations do not effectively figure in the decision to take the trip.

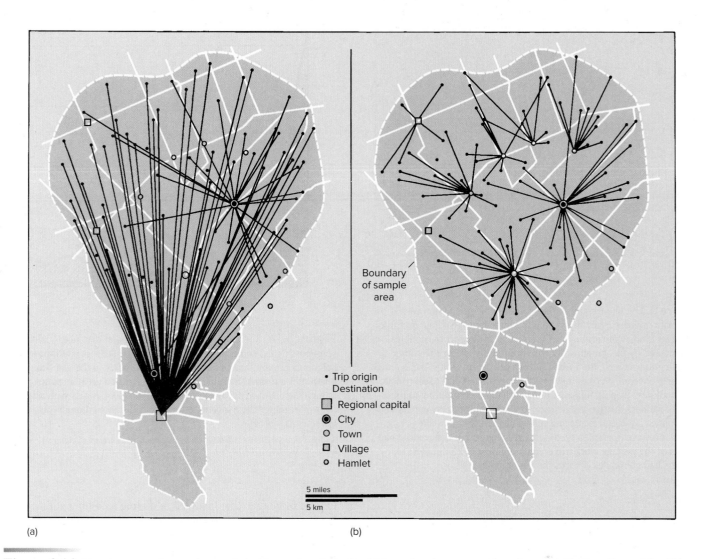

Figure 3.14 Travel patterns for purchases of clothing and yard goods of (*a*) rural cash-economy Canadians and (*b*) Canadians of the Old Order Mennonite sect. These strikingly different travel behaviors mapped many years ago in central Canada demonstrate the great differences that may exist in the action spaces of different culture groups occupying the same territory. At that time, "modern" rural Canadians, owning cars and wishing to take advantage of the variety of goods offered in the more distant regional capital, were willing and able to travel longer distances than were neighboring people of a traditionalist culture who had different mobility and whose different demands in clothing and other consumer goods were, by preference or necessity, satisfied in nearby small settlements. Unpublished studies suggest that similar contrasts in mobility and purchase travel patterns currently exist between buggy-using Old Order Amish (see Figure 7.2 in Chapter 7) and their car-driving neighbors.

Source: Robert A. Murdie, "Cultural differences in consumer travel," Economic Geography *41, no. 3 (Worcester, MA: Clark University, 1965).*

significance to an individual (**Figure 3.14**). A small child, for example, will make many trips up and down the block but is inhibited by parental admonitions from crossing the street. Different but equally effective distance constraints control adult behavior.

The journey to work plays a decisive role in defining the activity space of most adults. Formerly restricted by walking distance or by the routes and schedules of mass transit systems, the critical distances of work trips have steadily increased in European and Anglo American cities as the private automobile figures more importantly in the

movement of workers (**Figure 3.15**). Daily or weekly shopping may be within the critical distance of an individual, and little thought may be given to the cost or the effort involved. That same individual, however, may relegate shopping for special goods to infrequent trips and carefully consider their cost and effort. The majority of our social contacts tend to be at short distance within our own neighborhoods or with friends who live relatively close at hand; longer social trips to visit relatives are less frequent. In all such trips, however, the distance decay function is clearly at work (**Figure 3.16**).

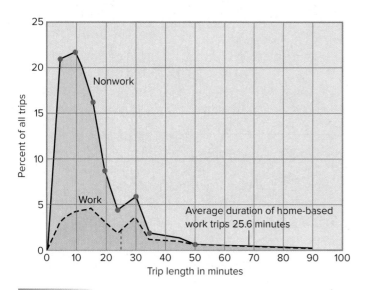

Figure 3.15 The frequency distribution of work and nonwork trip lengths in minutes in the seven-county Minneapolis–St. Paul metropolitan area. Studies in various metropolitan areas support the conclusions documented by this graph: work trips are usually longer than other recurring journeys. In the United States in the early 1990s, the average work trip covered 17.1 kilometers (10.6 miles), and half of all trips to work took less than 22 minutes; for suburbanites commuting to the central business district, the journey to work involved between 30 and 45 minutes. By 2000, increasing sprawl had lengthened average commuting distances and, also because of growing traffic congestion, had increased the average work trip commuting time to 25 minutes; many workers had commutes of more than 45 minutes.

Source: Metropolitan Council: The 2000 Travel Behavior Inventory.

Spatial Interaction and the Accumulation of Information

Critical distances, even for the same activity, are somewhat different for each person. The variables of life stage, mobility, and opportunity, together with an individual's interests and demands, help define how often and how far a person will travel. On the basis of these variables, we can make inferences about the amount of information that a person is likely to acquire about his or her activity space and the area beyond. The accumulation of information about the opportunities and rewards of spatial interaction helps increase and justify movement decisions.

For information flows, however, space has a somewhat different meaning than it does for the movement of commodities. Communication does not necessarily imply the time-consuming physical relocations of freight transportation, though in the case of letters and print media, it usually does. In modern telecommunications, the process of information flow may be very quick regardless of distance. The result is extensive space-time compression. A Bell System report tells us that in 1920, putting through a transcontinental telephone call took 14 minutes and eight operators and cost more than $15 for a 3-minute call. By 1940, the call completion time was reduced to less than 1½ minutes, and the cost fell to $4. In the 1960s, direct distance dialing allowed a transcontinental connection in less than 30 seconds, and electronic switching has now reduced the completion time

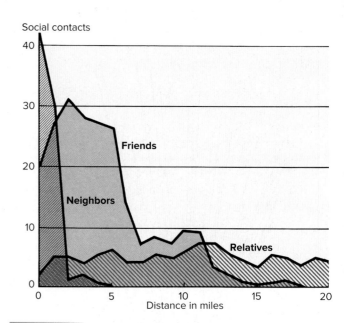

Figure 3.16 Social interaction as a function of distance. Visits with neighbors on the same street are frequent; they are less common with neighbors around the corner and diminish quickly to the vanishing point after a residential relocation. Friends exert a greater spatial pull, though the distance decay factor is clearly evident. Visits with relatives offer the greatest incentive for longer distance (though relatively infrequent) journeys.

Source: Frederick P. Stutz, "Distance and network effects on urban social travel fields," Economic Geography 49(2) (Worcester, Mass.: Clark University, 1973), p. 139.

to little more than that involved in dialing a number and answering a phone. The extra price of long-distance conversation essentially disappeared with the advent of voice communication over the Internet early in this century.

The Internet and communication satellites have made worldwide personal and mass communication almost immediate and data transfers a matter of moments. The same technologies that have reduced the time separation of communications have tended to reduce their cost separation. Domestic mail, which once charged a distance-based postage, is now carried nationwide or across town for the same price. In the modern world, transferability is no longer a very significant consideration in information flows. A speculative view of the future suggests that as distance ceases to be much of a determinant of the cost or speed of communication, the spatial structure of economic and social decision making may be fundamentally altered. Determinations about where people live and work, the role of cities and other existing command centers, flows of domestic and international trade, constraints on human mobility, and even the concepts and impacts of national boundaries may fundamentally change with new and unanticipated consequences for patterns of spatial interaction.

Information Flows

Spatially significant information flows are of two types: individual (person-to-person) exchanges and mass (source-to-area) communication. A further subdivision into formal and informal interchange recognizes, in the former, the need for an interposed channel (radio, press, postal service, or telephone, for example) to convey messages. Informal communication requires no such institutionalized message carrier.

Short-range informal individual communication is as old as humankind itself. Contacts and exchanges between individuals and within small groups tend to increase as the complexity of social organization increases, as the size and importance of the population center grow, and as the range of interests and associations of the communicating person expands. Each individual develops a **personal communication field,** the informational counterpart of that person's activity space. Its size and shape are defined by the individual's contacts in work, recreation, shopping, school, or other regular activities. Those activities, as we have seen, are functions of the age, sex, education, employment, income, and so on of each person. An idealized personal communication field is suggested in **Figure 3.17**.

Each interpersonal exchange constitutes a link in the individual's personal communication field. Each person, in turn, is a node in the communication field of those with whom he or she makes or maintains contact. The total number of such separate informal networks essentially equals the total count of people alive. Despite the number of those networks, all people, in theory, are interconnected by multiple shared nodes (**Figure 3.18**). One debated claim suggested that through such interconnections, no person in the United States is more than six links removed from any other person, no matter where located or how unlikely the association.

Mass communication is the formal, structured transmission of information in essentially a one-way flow between single points of origin and broad areas of reception. There are few transmitters and many receivers. The mass media are by nature "space filling." From single origin points, they address their messages by print, radio, or television to potential receivers within a defined area. The number and location of disseminating points, therefore, are related to their spatial coverage characteristics, to the minimum size of area and population necessary for their support, and to the capability of the potential audiences to receive their message. The coverage area is determined both by the nature of the medium and by the corporate intent of the agency.

There are no inherent spatial restrictions on the dissemination of printed materials, though of course limitations and restrictions may be imposed by obscenity laws, religious prohibitions, restrictions in some countries on certain forms of political speech, and the like. And not everyone has access to bookstores or libraries or funds to buy printed material, and not everyone can read. Unlike the distance limitations on the transmission of AM or FM radio waves, however, these restrictions are independent of the area over which printed material could be physically distributed and made available.

In the United States, much book and national magazine publishing is localized in metropolitan New York City, as have the services supplying news and features for sale to the print media located there and elsewhere in the country. Paris, Buenos Aires, Moscow, London—indeed, the major metropolises and/or capital cities of other countries—show the same spatial concentration. Regional journals emanate from regional capitals, and major metropolitan newspapers, though serving primarily their home markets, are distributed over (or produce special editions for distribution within) tributary areas whose size and shape depend on the intensity of competition from other metropolises. A spatial information hierarchy has thus emerged.

Hierarchies are also reflected in the market-size requirements for different levels of media offerings. National and international organizations are required to expedite information flows (and, perhaps, to control their content), but market demand is heavily weighted in favor of regional and local coverage. In the electronic media, the result has been national networks with local affiliates acting as the gatekeepers of network offerings and adding to them locally originating programs and news content. A similar market subdivision is represented by the regional editions of national newspapers and magazines.

The technological ability to fill space with messages from different mass media is useless if receiving audiences do not exist. In illiterate societies, publications cannot inform or influence. Unless the appropriate receivers are widely available, television and radio broadcasts are a waste of resources. Perhaps no invention in history has done more to weld isolated individuals and purely person-to-person communicators into national societies exposed to centralized information flows than has the low-cost transistor radio. Its battery-powered transportability converts the remotest village and the most isolated individual into a receiving node of entertainment, information, and political messages. The direct satellite broadcast of television programs to community antennae or communal sets brings that mass medium to remote areas of Arctic Canada, India, Indonesia, and other world areas able to invest in the technology but as yet unserved by ground stations.

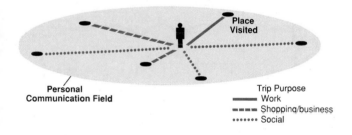

Figure 3.17 A personal communication field is determined by individual spatial patterns of communication related to work, shopping, business trips, social visits, and so on.

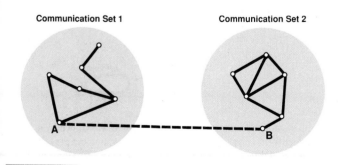

Figure 3.18 Separate population sets (groups) are interconnected by the links between individuals. If link A–B exists, everyone in the two sets is linked.

Information and Cognition

Human spatial interaction, as we have seen, is conditioned by a number of factors. Complementarity, transferability, and intervening opportunities help pattern the movement of commodities and peoples. Flows between points and over area are influenced by distance decay and partially explained by gravity and potential models. Individuals in their daily affairs operate in activity spaces that are partly determined by stage in life, mobility, and a variety of socioeconomic characteristics. In every instance of spatial interaction, however, decisions are based on people's beliefs about the opportunity or feasibility of movement, exchange, and the satisfaction of needs and desires, not the objective truth about these things.

More precisely, our actions and decisions are based on our cognition—the awareness that we have, as individuals, of home and distant places, and the beliefs that we hold about their properties. Geographers often refer to such cognitions about Earth locations with the term **place perception,** which is our beliefs, impressions, and feelings—rational or irrational, consciously realized or not—about the natural and cultural characteristics of an area and about its opportunity structure. (Used in this way, geographers are treating the term *perception* as synonymous with cognition.) Whether our views accord with those of others or accurately reflect the "real" world seen in objective terms is not all that matters. Our cognitions are important because the decisions that we make about how to spend our time or about what actions to take in space are not based directly on reality but on our assumptions and impressions of reality.

Cognition of Environment

Psychologists and geographers are interested in determining how we arrive at our cognitions of place and environment both within and beyond our normal activity space. The images we form firsthand of our home territory have been in part reviewed in the discussion of mental maps in Chapter 1. The cognitions we have about more distant places are less directly derived (**Figure 3.19**). In technologically advanced societies, television and radio, magazines and newspapers, Web sites, books and lectures, travel brochures, and hearsay all combine to help us develop a mental picture of unfamiliar places and of the interaction opportunities that they may contain. Again, however, the most effectively transmitted information seems to come from word-of-mouth reports. These may be in the form of letters or visits from relatives, friends, and associates who supply information that helps us develop ideas about relatively unknown areas.

There are, of course, barriers to the flow of information, including that of distance decay. Our knowledge of close places is greater than our knowledge of distant points; our contacts with nearby persons theoretically yield more information than we receive from afar.

Yet in crowded areas with maximum interaction potential, people commonly set psychological barriers around themselves so that only a limited number of those possible interactions and information exchanges actually occur. We raise barriers against information overload and to preserve a sense of privacy that permits the filtering out of information that does not directly affect us. There are obvious barriers to long-distance information flows as well, such as time and money costs, mountains, oceans, rivers, and differing religions, languages, ideologies, and political systems.

Barriers to information flow give rise to what we earlier (Section 3.1) called *direction bias*. In the present usage, this implies a tendency to have greater knowledge of places in some directions than in others. Not having friends or relatives in one part of a country may represent a barrier to individuals, so interest in and knowledge of the area beyond the "unknown" region are low. In the United States, both northerners and southerners tend to be less well informed about each other's areas than about the western part of the country. Traditional communication lines in the United States follow an east-west rather than a north-south direction, the result of early migration patterns, business connections, and the pattern of the development of major cities. In Russia, directional bias favors a north-south information flow within the European part of the country and less familiarity with areas far to the east. Within Siberia, however, east-west flows dominate.

When information about a place is sketchy, blurred pictures develop. These influence the impressions that we have of places and cannot be discounted. Many important decisions are made on the basis of incomplete information or biased reports, such as decisions to visit or not, to migrate or not, to hate or not, even to make war or

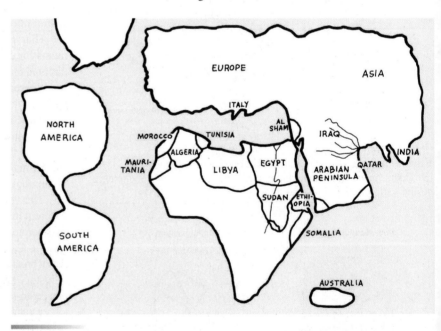

Figure 3.19 A Palestinian student's view of the world. This map was drawn by a Palestinian high school student from Gaza, and it reflects the instruction and classroom impressions that the student has received. The Gaza curriculum conforms to the Egyptian national standards and thus is influenced by the importance of the Nile River and pan-Arabism. Al Sham is the old, but still used, name for the area that includes Syria, Lebanon, and Palestine. The map might be quite different in emphasis if the Gaza school curriculum were designed by Palestinians or if it had been drawn by an Israeli student.

not. Awareness of places is usually accompanied by opinions about them, but there is no necessary relationship between the depth of knowledge and the beliefs held (see the discussion of mental maps and place stereotypes in Chapter 1). In general, the more familiar we are with a locale, the more sound the factual basis of our mental image of it will be. But individuals form firm impressions of places totally unknown to them personally, impressions that may be very inaccurate, and these may color interaction decisions.

One way to determine how individuals envisage home or distant places is to ask them what they think of different locales. For instance, they may be asked to rate places according to desirability—perhaps residential desirability—or to make a list of the 10 best and the 10 worst cities in their country of residence. Certain regularities appear in such inquiries. **Figure 3.20** presents some residential desirability data elicited from college students in three provinces of Canada. These and comparable preference maps derived from studies conducted by researchers in many countries suggest that people generally like the place where they are from but also like some places far away that are widely popular in their culture. For example, students from colleges around the United States tend to rate West Coast states such as California highly, but unless they are from states in the Southeast or the Plains Region, they tend to rate those states poorly. Individuals tend to be indifferent to unfamiliar places and areas and to dislike those that have competing interests (such as distasteful political and military activities or conflicting economic concerns) or a physical environment believed to be unpleasant.

On the other hand, places believed to have superior climates or landscape amenities are rated highly in preference map studies and favored in tourism and migration decisions. Holiday tours to Spain, the south of France, and the Mediterranean islands are heavily booked by the British seeking to escape their damp, cloudy climate. A U.S. Census Bureau study indicates that *climate* is, after work and family proximity, the most often reported reason for interstate moves by adults of all ages. International studies reveal a similar migration motivation based not only on climate, but also on concepts of natural beauty and amenities.

Perception of Natural Hazards

Less certain is the negative impact on spatial interaction or relocation decisions of assessments of **natural hazards**. Natural hazards are elements, processes, or events in the environment

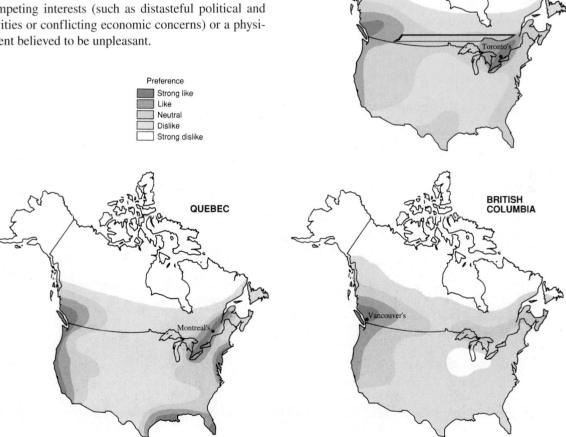

Figure 3.20 Residential preferences of Canadians. These maps show the residential preference of a sampled group of Canadians from the provinces of British Columbia, Ontario, and Quebec, respectively. Note that each group of respondents prefers its own area, but all like the Canadian and U.S. west coasts.

Source: Herbert A. Whitney, "Preferred locations in North America: Canadians, clues, and conjectures," Journal of Geography *83, no. 5. (Indiana, PA: National Council for Geographic Education, 1984): 222.*

that can cause harm to humans. Although the term implies that these hazards are "natural," human adaptation and location decisions always play a major role in determining how disastrous the results are. For example, where we decide to build homes and businesses, and how we construct them, has a major impact on how much destruction is caused by an earthquake.

Mental images of home areas do not generally include an acknowledgment of potential natural dangers as an overriding concern. The cyclone that struck the delta area of Bangladesh on November 12, 1970, left at least 500,000 people dead, yet after the disaster, the movement of people into the area swelled the population above precyclone levels—a resettlement repeated after other, more recent cyclones. The July 28, 1976, earthquake in the Tangshan area of China devastated a major urban industrial complex, with casualties estimated at about a quarter-million, and between 50,000 and 100,000 city dwellers and villagers reportedly perished during and after the January 2001 quake in Gujarat state of western India. In both cases, rebuilding began almost immediately, as it usually does following earthquake damage (**Figure 3.21**) or after the devastation of earthquake-induced tsunamis like the December 2004 inundation of the Indonesian, Thai, and Indian coasts. The human response to even such major and exceptional natural hazards is duplicated by a general tendency to discount dangers from more common hazard occurrences. Johnstown, Pennsylvania, has suffered recurrent floods, and yet its residents rebuild; violent storms like Hurricane Katrina recurrently strike the Gulf and East Coasts of the United States (**Figure 3.22**), and people remain or return. Californians contemplating a move to Kansas may be concerned about tornadoes there but be unconcerned about earthquake dangers at home.

Why do people choose to settle in areas of high-consequence hazards in spite of the potential threat to their lives and property? Why do hundreds of thousands of people live along the San Andreas Fault in California, build houses in Pacific coastal areas known to experience severe erosion during storms, return to flood-prone river valleys in Europe or Asia, or avalanche-threatened Andean valleys? What is it that makes the risk worth taking? Ignorance of natural hazard danger is not necessarily a consideration. People in seismically active regions of the United States and Europe, at least, do believe that damaging earthquakes are a possibility in their districts but, research indicates, are reluctant to do anything about the risk. Similar awareness and reticence accompanies other low-incidence/high-consequence natural dangers. Less than one-tenth of 1 percent of respondents to a federal survey gave "natural disaster" as the reason for their interstate residential move. Geographers have long wondered about these residential decisions by people in the face of repeated hazards. Are they paradoxical, foolish, or, in some way, rational? There are many reasons why natural hazard risk does not deter settlement or adversely affect space-behavioral decisions. Of importance, of course, is the persistent belief that the likelihood of an earthquake or a flood or other natural calamity is sufficiently remote so that it is not reasonable or pressing to modify behavior because of it. People

Figure 3.21 Destruction from the San Francisco earthquake and fire. The first shock struck San Francisco early on the morning of April 18, 1906, damaging the city's water system. Fire broke out and raged for three days. It was finally stopped by dynamiting buildings in its path. When it was over, some 700 people were dead or missing, and 25,000 buildings had been destroyed. Locally, the event is usually referred to as the Great Fire of 1906, suggesting a denial of the natural hazard in favor of assigning blame to correctable human error. Post-destruction reconstruction began at once. Rebuilding following earthquake damage is the general rule.

©*Bettmann/Getty Images*

80   Chapter 3   Spatial Interaction and Spatial Behavior

Figure 3.22 People waiting to be rescued from a New Orleans roof-top following Hurricane Katrina's assault in late August of 2005. More than 1,600 died, hundreds of thousands were left homeless, and tens of billions of dollars of damage were incurred from the storm, which was immediately followed by government and private efforts at recovery and rebuilding.

Source: Jocelyn Augustino/FEMA

Table 3.1

Common Responses to the Uncertainty of Natural Hazards

Eliminate the Hazard

Deny or Denigrate Its Existence	Deny or Denigrate Its Recurrence
"We have no floods here, only high water."	"Lightning never strikes twice in the same place."
"It can't happen here."	"It's a freak of nature."

Eliminate the Uncertainty

Make It Determinate and Knowable	Transfer Uncertainty to a Higher Power
"Seven years of great plenty.... After them seven years of famine."	"It's in the hands of God."
"Floods come every five years."	"The government is taking care of it."

1. Source: *Burton and Kates , "The perception of natural hazards in resource management,"* Natural Resources Journal *435(3) (1964), University of New Mexico School of Law, Albuquerque, NM.*

are influenced by their natural optimism (which some might call willful ignoring of real risks) and the predictive uncertainty about timing or severity of a rare but calamitous event. Past experiences in high-hazard areas might be relevant. If people have not suffered much damage in the past, they may be optimistic about the future. If, on the other hand, past damage has been great, they may think that the probability of repetition in the future is low, which often represents fallacious thinking (Table 3.1). Lightning may well be *more* likely to strike in the same place, because the first strike we hear about is evidence that the place is attractive to lightning (think of a metal tower on a hill).

Perception of place as attractive or desirable may be quite divorced from any understanding of its hazard potential. Attachment to locale or region may be an expression of emotion and economic or cultural attraction, not just a rational assessment of risk. The culture hearths of antiquity discussed in Chapter 2 and shown on Figure 2.15 were for the most part sited in flood-prone river valleys; their enduring attraction was undiminished by that potential danger. The home area, whatever disadvantages an outside observer may discern, exerts a force of attachment to place and identification with place that is not easily dismissed or ignored.

Indeed, high-hazard areas are often sought out because they possess desirable topography or scenic views, as do, for instance, coastal areas subject to storm damage. Once people have purchased property in a known hazard area, they may be unable to sell it for a reasonable price even if they so desire. They think that they have no choice but to remain and protect their investment. The cultural hazard—loss of livelihood and investment—appears more serious than whatever natural hazards there may be. Carried further, it has been observed that spatial adjustment to perceived natural hazards is a luxury not affordable to impoverished people in general or to the urban and rural poor of Third World countries in particular. Forced by population growth and economic necessity to exert ever-greater

pressures upon fragile environments or to occupy at higher densities hazardous hillside and floodplain slums, their margin of safety in the face of both chronic and low-probability hazards is minimal to nonexistent (**Figure 3.23**). In sum, these considerations suggest that there are great economic, cultural, and ideological benefits to living in hazardous places, which when taken together, apparently outweigh the risks to most people.

Figure 3.23 Many of the poor of Rio de Janeiro, Brazil, occupy steep hillside locations above the reach of sewer, water, and power lines that keep the more affluent at lower elevations. Frequent heavy rains cause mudflows from the saturated hillsides that wipe away the shacks and shelters that insecurely cling to them, and deposit the homes and hopes of the poor in the richer neighborhoods below.

©VANDERLE ALMEIDA/AFP/Getty Images

Migration

When continental glaciers began their retreat some 11,000 years ago, the activity and awareness spaces of Stone Age humans were limited. As a result of pressures of numbers, need for food, changes in climate, and other inducements, those spaces were collectively enlarged to encompass most of the terrestrial world. *Migration*—the permanent or planned long-term relocation of residential place—has been one of the enduring themes of human history. It has contributed to the evolution of separate cultures, to the diffusion of those cultures and their components by interchange and communication, and to the frequently complex mix of peoples and cultures found in different areas of the world. Indeed, it has been a major force in shaping the world as it is today and continues to be an important force in ongoing world change.

Massive movements of people within countries, across national borders, and among continents have emerged as a pressing concern of recent decades. They affect national economic structures, determine population density and distribution patterns, alter traditional ethnic, linguistic, and religious mixtures, and inflame national debates and international tensions. Because migration patterns and conflicts touch so many aspects of social and economic relations and have become so important a part of current human geographic realities, their specific impact is a significant aspect of several of our topical concerns. Portions of the story of migration have been touched on already in Chapter 2; other elements of it are part of later discussions of population (Chapter 4), ethnicity (Chapter 6), economic development (Chapter 10), urbanization (Chapter 11), and international political relations (Chapter 12). Because voluntary migration is a near-universal expression of spatial assessment and interaction, reviewing its behavioral basis now will give us common ground for understanding its impacts in other contexts later.

Migration embodies all the principles of spatial interaction and space relations that have already been discussed in this chapter. Complementarity, transferability, and intervening opportunities and barriers all play a role. Space information and place perception are important, as are the sociocultural and economic characteristics of the migrants and the distance relationships between their original and prospective locations of settlement. In less abstract terms, group and individual migration decisions may express real-life responses to poverty, rapid population growth, environmental deterioration, or international and civil conflict or war. In its current troubling dimensions, migration may be as much a strategy for survival as an unforced but reasoned response to economic and social opportunity.

Naturally, the length of a specific move and its degree of disruption of established activity space patterns raise distinctions important in the study of migration. A change of residence from the central city to the suburbs certainly changes both residence and activity space of schoolchildren and of adults in many of their nonworking activities, but the working adults may still retain the city—indeed, the same place of employment there—as an activity space. College students frequently change dorm rooms or rental homes during their time at school, even though they attend classes at the same campus and socialize with most of the same friends. On the other hand, the immigration from Europe to the United States and the massive farm-to-city movements of rural Americans late in the 19th and early in the 20th centuries clearly meant a total change of all aspects of behavioral patterns.

Principal Migration Patterns

Migration flows can be discussed at different scales, from massive intercontinental torrents to individual decisions to move to a new house or apartment within the same metropolitan area. Geographers broadly distinguish **total displacement migrations,** wherein migrants travel so far that they have completely new activity spaces that do not overlap at all with their former home ranges, from **partial displacement migrations,** local moves wherein migrants move to a new residence nearby, with new activity spaces overlapping some with their former home ranges. As discussed more below, such partial displacement moves are actually quite common, an expression of distance decay in migration behavior. At each scale, although the underlying controls on spatial behavior remain constant, the immediate motivating factors influencing the spatial interaction are typically different, with varying impacts on population patterns and cultural landscapes.

At the broadest scale, *intercontinental* movements range from the earliest peopling of the habitable world to the most recent flight of Asian or African refugees to countries of Europe or the Western Hemisphere. The population structure of the United States, Canada, Australia, New Zealand, and Argentina, Brazil, and other South American countries—as Chapter 4 suggests—is a reflection and result of massive intercontinental flows of immigrants that began as a trickle during the 16th and 17th centuries and reached a flood during the 19th and early 20th (Figure 4.20). Later in the 20th century, World War II (1939–1945) and its immediate aftermath involved more than 25 million permanent population relocations, all of them international, but not all intercontinental.

Intracontinental and *interregional* migrations involve movements between countries (*external migrations*) and within countries (*internal migrations*), most commonly in response to individual and group assessments of improved economic prospects, but often reflecting flight from difficult or dangerous environmental, military, economic, or political conditions. The millions of **refugees** leaving their homelands following the dissolution of Eastern European communist states, including the former USSR and Yugoslavia, exemplify that kind of flight. Between 1980 and 2005, Europe received some 23 million newcomers, often refugees, who joined the 15 million labor migrants ("guest workers") already in West European countries by the early 1990s **(Figure 3.24)**. North America has its counterparts in the hundreds of thousands of immigrants coming (many illegally) to the United States each year from Mexico, Central America, and the Caribbean region, particularly during the 1990s and 2000s.

The Hauns, whose westward trek opened this chapter, were part of a massive 19th-century regional shift of Americans that continues today **(Figure 3.25)**. Russia experienced a similar, though eastward, flow of people in the 20th century. In 2007, nearly 200 million people—roughly 1 of every 33 then alive—lived outside the country of their birth, and migration had become a world social, economic, and political issue of first priority.

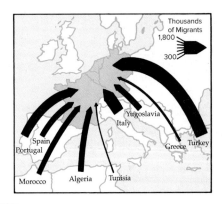

Figure 3.24 International "guest worker" flows to Western Europe. Labor shortages in expanding Western European economies beginning in the 1960s offered job opportunities to workers immigrating under labor contract from Eastern and Southern Europe and North Africa. Economic stagnation and domestic unemployment halted foreign worker contracting in Germany, France, Belgium, the Netherlands, and Switzerland in the later 1980s and 1990s, but continuing immigration raised the share of foreign workers in the labor force to 20 percent in Switzerland, 10 percent in Austria, and 9.5 percent in Germany by 2000.

Source: Data from Gunther Glebe and John O'Loughlin, eds., "Foreign Minorities in Continental European Cities," Erdkundliches Wissen 84 (Wiesbaden, Germany: Franz Steiner Verlag, 1987).

In the 20th century, nearly all countries experienced a great movement of peoples from agricultural areas to the cities, continuing a pattern of *rural-to-urban* migration that first became prominent during the 18th- and 19th-century Industrial Revolution in advanced economies and now is even more massive than international migrant flows. The rural-to-urban migration going on in China is especially remarkable, as more than 300 million people (about equal to the entire U.S. population!) are predicted to migrate

in this way by 2020. Rapid increases in impoverished rural populations of developing countries put increasing and unsustainable pressures on land, fuel, and water in the countryside. Landlessness and hunger, as well as the loss of social cohesion that growing competition for declining resources induces, help force migration to cities. As a result, although the rate of urban growth is decreasing in the more developed countries, urbanization in the developing world continues apace, as will be discussed more fully in Chapter 11.

Motivations to Migrate

Migrations may be forced or voluntary or, in many instances, reluctant relocations imposed on the migrants by circumstances. Put another way, all migrations can be placed on a continuum of motivation from being strongly forced or compelled to being entirely voluntary, with all possible degrees of motivation in between.

In **forced migrations,** the relocation decision is made solely by people other than the migrants themselves (**Figure 3.26**). An estimated 10 to 12 million Africans were forcibly transferred as slaves to the Western Hemisphere from the late 16th to early 19th centuries. Half or more were destined for the Caribbean and most of the remainder for Central and South America, though nearly a million arrived in the United States. Australia owed its earliest European settlement to convicts transported after the 1780s to the British penal colony established in southeastern Australia (New South Wales). More recent involuntary migrants include millions of Soviet citizens forcibly relocated from countryside to cities and from the western areas to labor camps in Siberia and the Russian Far East beginning in the late 1920s. During the 1980s and 1990s, many refugee destination countries in Africa, Europe, and Asia expelled immigrants or encouraged or forced the repatriation of foreign nationals within their borders.

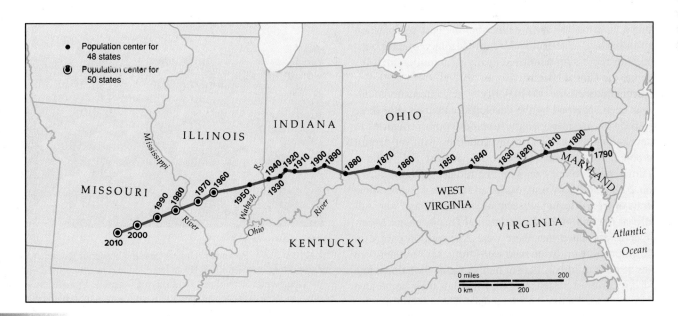

Figure 3.25 Westward shift of population, 1790–2010. More than 200 years of western migration and population growth are recorded by the changing U.S. center of population. (The *center of population* is that point at which a rigid map of the United States would balance, reflecting the identical weights of all residents in their location on the census date.) The westward movement was rapid for the first 100 years of census history and slowed between 1890 and 1950. Some of the post-1950 acceleration reflects population growth in the Sunbelt. However, the large shift in 1960 as compared to 1950 also reflects the geographic pull on the center of population exerted by the admission of Alaska and Hawaii to statehood in 1959.

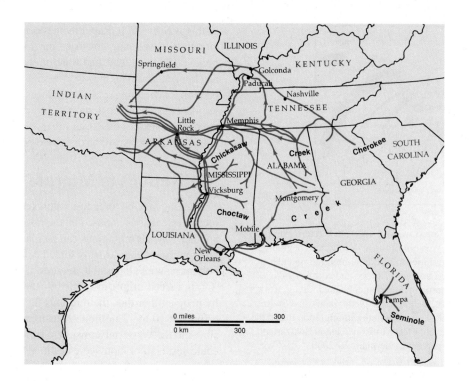

Figure 3.26 Forced migrations: The Five Civilized Tribes. Between 1825 and 1840, some 100,000 southeastern Native Americans were removed from their homelands and transferred by the Army across the Mississippi River to "Indian Territory" in present-day Oklahoma. By far, the largest number consisted of members of the "Five Civilized Tribes" of the South: Cherokees, Choctaws, Chickasaws, Creeks, and Seminoles. Settled, Christianized, literate small-farmers, their forced eviction and arduous journey—particularly along what the Cherokees named their "Trail of Tears" in the harsh winter of 1837–1838—resulted in much suffering and death.

Less than fully voluntary migration—**reluctant relocation**—of some 8 million Indonesians has taken place under an aggressive governmental campaign begun in 1969 to move people from densely settled Java (roughly 775 per square kilometer or 2,000 people per square mile) to other islands and territories of the country, in what has been called the "biggest colonization program in history." International refugees from war and political turmoil or repression numbered nearly 13.5 million in 2005, according to the United Nations.

In recent decades, the vast majority of new international migrants has been absorbed by the developed countries. But there has also been an increasing flight from developing countries to other developing regions, and many countries with the largest refugee populations are among the world's poorest. Sub-Saharan Africa alone houses more than 3 million refugees (**Figure 3.27**). Between 2003 and 2007, Iran, Syria, and Jordan became home to millions of Iraqis fleeing persecution, terrorism, and war. Additionally, at the end of 2005, the Internal Displacement Monitoring Centre estimated that there were 25.3 million people in some 40 countries worldwide who were "refugees" in their own countries as a result of conflicts or human rights violations; in a search for security or sustenance, they had left their home area but not crossed an international boundary. The total did not include those additional millions internally displaced by environmental disasters such as the southeast Asian tsunami in 2004 or Hurricane Katrina in 2005.

Poverty is the great motivator. Some 30 percent of the world's population—nearly 2 billion persons—have less than $1 per day of income. In addition, many are victims of drought, floods, other natural catastrophes, or wars and terrorism. Poverty in developing countries is greatest in the countryside; rural areas

Figure 3.27 Rwandan refugees near the border of Rwanda and Zaire. More than 1 million Rwandans fled into neighboring Zaire (now, the Democratic Republic of the Congo), Tanzania, Uganda, and Burundi in 1994 to escape an interethnic civil war in their home country and the genocide and retribution that killed at least 750,000 people. Early in the 21st century, more than 14 million Africans remained uprooted (that is, internally displaced and refugees combined). Fleeing war, repression, and famine, millions of people in developing nations have become reluctant migrants from their homelands.

©GYSEMBERGH Benoit/Getty Images

are home to around 750 million of the world's poorest people. Of these, some 20 to 30 million move each year to towns and cities, many as "environmental refugees" abandoning land so eroded or exhausted it can no longer support them. In the cities, they join the 40 percent or more of the labor force that is unemployed or underemployed in their home country and seek legal or illegal entry into more promising economies of the developed world. All, rural or urban, respond to the same basic forces—the push of poverty and the pull of perceived or hoped-for opportunity.

Those motivating forces are controlling in much of the international flow of illegal migrants whose economic condition in their homelands, they feel, is so intolerable that to seek employment they risk their lives in flight by unsafe boat and raft or through forbidding natural boundary barriers as involuntary but unforced migrants. Without immigration papers or legal status, subject to arrest and deportation or worse, illegal immigrants able to find work and income satisfy some of their migration objectives by sending money home to ensure their families' survival. Immigrants from poor countries, the World Bank estimated, sent more than 260 billion traceable dollars home in 2006; money sent through informal channels increased that total by as much as 50 percent. The traceable **remittance** amount alone in 2005 was on a par with the total of foreign direct investment in developing countries and twice the value of foreign aid. The estimated 17 million Latin American immigrants (again, not all illegal) in the United States sent an estimated $45 billion a year in legal and illegal remittances to their home countries in 2006 ($19 billion to Mexico alone). For some Latin American countries, those remittances account for about 20 percent of the money circulating in their economies.

The great majority of migratory movements, however, are **voluntary migrations** (volitional), representing individual response to the factors influencing all spatial interaction decisions. At root, migrations take place because the migrants believe that their opportunities and life circumstances will be better at their destination than they are at their present location.

Controls on Migration

Economic considerations crystallize most migration decisions, though nomads fleeing the famine and spreading deserts of the Sahel obviously are impelled by different economic imperatives than is the executive considering a job transfer to Montreal or the resident of Appalachia seeking factory employment in the city. Among the aging, affluent populations of highly developed countries, retirement amenities figure importantly in beliefs about the residential attractiveness of areas. Educational opportunities, major life events such as getting married, moving closer to family members, and environmental attractions or repulsions are but a few other possible migration motivations. Migration theorists attribute international economic migrations to a series of often overlapping mechanisms. Differentials in wages and job opportunities between home and destination countries are perhaps the major driving force in such individual migration decisions. Those differentials are in part rooted in a built-in demand for workers at the bottom of the labor hierarchy in more prosperous developed countries whose own workers disdain low-income,

menial jobs with poor wages and benefits. Migrants are available to fill those jobs, some argue, because advanced economies make industrial investment in developing or colonial economies to take advantage of lower labor costs there. New factories inevitably disturb existing economies, employ primarily short-term female workers, and leave a residue of unemployed males available and prone to migrate in search of opportunity. If successful, international economic migrants, male or female, help diversify sources of family income through their remittances from abroad, a form of household security that in itself helps motivate some international economic migration.

Negative home conditions that impel the decision to migrate are called **push factors.** They might include loss of job, lack of professional opportunity, overcrowding or slum clearance, or a variety of other influences, including poverty, war, violent crime, and famine. The presumed positive attractions of the migration destination are known as **pull factors.** They include all the attractive attributes believed to exist at the new location—safety and food, perhaps, or job opportunities, better climate, lower taxes, quality schools, more room, and so forth. Very often, migration is a result of both push and pull factors. But it is important to recognize that it is the migrant's beliefs about the areal pattern of opportunities and want satisfaction that is important here, whether or not those beliefs are supported by objective reality.

The concept of place utility helps us to understand the decision-making process that potential voluntary migrants undergo. **Place utility** is the measure of an individual's satisfaction with a given residential location. The decision to migrate is a reflection of the appraisal—the perception—by the prospective migrant of the current homesite compared to other sites of which something is known or hoped for. In the evaluation of comparative place utility, the decision maker considers not only the believed value of the present location, but also the expected place utility of potential destinations. Seen in this way, we can understand that all migrations not strictly forced are a combined response to push factors at the current location relative to the pull factors of potential new residential sites.

Those evaluations are matched with the individual's aspiration level, that is, the level of accomplishment or ambition that the person sees for herself or himself. Aspirations tend to be adjusted to what one considers attainable. If one finds present circumstances satisfactory, then **spatial search** behavior—the process by which locational alternatives are evaluated—is not initiated. If, on the other hand, dissatisfaction with the home location is felt, then a utility is assigned to each of the possible migration locations. The utility is based on past or expected future rewards at various sites. Because new places are unfamiliar to the searcher, the information received about them acts as a substitute for the personal experience of the homesite. Decision makers can do no more than sample information about place alternatives and, of course, there may be errors in both information and interpretation. Ultimately, these anticipated utilities depend on beliefs—place perceptions—about the places being considered and on the motivations of potential migrants that prompt them to consider long-distance migration, or even relocation of residence within the local area. In the latter instance, of course, the spatial search usually involves actual site visits in evaluating the potential move (**Figure 3.28**). Of course, even actual visits cannot generally show people everything important there is to know about a place being considered for a new home (just as one date does not usually tell us whom to marry!).

One goal of the potential migrant is to avoid physically dangerous or economically unprofitable outcomes in the final migration decision. Place utility evaluation, therefore, requires assessments not only of hoped-for pull factors of new sites, but also of the potentially negative economic and social reception the migrant might experience at those sites. An example of that observation can be seen in the case of the large numbers of young Mexicans and Central Americans who have migrated both legally and illegally to the United States (**Figure 3.29**). Faced with poverty, crime, and overpopulation at home, they regard the place utility in Mexico as minimal. With a willingness to work, they learn from friends and relatives of job opportunities north of the border and, hoping for success or even wealth, quickly place high utility on relocation to the United States. Many know that dangerous risks are involved in entering the country illegally, but even legal immigrants face legal restrictions or rejections that are advocated or designed to reduce the pull attractions of the United States (see the feature "Porous Borders").

Another migrant goal is to reduce uncertainty about the move and the new residential destination. That objective may be achieved either through a series of transitional relocation stages or when the migrant follows the example of known predecessors. **Step migration** involves the place transition from, for example, rural to central city residence through a series of less extreme locational changes—from farm to small town to suburb and, finally, to the major central city itself. **Chain migration** assures that the mover is part of an established migrant flow from a common origin to a prepared destination. An advance group of migrants, having established itself in a new home area, is followed by second and subsequent migrations originating in the same home district and frequently united by kinship or friendship ties. Public and private services for legal migrants and informal service networks for undocumented or illegal migrants become established and contribute to the continuation or expansion of the chain migration flow. Ethnic and foreign-born enclaves in major cities and rural areas in a number of countries are the immediate result, as we shall see more fully in Chapter 6.

Sometimes the chain migration is specific to occupational groups. For example, nearly all newspaper vendors in New Delhi, in the north of India, come from one small district in Tamil Nadu, in the south of India. Most construction workers in New Delhi come either from Orissa, in the east of India, or Rajasthan, in the northwest. The diamond trade of Mumbai, India, is dominated by a network of about 250 related families who come from a small town several hundred miles to the north. Many members of particular small towns in Mississippi ended up in the same neighborhood in Chicago as part of the great internal migration of African Americans from the rural South to northern and western cities during the mid-twentieth century.

Certainly, not all immigrants stay permanently at their first destination. Of the some 80 million newcomers to the United States between 1900 and 1980, some 10 million returned to their homelands or moved to another country. Estimates for Canada indicate that perhaps 40 of each 100 immigrants eventually leave, and about 25 percent of newcomers to Australia also depart permanently. Therefore, a corollary of all out-migration flows is **counter** (or **return**) **migration,** the likelihood that as many as 25 percent of all migrants will return to their place of origin.

Within the United States, return migration—defined as moving back to one's state of birth—makes up about 20 percent of all domestic moves. That figure varies dramatically among states. More than a third of recent in-migrants to West Virginia, for example, were returnees—as were more than 25 percent of those moving to Pennsylvania, Alabama, Iowa, and a few other states. Such widely different states as New Hampshire, Maryland, California, Florida, Wyoming, and Alaska were among the several that found returnees were fewer than 10 percent of their in-migrants. Interviews suggest that states deemed attractive draw new migrants in large numbers, whereas those with high proportions of returnees in the migrant stream are not believed to be desirable destinations by other than former residents. The preference maps we looked at above reveal these sorts of place preferences.

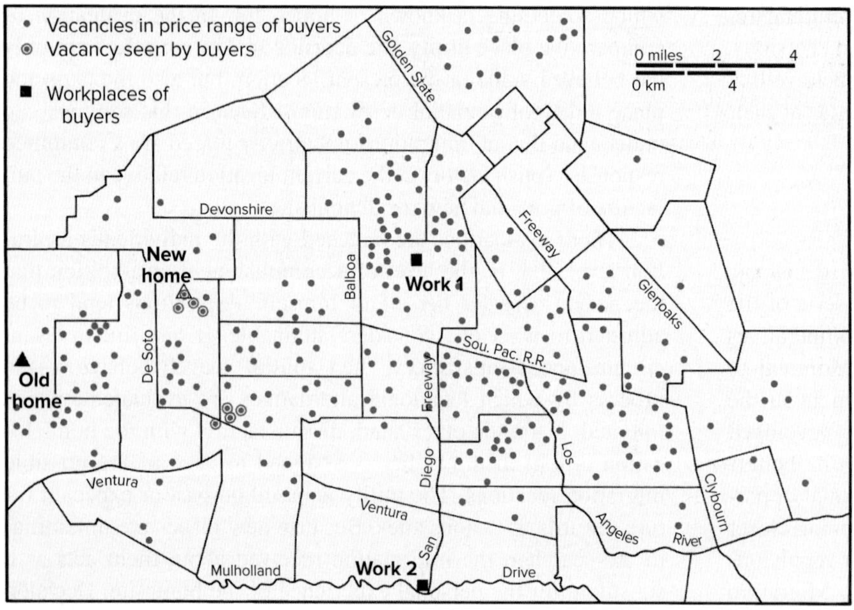

Figure 3.28 An example of a residential spatial search. The dots represent the house vacancies in the price range of a sample family. Note (1) the relationship of the new house location to the workplaces of the married couple, (2) the relationship of the old house location to the chosen new homesite; and (3) the limited total area of the spatial search. This example from the San Fernando Valley area of Los Angeles is typical of intraurban moves.

Source: J. O. Huff, Annals of the Association of American Geographers, *Vol. 76, 217 -221. Association of American Geographers, 1986.*

(a)

(b)

Figure 3.29 *(a)* Unauthorized Mexican immigrants running from the U.S. Border Patrol. *(b)* Workers build portions of a barrier fence on the U.S.–Mexico border, dividing Calexico, California, from Mexicali, Baja California.

Source: (b) Image released by the United States Air Force with the ID 061003-F-1726H-004.

(a) ©Per-Anders Pettersson/Hulton Archive/Getty Images; (b) Source: DoD photo by Staff Sgt. Dan Heaton, U.S. Air Force

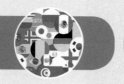

Geography and Citizenship

Porous Borders

Many residents of the U.S. were born in another country—about 13 to 14 percent of the United States population as of 2016. The United States has been the most popular destination for international migrants for at least half a century (and quite a few before then as well); about 20 percent of the world's current international migrants live there. But migrants can enter a country legally—with a passport, visa, working permit, or other authorization—or illegally, without authorization (also referred to as *undocumented*). And some immigrants arrive claiming the right of political asylum but actually seeking economic opportunity. Others enter the country legally but on a temporary basis (as a student or tourist, for example) but then remain after their official departure date. The U.S. Department of Homeland Security estimates that between one-fourth and one-third of those residing without authorization in the United States entered the country legally but then overstayed their visas.

It is impossible to determine the precise number of people residing without authorization in the United States; this population is very difficult to count accurately. At this time, the Census attempts to count everyone living in the United States but does not attempt to determine residents' legal status (the Trump administration proposes to add this question to the 2020 Census). The number of people residing without authorization changes on a daily basis, and most of these people are obviously not eager to make sure government agencies know they reside here. A widely accepted figure (from the Pew Research Center and other sources) estimates the number of people living without authorization in the United States in the year 2017 at about 11 million. This is much higher than the 3½ million estimated to have resided illegally in the United States in 1990, but actually a decline from the more than 12 million believed to have been here in 2007, perhaps the peak year in American history (the United States had virtually an open border until the 1920s). The main cause for the decline during the last 10 years has apparently been the large economic recession starting in 2008.

About 52 percent of unauthorized immigrants in the United States came from Mexico,

another 21 percent from other Central American countries and the Caribbean, 12 percent from Asia, 6 percent from South America, 3 percent from Africa, and the rest from Canada, Europe, and Oceania. Although, historically, a substantial majority of people migrating to the United States without authorization came from Mexico, the source of unauthorized immigrants has changed considerably during the last decade. During this time, there has actually been net out-migration of Mexican people here without authorization—more people here illegally from Mexico have left the United States (most returning to Mexico) than have entered. Those coming from other countries have increased during this time, however, in some sense "making up" for the decrease in Mexican immigrants. There has especially been a big increase in people coming without authorization from Central America south of Mexico (who generally travel through Mexico to get to the United States) and from Asia, particularly China and India (the countries with the two largest populations in the world).

More than half of unauthorized immigrants reside in four states: California (25 percent), Texas (11 percent), New York (10 percent), and Florida (6 percent). They are clustered in large urban areas, with just 20 large metropolitan areas (led by New York, Los Angeles, and Houston) accounting for something like 60 percent of the unauthorized residents. But they reside in every state, and in both rural and urban areas. The states with the smallest unauthorized populations are Montana, North Dakota, and West Virginia, each with less than 5,000 such residents.

Attitudes against unauthorized immigrants that some have expressed have been directed mainly against those from Mexico and other Latin American countries, most of whom are unskilled workers. Once in the United States, Latin American immigrants have found work in agricultural fields, animal slaughtering and meatpacking facilities, construction, hotels, and restaurants. Many work in private residences as maids, nannies, and gardeners, and, hence, citizens of wealthy enclaves often express stronger support for the rights of unauthorized immigrants to stay in the United States The demand for their labor is great enough that male immigrants here without authorization have a higher labor-force participation rate than male

immigrants here legally (92 percent versus 85 percent), who, in turn, participate at a higher rate than males who are native-born (81 percent). Interestingly, the pattern is reversed for females. Females here without authorization have lower labor-force participation rates than female immigrants here legally (61 percent versus 64 percent), who themselves participate at a lower rate than native-born females (72 percent).

Public opinion is distinctly split about what the United States should do concerning both the continuing streams of unauthorized immigrants and those who are here already. Many people are concerned about the loss of employment opportunities or depression in wages they believe immigrants cause, as well as the increased strain they put on housing stocks, traffic congestion, emergency room visits, and public schools, the latter particularly considering the challenges of providing public education for students who do not speak English well. There is concern that illegal immigration opens our country to disease, drug smuggling, and terrorism (concern about disease is a very traditional and long-standing anxiety about foreign immigration). Some people, including many legal immigrants, just feel that it is wrong for people to enter without the long and rather arduous process of applying to enter and reside legally. In contrast, other people believe that the United States acts compassionately by accepting new residents who want only to work, provide for their families, and escape from poverty and drug violence. Some supporters claim that immigrants increase the strength of the U.S. economy, through their consumption, production, and payment of taxes. Others argue that unauthorized immigrants keep the costs of agricultural products low and do jobs "Americans will not take" (a claim of the George W. Bush administration). They value the cultural diversity immigrants bring and do not believe that immigrants (authorized or not) are major contributors to disease and crime (evidence supports their view on these claims). Many supporters of legalizing unauthorized immigrants simply believe it is practically and morally unimaginable to propose deporting so many people from our country, in many cases returning them to harsh and dangerous conditions.

Thus, attitudes toward unauthorized immigration vary greatly, with multiple arguments on both sides of the issue. However, it appears that a clear majority of Americans favor some way of granting legal status to immigrants currently residing illegally in the United States In the 2015 *American Values Atlas* survey (Public Religion Research Institute), 62 percent of Americans agreed that unauthorized immigrants should be allowed to become citizens providing they meet certain requirements, and 15 percent agreed they should be allowed to become permanent legal residents but not citizens. Just 19 percent believed they should be identified and deported (4 percent offered no opinion).

Concern over the large number of unauthorized immigrants has been reflected in a number of actions in recent decades. Greater efforts have been made to deter unauthorized crossings along the 1,950-mile U.S.-Mexico border. This border is enforced with nearly 700 miles of fencing of different types (some of which is fairly easy to climb over, if you can get to that spot over rough terrain). Thousands of Border Patrol agents use automotive vehicles, helicopters, drones, night-vision cameras, and hidden electronic sensors to surveil the border. Apprehensions at the border vary year to year, in part because of changing emphasis by presidential administrations. During the Obama administration, the number was as high as 800,000 in 2010 and as low as 460,000 in 2015. In parts of Arizona and California, self-appointed Minutemen —groups of volunteer militia—patrol the border "to protect our country from a 40-year long invasion across our southern border with Mexico," as one vigilante put it. A cornerstone of President Trump's platform when running for office was a promise to build a tall barrier wall along the entire southern border. As of this writing, his administration continues to push Congress to provide initial funding for the wall. Interestingly, the Pew Research Center reports that as of early 2018, more than half (60 percent) of Americans are against having the government "substantially expand the wall along the U.S. border with Mexico". Democrats are strongly against it (6½ to 1 against), whereas Republicans strongly support such an expansion (3 to 1 in favor).

With respect to unauthorized immigrants who are already in the United States, many people advocate implementing a process of vetting and eventually granting them authorization

to be here. This has happened more than once before; millions of unauthorized residents managed to achieve legal status by taking advantage of government amnesties offered between 1984 and 2000. In 2012, the Obama administration launched the *Deferred Action for Childhood Arrivals*, or *DACA*, program. It offered two years of work authorization and relief from the threat of deportation to young adult immigrants who were brought into the United States without authorization by their parents or other adults before they were 16 years of age, as long as they had lived continuously in the United States, had gone to school, and had not been convicted of significant criminal behaviors or deemed a threat to public safety or national security. It has been estimated that more than a million people currently meet these criteria, with another half million nearly meeting them. Most of these applicants have applied for the program, a majority of whom have been granted DACA status. In September of 2017, however, the Trump administration rescinded DACA with a six-month wind-down. Public opinion does not support this much—the Pew Center report from early 2018 finds that 74 percent of Americans favor granting legal status to DACA candidates. Democrats and those leaning Democratic are at least 9 to 1 in favor of it, and although Republicans and those leaning Republican are much less in favor, they are still a bit more likely to support it than oppose it. A political compromise being debated by Congress and the president is to reinstate much of the DACA program in return for funding for the border wall. Clearly, the issue of unauthorized immigration continues to be prominent in political and policy debates in the United States.

Questions to Consider

1. Several hundred thousand undocumented immigrants enter the United States each year, and most find gainful employment, yet the country issues only 5,000 visas a year for unskilled foreigners seeking year-round work. Should the United States increase the number of visas available?

2. Making unauthorized crossings more difficult in California did not diminish the number of migrants making the journey north; it simply pushed coyotes, the people

who lead migrants across the border, into Arizona. Some people believe resourceful migrants will always find a way to get across the border. As one observer noted, "It's like putting rocks in a river—the water just goes around it." Do you think there is any way to seal the entire U.S.-Mexico border, or will immigration continue so long as the income gap between the United States and Latin America remains great? Could the United States reduce immigration pressures by improving the Mexican economy?

3. With regard to proposals for a temporary-worker program, the libertarian Cato Institute argues that when there is no immigration barrier, circular migration occurs, with migrant workers entering and leaving almost at will. It cites Puerto Rico as an example; many who move to the U.S. mainland stay for just a few years, and out-migration from the island is very low. A temporary worker program, on the other hand, encourages migrants to move north with their entire families, and those who are already in the United States stay for good because border crossings become more expensive and dangerous. If you were a member of Congress, would you be in favor of creating a guest-worker program? Why or why not?

4. It is often said that unauthorized immigrants perform jobs that "Americans won't take". Why do you think many immigrants are willing to work for low wages, often under poor working conditions? Would Americans take the jobs if they paid, say, $20 per hour and offered health care and other benefits?

5. What would happen if all states followed Arizona's lead in passing and enforcing laws targeting employers of undocumented workers? One in 10 or 11 workers in Arizona is unauthorized, and opponents of the legislation contend that the state has put its economy in jeopardy. The workers tend to be reliable, they fill necessary jobs, spend their wages in the communities in which they live and work, and most pay taxes because employers in the construction, hotel, restaurant, and other industries withhold taxes from paychecks.

6. Anyone born on U.S. soil (except the children of foreign diplomats or the like) automatically becomes a U.S. citizen, even if their parents entered the country illegally. Likewise, the immediate family members (spouses, children, parents) of anyone in the United States legally gets preferential legal admission to move to the United States This *family reunification* principle is an example of chain migration. Critics argue that these principles are inappropriate and responsible for excess immigration into the United States Argue for or against each of these principles.

7. Should unauthorized residents already here be given the opportunity to get worker permits and the possibility of eventual citizenship if they have no criminal record, pay a fine, and demonstrate that they are gainfully employed and have paid taxes?

8. Do you believe the federal government has an obligation to fully or partially reimburse state and local governments for the costs of education, medical care, incarceration, and other legal services for unauthorized immigrants? Why or why not?

9. Should the United States require citizens to have a national identification card and to present it to officials upon demand? If so, would all people who look like they may have been born abroad feel it necessary to carry proof of citizenship at all times?

Once established, origin and destination pairs of places tend to persist. Areas that dominate a locale's in- and out-migration patterns make up the **migration fields** of the place in question. As we would expect, areas near the point of origin comprise the largest part of the migration field (remember distance decay and the First Law), though larger cities more distantly located may also be prominent as the ultimate destination of hierarchical step migration (**Figure 3.30**). As **Figure 3.31** shows, some migration

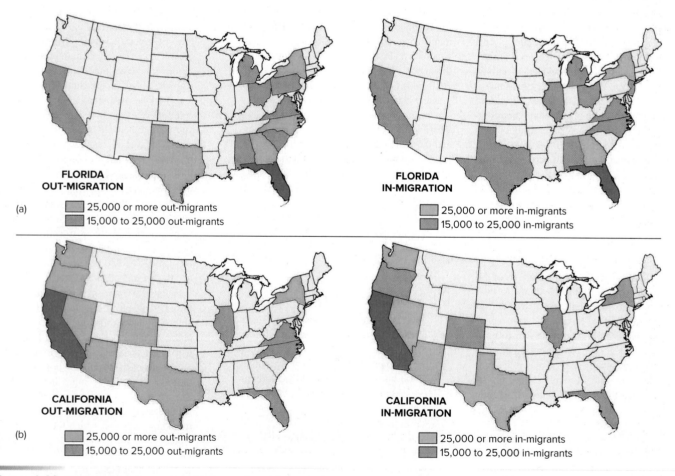

(a)

FLORIDA
OUT-MIGRATION

☐ 25,000 or more out-migrants
☐ 15,000 to 25,000 out-migrants

FLORIDA
IN-MIGRATION

☐ 25,000 or more in-migrants
☐ 15,000 to 25,000 in-migrants

(b)

CALIFORNIA
OUT-MIGRATION

☐ 25,000 or more out-migrants
☐ 15,000 to 25,000 out-migrants

CALIFORNIA
IN-MIGRATION

☐ 25,000 or more in-migrants
☐ 15,000 to 25,000 in-migrants

Figure 3.30 The migration fields of Florida and California in 2005–2010. (*a*) For Florida, nearby Georgia receives most out-migrants, but in-migrants originate in large numbers from the northeastern United States. (*b*) For California, the nearby western states receive large numbers of out-migrants, and there are fewer in-migrants from those states.

Source: U.S. Census Bureau.

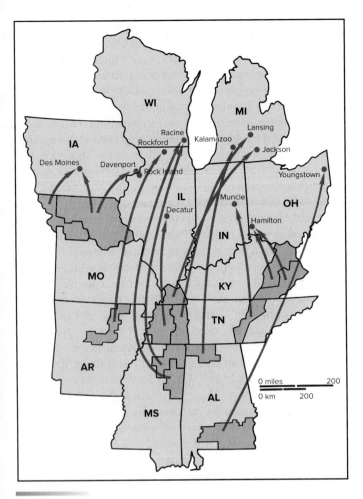

Figure 3.31 In the United States in the late 20th century, channelized migration flows from the rural south to midwestern cities of medium size. Distance is not the only determinant of flow direction. Through family and friendship links, the rural southern areas were tied to particular midwestern destinations.

Source: Proceedings of the Association of American Geographers, *C. C. Roseman, Vol. 3, p. 142.*

studies. Movers seek to minimize the friction of distance. In selecting between two potential destinations of equal merit, a migrant tends to choose the nearer as involving less effort and expense. And because information about distant areas may be less complete and less certain than knowledge about nearer localities, short moves are favored over long ones. Research indicates that determined migrants with specific destinations in mind are unlikely to be deterred by distance considerations. However, groups for whom push factors determine the migration decision more than specific destination pulls are likely to limit their migration distance in response to encountered apparent opportunities. For them, intervening opportunities affect locational decisions. The concept of hierarchical migration also helps explain some movement decisions. Individuals in domestic relocations tend to move up the level in the urban hierarchy, from small places to larger ones. Often, levels are skipped on the way up; only in periods of general economic decline is there considerable movement down the hierarchy. Because suburbs of large cities are considered part of the metropolitan area, the movement from a town to a suburb is considered a move up the hierarchy.

Observations such as these were summarized in the 1870s and 1880s as a series of "laws of migration" by E. G. Ravenstein (1834–1913). Among those that remain relevant are the following:

- Most migrants go only a short distance.
- Longer-distance migration favors big-city destinations.
- Most migration proceeds step by step.
- Most migration is rural to urban.
- Each migration flow produces a counterflow.
- Most migrants are adults; families are less likely to make international moves.
- Most international migrants are young males.

The latter two "laws" introduce the role of personal attributes (and attitudes) of migrants: their age, sex, education, and economic status. Migrants do not represent a cross section of the populace from which they come. Selectivity of movers is evident, and the selection shows some regional differences. In most societies, young adults are the most mobile (**Figure 3.32**). In the United States, mobility peaks among those in their twenties, especially the later twenties, and tends to decline thereafter. Among West African cross-border migrants, a World Bank study reveals, the age group 15–39 predominated. Young adults have weaker ties to their place of origin than older adults, being less likely to have careers or stable employment, mates, children, houses, and other material possessions. They are also generally healthier with more energy and endurance for the challenges of relocation.

Ravenstein's conclusion that young adult males are dominant in economically pushed international movement is less valid today than when first proposed. In reality, women and girls now comprise 40 to 60 percent of all international migrants worldwide (see the feature "Gender and Migration"). It is true that legal and illegal migrants to the United States from Mexico and Central America are primarily young men, as were first-generation "guest workers" in European cities. But population

fields reveal a distinctly channelized pattern of flow. The channels link areas that are in some way tied to one another by past migrations, by economic trade considerations, or some other affinity. The flow along them is greater than otherwise would be the case (such as predicted by a standard gravity model) but does not necessarily involve individuals with personal or family ties. The former streams of southern blacks and whites to northern cities, of Scandinavians to Minnesota and Wisconsin, and of U.S. retirees to Florida and Arizona or their European counterparts to Spain, Portugal, or the Mediterranean coast are all examples of **channelized migration.**

Voluntary migration is responsive to the other controls that influence all forms of spatial interaction. Push-pull factors may be equated with complementarity; costs (emotional and financial) of a residence relocation are expressions of transferability. Other things being equal, large cities exert a stronger migrant pull than do small towns, a reflection of the impact of the gravity model. We noted the influence of distance decay in migration

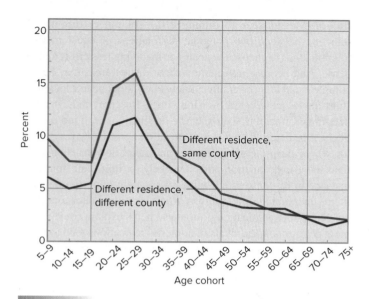

Figure 3.32 Percentage of 2004 population over 5 years of age with a different residence than in 2003. Young adults figure most prominently in both short- and long-distance moves in the United States, an age-related pattern of mobility that has remained constant over time. For the sample year shown, 28 percent of people in their 20s moved, whereas fewer than 5 percent of those 65 and older did so. Short-distance moves predominate; 58 percent of the 39 million total U.S. movers between March 2003 and March 2004, relocated within the same county and another 20 percent moved to another county in the same state. Some two-thirds of intracounty (mobility) moves in that year were made for housing-related reasons; long-distance moves (migration) were made for work-related (31 percent) and family (25 percent) reasons.

Source: U.S. Bureau of the Census.

projections for West European countries suggest that women will shortly make up the largest part of their foreign-born population, and in one-third of the countries of sub-Saharan Africa, including Burkina Faso, Swaziland, and Togo, the female share of foreign-born populations was as large as the male. Further, among rural to urban migrants in Latin America since the 1960s, women have been in the majority.

Female migrants are motivated primarily by economic pushes and pulls. Surveys of women migrants in Southeast Asia and Latin America indicate that 50 percent to 70 percent moved in search of employment and commonly first moved while in their teens. The proportion of young, single women is particularly high in rural-to-urban migration flows, reflecting their limited opportunities in increasingly overcrowded agricultural areas. To the push and pull factors normally associated with migration decisions are sometimes added family pressures that encourage young women with few employment opportunities to migrate as part of a household's survival strategy. In Latin America, the Philippines, and parts of Asia, emigration of young girls from large, landless families is more common than from smaller families or those with land rights. Their remittances of foreign earnings help maintain their parents and siblings at home.

An eighth internationally relevant observation may be added to those cited in Ravenstein's list: On average, emigrants tend to be relatively well educated. A British government study reveals that three-quarters of Africa's emigrants have higher (beyond high school) education, as do about half of Asia's and South America's. Of the more than 1 million Asian Indians living in the United States, more than three-quarters of those of working age have at least a bachelor's degree. The loss to home countries can be draining; about 30 percent of all highly educated West African Ghanaians and Sierra Leoneans live abroad. Outward migration of the educated affects developed countries as well as poorer developing states. It is claimed that from 1997 to 2002, between 15 percent and 40 percent of each year's graduating classes from Canadian colleges emigrated to the United States, while in Europe, half the mid-1990s' graduating physics classes at Bucharest University left the country.

For modern Americans, the decisions to migrate are more ordinary, but individually just as compelling. They appear to involve (1) major life events (e.g., getting married, having children, getting a divorce, etc.); (2) changes in the career course (getting a first job or a promotion, receiving a career transfer, seeking work in a new location, retiring, etc.); or (3) changes of residence associated with individual personality (**Figure 3.33**).

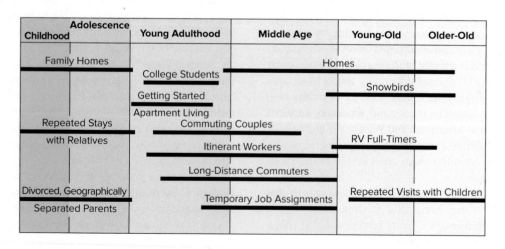

Figure 3.33 Examples of multiple residences by stage in life. Each horizontal line represents a period of time in a possible new residence.

Source: From K. McHugh, T. Hogan, and S. Happel, "Multiple residence and cyclical migration," The Professional Geographer 47(3), Figure 1, p. 253. Association of American Geographers, 1995.

Gender and Migration

Gender is involved in migration at every level. In a household or family, women and men are likely to play different roles regarding decisions or responsibilities for activities such as child care. These differences, and the inequalities that underlie them, help determine who decides whether the household moves, which household members migrate, and the destination for the move. Outside the household, societal norms about women's mobility and independence often restrict their ability to migrate.

The economies of sending and receiving areas play a role as well. If jobs are available for women in the receiving area, women have an incentive to migrate, and families are more likely to encourage the migration of women as necessary and beneficial. Thousands of women from East and Southeast Asia have migrated to the oil-rich countries of the Middle East, for example, to take service jobs.

The impact of migration is also likely to be different for women and men. Moving to a new economic or social setting can affect the regular relationships and processes that occur within a household or family. In some cases, women might remain subordinate to the men in their families. A study of Greek-Cypriot immigrant women in London and of

Turkish immigrant women in the Netherlands found that although these women were working for wages in their new societies, these new economic roles did not affect their subordinate standing in the family in any fundamental way.

In other situations, however, migration can give women more power in the family. In former Zaire (now the Democratic Republic of the Congo), women in rural areas moved to towns to take advantage of job opportunities there, and gain independence from men in the process.

One of the keys to understanding the role of gender in migration is to disentangle household decision-making processes. Many researchers see migration as a family decision or strategy, but some members will benefit more than others from those decisions.

For many years, men predominated in the migration streams flowing from Mexico to the United States. Women played an important role in this migration stream even when they remained in Mexico. Mexican women influenced the migration decisions of other family members; they married migrants to gain the benefits from and opportunity for migration; and they resisted or accepted the new roles in their families that migration created.

In the 1980s, Mexican women began to migrate to the United States in increasing numbers. Economic crises in Mexico and an increase in the number of jobs available for women in the United States, especially in factories, domestic service, and service industries, have changed the backdrop of individual migration decisions. Now, women often initiate family moves or resettlement efforts.

Mexican women have begun to build their own migration networks, which are key to successful migration and resettlement in the United States. Networks provide migrants with information about jobs and places to live and have enabled many Mexican women to make independent decisions about migrating.

In immigrant communities in the United States, women are often the vital links to social institution services and to other immigrants. Thus, women have been instrumental in the way that Mexican immigrants have settled and become integrated into new communities.

Nancy E. Riley, "Gender, Power, and Population Change," Population Bulletin 52, no. 1 (1997): 32-33. Reproduced by permission.

Work-related relocations are most important in U.S. long-distance (intercounty) migrations, and in both intra- and interstate relocations, more migrants move down the urban hierarchy—that is, from larger to smaller centers—than vice versa. Some observers suggest that pattern of deconcentration reflects modern transportation and communication technologies, more and younger retirees, the cost of housing, and the attractions of amenity-rich smaller places. Some people, of course, simply seem to move often for no discernible reason, whereas others settle into a community permanently. For other developed countries, a somewhat different set of summary migration factors may be present.

Globalization

We have seen how the cost of communication affects the degree of spatial interaction. Since the 1980s, the Internet and relatively low transportation costs have made it increasingly easy to buy goods from abroad and to travel throughout the

world. There has been simultaneously a strong international movement to reduce barriers to trade and to foreign investment and ownership. For example, the creation of the European Union (EU) has dismantled restrictive national borders there, and its monetary unit, the euro, makes possible financial transactions in a single currency over a massive multinational common market.

Integration and interdependence characterize globalization and affect economic, political, and cultural patterns across the world. The unification of Eastern and Western Europe or the creation of regional trade alliances such as the North American Free Trade Agreement (NAFTA) in North America or Mercosur in South America is as much a function of the revolution in communication and computer technology as it is of the will of political or financial leaders. Low-cost, high-speed computers, communication satellites, fiber-optic networks, and the Internet are the main technologies of the revolution, with robotics, microelectronics, e-mail, cell phones, and more making their contributions.

The fact that a consumer in Italy can order a book from Amazon.com or clothes from Land's End, obtain news from CNN, or make an investment through the London Stock Exchange while talking on a cell phone to a colleague in Tokyo is revolutionary and proof that globalization brings about greater world integration and spatial interaction (**Figure 3.34**). The Web browsers built into even basic cell phones enable large numbers of people to be integrated into the global community. The cell phone's capacity for immediate voice, instant message, and Internet data transfer has made it an agent of universal globalization in the early 21st century.

International banking is a financial case in point with nearly instantaneous movements across borders of billions of dollars in response to changing foreign exchange values and investment opportunities. Split-second changes in all the interconnected markets are certain. Within minutes of the September 11, 2001, attacks on the World Trade Center, stock markets everywhere went down as investors sensed that the international marketplace was in danger of losing stability. Internationalization of finance is also demonstrated by the immense sums in foreign investments held by citizens of all countries. American ownership of foreign stocks and bonds directly or through mutual funds and pension plans, for example, tie citizens of this country to the economic institutions of distant areas; at the same time, people outside the United States have significant holdings in U.S. companies and in U.S. treasury bonds. The U.S. subprime lending crisis and housing price bubble collapse of 2008 led to financial crises and an economic slowdown around the globe.

Transnational corporations (TNCs), discussed more fully in Chapter 9, are important forces driving the globalization of the world economy. With headquarters in one country and subsidiary companies, factories, warehouses, laboratories, and so on in several others, some 65,000 transnational corporations control several hundred thousand affiliates worldwide and sell their multitude of products on the international market. TNCs, by some estimates controlling about one-third of the world's productive assets, exploit the large differential in wage rates around the world to keep production costs low, not only decentralizing manufacturing and other business activities internationally but also diffusing the infrastructure and technology of modern business and industry to formerly underdeveloped regions, integrating them more fully and competitively into the global economy. At the same time, many employment opportunities decrease for citizens of the developed countries where most of the owners and stockholders of the TNCs reside. It is interesting to keep in mind that in spite of globalization, the areal differentiation in wages and employment opportunities is proof that the world has not yet been made completely homogeneous by technologies.

The internationalization of popular culture is more apparent to most of the world's people than is the less visible globalization of commerce and industry. In widely different culture realms, teenagers wear Yankee baseball caps, Gap shirts, Levis, and Reeboks; eat at McDonald's; drink coffee at Starbucks; and listen to pop music on their iPod. The culture they embrace is largely Western in origin and chiefly American. U.S. movies, television shows, video games, software, music, food, and fashion are marketed worldwide. They influence the beliefs, tastes, and aspirations of people in virtually every country, though their effect is most pronounced on young people. Like the globalization of finance, industry, and commerce, this internationalization of popular culture is further evidence of the transformative nature and impact of modern spatial behavior and interaction.

Figure 3.34 The old and the new: A traditional gondolier in Venice, Italy, conducting business on a cell phone.

©Arthur Getis

SUMMARY

Spatial interaction is the dynamic evidence of a world marked by large differences among places and of the interdependence among geographic locations. The term refers to the movement of goods, information, people, ideas—indeed, of every facet of economy and society—between one place and another. It includes the daily spatial activities of individuals and the collective patterns of their short- and long-distance behavior in space. The principles and constraints that unite, define, and control spatial behavior in this sense constitute an essential organizing focus for the study of human geographic patterns on the Earth.

We have seen that whatever the type of spatial behavior or flow, a limited number of recurring mechanisms of guidance and control are encountered. Three underlying bases for spatial interaction are complementarity, which encourages flows between areas by balancing supply with demand or satisfying need with opportunity; transferability, which affects movement decisions by introducing cost, effort, and time considerations; and intervening opportunities, which suggests that costs of overcoming distance may be reduced by finding closer alternate points where needs can be satisfied. The flows of commodities, ideas, and people governed by these interaction factors are interdependent and additive. Flows of commodities establish and reinforce traffic patterns, for example, and also channelize the movement of information and people.

Those flows and interactions may further be understood by the application of uniform models to all forms of spatial interaction from interregional commodity exchanges to an individual's daily pattern of movement. Distance decay tells us of the inevitable decline of interaction with increasing distance. The gravity model suggests that major centers of activity can exert interaction pulls that partly compensate for distance decay. Recognition of movement biases explains why spatial interaction in the objective world may deviate from that proposed by abstract models.

Humans in their individual and collective short- and long- distance movements are responsive to these impersonal spatial controls. Their spatial behaviors are also influenced by their separate circumstances. Each has an activity and awareness space reflective of individual socioeconomic and life-stage conditions. Each differs in mobility. Each has unique wants and needs and beliefs about their satisfaction. Human response to distance decay is expressed in a controlling critical distance beyond which the frequency of interaction quickly declines. That decline is partly conditioned by unfamiliarity with distant points outside normal activity space. Beliefs about home and distant territory therefore color interaction flows and space evaluations. In turn, those beliefs, no matter how accurate, underlie travel and migration decisions, part of the continuing spatial diffusion and interaction of people. It is to people and their patterns of distribution and regional growth and change that we turn our attention in the following chapter.

KEY WORDS

activity space	law of retail gravitation	refugee
attitude	link	Reilly's Breaking-Point Law
awareness space	migration	reluctant relocation
barrier	migration field	remittance
behavior	mobility	space-time compression
behavioral approach	movement bias	space-time path
chain migration	natural hazard	space-time prism
channelized migration	network	spatial interaction
cognition	network bias	spatial search
complementarity	node	step migration
counter (return) migration	partial displacement migration	temporary travel
critical distance	personal communication field	territoriality
distance decay	personal space	time geography
First Law of Geography	place perception	total displacement migration
forced migration	place utility	transferability
friction of distance	potential model	voluntary migration
gravity model	pull factor	
intervening opportunity	push factor	

FOR REVIEW

1. What is meant by spatial interaction? What are the three fundamental conditions governing all forms of spatial interaction? What is the distinctive impact or importance of each of the conditions?

2. What variations in distance decay curves might you expect if you were to plot shipments of ready-mixed concrete, potato chips, and computer parts? What do these respective curves tell us about transferability?

3. What is activity space? What factors affect the areal extent of an individual's activity space?

4. On a piece of paper, and following the model of Figure 3.10, plot your space-time path for your movements on a typical class day. What alterations in your established movement habits might be necessary (or become possible) if (a) instead of walking, you rode a bike? (b) Instead of biking, you drove a car? (c) Instead of driving, you had to use the bus?

5. What does the idea that transportation and communication are space-adjusting imply? In what ways has technology affected the "space adjustment" in commodity flows? In information flows?

6. Recall the places that you have visited in the past week. In your movements, were the rules of distance decay and critical distance operative? What variables affect your critical distances?

7. What considerations influence the decision to migrate? How do perceptions of place utility induce or inhibit migration?

8. What is a migration field? Some migration fields show a channelized flow of people. Select a particular channelized migration flow (such as the movement of Scandinavians to Michigan, Wisconsin, and Minnesota, or people from the Great Plains to California, or southern blacks to the North) and speculate why a channelized flow developed.

KEY CONCEPTS REVIEW

1. **What are the three bases for all spatial interaction?** Section 3.1

   Spatial interaction reflects areal differences and is controlled by three "flow-determining" factors. Complementarity implies a local supply of an item for which effective demand exists elsewhere. Transferability expresses the costs of movement from source of supply to locale of demand. An intervening opportunity serves to reduce flows of goods between two points by presenting nearer or cheaper sources.

2. **How is the likelihood of spatial interaction probability measured?** Section 3.1

   The probability of aggregate spatial movements and interactions may be assessed by the application of established models. Distance decay is the decline of interaction with increase in separation; the gravity model tells us that distance decay can in part be overcome by the enhanced attraction of larger centers of activity; and movement bias helps explain interaction flows contrary to model predictions.

3. **What are the forms, attributes, and controls of human spatial behavior?** Section 3.3–3.5

   While humans react to distance, time, and cost considerations of spatial movement, their spatial behavior is also affected by separate conditions of activity and awareness space, of individual economic and life-stage circumstances, by degree of mobility, and by unique beliefs of wants and needs.

4. **What roles do information and cognition play in human spatial actions?** Section 3.5

   Humans base their decisions about the opportunity or feasibility of spatial movements, exchanges, or want satisfactions on place perceptions (cognitions). These condition the feelings we have about physical and cultural characteristics of areas, the opportunities they possess, and their degree of attractiveness. Those cognitions may not accurately reflect reality or be well-informed in general. Distant places are less known than nearby ones, for example, and real natural hazards of areas may be mentally minimized through familiarity or rationalization.

5. **What kinds of migration movements can be recognized and what influences their occurrence?** Section 3.5

   Migration means the permanent relocation of residence and activity space. It is subject to all the principles of spatial interaction and behavior and represents both a survival strategy for threatened people and a reasoned response to perceptions of opportunity. Migration has been important throughout human history and occurs at separate scales, from intercontinental to regional and local. Negative home location conditions (push factors) coupled with believed positive destination attractions (pull factors) are important, as are age and sex of migrants and the spatial search they conduct. Step and chain migration and return migratory flows all affect the patterns and volume of flows.

POPULATION:
World Patterns, Regional Trends

The Djemaa el-Fna open-air market in the old quarter of Marrakesh, Morocco, is among the liveliest of markets in North Africa.
©Goodshoot/Getty Images

Key Concepts

4.1-4.2 Data and terminology used by population geographers: the meaning and purpose of population cohorts, population pyramids, fertility and mortality rates, and other measures.

4.3-4.4 Understanding demographic change using the demographic transition model and forecasting population change using the demographic equation.

4.5-4.6 World population distribution and densities.

4.7-4.8 Population projections, policies, and prospects: forecasting the future.

"*Zero, possibly even negative [population] growth*" *was the 1972 slogan proposed by the prime minister of Singapore, an island country in Southeast Asia. His nation's population, which stood at 1 million at the end of World War II (1945), had doubled by the mid-1960s. To avoid the overpopulation he foresaw, the government decreed "Boy or girl, two is enough" and refused maternity leaves and access to health insurance for third or subsequent births. Abortion and sterilization were legalized, and children born fourth or later in a family were to be discriminated against in school admissions policy. In response, by the mid-1980s birth rates had fallen to below the level necessary to replace the population, and abortions were terminating more than one-third of all pregnancies.*

"At least two. Better three. Four if you can afford it" was the national slogan proposed by that same prime minister in 1986, reflecting fears that the earlier campaign had gone too far. Gone was concern that overpopulation would doom the country to perpetual poverty. Instead, Prime Minister Lee Kuan Yew was moved to worry that population limitation would deprive Singapore of economic growth potential and the youthful, educated workforce needed to support its aging population. His 1990 national budget provided for sizable long-term tax rebates for second children born to mothers younger than 28. Not certain that financial inducements alone would suffice to increase population, the Singapore government annually renewed its offer to take 100,000 Hong Kong Chinese who might choose to leave when China took over that territory in 1997. By 2018, Singapore was among the richest countries in the world. It also had one of the lowest fertility rates in the world. Population decline was avoided, however, by carefully increasing immigration rates to make up for the lack of births.

The policy reversals in Singapore reflect an inflexible fact of population: the structure of the present controls the content of the future. The size, characteristics, growth trends, and migrations of today's populations help shape the well-being of peoples yet unborn. The numbers, age, and sex distribution of people; patterns and trends in their fertility and mortality; and their density of settlement and rate of growth both affect and are affected by the social, political, and economic organization of a society. Through population analysis, we can understand the relationship between population and resources, evaluate national and international population policies, and make reasoned forecasts of what the future may bring.

Population geography provides the background concepts and theories to understand and forecast the size, composition, and distribution of the human population. It differs from **demography,** the statistical study of human population, in its concern with *spatial* analysis—location, density, pattern, and relationship to the physical environment. Regional natural resources, standard of living, food supply, and conditions of health and well-being are basic to geography's population concerns. In addition, they are fundamental expressions of the human–environmental relationships that are one of the core themes of human geography.

4.1 Population Growth

Sometime in 2017, a human birth raised the Earth's population to 7.5 billion people. In 1999, the count reached 6 billion. In 2017, the world was adding an additional 83 million people annually,

or some 230,000 per day. By contrast, it took from the beginning of human history to about the year 1800 to reach 1 billion, and another 130 years to add the second billion. However, the annual rate of population increase has slowed from a peak value of 2.1 percent in 1962 to 1.1 percent in 2017. Even with the slowing rate of population growth, the United Nations in 2017 projected that the world would likely be home to 9.8 billion inhabitants in 2050. Projections of world population contain uncertainty that increases as they extend further into the future. The U.N. projections for the year 2100 range from 9.5 to 13.3 billion, with a best projection of 11.2 billion. U.N. projections for 2300 range from 2.3 to 36 billion.

All demographic forecasts agree, however, that essentially all of any future growth will occur in countries now considered *developing* (**Figure 4.1**), with especially rapid growth in the 47 least-developed states. The world's 10 most populous countries are mostly found in the developing regions of the world, and that trend will become even more pronounced by 2050 (**Table 4.1**). We will return to these projections and to the difficulties and disagreements inherent in making them later in this chapter.

Just what is implied by numbers in the millions and billions? With what can we compare the 2017 population of Trinidad and Tobago in the Caribbean (about 1.4 million) or of China (about 1.4 billion)? It is difficult to appreciate a number as vast as 1 million or 1 billion, and the great distinction between them. Here is an example offered by the Population Reference Bureau to help visualize the immensity of these numbers.

- A 2.5-centimeter (1-inch) stack of U.S. paper currency contains 233 bills. If you had a *million* dollars in thousand-dollar bills, the stack would be 11 centimeters (4.3 inches) high. If you had a *billion* dollars in thousand-dollar bills, your pile of money would reach 109 meters (358 feet)—about the length of a football field.

Table 4.1

World's Most Populous Countries, 2017 and 2050

2017		2050	
Country	Population (millions)	Country	Population (millions)
China	1,387	India	1,676
India	1,353	China	1,343
United States	325	Nigeria	411
Indonesia	264	United States	397
Brazil	208	Indonesia	322
Pakistan	199	Pakistan	311
Nigeria	191	Brazil	231
Bangladesh	165	Congo Dem. Rep.	216
Russia	117	Bangladesh	202
Mexico	129	Ethiopia	191

Source: *Population Reference Bureau, 2017.*

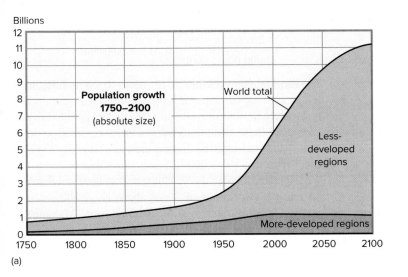

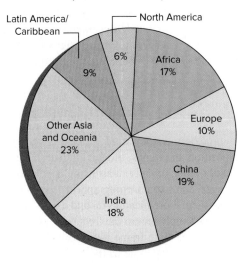

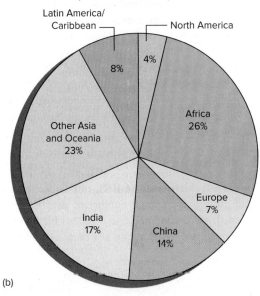

Figure 4.1 World population numbers and projections. (*a*) World population began explosive expansion after World War II ended in 1945. Numbers in more developed regions will remain stable or decline during this century due to low fertility rates. However, higher immigration and higher fertility among immigrants are projected to increase the U.S. population by more than 20 percent between 2017 and 2050, and large-volume immigration into Europe could alter its projected population decline. Between 2000 and 2100, nearly all population growth is projected to take place in 58 high-fertility countries. The high-fertility countries are all classified as *less developed*, and 39 are located in Africa, 9 in Asia, 6 in Oceania, and 4 in Latin America/Caribbean. (*b*) Africa is forecast to grow faster than the other world regions, increasing its share of the world's population substantially. China and Europe will decline slightly in total population and see a decline in their share of world population.

Sources: (a) Estimates from Population Reference Bureau and United Nations Population Fund; (b) Based on United Nations and U.S. Bureau of the Census data and projections.

The implications of the present numbers and the potential increases in population are of vital current social, political, and ecological concern. Population numbers were much smaller some 12,000 years ago when continental glaciers began their retreat, people spread to formerly unoccupied portions of the globe, and human experimentation with food sources initiated the Agricultural Revolution. The 5 or 10 million people who then constituted all of humanity obviously had considerable potential to expand their numbers. In retrospect, we see that the natural resource base of the Earth had a population-supporting capacity far in excess of the pressures exerted on it by early hunting and gathering groups.

Some observers maintain that despite a large and growing world population, the adaptive and exploitive ingenuity of humans will generate solutions to resource shortages. To them, population growth means more pressure for innovations and a larger pool of human talent to generate those innovations.

Others, however, frightened by the resource demands of a growing world population that had already expanded fourfold—from 1.6 billion to 6.1 billion—in the century from 1900 to 2000, compare the Earth to a lifeboat or spaceship with an ever-increasing number of passengers. They point to recurring problems of malnutrition and starvation (though these are realistically more a matter of failures of distribution than of inability to produce enough food worldwide). They cite global climate change, air and water pollution, the loss of forest and farmland, rising prices of many minerals and fossil fuels, and other evidences of strains on world resources as signs that the world population has reached the Earth's physical capacity.

On a worldwide basis, populations grow only one way: The number of births in a given period exceeds the number of deaths. Ignoring for the moment regional population changes resulting from migration, we can conclude that humans have been highly successful in overcoming natural controls on their numerical growth. In contrast, current estimates of slowing population growth and eventual stability clearly indicate that humans by their individual and collective decisions may effectively limit growth and control global population numbers. The implications of these observations will become clearer after we define some terms important in the study of population and explore their significance.

4.2 Some Population Definitions

Demographers employ a wide range of measures of population composition and trends, though all their calculations start with a count of events: of individuals in the population, of births, deaths, marriages, and so on. Demographers convert those counts to *rates* to make them more meaningful and useful in population analysis.

Rates simply record the frequency of occurrence of an event during a given time frame for a designated population—for example, the marriage rate as the number of marriages performed per 1,000 population in the United States last year. Demographers also place populations into *cohort* groups as they calculate birth rates, death rates, and so on. A **cohort** is a population group unified by a specified temporal characteristic—the age cohort of 0–4 years, perhaps, or the college class of 2025 (**Figure 4.2**). Basic values and rates useful in the analysis of world population and population trends have been reprinted with the permission of the Population Reference Bureau as Appendix B to this book. Comparing values in Appendix B and studying the choropleth maps in this chapter will help illustrate the discussion that follows.

Birth Rates

The **crude birth rate (CBR),** often referred to simply as the *birth rate,* is the annual number of live births per 1,000 population. It is "crude" because it relates births to total population without regard to the age or sex composition of that population.

Figure 4.2 Whatever their differences may be by race, sex, or ethnicity, these babies will forever be clustered demographically into a single birth cohort.

©Diane Macdonald/Stockbyte/Getty Images

A country with a population of 2 million and with 40,000 births a year would have a crude birth rate of 20 per 1,000.

$$\frac{40,000}{2,000,000} = \frac{20}{1,000} = 20 \text{ per thousand}$$

The birth rate of a country is, of course, strongly influenced by the age and sex structure of its population, the customs and family size expectations of its inhabitants, and its population policies. Because these conditions vary widely, recorded national birth rates vary—as of 2017, from a high of 48 in Niger in sub-Saharan Africa to 8 per 1,000 in Hong Kong, Japan, and South Korea in Asia and Greece, Italy, Monaco, Portugal, and San Marino in Europe. Birth rates of 30 or above per 1,000 are considered *high* and are found in sub-Saharan Africa, Afghanistan, Egypt, Iraq, the Palestinian Territory, Timor-Leste, and Yemen. In these countries, poverty is widespread and a high proportion of the female population is young. In many of them, birth rates may be significantly higher than official records indicate.

Birth rates of less than 18 per 1,000 are reckoned *low* and are characteristic of industrialized, urbanized regions. Overall, the Caribbean, East Asia, Europe, North America, Oceania, and South America have low birth rates. In recent years, low birth rates have been observed in an increasing number of developing states. Some of these, such as China (see "China's Way—and Others"), have adopted effective government-led population programs. In others, changed cultural norms have reduced desired family size. *Transitional* birth rates (between 18 and 30 per 1,000) characterize developing and newly industrializing regions such as Central America, Central Asia, South Asia, Western Asia, Southern Africa, and Northern Africa.

As the recent population histories of Singapore and China indicate, birth rates are subject to change. The transition to low birth rates in more developed countries is usually ascribed to industrialization, urbanization, and, in recent years, aging populations. While restrictive family planning policies in China rapidly reduced the birth rate from more than 33 per 1,000 in 1970 to 18 per 1,000 in 1986, industrializing Japan experienced a comparable 15-point decline in the decade 1948–1958 with little governmental intervention. Indeed, the stage of economic development appears closely related to variations in birth rates among countries, although rigorous testing of this relationship proves it to be imperfect (**Figure 4.3**). As a group, the more developed states of the world showed a crude birth rate of 11 per 1,000 in 2017; less-developed countries (excluding China) registered about 24 per 1,000 (down from 35 in 1990).

Technological developments such as the birth control pill have played a major role in declining birth rates. However, the sociological and ideological subsystems of culture are also important. Religious and political beliefs can also influence birth rates. In a number of different religions, more devout individuals tend to have more children. The convictions of many Roman Catholics and Muslims that their religion forbids the use of artificial birth control techniques often led to high birth rates among believers. However, predominantly Catholic Italy has one of the world's lowest birth rates. Islam itself does not prohibit contraception, and birth rates vary widely across Muslim countries. Regional variations in projected percentage contributions to world population growth are summarized in **Figure 4.4**.

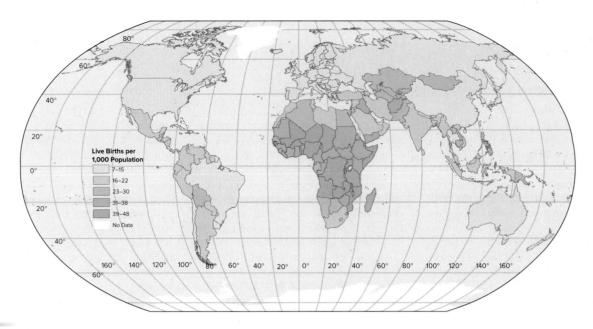

Figure 4.3 Crude birth rate, 2017. Africa stands out for its significantly higher birth rates. The map suggests a degree of precision that is misleading in the absence of reliable, universal registration of births. The pattern shown serves, however, as a generally useful summary of comparative reproduction patterns.

Source: Data from Population Reference Bureau, 2017.

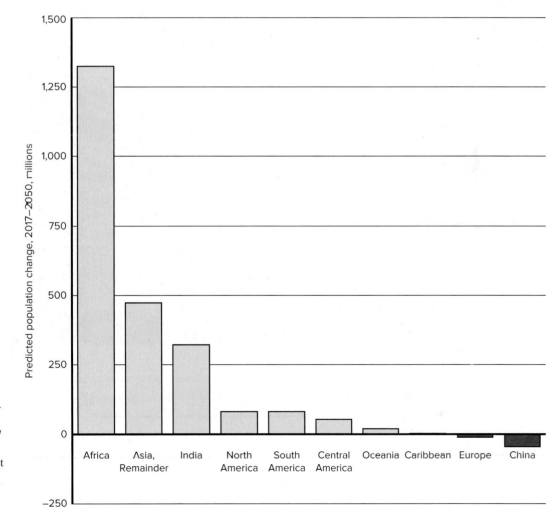

Figure 4.4 Projected contributions to world population growth by region, 2017 to 2050. Africa is projected to contribute more than half of the world's population growth in the period 2017 to 2050. India will surpass China as the world's most populous country as it continues to grow. Meanwhile, between 2017 and 2050, China's population will level off and begin a modest decline.

Source: Calculated from Population Reference Bureau, 2017.

China's Way—and Others

An ever larger population is "a good thing," Chairman Mao announced in 1965 when China's birth rate was 37 per 1,000 and population totaled 717 million. At Mao's death in 1976, numbers reached 852 million, though the birth rate then had dropped to 25. During the 1970s, China introduced a well-publicized campaign advocating the "two-child family" and providing services, including abortions, supporting that program. It was a bold attempt to create a more prosperous and powerful China while avoiding the ecological problems forecast by neo-Malthusians. In response, China's birth rate dropped to 19.5 per 1,000 by the late 1970s.

"One couple, one child" became the slogan of a more stringent population control drive launched in 1979, backed by both incentives and penalties to assure its success in China's tightly controlled society. Late marriages were encouraged; free contraceptives, cash awards, abortions, and sterilizations were provided. Penalties, including steep fines, were levied for second births. At the campaign's height in 1983, the government ordered the sterilization of either husband or wife for couples with more than one child. Infanticide—particularly the abandonment or murder of female babies—was a reported means both of conforming to a one-child limit and of increasing the chances that the one child would be male.

The one-child policy was relaxed in 1984 to permit two-child limits in rural areas where the majority of the Chinese population still resided. In contrast, newly prosperous urbanites have voluntarily reduced their fertility to well below replacement levels, with childless couples increasingly common.

Nationally, the one-child policy was so successful that when it was officially ended in 2015, an estimated 400 million births had been averted. Indeed, demographers and government officials express serious concerns that population aging and decline, not growth, are the next problems to be confronted. Projections suggest that by 2027, because of lowered fertility rates, China's population numbers will actually start falling. The country is already beginning to face a pressing social problem: a declining proportion of working-age persons and an absence of an adequate welfare network to care for a rapidly growing number of senior citizens.

China and India are the two most populous countries in the world, home to more than one-third of the global population. Back in 1950, both countries had total fertility rates of about 6 children per woman. China had 560 million residents, compared to 370 million in India. But India is predicted to surpass China by 2025 to become the most populous country on Earth. China and India's different approaches to population policy offer a contrast with great global significance. India was one of the first countries to implement a national family planning program. It has long subsidized birth control devices, legalized abortion in 1972, and has performed more surgical sterilizations than any other country. In the mid-1970s during the Indian emergency, Indira Ghandi's government implemented draconian population control measures including mass sterilization camps and even forced sterilizations.

Reactions to abuses during the Indian emergency led to a rejection of coercive population policies. Instead, India relies on education and advertising campaigns. The success of the programs has been limited and India's population continues to grow while China's plateaus.

Concerned with their own increasing numbers, many developing countries have introduced their own programs of family planning, stressing access to contraception and sterilization. International agencies have encouraged these programs, buoyed by such success as the 21 percent fall in fertility rates in Bangladesh from 1970 to 1990 as the proportion of married women using contraceptives rose from 3 percent to 40 percent under intensive family planning encouragement.

With some convincing evidence, improved women's education has been proposed as a surer way to reduce births than either encouraged contraception or China's coercive efforts. Studies from individual countries indicate that 1 year of female schooling can reduce the fertility rate by between 5 percent and 10 percent.

Figure 4A Chinese Family planning poster on advertising billboard.
©Alasdair Drysdale

Fertility Rates

Crude birth rates display such regional variability because of differences in age and sex composition and/or disparities in births among those of reproductive-age. The rate is "crude" because its denominator contains persons who have no chance at all of giving birth—males, young girls, and older women. For example, we would expect a very low crude birth rate in a community with a high percentage of elderly. The **total fertility rate (TFR)** is a more refined and thus more accurate measure for showing the

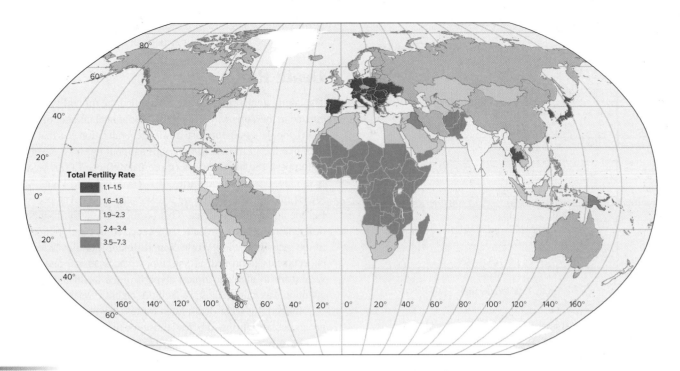

Figure 4.5 The total fertility rate (TFR) indicates the average number of children that would be born to each woman if, during her childbearing years, she bore children at the same rate as women of those ages actually did in a given year. Unlike the crude birth rate, the TFR is not affected by the age distribution of a population and thus is a more direct measure of fertility. Depending on mortality conditions, a TFR of 2.1 to 2.3 or more children per woman is considered the "replacement level," at which a population will eventually stabilize. Total fertility rates illustrate the demographic divide between the high fertility countries of sub-Saharan Africa and the low fertility countries of Europe and East Asia.

Source: Data from Population Reference Bureau, 2017.

rate and probability of reproduction among fertile females, the only segment of population capable of bearing children.

The TFR (**Figure 4.5**) tells us the average number of children that would be born to each woman if, during her childbearing years, she bore children at the current year's rate for women that age. Thus, a TFR of 3 means that the average woman in a population would be expected to have three births in her lifetime. The fertility rate is not affected by the share of women in their childbearing years and best summarizes the expected reproductive behavior of women. Thus, it is a more useful and more reliable figure for comparative and predictive purposes than the crude birth rate.

Although a TFR of 2.0 would seem sufficient exactly to replace present population (one baby to replace each parent), in reality replacement levels are reached only with TFRs of 2.1 to 2.3 or more. The fractions over 2.0 are required to compensate for mortalities that occur before women complete their childbearing years. The concept of *replacement level fertility* is useful here. It marks the level of fertility at which each successive generation of women produces exactly enough children to ensure that the same number of women survive to have offspring themselves. In general, then, the higher the level of mortality in a population, the higher the replacement level of fertility will be. For Mozambique early in the 21st century, the replacement level fertility was 3.4 children per woman.

On a worldwide basis, the TFR in 2017 was 2.5, down from 3.6 a quarter century earlier. The more developed countries recorded a 1.6 TFR in 2017. That decrease has been dwarfed by the rapid changes in reproductive behavior in much of the developing world. Since 1960, the average TFR in the less-developed world has fallen by half from the traditional 6.0 or more to 2.6 in 2017. That

dramatic decline reflects the fact that women and men in developing countries are marrying later and having fewer children, following the pattern earlier set in the developed world. There has been, as well, a great increase in family planning and contraceptive use. In 2017, the United Nations reported that 85 percent of the global demand for family planning and contraceptives had been met.

The recent fertility declines in developing states have been more rapid and widespread than anyone expected. The TFRs for so many of them have dropped so dramatically since the early 1960s (**Figure 4.6**), that the fantastic world population projections issued then are now generally discounted and rejected. Indeed, in 2017, half of the global population lived in countries with fertility rates of 2.1 or less. China's decrease from a TFR of 5.9 births per woman in the period 1960–1965 to officially about 1.8 in 2017 and comparable drops in TFRs of Bangladesh, Brazil, Mozambique, and other states demonstrate that fertility reflects cultural values, not biological imperatives.

In fact, demographers have long assumed that recently observed developing country—and therefore global—fertility rate declines to the replacement level would continue and in the long run lead to stable population numbers. However, nothing in logic or history requires population stability at any level. Indeed, the experience of sub-replacement fertility in most developed countries suggests the real possibility of absolute decline.

World regional and national fertility rates reported in Appendix B and other sources are summaries that conceal significant variations between population groups. India's published TFR of 2.3 is a pooled number that hides the variation between a low value of 1.4 in the state of Goa and 3.4 in Bihar. The U.S. 2015 national average

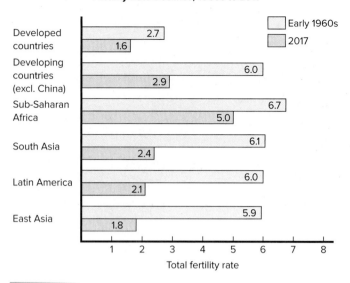

Fertility Rate Declines, 1960s to 2017

Legend:
- Early 1960s
- 2017

Region	Early 1960s	2017
Developed countries	2.7	1.6
Developing countries (excl. China)	6.0	2.9
Sub-Saharan Africa	6.7	5.0
South Asia	6.1	2.4
Latin America	6.0	2.1
East Asia	5.9	1.8

Total fertility rate

Figure 4.6 Differential fertility rate declines. Fertility has declined most rapidly in Latin America and Asia and much more slowly in sub-Saharan Africa. In 88 countries, the 2017 fertility rate was below the replacement level. The lowest fertility rates are found in Eastern and Southern Europe and in the more developed countries of Asia. The United States was long thought to be unique among developed states with its near-replacement fertility rates. However, it too has dropped below replacement levels with a 2017 TFR of 1.8.

Sources: Population Reference Bureau, 2017.

fertility rate of 1.8 does not reveal that the TFR for Hispanics was 2.1, compared to only 1.3 for American Indians and Alaska Natives.

Death Rates

The **crude death rate (CDR),** also called the **mortality rate,** is calculated in the same way as the CBR: the annual number of events per 1,000 population. In the past, a valid generalization was that the death rate, like the birth rate, varied with national levels of development. In the past, the highest rates (more than 20 per 1,000) were found in the less-developed countries of Africa, Asia, and Latin America; lowest rates (less than 10) were associated with developed states of Europe and North America. That correlation mostly disappeared as dramatic reductions in death rates occurred in developing countries in the years following World War II. Infant mortality rates and life expectancies improved as antibiotics, vaccinations, and pesticides were made available in almost all parts of the world and as increased attention was paid to sanitary facilities and safe water supplies.

Indeed, by 1994 death rates for less-developed countries as a group actually dropped below those for the more developed states and have remained lower since (**Figure 4.7**). Less improvement has been seen in reducing maternal mortality rates (see the feature "The Risks of Motherhood"). Like crude birth rates, death rates are affected by a population's age distribution. Countries with a high proportion of elderly people, such as Japan, have higher death rates than those with a high proportion of young people, such as Mexico, despite differences in living standards. The pronounced youthfulness of populations in developing countries, as much as

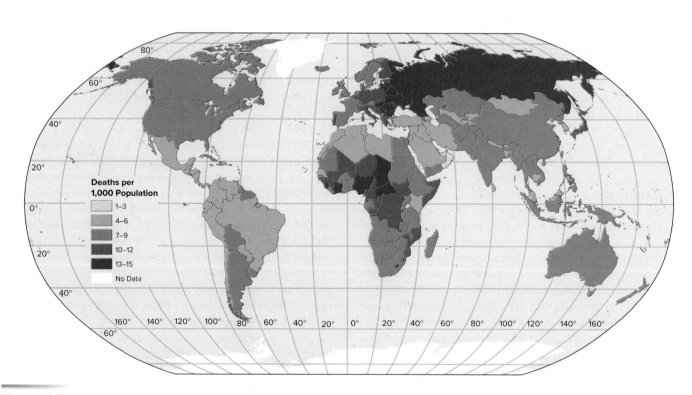

Figure 4.7 Crude death rates (CDRs) show less worldwide variability than do the birth rates displayed in Figure 4.3. The widespread availability of at least minimal health protection measures and a generally youthful population in the developing countries yield death rates frequently lower than those recorded in "old age" Europe. With its youthful population, Mexico has a lower death rate than the United States.

Source: Data from Population Reference Bureau, 2017.

The Risks of Motherhood

One of the most glaring global inequities is the maternal mortality ratio—maternal deaths per 100,000 live births. According to the World Health Organization (WHO), approximately 830 women die each day from preventable causes related to pregnancy and childbirth, leaving hundreds of thousands of children motherless.

As shown in the chart, the geography of maternal mortality is highly uneven; 99 percent of maternal deaths take place in less-developed states and two-thirds take place in sub-Saharan Africa. Pregnancy complications, childbirth, and unsafe abortions are the leading slayers of women of reproductive age throughout the developing world. According to 2015 data, the lifetime maternal mortality risk is 1 in 17 in Sierra Leone, versus 1 in 24,000 in Greece. Countries with extraordinarily high maternal death ratios are found in war-torn or politically unstable areas of Africa and Asia. In regions with high HIV prevalence, AIDS indirectly contributes to higher maternal death rates.

The vast majority of maternal deaths in the developing world are preventable. Most result from causes rooted in the social, cultural, and economic barriers confronting the world's poor: malnutrition, lack of access to prenatal health care, and unavailability of trained medical assistance, medications, or blood transfusions at birth. Education of women and girls is an important element in reducing maternal mortalities. Education empowers women to challenge practices that endanger their health.

Given the immense human suffering associated with preventable maternal deaths, the United Nations included improving maternal health among its eight Millennium Development Goals (MDGs). Specifically, the goal called for reducing the maternal mortality rate by three-quarters between 1990 and 2015. Substantial progress was made in all world regions with a 44 percent global drop in maternal deaths. The most dramatic improvements were in South Asia, which saw its rate drop from 590 to 176 deaths per 100,000 births. Meanwhile, sub-Saharan Africa lowered its rate from 870 to 546, yet still has the furthest to go. In 2016, the UN issued its Sustainable Development Goals calling for reducing maternal mortalities below 70 deaths per 100,000 live births by 2030.

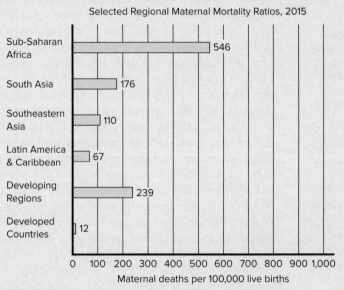

Selected Regional Maternal Mortality Ratios, 2015

Figure 4B Maternal deaths per 100,000 live births by regions.

Source: *Graph data from* Trends in Maternal Mortality: 1990 to 2015. *Geneva, Switzerland: WHO Press, 2015*

improvements in sanitary and health conditions, is an important factor in the recently reduced mortality rates of those areas.

To overcome that lack of comparability, death rates can be calculated for specific age groups. The **infant mortality rate**, for example, is the ratio of deaths of infants aged 1 year or younger per 1,000 live births:

$$\frac{\text{deaths age 1 year or younger}}{1,000 \text{ live births}}$$

Infant mortality rates are significant because it is at these ages that the greatest declines in mortality have occurred, largely as a result of better health services, clean water, and sanitation. The drop in infant mortality accounts for much of the decline in the overall death rate in the last few decades, for mortality during the first year of life is usually greater than in any other year.

Two centuries ago, it was not uncommon for 200–300 infants per 1,000 to die in their first year. Today, rates are in the single digits in developed countries and 32 for the world as a whole (**Figure 4.8**). Still, striking world regional and national variations remain. The highest infant mortality rates are in war-torn or politically unstable countries in Africa. For all of Africa, infant mortality rates are more than 50 per 1,000. In contrast, infant mortality rates in more developed countries are more uniformly in the 2–7 range. Infant mortality rates are not solely a matter of the economic status of a country. For example, the U.S. infant mortality rate exceeds that of many countries and even Cuba and Iran. Nor are rates uniform within a country. In the United States, African Americans have an infant mortality rate much higher than that of Asians, whites, or Hispanics.

Modern medicine and sanitation have increased life expectancy and altered age-old relationships between birth and death rates. In the early 1950s, only five countries, all in northern Europe, had life expectancies at birth of more than 70 years. By 2015, the average life expectancy at birth for the global population had risen to 71.4 years. The leading causes of death are now non-communicable diseases such as heart disease and stroke. The availability of

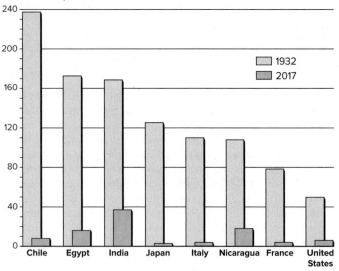

Infant deaths per 1,000 live births

Figure 4.8 Infant mortality rates for selected countries. Dramatic declines in the rate have occurred in all countries. Nevertheless, the decreases have been proportionately greatest in the urbanized, industrialized countries, where sanitation, safe water, and quality health care are widely available.

Sources: Data from U.S. Bureau of the Census and Population Reference Bureau.

clean water and sanitation varies regionally, and the least developed countries have least benefited from them. In low-income countries, leading causes of death also include communicable diseases such as malaria, tuberculosis, and water-borne diseases caused by a lack of sanitation, such as cholera, dysentery, and typhoid.

Starting in the 1950s, global life expectancies increased by about 3 years per decade. But in the 1990s, HIV/AIDS undermined global improvements in life expectancies. AIDS is the fifth most common cause of death in low-income countries and is forecast to surpass the Black Death of the 14th century—which caused an estimated 25 million deaths in Europe and 13 million in China—as history's worst-ever epidemic. According to a report by UNAIDS, AIDS-related illnesses had killed 35 million people by 2017. The United Nations estimated 37 million people were HIV positive in 2017. Some 70 percent of those infected live in sub-Saharan Africa, where HIV (originally a disease of monkeys) first established itself as a virulent human epidemic strain in the 1950s.

Southern and eastern Africa have the highest rates of HIV infection in the world. Four countries in southern Africa have adult HIV infection rates above 15 percent: Swaziland, Lesotho, Botswana, and South Africa. In southern and eastern Africa, HIV/AIDS has reduced life expectancies and slowed economic growth as parents and workers in the prime of life become ill and die prematurely. More than 10 million children in sub-Saharan Africa have lost one or both parents to HIV/AIDS. Botswana is notable for its strong national response to HIV and its provision of free antiretroviral treatment (ART) to all people with HIV. The use of ART has reduced the number of deaths to AIDS-related illnesses and reduced transmission of HIV, especially from mothers to their newborn children. The availability of expensive ART medications has been made possible by international donor and aid agencies. The gains due to ART programs, however, are fragile and vulnerable to shifting priorities at donor agencies. In addition to funding challenges, an effective response to HIV must

overcome gender inequalities which result in higher infection rates among women, lack of HIV testing, lack of education, and the social stigma associated with HIV status. In Kenya, the number of annual new HIV cases declined from about 250,000 in the early 1990s to about 100,000 in 2013. To achieve the goal of lowering new infections to near zero by 2030, the Kenyan Ministry of Health developed a comprehensive HIV prevention strategy. Central to that strategy was geographical analysis of HIV incidence and diffusion rates across Kenya. Based on a finding that 65 percent of new infections occurred in a cluster of nine counties that were home to just 23 percent of Kenya's population, the national HIV prevention strategy targets high priority geographic areas.

Population Pyramids

A **population pyramid** is a powerful means of visualizing and comparing a population's age and sex composition. The term *pyramid* describes the diagram's shape for many countries in the 1800s, when the display was created: a broad base of younger age groups and a progressive narrowing toward the apex as older populations were thinned by death. Now many different shapes are formed, each reflecting a different population history (**Figure 4.9**), and some suggest *population profile* is a more appropriate label. By showing the size of different age cohorts, the pyramids highlight the impact of "baby booms," population-reducing wars, birth rate reductions, and external migrations.

A rapidly growing country such as Nigeria has most people in the lowest age cohorts; the percentage in older age groups declines successively, yielding a classic pyramid shape with stepped sides. Typically, life expectancy is reduced in older cohorts of less-developed countries, so that for Nigeria, the proportion of the population in older age groups is lower than in, for example, New Zealand. Female life expectancy and mortality rates may also be affected by cultural rather than economic developmental causes (see the feature "Millions of Women Are Missing"). In New Zealand, a wealthy country with a very slow rate of growth, the population is nearly equally divided among the age groups, giving a "pyramid" with almost vertical sides. Below-replacement fertility rates create a pyramid with smaller cohorts at the bottom, as in the case of Japan. Among older cohorts, as Japan shows, there may be more women than men because of the greater life expectancy of the former. The impacts of war, as Russia's 2018 pyramid vividly demonstrated, were evident in that country's depleted age cohorts and male-female disparities. The sharp contrasts between the pyramids for Japan and Nigeria are representative of the differing population concerns of the developing and developed regions of the world. Within countries, people sort themselves out geographically in ways that are clearly revealed in the population pyramids for different communities (**Figure 4.10**).

The population profile provides a quickly visualized demographic picture of practical and predictive value. For example, the percentage of a country's population in each age group strongly influences demand for goods and services within that national economy. A country with a high proportion of young has a high demand for educational facilities and certain types of health delivery services. In addition, of course, a large portion of the population is too young to be employed (**Figures 4.10** and **4.11**). On the other hand, a population with a high percentage of elderly people is burdened with high costs for pensions and health care, and these

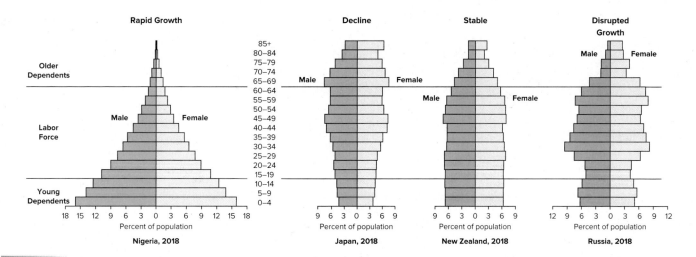

Figure 4.9 Four patterns of population structure. These diagrams show that population "pyramids" assume many shapes. The age distribution of populations reflects the past, records the present, and foretells the future. In countries such as Nigeria, social costs related to the young are important and economic expansion is vital to provide employment for new entrants in the labor force. Japan's negative growth means a future with fewer workers to support a growing demand for social services for the elderly. The 2018 pyramid for Russia records the sharp decline in births after the collapse of the Soviet Union in 1991 and during World War II as a "pinching" of the 0–24 and 70–74 cohorts, and showed in the large deficits of men above age 70 the sharp reductions in Russian male longevity.

Sources: The 2018 pyramids for Nigeria, Japan, New Zealand, and Russia: U.S. Bureau of the Census, International Data Bases.

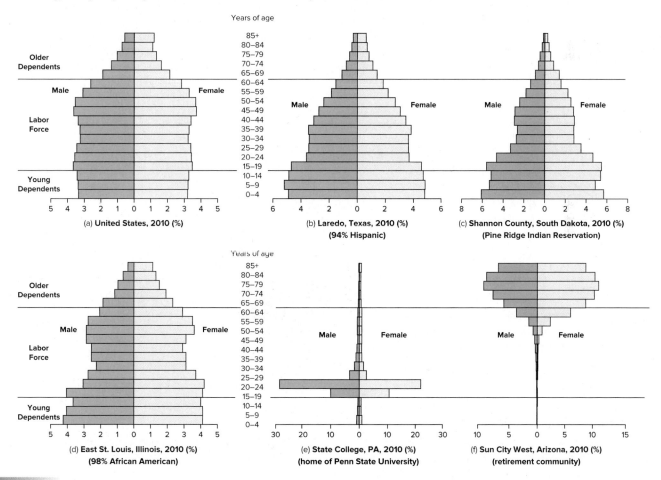

Figure 4.10 Population pyramids for different communities. The 2010 pyramids show the dramatic differences in the population structure across the United States. Laredo, Texas, on the U.S.-Mexico border, and Shannon County, South Dakota, on the Pine Ridge Indian Reservation, both display youthful populations with high birth rates. The pyramid for Shannon County also shows missing young adults and middle-aged adults who have moved away to urban areas. The pyramid for East St. Louis, Illinois shows the negative effects of high rates of incarceration and lower life expectancies for African American males. College towns and retirement communities create age-segregated places.

Source: U.S. Bureau of the Census, 2010.

Millions of Women Are Missing

Worldwide, according to one UN estimate, about 117 million women are demographically missing, victims of nothing more than their sex. Their absence is the result of sex-selective abortions, female infanticide, abandonment, neglect, and violence in countries where boys are favored.

The absence of millions of women is a tragic example of the interrelationship between the three subsystems of culture: technology (artifacts), social relationships (sociofacts), and belief systems (mentifacts). In some Asian cultures, the combination of a deeply ingrained cultural preference for boys and widespread availability of ultrasound technology has fueled the growth of sex-selective abortions. Traditionally in these societies, a married woman will care for her husband's parents in their old age. This practice has created the widespread perception that having a girl is an economic burden, while the birth of a boy is a blessing. Because higher-income families have greater access to ultrasound

tests, they are more likely to abort a female fetus, even against government directives.

The evidence for the missing women starts with a comparision against a normal **sex ratio:** Between 103 and 106 males are conceived and born for every 100 females. Normally, girls are hardier and more resistant to disease than boys, and in populations where the sexes are treated equally in matters of nutrition and health care, the number of males and females tends to equalize as they age. However, in many Asian and southeastern European countries, the ratio of males to females has been rising since the introduction of prenatal sex selection. China has the most dramatic imbalances with a male-to-female sex ratio of 117 males for every 100 females for cohorts under 20 years of age. In small pockets of rural China and India, sex ratios as high as 150 have been reported. China's male-to-female sex ratio at birth increased from a normal ratio in the early 1980s to the distorted ratios witnessed today.

Sex ratio deviations are most striking for second and subsequent births. In South Korea,

and Armenia, for example, the figures for first-child sex ratios are near normal, but rise to 142 boys per 100 girls for a third Korean child and to 177 per 100 for a third Armenian child.

Concern has been expressed over the lack of potential marriage partners for males reaching adulthood in India and China. The sex imbalance may exacerbate kidnapping and sex trafficking of women and girls.

But not all poor countries show the same disparities. In sub-Saharan Africa, where poverty and disease are perhaps more prevalent than on any other continent, but where there is no tradition of deadly violence against women, there are 102 females for every 100 males, and in Latin America and the Caribbean, there are equal numbers of males and females. To a lesser extent, the sex ratio disparities due to sex-selective abortions are also found among East Asian and South Asian immigrant communities in North America. Cultural norms and practices, not poverty or underdevelopment, seem to determine the fate and swell the numbers of the world's millions of missing women.

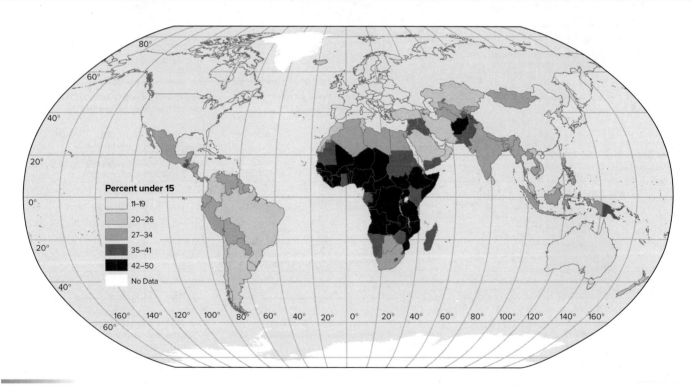

Figure 4.11 Percentage of population under 15 years of age. A high proportion of a country's population under 15 increases the dependency ratio of that state and promises future population growth as the youthful cohorts enter childbearing years.

Source: Data from Population Reference Bureau, 2017.

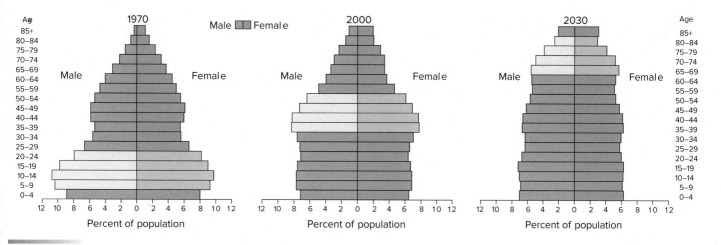

Figure 4.12 The progression of the "boomers"—the baby boom cohort born between 1946 and 1964—through the U.S. population pyramid has been associated with changing American lifestyles and expenditure patterns. In 1970, national priorities focused on the needs of children and young adults, such as building schools and expanding universities. At the turn of the 21st century, boomers formed the largest share of the working-age adult population, and their wants and spending patterns shaped the national culture and economy. By 2030, the pyramid foretells, their desires and support needs—now for retirement facilities and old-age care—will again be central concerns.

Source: Redrawn from Christine L. Himes, "Elderly Americans," Population Bulletin 56, no. 4 (December 2001), Fig. 1.

people must be supported by a smaller working-age population. Thus, as the profile of a national population changes, differing demands are placed on a country's social and economic systems (**Figure 4.12**). The **dependency ratio** is a simple measure of the number of economic dependents, old or young, that each 100 people in the productive years (usually, 15–64) must support. Population pyramids give visual evidence of that ratio.

Population pyramids also foretell future problems resulting from present population policies or practices. The strict family-size rules and widespread preferences for sons in China, for example, skews the pyramid in favor of males. On current evidence, about 1 million excess males a year enter an imbalanced marriage market in China. The 40 million bachelors China is likely to have in 2020, unconnected to society by wives and children, may pose threats to social order and, perhaps, national stability not foreseen or planned when family control programs were put in place, but clearly suggested by population pyramid distortions.

Natural Increase and Doubling Times

Knowledge of a country's sex and age distributions also enables demographers to forecast its future population levels, though the reliability of projections decreases with increasing length of forecast (**Figure 4.13**). Thus, a country with a high proportion of young people will experience a high rate of natural increase unless there is a very high mortality rate among infants and juveniles or fertility and birth rates change materially. The **rate of natural increase** of a population is derived by subtracting the crude death rate from the crude birth rate. *Natural* means that increases or decreases due to migration are not included. If a country had a birth rate of 22 per 1,000 and a death rate of 12 per 1,000 for a given year, the rate of natural increase would be 10 per 1,000. This rate is usually expressed as a percentage, that is, as a rate per 100 rather than per 1,000. In the example given, the annual increase would be 1 percent.

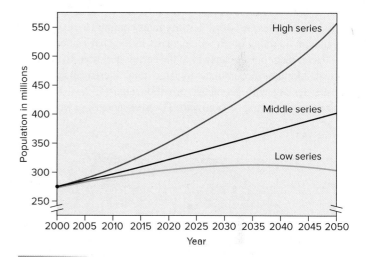

Figure 4.13 Possible population futures for the United States. Population projections account for age-specific birth and death rates for each ethnic and racial group as well as future changes in fertility and immigration rates. Longer term forecasts often diverge widely because of the effect of slightly different assumptions about fertility, death rates, or immigration.

Source: U.S. Bureau of the Census.

The rate of increase can be related to the time it takes for a population to double if the present growth rate remains constant—that is, the **doubling time.** Table 4.2 shows that it would take 70 years for a population with a rate of increase of 1 percent (approximately the rate of growth of South America in 2017) to double. A 2 percent rate of increase—recorded in 2017 by Namibia—means that the population would double in only 35 years. (Population doubling time can be roughly estimated by applying the Rule of 70, which works for exponential growth and simply involves dividing 70 by the growth rate.) How could adding only 20 people per 1,000 cause a population to grow so quickly? The principle is the same as that used to compound interest in a bank.

Table 4.2

Doubling Time in Years at Different Rates of Increase

Annual Percentage Increase	Doubling Time (Years)
0.5	140
1.0	70
2.0	35
3.0	24
4.0	18
5.0	14
10.0	7

Table 4.3

Population Growth and Approximate Doubling Times Since C.E. 1

Year	Estimated Population	Time to Double
1	250 million	
1650	500 million	1650
1804	1 billion	154
1927	2 billion	123
1974	4 billion	47
World population may reach:		
2024	8 billion	50[a]

[a]No current projections contemplate a further doubling to 16 billion people.
Source: *United Nations Population Division.*

Until recently, for the world as a whole, the rates of increase have risen over the span of human history. Therefore, the doubling time has decreased (Table 4.3). Growth rates vary regionally, and in countries with high rates of increase (Figure 4.14), the doubling time is less than the 60 years projected for the world as a whole at 2017 growth rates. As world fertility rates continue to decline, the population doubling times will increase, as they have since 1990.

Here, then, lies the answer to the earlier question. Even small annual additions accumulate to large total increments because we are dealing with geometric or exponential (1, 2, 4, 8) rather than arithmetic (1, 2, 3, 4) growth. The ever-increasing world base

population has reached such a size that each additional doubling results in an astronomical increase in the total. A simple mental exercise suggests the inevitable consequences of such doubling, or **J-curve,** growth. Take a very large sheet of the thinnest paper you can find and fold it in half. Fold it in half again. After seven or eight folds, the sheet will have become as thick as a book—too thick for further folding by hand. If you could make 20 folds, the

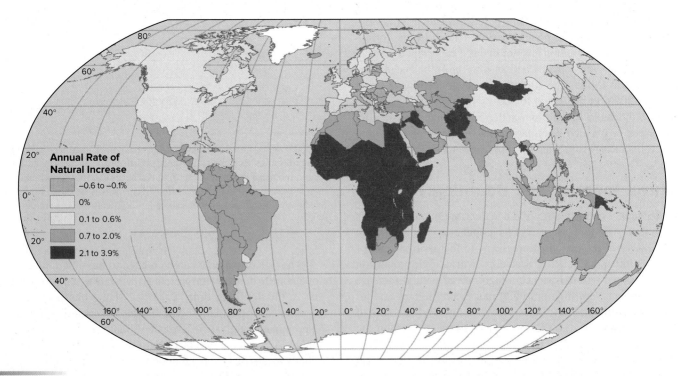

Figure 4.14 Annual rates of natural increase. The world's 2017 rate of natural increase (1.2 percent) would mean a doubling of population in 60 years. Because many demographers now anticipate world population will stabilize at around 11.2 billion (in about C.E. 2100) and perhaps actually decline after that, the "doubling" implication and time frame of current rates of natural increase reflect mathematical, not realistic, projections. Many individual continents and countries, of course, deviate widely from the global average rate of growth and have vastly different potential doubling times. Africa as a whole has the highest rates of increase, followed by southern and western Asia. Europe as a whole (including Russia) has a natural increase rate of 0.0.

Source: Data from Population Reference Bureau, 2017.

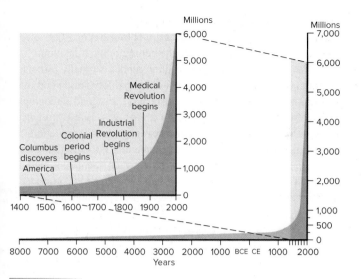

Figure 4.15 World population growth 8000 B.C.E. to C.E. 2000. Notice that the bend in the J-curve begins in the mid-1700s when industrialization started to provide new means to support the population growth through improvements in agriculture and food supply. Improvements in medical science, sanitation, and nutrition reduced death rates in the late 19th and early 20th centuries in the industrializing countries.

stack would be nearly as high as a football field is long. From then on, the results of further doubling are astounding. At 40 folds, the stack would be well on the way to the moon, and at 70, it would reach twice as far as the distance to the nearest star. Rounding the bend on the J-curve, which **Figure 4.15** suggests the world population did around 1900, fostered dire predictions of inevitable unsupportable pressures on the planet's population support capabilities.

Today, rates of natural increase in developed countries, particularly in Europe, are approaching zero or even negative values.

But individual country growth is also dependent on patterns of immigration and emigration. For example, Canada has a 0.3 percent rate of natural increase and a doubling time of 233 years; but with its high rates of immigration, it had, however, an *overall* growth rate of 1.2 percent with a doubling time of 58 years.

With replacement or below-replacement fertility rates throughout the developed world and declining fertility rates in most of the developing world, doubt is cast on the applicability of long-term doubling time projections. Although the doubling times are easier to understand than growth rates, they can be very misleading because they are based on the dubious assumption that present growth rates will continue indefinitely.

4.3 The Demographic Transition Model

Exponential population growth cannot continue indefinitely on a finite planet. Some form of braking mechanism must operate to control population growth. If voluntary population limitation is not undertaken, involuntary controls such as famine, disease, or resource wars may occur.

One attempt to summarize the observed voluntary population control that has accompanied economic development and modernization is the **demographic transition** model. The model traces the changing levels of human fertility and mortality associated with industrialization, health care improvements, urbanization, and changing cultural attitudes regarding childbearing. As societies move through the model's stages, high birth and death rates are replaced by low rates (**Figure 4.16**). During the intermediate stages of the model, populations grow rapidly before stabilizing in the final stage.

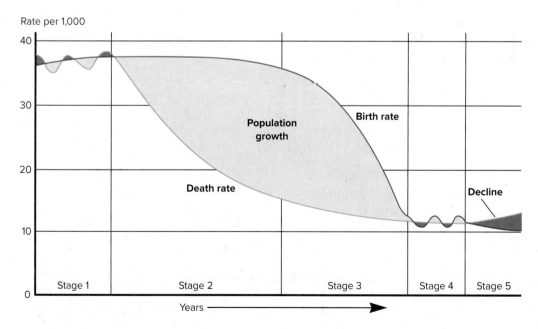

Figure 4.16 Stages in the demographic transition. During the first stage, birth and death rates are both high, and population grows slowly. When, during stage 2, the death rate drops while the birth rate remains high, there is a rapid increase in numbers. During the third stage, birth rates decline, and population growth is less rapid. The fourth stage is marked by low birth and death rates and, consequently, by stabilization of the population. The negative growth rates of many developed countries and the falling birth rates in other regions suggest that a fifth stage, one of population decline, is regionally—and ultimately worldwide—a logical extension of the transition model.

The *first stage* of the demographic transition model is characterized by high birth and high (but fluctuating) death rates. Birth rates and death rates are similar, and population grows slowly. Demographers think that it took from approximately 1 C.E. to 1500 C.E. for the population to increase from 250 million to 500 million, a doubling time of a millennium and a half. Growth was not steady, of course. There were periods of regional expansion that were usually offset by sometimes catastrophic decline. Wars, epidemics, poor harvests, and natural disasters took heavy tolls. For example, the bubonic plague (the Black Death), which swept across Europe in the 14th century, is estimated to have killed between one-third and one-half of the population of that continent and epidemic diseases brought by Europeans to the Western Hemisphere are believed to have reduced New World native populations by 95 percent within a century or two of first contact. The high birth rates, high death rates, and slow population growth of the first stage describe all of human history until about 1750 C.E., when Western Europe entered stage two of the transition. Over the past 150 years, the decline in death rates that signaled the transition to stage two diffused around the globe. Today, death rates in even the poorest countries are below 20 per 1,000 and no country remains in stage one of the demographic transition.

The *second stage* of the transition model is usually associated with the consequences of Europe's Industrial Revolution and modernization. Rapid population growth during the second demographic stage results as birth rates outstrip death rates. Life expectancies increase dramatically due to advances in medicine, sanitation, agricultural productivity, and food distribution. Urbanization during the second stage provides the stage on which sanitation, medical, and food distribution improvements are concentrated (**Figure 4.17**). Birth rates do not fall as soon as death rates; ingrained cultural beliefs and social relationships—mentifacts and sociofacts—change more slowly than technologies. In most traditional societies, large families are valued for their social and economic advantages. Children are the focus of activities and rituals by which culture is transmitted, and in low-income families, they contribute economically by starting work at an early age, especially on farms or family businesses, and by supporting their parents in old age.

Figure 4.17 London, England, in the late 19th century. A modernizing Europe experienced improved living conditions and declining death rates during that century of progress.

Source: Alfred Gilbert/Library of Congress Prints and Photographs Division [LC-DIG-ppmsc-08577].

The *third stage* of the demographic transition occurs when birth rates decline as people begin to control family size. In urbanized, industrialized cultures, raising children is expensive. In fact, such cultures may view children as economic liabilities rather than assets. The elderly rely on government pension programs rather than their children. In addition, declining childhood mortality rates make parents feel that they can have fewer children, knowing that most of them are likely to survive. When the birth rate falls and the death rate remains low, the rate of population increase slows.

The classic demographic transition model ends with a fourth and final stage characterized by very low, nearly equal birth and death rates. At this point, a condition of zero population growth approaches as natural increase rates drop to zero. With longer life expectancies and fewer births, a significant aging of the population accompanies this stage.

In some countries that have completed the demographic transition, death rates equal or exceed birth rates and populations are actually declining. This extension of the fourth stage into a fifth stage of population decrease has so far been largely confined to the rich, industrialized world—notably Europe and Japan—but increasingly promises to affect much of the rest of the world as well. The dramatic decline in fertility that has been recorded in almost all countries since the 1980s means that a majority of the world's population resides in areas where the only significant population growth is from **demographic momentum** (see the section "Population Prospects," later in this chapter).

The Western Experience

The original transition model was devised to describe the experience of northwest European countries as they transitioned from rural-agrarian societies to urban-industrial ones. It may not accurately predict the course of events for all developing countries. In Europe, church and municipal records, some dating from the 16th century, show that people tended to marry late, if at all. In England before the Industrial Revolution, as many as half of all women in the 15–50 age cohort were unmarried. Infant mortality was high, life expectancy low. With the coming of industrialization in the 18th and 19th centuries, immediate factory wages instead of long apprenticeship programs permitted earlier marriage and more children. Because improvements in sanitation and health came only slowly, death rates remained high. Around 1800, 25 percent of Swedish infants died before their first birthday. Population growth rates remained below 1 percent per year in France throughout the 19th century.

Beginning about 1860, first death rates and then birth rates began their significant, though gradual, decline. This "mortality revolution" came first, as an *epidemiological transition:* formerly fatal epidemic diseases became endemic (that is, essentially continual within a population). As people developed partial immunities, mortality rates declined. Improvements in livestock raising, crop rotation, fertilizer use, and new crops from overseas colonies (the potato was an early example) raised the level of health of the European population in general.

At the same time, sewage systems and sanitary water supplies became common in larger cities, reducing the frequency of water-borne illnesses such as cholera and typhoid (**Figure 4.18**). Deaths due to infectious, parasitic, and respiratory diseases and to malnutrition declined, while those related to chronic illnesses associated with an aging population increased. Western Europe passed from a first stage, the "Age of Pestilence and Famine," to the "Age of Degenerative and Human-Origin Diseases." However, recent increases in drug- and antibiotic-resistant diseases, pesticide resistance of disease-carrying insects, and such new scourges

Figure 4.18 Pure piped water replacing individual or neighborhood wells, and sewers and waste treatment plants instead of privies, became increasingly common in urban Europe and North America during the 19th century. Their modern successors, such as the Las Vegas, Nevada, treatment plant shown here, helped complete the *epidemiologic transition* in developed countries. Lynn Betts, USDA, Natural Resources Conservation Center

Source: Lynn Betts, USDA Natural Resources Conservation Service.

of both the less-developed and more-developed countries as HIV cast doubt on the stability of that ultimate stage (see the feature "Our Delicate State of Health"). Nevertheless, even old and new scourges such as malaria, tuberculosis, and Ebola are unlikely to have decisive demographic consequences on the global scale.

In Europe, the striking reduction in death rates was echoed by similar declines in birth rates as societies began to alter their traditional concepts of ideal family size. In cities, child labor laws and mandatory schooling meant that children no longer were important contributors to family economies. As "poor-relief" legislation and other forms of social welfare substituted for family support structures, the value of children as a social safety net declined. Family consumption patterns altered as the Industrial Revolution made widely available consumer goods that once were considered luxuries. For some, children came to be seen as hindrance rather than aid to social mobility, lifestyle improvement, and self-expression. Perhaps most important were changes in the education levels, career opportunities, and social status of women that spread the conviction that control over childbearing was within their power and to their benefit.

A Divided World, A Converging World

The dramatic decline in mortality that had emerged gradually throughout the European world diffused with startling speed to developing countries after 1950. The increasing availability in developing countries of Western technologies of medicine and public health, including antibiotics, insecticides, sanitation,

immunization, infant and child health care, and the eradication of smallpox, dramatically increased life expectancies. Such imported technologies and treatments accomplished in a few years what it took Europe 50 to 100 years to experience. Sri Lanka, for example, sprayed extensively with DDT to combat malaria; life expectancy there jumped from 44 years in 1946 to 60 only eight years later. With similar public health programs, India also experienced a steady reduction in its death rate after 1947. Simultaneously, with international sponsorship, food aid cut the death toll of developing states during drought and other disasters. Thus, the second stage of the demographic transition—declining death rates accompanied by continuing high birth rates—diffused worldwide.

Many countries in sub-Saharan Africa display the characteristics of the second stage in the model. Ghana, with a birth rate of 32 and a death rate of 8, is typical. The annual rate of increase for Ghana is 2.4 percent, and its population will double in about 30 years. Such rates, of course, do not mean that the full impact of modernization has been worldwide; they do mean that the underdeveloped societies have been beneficiaries of the life preservation techniques it generated.

Birth rate declines, of course, depend in part on technology, but even more important is social acceptance of the idea of fewer children and smaller families. That acceptance has diffused broadly but unevenly worldwide. The steep declines in birth rates in developing countries indicate that most of them have moved into the third or fourth stages of the demographic transition (**Figure 4.19**). In 1984, only 18 percent of world population lived

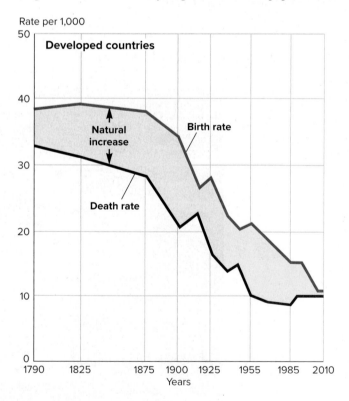

Figure 4.19 World birth and death rates to 2010. The "population explosion" after World War II (1939–1945) reflected the effects of drastically reduced death rates in developing countries without simultaneous and compensating reductions in births. Today, however, three interrelated trends appear in many developing countries: (1) fertility has overall dropped further and faster than was predicted; (2) contraceptive acceptance and use has increased markedly; and (3) age at marriage is rising. In consequence, the demographic transition had been compressed from a century to a generation in some developing states.

Source: Revised and redrawn from Elaine M. Murphy, World Population: Toward the Next Century, revised ed. *(Washington, D.C.: Population Reference Bureau, 1989) and Population Reference Bureau, annual World Population Data Sheet.*

Our Delicate State of Health

Death rates have plummeted, and the benefits of modern medicines, vaccines, antibiotics, and sanitary practices have enhanced both the quality and length of life worldwide. This epidemiological transition is the dominant story in global health. Far from being won, however, the struggle against infectious diseases is growing in intensity and is, perhaps, unwinnable. Almost a century after the discovery of antibiotics, both old and new disease-causing microorganisms are emerging and spreading all over the world. Infectious and parasitic diseases kill about 15 million people each year.

The five leading infectious killers are, in order, acute respiratory infections such as pneumonia, diarrheal diseases, HIV/AIDS, tuberculosis, and malaria. The incidence of infection, of course, is much greater than the occurrence of deaths. Nearly 30 percent of the world's people, for example, are infected with the bacterium that causes tuberculosis, but only 2 to 3 million are killed by the disease each year. More than 500 million people are infected with such tropical diseases as malaria, sleeping sickness, schistosomiasis, and river blindness, with perhaps 3 million annual deaths. Newer pathogens are constantly appearing, such as those causing Lassa fever, Rift Valley fever, Ebola fever, hantavirus pulmonary syndrome, West Nile encephalitis, hepatitis C, and severe acute respiratory syndrome (SARS). In fact, at least 30 previously unknown infectious diseases have appeared since the mid-1970s.

The 1918–1919 Spanish Flu pandemic demonstrates how diseases can diffuse globally with devastating effect. Striking as World War I was drawing to an end, the epidemic's spread was likely aided by massive troop movements as it diffused along shipping routes. The influenza virus struck down the young and healthy and

Figure 4C Workers practicing proper hand washing techniques at a health clinic in Guinea during the 2014 Ebola outbreak.

Source: Centers for Disease Control and Prevention/Conne Ward-Cameron

killed an estimated 20 to 50 million people. The 2014–2015 Ebola outbreak emerged in rural areas and spread through Guinea, Sierra Leone, and Liberia along roadways leading to the capital cities on the coast. Diffusion was slower in the Democratic Republic of Congo where settlements are more isolated.

The spread and virulence of infectious diseases are linked to the dramatic changes so rapidly occurring in the Earth's physical and social environments. Climate change permits temperature-restricted pathogens to invade new areas and claim new victims. Deforestation, wetland drainage, and other human-induced alterations to the physical environment disturb ecosystems and simultaneously disrupt the natural system of controls that keep infectious diseases in check.

Rapid population growth and explosive urbanization, increasing global tourism, population-dislocating wars and migrations, and expanding world trade all increase the diffusion of diseases. Some disease-causing microbes have expanded their range as previously isolated areas are opened up by new roads and air routes.

The most effective weapons in the battle against epidemics are already known. They include improved health education; disease prevention and surveillance; research on disease vectors and incidence areas (using GIS and spatial analysis); drug therapy; mosquito control programs; clean water supplies; and distribution of such simple and cheap remedies as childhood immunizations and oral rehydration therapy. All, however, require expanded investment and attention to those spreading infectious diseases.

---

in countries with fertility rates at or below replacement levels (that is, countries that had completed the demographic transition). Today, about half live in such countries, and it is increasingly difficult to distinguish between developed and developing societies on the basis of their fertility rates. Those rates in many Indian states (Kerala and Tamil Nadu, for example) and in such countries as Sri Lanka, Thailand, South Korea, and China are below those of the United States and some European countries. Significant decreases to near the replacement level have also occurred

in the space of a single generation in many other Asian and Latin American states with high recent rates of economic growth. Increasingly, it appears, low fertility is becoming a feature of both rich and poor, developed and developing states.

The demographic transition model assumes an inevitable course of events from the high birth and death rates of premodern (nonindustrialized) societies to the low and stable rates of advanced (industrialized) countries. The model fails to anticipate, however, that the European experience would not be matched by all developing countries. Some developing societies remain in the second stage of the model, unable to realize the economic gains and social changes necessary to progress to the third stage of falling birth rates. Thus, despite a substantial convergence of fertility rates, many observers point to a continuing and growing demographic divide. On one side of the divide are the low-fertility countries that are home to 46 percent of the world's population. The low fertility rates of these mostly wealthy countries guarantee future population decline and rapid aging unless they attract significant immigration. On the other side of the divide are a group of high-fertility countries that account for just 18 percent of the world's population but are projected to triple in size and increase their global population share to 42 percent by 2100. Nearly all of the high-fertility countries are included on the United Nation's list of least-developed countries. Most of them are in sub-Saharan Africa, and all suffer from low per capita income, illiteracy, low standards of living, and inadequate health facilities and care.

The established patterns of both high- and low-fertility regions tend to be self-reinforcing. Low growth permits the expansion of personal income and the accumulation of capital to enhance the quality of life and make large families less attractive or essential. In contrast, in high-fertility regions, population growth requires spending on social services that otherwise might have been invested to promote economic expansion. Growing populations place ever greater demands on limited natural resources. As the environmental base deteriorates, productivity declines, undermining the economic progress on which the demographic transition depends (see the feature "International Population Policies"). The vastly different future prospects for personal and national prosperity between the high-fertility countries and the rest of the world make the demographic divide a matter of continuing international concern.

4.4 The Demographic Equation

Change in a region's population stems from three basic life events: people are born, they migrate, and they die. Migration involves the long-distance movement of people from one residence to another. When that relocation occurs across political boundaries, it affects the population structure of both the origin and destination regions. The **demographic equation** summarizes the contribution made to regional population change over time by the combination of *natural change* (births minus deaths) and **net migration** (difference between in-migration and out-migration). **Zero population growth (ZPG)** is the condition for individual countries when births plus immigration equals deaths plus emigration. On a global scale, of course, all population change is accounted for by natural change. The impact of migration on the demographic equation often increases as the population size of the areal unit decreases.

$$P_f = P_i + \text{Births} + \text{In-Migration} - \text{Deaths} - \text{Out-Migration}$$

where:

P_f = Final Population

P_i = Initial Population

Population Relocation

In the past, emigration proved an important device for relieving the pressures of rapid population growth during the demographic transition for many European countries (**Figure 4.20**). For example, in one 90-year span, 45 percent of the natural increase in the population of the British Isles emigrated, and between 1846 and 1935, some 60 million Europeans of all nationalities left that continent. Today, the major population relocations connect growing countries going through the second and third stages of the demographic transition with aging societies in the fourth stage. Yet, despite recent massive movements of economic and political refugees across Asian, African, and Latin American boundaries, emigration today provides no comparable relief valve for developing countries. Total population numbers are too great to be much affected by migrations of even millions of people. In only a few countries—Afghanistan, Cuba, El Salvador, Haiti, and Syria, for example—have as many as 10 percent of the population emigrated in recent decades.

Net migration, the balance between in-migration and out-migration, is positive for the developed countries as a group, and negative for the less-developed countries. Net migration rates are modest, but the direction of movement reflects the demand for additional workers in high-income countries that have completed the demographic transition and the ample supply of willing workers in rapidly growing, low-income countries. The highest negative (net out-migration) rates are found in war-torn Syria. The highest positive (net in-migration) rates are found in oil-producing states in the Near East.

Immigration Impacts

Where cross-border movements are massive enough, migration may have a pronounced impact on the demographic equation and result in significant changes in the population structures of both the origin and destination regions. Past European and African migrations, for example, not only altered but substantially created the population structures of new, sparsely inhabited lands of colonization in the Western Hemisphere and Australasia. In some decades of the late 18th and early 19th centuries, 30 percent to more than 40 percent of population increase in the United States was accounted for by immigration. Similarly, eastward-moving Slavs colonized under populated Siberia and overwhelmed native peoples.

Migrants are rarely a representative cross section of the population group they leave, and they add an unbalanced age and sex component to the group they join. A recurrent research observation is that emigrant groups are heavily skewed in favor of young singles. Although males traditionally far exceeded females in international flows, under some circumstances females outnumber males among transborder migrants.

The receiving country will have its population structure altered by an outside increase in its younger age cohorts. The results are both immediate in a modified population pyramid, and potential in future impact on birth rates. The origin area will have lost a portion of its young, active members in their child-bearing years. The outmigration may distort its young adult sex ratios,

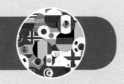

Geography and Citizenship

International Population Policies

After a sometimes rancorous nine-day meeting in Cairo in September 1994, the U.N. International Conference on Population and Development endorsed a strategy for stabilizing the world's population at 7.27 billion by no later than 2015. The 20-year program of action accepted by 179 signatory countries sought to avoid the environmental consequences of excessive population growth. Its proposals were therefore linked to discussions and decisions of the U.N. Conference on Environment and Development held in Rio de Janeiro in June 1992.

The Cairo plan abandoned several decades of top-down governmental programs that promoted "population control" (a phrase avoided by the conference) based on targets and quotas and, instead, embraced for the first time policies giving women greater control over their lives, greater economic equality and opportunity, and a greater voice in reproduction decisions. It recognized that limiting population growth depends on programs that lead women to want fewer children and make them partners in economic development. In that recognition, the Conference accepted the documented link between increased educational access and economic opportunity for women and falling birth rates and smaller families. Earlier population conferences— 1974 in Bucharest and 1984 in Mexico City—did not fully address these issues of equality, opportunity, education, and political rights; their adopted goals failed to achieve hoped-for changes in birth rates in large part because women in many traditional societies had no power to enforce contraception and feared their other alternative, sterilization.

The earlier conferences carefully avoided or specifically excluded abortion as an acceptable family planning method. It was the open discussion of abortion in Cairo that prompted religious objections by the Vatican and many Muslim and Latin American states to the inclusion of legal abortion as part of health care. Although the final text of the conference declaration did not promote any universal right to abortion and excluded it as a means of family planning, some delegations still registered reservations to its wording on both sex and abortion. At conference close, however, the Vatican endorsed the declaration's underlying principles, including the family as "the basic unit of society" and the need to stimulate economic growth, and to promote "gender equality, equity, and the empowerment of women."

In follow-up conferences, the UN reported progress toward reaching the Cairo goals. The consensus was that much remained to be done to broaden sex education and family planning programs for the poorest population groups, to strengthen laws ending discrimination against women, and to encourage donor countries to meet their agreed-on contributions to the program.

Since 2012, one of the world's wealthiest private foundations, the Bill and Melinda Gates Foundation, has prioritized access to birth control and contributed more than $1 billion to providing contraception in developing countries. Declining fertility in most developing countries suggests that the combined efforts of government and the private sector are working. Some demographers and many women's health organizations interpret the falling fertility rates as the expected result of women assuming greater control over their economic and reproductive lives. The director of the U.N. Population Division noted: "A woman in a village making a decision to have one or two or at most three children is a small decision in itself. But . . . compounded by millions and millions . . . of women in India and Brazil and Egypt, it has global consequences."

That women are making those decisions, reflects important cultural factors emerging since Cairo. Satellite television brings contraceptive information to even remote villages and shows programs of small, apparently happy families that viewers want to emulate. Increasing urbanization reduces some traditional family controls on women and makes contraceptives easier to find, and declining infant mortality makes mothers more confident their babies will survive. Perhaps most important, population experts assert, is the dramatic increase in most developing states in female school attendance and corresponding reductions in the illiteracy rates of girls and young women who will soon be making fertility decisions.

Questions to Consider

1. Do you think it is appropriate for international aid agencies to promote policies affecting such personal or national concerns as reproduction and family planning? Why or why not?

2. Do you think that current international concerns over population growth, development, and the environment are sufficiently pressing to override long-enduring cultural norms and religious practices in traditional societies? Why or why not?

3. Many environmentalists see the world as a finite system unable to support ever-increasing populations; to exceed its limits would cause frightful environmental damage and global misery. Many economists counter that free markets will keep supplies of needed commodities in line with growing demand and that science will, as necessary, supply technological fixes in the form of substitutes or expansion of production. In light of such diametrically opposed views of population growth consequences, is it appropriate or wise to base international programs solely on one of them? Why or why not?

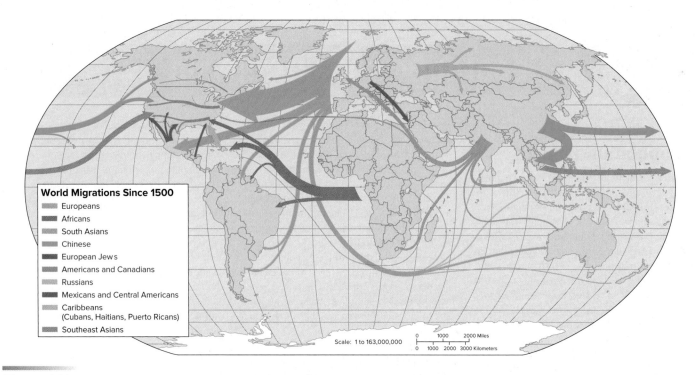

Figure 4.20 Principal migrations since 1500. The arrows show the direction and relative magnitude of the major voluntary and forced international migration since about 1500. More recently, there have been significant population movements into the United Kingdom, France, Belgium, and the Netherlands from their former colonies and from less-prosperous countries within the EU.

Source: Shaded zones after Daniel Noin, Géographie de la Population *(Paris: Masson, 1979), p. 85.*

and it certainly will contribute to statistical aging of its population. The destination country will likely experience increased births associated with the youthful newcomers and, in general, have its average age reduced.

4.5 World Population Distribution

The world's population is not uniformly distributed over the Earth. The most striking feature of the world population distribution map (**Figure 4.21**) is the very unevenness of the pattern. Some land areas are nearly uninhabited, others are sparsely settled, and still others contain dense agglomerations of people. Until about 2007, rural folk outnumbered urban people. After 2007, however, urbanites will remain dominant with cities capturing most of the world's population growth.

Earth regions of apparently very similar physical makeup show quite different population numbers and densities, perhaps the result of differently timed settlement or of settlement by different cultural groups. Northern and Western Europe, for example, inhabited thousands of years before North America, contain as many people as the United States on 70 percent less land; the present heterogeneous population of the Western Hemisphere is vastly more dense than was that of earlier Native Americans.

We can draw certain generalizations from the uneven distribution of population shown in Figure 4.21. First, 88 percent of all people live north of the equator and two-thirds of the total dwell in the midlatitudes between 20° and 60° North (**Figure 4.22**). Second, a large majority of the world's inhabitants occupy only a small part of its land surface. More than half the people live on about 5 percent of the land. Third, people congregate in lowland areas; their numbers decrease sharply with increases in elevation. Temperature, length of growing season, slope and erosion problems, even oxygen reductions at very high altitudes, all appear to limit the habitability of higher elevations. One estimate is that 34 percent of all people live below 100 meters (330 ft), a zone containing about 17 percent of total land area.

Fourth, although low-lying areas are preferred settlement locations, not all such areas are equally favored. Continental margins have attracted the densest settlement. On average, density in coastal areas is twice the world's average population density. Low temperatures and infertile soils of the extensive Arctic coastal lowlands of the Northern Hemisphere have restricted settlement there. Mountainous or desert coasts are sparsely occupied at any latitude, and some tropical lowlands and river valleys that are marshy, forested, and disease-infested are also unevenly settled.

The world contains four great clusters of population, listed in order of decreasing size: South Asia, East Asia, Europe, and the northeastern United States/southeastern Canada. The *South Asia* cluster is composed primarily of countries associated with the Indian subcontinent—Bangladesh, India, Nepal, Pakistan, and the island state of Sri Lanka—and is the most populous cluster, containing a fourth of the world's people. The *East Asia* zone, which includes China, Japan, North Korea, South Korea,

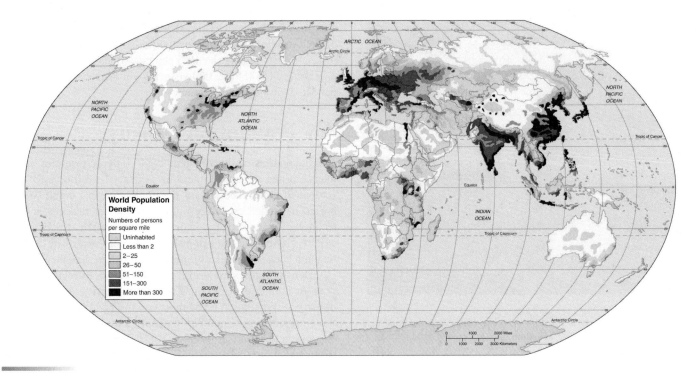

Figure 4.21 World population density.

and Taiwan, is the second-largest in population. Together, the South and the East Asian concentrations are home to nearly one-half the world's people.

Europe—southern, western, and eastern through Ukraine and much of European Russia—is the third extensive world population concentration, about 10 percent of its inhabitants. Much smaller in extent and total numbers is the cluster in the *northeastern United States/southeastern Canada*. Other smaller but pronounced concentrations are found around the globe: on the island of Java in Indonesia and along the Nile River in Egypt.

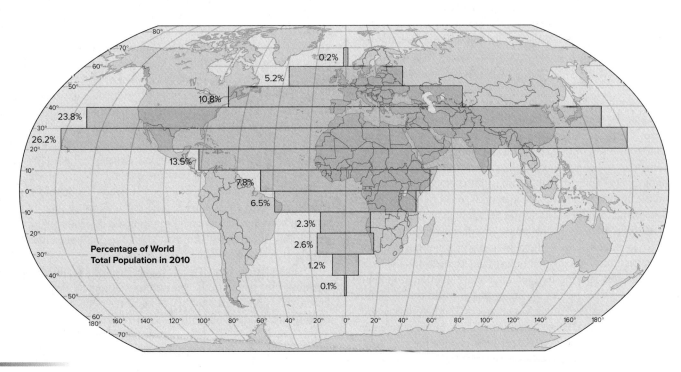

Figure 4.22 The population dominance of the Northern Hemisphere is strikingly evident from this bar chart. Only 12 percent of the world's people live south of the equator—not because the Southern Hemisphere is underpopulated, but because it is mainly water. The median latitude of the world's population is at 26° North, roughly that of Miami, Florida.

Data Source: Center for International Earth Science Information Network (CIESIN), Columbia University. 2005. Gridded Population of the World: Future Estimates (2011).

Figure 4.23 Terracing of hillsides is one device to extend a naturally limited productive area. The technique is effectively used here on the island of Bali, Indonesia.

©Photodisc/Houghton Mifflin Harcourt/Getty Images

The term **ecumene** is applied to permanently inhabited areas of the Earth's surface. The ancient Greeks used the word, derived from their verb "to inhabit," to describe their known, inhabited world between what they believed to be the unpopulated, searing hot "torrid" tropical zone and the permanently frozen northern "frigid" zone. Clearly, natural conditions are less restrictive than Greek geographers believed. Both ancient and modern technologies have rendered habitable areas that natural conditions make forbidding. Irrigation, terracing, diking, and draining are among the methods devised to extend the ecumene locally (**Figure 4.23**).

Although the ancient Greeks were wrong in labeling the tropics uninhabitable, the observation that some places are uninhabitable appears remarkably astute. The **nonecumene,** the uninhabited or very sparsely occupied zone, includes the permanent ice caps of the Far North and Antarctica and large segments of the tundra and coniferous forest of northern Asia and North America. But the nonecumene is not continuous, as the ancients supposed. It is discontinuously encountered in all portions of the globe and includes parts of the tropical rain forests of equatorial zones, deserts of both the Northern and Southern Hemispheres, and high mountain areas.

Even parts of these unoccupied or sparsely occupied districts have localized dense settlement nodes or zones based on irrigation agriculture, mining and industrial activities, and the like. Perhaps the most striking case of settlement in an environment elsewhere considered part of the nonecumene world is that of the dense population in the Andes Mountains of South America and the plateau of Mexico. Here Native Americans found temperate conditions away from the dry coast regions and the hot, wet Amazon basin. The fertile high basins have served a large population for more than a thousand years.

Even with these locally important exceptions, the nonecumene portion of the Earth is extensive. Some 35 to 40 percent of all the world's land surface is inhospitable and without

significant settlement. This is, admittedly, a smaller proportion of the Earth than would have qualified as uninhabited in ancient times, or even during the 19th century. Since the end of the Ice Age some 11,000 to 12,000 years ago, humans have steadily expanded their areas of settlement.

4.6 Population Density

The boundaries of the ecumene could only be extended as humans learned to support themselves from the resources of new settlement areas. The size of the population that could be supported depended on the resource potential of the area and the technologies and social organization possessed by the occupying culture group. The term **population density** expresses the relationship between number of inhabitants and the area they occupy.

Density figures are useful representations of regional variations of human distribution. The **crude density,** or **arithmetic density,** of population is the most common and simplistic variation. It calculates the number of people per unit area of land, usually within the boundaries of a political entity. All that is required is information on total population and total land area, both commonly available for national or other political units. The figure can, however, be misleading and may obscure as much as it reveals. The calculation is a national average, and does not reveal whether a country's population is evenly distributed or contains extensive regions that are only sparsely populated (**Figure 4.24**) alongside intensively settled districts. In general, the larger the political unit for which crude or arithmetic population density is calculated, the less useful is the figure.

Its usefulness is improved if the area in question can be subdivided into comparable regions or units. For example, it is more revealing to know that in 2010, New Jersey had a density of 467 and Wyoming of 2 persons per square kilometer (1,196 and 5.8 per sq mi) of land area than to know only that the figure for the United States was 32 per square kilometer (99.6 per sq mi).

Another potential refinement of crude density relates population not simply to total area of national territory but to that area of a country that is or may be cultivated; that is, to *arable* land. When total population is divided by arable land area alone, the resulting figure is the **physiological density,** which provides a measure of the population pressure exerted on agricultural land. Table 4.4 shows striking contrasts between crude and physiological densities of countries. The calculation of physiological density, however, has its own limitations. It depends on uncertain definitions of arable and cultivated land and assumes that all arable land is equally productive.

Agricultural density is another useful variant. It simply excludes city populations from the physiological density calculation and reports the number of rural residents per unit of agriculturally productive land. It is, therefore, an estimate of the number of farmers per unit area and offers insights into the type of agriculture practiced in a country.

Overpopulation

After comparing population densities, it is common to draw conclusions about overcrowding or overpopulation. However, it is

Figure 4.24 Tundra vegetation and landscape in the Alaska National Wildlife Refuge. While this area may contain underground petroleum deposits, it is part of the *nonecumene*—sparsely populated portions of the Earth's surface.

Source: Greg Weiler/U.S. Fish & Wildlife Service.

wise to remember that labeling a place overcrowded involves making value judgments regarding the amount of space needed per person. To someone from Wyoming, New York City might feel overcrowded, while to a New Yorker, Wyoming is likely to feel underpopulated. **Overpopulation** refers to the situation in which an environment or territory cannot support its present population and involves many factors other than just population density. Overpopulation is a reflection, not of numbers per unit area but of

Table 4.4

Comparative Densities for Selected Countries

Country	Crude Density		Physiological Density[a]		Agricultural Density[b]	
	sq mi	km²	sq mi	km²	sq mi	km²
Argentina	42	16	293	113	23	9
Australia	8	3	138	53	14	5
Bangladesh	3,276	1,265	5,497	2,122	3,573	1,380
Canada	10	4	218	84	39	15
China	383	148	3,012	1,163	1,295	500
Egypt	243	94	8,379	3,235	4,776	1,844
India	1,178	455	2,240	865	1,501	579
Japan	900	348	7,826	3,022	470	181
Nigeria	543	210	1,455	562	742	287
United Kingdom	709	274	2,858	1,103	486	188
United States	92	36	555	214	105	41

[a]Total population divided by area of agricultural land.
[b]Rural population divided by area of agricultural land.
Rounding may produce apparent conversion discrepancies.
Sources: *World Bank,* World Development Indicators 2017; *and Population Reference Bureau,* World Population Data Sheet, 2017. Calculations by authors.

the **carrying capacity** of the land, the prevailing agricultural technology, and the ability to afford imported food. A region devoted to energy-intensive commercial agriculture that makes heavy use of irrigation, fertilizers, and biocides can support more people at a higher level of living than one engaged in the slash-and-burn agriculture that will be described in Chapter 8. An industrial society that takes advantage of resources such as coal and iron ore and has access to imported food will not feel population pressure at the same density level as a country with rudimentary technology. In fact, we should be careful in borrowing the concept of carrying capacity from ecology because compared to other animals, humans are distinguished by their ability to adopt new production technologies, to increase their appetite for consumption far beyond biological requirements, and to trade resources globally.

In a world of growing global interdependence, high physiological densities alone do not indicate overpopulation. Few countries are agriculturally self-sufficient. Japan has a very high physiological density, as Table 4.4 indicates, and produces only about 40 percent of the calories that its population consumes. Nonetheless Japan ranks high on all indicators of national well-being and prosperity. For countries such as Japan, South Korea, Malaysia, and Taiwan—all of which currently import the majority of the grain that they consume—a sudden cessation of the international trade that permits the exchange of industrial products for imported food and raw materials would be disastrous. Domestic food production could not maintain the dietary levels now enjoyed by their populations and they, more starkly than many underdeveloped countries, would be *overpopulated*.

One measure of overpopulation might be a lack of **food security.** Food security means having access to safe and nutritious food supplies sufficient to meet individual dietary needs in accord with cultural preferences. Unfortunately, dietary insufficiencies—with long-term adverse implications for life expectancy, physical vigor, and mental development—are most likely to be encountered in developing countries, where much of the population is young and vulnerable (Figure 4.11). Over the past decades, sub-Saharan Africa saw its population grow faster than food production, widening the population-food gap and increasing reliance on imports. For poor countries, reliance on food imports means vulnerability to price volatility in international markets and rising rates of undernourishment whenever food prices are high. Although it may be tempting to conclude that a country with a high rate of undernourishment is overpopulated, the underlying causes are found in the interaction of poverty and population pressure, as discussed in Chapter 10.

It is difficult to draw meaningful conclusions about the relationship between population density and levels of development. As a group, the less-developed countries have higher densities than the more-developed countries. Densities in Australia, Canada, New Zealand, Norway, and the United States, where there is a great deal of unused and unsettled land, are considerably lower than those in Bangladesh, where most land is arable and which, with some 1,265 people per square kilometer (3,280 per square mile), is the most densely populated non-island state in the world. However, counterexamples abound. Mongolia, a sizable state between China and Siberian Russia, has 2 persons per square kilometer (5 per square mile), while Singapore, a highly urbanized island country in Southeast Asia has 7,930 persons per square kilometer (20,500 per square mile). Incomes are about 8 times higher in Singapore than in Mongolia. Many African countries have low population densities and low levels of income, whereas Japan combines both high densities and wealth.

4.7 Population Data and Projections

Population geographers, demographers, planners, governmental officials, and a host of others rely on detailed population data to make their assessments of present national and world population patterns and to estimate future conditions. Birth rates and death rates, rates of fertility and the age and sex composition of the population are all necessary ingredients for their work. In some countries, using the values for demographic variables specific to individual regions or ethnic groups is essential to making accurate projections.

Population Data

The data that demographers use come primarily from the United Nations Statistical Office, the World Bank, the Population Reference Bureau, and ultimately, from national censuses and sample surveys. Unfortunately, the data are far from perfect. A national census is a massive undertaking. In developing regions, isolation and poor transportation, insufficiency of funds and trained census personnel, high rates of illiteracy limiting the type of questions that can be asked, and populations suspicious of government data collectors serve to restrict the frequency, coverage, and accuracy of population reports.

However derived, detailed data are published by the major reporting agencies for all national units even when those figures are of poor quality. For years, data on the total population, birth and death rates, and other vital statistics for Somalia were regularly reported and annually revised. The fact was, however, that Somalia had never had a census and had no system whatsoever for recording births. Seemingly precise data were regularly reported as well for Ethiopia. When that country had its first-ever census in 1985, at least one data source had to drop its estimate of the country's birth rate by 15 percent and increase its figure for Ethiopia's total population by more than 20 percent. And a disputed 1991 census of Nigeria officially reported a population of 89 million, still the largest in Africa but far below the then generally accepted and widely cited estimates of between 105 and 115 million Nigerians. The 2006 census in Nigeria was surrounded by protests, boycotts, and fraud charges despite leaving out questions about religious and ethnic identity that were controversial in a country with more than 250 ethnic groups and a population nearly evenly divided between Muslims and Christians.

Fortunately, census coverage is improving. Almost every country has now had at least one census of its population and most have been subjected to periodic sample surveys (**Figure 4.25**). However, only about 10 percent of the developing world's population live in countries with anything approaching complete systems for registering births and deaths. Sub-Saharan Africa has the

Figure 4.25 To encourage complete participation in the 2000 census, the U.S. government created this billboard and other advertisements aimed at Hispanic residents.

©The McGraw-Hill Education/Barry Barker, photographer

highest percentage of unregistered births (65 percent), according to UNICEF. Apparently, deaths are even less completely reported than births throughout Asia. And whatever the deficiencies of Asian states, African statistics are still less complete and reliable. It is, of course, on just these basic birth and death data that projections about population growth and composition are founded.

Population Projections

For all their inadequacies and imprecisions, current data reported for country units form the basis of **population projections,** estimates of future population size, age, and sex composition based on current data. Projections are not forecasts, and demographers are not the social science equivalent of meteorologists. Weather forecasters work with a myriad of accurate observations applied against a known, tested model of the atmosphere. The demographer, in contrast, works with sparse, imprecise, out-of-date, and missing data applied to human actions that will be unpredictably responsive to stimuli not yet evident.

Population projections, therefore, are based on assumptions for the future applied to current data that are, themselves, frequently imperfect. Because projections are not predictions, they can never be wrong. They are simply the inevitable result of calculations about fertility, mortality, and migration rates applied to each age cohort of a population now living, and the making of birth rate, survival, and migration assumptions about cohorts yet unborn. Of course, the perfectly valid *projections* of future population size and structure resulting from those calculations may be dead wrong as *predictions*.

Because those projections are invariably treated as scientific expectations by a public that ignores their underlying qualifying assumptions, agencies such as the UN that estimate the population of, say, Africa in the year 2050, do so by not one but by three or more projections: high, medium, and low, for example. For areas as large as Africa, a medium projection is assumed to benefit from compensating errors and statistically predictable behaviors of very large populations. For individual African countries and smaller

populations, the medium projection may be much less satisfying. The usual tendency in projections is to assume that something like current conditions will be applicable in the future.

Projections for the world population in 2050 do not vary by much because many of those expected to be living in 2050 are already alive. Projections for 2010 differ significantly because small variations in birth rates can make a large difference in the long term.

Population Controls

As the number of humans and the extent of the ecumene have expanded, attention has turned to the possibility of overpopulation and the need for population control (**Figure 4.26**). Ancient Chinese thinker Confucius warned against rapid population growth, and ancient Greek philosophers Plato and Aristotle gave careful thought to the ideal population size for a city-state. All population projections include an assumption that at some point in time, fertility rates will stabilize at replacement levels and population growth will cease. Otherwise, future numbers become unthinkably large. At present growth rates, there would be 1 trillion people living on the Earth four centuries from now. Although there is reasonable debate whether the world is now overpopulated and what its optimum or maximum sustainable populations might be, totals in the trillions are beyond our imagination.

Population pressures do not come from the amount of physical space that humans occupy. For example, it has been calculated that the entire human race could easily be accommodated within the boundaries of the state of Delaware. The problems stem from the energy, food, water, and other resources necessary to support the population and absorb its waste products. Past technological changes have increased the carrying capacity of the Earth, allowing humans to achieve higher plateau populations (**Figure 4.27**). Nonetheless, population will have to stop increasing at some point. The demographic transition model provides the reassuring message that the decreases in birth rates that accompany modernization will automatically bring about a condition of zero population growth. However, the presence of a significant demographic divide between the nearly stable populations of the developed countries and rapidly growing populations in developing regions lead some observers to question the model's universality. According to these observers, completion of the demographic transition in the least-developed countries will not take place soon enough, if at all. If the demographic transition is not completed soon enough, environmental limits will appear in more dramatic fashion.

Thomas Robert **Malthus** (1766–1834), an English clergyman and economist, published *An Essay on the Principle of Population* in 1798, which set the framework for ongoing debates on population and resources. Prior to Malthus, most European economic and political thinkers were pronatalist—that is, they supported population growth to increase the number of workers and military might of a state. According to Malthus, the biological potential for population growth outstripped the potential for increasing food supplies to meet human subsistence needs. Malthus wrote:

> Taking the populations of the world at any number, a thousand million, for instance, the human species would increase in the ratio of −1, 2, 4, 8, 16, 32, 64, 128, 256, 512, & c. and subsistence as −1, 2, 3, 4, 5, 6, 7, 8, 9, 10, & c. (9)

Figure 4.26 Protesters in India calling for increased government action to reduce the country's high fertility rate.
©Hindustan Times/Getty Images

In essence, Malthus argued that if unchecked, human population would increase at a geometric rate while food supplies expanded at an arithmetic rate. If humans did not restrain their reproductive capacity with "private" means of moral restraint such as late marriage, or celibacy, nature would enact "destructive" checks on overpopulation:

> The power of population is so superior to the power in the earth to produce subsistence for man that premature death must in some shape or other visit the human race. The vices of mankind are active and able ministers of depopulation. . . . But should they fail in the war of extermination, sickly seasons, epidemics, pestilence, and plague, advance in terrific array, and sweep off their thousands and ten thousands. Should success be still incomplete, gigantic inevitable famine stalks in the rear, and with one mighty blow, levels the population with the food of the world. (44)

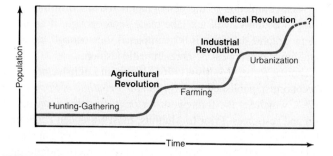

Figure 4.27 The steadily higher plateau populations achieved by humans are evidence of their ability to increase the carrying capacity of the land through technological advance. Each new plateau represents the conversion of the J-curve into an **S-curve**. *Medical revolution* implies the range of modern sanitary and public health technologies and disease preventative and curative advances that reduced morbidity and mortality rates.

Allied with the property-owning elite, Malthus was pessimistic about the lower social classes' ability to restrain their reproduction. He argued against charity because alleviating the suffering of the hungry would merely encourage them to increase in number, while discouraging thrift and hard work. Malthus was not alone in drawing pessimistic conclusions about population growth and resources; Hung Liangchi of China wrote in the 19th century that "Within a hundred years or so, the population can increase from fivefold to twentyfold, while the means of subsistence . . . can increase only from three to five times."

Malthus did not have the benefit of witnessing the demographic transition that was just getting underway or more recent advances in contraceptives and food production technology. Yet that did not stop the revival of his ideas after World War II, when many developing countries entered the second stage of the demographic transition. **Neo-Malthusians,** most notably Paul **Ehrlich,** a Stanford University biologist and author of the best-seller *The Population Bomb,* updated Malthus's arguments for the 20th century. As a population biologist, Ehrlich was familiar with animal studies that showed a steep J-curve of population growth followed by dieback when numbers exceeded the carrying capacity of the environment. Sounding the alarm in 1968, Ehrlich wrote:

> The battle to save humanity is over. In the 1970s and 1980s hundreds of millions of people will starve to death despite any crash programs embarked upon now. At this late date, nothing can be done to prevent a substantial increase in the world death rate, although many lives could be saved through dramatic programs to "stretch" the carrying capacity of the earth by increasing food production and providing for more equitable distribution of whatever food is available. But these programs will only provide a stay of execution unless they are accompanied by determined and successful efforts at population control (1968, xi).

Neo-Malthusianism gained popularity among environmentalists and international development specialists who viewed rapid population growth as damaging the environment and diverting scarce resources from capital investment. In order to lift living standards, the existing national efforts to lower mortality rates had to be balanced by governmental programs to reduce birth rates. Thus, neo-Malthusian thinking became the basis for national and international programs of population control. These programs were promoted around the world by the United Nations and non-governmental organizations and were especially influential in Asian countries such as China and India.

Arrayed against the pessimism of Malthus and the neo-Malthusians are a number of competing perspectives, most notably the Marxist critique of Malthus and the economic theories of Esther Boserup and Julian Simon. **Karl Marx** rejected the Malthusian interpretation of poverty, arguing that what appeared to be overpopulation was actually the unemployed surplus labor population needed by the capitalist system. Danish economist Esther Boserup developed the **Boserup thesis,** on the basis of detailed historical and field studies, which argues that past agricultural improvements occurred as a result of population pressure. In order to feed more people, farmers developed new ways to use their land and labor more intensively. In other words, population growth was a stimulus, not a deterrent, to development. American economist **Julian Simon** went further, developing an optimistic cornucopian perspective on population growth. Simon argued that resources do not exist in nature but are created by human ingenuity, which is the world's ultimate resource base. For example, a resource such as oil was just a black gooey substance until humans discovered ways of refining it and capturing its energy content. For cornucopians, more people means more scientists and inventors. Since the time of Malthus, they observed, the world's population had grown from 900 million to more than 7 billion without the predicted dire consequences—proof that Malthus failed to recognize the importance of technology in raising the carrying capacity of the Earth. Still higher population numbers, they suggest, are sustainable, perhaps even with improved standards of living for all.

An intermediate view admits that products of human ingenuity, such as the Green Revolution (discussed in Chapter 8), allowed food production to keep pace with rapid population growth but warns that continued gains in food production technology are not guaranteed. Both complacency and inadequate research support have hindered continuing progress in recent years. And even if further advances are made, they observe, not all countries or regions have the social and political will or capacity to take advantage of them. Those that do not will fail to keep pace with the needs of their populace and will sink into varying degrees of poverty and environmental decay, creating national and regional—though not necessarily global—crises.

Drawing upon both nationalistic and Marxist concepts, many less-developed countries rejected the neo-Malthusian population control programs promoted by Western states (See the feature "International Population Policies"). They maintained that remnant colonial era social, economic, and class structures, rather than population increase, hindered their development. Some government leaders recognized a link between population size and power and pursued pronatalist policies that encouraged childbearing, as did Mao's China during the 1950s and early 1960s. Africa and the Middle East have been generally less receptive to neo-Malthusian

arguments because of established cultural preferences for large families. Islamic fundamentalist resistance to birth restrictions has had an influence in the Near East and North Africa, although the Muslim theocracy in Iran has endorsed a range of contraceptive options and supports an aggressive family planning program. Predominantly Roman Catholic countries in southern Europe, Central America, and South America have witnessed substantial fertility declines despite the official pronatalist policies of the church.

4.8 Population Prospects

Regardless of population philosophies, theories, or cultural norms, the fact remains that many or most developing countries are showing significantly declining population growth rates. Global fertility and birth rates are falling to an extent not anticipated by pessimistic Malthusians and at a pace that suggests a peaking of world population numbers sooner—and at lower totals—than previously projected and possible decline in the developed world. In all world regions, steady and continuous fertility declines have been recorded over the past years, reducing fertility from global 5-children-per-woman levels in the early 1950s to 2.5 per woman today. Most continuing population growth in developed countries is due to momentum from the past, and aging will be an inevitable consequence of the recent changes in fertility patterns.

Population Implosion in the Developed World?

For much of the last half of the 20th century, demographers and economists focused on a "population explosion" and its implied threat of a world with too many people and too few resources of food and minerals to sustain them. In the 21st century, those fears were replaced by a new prediction of world regions with too few rather than too many people.

That possibility was suggested by two related trends. The first became apparent by 1970 when total fertility rates (TFRs) of 19 countries, almost all of them in Europe, had fallen below the **replacement level**—the level of fertility at which populations replace themselves—of 2.1. Simultaneously, Europe's population pyramid began to become noticeably distorted, with a smaller proportion of young and a growing share of middle-aged and retirement-age inhabitants. The decrease in native working-age cohorts had already, by 1970, encouraged the influx of non-European "guest workers" whose labor was needed to maintain economic growth and to sustain the generous security provisions guaranteed to what was becoming the oldest population of any continent.

Many countries of Western and Eastern Europe sought to reverse their birth rate declines by adopting pronatalist policies. The communist states of the East rewarded pregnancies and births with generous family allowances, free medical and hospital care, extended maternity leaves, and child care. France, Italy, the Scandinavian countries, and others gave similar bonuses or awards for first, second, and later births. Despite those inducements, however, reproduction rates continued to fall. By the early 21st century, every European country and territory had fertility rates below replacement levels. The personal decisions that produced sub-replacement fertility were influenced by cultural changes, increased educational levels of women, opportunities for women to work in challenging careers, and the increasing cost of rearing children. Cultural expectations about ideal family size have shifted downward, and an increasing number of adults are choosing personal pursuits over family obligations. The effect on national growth prospects has been striking. Every country in eastern Europe and most in southern Europe are projected to shrink in population by 2050. Germany's willingness to accept more than one million Syrian refugees in 2015 to 2016 was strongly influenced by the country's prospect of population decline. Europe as a whole is forecast to shrink in population by mid-century. "In demographic terms," France's prime minister remarked, "Europe is vanishing."

Europe's experience has been echoed in other societies with advanced economies. By the late 20th century, Canada, Australia, New Zealand, Japan, Taiwan, South Korea, Singapore, and other older and newly industrializing countries (NICs) registered fertility rates below the replacement level. As they have for Europe, simple projections foretold their aging and declining population. Japan's elderly population outnumbers its youth population more than twofold and its total population is projected to shrink by 20 percent by 2050.

The second trend indicating to many that world population numbers could stabilize during the lifetimes of today's college cohort is a simple extension of the first: TFRs are being reduced to or below the replacement levels in countries at all stages of economic development in all parts of the world. Exceptions to the trend are found in sub-Saharan Africa, and in some areas of South, Central, and West Asia; but even in those regions, fertility rates have been decreasing in recent years. "Powerful globalizing forces [are] at work pushing toward fertility reduction everywhere," was an observation of the French National Institute of Demographic Studies.

Achievement of zero population growth has social and economic consequences not always perceived by its advocates. These inevitably include an increasing proportion of older citizens, fewer young people, a rise in the median age of the population, and a growing old-age dependency ratio with ever-increasing pension and social service costs borne by a shrinking labor force.

Momentum

Reducing fertility levels to the replacement level of about 2.1 births per woman does not mean an immediate end to population growth. Because of the age composition of many societies, births will exceed deaths even as fertility rates per woman decline. The reason is to be found in **population** (or **demographic**) **momentum,** which is similar to a car continuing to coast for some time after the driver has lifted their foot from the accelerator pedal. The key to momentum is the age structure of a country's population.

When a high proportion of the population is young, the product of past high fertility rates, larger numbers enter the childbearing age each year; that is the case for major parts of the world today. The populations of developing countries are far younger than those of the established industrially developed regions (see Figure 4.11), with more than 40 percent in Africa below the age of 15. The fertility of these young people has yet to be realized. A population with a greater number of young people tends to

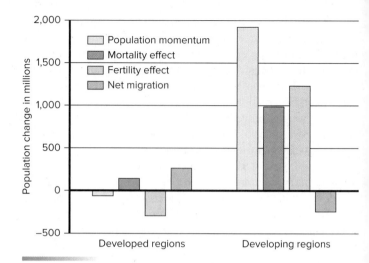

Figure 4.28 Components of population change, 2010–2100. The most dramatic population changes will take place in developing regions where momentum is the largest component of growth. Population momentum represents the lingering effects of past high fertility. Worldwide declines in total fertility rates will not immediately be reflected by equivalent declines in the growth of population. Because of past high fertility, the numbers of women in their childbearing years are increasing both absolutely and relative to the rest of the population.

Source: United Nations Population Division, 2013.

grow rapidly regardless of fertility rates. The results will continue to be felt until the youthful cohorts mature and work their way through the population pyramid (**Figure 4.28**).

Inevitably, while this is happening, even the most stringent national policies limiting growth cannot stop it entirely. A country with a large youth population will experience large numerical increases despite declining birth rates. Indeed, the higher fertility was to begin with and the sharper its drop to low levels, the greater will be the role of momentum even after rates drop below replacement. For example, Iran has recently implemented strong population policies that have lowered fertility rates to 1.9, well below replacement values. However, due to momentum, the population is projected to grow by 12 million by 2050. Population momentum also works in reverse. Low fertility rates create a top-heavy population pyramid containing momentum for future population decline.

Aging

Eventually, of course, young populations grow older, and even the youthful developing countries are beginning to face the consequences of that reality. The problems of a rapidly aging population that already confront the industrialized economies are now being realized in the developing world as well. Throughout human history, young people outnumbered the elderly; sometime between 2015 and 2020, persons 65 years and older surpassed those under 5 years of age. Europe is the oldest world region and Africa is the youngest. Japan leads the world in the percentage of its population aged 65 and older. Italy and Germany are close behind.

The progression toward older populations is considered irreversible, the result of longer life expectancies and the now-global demographic transition from high to low levels of fertility and mortality. The youthful majorities of the past are unlikely to occur again.

In the developing world, older persons are projected to make up 20 percent of the population by 2050 in contrast to the 8 percent over age 60 there in 2000. Because the demographic transition took place more quickly, the pace of aging will be much faster in the developing countries. Thus, they will have less time than the developed world did to adjust to the consequences of that aging. And those consequences will be experienced at lower levels of personal and national income and economic strength.

In both rich and poor states, the working-age populations will face increasing burdens and obligations. The potential support ratio, or PSR (the number of persons aged 15–64 years per one citizen aged 65 or older), has steadily fallen. Between 1950 and 2000, it dropped from 12 to 9 workers for each older person; by mid-century, the PSR is projected to drop to 4. The implications for Social Security programs, private pension plans, health insurance, government finances, and social support obligations are ominous. Exacerbating the strain on resources, the sharp increase in the oldest population (80 years or older) will require greater expenditures for health and long-term care. We can expect strained government budgets to become normal. The consequences of population aging appear most intractable for the world's poorest developing states that generally lack health, income, housing, and social service support systems adequate to the needs of their older citizens. To the social and economic concerns of their present population momentum, therefore, developing countries must add the consequences of future aging (**Figure 4.29**).

Figure 4.29 These senior residents of a Moroccan nursing home are part of the rapidly aging population of many developing countries. Worldwide, the over-60 cohort will number some 22 percent of total population by 2050 and be larger than the number of children less than 15 years of age. By 2020, a third of Singapore citizens will be 55 or older, and China will have as large a share of its population over 60—about one in four—as will Europe. Already, the numbers of old people in the world's poorer countries are beginning to dwarf those in the rich world. There are nearly twice as many persons over 60 in developing countries as in the advanced ones, but most are without the old age assistance and welfare programs that developed countries have put in place.
©*Nathan Benn/Corbis Historical/Getty Images*

SUMMARY

Birth, death, and migration rates are important in understanding the size, composition, distribution, and spatial trends of population. Recent "explosive" increases in human numbers and the prospects of continuing population expansion may be traced to sharp reductions in death rates, increases in longevity, and the impact of demographic momentum on a youthful population largely concentrated in the developing world. Control of population growth historically was accomplished through a demographic transition first experienced in European societies that adjusted their fertility rates downward as death rates fell and life expectancies increased. The introduction of advanced technologies of preventive and curative medicine, pesticides, and famine relief have reduced mortality rates in developing countries pushing developing regions into the demographic transition.

Even with the advent of more widespread fertility declines, the 7.5 billion human beings present today will still likely grow to perhaps 9.8 billion by the middle of the 21st century. That growth is largely unavoidable because of the size and youth of populations in developing countries. Eventually, a new balance between population and carrying capacity will be reached, as it has always been following past periods of rapid population increase.

People are unevenly distributed over the Earth. The ecumene, or permanently inhabited portion of the globe, is discontinuous and marked by pronounced differences in population concentrations and numbers. South Asia, East Asia, Europe, and northeastern United States/southeastern Canada represent the world's greatest population clusters, though smaller areas of great density are found in other regions and continents.

As respected geographer Glenn Trewartha once commented, "population is the point of reference from which all other elements [of geography] are observed." Certainly, population geography is the essential starting point for studying the human impact on the environment. But human populations are not merely collections of numerical units; nor are they to be understood solely through statistical analysis. Societies are distinguished not just by abstract statistical data, but by experiences, beliefs, and aspirations that collectively constitute what social scientists call *culture*. It is to that fundamental human diversity that we next turn our attention.

KEY WORDS

FOR REVIEW

1. How do the *crude birth rate* and the *fertility rate* differ? Which measure is the more accurate statement of the amount of reproduction occurring in a population?

2. How is the *crude death rate* calculated? What factors account for the worldwide decline in death rates since 1945?

3. How is a *population pyramid* constructed? What shape of "pyramid" reflects the structure of a rapidly growing country? Of a population with a slow rate of growth? What can we tell about future population numbers from those shapes?

4. What variations do we discern in the spatial pattern of the *rate of natural increase* and, consequently, of population growth? What rate of natural increase would double population in 35 years?

5. How are population numbers projected from present conditions? Are projections the same as predictions? If not, in what ways do they differ?

6. Describe the stages in the *demographic transition*. Where has the final stage of the transition been achieved? Why do some analysts doubt the applicability of the demographic transition to all parts of the world?

7. Contrast *crude population density, physiological density,* and *agricultural density.* For what differing purposes might each be useful? How is *carrying capacity* related to the concept of density?

8. What was Malthus's underlying assumption concerning the relationship between population growth and food supply? In what ways do the arguments of *neo-Malthusians* differ from the original doctrine? What governmental policies are implicit in *neo-Malthusianism?*

9. Why is *population momentum* a matter of interest in population projections? In which world areas are the implications of demographic momentum most serious in calculating population growth, stability, or decline?

KEY CONCEPTS REVIEW

1. **What are some basic terms and measures used by population geographers?** Section 4.1–4.2

   A *cohort* is a population group, usually an age group, treated as a unit. *Rates* record the frequency of occurrence of an event over a given unit of time. Rates are used to trace a wide range of population features and trends: births, deaths, fertility, infant or maternal mortality, natural increase, and others. Those rates tell us both the present circumstances and likely prospects for national, country group, or world population structures. Population pyramids give visual evidence of the current age and sex cohort structure of countries or country groupings.

2. **What are meant and measured by the demographic transition model and the demographic equation?** Section 4.3–4.4

   The *demographic transition* model traces the presumed relationship between population growth and economic development. In Western countries, the transition model historically displayed four stages:
   (a) high birth and death rates; (b) high birth and declining death rates; (c) declining births and reduced growth rates; and (d) low birth and death rates. A fifth stage of population decline is observed for some aging societies. The transition model has been observed to be not fully applicable to all developing states. The *demographic equation* attempts to incorporate cross-border population migration into projections of national population trends.

3. **What descriptive generalizations can be made about world population distributions and densities?**
Section 4.5–4.6

World population is primarily concentrated north of the equator, in lower (below 200 meters) elevations, along continental margins. Major world population clusters include *East Asia* with 25 percent of the total, *South Asia* with more than 20 percent, *Europe* and *northeastern United States/southeastern Canada* with significant but lesser shares of world population. Other smaller but pronounced concentrations are found discontinuously on all continents. Within the permanently inhabited areas—the *ecumene*—population densities vary greatly. Highest densities are found in cities; almost one-half of the world's people are urban residents now and the vast majority of world population growth over the first quarter of the 21st century will occur in cities of the developing world.

4. **What are population projections, and how are they affected by various controls on population growth?**
Section 4.7–4.8

Population projections are merely calculations of the future size, age, and sex composition of regional, national, or world populations; they are based on current data and manipulated by varying assumptions about the future. As simple calculations, projections cannot be wrong. They may, however, totally misrepresent what actually will occur because of faulty current data or erroneous assumptions used in their calculation. They may also be invalid because of unanticipated self-imposed or external brakes on population growth, such as changing family size desires or limits on areal carrying capacity that slow or halt current growth trends. Even with such growth limitations, however, population prospects are always influenced greatly by *demographic momentum,* the inevitable growth in numbers promised by the high proportion of younger cohorts yet to enter childbearing years in the developing world, and by the consequences of global population aging.

CHAPTER 5

LANGUAGE AND RELIGION:
Mosaics of Culture

Ethiopia's 1,600-year-old Coptic Christian Meskel Festival marks the finding of the true cross on which Christ was crucified.
©Anadolu Agency/Getty Images

Key Concepts

Language

5.1 The classification, spread, and distribution of the world's languages; the nature of language change; language standards and variants, from dialects to official tongues

5.2–5.3 Language as source of cultural identity and landscape relic

Religion

5.4 The cultural significance and role of religion

5.5 How world religions are classified and distributed

5.6 The origins, nature, and diffusions of principal world religions

*W*hen God saw [humans become arrogant], he thought of something to bring confusion to their heads: he gave the people a very heavy sleep. They slept for a very, very long time. They slept for so long that they forgot the language they had used to speak. When they eventually woke up from their sleep, each man went his own way, speaking his own tongue. None of them could understand the language of the other any more. That is how people dispersed all over the world. Each man would walk his way and speak his own language and another would go his way and speak in his own language. . . .

God has forbidden me to speak Arabic. I asked God, "Why don't I speak Arabic?" and He said, "If you speak Arabic, you will turn into a bad man." I said, "There is something good in Arabic!" And He said, "No, there is nothing good in it! . . ."

Here, I slaughter a bull and I call [the Muslim] to share my meat. I say, "Let us share our meat." But he refuses the meat I slaughter because he says it is not slaughtered in a Muslim way. If he cannot accept the way I slaughter my meat, how can we be relatives? Why does he despise our food? So, let us eat our meat alone. . . . Why, they insult us, they combine contempt for our black skin with pride in their religion. As for us, we have our own ancestors and our own spirits; the spirits of the Rek, the spirits of the Twic, we have not combined our spirits with their spirits. The spirit of the black man is different. Our spirit has not combined with theirs.[1]

In this chapter, we examine two prominent threads in the tapestry of cultural diversity—language and religion. Language and religion are basic components of cultures, the learned ways of life of different human communities. They help identify who and what we are and clearly place us within larger communities of persons with similar characteristics. At the same time, as the words of Chief Makuei suggest, they separate and divide peoples of different tongues and faiths. Language and religion are mentifacts, components of the ideological subsystem of culture that help shape the belief system of a society and transmit it to succeeding generations. Both within and between cultures, language and religion are fundamental strands in the complex web of culture, serving to shape and to distinguish people and groups.

They are ever-changing strands. Languages evolve in place, responding to the dynamics of social and economic change and spatial interaction in a closely integrated world. New artifacts and sociofacts demand new words like *e-shopping* and *co-parenting,* which entered the *Oxford English Dictionary* in 2018. Languages diffuse in space, carried by streams of migrants, colonizers, and conquerors. They may be rigorously defended and preserved as essential elements of cultural identity, or they may be abandoned in the search for acceptance into a new society. Religions, too, are dynamic, sweeping across national, linguistic, and cultural boundaries by conversion, conviction, and conquest. Their broad spatial patterns—distinctive culture regions in their own right— are also fundamental in defining the culture realms outlined in Figure 2.4, while at a different scale, religious differences may

Figure 5.1 This small town welcome sign offers evidence of religious diversity in the United States. However, the sign details only a few Christian congregations. In reality, the United States has become the most religiously diverse country in the world, with essentially all of the world's faiths represented within its borders.
©*Joe Sohm/Visions of America/Universal Images Group/Getty Images*

contribute to the cultural diversity and richness within the countries of the world (**Figure 5.1**).

5.1 Classification of Languages

On a clear, dark night, the unaided eye can distinguish between 4,000 and 6,000 stars, a number comparable to some estimates of the probable total number of the world's languages. In reality, no precise figure is possible, for even today in Africa, Latin America, New Guinea, and elsewhere, linguists race to identify and classify the tongues spoken by isolated peoples before some disappear.

In the broadest sense, **language** is any systematic method of communicating ideas, attitudes, or intent through the use of mutually understood signs, sounds, or gestures. For our geographic purposes, we may define language as an organized system of spoken words by which people communicate with one another with mutual comprehension. But such a definition fails to recognize the gradations among languages or to grasp the varying degrees of mutual comprehension between two or more of them. The language commonly called *Chinese,* for example, is more properly seen as a group of distinct but related languages— Mandarin, Cantonese, Hakka, and others—that are as different from one another as are such comparably related European languages as Spanish, Italian, French, and Romanian. *Chinese* has uniformity only in the fact that all of the varied Chinese languages are written alike. No matter how it is pronounced, the same symbol for *house* or for *rice,* for example, is recognized by all literate speakers of any Chinese language variant (**Figure 5.2**). Again, the language known as *Arabic* represents a number of related but distinct tongues, so Arabic spoken in Morocco differs from Palestinian Arabic, roughly as Portuguese differs from Italian.

Languages differ greatly in their relative importance, if "importance" can be taken to mean the number of people using them. More than half of the world's inhabitants are native speakers of just eight of its thousands of tongues, and at least half regularly use

---

[1]Source: The words of Chief Makuei Bilkuei of the Dinka, a Nilotic people of the southern Sudan. His comments are directed at the attempts to unite into a single people the Arabic Muslims of the north of the Republic of the Sudan with his and other black, Luo-speaking animist and Christian people of the country's southern areas. Recorded by Francis Mading Deng, *Africans of Two Worlds: The Dinka in Afro-Arab Sudan.*

The Geography of Language

Forever changing and evolving, language in spoken or written form makes possible the cooperative efforts, group understandings, and shared behavior that is central to culture. Language is the most important medium by which culture is transmitted. It is what enables parents to teach their children about the world they live in and what they must do to become functioning members of society. Some argue that the language of a society structures the perceptions of its speakers. By the words that it contains and the concepts that it can formulate, language is said to influence the attitudes, understandings, and responses of the society to which it belongs.

If that conclusion be true, one aspect of cultural heterogeneity may be easily understood. The nearly 8 billion people on Earth speak many thousands of different languages. Knowing that more than 2,100 languages and language variants are spoken in Africa (though 85 percent of Africans speak

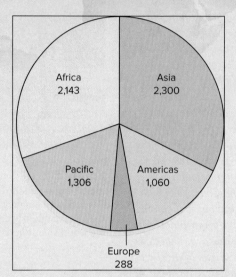

Figure 5A World distribution of 7,100 languages still spoken today, one-third are found in Asia, 30 percent in Africa, 18 percent in the Pacific area, 15 percent in the Americas, and 3 percent in Europe. Linguists' estimates of the number of languages ever spoken on Earth range from 31,000 to as many as 300,000 or more. Assuming the lower estimate (or even one considerably smaller), dead languages far outnumber the living. Approximately 10 percent of the world's languages are classified as moribund or nearly extinct because only the grandparent generation uses the language.

Source: Estimates based on Ethnologue, 21st edition, 2018.

one or more variants of 15 core languages) gives us a clearer appreciation of the political and social divisions in that continent. Europe alone has some 288 languages and dialects

(Figure 5A). Language is a hallmark of cultural diversity, an often fiercely defended symbol of cultural identity that helps to distinguish the world's diverse social groups.

---

or have competence in just four of them. That restricted language dominance reflects the reality that the world's linguistic diversity is rapidly shrinking. Of the at most 6,900 tongues still remaining, more than half are no longer being learned by children or used in everyday life and are endangered. One estimate anticipates that no more than 600 of the world's current living languages will still be in existence in 2100. Table 5.1 lists those languages currently spoken as a primary tongue by 90 million or more people. At the other end of the scale are a number of rapidly declining languages whose speakers number in the hundreds or, at most, the few thousands.

The diversity of languages is simplified when we classify them into families. A language family is a group of languages descended from a single, earlier tongue. By varying estimates, from at least 30 to perhaps 100 such families of languages are found worldwide. The families, in turn, may be subdivided into subfamilies, branches, or groups of more closely related tongues. Some 2,000 years ago, Latin was the common language spoken throughout the Roman Empire. The fall of the empire in the 5th

Table 5.1

First Languages Spoken by 100 Million or More People as of 2018

Language	Millions of Speakers	Total Countries
Chinese	1,299	38
Spanish	442	31
English	378	118
Arabic[c]	315	58
Hindi/Urdu[b] (India, Pakistan)	260	4
Bengali (Bangladesh/India)	243	4
Portuguese	223	15
Russian/Belorussian	154	18
Japanese	128	2
Lahnda (Pakistan)	119	6

[a]The official dialect of Mandarin is spoken by an estimated 909 million.
[b]Hindi and Urdu are basically the same language: Hindustani. Written in the Devangari script, it is called *Hindi*, the official language of India; in the Arabic script, it is called *Urdu*, the official language of Pakistan.
[c]The figure given includes speakers of the many often mutually unintelligible versions of colloquial Arabic. Classical or literary Arabic, the language of the Koran, is uniform and standardized but restricted to formal usage as a spoken tongue. Because of its religious association, Arabic is a second language for many inhabitants of Muslim countries with other native tongues.
Sources: *Based on data from* Ethnologue: Languages of the World, 21st edition, 2018.

HOUSE   RICE   TREE

Figure 5.2 All literate Chinese, no matter which of the many languages of China they speak, recognize the same ideographs for *house, rice,* and *tree.*

century CE broke the unity of Europe, and regional variants of Latin began to develop in isolation. In the course of the next several centuries, these Latin derivatives, changing and developing as all languages do, emerged as the individual Romance languages—Italian, Spanish, French, Portuguese, and Romanian—of modern Europe and of the world colonized by their speakers. Catalan, Sardinian, Provençal, and a few other spatially restricted tongues are also part of the Romance language group.

Family relationship between languages can be recognized through similarities in their vocabulary and grammar. By tracing regularities of sound changes in different languages back through time, linguists are able to reconstruct earlier forms of words and, eventually, determine a word's original form before it underwent alteration and divergence. Such a reconstructed earlier form is said to belong to a **protolanguage.** In the case of the Romance languages, of course, the well-known ancestral tongue was Latin, which needs no such reconstruction. Its root relationship to the Romance languages is suggested by modern variants of *panis,* the Latin word for "bread": *pane* (Italian), *pain* (French), *pan* (Spanish), *pão* (Portuguese), *pâine* (Romanian). In other language families, similar word relationships are less confidently traced to their protolanguage roots. For example, the Germanic languages, including English, German, Dutch, and the Scandinavian tongues, are related descendants of a less well-known proto-Germanic language spoken by peoples who lived in southern Scandinavia and along the North Sea and Baltic coasts from the Netherlands to western Poland. The classification of languages by origin and historical relationship is called a *genetic classification.*

Further tracing of language roots tells us that the Romance and the Germanic languages are individual branches of an even more extensive family of related languages derived from proto-Indo-European, or simply Indo-European. Of the principal recognized language clusters of the world, the Indo-European family is the largest, embracing most of the languages of Europe and a large part of Asia, and the introduced—not the native—languages of the Americas (**Figure 5.3**). All told, languages in the Indo-European family are spoken by about half the world's peoples.

By recognizing similar words in most Indo-European tongues, linguists deduce that the Indo-European people—originally hunters and fishers but later switching to agriculture—developed somewhere in eastern Europe or the Ukrainian steppes about 5,000 years ago (though some conclude that central Turkey was the more likely site of origin and that the ancestral tongue existed 8,700 to 10,000 or more years ago). By at least 2500 BCE, their society apparently fragmented; they left the homeland, carrying segments of the parent culture in different directions. Some migrated into Greece, others settled in Italy, still others crossed central and western Europe, ultimately reaching the British Isles. Another group headed into the

Russian forest lands, and still another branch crossed Iran and Afghanistan, eventually to reach India. Wherever this remarkable people settled, they appear to have dominated local populations and imposed their language on them. For example, the word for *sheep* is *avis* in Lithuanian, *ovis* in Latin, *avis* in Sanskrit (the language of ancient India), and *hawi* in the tongue used in Homer's Troy. Modern English retains its version in the word *ewe.* All, linguists infer, derive from an ancestral word, *owis* in Indo-European. Similar relationships and histories can be traced for other protolanguages.

World Pattern of Languages

The present world distribution of major language families (**Figure 5.4**) records not only the migrations and conquests of our linguistic ancestors but also the continuing dynamic pattern of recent human movement, settlement, and colonizations. Indo-European languages have been carried far beyond their Eurasian homelands from the 16th century onward by western European colonizers in the Americas, Africa, Asia, and Australasia. In the process of linguistic imposition and adoption, innumerable indigenous languages and language groups have been modified or totally lost. Most of the

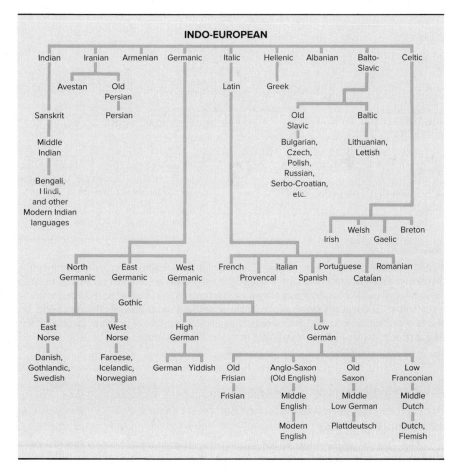

Figure 5.3 The Indo-European linguistic family tree. Euskara (Basque), Estonian, Finnish, Hungarian, Maltese, and Lappish are the only European languages not in the Indo-European family. (See also Figure 5.7.)

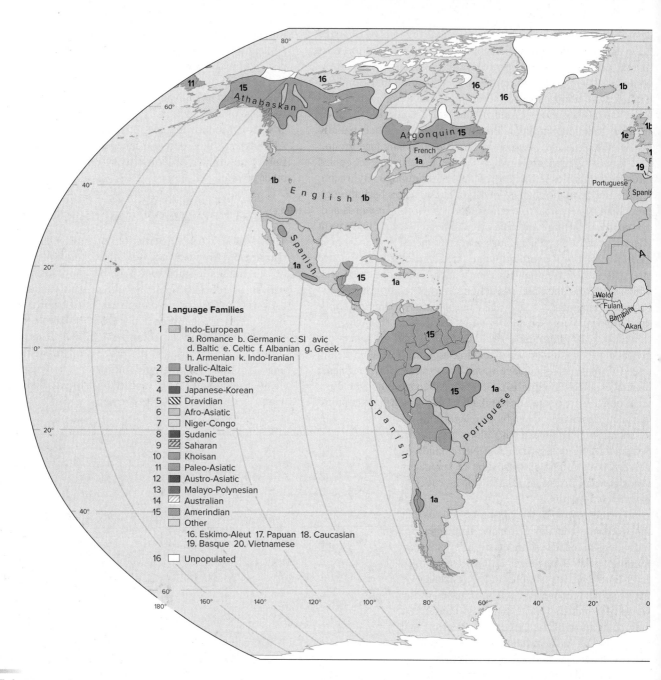

Figure 5.4 World language families. Language families are groups of individual tongues that had a common but remote ancestor. By suggesting that the area assigned to a language or language family uses that tongue exclusively, the map pattern conceals important linguistic detail. Many countries and regions have local languages spoken in territories too small to be recorded at this scale. The map also fails to report that the population in many regions is fluent in more than one language or that a second language serves as the necessary vehicle of commerce, education, or government. Nor is important information given about the number of speakers of different languages; the fact that there are more speakers of English in India or Africa than in Australia is not even hinted at by a map at this scale.

Note that some linguistic boundaries match political boundaries, and others do not.

estimated 1,000 to 2,000 Amerindian tongues of the Western Hemisphere disappeared in the face of European conquest and settlement (**Figure 5.5**).

The Slavic expansion eastward across Siberia beginning in the 16th century obliterated most of the Paleo-Asiatic languages there. Similar loss occurred in Eskimo and Aleut language areas. Large linguistically distinctive areas comprise the northern reaches of both Asia and America

(see Figure 5.4). Their sparse populations are losing the mapped languages as the indigenous people adopt the tongues of the majority cultures of which they have been forcibly made a part. In the Southern Hemisphere, the several hundred original Australian languages also loom large spatially on the map but have at most 50,000 speakers, exclusively Australian aborigines. Numerically and effectively, English dominates that continent.

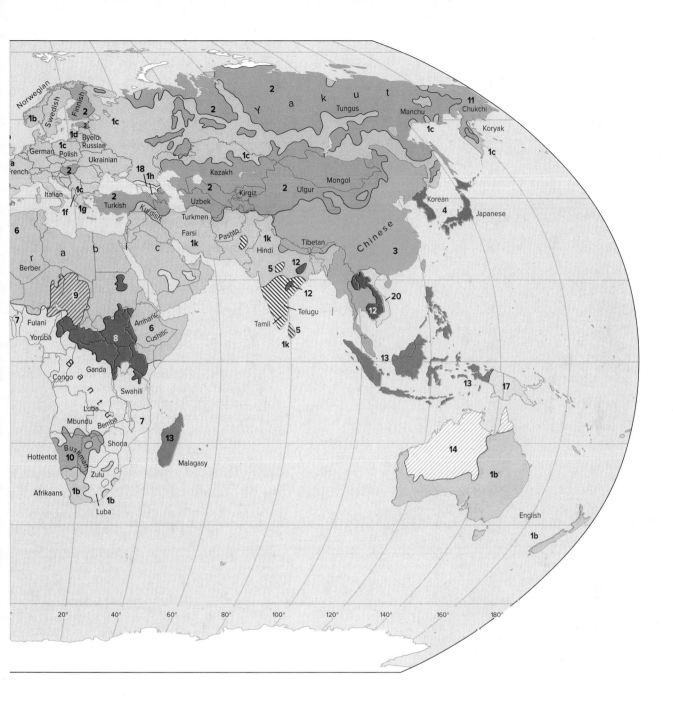

Examples of linguistic conquest by non-Europeans also abound. In Southeast Asia, formerly extensive areas identified with different members of the Austro-Asiatic language family have been reduced through conquest and absorption by Sino-Tibetan (Chinese, Thai, Burmese, and Lao, principally) expansion. Arabic—originally a minor Afro-Asiatic language of the Arabian Peninsula—was dispersed by the explosive spread of Islam through much of North Africa and southwestern Asia, where it largely replaced a host of other locally variant tongues and became the official or the dominant language of more than 20 countries and more than 300 million people. The more than 300 Bantu languages found south of the "Bantu line" in sub-Saharan Africa are variants of a proto-Bantu carried by an expanding, culturally advanced population that displaced more primitive predecessors (**Figure 5.6**).

Language Diffusion

Language diffusion represents the increase or relocation over time in the geographic area within which a language is spoken. The Bantu of Africa or the English-speaking settlers of North America displaced preexisting populations and replaced the languages previously spoken in the areas of penetration. Therefore, we find one explanation of the spread of language families to new areas of occurrence in massive population relocations such as those accompanying the colonization of the Americas or of Australia. That is, languages may spread through migration in a process called *relocation diffusion.*

Latin, however, replaced earlier Celtic languages in western Europe not by force of numbers—Roman legionnaires, administrators, and settlers never represented a majority population—but

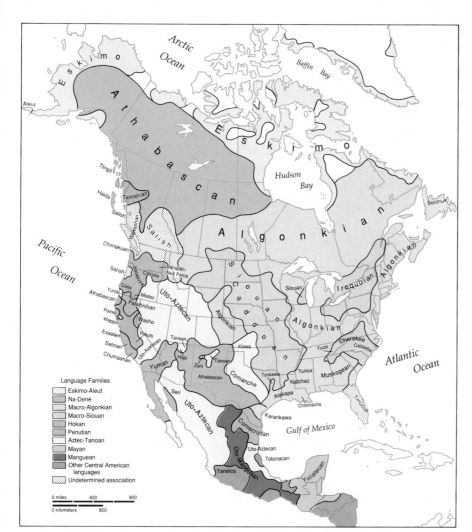

Figure 5.5 Amerindian language families of North America. As many as 300 different North American and more than 70 Mesoamerican tongues were spoken at the time of first European contact. The map summarizes the traditional view that these were grouped into 9 or 10 language families in North America, as many as 5 in Mesoamerica, and another 10 or so in South America. More recent research, however, suggests close genetic relationships between Native American tongues, clustering them into just three families: Eskimo-Aleut in the extreme north and Greenland; Na-Dené in Canada and the U.S. Southwest, and Amerind elsewhere in the hemisphere. Because each family has closer affinities with Asian language groups than with one another, it is suggested that each corresponds to a separate wave of Asian migration to the Americas: the first giving rise to the Amerind family, the second to the Na-Dené, and the last to the Eskimo-Aleut. Many Amerindian tongues have become extinct; others are still known only to very small groups of mostly elderly speakers.

Source: Data from various sources, including C. F. and F. M. Voegelin, Map of North American Indian Languages (Seattle: University of Washington Press, 1986).

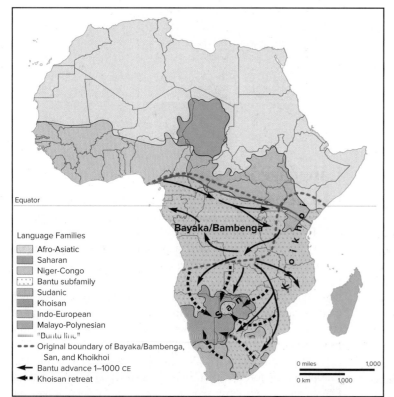

Figure 5.6 Bantu advance, Khoisan retreat in Africa. Linguistic evidence suggests that proto-Bantu speakers originated in the region of the Cameroon-Nigeria border, spread eastward across the southern Sudan, then turned southward to Central Africa. From there, they dispersed slowly eastward, westward, and, against slight resistance, southward. The earlier Khoisan-speaking occupants of sub-Saharan Africa were no match against the advancing metal-using Bantu agriculturalists. The Bayaka/Bambenga, adopting a Bantu tongue, retreated deep into the forests; San and Khoikhoi retained their distinctive Khoisan "click" language but were forced out of forests and grasslands into the dry steppes and deserts of the southwest.

by the gradual abandonment of their former languages by native populations brought under the influence and control of the Roman Empire and, later, of the Western Christian church. Adoption rather than eviction of language was the rule followed in perhaps the majority of historical and contemporary instances of language spread. Knowledge and use of the language of a dominating culture may be seen as a necessity when that language is the medium of commerce, law, civilization, and personal prestige. Usually, those who are in or aspire to positions of importance are the first to adopt the new language of control and prestige. Later, through schooling, daily contact, and business or social necessity, other, lower social strata of society may gradually be absorbed into the expanding pool of language adopters. It was on that basis, not through numerical superiority, that Indo-European tongues were dispersed throughout Europe and to distant India, Iran, and Armenia. Likewise, Arabic became widespread in western Asia and North Africa not through massive population relocations but through conquest, religious conversion, and dominating culture. Thus, languages may spread through expansion diffusion as they acquire new speakers.

Hierarchical diffusion of an official or prestigious language within urban centers and centers of power has occurred in many societies. In India during the 19th century, the English established an administrative and judicial system that put a very high premium on their language as the sole medium of education, administration, trade, and commerce. Proficiency in it was the hallmark of the cultured and educated person (as knowledge of Sanskrit and Persian had been under previous conquerors of India). English, French, Dutch, Portuguese, and other languages introduced during the age of empire retain a position of prestige and even status as the official language in multilingual societies, even after independence has been achieved by former colonies. In Uganda and other former British possessions in Africa, a stranger may be addressed in English by one who wishes to display his or her education and social status, though standard Swahili, a second language for many different culture groups, may be chosen if certainty of communication is more important than prestige.

As a diffusion process, language spread may be impeded by barriers or promoted by their absence. Cultural barriers may retard or prevent language adoption. Speakers of Greek resisted centuries of Turkish rule of their homeland, and the language remained a focus of cultural identity under foreign domination. Breton, Catalan, Gaelic, Welsh, and other localized languages of Europe remain symbols of ethnic separateness from dominant national cultures.

Physical barriers to language spread have also left their mark (see Figure 5.4). Migrants or invaders follow paths of least topographic resistance and disperse most widely where access is easiest. Once past the barrier of the Pamirs and the Hindu Kush mountains, Indo-European tongues spread rapidly through the Indus and Ganges river lowlands of the Indian subcontinent but made no headway in the mountainous northern and eastern border zones. The Pyrenees Mountains serve as a linguistic barrier separating France and Spain. They also house the Basques, who speak the only language—Euskara in their tongue—in southwestern Europe that survives from pre-Indo-European times (**Figure 5.7**). Similarly, the Caucasus Mountains between the Black and Caspian seas separate the Slavic speakers to the north and the areas of Ural-Altaic languages to the south. At the same time, in their rugged mountains,

Figure 5.7 In their mountainous homeland, the Basques have maintained a linguistic uniqueness despite more than 2,000 years of encirclement by dominant lowland speakers of Latin or Romance languages. This sign—thanking travelers for their visit and wishing them a good trip home—gives its message in both Spanish and the Basque language, Euskara.
©Mark Antman/The Image Works

they contain an extraordinary mixture of languages, many unique to single valleys or villages, lumped together spatially into a separate Caucasian language family.

Language Change

Migration, segregation, and isolation give rise to separate, mutually unintelligible languages. Changes occur naturally in word meaning, pronunciation, vocabulary, and syntax (the way words are put together in phrases and sentences). Because they are gradual, minor, and readily adopted, such changes tend to go unnoticed. Yet, cumulatively, they can result in language change so great that over the course of centuries, an essentially new language has been created. The English of 17th-century Shakespearean writings or the King James Bible (1611) sounds stilted to our ears. Few of us can easily read Chaucer's 14th-century *Canterbury Tales,* and the 8th-century *Beowulf* is practically unintelligible.

Change may be gradual and cumulative, with each generation deviating in small degree from the speech patterns and vocabulary of its parents, or it may be massive and abrupt. English gained about 10,000 new words from the Norman-French of the 11th-century Norman conquerors. In some 70 years (1558–1625) of literary and linguistic creativity during the reigns of Elizabeth I and James I, an estimated 12,000 words—often borrowed from Latin or Greek (such as the word *geography*)—were introduced.

Discovery and colonization of new lands and continents in the 16th and 17th centuries greatly expanded English as new foods, vegetation, animals, and artifacts were encountered and adopted along with their existing aboriginal American, Australian, or African names. The Indian languages of the Americas alone brought more than 200 relatively common daily words to English,

80 or more from the North American native tongues and the rest from the Caribbean and Central and South America. More than 2,000 more specialized or localized words were also added. *Moose, raccoon, skunk, maize, squash, succotash, igloo, toboggan, hurricane, blizzard, hickory, pecan,* and a host of other terms were taken directly into English; others were adopted secondhand from Spanish variants of South American native words: *cigar, potato, chocolate, tomato, tobacco, hammock.* More recently, scientific developments have enriched and expanded vocabularies by adding many words of Greek and Latin derivation. Introduction of new technologies requires new terms such as *Internet* and *cyberspace,* many of which have been adopted by other languages.

The Story of English

English itself is a product of change and diffusion, an offspring of proto-Germanic (see Figure 5.3) descending through the dialects brought to England in the 5th and 6th centuries by conquering Danish and North German Frisians, Jutes, Angles, and Saxons. Earlier Celtic-speaking inhabitants found refuge in the north and west of Britain and in the rugged uplands of what are now Scotland and Wales. Each of the transplanted tongues established its own area of dominance, but the West Saxon dialect of southern England emerged in the 9th and 10th centuries as Standard Old English on the strength of its literary richness.

It lost its supremacy after the Norman Conquest of 1066, as the center of learning and culture shifted northeastward from Winchester to London, and French rather than English became the language of the nobility and the government. When the tie between France and England was severed after the loss of Normandy (1204), French fell into disfavor and English again became the dominant tongue, although now as the French-enriched Middle English used by Geoffrey Chaucer and mandated as the official language of the law courts by the Statute of Pleading (1362). During the 15th and 16th centuries, English as spoken in London emerged as the basic form of Early Modern English.

During the 18th century, attempts to standardize and codify the rules of English were unsuccessful. But Samuel Johnson's *A Dictionary of the English Language* (published 1755)—based on the cultured language of contemporary London and the examples of authors such as Shakespeare—helped establish norms of form and usage. A worldwide diffusion of the language resulted as English colonists carried it as settlers to the Western Hemisphere and Australasia; through merchants, conquest, or territorial claim, it established footholds in Africa and Asia. In that spatial diffusion, English was further enriched by its contacts with other languages. By becoming the accepted language of commerce and science, it contributed, in turn, to the common vocabularies of other tongues (see the feature "Language Exchange").

Within some 400 years, English has developed from a localized language of 7 million islanders off the European coast to a truly international language with some 330 million native speakers, and more than 1 billion total speakers worldwide. English is the most commonly used of the three working languages of the European Union (EU) and is an official language in some 60 countries (**Figure 5.8**), far exceeding French (32), Arabic (25), and Spanish

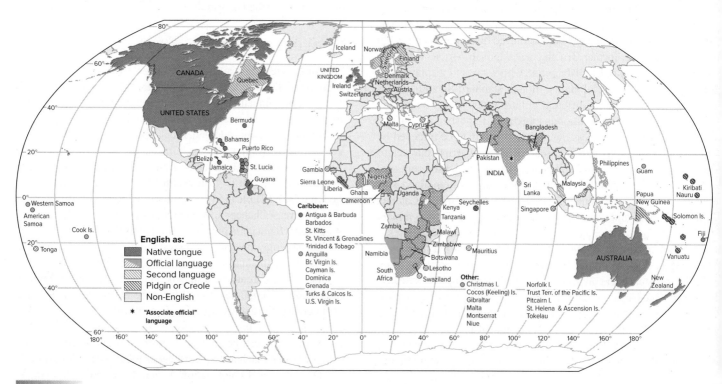

Figure 5.8 International English. In worldwide diffusion and acceptance, English has no past or present rival. English is the sole or joint official language of more nations and territories, some too small to be shown here, than any other tongue. It also serves as the effective unofficial language of administration in other multilingual countries with different formal official languages. "English as a second language" is indicated for countries with near-universal or mandatory English instruction in public schools. Not evident on this map is the full extent of English penetration of continental Europe, where more than 80 percent of secondary school students (and 92 percent of those of EU states) study it as a second language.

Language Exchange

The English language in highly indebted to other tongues. The geographer's vocabulary is full of words derived from Greek or Latin. From the Greek words *topos* (place) and *kosmos* (world) we get the English words *topography* and *cosmopolitan*. Where would the geographer be without these Latin-derived terms: *latitude, longitude, propinquity, province,* and *village*? From French, the English language borrowed words of culture: *boutique, buffet, chaise lounge, etiquette, feminine, lingerie, masculine,* and *naive*. From Arabic, the English language picked up *alcohol, coffee,* and *masquerade,* from German, angst, *hinterland* and *kindergarten,* and from Spanish, *machete*. In South Asia, the English encountered new household items and Hindi/Urdu words to describe them: *bungalow, cot, dungarees, pajamas, shampoo,* and *veranda*. They also picked up new religious ideas and words in South Asia: *avatar, guru, pundit, nirvana,* and *yoga*.

As a lingua franca, English words are constantly being borrowed and modified by other languages. Recent English exports include *airport, jeans, know-how,* and *sex-appeal*. In France, young people listen to the popular music of *les rappeurs*. German has borrowed *das Bodybuilding, der Computer,* and das *Mobiltelefon*. In South Africa, a Zulu-speaking university student might wear *izingilazi* (eyeglasses) while driving her *imotokali* (motor car) or working at her *ikhomp'yutha* (computer). Some words go full circle and pick up new meanings. From the English word *animation,* the Japanese got their word *animēshon,* which in turn was borrowed back into English and shortened to *anime*.

---

(21), the other leading current international languages. Global integration requires a common language, and English has taken on that role, being used in international air traffic control, diplomacy, two-thirds of scientific publishing, and most international academic conferences, While the dominance of English on the Internet is diminishing, it still leads with a 25 percent share of Internet users, followed by Mandarin at 19 percent and Spanish at 8 percent, No other language in history has assumed so important a role on the world scene.

Standard and Variant Languages

People who speak a common language such as English are members of a speech community, but membership does not necessarily imply linguistic uniformity. A speech community usually possesses both a standard language—comprising the accepted community norms of syntax, vocabulary, and pronunciation—and a number of more or less distinctive dialects reflecting the ordinary speech of regional, social, professional, or other subdivisions of the general population.

Standard Language

A dialect may become the standard language through connection with the speech of the most prestigious and most powerful members of the community. A rich literary tradition may help establish its primacy, and its adoption as the accepted written and spoken norm in administration, economic life, and education will solidify its position, eliminating deviant, nonstandard forms. The dialect that emerges as the basis of a country's standard language is often the one identified with its capital or center of power at the time of national development. Standard French is based on the dialect of the Paris region, a variant that assumed dominance in the latter half of the 12th century and was made the only official language in 1539. Castilian Spanish became the standard after 1492 with the Castile-led reconquest of Spain from the Moors and the export of the dialect to the Americas during the 16th century. Its present form, however, is a modified version associated not with Castile but with Madrid, the modern capital of Spain. Standard Russian is identified with the speech patterns of the former capital, St. Petersburg, and Moscow, the current capital. Modern Standard Chinese is based on the Mandarin dialect of Beijing. In England, Received Pronunciation— "Oxford English," the speech of educated residents of London and southeastern England and used by the British Broadcasting Corporation (BBC)—was until recently the accepted standard. It is now being modified, or replaced by a dialect called "Estuary English," which refers to the region around the lower River Thames in southeastern England.

Forces other than politics may affect language standardization. In its spoken form, Standard German is based on norms established and accepted in the theater, the universities, public speeches, and radio and television. The Classical or Literary Arabic of the Koran became the established norm from the Indian to the Atlantic Ocean. Standard Italian was derived from the Florentine dialect of the 13th and 14th centuries, which became widespread as the language of literature and economy.

In many societies, the official or unofficial standard language is not the dialect of home or daily life, and populations in effect have two languages. One is their regional dialect, which they employ with friends, at home, and in local community contacts; the other is the standard language used in more formal situations. In some cases, the contrast is great; regional variants of Arabic may be mutually unintelligible. Most Italians encounter Standard Italian for the first time in primary school. In India, the several totally distinct official regional languages are used in writing and taught in school but have no direct relationship to local speech; citizens must be bilingual to communicate with government officials who know only the regional language but not the local dialect.

Dialects

Just as no two individuals speak exactly the same, all but the smallest and most closely knit speech communities display recognizable speech variants called **dialects.** Simply ordering a carbonated beverage in different parts of the United States offers plenty of proof of regional speech variation (**Figure 5.9**). Vocabulary, pronunciation, rhythm, and the speed at which the language is spoken may set groups of speakers apart from one another and, to a trained observer, clearly mark the origin of the speaker. In George Bernard Shaw's play *Pygmalion,* on which the musical *My Fair Lady* was based, Henry Higgins—a professor of phonetics—is able to identify the London neighborhood of origin of a flower girl by listening to her vocabulary and accent. Sometimes, such variants are totally acceptable; in others, they mark the speaker as a social, cultural, or regional "outsider" or "inferior." In the song "Why Can't the English?" Professor Higgins proclaims, "An Englishman's way of speaking absolutely classifies him, the moment he begins to talk he makes some other Englishman despise him."

Shaw's play shows us dialects coexisting in space. Cockney and cultured English share the streets of London; Black English and Standard American are heard in the same schoolyards throughout the United States. In many societies, dialects denote social class and educational level. Speakers of higher socioeconomic status or educational achievement are most likely to follow the norms of their standard language; less-educated or lower-status persons or groups consciously distinguishing themselves from the mainstream culture are more likely to use the **vernacular**—nonstandard language or dialect native to the locale or adopted by the social group. In some instances, however, as in Germany and German-Switzerland, local dialects are preserved and prized as badges of regional identity.

Different dialects may be part of the speech patterns of the same person. Professionals discussing, for example, medical, legal, financial, or scientific matters with their peers employ vocabularies and formal modes of address and sentence structure that are quickly changed to informal colloquial speech when the conversation shifts to sports, vacations, or personal anecdotes. Even gender may be the basis for linguistic differences.

More commonly, we think of dialects in spatial terms. Speech is a geographic variable; each locale is apt to have its own, perhaps slight, language differences from neighboring places. Such differences in pronunciation, vocabulary, word meanings, and other language characteristics tend to accumulate with increasing distance. When they are mapped, they help define the **linguistic geography**—the character and spatial pattern of dialects and languages—of a speech community.

Every dialect feature has a territorial extent. The outer limit of its occurrence is a boundary line called an **isogloss** (the term *isophone* is used if the areal variant is marked by difference in sound rather than word choice). Each isogloss is a distinct entity, but taken together, isoglosses give clear map evidence of dialect regions that in turn may reflect paths and barriers for spatial interaction, long-established political borders, or past migration flows and diffusion of word choice and pronunciation.

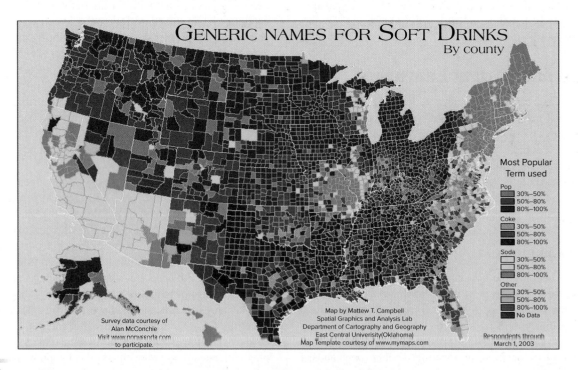

Figure 5.9 The generic term for a carbonated beverage varies regionally across the United States, from soda to pop to coke. Maps such as this visually record geographic variations in word usage or accent or pronunciation. Despite the presumed influence of national radio and television programs in promoting a "general" or "standard" American word usage, regional variations persist.

Source: M. Campbell and G. Plumb, Web Atlas of Oklahoma, East Central University.

Geographic or **regional dialects** may be recognized at different scales. On the world scene, for example, British, American, Indian, and Australian English are all acknowledged distinctive dialects of the same evolving language (see the feature "World Englishes"). Regionally, in Britain alone, one can recognize Southern British English, Northern British English, and Scottish English, each containing several more localized variants. Japanese has three recognized dialect groups; and China's Han ethnic group—making up more than 90 percent of the population of a country whose official language is Standard Mandarin—speak as many as 1,500 dialects, many almost entirely mutually incomprehensible.

Indeed, all long-established speech communities show their own structure of geographic dialects whose number and diversity tend to increase in areas longest settled and most fragmented and isolated. For example, the local speech of Newfoundland—isolated off the Atlantic coast of mainland Canada—retains much of the 17th-century flavor of the four West Counties of England from which the overwhelming majority of its settlers

came. Yet the isolation and lack of cultural mixing of the islanders have not led to a general Newfoundland "dialect"; settlement was coastal and in the form of isolated villages in each of the many bays and indentations. There developed from that isolation and the passage of time nearly as many dialects as there are bay settlements, with each dialect separately differing from Standard English in accent, vocabulary, sounds, and syntax. Isolation has led to comparable linguistic variation among the 47,000 inhabitants of the 18 Faeroe Islands between Iceland and Scotland; their Faeroese tongue has 10 dialects.

Dialects in the United States Mainland North America had a more diversified colonization than did Newfoundland, and its more mobile settlers mixed and carried linguistic influences away from the coast into the continental interior. Nonetheless, as early as the 18th century, three distinctive dialect regions had emerged along the Atlantic coast of the United States (**Figure 5.10**) and are evident in the linguistic geography of North America to the present day.

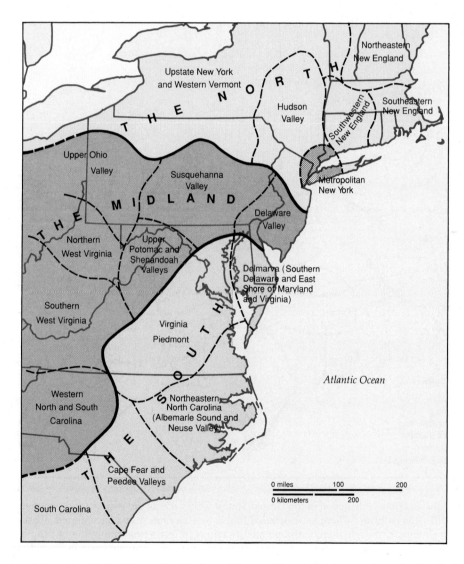

Figure 5.10 Dialect areas of the eastern United States. The Northern dialect and its subdivisions are found in New England and adjacent Canada (the international boundary has little effect on dialect borders in North America), extending southward to a secondary dialect area centered on New York City. Midland speech is found along the Atlantic Coast only from central New Jersey southward to central Delaware, but it spreads much more extensively across the interior of the United States and Canada. The Southern dialect dominates the East Coast from Chesapeake Bay South.

Source: Hans Kurath, A Word Geography of the Eastern United States *(Ann Arbor: University of Michigan Press, 1949).*

With the extension of settlement after the Revolutionary War, each of the dialect regions expanded inland. Speakers of the Northern dialect moved along the Erie Canal and the Great Lakes. Midland speakers from Pennsylvania traveled down the Ohio River, and the related Upland Southern dialect moved through the mountain gaps into Kentucky and Tennessee. The Coastal Southern dialect was less mobile, held to the east by plantation prosperity and the long resistance to displacement exerted by the Cherokees and the other Civilized Tribes (**Figure 5.11**).

Once across the Appalachian barrier, the diffusion paths of the Northern dialect were fragmented and blocked by the time they reached the Upper Mississippi. Upland Southern speakers spread out rapidly: northward into the old Northwest Territory, west into Arkansas and Missouri, and south into the Gulf Coast states. But the Civil War and its aftermath halted further major westward movements of the southern dialects. The Midland dialect, apparently so restricted along the Eastern Seaboard, became, almost by default, the basic form for much of the interior and West of the United States. It was altered and enriched there by contact with the Northern and Southern dialects, by additions from Native American languages, by contact with Spanish culture in the Southwest, and by contributions from the great non-English immigrations of the late 19th and early 20th centuries. Naturally, dialect subregions are found in the West, but their boundary lines—so clear in the eastern interior—become less distinct from the Plains States to the Pacific.

In areas with strong infusions of recently arrived Hispanic, Asian, and other immigrant groups, language mixing tends to accelerate language change as more and different non-English

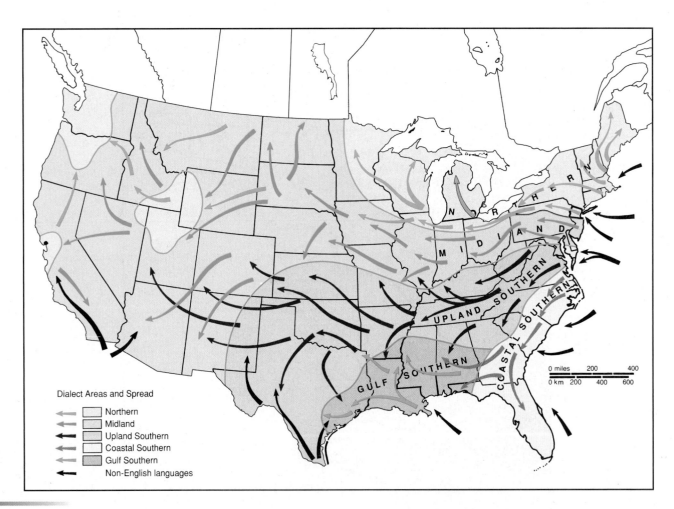

Figure 5.11 Speech regions and dialect diffusion in the United States. This generalized map is most accurate for the Eastern Seaboard and the easternmost diffusion pathways where most detailed linguistic study has been concentrated. West of the Mississippi River, the Midland dialect becomes dominant, though altered through modifications reflecting intermingling of peoples and speech patterns. Northern speech characteristics are still clearly evident in the San Francisco Bay area, brought there in the middle of the 19th century by migrants coming by sea around Cape Horn. Northerners were also prominent among the travelers of the Oregon Trail and the California Gold Rush.

Sources: Based on Raven I. McDavid, Jr. "The Dialects of American English," in W. Nelson Francis, The Structure of American English *(New York: Ronald Press, 1958); "Regional Dialects in the United States,"* Webster's New World Dictionary, *2nd College Edition (New York: Simon and Schuster, 1980); and Gordon R. Wood,* Vocabulary Change *(Carbondale: Southern Illinois University Press, 1971), Map 83, p. 358.*

World Englishes

Non-native speakers of English far outnumber those for whom English is the first language. Most of the more than 1.5 billion people who speak and understand at least some English as a second language live in Asia; they are appropriating the language and remaking it in regionally distinctive fashions to suit their own cultures, linguistic backgrounds, and needs.

It is inevitable that widely spoken languages separated by distance, isolation, and cultural differences will fragment into dialects that, in turn, evolve into new languages. Latin splintered into French, Spanish, Italian, and other Romance languages. English is similarly experiencing that sort of regional differentiation, shaped by the variant life-worlds of its far-flung community of speakers, and following the same path to mutual unintelligibility. Although Standard English may be one of or the sole official language of their countries of birth, millions of people around the world claiming proficiency in English or English as their national language cannot understand one another. Even teachers of English from India, Malaysia, Nigeria, or the Philippines, for example, may not be able to communicate in their supposedly common tongue—and find the cockney English of London utterly alien.

The splintering of spoken English is a fact of linguistic life and its offspring—called "World Englishes" by linguists—defy frequent attempts by governments to encourage adherence to international standards.

Singlish (Singapore English) and Taglish (a mixture of English and Tagalog, the dominant language of the Philippines) are commonly cited examples of the multiplying World Englishes, but equally distinctive regional variants have emerged in India, Malaysia, Hong Kong, Nigeria, the Caribbean, and elsewhere. One linguist suggests that beyond an "inner circle" of states where English is the first and native language—for example, Canada, Australia, United States—lies an "outer circle" where English is a second language (Bangladesh, Ghana, India, Kenya, Pakistan, Zambia, and many others) and where the regionally distinctive World Englishes are most obviously developing. Even farther out is an "expanding circle" of such states as China, Egypt, Korea, Nepal, Saudi Arabia, and others where English is a foreign language and distinctive local variants in common usage have not yet developed.

Each of the emerging varieties of English is, of course, "correct," for each represents a functioning mode of communication with mutual comprehension between its speakers. Each also represents a growing national cultural confidence and pride in the particular characteristics of the local varieties of English, and each regional variant is strengthened by local teachers who do not themselves have a good command of the standard language. Conceivably, these factors may mean that English will fragment into scores or hundreds of mutually unintelligible tongues. But equally conceivably, the worldwide influence of globalized business contacts, the Internet, worldwide American radio and television broadcasts, near-mandatory use of English in scientific publication, and the like will mean a future English more homogeneous and, perhaps, more influenced and standardized by American usage.

Most likely, observers of World Englishes suggest, both divergence and convergence will take place. While use of English as the major language of communication worldwide is a fact in international politics, business, education, and the media, increasingly, speakers of English learn two "dialects"—one of their own community and culture and one in the international context. While the constant modern world electronic and literary interaction between the variant regional Englishes make it likely that the common language will remain universally intelligible, it also seems probable that mutually incomprehensible forms of English will become entrenched as the language is taught, learned, and used in world areas far removed from contact with first-language speakers and with vibrant local economies and cultures independent of the Standard English community. "Our only revenge," said a French official, deploring the declining role of French within the EU, "is that the English language is being killed by all these foreigners speaking it so badly."

---

words enter the general vocabulary of all Americans. In many cases, those infusions create or perpetuate pockets of linguistically unassimilated peoples whose urban neighborhoods in shops, signage, and common speech bear little resemblance to the majority Anglo communities of the larger metropolitan area. Even as immigrant groups learn and adopt English, there is an inevitable retention of familiar words and phrases and, for many, the unstructured intermixture of old and new tongues into such hybrids as "Spanglish" among Latin and Central American immigrants, for example, or "Runglish" among the thousands of Russian immigrants of the Brighton Beach district of New York City.

Local dialects and accents do not display predictable patterns of consistency or change. Regional differences still persist in the face of the presumed leveling effects of the mass media. The distinct evidence of increasing contrasts between the speech patterns and accents of Chicago, New York, Birmingham, St. Louis, and other cities is countered by reports of decreasing local dialect pronunciations in large southern cities such as Dallas and Atlanta that have experienced major influxes of Northerners. Other studies find that some regional accents are fading in small towns and rural areas, presumably because mass media standardization is more influential than local dialect reinforcement as local populations decline and physical and social mobility increase.

Pidgins and Creoles

Language is rarely a total barrier in communication among peoples, even those whose native tongues are mutually incomprehensible. **Bilingualism** or **multilingualism** may permit skilled linguists to communicate in a jointly understood third language, but long-term contact may require the creation of new language—a pidgin—learned by both parties. In the past 400 years, more than 100 new languages have been created out of the global mixing of peoples and cultures. A pidgin is an amalgam of languages, usually a simplified form of one, such as English or French, with words borrowed from another one, perhaps a non-European local language. In its original form, a pidgin is not the mother tongue of any of its speakers; it is a second language for everyone who uses it, a language generally restricted to such specific functions as commerce, administration, or work supervision. For example, given the variety of languages spoken among the some 270 ethnic groups of the Democratic Republic of the Congo, a special tongue called Lingala, a hybrid of Congolese dialects and French, emerged to permit, among other things, communication with soldiers drawn from across the country.

Pidgins are initially characterized by a highly simplified grammatical structure and a limited vocabulary, adequate to express basic ideas but not complex concepts. For example, fanagalo, a pidgin created earlier in South Africa's gold mines to allow spoken communication between workers of different tribes and nationalities and between workers and Afrikaner bosses, is being largely abandoned. Since the mid-1990s, workers have increasingly been schooled in basic English as fanagalo—which lacks the vocabulary to describe how to operate new, automated mining machinery and programmable winches with their multiple sensors and warnings in English—became less useful. In South America, when the Portuguese arrived five centuries ago, the challenge of communicating with conquered native peoples speaking more than 700 languages led Jesuit priests to concoct a pidgin mixture of Indian, Portuguese, and African words they called "lingua geral," or the "general language." As a living language, lingua geral gradually died out in most of Brazil but has been retained and adopted as an element of their cultural identity by some isolated Indian groups that have lost their own original mother tongue. If a pidgin becomes the first language of a group of speakers—who may have lost their former native tongue through disuse—a **creole** language has evolved. In their development, creoles invariably acquire a more complex grammatical structure and enhanced vocabulary.

Creole languages have proved useful integrative tools in linguistically diverse areas; several have become symbols of nationhood. Swahili, a pidgin formed from a number of Bantu dialects with major vocabulary additions from Arabic, originated in the coastal areas of East Africa and spread inland first by Arab ivory and slave caravans and later by trade during the period of English and German colonial rules. When Kenya and Tanzania gained independence, they made Swahili the national language of administration and education. Other examples of creolization are Afrikaans (a pidginized form of 17th-century Dutch used in the Republic of South Africa); Haitian Creole (the language of Haiti, derived from the pidginized French used in the slave trade); and Bazaar Malay (a pidginized form of the Malay language, a version of which is the official national language of Indonesia).

Lingua Franca

A **lingua franca** is an established language used habitually for communication by people whose native tongues are mutually incomprehensible. For them, it is a second language, one learned in addition to the native tongue. Lingua franca, literally "Frankish tongue," was named from the dialect of France adopted as their common tongue by the Crusaders assaulting the Muslims of the Holy Land. Later, it endured as a language of trade and travel in the eastern Mediterranean, useful as a single tongue shared in a linguistically diverse region.

Between 300 BCE and 500 CE, the Mediterranean world was unified by Common Greek. Later, Latin became a lingua franca, the language of empire and, until replaced by the vernacular European tongues, of the Church, government, scholarship, and the law. Outside the European sphere, Aramaic served the role from the 5th century BCE to the 4th century CE in the Near East and Egypt; Arabic followed Muslim conquest as the unifying language of that international religion after the 7th century. Mandarin Chinese and Hindi in India have a lingua franca role in their linguistically diverse countries. The immense linguistic diversity of Africa has made regional lingua francas such as Swahili necessary (**Figure 5.12**), and in a world of increased spatial interaction, English increasingly serves as the lingua franca of globalization.

Official Languages

Governments may designate a single tongue as a country's official **language,** the required language of instruction in the schools and universities, government business, the courts, and other official and semiofficial public and private activities. In societies in which two or more languages are in common use (multilingualism), such an official language may serve as the approved national lingua franca, guaranteeing communication among all citizens of differing native tongues.

In many immigrant societies, such as the United States, only one of the many spoken languages may have implicit or official government sanction. Many Americans are surprised to discover, however, that English is not the official language of the United States. Nowhere does the Constitution provide for an official language, and no federal law specifies one. The country was built by a great diversity of immigrants and in 2016, 21 percent of U.S. households spoke a language other than English at home. A majority of states offer driving tests in foreign languages, multilingual ballots are provided in many jurisdictions, and many school districts offer bilingual teaching. There have been a number of failed attempts to add a Constitutional amendment making English the official national

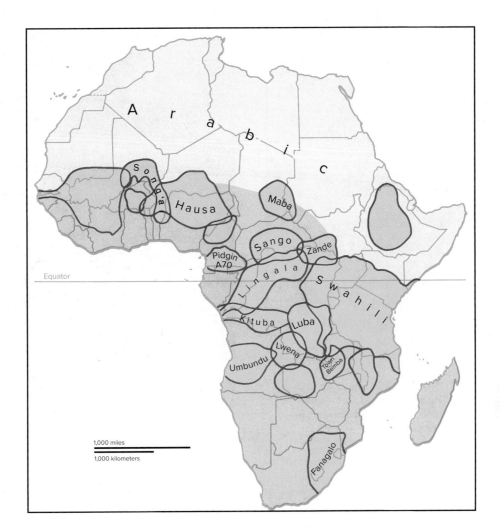

Figure 5.12 Lingua francas of Africa. The importance and extent of competing lingua francas in sub-Saharan Africa change over time, reflecting the spread of populations and the relative economic or political stature of speakers of different languages. In many areas, an individual may employ different lingua francas, depending on activity: dealing with officials, trading in the marketplace, conversing with strangers. Among the elite in all areas, the preferred lingua franca is apt to be a European language. When a European tongue is an official language (Figure 5.13) or the language of school instruction, its use as a lingua franca is more widespread. Throughout northern Africa, Arabic is the usual lingua franca for all purposes.

Source: Adapted from Bernd Heine, Status and Use of African Lingua Francas *(Munich: Weltforum Verlag; and New York: Humanities Press, 1970).*

language of the United States. In general, however, immigrants have been eager to learn English as they enter mainstream American life.

Nearly every country in linguistically complex sub-Saharan Africa has selected a European language—usually that of their former colonial governors—as an official language (**Figure 5.13**), only rarely designating a native language or creole as an alternate official tongue. Indeed, less than 10 percent of the population of sub-Saharan Africa live in countries with any indigenous African tongue given official status. Nigeria has some 350 clearly different languages and is dominated by three of them: Hausa, Yoruba, and Ibo. English is not a native tongue for any Nigerian, yet throughout the country, English is the sole language of instruction and the sole official

language. Effectively, all Nigerians must learn a foreign language before they can enter the mainstream of national life. Most Pacific Ocean countries, including the Philippines (with between 80 and 110 Malayo-Polynesian languages) and Papua New Guinea (with more than 850 distinct Papuan tongues), have a European language as at least one of their official tongues.

Increasingly, the diffusion of popular English words and phrases in everyday speech, press, and television threatens the "purity" of other European languages. So common has the adoption of phrases such as *Das Laptop* and *le hot dog* that some nearly new language variants are now recognized: franglais in France and Denglisch in Germany are the best-known examples. Both have spurred resistance movements

Figure 5.13 Europe in Africa through official languages. Both the linguistic complexity of sub-Saharan Africa and the colonial histories of its present political units are implicit in the designation of a European language as the sole or joint "official" language of the different countries.

Figure 5.14 French/English bilingual stop sign in Ottawa, Ontario. Although stop is considered an acceptable French word by the Quebec Board of the French language, a variety of bilingual combinations can be observed on stop signs in Canada, including some featuring French or English alongside an indigenous language.

©Mark Bjelland

from officially sanctioned language monitors of, respectively, the French Academy and the Institute for the German Language. Japan, Latvia, Poland, and Spain are among the states seeking to preserve the purity of their official languages from contamination by English or other borrowed foreign words.

In some countries, multilingualism has official recognition through designation of more than a single state language. Canada and Finland, for example, have two official languages (bilingualism), reflecting rough equality in numbers or influence of separate linguistic populations comprising a single country. In a few multilingual countries, more than two official languages have been designated. Bolivia and Belgium have three official tongues, and Singapore has four. South Africa's constitution designates 11 official languages, and India gives official status to 18 languages at the regional, though not at the national, level.

Multilingualism may reflect significant cultural and spatial divisions within a country. In Canada, the Official Languages Act of 1985 accorded French and English equal status as official languages of Parliament and government throughout the nation (**Figure 5.14**). French-speakers are concentrated in the province of Quebec, however, and constitute a culturally distinct population sharply divergent from the English-speaking majority of other parts of Canada (**Figure 5.15**). Within sections of Canada, even greater linguistic diversity is recognized; the legislature of the Northwest Territories, for example, has eight official languages—six native, plus English and French.

Few countries remain purely **monolingual.** Past and recent movements of peoples as colonists, refugees, or migrants have assured that most of the world's countries contain linguistically mixed populations. Maintenance of native languages among such populations is not assured, of course. Where numbers are small or pressures for integration into the dominant culture are strong, immigrant and aboriginal (native) linguistic minorities tend to adopt the majority or official language for all purposes. On the other hand, isolation and relatively large numbers of speakers may serve to preserve native tongues. In Canada, for example, aboriginal languages with large populations of speakers—Cree, Ojibwe, and Inuktitut—are well maintained in their areas of concentration (respectively, northern Quebec, the northern prairies, and Nunavut). In contrast, much smaller language groups in southern Canada have a much lower ratio of retention among native speakers.

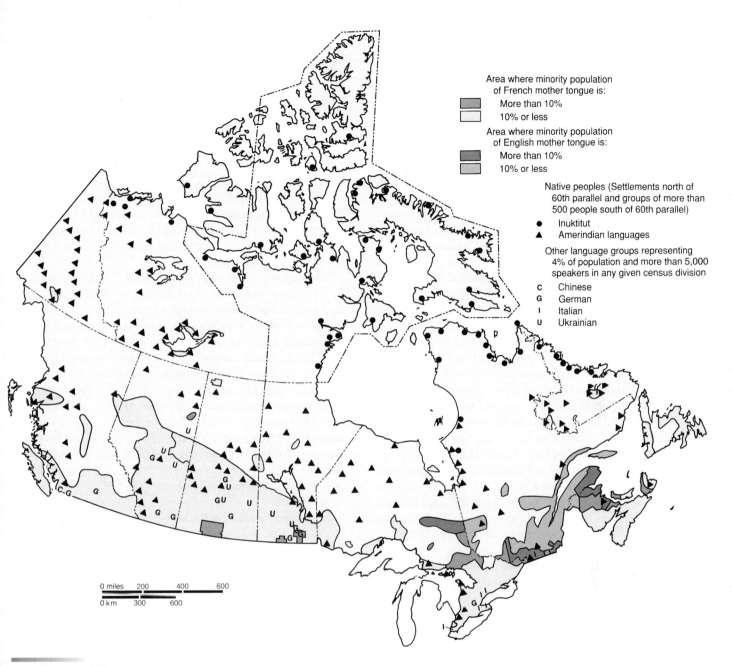

Figure 5.15 Bilingualism and diversity in Canada. The map shows areas of Canada that have a minimum of 5,000 inhabitants and include a minority population identified with an official language.

Source: Commissioner of Official Languages, Government of Canada.

5.2 Language, Territoriality, and Cultural Identity

The designation of more than one official language does not always satisfy the ambitions of linguistically distinct groups for recognition and autonomy. Language is a defining characteristic of ethnic and cultural identity. Languages contain unique expressions of a culture's heritage. Thus, many groups from around the world believe that losing their linguistic identity would be the worst and final evidence of discrimination and subjugation.

Language has often been the focus of separatist movements among spatially distinct, minority linguistic groups.

In Europe, France, Spain, Britain—and Yugoslavia and the Soviet Union before their dismemberment—experienced such language "revolts" and acknowledged, sometimes belatedly, the local concerns that they express. After 1970, France dropped its ban on teaching regional tongues and Spain relaxed its earlier total rejection of Basque and Catalan as regional languages, recognizing Catalan as a co-official language in its home region in northeastern Spain. In the United Kingdom, parliamentary debates concerning greater regional autonomy have resulted in bilingual road and informational signs in Wales, a publicly supported Welsh-language television channel, and compulsory teaching of Welsh in all schools in Wales.

In fact, throughout Europe nonofficial native regional languages have increasingly not only been tolerated, but encouraged as a buffer against the loss of regional institutions and traditions threatened by a multinational "superstate" under the EU. The Council of Europe, a 41-nation organization promoting democracy and human rights, has adopted a charter pledging encouragement of the use of indigenous languages in schools, the media, and public life. The language charter acknowledges that cultural diversity is part of Europe's wealth and heritage and that its retention strengthens, not weakens, the separate states of the continent and the larger European culture realm as a whole. In North America, the designation of French as the official language of the Canadian province of Quebec reinforces and sustains the distinctive cultural and territorial identity that is so important to the *Québecois*. Quebec's language laws enforce a French appearance to the landscape, requiring that billboards and commercial signs give priority to the French language.

Many other world regions have continuing conflict based on language. Language has long been a divisive issue in South Asia, for example, leading to wars in Pakistan and Sri Lanka and periodic demands for secession from India by southern states such as Tamil Nadu, where the Dravidian Tamil language is defended as an ancient tongue as worthy of respect as the Indo-European official language, Hindi. In Russia and several other successor states of the former USSR (which housed some 200 languages and dialects), linguistic diversity forms part of the justification for local separatist movements, as it did in the division of Czechoslovakia into Czech- and Slovak-speaking successor states and in the violent dismemberment of former Yugoslavia.

Even more than in its role in ethnic identity and separatism, language embodies the culture complex of a people, reflecting environment, experience, and shared understandings. Arabic has 80 words related to the camel, an animal on which a regional culture relied for food, transport, and labor, and Japanese contains more than 20 words for various types of rice. Russian is rich in terms for ice and snow, indicative of the prevailing climate of its linguistic cradle; Hawaiians reportedly have 108 words for sweet potato, 65 for fishing net, and 47 for banana. The 15,000 tributaries and subtributaries of the Amazon River have obliged the Brazilians to enrich Portuguese with words that go beyond *river*. Among them are *paraná* (a stream that leaves and reenters the same river), *igarapé* (an offshoot that runs until it dries up), and *furo* (a waterway that connects two rivers).

Most—perhaps all—cultures display subtle or pronounced differences in ways males and females use language. Most have to do with vocabulary and with grammatical forms peculiar to individual cultures. For example, among the Caribs of the Caribbean, the Zulu of Africa, and other groups, men have words that women through custom or taboo are not permitted to use, and women have words and phrases that the men never use "or they would be laughed to scorn," an informant reports. The greater and more inflexible the difference in the social roles of men and women in a particular culture, the

greater and more rigid are the observed linguistic differences between the sexes.

5.3 Language on the Landscape: Toponymy

Toponyms—place names—are language on the landscape, the record of past inhabitants whose naming practices endure, perhaps corrupted and disguised, as reminders of their existence. Toponymy is the study of place names, a special interest of cultural geography. It is also a powerful tool of historical geography, for place names remain a part of the cultural landscape long after the name givers have passed from the scene.

In England, for example, place names ending in *chester* (as in Winchester and Manchester) evolved from the Latin *castra,* meaning "camp." Common Anglo-Saxon suffixes for tribal and family settlements were *ing* (people or family) and *ham* (hamlet or, perhaps, meadow), as in Birmingham or Gillingham. Norse and Danish settlers contributed place names ending in *thwaite* ("meadow") and others denoting such landscape features as *fell* (an uncultivated hill) and *beck* (a small brook). The Celts, present in Europe for more than 1,000 years before Roman times, left their tribal names in corrupted form on territories and settlements taken over by their successors. The Arabs, sweeping out from Arabia across North Africa and into Iberia, left their imprint in place names to mark their conquest and control. *Cairo* means "victorious," *Sudan* is "the land of the blacks," and *Sahara* is "wasteland" or "wilderness." In Spain, a corrupted version of the Arabic *wadi*, "watercourse," is found in *Guadalajara and Guadalquivir.*

In the New World, many groups placed names on landscape features and new settlements. In doing so, they remembered their homes and homelands, honored their monarchs and heroes, borrowed and mispronounced from rivals, followed fads, recalled the Bible and Classical Greek and Roman places, and adopted and distorted Amerindian names. Homelands were recalled in New England, New France, or New Holland; settlers' hometown memories brought Boston, New Bern, New Rochelle, and Cardiff from England, Switzerland, France, and Wales. Monarchs were remembered in Virginia for the Virgin Queen Elizabeth, Carolina for one English king, Georgia for another, and Louisiana for a king of France. Washington, D.C.; Jackson, cities in Mississippi and Michigan; Austin, Texas; and Lincoln, Illinois memorialized heroes and leaders. Names given by the Dutch in New York were often distorted by the English; Breukelyn, Vlissingen, and Haarlem became Brooklyn, Flushing, and Harlem. French names underwent similar twisting or translation, and Spanish names were adopted, altered, or, later, put into such bilingual combinations as Hermosa Beach. Amerindian tribal names—Yenrish, Maha, and Kansa— were modified, first by French and later by English speakers—to Erie, Omaha, and Kansas. A faddish "Classical Revival" after the Revolution gave us Troy, Athens, Rome, Sparta, and other ancient town names and later spread them across the country. Bethlehem, Ephrata, Nazareth, and Salem came from the Bible. Popular place names diffused westward across the United States frontier (**Figure 5.16**).

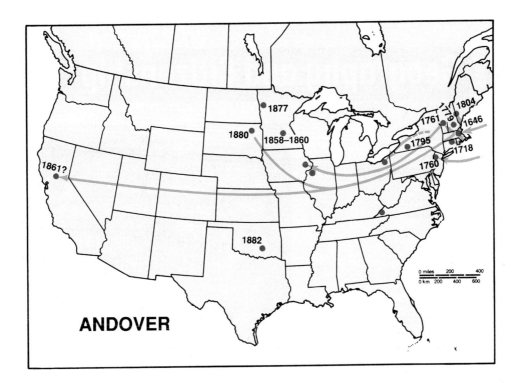

ANDOVER

Figure 5.16 Migrant Andover. Place names in a new land tend to be transportable, carried to new locales by migrating town founders. They are a reminder of the cultural origins and diffusion paths of settlers. Andover, a town name from England, was brought to New England in 1646 and later carried westward.

Source: American Name Society.

Place names, whatever their language of origin, frequently consist of two parts: generic (classifying) and specific (modifying or particular). *Big River* in English is translated into *Rio Grande* in Spanish, *Mississippi* in Algonquin, and *Ta Ho* in Chinese. The *order* of generic and specific, however, may alter among languages and give a clue to the group originally bestowing the place name. In English, the specific usually comes first: *Hudson River, Bunker Hill, Long Island,* and so on. When, in the United States, we find *River Rouge* or *Isle Royale,* we also find evidence of French settlement—the French reverse the naming order. Some generic names can be used to trace the migration paths across the United States of the three Eastern dialect groups (see Figure 5.10). Northern dialect settlers tended to carry with them their habit of naming a community and calling its later neighbors by the same name modified by direction—Lansing and East Lansing, for example. *Brook* is found in the New England settlement area, *run* is from the Midland dialect, and *bayou* and *branch* are from the Southern area.

European colonists and their descendants gave new place names to a physical landscape already extensively named by indigenous peoples, effectively erasing signs of indigenous presence. In other cases, names were adopted, but often shortened, altered, or—certainly—mispronounced. The vast territory that local Amerindians called *Mesconsing,* meaning "the long river," was recorded by Lewis and Clark as *Quisconsing,* later to be further distorted into *Wisconsin. Milwaukee* and *Winnipeg,*

Potomac and *Niagara, Adirondack, Chesapeake, Shenandoah,* and *Yukon;* the names of 27 of the 50 United States; and the present identity of thousands of North American places and features, large and small, had their origin in Native American languages.

In northern Canada, Indian and Inuit (Eskimo) place names are returning. The town of Frobisher Bay has reverted to its Eskimo name *Iqaluit* ("place of the fish"); Resolute Bay has become *Kaujuitok* ("place where the sun never rises") in Inuktitut, the lingua franca of the Canadian Eskimos; the Jean Marie River has returned to *Tthedzehk'edeli* ("river that flows over clay"), its earlier Slavey name. These and other official name changes reflect a decision that community preference will be the standard for all place names, no matter how entrenched European versions might be.

Decisions to rename places recognize the importance of language as a powerful unifying thread in peoples' culture complexes (see the feature "Changing Toponyms."). Language may serve as a key marker of ethnicity and a fiercely defended symbol of the history and individuality of a distinctive social group. Hispanic Americans demand the right of instruction in their own language, and Basques wage civil war to achieve a linguistically based separatism. Indian states were adjusted to coincide with language boundaries, and the Polish National Catholic Church was created in America, not Poland, to preserve Polish language and culture in an alien environment.

Geography and Citizenship

Changing Toponyms

Toponyms are powerful. The act of naming a place is to define its identity, at least in part. Toponyms demonstrate the power of the person or country who assigned the name and the legitimacy of the person for whom the place was named. Local governments have discovered that they can raise revenues by selling the naming rights to important, publicly owned facilities such as stadiums, public buildings, and transit stations. In Dubai, in the United Arab Emirates, the names of several transit stations have been sold. Instead of being named for the streets or neighborhood where they are located, the stations carry corporate names such as First Gulf Bank station. The names of sports stadium once reflected the name of the home team such as Dodger Stadium or their location such as Riverfront Stadium in Cincinnati and Fenway Park in Boston. More recently, stadium names have become market commodities, reflecting corporate control over popular culture and the cultural landscape. For example, Three Rivers Stadium in Pittsburgh was replaced by PNC Park, named for a bank.

Toponyms reflect political power. Grosse Frankfurter Strasse, one of the most important streets in Berlin, Germany, was renamed Stalinallee by the communists in 1949, and then renamed Karl Marx Allee in 1961 when Stalin fell out of fashion. Proposals to revert to the original street name have been discussed. Colonial powers imposed a new language and often erased signs of indigenous culture. The post-colonial era has been marked by a new wave of name changes. In India, rising post-colonial pride resulted in decisions to undo anglicized city names: Bombay changed to *Mumbai*, Madras to *Chennai*, and Calcutta to *Kolkata*.

The United States Board on Geographic Names is charged with designating official names of geographic features. Recently, it has received many proposals to eliminate derogatory or racist names and recover indigenous names. Dozens of toponyms using the racial epithet *squaw* have been changed. Utah's Negro Bill Canyon was renamed Grandstaff Canyon. The most

Figure 5B The name of Minneapolis' Lake Calhoun changed to Bde Maka Ska. ©*Mark Bjelland*

significant recent action was the official name change of North America's highest mountain from Mount McKinley to *Denali*. The name *Mount McKinley* was given to commemorate President McKinley who was assassinated in 1901 but had never travelled to Alaska and had no meaningful connections to the mountain. The name *Denali* means "the tall one" in the indigenous Athabascan language and had been used for centuries. Thus, in keeping with the wishes of Alaskans who had been petitioning for the change since the 1970s, the name of the mountain was officially changed in 2015.

Minneapolis' Lake Calhoun was named in honor of Secretary of War John C. Calhoun who commissioned the 1817 army survey that first mapped the lake. Calhoun later served as Vice President of the United States where he was an outspoken defender of slavery. Almost two centuries after the lake was named in his honor, public pressure to eliminate symbols of racism began to focus on Lake Calhoun, even though most Minneapolis residents had no idea where the name came from. Activists succeeded in convincing the City of Minneapolis to revert to the original indigenous Dakota name—*Bde Maka Ska* which means "white earth lake."

Considering the Issues

1. Have any stadiums or other public facilities in your city or sold their naming rights? What do the new names suggest about who holds cultural and economic power? Where would you draw the line in allowing public facilities and landscape features to have their naming rights sold to the highest bidder?

2. Birch Creek in Alaska recently reverted to its indigenous name K'iidòotinjik River. Some detractors complain that non-English names are difficult to spell and pronounce. Is maintaining language diversity and indigenous cultural identity worth the inconvenience to English speakers who must learn new, unfamiliar names?

3. Minneapolis' Lake Calhoun Parkway is home to many apartments, houses, and businesses. Is eliminating a symbol of racism worth the costs of changing road signs, maps, and street addresses?

4. Some defenders of the status quo argue that keeping old toponyms is an important part of remembering the past in order to avoid repeating it, even if that past was racist, sexist, oppressive, or violent. Do you think it is important to keep offensive toponyms in order to remember the past or is eliminating those toponyms an important step in overcoming the evils of the past?

5.4 Religion and Culture

Unlike language, which is an attribute of all people, religion varies in its cultural role—dominating some societies, unimportant or denied totally in others. All societies have belief systems—common understandings, expectations, and objects held in high regard—that unite their members and set them off from other culture groups. Such a value system is termed a **religion** when it involves practices of formal or informal worship and addresses questions of meaning and ultimate significance. In a more inclusive sense, religion may be viewed as a unified system of beliefs and practices that join all those who adhere to them into a single moral community. Religion can intimately affect all facets of a culture. Religious belief is by definition an element of the ideological subsystem; formalized and organized religion is an institutional expression of the sociological subsystem. And religious beliefs strongly influence attitudes toward the use and development of the technological subsystem.

Nonreligious belief and value systems exist—humanism or Marxism, for example—that are just as binding on the societies that espouse them as are more traditional religious beliefs. Even societies that largely reject religion—that are officially atheistic or secular—may be strongly influenced by traditional values and customs set by predecessor religions, in days of work and rest, for example, or in legal principles.

Because religions are formalized views about the relation of the individual to this world and to the sacred, each carries a distinct conception of the meaning and value of this life, and most contain rules for living. These rules for living become interwoven with the traditions of a culture. For example, the Muslim observance of daily prayers led to a defining visual feature of Muslim countries—minaret towers on mosques from which the call to prayer is issued five times each day (see **Figure 5.25**). Economic patterns may be intertwined with past or present religious beliefs. Traditional restrictions on food and drink may affect the kinds of animals that are raised or avoided, the crops that are grown, and the importance of those crops in the daily diet. Occupational assignment in the Hindu caste system is in part religiously supported. Religious beliefs and practices may justify social inequalities or work toward their elimination through social services and activism.

In many countries, there is a state religion—that is, religious and political structures are intertwined. Islam is the most common official state religion. By their official names, the Islamic Republic of Iran and the Islamic Republic of Pakistan proclaim their close links between religion and government. Despite declining support, many European countries still have an official state Christian church. Buddhism has been the state religion in Myanmar, Laos, and Thailand. Despite Indonesia's overwhelming Muslim majority, that country sought and formerly found domestic harmony by recognizing five official religions and a state ideology—*pancasila*—whose first tenet is belief in one god.

The landscape imprint of religion is sometimes obvious, sometimes subtle (see feature "Religious Attire in Secular Spaces"). The structures of religious worship—temples, churches, mosques, stupas, or cathedrals—landscape symbols such as shrines or statues, and such associated land uses as monasteries may lend a regionally distinctive cultural character to an area. "Landscapes of death" may also be visible regional variables, for different religions and cultures dispose of their dead in different manners. Cemeteries are significant and reserved land uses among Christians, Jews, and Muslims who typically bury their deceased with headstones or other markers and monuments to mark graves. Egyptian pyramids or elaborate mausoleums like the Taj Mahal are more grandiose structures of entombment and remembrance. On the other hand, Hindus and Buddhists have traditionally cremated their dead and scattered their ashes, leaving no landscape evidence or imprint.

Some religions make a subtle cultural stamp on the landscape through recognition of particular natural or cultural features as sacred. These **sacred places** are infused with special religious significance and are sites of reverence, fear, pilgrimage, or worship. Grottos, lakes, single trees or groves, rivers such as the Ganges or Jordan, or special mountains or hills, such as Mount Ararat or Mount Fuji, are examples of sacred places unique to specific religions. In some cases, places such as the Old City of Jerusalem are sacred to more than one religion (**Figure 5.17**).

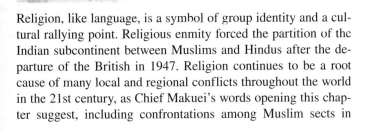

patterns of religion

Religion, like language, is a symbol of group identity and a cultural rallying point. Religious enmity forced the partition of the Indian subcontinent between Muslims and Hindus after the departure of the British in 1947. Religion continues to be a root cause of many local and regional conflicts throughout the world in the 21st century, as Chief Makuei's words opening this chapter suggest, including confrontations among Muslim sects in Lebanon, Iran, Iraq, and Syria; Muslims and Jews in Palestine; Christians and Muslims in the Philippines and Nigeria, Buddhists and Muslims in Myanmar; and Buddhists, Hindus, and Muslims in Sri Lanka. More peacefully, in the name of their beliefs, American Amish, Hutterite, Shaker, and other religious communities have isolated themselves from the secular world and pursued their own ways of life.

Figure 5.17 The Old City of Jerusalem in Israel contains sites sacred to Jews, Christians, and Muslims. In the foreground, Jews gather to pray at the Western Wall, also known as the Wailing Wall, a remnant of the ancient Jewish temple which was the center of Jewish religious life before it was destroyed by the Romans in 70 CE. For Christians, Jerusalem is sacred because Jesus was crucified just outside the city walls. During the Middle Ages, Christians fought the Crusades against Muslims over control of Jerusalem. The dome shown at the top of the photo is part of al-Haram al-Shariff (the Noble Sanctuary), a site sacred to Muslims because it marks where Muhammed is believed to have ascended to heaven.

©*McGraw-Hill Education/Mike Camille, photographer*

Classification of Religion

Religions are cultural innovations. They may be unique to a single culture group, closely related to the faiths professed in nearby areas, or derived from belief systems in a distant location. Although interconnections among religions can often be discerned—as Christianity and Islam both trace descent from Judaism—family groupings are not as useful to us in classifying religions as they were in studying languages. A distinction between **monotheism,** belief in a single deity, and **polytheism,** belief in many gods, is frequent, but less important to understanding the geography of religion. Simple territorial categories have been offered recognizing origin areas of religions: Western versus Eastern, for example, or African, Far Eastern, or Indian. With proper detail such distinctions may inform us where particular religions had their roots but do not reveal their courses of development, paths of diffusion, or current distributions.

The geographer's approach to religion is different from that of theologians or historians. We are not so concerned with specific doctrines or important figures. Rather, we are interested in religions' patterns and processes of diffusion, with the spatial distributions they have achieved, and with the impact and imprint of different religious systems on the landscape. To explain

their patterns of spatial diffusion, geographers have found it useful to categorize religions as *universalizing,* ethnic, or *tribal (traditional).*

Christianity, Islam, and Buddhism are the major **universalizing religions,** faiths that claim applicability to all humans and proselytize; that is, they seek to transmit their beliefs through missionary work and conversion. Membership in universalizing religions is open to anyone who chooses to make some sort of symbolic commitment, such as the initiation ritual of baptism for Christians. No one is excluded because of nationality, ethnicity, or previous religious belief.

Ethnic religions have strong territorial and cultural group identification. One usually becomes a member of an ethnic religion by birth or by adoption of a complex lifestyle and cultural identity, not by simple declaration of faith. These religions do not usually proselytize, and their members form distinctive closed communities identified with a particular ethnic group or political unit. An ethnic religion—for example, Judaism, Indian Hinduism, or Japanese Shinto—is an integral element of a specific culture; to be part of the religion is to be immersed in the totality of the culture.

Tribal, or **traditional religions,** are special forms of ethnic religions distinguished by their small size, their unique identity

Religious Attire in Secular Spaces

Religious believers express their faith in their daily lives at multiple geographic scales. While geographic patterns of religion are often mapped at national or global scales, finer geographic scales such as the home and the body are also important. For example, at the scale of the body, distinctive dress, dietary practices, and sexual mores can be ways of expressing one's religious devotion. Distinctive patterns of dress can express religious values such as modesty, demonstrate an adherent's religious devotion, or highlight a group's separation from the wider society. Orthodox Jews have adopted hats, beards, and dark coats for men and long sleeves and head coverings for women. Many Sikh men demonstrate their faith publicly through their personal appearance, keeping their hair uncut and wrapped in a turban. Muslim men often wear beards, following Muhammed's concern that believers distinguish themselves from nonbelievers. The Amish are a pacifist Christian group whose men wear beards without moustaches because of a historic association between moustaches and the German military.

Islamic attire is a powerful symbol of religious and national identities. The wearing of veils by Muslim women is traced to passages in the Quran calling for a woman's beauty to be disguised outside the home. Veiling practices vary from the headscarf or *hijab*, to the full face veil, or *niqab,* to the full body covering of the traditional Afghan *burqa*. The Muslim theocracy of Iran requires women to wear the hijab in public while in Saudi Arabia women are required to wear the hijab and a long black cloak or *abaya*.

Critics from secular societies often decry the control of women's bodies by laws mandating veiling. However, secular societies that celebrate personal freedoms have sometimes implemented strict bans on veiling. In Turkey, secular governments have treated veiling as a backward practice and for decades banned the wearing of Islamic headscarves in public office buildings and universities. Several western European countries with growing Muslim minority populations have passed laws restricting veiling in public spaces. European politicians have expressed concern that the veil symbolizes female subservience and the lack of cultural assimilation of new immigrants. In 2004, France banned the wearing of conspicuous religious symbols in public schools, arguing that schools were to be religiously neutral secular spaces. Banned were the hijab, large Christian crosses, and Jewish skullcaps. In 2011, France and Belgium implemented bans on the niqab in public spaces. Under France's ban, the niqab could be worn only in the private home, in a house of worship, or as a passenger in a private car. A Dutch law restricts the niqab on public transportation and in public spaces.

Geographers have observed that veiling is a dynamic spatial practice. Muslim women vary their veiling practices depending upon their geographic location. Further, the act of wearing a veil takes on different meanings at school, at work, in the street, at home, or at mosque. A thriving global Islamic fashion industry has emerged to produce its own fashion magazines and stylish hijab and abaya designs. Thus, for many European Muslims, wearing a veil is a personal choice that may express religious devotion, make a political statement, or merely display self confidence in one's identity as a modern, European Muslim woman.

---

with localized culture groups not yet fully absorbed into modern society, and their close ties to nature. **Animism** is the name given to their belief that life exists in all objects, from rocks and trees to lakes and mountains, or that such inanimate objects are the abode of the dead, spirits, or gods. **Shamanism** is a form of tribal religion that involves community acceptance of a *shaman,* a religious leader, healer, and worker of magic who, through special powers, can intercede with and interpret the spirit world.

5.5 Patterns and Flows

The nature of the different classes of religions is reflected in their distributions over the world and in their number of adherents. Universalizing religions tend to be expansionary, carrying their message to new peoples and lands. Ethnic religions, unless their adherents are dispersed, tend to be regionally confined or to expand only slowly and over long periods. Tribal religions tend to contract spatially as their adherents are incorporated into modern society and converted by proselytizing faiths.

As we expect in human geography, the map records only the latest stage of a constantly changing cultural reality. While established religious institutions tend to be conservative and resistant to change, religion is a dynamic culture trait. Personal and collective beliefs may alter in response to developing individual and societal needs and challenges. Religions may be imposed by conquest, adopted by conversion, and recede or persevere in the face of surrounding pluralism, hostility or indifference.

The World Pattern

The world map of religious affiliation offers a global view of the spatial distribution of major religions (**Figure 5.18**).

More than half of the world's population adheres to one of the major universalizing religions: Christianity, Islam, or Buddhism. Of these three, Figure 5.18 indicates, Christianity and Islam are most widespread; Buddhism is largely an Asian religion. Hinduism, the largest ethnic faith, is essentially confined to the Indian subcontinent, showing the spatial restriction characteristic of most ethnic and traditional religions even when found outside of their homeland area. Small Hindu emigrant communities in Africa, southeastern Asia, Britain, or the United States, for example, tend to remain isolated even in densely crowded urban areas. Although it is not localized beyond Israel, Judaism is also included among the ethnic religions because of its identification with a particular people and cultural tradition.

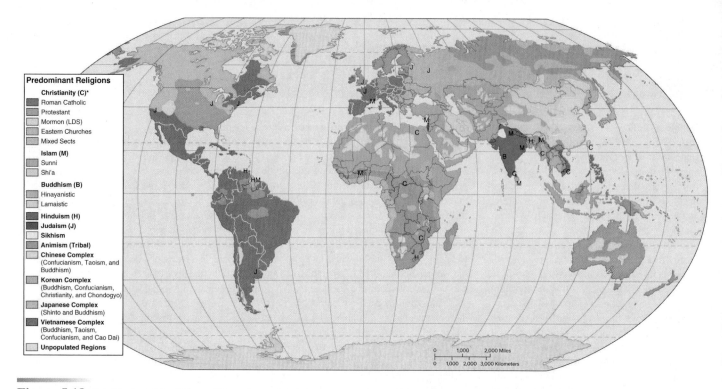

Figure 5.18 Principal world religions. The assignment of areas to a single religion category conceals a growing intermixture of faiths. The capital letters indicate pockets of locally important groups of minority religions. In European and other western countries that have experienced recent major immigration flows, those influxes are altering the effective, if not the numerical, religious balance. In nominally Christian, Catholic France, for example, low churchgoing rates suggest that now more Muslims than practicing Catholics reside there and, considering birth rate differentials, that someday Islam may be the country's predominant religion (as measured by the number of practicing adherents). Secularism—rejection of religious belief—is common in many countries but is not indicated on this map.

Extensive areas of the world are peopled by those who practice tribal or traditional religions, often in concert with the universalizing religions to which they have been outwardly converted. Tribal religions are found principally among peoples who have not yet been fully absorbed into modern cultures and economies or who are on the margins of more populous and advanced societies. Although the areas assigned to tribal religions in Figure 5.18 are significant, the number of adherents is small and declining. **Table 5.2** presents a reasonable ranking of the major religions of the world by estimated number of adherents. The list is not exhaustive; the *World Christian Encyclopedia* tabulates 10,000 distinct religions, including nearly 12,000 Christian denominations.

Frequently, members of a particular religion show areal concentration within a country. Thus, in urban Northern Ireland, Protestants and Catholics reside in separate areas whose boundaries are clearly understood. The "Green Line" in Beirut, Lebanon, marked a guarded border between the Christian East and the Muslim West sides of the city, while within the country as a whole, regional concentrations of adherents of different faiths and sects are clearly recognized.

The world map misses finer-scale detail and the reality of religious diversity in a given location. For example, the map identifies western Canada as mainly Protestant Christian, while a trip to the Vancouver suburb of Richmond, British Columbia, would complicate that picture. Lined up next to one another on a short stretch of a single street, one finds a Hindu temple, a Sikh temple, a Buddhist temple, a Buddhist monastery, an Islamic mosque, several

Table 5.2

Major Religions Ranked by Estimated Number of Adherents

Religion	Number of Adherents (millions)
Christianity	2,112
Islam	1,555
Hinduism	1,017
Nonreligious/Secular/Agnostic/Atheist	788
Syncretic religions	590
Buddhism	485
Tribal/animist religions	174
Shinto	108
Other religions	27
Sikhism	24
Judaism	14
Taoism	9.9
Jainism	5.1
Baha'i	4.8
Confucianism	4.0
Zoroastrianism	1.8

Source: Z. Maoz and E. A. Henderson. 2013. World Religion Project: Global Religion Dataset, 1945–2010.

The term *fundamentalism* entered the social science vocabulary in the late 20th century to describe conservative, reactionary religious movements. Originally, it designated an early 20th century American Christian movement embracing both traditional religious orthodoxy and ethical precepts. More recently, fundamentalism has become a generic description for all religious movements that seek to regain and publicly institutionalize traditional social and cultural values that are usually rooted in the teachings of a sacred text.

Springing from rejection of the secularist tendencies of modernity, fundamentalism is now found in every dominant religion wherever a western-style society has developed, including Christianity, Islam, Hinduism, Judaism, Sikhism, Buddhism, Confucianism, and Zoroastrianism. Fundamentalism is, therefore, a reaction to the modern world; it seeks to counteract cultural changes that undermine religious faith and traditional religious values. The near-universality of fundamentalist movements is seen by some as another expression of a widespread rebellion against secular globalization.

Fundamentalists place a high priority on doctrinal conformity. Further, they are convinced of the correctness of their beliefs To some observers, therefore, fundamentalism is by its nature undemocratic, and states controlled by fundamentalist regimes combining politics and religion, of necessity, stifle debate and punish dissent. In the modern world, that rigidity seems most apparent in Islam where, it is claimed, "all Muslims believe in the absolute inerrancy of the Quran . . ." (*The Islamic Herald,* April 1995) and several countries—for example, the Islamic Republic of Iran and the Islamic Republic of Pakistan—proclaim by official name their administrative commitment to religious control.

In most of the modern world, however, such commitment is not overt or official, and fundamentalists often believe that they and their religious convictions are under threat. They view modern secular society undermining the true faith and religious verities. Initially, therefore, a fundamentalist movement exhorts its followers to ardent prayer, ascetic practices, and physical or military training.

If it is unable to peacefully impose its beliefs on others, some fundamentalist groups justify other more extreme actions against perceived oppressors. Initial protests and nonviolent actions may escalate to outright terrorism. That escalation advances when inflexible fundamentalism is combined with the unending poverty and political impotence felt in many—particularly Middle Eastern—societies today. When an external culture or power—commonly a demonized United States—is seen as the source of cultural decay, and economic and military domination some fundamentalists have been able to justify international terrorism.

Protestant Christian Chinese churches, a Muslim high school, and a Christian high school. Further research would reveal that a high percentage of Vancouver residents claim no religious adherence at all. Religious diversity within countries may reflect the degree of toleration that a majority culture affords minority religions. In predominantly Muslim Indonesia (55 percent to 88 percent of the population, depending on the definition), Christian Bataks, Hindu Balinese, and Muslim Javanese for many years lived in peaceful coexistence. By contrast, the fundamentalist Islamic regime in Iran has persecuted and executed those of the Baha'i faith. The world map of religion has other limitations. Data on religious affiliation are imprecise because most nations do not have religious censuses, and religious groups report their membership differently. When communism was supreme in the former Soviet Union and Eastern European countries, official atheism dissuaded many from openly professing or practicing any religion; in nominally Christian Europe and North America, many who are counted as Christians are not active church members and may have renounced religion altogether.

5.6 The Principal Religions

Each of the major religions has its own mix of cultural values and expressions, its own pattern of innovation and spatial diffusion (**Figure 5.19**), and its unique impact on the cultural landscape. Together, they contribute importantly to the worldwide pattern of cultural diversity.

Judaism

We begin our review of world religions with *Judaism,* whose belief in a single God laid the foundation for both Christianity and Islam. Unlike its universalizing offspring, Judaism is closely identified with a single ethnic group. It emerged some 3,000 to 4,000 years ago in the Near East, one of the ancient culture hearth regions (see Figure 2.15). Early Near Eastern civilizations, including those of Sumeria, Babylonia, and Assyria, developed writing, codified laws, and formalized polytheistic religions featuring rituals of sacrifice and celebrations of the cycle of seasons.

Judaism was different. The Israelites' conviction that they were a chosen people, bound with God through a covenant of mutual loyalty and guided by complex formal rules of behavior, set them apart from other peoples of the Near East. Theirs became a distinctively *ethnic* religion, the determining factors of which are descent from Israel (the patriarch Jacob), the Torah (law and scripture), and the traditions of the culture and the faith. Early military success gave the Jews a sense of territorial and political identity to supplement their religious self-awareness. Later conquest by a series of empires led to their dispersion (*diaspora*) across much of the Mediterranean world and farther east into Asia by 500 CE (**Figure 5.20**).

Alternately tolerated and persecuted in Christian Europe, occasionally expelled from countries, and often spatially isolated in special residential quarters (ghettos), Jews managed

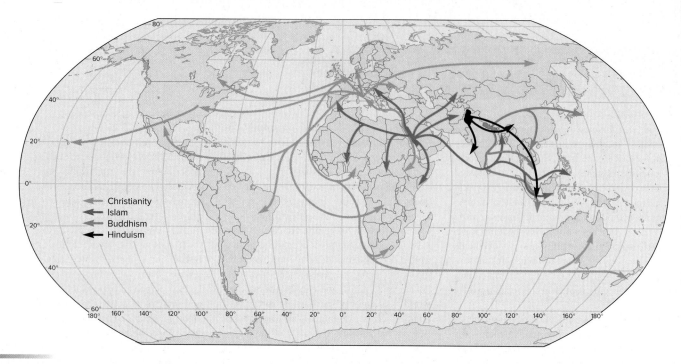

Figure 5.19 Innovation areas and diffusion routes of major world religions. The monotheistic (single deity) faiths of Judaism, Christianity, and Islam arose in southwestern Asia, the first two in Palestine in the eastern Mediterranean region and the last in western Arabia near the Red Sea. Hinduism and Buddhism originated within a confined hearth region in the northern part of the Indian subcontinent. Their rates, extent, and directions of diffusions are suggested here and detailed on later maps.

to retain their faith. Two separate branches of Judaism developed in Europe during the Middle Ages. The Sephardim were originally based in the Iberian Peninsula and expelled from there in the late 15th century; with ties to North African and Babylonian Jews, they retained their native Judeo-Spanish language (Ladino) and culture. Between the 13th and 16th centuries, the Ashkenazim, seeking refuge from intolerable persecution in western and central Europe, settled in Poland, Lithuania, and Russia (Figure 5.20). It was from Eastern Europe that many of the Jewish immigrants to the United States came during the later 19th and early 20th centuries, though German-speaking areas of central Europe were also important source regions.

The Holocaust, which resulted in the murder of perhaps one-third of the world's Jewish population, fell most heavily upon the Ashkenazim. The establishment of the state of Israel in 1948 was a fulfillment of the goal of *Zionism,* the belief in the need to create an autonomous Jewish state in Palestine. It represented a determination that Jews not lose their identity by absorption into alien cultures and societies. The new state represented a reversal of the preceding 2,000-year history of dispersal and relocation diffusion. Israel became largely a country of immigrants, an ancient homeland again identified with a distinctive people and an ethnic religion. The Sephardim and Ashkenazim are present in roughly equal numbers In Israel. Of the world's 14 million Jews, about 7 million live in Israel and 6 million live in North America, where they are mostly found in the larger urban areas.

Judaism's imprint on the cultural landscape has been subtle and unobtrusive. The Jewish community reserves space for the practice of communal burial; the spread of the cultivated citron in the Mediterranean area during Roman times has been traced to Jewish ritual needs; and the religious use of grape wine assured the cultivation of the vine in their areas of settlement. The synagogue as place of worship can take on a wide variety of architectural styles. The essential for a religious service is a community of at least 10 adult males, not a specific structure. Orthodox Jews are a subgroup that adheres to a stricter set of beliefs and practices, one of which forbids driving a car on the Sabbath—the day of worship. This simple rule means that Orthodox Jews tend to live close together in cities. Orthodox Jews often construct *eruvin,* simple wire enclosures strung from utility poles or streetlights that define a home territory for purposes of obeying prohibitions against carrying objects outside the home on the Sabbath. Many non-Jews are surprised to discover they live within an *eruv.*

Christianity

Christianity had its origin in the life and teachings of Jesus, a Jewish preacher of the 1st century of the Common Era, whom his followers believed was the Christ, the savior promised in the Jewish Scriptures. Christians understand their religion as fulfilling the promises of Judaism and extending them to all humankind rather than to just a chosen people.

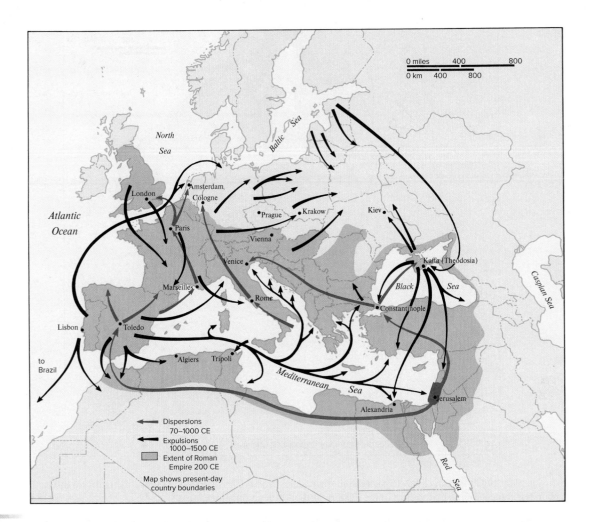

Figure 5.20 Jewish dispersions, 70–1500 CE. A revolt against Roman rule in 66 CE was followed by the destruction of the Jewish Temple four years later and an imperial decision to Romanize the city of Jerusalem. Judaism spread from the hearth region by *relocation diffusion,* carried by its adherents dispersing from their homeland to Europe, Africa, and eventually in great numbers to the Western Hemisphere. Although Jews established themselves and their religion in new lands, they did not lose their sense of cultural identity and did not seek to attract converts to their faith.

As a universal religion, Christianity spread quickly in both the eastern and western parts of the Roman Empire, carried by missionaries to major cities and ports along the excellent system of Roman roads and sea lanes (**Figure 5.21**). *Expansion diffusion* followed the establishment of missions and colonies of converts in locations distant from the hearth region. Important among them were the urban areas that became administrative seats of the new religion. For the Western Church, Rome was the principal center for dispersal, through *hierarchical diffusion,* to provincial capitals and smaller Roman settlements of Europe. From those nodes and from monasteries established in pagan rural areas, *contagious diffusion* disseminated Christianity throughout the continent. The acceptance of Christianity as the state religion of the empire by the Emperor Constantine in 313 CE was also an expression of hierarchical diffusion of great importance in establishing the faith throughout the Roman world. Finally, and much later, *relocation diffusion,* missionary efforts, and in Spanish colonies, forced conversion of Native Americans brought the faith to the New World with European settlers (see Figure 5.18).

The dissolution of the Roman Empire into a western and an eastern half after the fall of Rome also divided Christianity. The

Western Church, based in Rome, was one of the very few stabilizing and civilizing forces uniting western Europe. Its bishops became the civil as well as ecclesiastical authorities over vast areas devoid of other effective government. Parish churches were the focus of rural and urban life, and the cathedrals replaced Roman monuments and temples as the symbols of the social order (**Figure 5.22**). Everywhere, the Roman Catholic Church and its ecclesiastical hierarchy were dominant.

Secular imperial control endured in the eastern empire, whose capital was Constantinople. Thriving under its protection, the Eastern Church expanded into the Balkans, Eastern Europe, Russia, and the Near East. The fall of the eastern empire to the Turks in the 15th century opened Eastern Europe temporarily to Islam, though the Eastern Orthodox Church (the direct descendant of the Byzantine state church) remains, in its various ethnic branches, a major component of Christianity.

The Protestant Reformation of the 15th and 16th centuries split the church in the west, leaving Roman Catholicism supreme in southern Europe but installing a variety of Protestant denominations and national churches in western and northern Europe. The split was reflected in the subsequent worldwide dispersion of

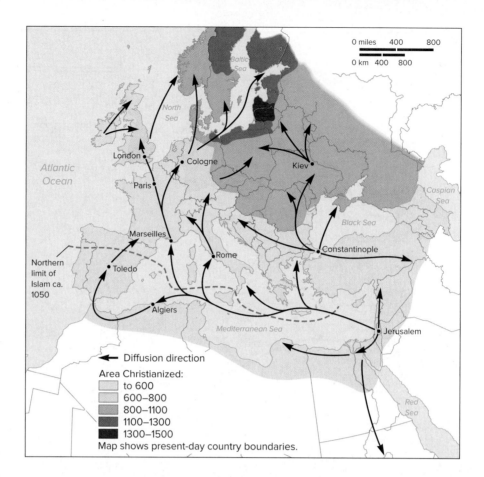

Figure 5.21 Diffusion paths of Christianity, 100–1500 CE. Routes and dates are for Christianity as a composite faith. No distinction is made between the Western church and the various subdivisions of the Eastern Orthodox denominations.

Figure 5.22 The building of Notre Dame Cathedral of Paris, France, begun in 1163, took more than 100 years to complete. Perhaps the best known of the French Gothic churches, it was part of the great period of cathedral construction in Western Europe during the late 12th and the 13th centuries. Between 1170 and 1270, some 80 cathedrals were constructed in France alone. The cathedrals were located in the center of major cities; their plazas were the sites of markets, public meetings, morality plays, and religious ceremonies. They were the focus of public and private life and the symbol not only of the faith, but of the pride and prosperity of the towns and regions that erected them.

©Carl & Ann Purcell/Corbis/Getty Images

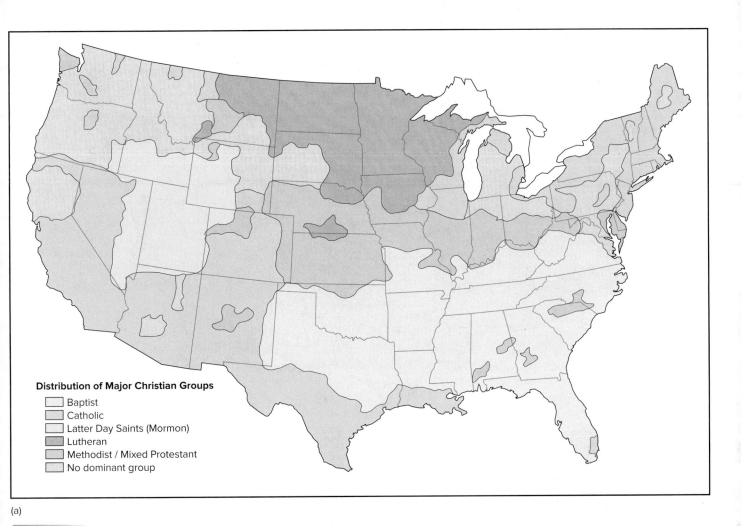

Distribution of Major Christian Groups
- Baptist
- Catholic
- Latter Day Saints (Mormon)
- Lutheran
- Methodist / Mixed Protestant
- No dominant group

(a)

Figure 5.23 (*a*) Religious affiliation in the conterminous United States. The greatly generalized areas of religious dominance shown conceal the reality of immense diversity of church affiliations throughout the United States. *Major* simply means that the indicated category had a higher percentage response than any other affiliation; in practically no case was that as much as 50 percent.

Christianity. Catholic Spain and Portugal colonized Latin America, taking both their languages and the Roman church to that area (see Figure 5.19), as they did to colonial outposts in the Philippines, India, and Africa. Catholic France colonized Quebec in North America. Protestants, many of them fleeing Catholic or repressive Protestant state churches, were primary early settlers of the United States, Canada, Australia, New Zealand, and South Africa.

In Africa and Asia, both Protestant and Catholic missionaries attempted to convert nonbelievers. Both achieved success in sub-Saharan Africa, though traditional religions persist and are common in areas that are shown on Figure 5.18 as predominantly Christian. Neither was particularly successful in China, Japan, or India, where strong ethnic religious cultural systems were less permeable to the diffusion of the Christian faith. Although it still accounts for nearly one-third of the world's population and is territorially the most extensive belief system, Christianity has declined in some regions where it once was strongest. It is no longer numerically important in or near its original hearth. Nor is it any longer dominated by Western adherents. In 1900, 93 percent of all Christians lived in Europe and the Americas; in 2010, 37 percent of an estimated 2.1 billion total lived in Africa and Asia.

Regions and Landscapes of Christianity

All of the principal world religions have experienced theological, doctrinal, or political divisions; frequently these have spatial expression. In Christianity, the early split between the Western and Eastern Churches was initially unrelated to dogma but nonetheless resulted in a territorial separation still evident on the world map. The later subdivision of the Western Church into Roman Catholic and Protestant branches gave a more intricate spatial patterning in Western Europe that can be only generally suggested at the scale shown in Figure 5.18. Still more intermixed are the areal segregations and concentrations that have resulted from the denominational subdivisions of Protestantism.

In the United States and Canada, the beliefs and practices of various immigrant groups and the innovations of domestic congregations have created a particularly varied spatial pattern, though intermingling rather than rigid territorial division is the norm (see Figure 5.1). One observer has suggested a pattern of "religious regions" of the country (**Figure 5.23**) that reflects a larger cultural regionalization of the United States.

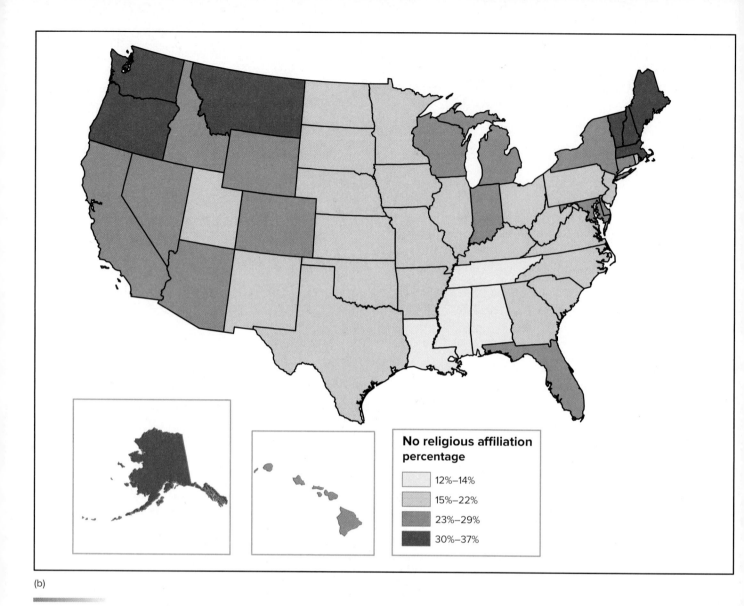

No religious affiliation percentage

- 12%–14%
- 15%–22%
- 23%–29%
- 30%–37%

(b)

Figure 5.23 (*Continued*) (*b*) Percentage reporting "No Religious Affiliation" in 2014. The number of persons claiming no religious affiliation has increased significantly in recent decades. It is most pronounced in New England and the Northwest, both regions with smaller minority populations.

Sources: (a) Based on data or maps from the 2001 "American Religious Identity Survey" by the Graduate School at City University of New York: Religious Denomination Maps Prepared by Ingolf Vogeler of the University of Wisconsin, EauClaire, based on data compiled by the Roper Center for Public Research; and Churches and Church Membership in the United States (Atlanta, Georgia: Glenmary Research Center, 1992). (b) Based on Pew Research Center, Religious Landscape Study, 2014.

Strongly French-, Irish-, and Portuguese-Catholic New England, the Hispanic-Catholic Southwest, and the French-Catholic vicinity of New Orleans (evident in Figure 5.23a) are commonly recognized regional subdivisions of the United States. Each has a cultural identity that includes, but is not limited to, its dominant religion. The Mormon (more properly, Church of Jesus Christ of Latter-Day Saints, or LDS) culture region centered on Utah is a prominent feature. The Baptist presence in the South and that of German and Scandinavian Lutherans in the Upper Midwest (see Figure 5.23a) help determine the boundaries of other distinctive religious regions. The zone of cultural mixing across the center of the country from the Middle Atlantic states to the western LDS region—so evident in the linguistic geography of the United States (see Figure 5.11)—is again apparent on the map of religious affiliation. No single church or denomination dominates, which

is also a characteristic of the Far Western zone. The northeastern and northwestern parts of the United States are home to the highest proportion of people claiming no religious identification. In Canada, the three top Christian groups are Roman Catholic, the United Church of Canada, and Anglican, together comprising 60 percent of the country's population. The "No Religion" category ranks second with 16 percent. Muslims comprise about 2 percent, and Hindus, Sikhs, and Buddhist are each about 1 percent.

The mark of Christianity on the cultural landscape has been prominent and enduring. In pre-Reformation Catholic Europe, the parish church formed the center of life for small neighborhoods of every town, and the village church was the centerpiece of every rural community. In York, England, with a population of 11,000 in the 14th century, there were 45 parish churches, one for each 250 inhabitants. In addition, the central cathedral served simultaneously as a glorification of God, a symbol of piety, and the focus of religious and secular life. The Spanish Laws of the Indies (1573) perpetuated that landscape dominance in the New World, decreeing that all Spanish American settlements should have a church or cathedral on a central plaza (**Figure 5.24a**).

While in Europe and Latin America a single dominant central church was the rule, North American Protestantism placed less importance on the church edifice as a monument and urban symbol. The structures of the principal denominations of colonial New England were, as a rule, clustered in the village center, and that centrality remained a characteristic of small-town America to the present day (Figure 5.24b). Church architecture in North America is highly diverse, ranging from stone or brick revivals of European styles to modern or nondescript styles. In earlier periods, churches were often adjoined by a cemetery, for Christians—in common with Muslims and Jews—practice burial in areas reserved for the dead. In Christian countries in particular, the cemetery—whether connected to the church or separate from it—has traditionally been a significant land use within urban areas. Frequently, the separate cemetery, originally on the outskirts of the community, becomes surrounded by urban growth distorting or blocking the growth of the city.

Islam

Islam—the word means "submission" (to the will of God)—springs from the same Judaic roots as Christianity and embodies similar monotheistic beliefs. Muhammed is revered as the prophet of Allah (God), succeeding and completing the work of earlier prophets of Judaism and Christianity, including Moses, David, and Jesus. The Koran, the holy book of the Muslims, contains not only rules of worship and doctrine, but also instructions on the conduct of human affairs. All Muslims are expected to observe the five pillars of the faith: (1) repeated saying of the basic creed; (2) prayers five times daily at appointed times; (3) a month of daytime fasting during Ramadan; (4) almsgiving; and, (5) if possible, a pilgrimage to Mecca. Two of the five pillars of Islam are explicitly geographical: prayers are done facing Mecca and the pilgrimage to the sacred city of Mecca is among the world's greatest gatherings (**Figure 5.25**). To pray facing Mecca, the direction known as Qiblah, has traditionally required Muslim scholars to calculate a great circle route. Thus, for example, most North American Muslims pray facing northeast because that is the shortest route to Mecca (try it by stretching a string from North America to Saudi Arabia on a globe).

Islam unites the faithful into a brotherhood that crosses boundaries of race, ethnicity, language, and social status. That law of brotherhood served to unify an Arab world sorely divided by tribes, social ranks, and multiple local deities. Muhammed was a resident of Mecca but fled in 622 CE to Medina, where the Prophet proclaimed a constitution and announced the universal mission of the Islamic community. By the time of Muhammed's death in 632 CE, all of Arabia had joined Islam. The new religion swept quickly by expansion diffusion outward from that source region over most of central

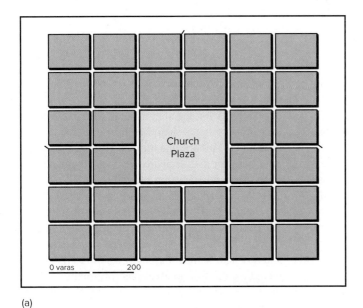

(a)

(b)

Figure 5.24 In Christian societies, the church assumes a prominent central position in the cultural landscape. (*a*) By royal decree, Spanish-planned settlements in the New World were to focus on a cathedral and a plaza centered within a gridiron street system. (*b*) Individually less imposing than the central cathedral of Catholic areas, the multiple Protestant churches common in small and large U.S. and Canadian towns collectively constitute an important land use, frequently seeking or claiming space in the center of the community. This Protestant church in Minnesota was built by Swedish Lutheran immigrants. The church's steeple is the tallest and most prominent structure in the village while the church yard serves as the place of burial for the dead.

(b) ©Mark Bjelland

Figure 5.25 Worshipers gathered during hajj, the annual pilgrimage to Mecca. The black structure is the Ka'ba, the symbol of God's oneness and of the unity of God and humans. All able-bodied Muslims are expected to make a pilgrimage to Mecca once in their lifetime, if possible. As many as 3 million Muslims converge on Mecca for the *hajj*.
©AHMAD FAIZAL YAHYA/Shutterstock

Asia and, at the expense of Hinduism, into northern India (**Figure 5.26**).

The advance westward was particularly rapid and inclusive in North Africa. In Western Europe, 700 years of Muslim rule in much of Spain were ended by Christian reconquest in 1492. In Eastern Europe, conversions made under an expansionary Ottoman Empire are reflected in Muslim components in Bosnia and Kosovo, regions of former Yugoslavia, in Bulgaria, and in the 70 percent Muslim population of Albania. Later, by *relocation diffusion,* Islam was dispersed into Indonesia, southern Africa, and the Western Hemisphere. Muslims now form the majority population in 49 countries.

Asia has the largest absolute number of Muslims while Africa has the highest proportion of Muslims among its population. Through immigration, high birth rates, and conversions, Islam is growing rapidly in Europe and North America. Although second to Christianity in absolute numbers, it is growing faster than any other major world religion due to high rates of natural increase. Islam, with an estimated 1.6 billion adherents worldwide, has been a prominent element in recent and current political affairs. Sectarian divisions fueled the 1980–1988 war between Iran and Iraq as well as recent internal conflicts in Iraq and Syria. Afghan mujahideen— "holy warriors"—found inspiration in their faith to resist Soviet

occupation of their country, and Chechens drew strength from Islam in resisting the Russian assaults on their Caucasian homeland during the 1990s and after. Islamic fundamentalism led to the 1979 overthrow of Iran's shah. Muslim separatism is a recurring theme in Philippine affairs, and militant groups seek establishment of religiously rather than secularly based governments in several Muslim states. Extremist Muslim militants carried out the September 11, 2001, World Trade Center attack and more recent acts of terrorism.

Islam initially united a series of separate tribes and groups, but disagreements over the succession of leadership after the Prophet led to a division between two sectarian groups, Sunnis and Shi'ites. Sunnis, the majority (80 to 85 percent of Muslims) recognize the first four caliphs (originally, "successor" and later the title of the religious and civil head of the Muslim state) as Muhammed's rightful successors. The Shi'ites reject the legitimacy of the first three and believe that Muslim leadership rightly belonged to the fourth caliph, the Prophet's son-in-law, Ali, and his descendants. Sunnis constitute the majority of Muslims in all countries except Iran, Iraq, Bahrain, and perhaps Yemen.

The mosque—place of worship, community clubhouse, meeting hall, and school—is the focal point of Islamic communal life and the primary imprint of the religion on the cultural landscape. Its principal purpose is to accommodate the Friday

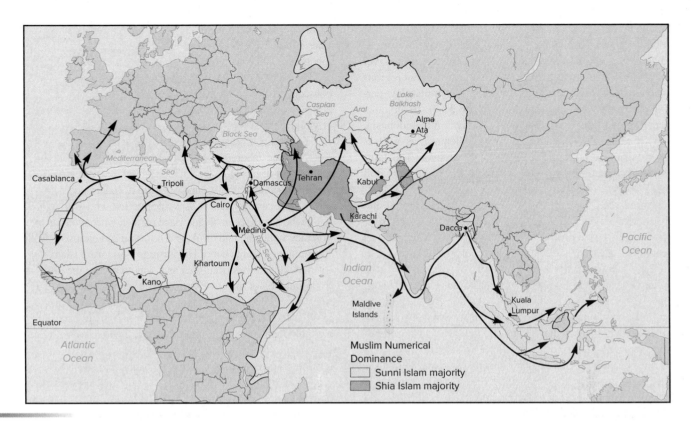

Figure 5.26 Spread and extent of Islam. Islam predominates in 49 countries along a band across northern Africa to central Asia, and the northern part of the Indian subcontinent. Still farther east, Indonesia has the largest Muslim population of any country. Islam's greatest development is in Asia, where it is second only to Hinduism, and in Africa, where it trails only Christianity. Current Islamic expansion is particularly rapid in the Southern Hemisphere.

prayer service mandatory for all male Muslims. United by the use of Arabic for prayers and readings, many mosques in North America draw worshippers from diverse immigrant groups and multiple native tongues. The mosque is a sacred space with rules pertaining to proper conduct and dress. It is typically a gendered space with certain spaces reserved for men and others for women. Rooms are usually oriented so that worshippers face Mecca along a precisely calculated great circle route. The distinctive mosque architecture found throughout the world of Islam draws upon Roman, Byzantine, and Indian design elements. With its perfectly proportioned, frequently gilded or tiled domes, geometric artwork, graceful, soaring towers and minarets (from which the faithful are called to prayer), and delicately wrought parapets and cupolas, the carefully tended mosque is frequently the community's most elaborate and imposing structure (**Figure 5.27**).

Hinduism

Hinduism is the world's oldest major religion. Though it has no datable founding event or initial prophet, some evidence traces its origin back 4,000 or more years. Hinduism is a polytheistic religion woven into an intricate web of philosophical, social, economic, and artistic elements that comprise a distinctive Indian civilization. Its estimated 800 million adherents are largely confined to India, where it claims 80 percent of the population.

Figure 5.27 The common architectural features of the mosque, with its dome and soaring minaret towers, make it an unmistakable landscape evidence of the presence of Islam in any local culture. The visual dominance and symbolism of minarets has drawn attention to new mosques and provoked controversies about the place of Islam in western societies. Some newer mosques in Europe and North America have forgone the minaret towers to better blend in. The Blue Mosque in Istanbul, Turkey, would not be out of place architecturally in Muslim Malaysia or Indonesia.

©Glen Allison/Photodisc/Getty Images

Hinduism derives its name from its cradle area in the valley of the Indus River. From that district of present-day Pakistan, it spread by *contagious diffusion* eastward down the Ganges River and southward throughout the subcontinent and adjacent regions by amalgamating, absorbing, and eventually supplanting earlier native religions and customs (see Figure 5.19). Its practice eventually spread throughout southeastern Asia, into Indonesia, Malaysia, Cambodia, Thailand, Laos, and Vietnam, as well as into neighboring Myanmar (Burma) and Sri Lanka. The largest Hindu temple complex is in Cambodia, not India, and Bali remains a Hindu pocket in dominantly Islamic Indonesia. Hinduism's more recent growing presence in Western Europe and North America reflects a *relocation diffusion* of its adherents.

Hinduism is based on the concepts of reincarnation and passage from one state of existence to another in an unending cycle of birth and death in which all living things are caught. One's position in this life is determined by one's *karma,* or deeds and conduct in previous lives. That conduct dictates the condition and the being—plant, animal, or human—into which a soul is reborn. All creatures are ranked, with humans at the top of the ladder. But humans themselves are ranked, and the social caste into which an individual is born is an indication of that person's spiritual status. The goal of existence is to move up the hierarchy, eventually to be liberated from the cycle of rebirth and death and to achieve salvation and eternal peace through union with the Brahman, the universal soul.

The **caste** (meaning "birth") structure of society is an expression of the eternal transmigration of souls. For the Hindu, the primary aim of this life is to conform to prescribed social and ritual duties and to the rules of conduct for the assigned caste and profession. Those requirements comprise that individual's *dharma*—law and duties. To violate them upsets the balance of society and nature and yields undesirable consequences. Careful observance improves the chance of promotion at the next rebirth. Traditionally, each craft or profession is the property of a particular caste: Brahmins (scholar-priests), Kshatriyas (warrior-landowners), Vaishyas (businessmen, farmers, herdsmen), Sudras (servants and laborers). Dalits, or *untouchables* for whom the most menial and distasteful tasks were reserved, and backwoods tribes—together accounting for around one-fifth of India's population—stand outside the caste system. Caste rules define who you can socialize with, who is an acceptable marriage partner, where you can live, what you may wear, eat, and drink, and how you can earn your livelihood. Conversion of Dalits out of Hinduism to Buddhism, Islam, Christianity, or Sikkhism has been seen as a way to escape the prejudice and discrimination of the caste system. As a secular democracy, religious freedom and protection against caste-based discrimination are written into the Indian constitution. However, well-publicized mass conversions to Buddhism and Christianity have provoked a strong response from Hindu fundamentalists who have successfully fought for legal restrictions on conversions in some Indian states.

The practice of Hinduism is rich with rites and ceremonies, festivals and feasts, processions and ritual gatherings of literally millions of celebrants. Pilgrimages to holy rivers and sacred places are thought to secure deliverance from sin or pollution and to preserve religious worth (**Figure 5.28**). In what is perhaps

Figure 5.28 Pilgrims at dawn worship in the Ganges River at Varanasi (Banares), India, one of the seven most sacred Hindu cities and the reputed Earthly capital of Siva, the Hindu god of destruction and regeneration. Hindus believe that to die in Varanasi means release from the cycle of rebirth and permits entrance into heaven.

©*PORTERFIELD- CHICKERING/Science Source*

the largest periodic gathering of humans in the world, millions of Hindus of all castes, classes, and sects gather about once every 12 years for a ritual washing away of sins in the Ganges River near Allahabad. Worship in the temples and shrines and making offerings to secure merit from the gods are requirements for Hindus. The doctrine of ahimsa—also fundamental in Buddhism—instructs Hindus to refrain from harming any living being.

Temples and shrines are everywhere. Within them, innumerable icons of deities such as Vishnu, Shiva, and Ganesha are enshrined, the objects of veneration, offerings, and daily care. All temples have a circular spire as a reminder that the sky is the real dwelling place of the god who resides only temporarily within the temple (**Figure 5.29**). The temples, shrines, daily rituals and worship, numerous specially garbed or marked holy men and ascetics, and the ever-present sacred animals mark the cultural landscape of Hindu societies—a landscape infused with religious symbols and sights that are part of a total cultural experience.

Numerous reform movements have derived from Hinduism over the centuries, some of which have endured to the present day as major religions on a regional or world scale. Jainism, begun in the 6th century BCE as a revolt against the authority of the early Hindu doctrines, rejects caste distinctions and modifies concepts of karma and transmigration of souls; it counts perhaps 5 million adherents. Combining elements of Hinduism and Islam, *Sikhism* developed in the Punjab area of northwestern India in the late 15th century CE. Sikhism is an ethnic religion with an estimated 24 million adherents. The great majority of Sikhs live in India, mostly in the Punjab, though substantial numbers have settled in the United Kingdom and Canada.

Buddhism

The largest and most influential of the dissident movements within Hinduism has been Buddhism, a universalizing faith founded in the 6th century BCE in northern India by Siddhartha Gautama, the Buddha (*Enlightened One*). The Buddha's teachings were more a moral philosophy that offered an explanation for evil and human suffering than a formal religion. He viewed the road to enlightenment and salvation to lie in understanding the "four noble truths": existence involves suffering; suffering is the result of desire; pain ceases when desire is destroyed; the destruction of desire comes through knowledge of correct behavior and correct thoughts. The Buddha's message was open to all castes, raising Buddhism from a philosophy to a universalizing religion.

Contagious diffusion spread the belief system throughout India, where it was made the state religion in the 3rd century BCE. It was carried elsewhere into Asia by missionaries, monks, and merchants. While expanding abroad, Buddhism began to decline at home as early as the 4th century CE, slowly but irreversibly reabsorbed into a revived Hinduism. By the 8th century, its dominance in northern India was broken by conversions to Islam; by the 15th century, it had essentially disappeared from the Indian subcontinent.

Present-day spatial patterns of Buddhist adherence reflect the schools of thought, or *vehicles*, that were dominant during

Figure 5.29 The Hindu temple complex at Belur, Karnataka in southern India. The creation of temples and the images that they house has been a principal outlet of Indian artistry for more than 3,000 years. At the village level, the structure may be simple, containing only the windowless central cell housing the divine image, a surmounting spire, and the temple porch or stoop to protect the doorway of the cell. The great temples, of immense size, are ornate extensions of the same basic design.

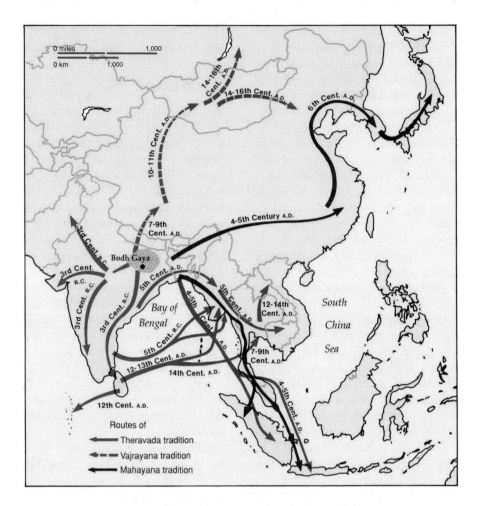

Figure 5.30 Diffusion paths, times, and "vehicles" of Buddhism.

different periods of dispersion (**Figure 5.30**). Earliest, most conservative, and closest to the origins of Buddhism was *Theravada* (Vehicle of the Elders) Buddhism, which was implanted in Sri Lanka and Southeast Asia beginning in the 3rd century BCE. Its emphasis is on personal salvation through the four noble truths; it mandates a portion of life to be spent as monk or nun.

Mahayana (Greater Vehicle) was the dominant tradition when Buddhism was accepted into East Asia—China, Korea, and Japan—in the 4th century CE and later. Itself subdivided and diversified, Mahayana Buddhism considers the Buddha divine and, along with other deities, a savior for all who are truly devout. It emphasizes meditation (contemplative Zen Buddhism is a variant form), does not require service in monasteries, and tends to be more polytheistic and ritualistic than does Theravada Buddhism.

Vajrayana (the Diamond Vehicle) was dominant when the conversion of Tibet and neighboring northern areas began, first in the 7th century and again during the 10th and 11th centuries as a revived Lamaist tradition. That tradition originally stressed self-discipline and conversion through meditation and the study of philosophy, but it later became more formally monastic and ritualistic, elevating the Dalai Lama as the reincarnated Buddha, who became both spiritual and temporal ruler. Before Chinese conquest and the flight of the Dalai Lama in 1959, as many as

one out of four or five Tibetan males was a monk whose celibacy helped keep population numbers stable. Tibetan Buddhism was further dispersed, beginning in the 14th century, to Mongolia, northern China, and parts of southern Russia.

Buddhism imprints its presence vividly on the cultural landscape. Buddha images in stylized human form began to appear in the 1st century CE and are common in painting and sculpture throughout the Buddhist world. Equally widespread are the three main types of buildings and monuments: the *stupa,* a commemorative shrine; the temple or pagoda enshrining an image or relic of the Buddha; and the monastery, some of them the size of small cities (**Figure 5.31**). Common, too, is the *bodhi* (or *bo*) tree, a fig tree of great size and longevity. Buddha is said to have received enlightenment seated under one of them at Bodh Gaya, India, and specimens have been planted and tended as an act of reverence and symbol of the faith throughout Buddhist Asia. Sacred places for Buddhists are largely associated with key events in the life of the Buddha—many of which are in areas that are no longer predominantly Buddhist.

Buddhism has suffered greatly in Asian lands that came under communist control: Inner and Outer Mongolia, Tibet, North Korea, China, and parts of Southeast Asia. Communist governments abolished the traditional rights and privileges of the monasteries. In those states, Buddhist religious buildings were

Figure 5.31 The golden stupas of the Swedagon pagoda, Yangon, Myanmar (formerly known as Rangoon, Burma).

©Medioimages/Photodisc/Getty Images

taken over by governments and converted into museums or other secular uses, abandoned, or destroyed. As a consequence, the number of Buddhists is uncertain, with world totals estimated to be 485 million.

East Asian Ethnic Religions

When Buddhism reached China from the south some 1,500 to 2,000 years ago and was carried to Japan from Korea in the 7th century, it encountered and later amalgamated with already well-established belief systems. The Far Eastern ethnic religions are **syncretisms** (combinations of different beliefs and practices). In China, the union was with Confucianism and Taoism, themselves becoming intermingled by the time of Buddhism's arrival. In Japan, it was with Shinto, a polytheistic animism and shamanism.

Chinese belief systems address more ethical or philosophical questions than religious ones, having little interest in the hereafter. Confucius (K'ung Fu-tzu), a compiler of traditional wisdom who lived about the same time as the Buddha, emphasized the importance of proper conduct—between ruler and subjects and

among family members. The family was extolled as the nucleus of the state, and children's respect for their parents was the loftiest of virtues. There are no places of worship or clergy in **Confucianism,** though its founder believed in a Heaven seen in naturalistic terms, and the Chinese custom of ancestor worship as a mark of gratitude and respect was encouraged. After his death, the custom was expanded to include worship of Confucius himself in temples erected for that purpose. That worship became the official state religion in the 2nd century BCE, and for some 2,000 years—until the start of the 20th century CE—Confucianism, with its emphasis on ethics and morality rooted in Chinese traditional wisdom, formed the basis of the belief system of China.

It was joined by, or blended with, **Taoism,** an ideology that according to legend was first taught by Lao-tsu in the 6th century BCE. Its central theme is *Tao,* the Way, a philosophy teaching that eternal happiness lies in total identification with nature and deploring passion, unnecessary invention, unneeded knowledge, and government interference in the simple life of individuals. Beginning in the 1st century CE, this philosophical naturalism was coupled with a religious Taoism involving deities, spirits, magic, temples, and priests. Buddhism, stripped by Chinese pragmatism of much of its Indian otherworldliness and defining a *nirvana* achievable in this life, was easily accepted as a companion to these traditional Chinese belief systems. Along with Confucianism and Taoism, Buddhism became one of the honored Three Teachings, and to the average person, there was no distinction in meaning or importance between a Confucian temple, Taoist shrine, or Buddhist stupa.

Buddhism also joined and influenced Japanese **Shinto,** the traditional religion of Japan that developed out of nature and ancestor worship. Shinto—the Way of the Gods—is a set of customs and rituals rather than an ethical or moral system. It observes a complex set of deities, including deified emperors, family spirits, and the divinities residing in rivers, trees, certain animals, mountains, and, particularly, the sun and moon. Buddhism eventually intertwined with traditional Shinto with Buddhist deities seen as Japanese gods in a different form. More recently, Shinto divested itself of many Buddhist influences and became, under the reign of the Emperor Meiji (1868–1912), the official state religion, emphasizing loyalty to the emperor. The centers of worship are the numerous shrines and temples in which the gods are believed to dwell and which are approached through ceremonial *torii,* or gateway arches (**Figure 5.32**).

Secularism

One cannot assume that all people within a mapped religious region are adherents of the designated faith, or that membership in a religious community means active participation in its belief system. **Secularism,** a weakening of the influence of religion and an indifference to or rejection of religious belief, is an increasing part of many modern societies. The population of nonreligious persons is most pronounced in the industrialized nations and those now or formerly under communist regimes. In Britain, for example, about half the population attended church regularly in 1851 but that figure is now about 8 percent. While the Church of England's parish churches are an enduring feature of the rural landscape, vacant churches are also a common landscape feature,

Figure 5.32 Floating torii gate at Itsukushima Shrine on Miyajima Island, Japan.

©GeoStock/Photodisc/Getty Images

especially in urban areas. It is not uncommon to see once-grand churches converted to arts facilities, stores, or restaurants, or simply boarded up (**Figure 5.33**). Two-thirds of the French describe themselves as Catholic, but less than 5 percent regularly go to church. Even in devoutly Roman Catholic South American states, church attendance rates of less than 20 percent in most countries attest to the rise of at least informal secularism. Estimates put the world number of the nonreligious at 1.0 billion. In societies undergoing secularization, responsibility for governance, education, health care, and social services is often transferred from religious bodies to nonreligious institutions. In its stronger form, secularism is openly anti-religious and may involve restricting the practice of religion and destroying places of worship. Official governmental policies of religious neutrality (as in the cases of the United States or India, for example) are, of course, distinct from personal secular or nonreligious beliefs.

Change and Diversity in the Geography of Religion

The map of principal world religions is continually changing as religions grow, diffuse, and recede. One of the most dramatic recent changes is the expansion of the universalizing religions of Christianity and Islam in areas of Africa once primarily associated with traditional religions. Although traditional African religions

have receded from much of the map, traditional religious practices such as fortune-telling or ceremonies honoring ancestral spirits still are widely practiced and are frequently blended with Islam or Christianity. The fault line between expanding Islam and Christianity runs through a number of countries, including Nigeria, Ethiopia, and Tanzania, and has triggered episodic violence. In the former unified Sudan, this religious fault line coincided with ethnic divides, which together led to the overwhelming vote in 2011 for independence for South Sudan.

While Europe is the continent most closely associated with Christianity, it has witnessed dramatic religious shifts in the past century. The most dramatic changes are the secularization of large portions of European society and the rise of new religions brought by immigrants, primarily Islam and Hinduism. In the 2011 census in the United Kingdom, 59 percent stated that they were Christian, 25 percent reported that they had no religion, 5 percent stated Muslim, and 1.5 percent stated Hindu. Immigrants, mostly from former colonies in developing regions, have added mosques, temples, and new storefront churches to the landscape.

One striking aspect of the North American religious landscape is the great number of different churches, synagogues, mosques, temples, gurdwaras, and related structures (**Figure 5.34**). In addition to the formal places of worship of the

Figure 5.33 A vacant, boarded-up church in Cardiff, Wales symbolizes the rise of secularism in Europe.

©Mark Bjelland

(a)

(c)

(b)

(d)

Figure 5.34 Diverse religious landscapes in the United States and Canada. (*a*) A Christian megachurch in the suburbs of Chicago, Illinois. Megachurches often resemble a combination of a shopping mall, convention center, and concert hall. Locations near highways and vast parking lots are usually essential to megachurch success. (*b*) The Thrangu Buddhist monastery on Number 5 Road in the suburb of Richmond, British Columbia, in Canada is built in a traditional Tibetan style. (*c*) Also on Number 5 Road in Richmond is this Islamic center, mosque, and Muslim school. The distinctive minarets are crowned with the traditional crescent moon. (*d*) Next door to the mosque is the Sikh Temple (Gurdwara) and Indian Cultural Centre.

(a,b,c,d) ©Mark Bjelland

more established groups, innumerable storefront places of worship—mainly associated with poorer neighborhoods, changing immigrant ethnic communities, and splinter Protestant sects—have become part of the American scene. Another recent feature on the North American religious landscape is the megachurch, defined as a congregation that draws more than 2,000 attendees in a typical weekend. There are an estimated 1,250 megachurches in the United States and Canada. Megachurches, often located in newer suburban areas, feature massive parking lots and architecture more typical of convention centers, sports arenas, or shopping malls (not surprisingly, because many meet in converted buildings once used for those purposes). Also, distinctively (though not exclusively) American is the proliferation of religious and denominational signage (see **Figure 5.1**) on city buildings, storefronts, or highway billboards. Compared with more secularized Europe, the religious diversity and vitality of the United States is remarkable. Some scholars attribute this vitality to the lack of an established state religion and to the successive waves of immigrants, most of whom are religious and have found that creating a religious congregation of their own is a way to be American while preserving their ethnic identity.

SUMMARY

Language and religion are basic threads in the web of culture. They are essential elements of cultural identity. A person's speech offers clues as to where they are from, their social status, and their education level. In some societies, religion may serve as a similar identifier. Both language and religion are mentifacts, parts of the ideological subsystem of culture; both are transmitters of culture as well as markers. Both have distinctive spatial patterns—reflecting past and present processes of spatial interaction and diffusion—that are basic to the identity of world culture realms.

Languages may be grouped genetically—by origin and historical development—but the world distribution of language families depends as much on the movement of peoples and histories of conquest and colonization as it does on patterns of linguistic evolution. As different language groups come into contact, pidgin and creole languages may emerge. With increased spatial interaction and global integration, English has become a new lingua franca, prompting efforts to preserve local languages. Toponymy, the study of place names, helps document the historical and cultural geography of a region.

Religion varies in its impact and influence on culture groups. Some societies are dominated in all aspects by their controlling religious belief: Hindu India, for example, or Islamic Iran. Where religious beliefs are strongly held, they can unite a society of adherents and divide nations and peoples holding divergent faiths. Religions may be classified as ethnic religions such as Judaism and Hinduism or universalizing such as Buddhism, Christianity, and Islam. The spatial distribution of religions reflects past and present patterns of migration, conquest, and diffusion, essential dynamics of cultural geography.

Although distinct, language and religion are not totally unrelated. Religion can influence the spread of languages to new peoples and areas, as Arabic, the language of the Koran, was spread by conquering armies of Muslims. Religion may conserve as well as disperse language. Yiddish remains the language of worship in Hasidic Jewish communities; many U.S. and Canadian Lutheran churches held services in German, Swedish, or other Scandinavian languages until World War II. Until the 1960s, Latin was the language of liturgy in the Roman Catholic Church and Sanskrit remains the language of the Vedas, sacred in Hinduism. Sacred texts may demand the introduction of an alphabet to nonliterate societies: the Roman alphabet follows Christian missionaries, Arabic script accompanies Islam. The Cyrillic alphabet of Eastern Europe was developed by missionaries. The tie between language and religion is not inevitable. The French imposed their language (but not their religion) on Algeria; Spanish Catholicism (but not the Spanish language) became dominant in the Philippines.

Language and religion are important components of ethnic identities that distinguish and isolate groups from the larger population within which they reside. Our attention turns next, in Chapter 6, to the concept and patterns of ethnicity, a distinctive piece in the mosaic of human culture.

KEY WORDS

animism	Judaism	sacred places
bilingualism	language	secularism
Buddhism	language family	shamanism
caste	lingua franca	Shinto
Christianity	linguistic geography	speech community
Confucianism	monolingual	standard language
creole	monotheism	syncretism
dialect	multilingualism	Taoism
ethnic religion	official language	toponyms
geographic (regional) dialect	pidgin	toponymy
Hinduism	polytheism	tribal (traditional) religion
Islam	protolanguage	universalizing religion
isogloss	religion	vernacular

FOR REVIEW

1. Why might one consider language the most important differentiating element of a culture?

2. In what way can religion affect other cultural traits of a society? In what cultures or societies does religion appear to exert a growing influence? In what cultures does the importance of religion appear to be declining? What might be the broader social or economic consequences of those changes?

3. In what way does the concept of *protolanguage* help us in linguistic classification? What is meant by *language family?* Is *genetic* classification of language an unfailing guide to spatial patterns of languages? Why or why not?

4. What do toponyms tell us about the history and values of a place? What do recent renaming efforts tell us about the changing cultural identities?

5. In what ways do *isoglosses* and the study of *dialects* help us understand other human geographic patterns?

6. Cite examples that indicate the significance of religion in the internal and foreign relations of nations.

7. How does the classification of religions as *universalizing, ethnic,* or *tribal* help us to understand their patterns of distribution and spatial diffusion?

8. What connection, if any, do you see among language, religion, and intergroup conflict in the contemporary world?

KEY CONCEPTS REVIEW

Language

1. **How are the world's languages classified and distributed?** Section 5.1

   The 7,100 languages spoken today may be grouped within a limited number of language families that trace their origins to common protolanguages. The present distribution of tongues reflects the current stage of continuing past and recent dispersion of their speakers and their adoption by new users. Languages change through isolation, migration, and the passage of time.

2. **What are standard languages and what kinds of variants can be observed?** Section 5.1

   All speakers of a given language are members of its speech community, but not all use the language uniformly. The standard language is that form of speech that has received official sanction or acceptance as the "proper" form of grammar and pronunciation. Dialects, regional and social, represent nonstandard or vernacular variants of the common tongue. A pidgin is a created, composite, simple language designed to promote exchange among speakers of different tongues. When evolved into a complex native language of a people, the pidgin has become a creole. Governments may designate one or more official state languages (including, perhaps, a creole such as Swahili).

3. **How does language serve as a cultural identifier and landscape artifact?** Section 5.2–5.3

Language is a mentifact, a part of the ideological subsystem of culture. It is, therefore, inseparable from group identity and self-awareness. Language may also be divisive, creating rifts within multilingual societies when linguistic minorities seek recognition or separatism. Toponyms (place names) record the perceptions and values of past and present occupants or explorers. Toponymy in tracing that record becomes a valuable tool of historical cultural geography.

Religion

4. **What is the cultural role of religion?** Section 5.4

Like language, religion is a basic component of culture, a mentifact that serves as a cultural rallying point. Frequently, religious beliefs divide and alienate different groups within and among societies. Past and present belief systems of a culture may influence its legal norms, dietary customs, social relations, economic patterns, and landscape imprints.

5. **How are religions classified and distributed?** Section 5.5

Distinguishing among universalizing, ethnic, and traditional religions helps understand diffusion processes. The spatial patterns of world religions reflect their origin areas, migrations, conquests, and the converts they have attracted in home and distant areas.

6. **What are the principal world religions and how are they distinguished in patterns of innovation, diffusion, and landscape imprint?** Section 5.6

The text briefly traces those differing origins, spreads, and cultural landscape impacts of Judaism, Christianity, Islam, Hinduism, Buddhism, and certain East Asian ethnic religions.

ETHNIC GEOGRAPHY:

Threads of Diversity

Chinatown in San Francisco, California is a Chinese ethnic enclave that dates to 1848.

©Spondylolithesis/iStock/Getty Images

Key Concepts

6.1–6.2 Race, ethnicity, ethnic diversity, and the changing immigration streams to multiethnic cities in the United States and Canada

6.3 Acculturation and the persistence of ethnic clusters and identities in the United States and Canada and elsewhere

6.4–6.5 Urban ethnic diversity and patterns of segregation

6.6 The landscape impacts of ethnic diversity

*W*e must not forget that these men and women who file through the narrow gates at Ellis Island, hopeful, confused, with bundles of misconceptions as heavy as the great sacks upon their backs—we must not forget that these simple, rough-handed people are the ancestors of our descendants, the fathers and mothers of our children.

So it has been from the beginning. For a century, a swelling human stream has poured across the ocean, fleeing from poverty in Europe to a chance at a new life in America. English, Welsh, Scotch, Irish; German, Swede, Norwegian, Dane; Jew, Italian, Bohemian, Serb; Syrian, Hungarian, Pole, Greek—one race after another has knocked at our doors, been given admittance, has married us and begot our children.... A few hours, and the stain of travel has left the immigrant's cheek; a few years, and he loses the odor of alien soils; a generation or two, and these outlanders are irrevocably our race, our nation, our stock.[1]

The United States and Canada are cultural composites—as increasingly are most of the countries of the world. North America's peoples include aborigine and immigrant, native-born and new arrival. Had this chapter's introductory passage been written in the 21st century rather than early in the 20th, the list of foreign origins would have been lengthened to include many Latin American, African, and Asian countries as well.

The majority of the world's societies, even those that outwardly seem most homogeneous, house distinctive **ethnic groups,** populations that feel themselves bound together by a common origin and set off from other groups by ties of culture, race, religion, language, or nationality. Ethnic diversity is a near-universal part of human geographic patterns; the approximately 200 independent countries are home to at least 5,000 ethnic groups. The factors driving globalization, such as the growth of transnational corporations, relaxed border restrictions, low-cost travel, and high-speed global communications are all encouraging greater movement and ethnic mixing. In response to labor shortages, European Union (EU) countries increasingly welcome workers from other EU states, as well as African and Asian immigrants and guest workers, effectively making their societies multiethnic. Refugees and job-seekers are found in alien lands throughout both hemispheres (**Figure 6.1**). Cross-border movements and refugee resettlements in West Asia, Southeast Asia, and Africa are prominent current events. European colonialism created pluralistic societies in tropical lands through the introduction of both ruling elites and, frequently, nonindigenous laborers. Polyethnic Russia, Afghanistan, China, India, and most African countries have native—rather than immigrant—populations more characterized by racial and cultural diversity than by uniformity. Tricultural Belgium has a nearly split personality in matters political and social. The idea of an ethnically pure nation-state is mostly obsolete.

Like linguistic and religious differences within societies, such population interminglings are masked by the "culture realms" shown in Figure 2.4, but at a finer scale, they are important threads in the cultural-geographic tapestry of our world. The multiple movements, diffusions, migrations, and mixings of

Figure 6.1 Hispanic students from San Diego, California protest legislation that would increase penalties for undocumented immigrants. Immigration policies generate heated political debates with serious consequences for affected persons.

©McGraw-Hill Education/John Flournoy, photographer

peoples of different origins are the subject of **ethnic geography.** Its concerns are those of spatial distributions and interactions of ethnic groups, of the cultural characteristics and influences underlying them, and of how the built environment reflects the imprint of various ethnic groups.

Culture, we saw in Chapter 2, is the composite of traits making up the way of life of a human group—collective beliefs, symbols, values, forms of behavior, and complexes of such nonmaterial and material traits as social customs, language, religion, food habits, tools, structures, and more. Culture is learned; it characterizes the group and distinguishes it from all other groups that have collectively created and transmitted to its children still other "ways of life." *Ethnicity* is simply the identifying term assigned to a large group of people who share the traits of a distinctive common culture. It is always based on a clear understanding by members of a group that they are fundamentally different from others who do not share their distinguishing characteristics or cultural heritage.

Ethnicity is not, by itself, a spatial concept. However, ethnic groups are associated with clearly recognized territories—either larger homeland districts or smaller rural or urban enclaves—in which they are primary or exclusive occupants and upon which they have placed distinctive cultural imprints. Because territory and ethnicity are inseparable concepts, ethnicity exhibits important spatial patterns and is an important concern for the human geographer. Further, because ethnicity is often identified with a particular language and/or religion, consideration of ethnicity flows logically from the discussions of language and religion in Chapter 5.

Our examination of ethnic patterns will concentrate on the United States and Canada. Originally, this region was occupied by a multitude of distinctive Native American peoples, each with their own territory, culture, and language. Over time, these populations were overwhelmed and displaced by a wide spectrum of Old World ethnic groups. The United States and Canada provide

[1]From Walter E. Weyl, "The New Americans," *Harper's Magazine* 129: 615. Copyright 1914 Harper's Magazine Foundation, New York.

case studies of how distinctive immigrant culture groups partition urban and rural space and place their claims and imprints upon it. The experiences of these countries show the durability of ethnic distinctions even under conditions and national myths that emphasize intermixing and homogenization of population. Examples drawn from other countries and environments will serve to highlight ways in which generalizations based on the North American experience may be applied more broadly.

6.1 Ethnicity and Race

Each year on a weekend in May, New York City celebrates its ethnic diversity by closing off to all but pedestrian traffic a 1-mile stretch of street to conduct the Ninth Avenue International Food Festival. Along the reserved route from 42nd to 57th streets, a million or more New Yorkers come together to sample the foods, view the crafts, and take in the music and dance of the diverse ethnic groups represented among the citizens of the city. As a resident of the largest U.S. metropolis, each of the merchants and artists contributing one of the several hundred separate storefront, stall, or card-table displays of the festival becomes a member of the United States and Canada culture realm. Each, however, preserves a distinctive small-group identity within that larger collective "realm" (**Figure 6.2**).

The threads of diversity exhibited in the festival are expressions of **ethnicity,** a term derived from the Greek word *ethnos,* meaning a "people" or "nation." Ethnic groups are composed of individuals who share some prominent cultural traits or characteristics, some evident physical or social identifications setting them apart both from the majority population and from other distinctive minorities among whom they may live. An ethnic identity is recognized by both members of the group and by outsiders. No single trait denotes ethnicity. Group recognition may be based on language, religion, national origin, unique customs, a shared history, or—improperly—an ill-defined concept of race. Common unifying bonds of ethnicity are a shared ancestry and cultural heritage, distinctive traditions, territorial identification, and sense of community. The principal ethnic groups of the United States and Canada are shown in Table 6.1 and Table 6.4, respectively.

Race and ethnicity are frequently equated, but they are actually very different concepts. **Race** is an outdated categorization of humans based on outward physical characteristics such as skin color, hair texture, or eye color or shape. Although humans are all one species, there is obvious variation in our physical characteristics. The spread of human beings over the Earth and their occupation of different environments were accompanied by the development of variations in these visible characteristics, as well as internal differences such as blood composition or lactose intolerance. Physical differentiation among human groups is old and can reasonably be dated to the Paleolithic spread and isolation of population groups (occurring 100,000 to about 11,000 years ago). Geographic patterns of distinct combinations of physical traits emerged due to natural selection or adaptation, and genetic drift.

Natural selection favors the transmission of characteristics that enable humans to adapt to a particular environmental

Figure 6.2 The annual Ninth Avenue International Fair in New York City became one of the largest of its kind. Similar festivals celebrating America's ethnic diversity are found in cities and small towns across the country.
©ALEX QUESADA/AP Images

feature, such as climate. Studies have suggested a plausible relationship between solar radiation and skin color, and between temperature and body size. Dark skin indicates the presence of melanin, which protects against the penetration of damaging ultraviolet rays from the sun. Conversely, the production of vitamin D in the body, which is necessary to good health, is linked to the penetration of ultraviolet rays. In high latitudes, where winter days are short and the sun is low in the sky, light skin confers an adaptive advantage by allowing the production of vitamin D.

Genetic drift refers to a heritable trait that appears by chance in a group and is accentuated by inbreeding. If two populations are too spatially separated for much interaction to occur (*isolation*), a trait may develop in one but not in the other. Unlike natural selection, genetic drift differentiates populations in nonadaptive ways. Natural selection and genetic drift promote differentiation. Countering them is gene flow via interbreeding, which acts to homogenize neighboring populations. Opportunities for

Table 6.1

Leading U.S. Ancestries Reported, 2016

Ancestry	Number (millions)	Percentage of Total Population
German	45.9	15.2
Irish	33.1	10.4
English	24.4	7.7
Italian	17.2	5.4
Polish	9.3	2.9
French	8.2	2.6
Scottish	5.5	1.7
Norwegian	4.5	1.4
Dutch	4.2	1.3
Swedish	3.9	1.2
Scotch-Irish	3.0	1.0

Note: More than 20 million persons indicated "American" as their ancestry, almost 4 million reported "European" and more than 3 million reported "Sub-Saharan African." These reported ancestries did not include options for "African American" or "Hispanic.". The tabulation is based on self-identification of respondents, not on objective criteria. Many persons reported multiple ancestries and were tabulated by the Census Bureau under each claim.

Source: *U.S. Census Bureau, American Community Survey, 2012–2016, 5-Year Estimates.*

interbreeding, always part of the spread and intermingling of human populations, have accelerated with the growing mobility and migrations of people in the past few centuries.

Racial categorization is a scientifically outdated way of making sense of human variation. Focusing on visible physical characteristics, anthropologists in the 18th and 19th centuries created a variety of racial classification schemes, most of which derived from geographical variations of populations. Some anthropological studies at that time attempted to link physical traits with intellectual ability in order to construct racial hierarchies that were used to justify slavery, imperialism, immigration restrictions, anti-miscegenation laws, and eugenics. Contemporary biology has rejected racial categorization as a meaningful description of human variation. Skin color does not correspond to genetic closeness between "race" groups. Further, pure races do not exist, and DNA-based evidence shows that there is more variation within the so-called racial groups than there is between the groups.

Living in a society where racial categorization has been widespread, we may be tempted to group humans racially and attribute intellectual ability, athletic prowess, or negative characteristics to particular racial groups. This is problematic for many reasons, the most important being that there is only one race—the human race. Second, intellectual ability as measured on standardized tests is mostly a function of socioeconomic status. Finally, the athletic abilities displayed by top athletes are the

property of particular individuals, not a group trait, and, like intellectual ability, are strongly influenced by social factors.

Race has no meaningful application to any human characteristics that are culturally acquired. That is, race is *not* equivalent to ethnicity or nationality and has no bearing on differences in religion or language. There is no "Irish" or "Hispanic" race, for example. Such groupings are based on culture, not genes. Culture summarizes the way of life of a group of people, and members of the group may adopt it irrespective of their individual genetic heritage. Although races do not exist in a scientific, biological sense, race persists as an idea and basis for group identity, and racism—prejudice and discrimination based on racial categories—is very much alive.

If racial categorization was scientifically valid, the categories should be universal. But instead they vary widely from country to country, reflecting the unique history and geography of particular places. In the United Kingdom, the census asks about ethnic rather racial identity. The U.K. census subdivides the Asian category into Indian, Pakistani, Bangladeshi, Chinese and other Asian with no special category for Japanese or Koreans. In Brazil, the census asks respondents to identify their race-color as indigenous or one of four skin tones: white, black, yellow, or brown. As society's understanding of race and ethnicity changes, so do the official census categories. In 2000, the U.S. Census Bureau asked respondents to classify themselves into one of five racial categories and answer a separate question about Hispanic status (which is considered an ethnic category, not a racial category). For the 2010 census, people had their choice of 14 categories. Starting in 2000, respondents were allowed to identify as "Some Other Race" and as more than one race (Table 6.2).

Table 6.2

U.S. Population by Race and Hispanic Status, 2016

	Number (millions)	Percent of U.S. Population
Total Population	**318.6**	**100.0**
White, Non-Hispanic	197.4	62.0
Hispanic or Latino (of any race)	55.2	17.3
Black or African American	40.2	12.6
Asian	16.6	5.2
American Indian and Alaskan Native	2.6	0.8
Native Hawaiian and other Pacific Islander	0.6	0.2
Two or more races	9,7	3.1

Note: Race as reported reflects the self-identification of respondents. Numbers do not sum to 100% because of overlap between the Hispanic and non-White racial categories.

Source: *U.S. Bureau of the Census, 2012–2016 American Community Survey 5-Year Estimates.*

Ethnic Diversity and Separatism

Ethnocentrism is the term describing a tendency to evaluate other cultures against the standards of one's own. It implies the feeling that one's own ethnic group is superior. Ethnocentrism can divide multiethnic societies by establishing rivalries and provoking social and spatial discord and isolation. In addition, it can be an emotionally sustaining force, giving familiar values and support to the individual in strange and complex surroundings. The ethnic group maintains familiar cultural institutions and shares traditional food and music. More often than not, it provides the friends, spouses, business opportunities, and political identification of ethnic group members.

Territorial isolation strengthens ethnic separatism and assists individual groups to retain their identification. In Europe, Asia, and Africa, ethnicity and territorial identity are inseparable. Ethnic minorities are first and foremost associated with *homelands*. This is true of the Welsh, Bretons, and Basques of Western Europe (Figure 12.20); the Slovenes, Croatians, or Bosnians of Eastern Europe (Figure 6.6a); the non-Slavic "nationalities" of Russia; and the immense number of ethnic communities of South and Southeast Asia. These minorities have specific spatial identity even though they may not have political independence.

Where ethnic groups are intermixed and territorial boundaries imprecise—the former Yugoslavia (Figure 6.6a) is an example—or where a single state contains disparate, rival populations—such as in the case of many African and Asian (Figure 6.6b) countries—conflict among groups can be serious if peaceful relations or central governmental control break down. **Ethnic cleansing**, a polite term with grisly implications, has motivated brutal civil conflict in parts of the former Soviet Union and Eastern Europe and in several African and southeast Asian countries. The Holocaust slaughter of millions of Jews before and during World War II in Western and Eastern Europe was an extreme case of ethnic extermination, but comparable murderous assaults on racial or ethnic target populations are as old as human history. Such "cleansing" involves, through mass genocide, the violent elimination of a target ethnic group from a particular geographic area to achieve racial or cultural homogeneity and expanded settlement area for the perpetrating state or ethnic group. Its outcome is not only an alteration of the ethnic composition of the region in which the violence takes place, but of the ethnic mix in areas to which displaced victim populations have fled as refugees.

Few true homelands exist within the North American cultural mix. However, the "Chinatown" and "Little Italy" enclaves within North American cities have provided both the spatial refuge and the support systems essential to new arrivals in an alien culture realm. Asian and West Indian immigrants in London and other English cities and foreign guest workers—originally migrant and temporary laborers, usually male—that reside in Continental European communities assume similar spatial separation. While serving a support function, this segregation is as much the consequence of the housing market and of public and private restriction as it is simply of self-selection. In Southeast Asia, with the exception of Thailand, Chinese communities remain aloof from the majority culture not as a stage on the route to assimilation but as a permanent chosen isolation.

By retaining what is familiar of the old in a new land, ethnic enclaves have reduced cultural shock and have paved the way for the gradual process of adaptation that prepares both individuals and groups to operate effectively in the new, larger **host society**—the established, dominant group. The traditional ideal of the United States "melting pot," in which ethnic identity and division would be lost and full amalgamation of all minorities into a blended, composite majority culture would occur, was the expectation voiced in the chapter-opening quotation. For many ethnic groups, however, that ideal has not become a reality.

Recent decades have seen a resurgence of cultural pluralism and an increasing demand for ethnic autonomy not only in North America but also in multiethnic societies around the world (see the feature "Nations of Immigrants"). At least, recognition is sought for ethnicity as a justifiable basis for special treatment in the allocation of political power, the structure of the educational system, the toleration or encouragement of minority linguistic rights, and other evidences of group self-awareness and promotion. In some multiethnic societies, second- and third-generation descendants of immigrants, now seeking "roots" and identity, embrace the ethnicity that their forebears sought to deny. At the same time, **xenophobia**—deep-rooted and unreasonable fears of foreigners on the part of the host society—has led to calls for immigration restrictions or even violence toward outsiders.

6.2 Immigration Streams

The ethnic diversity found in the United States and Canada today is the product of continuous flows of immigrants representing, at different periods, movements to this continent from nearly all of the cultures and races of the world (**Figure 6.3**). For the United States, that movement took the form of three distinct immigrant waves, all of which, of course, followed much earlier Amerindian arrivals.

The first wave, lasting from pioneer settlement to about 1870, was made up of two different groups. One comprised white arrivals from western and northern Europe, with Great Britain, Ireland, and Germany best represented. Together, they established a majority society controlled by Protestant Anglo-Saxons and allied groups. The second group of first-wave immigrants was African slaves brought involuntarily to the New World, comprising nearly 20 percent of the U.S. population in 1790.

The wave of social changes that brought about rapid population growth and large-scale immigration diffused outward from the British Isles. That second immigrant wave, from 1870 to 1921, was heavily weighted in favor of eastern and southern Europeans and Scandinavians, who comprised the majority of new arrivals by the end of the 19th century. The second period ended with congressional adoption of a quota system in 1921 regulating both the numbers of individuals who would be

Figure 6.3 Although it was not opened until 1892, New York Harbor's Ellis Island—the country's first federal immigration facility—quickly became the symbol of all the migrant streams to the United States. By the time that it was closed in late 1954, it had processed 17 million immigrants. Today, their descendants number more than 100 million Americans. A major renovation project was launched in 1984 to restore Ellis Island as a national monument.

©*Ron Chapple Stock/Alamy Stock Photo*

accepted and the countries from which they could come. The quota system limited the number of new immigrants from a country to 2 percent of the number that were already present in 1890. The quotas dramatically slowed immigration by southern and eastern Europeans who were believed to be racially inferior by some supporters of the quotas. The quota system, plus a worldwide depression and World War II (1939–1945), greatly slowed immigration until a third-wave migration, rivaling the massive influx of the late 19th and early 20th centuries, was launched with the Immigration and Nationality Act of 1965. At that time, the old national quota system of immigrant regulation was replaced by one that was more welcoming to newcomers from Latin America, Asia, and Africa. Since then, more than 40 million legal immigrants have entered the United States, in addition to an estimated 12 million unauthorized (undocumented or illegal) immigrants. Quickly, Hispanics, particularly Mexicans, dominated the inflow and became the largest segment of new arrivals. The changing source areas of the newcomers are traced in Table 6.3 and **Figure 6.4**.

While the United States accepts the largest total number of immigrants of any country, and 13 percent of its population was born outside its borders, the proportion of the foreign-born population is even higher in Australia (28 percent) and Canada (22 percent). Canada experienced three quite different immigration streams (Table 6.4). Until 1760, most settlers came from France. After that date, the pattern abruptly altered as a flood of United Kingdom (English, Irish, and Scottish) immigrants arrived. Many came by way of the United States, fleeing, as

Table 6.3

Immigrants to the United States: Major Flows by Origin

Ethnic Groups	Time Period	Numbers in Millions (approximate)
Blacks	1650s–1800	1
Irish	1840s and 1850s	1.75
Germans	1840s–1880s	4
Scandinavians	1870s–1900s	1.5
Poles	1880s–1920s	1.25
East European Jews	1880s–1920s	2.5
Austro-Hungarians	1880s–1920s	4
Italians	1880s–1920s	4.75
Mexicans	1950s–Present	13
Cubans	1960s–Present	1.4
Asians	1960s–Present	9

Loyalists, to Canada during and after the American Revolutionary War. Others came directly from overseas. Another pronounced shift in arrival patterns occurred during the 20th century as the bulk of new immigrants came from Continental Europe

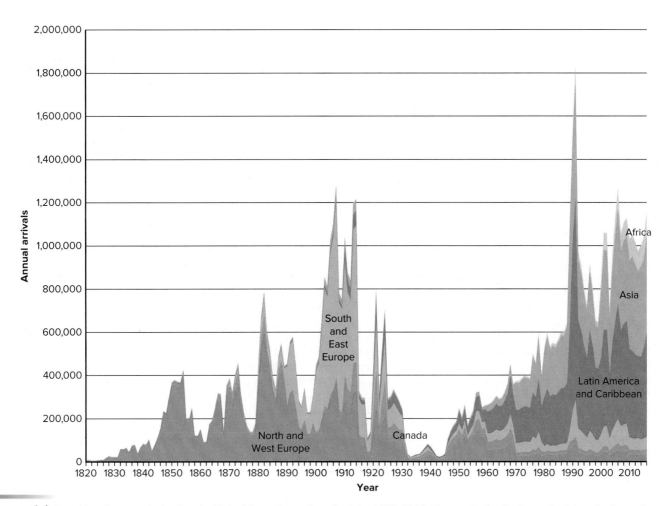

Figure 6.4 Legal immigrants admitted to the United States by region of origin, 1820–2016. The graph clearly shows the dramatic change in geographic origins of immigrants and the effect of immigration restrictions, the Great Depression, and World War II. After 1965, immigration restrictions based on national origin were shifted to priorities based on family reunification and needed skills and professions.

Sources: Data from United States Department of Homeland Security, Yearbook of Immigration Statistics, *annual.*

Table 6.4		
Canadian Population Ranked 2016		
Rank	**Ethnic Group**	**Total Number (millions)**
1	Canadian	11.1[1]
2	English	6.3
3	Scottish	4.8
4	French	4.7
5	Irish	4.6
6	German	3.3
7	Chinese	1.8
8	Italian	1.6
9	First Nations (North American Indian)	1.5
10	East Indian	1.4

[1]Includes both single ethnic origin and multiple ethnic origin responses.
Source: *Statistics Canada, 2017.*

and Asia. Immigration accounts for the majority of Canada's population growth.

The cultural diversity of the United States has increased as its immigration source regions have changed from the original European areas to Latin America and Asia, and both the number of visible and vocal ethnic communities and the number of regions housing significant minority populations have multiplied. Simultaneously, the proportion of foreign-born residents has increased in the U.S. population mix. In 1920, at the end of the period of the most active European immigration, more than 13 percent of the American population had been born in another country. That percentage declined each decade until a low of 4.8 percent foreign-born was reported in 1970 before rebounding in recent decades. As was the case during the 1920s, growing influxes from new immigrant source regions and, particularly, the large numbers of unauthorized entrants prompted movements to halt the flow and to preserve the ethnic status quo (see the feature "Porous Borders," Chapter 3). During the 2016 U.S. presidential election campaign, candidates debated restrictions on travel from specific countries, immigration restrictions, and construction of a border wall between Mexico and the United States.

Geography and Citizenship

Nations of Immigrants

Americans, steeped in the country's "melting pot" myth and heritage, are inclined to forget that many other countries are also "nations of immigrants" and that their numbers are dramatically increasing. In the United States, Canada, Australia, and New Zealand, early European colonists (and, later, immigrants from other continents) overwhelmed indigenous populations. In each, immigration has continued, contributing not only to national ethnic mixes diversity but maintaining or enlarging the proportion of the population that is foreign-born. In Australia and Canada, that proportion exceeds that of the United States.

In Latin America, foreign population domination of native peoples was and is less complete and uniform than in the United States and Canada. While in nearly all South and Central American states, European and other nonnative ethnic groups dominate the social and economic hierarchy, in a few they constitute only a minority of the total population. In Bolivia, for example, the vast majority (71 percent) pride themselves on their Native American descent, and Amerindians comprise between 25 percent and 55 percent of the populations of Bolivia, Guatemala, Peru, and Ecuador. Mestizo (mixed European and Amerindian ancestry) populations are the majority in many Latin American countries. But nonnative, largely European, ethnics make up essentially all—more than 94 percent—of the population of Argentina, Costa Rica, and southern Chile.

The original homelands of those immigrant groups are themselves increasingly becoming multiethnic, and several European countries are now home to as many or more of the foreign-born proportionately than is the United States. Some 25 percent of Switzerland's population, 13 percent of Germany's, and 17 percent of Sweden's are foreign-born, compared with 13 percent in the United States. Many came as immigrants and refugees fleeing unrest or poverty in post-communist Eastern Europe. Many are guest workers and their families who were earlier recruited in Turkey and North Africa; or they are immigrants from former colonial or overseas territories in Asia, Africa, and the Caribbean. More than a million refugees were admitted to Germany between 2015 and 2016.

The trend of ethnic mixing is certain to continue and accelerate. Cross-border movements of migrants and refugees in Africa, Asia, the Americas, and Europe are continuing common occurrences, reflecting growing incidences of ethnic strife, civil wars, famines, and economic hardships. But of even greater long-term influence are the growing disparities in population numbers and economic wealth between the older developed states and the developing world. The population of the world's poorer countries is growing twice as fast as Europe's did in the late 19th century, when that continent fed the massive immigration streams across the Atlantic. The current rich world, whose fertility rates are below replacement levels, will increasingly be a magnet for those from poorer countries where fertility rates are high. The economic and population pressures building in the developing world and the below-replacement fertility rates in developed countries ensure greater international and intercontinental migration and a rapid expansion in the numbers of "nations of immigrants."

Many of those developed host countries are beginning to resist that flow. Although the Universal Declaration of Human Rights declares individuals are to be free to move within or to leave their own countries, no right of admittance to any other country is conceded. Political asylum is often—but not necessarily—granted; refugees or migrants seeking economic opportunity or fleeing civil strife or starvation have no claims for acceptance. Increasingly, they are being turned away. Britain's vote in 2016 to leave the European Union was partly motivated by a desire for greater immigration controls.

Nor is Europe alone. Hong Kong ejects Vietnamese refugees; Congo orders Rwandans to return to their own country; India tries to stem the influx of Bangladeshis; the United States rejects "economic refugees" from Haiti. Algerians are increasingly resented in France as their numbers and cultural presence increase. Turks feel the enmity of a small but violent group of Germans, and East Indians and Africans find growing resistance among the Dutch. In many countries, policies of exclusion or restriction appear motivated by unacceptable influxes of specific racial, ethnic, or national groups.

Questions to Consider

1. Do you think all people everywhere should have a universal right to be admitted to the country of their choice? Why or why not?

2. Do you think it appropriate that destination states make a distinction between political and economic refugees? Why or why not?

3. Do you think it legitimate for countries to establish immigration quotas based on national origin or to classify certain potential immigrants as unacceptable or undesirable on the grounds that their national, racial, or religious origins are incompatible with the culture of the prospective host country? Why or why not?

6.3 Acculturation and Assimilation

In the United States and Canada, at least, the sheer volume of multiple immigration streams makes the concept of "minority" suspect when no single "majority" ethnic group exists (see Table 6.2). Indeed, high rates of immigration and subreplacement fertility rates among whites have placed the country on the verge of becoming a state with no racial—as well as no ethnic—majority. By the mid-21st century, the United States will be truly multiracial, with no group constituting more than 50 percent of the total population. Even now, American society is a composite of unity and diversity, with immigrants being both shaped by and shaping the larger community that they have joined.

Amalgamation theory is the formal term for the traditional "melting pot" concept of the merging of many immigrant ethnic heritages into a composite American mainstream. Dominant in the late 19th and early 20th centuries, amalgamation theory has more recently been rejected for a number of reasons. Recent experience in Western European countries and the United States and Canada indicates that immigrants strongly retain and defend their ethnic identities. On a practical level, ethnic distinctiveness is buttressed by the current ease—through radio, telephone, Internet, television, and rapid transportation—of communication and identification with the homeland societies of immigrants. More importantly, multiculturalism acknowledges the unique value of the world's diverse cultures. In an era of space-time compression, when the world seems to be getting smaller and the pace of change is accelerating, ethnic identities may offer reassuring stability. The old melting pot concept of the United States has largely dissolved, replaced with a greater emphasis on preserving the diverse cultural heritages of the country's many ethnic components.

Nonetheless, as we shall see, all immigrant groups find a controlling host group culture in place, with accustomed patterns of behavior and a dominant language for the workplace and government. The customs and practices of the host society have to be learned by newcomers if they are to be accepted. The process of **acculturation** is the adoption by the immigrants of the values, attitudes, ways of behavior, and speech of the receiving society. In the process, the ethnic group loses much of its separate cultural identity as it comes to accept the culture of the larger host community. It may, however, resist total absorption into the host society and proudly retain identifying features of its distinctive ethnic heritage: adherence to an ethnic worship center, celebration of traditional national or religious holidays with parades or festivals, and the like. To the extent that those ethnic retentions and identifications are long-lasting and characteristic of multiple ethnic groups, the presumed ideal of the melting pot is unattained, and a "salad bowl" ethnic mixture is the result.

Although acculturation most usually involves a minority group adopting the patterns of the dominant population, the process can be reciprocal. That is, the dominant group may also adopt at least some patterns and practices typical of new minority groups and become a "lumpy stew," in which the immigrant groups maintain their identities while both taking on the flavor of the host society and adding new flavor to the broader societal mix. New music and dance styles, ethnic foods, and a broadened selection of fruits and vegetables are familiar evidences of those immigrant contributions. For example, the most popular evening meal in the United Kingdom is curry—a dish brought by Indian and Pakistani immigrants.

Acculturation is a slow process for many immigrant individuals and groups, and the parent tongue may be retained as an identifying feature even after fashions of dress, food, and customary behavior have been substantially altered in the new environment. In 2016, one in five Americans above the age of 5 spoke a language other than English in the home; for almost two-thirds of them, that language was Spanish. In the light of recent immigration trends, we can assume that the number of people speaking a foreign language at home will only increase. The retention of the native tongue is encouraged rather than hindered by American civil rights regulations that give to new immigrants the right to bilingual education and (in some cases) special assistance in voting in their own language.

The language barrier that has made it difficult for foreign-born groups, past and present, to gain quick entrance to the labor force has encouraged their high rate of employment in small, family-held businesses. The consequence has been a continuing stimulus to the American economy and, through the creation of new neighborhood enterprises, the maintenance of the ethnic character of immigrant communities (**Figure 6.5**). The result has also been the gradual integration of the new arrivals into the economic and cultural mainstream of American society.

Figure 6.5 Immigrant neighborhoods exhibit a different mix of business than do established, older-majority neighborhoods. Food stores and specialty shops catering to the ethnic group predominate.

©Mark Bjelland

When an ethnic group can no longer be distinguished from the wider society, full **assimilation** has occurred. Full assimilation goes beyond acculturation; it implies integration into a common cultural life through shared experience, language, intermarriage, and sense of history. Assimilation of an ethnic group involves upward social and economic mobility, employment in a full range of occupations, establishment of social ties with members of the host society, and the adoption of prevailing attitudes and values. Employment segregation and intermarriage rates are important measures of assimilation. Assimilation is a two-way street. Not only does it require immigrant groups to absorb majority cultural values and practices, but it also demands that the majority society give full acceptance to members of the minority group and allow them to rise to positions of authority and power. Because where we live reflects our social status and influences our social ties and experiences, full assimilation is inevitably spatial. **Spatial assimilation** is measured by the degree of residential segregation that sets off the minority group from the larger general community. For most of the "old" (pre-1921 European) immigrants and their descendants, assimilation is complete. Most indicative of at least individual if not total group assimilation is election or appointment to high public office and business leadership positions. The 2008 election of President Barack Obama was an important marker of the assimilation of African Americans in the United States.

Assimilation is frequently partial, or segmented. Assimilation does not necessarily mean that ethnic consciousness or awareness of racial and cultural differences is lost. Evidence suggests that as ethnic minorities begin to achieve success and enter into mainstream social, political, and economic life, awareness of ethnic differences may be heightened. Frequently, ethnic identity may be most clearly experienced and expressed by those who can most successfully assimilate but who choose to promote group awareness and ethnic mobilization movements. That promotion is a reflection of pressures of American urban life and the realities of increased competition. Those pressures transform formerly isolated groups into recognized, self-assertive ethnic minorities pursuing goals and interests dependent on their position within the larger society.

While in the United States, it is usually expected that ethnic groups will undergo full assimilation, Canada established multiculturalism in the 1970s as a national policy. It was designed to reduce tensions between ethnic and language groups and to recognize that each thriving culture is an important part of the country's heritage. Since 1988, multiculturalism has been formalized by an act of the Canadian parliament and supervised by a separate government ministry. An example of its practical application can be seen in the way Toronto, Canada's largest and the world's most multicultural city routinely sends out property tax notices in six languages—English, French, Chinese, Italian, Greek, and Portuguese.

Australia, Canada, and the United States seek to incorporate their immigrant minorities into composite national societies. In other countries, quite different attitudes and circumstances may prevail when indigenous—not immigrant—minorities feel their cultures and territories being threatened. The Sinhalese comprise 75 percent of Sri Lanka's population, but the minority Tamils waged years of guerrilla warfare to defend what they see as majority threats

to their culture, rights, and property. In India, Kashmiri nationalists fight to separate their largely Muslim valley from the Hindu majority society. Expanding ethnic minorities made up nearly 8.5 percent of China's 2000 population total. Some, including Tibetans, Mongols, and Uighurs, face assimilation largely because of massive migrations of ethnic Chinese into their traditional homelands. And in many multiethnic African countries, single-party governments seek to impose a sense of national unity on populations whose primary loyalties are rooted in their tribes and regions and not the state that is composed of many tribes (see Figure 12.5). Across the world, conflicts among ethnic groups within states have proliferated in recent years. Armenia, Azerbaijan, Burma, Burundi, Ethiopia, Indonesia, Iraq, Russia, Rwanda, and the former Yugoslavia are others in a long list of countries where ethnic tensions have erupted into civil conflict.

Basques and Catalans of Spain and Corsicans, Bretons, and Normans of France have only recently seen their respective central governments relax strict prohibitions on teaching or using the languages that identified those ethnic groups. On the other hand, in Bulgaria, ethnic Turks, who unofficially comprise 10 percent of the total population, officially ceased to exist in 1984 (at least temporarily) when the government obliged Turkish speakers and Muslims to replace their Turkish and Islamic names with Bulgarian and Christian ones. The government also banned their language and strictly limited practice of their religion. The intent was to impose assimilation.

Elsewhere, ethnic minorities—including immigrant minorities—have grown into majority groups, posing the question of who will assimilate whom. Ethnic Fijians sought to resolve that issue by staging a coup to retain political power when the majority immigrant ethnic Indians came to power by election in 1987, and another in 2000 after the election of an ethnic-Indian prime minister. As these and innumerable other examples from all continents demonstrate, North American experiences and expectations have limited application to other societies differently constituted and motivated.

Areal Expressions of Ethnicity

Throughout much of the world, the close association between territory and ethnicity is well recognized and sometimes politically disruptive. Indigenous ethnic groups have developed over time in specific locations and, through ties of kinship, language, culture, religion, and shared history, have established themselves in their own and others' eyes as distinctive peoples with defined homeland areas. The boundaries of most countries encompass a number of racial or ethnic minorities, whose demands for special territorial recognition have increased rather than diminished with advances in economic development, education, and self-awareness (**Figure 6.6**).

The dissolution of the Soviet Union in 1991, for example, not only set free the 14 ethnically based union republics that formerly had been dominated by Russia and Russians, but also opened the way for many smaller ethnic groups—the Chechens of the northern Caucasus, for example—to seek recognition and greater local control from the majority populations, including Russians, within whose territory their homelands lay. In Asia, the

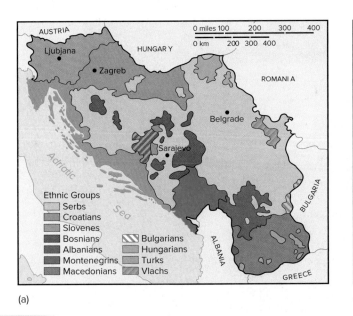

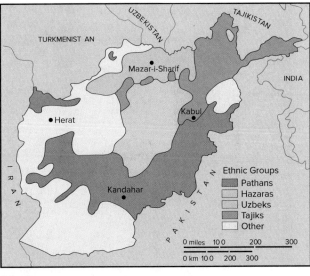

(a)　　　　　　　　　　　　　　　　　　　　　　　(b)

Figure 6.6 (*a*) Ethnicity in former Yugoslavia. Yugoslavia was formed after World War I (1914–1918) from a patchwork of Balkan states and territories, including the former kingdoms of Serbia and Montenegro, Bosnia-Herzegovina, Croatia-Slavonia, and Dalmatia. The authoritarian central government, created in 1945 and led by Josep Broz Tito, tried to forge a new Yugoslav ethnic identity but failed when in 1991, Serb minorities voted for regional independence. In response, Serb guerillas backed by the Serb-dominated Yugoslav military engaged in a policy of territorial seizure and "ethnic cleansing" to secure areas claimed as traditional Serb "homelands." Religious differences between Eastern Orthodox, Roman Catholic, and Muslim adherents compound the conflicts rooted in nationality. (*b*) Afghanistan houses Pathan, Tajik, Uzbek, and Hazara ethnic groups speaking Pashto, Dari Persian, Uzbek, and several minor languages, and split between majority Sunni and minority Shia Moslem believers. Ethnic and local warlord rivalries and regional guerilla resistance to the central government, supported by the North Atlantic Treaty Organization (NATO), contribute to national instability.

The Rising Tide of Nationalism

In recent decades, we have seen spreading ethnic self-assertion and demands for national independence and cultural purification of homeland territories. To some, these demands and the conflicts they frequently engender are the expected consequences of the decline of strong central governments and imperial controls. It has happened before. The collapse of the Roman and the Holy Roman empires were followed by the emergence of many new kingdoms and city-states during medieval and Renaissance Europe. The fall of Germany and the Austro-Hungarian Empire after World War I saw the creation of new ethnically based countries in Eastern Europe. The brief decline of post-czarist Russia permitted freedom for Finland, and, for 20 years, for Estonia, Latvia, and Lithuania. The disintegration of British, French, and Dutch colonial control after World War II resulted in new state formation in Africa, South and East Asia, and Oceania.

Few empires have collapsed as rapidly and completely as did that of the Soviet Union and its Eastern European satellites in the late 1980s and early 1990s. In the subsequent loss of strong central authority, the ethnic nationalisms that communist governments had for so long tried to suppress asserted themselves in independence movements. At one scale, the Commonwealth of Independent States and the republics of Estonia, Latvia, Lithuania, and Georgia emerged from the former Soviet Union. At a lesser territorial scale, ethnic animosities and assertions led to bloodshed in the Caucasian republics of the former USSR, in former Yugoslavia (see Figure 6.6a), in Moldova, and elsewhere, while Czechs and Slovaks agreed to peacefully go their separate ways at the start of 1993.

Democracies, too, risk disintegration or division along ethnic, tribal, or religious lines, at least before legal protections for minorities are firmly in place. Voter referenda on independence of Scotland from the United Kingdom and Catalonia from Spain are recent examples. African states with their multiple ethnic loyalties (see Figure 12.5 in Chapter 12) have frequently used those divisions to justify restricting political freedoms and continuing one-party rule. However, past and present ethnically inspired civil wars and regional revolts in Somalia, Ethiopia, Nigeria, Uganda, Liberia, Angola, Rwanda, Burundi, and elsewhere show the fragility of the political structure on that continent.

Figure 6.7 Although all of North America was once theirs alone, Native Americans have become now part of a larger cultural mix. In the United States, their areas of domination have been reduced to reservations found largely in the western half of the country and to the ethnic provinces shown in Figure 6.10. These are often areas to which Amerindian groups were relocated, not necessarily the territories occupied by their ancestors at the time of European colonization.

©Luc Novovitch/Alamy Stock Photo

Indian subcontinent was subdivided to create separate countries with primarily religious-territorial affiliations, and the country of India itself has adjusted the boundaries of its constituent states to accommodate linguistic-ethnic realities. Other continents and countries show a similar acceptance of the importance of ethnic territoriality in their administrative structure (see the feature "The Rising Tide of Nationalism").

With the exceptions of the Québécois (French Canadians) and some Native American groups, the United States and Canada lack the ethnic homelands that are so characteristic elsewhere in the world (**Figure 6.7**). The general absence of such claims is the result of the immigrant nature of American society. Even the Native American "homeland" reservations in the United States are dispersed, noncontiguous, and in large part artificial impositions.[2] In general, Native Americans were displaced from potentially productive agricultural lands and are today concentrated in the Arizona-New Mexico border region, Great Plains, and Oklahoma. The spatial pattern of ethnicity that has developed in North America is not based on absolute ethnic dominance but on interplay between a majority culture and, usually, several competing minority groups. It shows the enduring consequences of early settlement and the changing structure of a fluid, responsive, mobile North American society

Amerindians were never a single ethnic or cultural group and cannot be compared to a European national immigrant group in homogeneity. Arriving over many thousands of years, from many different origin points, with different languages, physical characteristics, customs, and skills, they are in no way comparable to a culturally uniform Irish or Slovak ethnic group arriving during the 19th century, or Salvadorans or Koreans during the 21st. Unlike most other minorities in the American melting pot, Amerindians have generally rejected the goal of full and complete assimilation into the national mainstream culture.

Charter Cultures

Except for the Québécois and Native Americans, no single ethnic minority homeland area exists in the United States and Canada today. However, a number of separate social and ethnic groups are of sufficient size and regional concentration to have put their impression on particular areas. Part of that imprint results from what the geographer Wilbur Zelinsky termed the "doctrine of **first effective settlement.**" That principle holds that

> Whenever an empty territory undergoes settlement, or an earlier population is dislodged by invaders, the specific characteristics of the first group able to effect a viable, self-perpetuating society are of crucial significance for the later social and cultural geography of the area, no matter how tiny the initial band of settlers may have been.[3]

[2]In Canada, a basic tenet of Aboriginal policy since 1993 has been the recognition of the inherent right of self-government under Section 35 of the Canadian Constitution. The territory of Nunavut, the central and eastern portion of the earlier Northwest Territories, is based largely on Inuit land claims and came into existence as a self-governing district in 1999.

[3]*The Cultural Geography of the United States.* Rev. ed. (Englewood Cliffs, N.J.: Prentice-Hall, 1992), p. 13.

On the North American stage, the English and their affiliates, although few in number, were the first effective entrants in the eastern United States, and they shared with the French that role in eastern Canada. Although the French were ousted from parts of Seaboard Canada, they retained their cultural and territorial dominance in Quebec province. In the United States, British immigrants (English, Welsh, Scottish, and Scotch-Irish) constituted the main portion of the new settlers in eastern Colonial America and retained their significance in the immigrant stream until after 1870.

The English, particularly, became the **charter group,** the dominant first arrivals establishing the cultural norms and standards against which other immigrant groups were measured. It is understandable, then, in the light of Zelinsky's "doctrine," that English became the national language; English common law became the foundation of the American legal system; British philosophers influenced the considerations and debates leading to the U.S. Constitution; English place names predominate in much of the country; and the influence of English literature and music remains strong. By their early arrival and initial dominance, the British established the majority culture of the United States and Canada; their enduring ethnic impact is felt even today.

Somewhat comparable to the British domination in the East is the Hispanic influence in the Southwest. Spanish and Mexican explorers established settlements in New Mexico a generation before the Pilgrims arrived at Plymouth Rock. Spanish-speaking El Paso and Santa Fe were prospering before Jamestown, Virginia, was founded in 1607. Although subsequently incorporated into an expanding "Anglo"-controlled cultural realm and dominated by it, the early established Hispanic culture, reinforced by continuing immigration, has proved enduringly effective. From Texas to California, Spanish-derived social, economic, legal, and cultural institutions and traditions remain an integral part of contemporary life—from language, art, folklore, and names on the land through Spanish water law to land ownership patterns reflecting Spanish tenure systems.

Ethnic Islands

Because the British already occupied much of the agricultural land of the East, other, later immigrant streams from Europe were forced to "leapfrog" those areas and seek settlement opportunities farther west. Those groups who arrived after most of the productive agricultural lands were claimed ended up settling in mining or manufacturing areas. The Germans of the Appalachian uplands, the Middle West, and Texas, Scandinavians in Minnesota and the Dakotas, the various Slavic groups farther west on the Plains, and Italians in California are examples of later arrivals occupying and becoming identified with different sections of the United States. Such areas of ethnic concentration are known as **ethnic islands,** the dispersed and rural counterparts of urban ethnic neighborhoods (**Figure 6.8**).

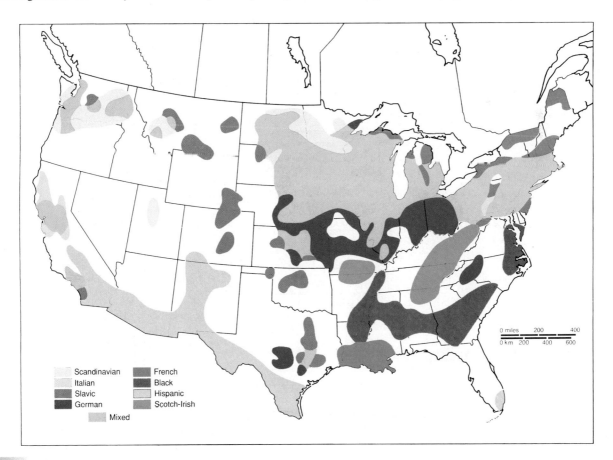

Figure 6.8 Ethnic islands in the United States.

Source: Russel Gerlach, Settlement Patterns in Missouri (Columbia: University of Missouri Press, 1986), p. 41.

Characterized usually by a strong sense of community, ethnic islands frequently placed their distinctive imprint on the rural landscape by retaining home-country barn and house styles and farmstead layouts, while their inhabitants may have retained their own language, manner of dress, and customs. With the passing of generations, rural ethnic identity has tended to diminish, and recent adaptations and dispersions have occurred. When long-enduring through spatial isolation or group determination, ethnic islands have tended to be considered landscape expressions of folk culture rather than purely ethnic culture; we shall return to them in that context in Chapter 7.

Similar concentrations of immigrant arrivals are found in Canada. Descendants of French and British immigrants dominate its ethnic structure, both occupying primary areas too large to be considered ethnic islands. Ethnic islands are most pronounced on the agricultural lands of the Western Prairie provinces, where Ukrainians are the third-largest group. The ethnic diversity of that central portion of Canada is suggested by **Figure 6.9**.

European immigrants arriving in the United States and Canada by the middle of the 19th century frequently took up tracts of rural land as groups rather than as individuals, assuring the creation of at least small ethnic islands. German and Ukrainian Mennonites in Manitoba and Saskatchewan, for example; Doukhobors in Saskatchewan; Mennonites in Alberta; Hutterites in South Dakota, Manitoba, Saskatchewan, and Alberta; the Pennsylvania Dutch (whose name is a corruption of *Deutsch,* or "German," their true nationality); Frisians in Illinois; and other ethnic groups settled as collectives. They sometimes acted on the advice and the land descriptions reported by advance agents sent out by the group. In most cases, sizable extents of rural territory received the imprint of a group of immigrants acting in concert. However, later in the century and in the less arable sections of the western United States, the disappearance of land available for homesteading and the changing nature of immigrant flows reduced the incidence of cluster settlement. Impoverished individuals rather than financially solid communities sought American refuge and found it in urban locations. While cluster migration created some ethnic concentrations of North America—in the Carolinas, Wisconsin, Kansas, Nebraska, and Oklahoma, for example—others evolved from the cumulative effect of

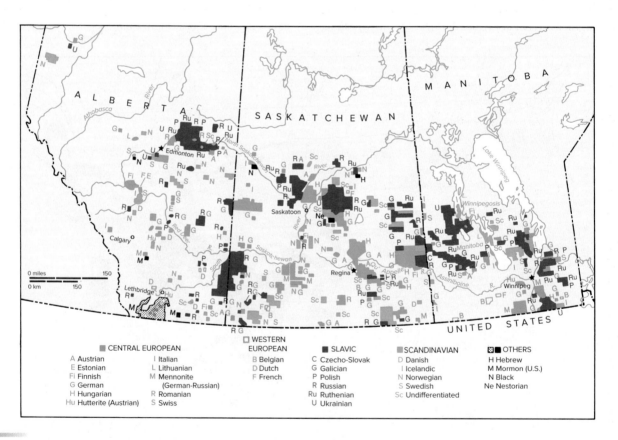

Figure 6.9 Ethnic diversity in the Prairie provinces of Canada. In 1991, 69 percent of all Canadians claimed some French or British ancestry. For the Prairie provinces, with their much greater ethnic mixture, only 15 percent declared any British or French descent. Immigrants comprise a larger share of Canadian population than they do of the U.S. population. Early in the 20th century, most newcomers were located in rural western Canada; and by 1921, about half the population of the Prairie provinces was foreign-born. Recent immigrants are mostly from Asia and concentrated in the three largest metropolitan centers of Toronto, Montreal, and Vancouver.

Source: D.G.G. Kerr, A Historical Atlas of Canada, *2nd ed., 1966. Thomas Nelson & Sons Ltd., 1966.*

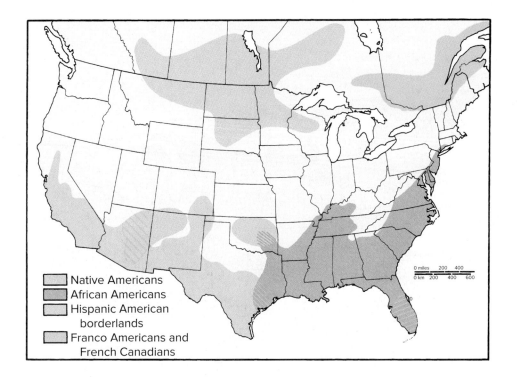

Native Americans
African Americans
Hispanic American
borderlands
Franco Americans and
French Canadians

Figure 6.10 Four North American ethnic groups and their provinces. Note how this generalized map differs from the more detailed picture of ethnic distributions shown in Figure 6.8.

chain migration—the assemblage in one area of the relatives, friends, or unconnected compatriots of the first arrivals, attracted both by favorable reports and by familiar presences in specific locales of the New World (see Section 3.5).

Some entire regions of North America—vastly larger than the distinctive ethnic islands—have become associated with larger ethnic or racial aggregations numbering in the thousands or millions. Such **ethnic provinces** include French Canadians in Quebec; African Americans in the southeast United States; Native Americans in Oklahoma, the Southwest, the Northern Plains, and Prairie provinces; and Hispanics in the southern border states of the western United States (**Figure 6.10**). The spatial distribution of Native Americans reflects a history of forced relocations onto reservations (**Figure 6.11**). The identification of distinctive communities with extensive regional units persists, even though ethnicity and race have not been fully reliable bases for dividing North America into regions. Cultural, ethnic, and racial mixing has been too complete to permit U.S. counterparts of Old World ethnic homelands to develop, even in the instance of the now-inappropriate association of African Americans with southern states.

The Black or African American Population

African Americans, involuntary immigrants to the continent, were nearly exclusively confined to rural areas of the South and Southeast prior to the Civil War (**Figure 6.12**). Even after emancipation, most remained on the land in the South. During the first two-thirds of the 20th century, however, those established patterns of southern rural residence and farm employment underwent profound changes. The decline of subsistence farming and share-cropping, the mechanization of southern agriculture, the demand for factory labor in northern cities starting with World War I (1914–1918), and the general urbanization of the American economy all induced African Americans to abandon the South in a "great migration" northward in search of manufacturing jobs and greater social equality.

Created by the U.S. Coast Survey using 1860 Census data, the map shown in Figure 6.12 was perhaps the first choropleth map depicting human geographic data. The map was useful in the Union's war efforts, convincing the public that slavery was the root cause of the war and showing that some sections of the South (such as Appalachia) had few slaves. Note the sharp difference in slave populations between the eastern and western sections of Virginia. That difference led to the secession of West Virginia from Virginia in 1862.

Between 1940 and 1970, more than 5 million African Americans left their homes in the South, in the largest internal ethnic migration ever experienced in the United States. A modest return migration of, particularly, middle-class African Americans that began in the 1970s picked up speed during the 1980s and gave evidence of being a reverse "great migration" in the past decades. That return movement, encouraged by an improving economic and racial environment in the South and by African Americans' enduring strong cultural and family ties, suggests a net inflow to the South of some 3 million African Americans between 1975 and 2010—more than half of the post-1940 out-migration. Prominent in that reverse flow are

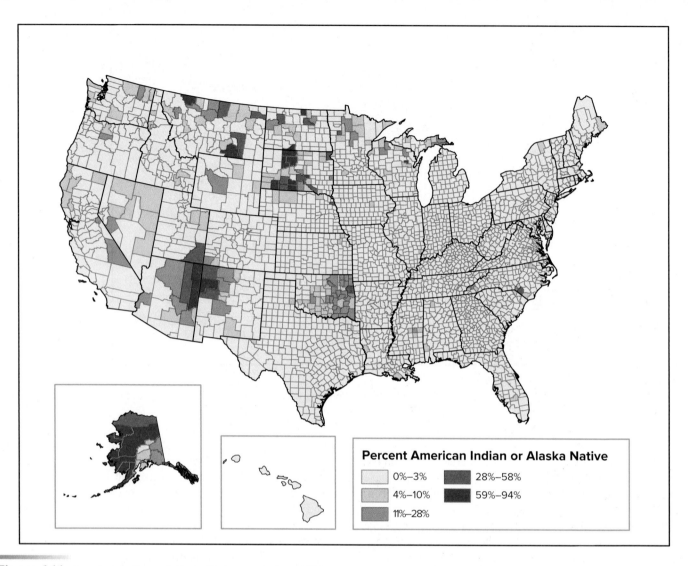

Figure 6.11 American Indian and Alaska Native populations, 2016.

African American professionals leaving such northern strongholds as Chicago, Detroit, and Philadelphia and settling in the suburbs of Atlanta, Charlotte, Dallas, Houston, Miami, and other Sunbelt metropolitan areas.

The growing African American population (about 13 percent of all Americans) is more urbanized than the general population and yet retains evidence of its rural roots in the former Confederacy states of the South (**Figure 6.13**). The South is home to more than half of the African American population.

Black Americans, like Asian Americans and Hispanics, have had thrust on them an assumed common ethnicity that does not, in fact, exist. Because of prominent physical or linguistic characteristics, quite dissimilar ethnic groups have been categorized by the white, English-speaking majority in ways totally at odds with the realities of their separate national origins or cultural inheritances. Although the U.S. Census Bureau makes some attempt to subdivide Asian ethnic groups—Chinese, Filipino, and Korean, for example—these are distinctions not necessarily recognized by members of the white majority. But even the Census Bureau, in its summary statistics, has treated "Black" and "Hispanic Origin" as catchall classifications that suggest ethnic uniformities where none necessarily exist.

In the case of African Americans, such categorizing is of decreasing relevance for two reasons. First, immigration has made the black population increasingly heterogeneous. Between 1970 and 2016, the share of foreign-born in the black community rose from a little more than 1 percent to about 10 percent. The immigrants originated in many countries of the Caribbean and Africa, with the largest percentages coming from Haiti, Jamaica, the Dominican Republic, Trinidad and Tobago, and Nigeria. Second, the earlier overwhelmingly rural Southern black community has become subdivided along socioeconomic rather than primarily regional lines, the result of its 20th-century spatial mobility encouraged by northern industrial job opportunities first apparent during World War I and continuing through the 1960s. Government intervention, which mandated and promoted racial equality, further opened the way for the creation of black urban middle and upper income and professional groups. Now, by separate experiences, African Americans have become as diversified as other ethnic or racial groups.

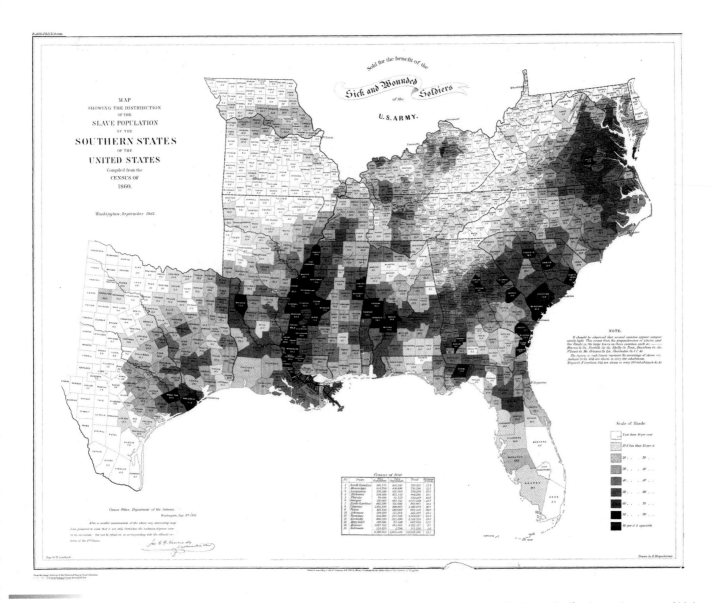

Figure 6.12 Slave population, 1860. The map shows the percentage of slaves in the total population, with darker shading indicating areas of high concentration such as South Carolina and the Mississippi Delta.

Source: Library of Congress, Geography and Map Division (g3861ecw0013200).

Hispanic Concentrations

Similarly, the members of the multiracial, multinational, and multicultural composite population lumped by the Census Bureau into the single category of "Hispanic, Latino, or Spanish origin" are not a homogeneous group. Indeed, it was the Census Bureau, not the group itself, that created the concept and distinct ethnic category of "Hispanics." Prior to 1980, no such composite group existed.

Hispanic Americans are a diverse group. By commonly used racial categories, they may also be white, black, or Native American; more than 50 percent of Hispanic Americans, in fact, report themselves to be white and more than one-third reported "some other race." Individually, they are highly diversified by country and culture of origin. Collectively, they also constitute the most rapidly growing minority component of U.S. residents—accounting for half of the country's population growth between

2000 and 2010, and surpassing African Americans as the largest minority, as Table 6.5 indicates. By 2016, the Hispanic population had grown to 57 million—18 percent of the U.S. population.

Mexican Americans account for about two-thirds of all Hispanic Americans (Table 6.6). Their highest concentrations are located in the southwestern states that constitute the ethnic province called the Hispanic American borderland (Figure 6.14). Beginning in the 1940s, the Mexican populations in the United States became increasingly urbanized and dispersed, losing their earlier primary identification as agricultural *braceros* (seasonal laborers) and as residents of the rural areas of Texas, New Mexico, and Arizona. California rapidly increased its Mexican American populations, as did the Midwest, particularly the chain of industrial cities near Chicago. Wherever they settled in the United States, Mexican immigrants represented a loss to their home country of a significant portion of its labor force and most educated residents.

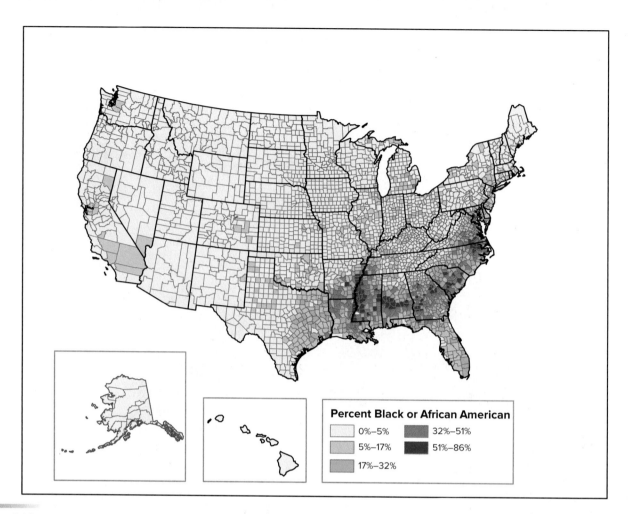

Figure 6.13 African American population, 2016. African Americans are particularly significant in the largely rural, relatively low-population states of the Southeast in a pattern reminiscent of their distribution in 1860.

Table 6.5

Actual and Projected U.S. Population Mix: 2000, 2016, 2030, and 2060

Population Group	Percent of Total			
	2000	2016	2030	2060
Non-Hispanic White (one race)	69.1	61.3	55.8	44.3
Hispanic or Latino	12.5	17.8	21.1	27.5
Black or African American (one race)	12.3	13.3	13.8	15.0
American Indian/Alaska Native (one race)	0.9	1.3	1.3	1.4
Asian/Pacific Islander (one race)	3.7	5.9	7.1	9.4
Two or More Races	2.4	2.6	3.6	6.2

Note: Black, Asian, and Native American categories exclude Hispanics, who may be of any race.

Source: *U.S. Bureau of the Census, Projected Race and Hispanic Origin, 2017–2060. Totals do not round to 100 percent because of "other race" category and because Hispanics may be of any race.*

Table 6.6

Composition of U.S. Hispanic Population, 2016

Hispanic Subgroup	Number (millions)	Percent
Mexican	35.1	63.6
Puerto Rican	5.3	9.6
Central American	5.0	9.1
South American	3.3	6.1
Cuban	2.1	3.8
Dominican	1.8	3.2
Other Hispanic origin[a]	2.6	4.7
Total Hispanic or Latino	**55.2**	**100**

a"Other Hispanics" includes those with origins in Spain or who identify themselves as "Hispanic," "Latino," "Spanish American," and so on.

Source: *U.S. Bureau of the Census, American Community Survey, 2012–2016 5-Year Estimates.*

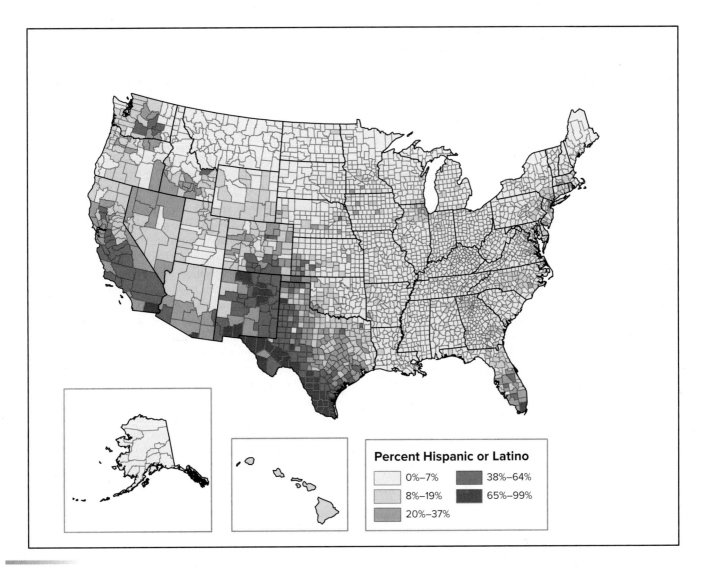

Figure 6.14 Hispanic American population, 2016. The highest concentrations of Hispanics are found in the Hispanic American borderland region, as well as areas of Cuban settlement in southern Florida. Significant Hispanic populations are also found in agricultural regions of the West, Midwest, and South.

Mexican Americans, representing a distinctive set of cultural characteristics, have been dispersing widely across the United States, though increases in the South and Midwest have been particularly noticeable. In similar fashion, immigrants from equally distinctive South, Central, and Caribbean American countries have been spreading out from their respective initial geographic concentrations. Puerto Ricans, already citizens, first localized in New York City, now the largest Puerto Rican city anywhere in numerical terms. Since 1940, however, when 88 percent of mainland Puerto Ricans were New Yorkers, there has been an outward dispersal primarily to other major metropolitan areas of the northeastern part of the country. The old industrial cities of New Jersey (Jersey City, Newark, Paterson, Passaic, and Hoboken); Philadelphia, Pennsylvania; Bridgeport and Stamford, Connecticut; the Massachusetts cities of Lowell, Lawrence, and Brockton; and Chicago and other central cities and industrial satellites of the Midwest have received the outflow. Miami and Dade County, Florida, play the

same magnet role for Cubans as New York City did for Puerto Ricans. The first large-scale movement of Cuban refugees from the Fidel Castro revolution occurred between 1959 and 1962. There followed a mixed period lasting until 1980, when emigration was alternately permitted and prohibited by the Cuban government. Suddenly and unexpectedly, in April 1980, a torrent of Cuban migration was released through the small port of Mariel. Although their flow was stopped after only five months, some 125,000 *Marielitos* fled from Cuba to the United States. A 1994 accord between the United States and Cuba allows for a steady migration of at least 20,000 Cubans each year, assuring strong Cuban presence in Florida, where most Cuban Americans reside, particularly in Miami's "Little Havana" community.

Early in the period of post-1959 Cuban influx, the federal government attempted a resettlement program to scatter the new arrivals around the United States. Some remnants of that program are still to be found in concentrations of Cubans in New York City, northern New Jersey, Chicago, and Los Angeles. The majority of early and late arrivals from Cuba, however, have settled in the Miami area. Immigrants from the Dominican Republic, many of them undocumented and difficult to trace, appear to be concentrating in the New York City area. Within that same city, Central and South Americans have congregated in the borough of Queens, with the South American contingent, particularly Colombians, settling in the Jackson Heights section. Elsewhere, Central American Hispanics also tend to cluster. Most Nicaraguans are found in the Miami area, most Hondurans in New Orleans. As noted, migrants from the Dominican Republic seek refuge in New York City; Salvadoran and Guatemalan migrants have dispersed themselves more widely, though they are particularly prominent in California.

Until recently, new arrivals tended to follow the paths of earlier arrivals from the home country. Chain migration and the security and support of an ethnically distinctive halfway community were as important for Hispanic immigrants as for their predecessors of earlier times and different cultures. However, the share of Hispanics living in states and counties with large concentrations of Hispanics has been slipping. The greater dispersion reflects spatial assimilation, as middle-class Hispanics following professional job opportunities throughout the country move to suburbs within and away from their former metropolitan concentrations. Also contributing to the dispersion are poorer, less-educated immigrants seeking jobs everywhere in construction, food processing plants, and service industries.

As the residential concentrations of the different Central American subgroups suggest, Hispanics as a whole are more urbanized than are non-Hispanic populations of the United States (**Figure 6.15**). Particularly the urbanized Hispanic population, it has been observed, appears confronted by two dominating but opposite trends. One is a drive toward conventional assimilation within American society. The other is consignment to a pattern of poverty, isolation, and, perhaps, cultural alienation from mainstream American life. Because of their numbers, which trend Hispanics follow will have significant consequences for American society as a whole.

To some observers, the very large and growing Mexican community poses a particular problem. Among other, earlier immigrant groups, they point out, fluency or even knowledge of the ancestral language was effectively lost by the third generation. Yet large majorities of second-generation Mexicans appear to emphasize the need for their children to be fluent in Spanish and to maintain close and continuing identification with Mexican culture in general. Because language, culture, and identity are intertwined, the fear has been that past and continuing Mexican immigration will turn America into a bilingual, bicultural, and therefore divided, country. Countering those fears, a SUNY-Albany study revealed that English not only is the language of choice among the majority of the children and grandchildren of Hispanic immigrants, but is increasing its appeal to them as they steadily move toward English monolingualism. This study found that nearly three-quarters of third-generation or later Hispanic children spoke English exclusively.

Figure 6.15 A proudly assertive street mural in the Boyle Heights, Los Angeles, *barrio*. Half of Los Angeles's population is Hispanic and overwhelmingly Mexican American. Their impact on the urban landscape—in choice of house colors, advertising signs, street vendors, and colorful wall paintings—is distinctive and pervasive.

©Stephanie Maze/Corbis Documentary/Getty Images

Asian American Contrasts

Since the Immigration Act of 1965 and its abolition of earlier exclusionary immigration limits, the Asian American population has grown from 1.5 million to 17 million in 2016; it is projected to grow to 24 million by 2030. Once largely U.S.-born and predominantly of Japanese and Chinese heritage, the Asian American population is now largely foreign-born and, through multiple national origins, is increasingly heterogeneous. Major sending countries include Korea, the Philippines, Vietnam, India, Thailand, and Pakistan, in addition to continuing arrivals from China. Although second to Hispanics in numbers of new arrivals, Asians comprise nearly one-third of the legal immigrant flow to the United States.

Their inflow was encouraged, first, by changes in immigration law that dropped the older national origin quotas and favored family reunification as an admission criterion. Educated Asians, taking advantage of professional preference categories in the immigration laws to move to the United States (or remain here on adjusted student visas), could become citizens after five years and send for immediate family and other relatives without restriction. They, in turn, after five years, could bring in other relatives. Chain migration was an important process. As a special case, the large number of Filipino Americans is related to U.S. control of the Philippines between 1899 and 1946. In the early part of the last century, Filipino workers were brought to Hawaii to work on sugar plantations, to California to labor on farms, or to Alaska to work in fish canneries. During World War II, Filipinos who served under the U.S. military were granted citizenship; immigration continues to be common today, especially for Filipino professionals.

Second, the wave of Southeast Asian refugees admitted during 1975–1980 under the Refugee Resettlement Program after the Vietnam War swelled the Asian numbers in the United States by more than 400,000, with 2.4 million more Asian immigrants admitted between 1980 and 1990. In 2011, about one-fourth of the U.S. foreign-born population were from Asia. Canada shows a similar increase in the immigrant flow from that continent.

Asia is a vast continent; successive periods of immigration have seen arrivals from many different parts of it, representing totally different ethnic groups and cultures. The major Asian American populations are detailed in Table 6.7, but even these groups are not homogeneous and cannot suggest the great diversity of other ethnic groups—Bangladeshi, Hmong, Karen, Nepalese, Sri Lankan, Mien, Indonesians of great variety, and many more—who have joined the American realm. Asian Americans as a whole are relatively concentrated in residence—far more so than the rest of the population. With the exception of Japanese Americans, most Asian Americans speak their native languages at home and maintain their distinctive ethnic cultures, values, and customs, suggesting only partial assimilation.

The highest concentrations of Asian Americans are found, as one would expect, in states bordering the Pacific (**Figure 6.16**). Japanese and Filipinos are particularly concentrated in Hawaii and the western states, where more than

Table 6.7

U.S. Leading Asian Populations by Ethnicity,[a] 2016

Ethnicity	Number (000s)	Percent of Asian American Total
Chinese, except Taiwanese	4,558	22.4%
Filipino	3,773	18.6%
Asian Indian	3,746	18.4%
Vietnamese	1,949	9.6%
Korean	1,796	8.8%
Japanese	1,414	7.0%
Pakistani	477	2.3%
Cambodian	318	1.6%
Hmong	289	1.4%
Thai	281	1.4%
Laotian	263	1.3%
Taiwanese	182	0.9%
Other Asian	587	2.9%
Total	**203,373**	**100**

[a]Ethnicity as reported by respondents, including claimed combination ethnicities.

Source: *U.S. Bureau of the Census 2012–2016 American Community Survey, 5-Year Estimates.*

half of the Chinese Americans are also found. In whatever part of the country they settled, Asian Americans (and Pacific Islanders) were drawn to metropolitan areas, where nearly all of them lived—more than half in suburban districts. For example, the Los Angeles, New York, and San Francisco metropolitan regions are home to one-third of the U.S. Asian population. Koreans and Filipinos are highly concentrated in the Los Angeles-Long Beach metropolitan area, and the Vietnamese in Orange County, south of Los Angeles. Although their metropolitan affinities have remained constant, the trend has been for greater dispersal around the country.

Immigrant Gateways and Clusters

Although new immigrants may ultimately seek residence in all parts of the United States, over the short term, immigrant concentrations rather than dispersals are the rule. Initially, most immigrants tend to settle near their points of entry (that is, nearest their country of origin) or in established immigrant communities. Six states—California, Texas, New York, Illinois, New Jersey, and Florida—are the most important immigrant gateways and have experienced the largest increases in their foreign-born populations. Those six states contain the country's three largest cities—New York, Los Angeles, and Chicago—and are the

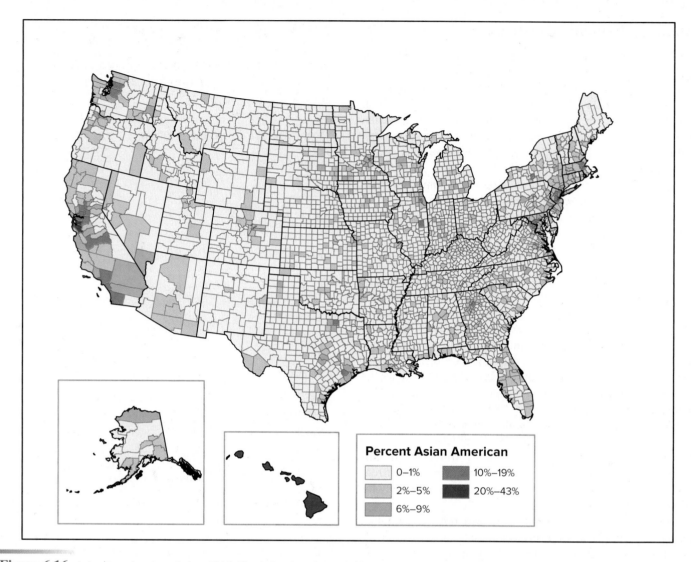

Figure 6.16 Asian American population, 2016. The Asian American population is concentrated in the West and in urban centers. California has the largest total number of Asian Americans, while Hawaii has the highest percentage of Asians in its population.

major points of entry from overseas and Latin America. As Ravenstein's laws of migration predict, the country's largest cities exhibit very high concentrations of new immigrants. New York City, for example, received one million immigrants in the 1990s and in 2016, 37 percent of its population had been born abroad. Other major cities attracting large numbers of new immigrants include San Francisco, Miami, Dallas, Houston, and Washington, D.C.

In Canada, immigrants are also concentrated in the largest gateway cities: Toronto, Montreal, and Vancouver. Of the immigrants arriving between 2006 and 2015, 61 percent settled in these three metropolitan areas, making them some of the world's most ethnically diverse places. These magnet cities contain established immigrant networks that offer social and economic support to new arrivals drawn to them by chain migration flows. Those attractions are not permanent and census evidence suggests that immigrant diffusion is occurring in areas where the existing labor supply does not satisfy needs for both low-skilled and technically trained workers.

Québec

The stamp of the Québécois (French Canadian) charter group on the ethnic province of French Canada is overwhelming. The province of Québec—with ethnic extensions into New Brunswick and northernmost Maine can be readily identified by its distinctive ethnic character. In language, religion, legal principles, system of land tenure, the arts, cuisine, philosophies of life, and urban and rural landscapes, Québec stands apart from the rest of Canada (**Figure 6.17**). Its uniqueness and self-assertion have won it special consideration and treatment within the political structure of the country.

Although the *Canadiens* of Québec were the charter group of eastern Canada and for some 200 years the controlling population, they numbered only some 65,000 when the Treaty of Paris ended the North American wars between the British and the French in 1763. That treaty, however, gave them control over three primary aspects of their culture and lives: language, religion, and land tenure. From these, they created their own distinctive and

Figure 6.17 The Château Frontenac hotel stands high above the lower older portion of Quebec City, where many streets show the architecture of French cities of the 18th century carried over to the urban heart of modern French Canada.

©*Perry Mastrovito/Stockbyte/Getty Images*

enduring ethnic province of some 1.5 million square kilometers (600,000 square miles) and 8 million people, more than 80 percent of whom have French as their native tongue (see Figure 5.15) and are at least nominally Roman Catholic. Québec City is the cultural heart of French Canada, though the bilingual Montreal metropolitan area with a population of 4.1 million is the largest center of Québec. The sense of cultural identity prevalent throughout French Canada imparted a spirit of nationalism not similarly expressed in other ethnic provinces of North America. Laws and guarantees recognizing and strengthening the position of French language and culture within the province assure the preservation of this distinctive North American cultural region, even if the movement for full political separation from the rest of Canada is never successful.

6.4 Urban Ethnic Diversity and Segregation

"Koreatown" and "Little Mogadishus" have joined the "Chinatowns," "Little Italys," and "Germantowns" of earlier eras as part of the American urban scene. The traditional practice of selective concentration of ethnic groups in their own well-defined subcommunities is evidence of the sharply defined social geography of urban America, in which ethnic neighborhoods have been a pronounced feature.

Protestant Anglo Americans created, from colonial times, the dominating host culture—the charter group—of urban North America. To that culture, the mass migrations of the 19th and early 20th centuries brought individuals and groups representative of different religious and ethnic backgrounds, including Irish Catholics, eastern European Jews, and members of every nationality, ethnic stock, and distinctive culture of central, eastern, and southern Europe. To them were added, both simultaneously and subsequently, newcomers from Asia and Latin America and such urbanizing rural Americans as Appalachian whites and Southern blacks.

Each newcomer sought to make a home within an urban environment established by the charter group. Frequently, new immigrants make their initial start in a new land by congregating within ethnic communities or neighborhoods. These are areas within the city where a particular culture group clusters, dominates, and which may serve as the core location from which diffusion or assimilation into the host society can occur. The rapidly urbanizing, industrializing society of 19th-century America became a mosaic of such ethnic enclaves. Their maintenance as distinctive social and spatial entities depended on the degree to which the assimilation of their population occurred. **Figure 6.18** and **Figure 6.19** show the ethnic concentrations that developed in Los Angeles and Chicago by the start of the 21st century. The increasing diversity of the immigrant stream and the multiplication of identified enclaves make comparable maps of older U.S. cities, such as New York, incredibly complex.

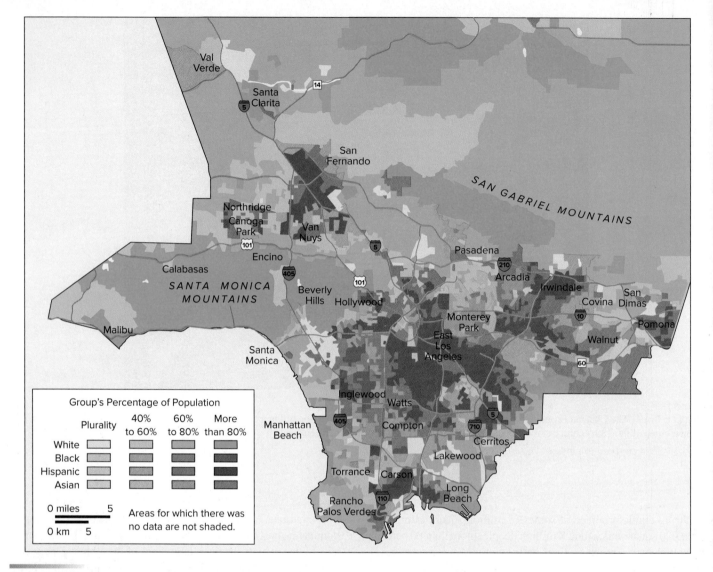

Figure 6.18 Racial/ethnic patterns in Los Angeles County, 2000, are greatly generalized on this map, which conceals much of the complex inter-mingling of different ethnic groups in several sections of Los Angeles city. However, the tendency of people to cluster in distinct neighborhoods by race and ethnicity is clearly evident.

Source: The New York Times, *March 30, 2001, p. A18.*

Immigrant neighborhoods are a measure of the **social distance** that separates the minority from the charter group. The greater the perceived differences between the two groups, the greater the social distance and the less likely the charter group is to easily accept or assimilate the newcomer. Consequently, the ethnic community will endure longer as a place both of immigrant refuge and of enforced segregation.

Segregation is a shorthand expression for the extent to which members of an ethnic group are not uniformly distributed in relation to the rest of the population. A commonly employed measure quantifying the degree to which a distinctive group is segregated is the segregation index or **residential dissimilarity index**. It indicates the degree to which the two component groups of a population are distributed differently across an urban region's neighborhoods, with values ranging from 0 (no segregation) to 100 (complete segregation). For

example, according to the 2010 Census, the index of dissimilarity in the Milwaukee metropolitan area was a very high 82, meaning that 82 percent of all blacks (or whites) would have to move to different neighborhoods before the two groups would be equally distributed across the metropolitan area. Evidence from cities throughout the world makes clear that most ethnic minorities tend to be sharply segregated from the charter group, and that segregation on racial or ethnic lines is usually greater than would be anticipated from the socioeconomic levels of the groups involved. Further, the degree of segregation varies among cities in the same country and among different ethnic mixes within each city.

Among the major racial and ethnic groups in U.S. cities, blacks are the most segregated and Asians the least, The most segregated cities for blacks are industrial cities in the Midwest and Northeast. Collectively, blacks, Hispanics, and Asians

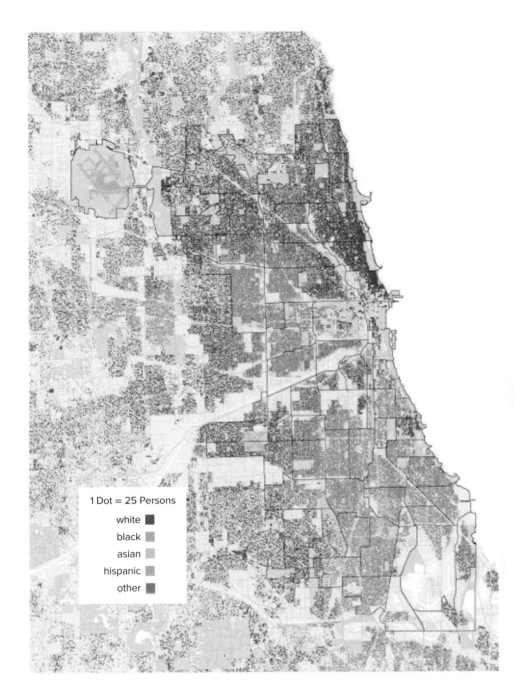

Figure 6.19 Racial/ethnic patterns in Chicago, 2000, are depicted using a dot distribution map. The social distance between groups is evident in their spatial separation.

lived in more integrated neighborhoods than did whites. Overall, although segregation remains high in America, for blacks, it steadily declined between 1970 and 2010. Of course, each world region and each country has its own patterns of national and urban immigration and immigrant residential patterns. Even when those population movements involve distinctive and contrasting ethnic groups, American models of spatial differentiation may not be applicable.

Foreign migrants to West European cities, for example, frequently do not have the same expectations of permanent residence and eventual amalgamation into the host society as their American counterparts. Many came under labor contracts with no initial legal assurance of permanent residence. Although many now have been joined by their families, they often find citizenship difficult to acquire. The Islamic populations from North Africa and Turkey tend to be more tightly grouped and defensive against the surrounding majority culture of western European cities than do African or south and east European Christian migrants. France, with some 5 million Muslim residents, most of them from North Africa, has tended to create bleak, outer suburban ghettoes in which immigrants remain largely isolated from mainstream French life.

Racial and ethnic divisions appear particularly deep and divisive in Britain. A British government report of 2001 claimed that in Britain, whites and ethnic minorities lead separate lives with no social or cultural contact and no sense of belonging to the same nation. Residential segregation in public housing and inner-city areas was compounded by deep social polarization. The nonwhite British population—largely Caribbean and Asian in origin—and the white majority, the report concluded, "operate on the basis of a series of parallel lives . . . that often do not seem to touch at any point," assigning blame for the situation on "communities choosing to live in separation rather than integration" (see the feature "The Caribbean Map in London"). The Home Secretary observed on the basis of the report that many "towns and cities lack any sense of civic identity or shared values." A similar total minority segregation is evident in the Sydney, Australia, suburb of Redfern, which houses an Aboriginal population that rarely ventures out to work or mingle in the surrounding white city and that white Australians avoid and ignore.

Spatial segregation is growing in the developing countries as well. Rapid urbanization in multiethnic India has resulted in cities of extreme social and cultural contrasts. Increasingly, Indian cities feature defined residential colonies segregated by village and caste origins of the immigrants. Chain migration has eased the influx of newcomers to specific new and old city areas; language, custom, religion, and tradition keep them confined. In Mumbai, for example, in Dharavi—considered the world's largest slum—Tamil, not Hindi, is spoken as the main language. Elsewhere, in Bangkok, Thailand, Burmese migrants are largely confined to the slum of Tlong Toey; the population of Hillbrow, a squatter slum in Johannesburg, South Africa, consists largely of Nigerian and French-speaking African immigrants; and the residents of the informal settlements of San José, Costa Rica, generally come from Nicaragua. International and domestic migration throughout ethnically diverse sub-Saharan Africa shows a repetitive pattern of residential segregation: the rural-to-urban population shift has created city neighborhoods defined on tribal and village lines. Worldwide in all continental and national urban contexts, the degree of immigrant segregation is at least in part conditioned by the degree of social distance felt between the newcomer population and the other immigrant and host societies among whom residential space is sought.

Constraints on assimilation and the extent of discrimination and segregation are greater for some minorities than for others. In general, the rate of assimilation of an ethnic minority by the host culture depends on two sets of controls: *external* controls, including attitudes toward the minority held by the charter group and other competing ethnic groups, and *internal* controls of group cohesiveness and defensiveness.

External Controls

When the majority culture or rival minorities perceive an ethnic group as threatening, the group tends to be spatially isolated by external "blocking" tactics designed to confine the rejected minority and to resist its "invasion" of neighborhoods. The more tightly knit the threatened group, the more adamant and overt are its resistance tactics. When confrontation measures (including, perhaps, housing market discrimination, threats, and vandalism) fail, the invasion of charter-group territory by the rejected minority proceeds until a critical percentage of newcomer housing occupancy is reached. That level, the **tipping point,** may precipitate a rapid exodus of the majority population. Invasion, followed by succession, then results in a new spatial pattern of ethnic dominance according to models of urban social geography developed for American cities and examined in Chapter 11.

Racial or ethnic discrimination in urban areas generally expresses itself in the relegation of the most recent, most alien, most rejected minority to the poorest available housing. That confinement has historically been reinforced by the concentration of the newest, least assimilated ethnic minorities at the low end of the job market. Distasteful, menial, low-paying service and factory employment unattractive to the charter group is available to those new arrivals, even when other occupational avenues may be closed. The dockworkers, street cleaners, slaughterhouse employees, and sweatshop garment workers of earlier America had and have their counterparts in other regions. In England, successive waves of West Indians and Commonwealth Asians took the posts of low-pay hotel and restaurant service workers, transit workers, refuse collectors, manual laborers, and the like; Turks in German cities and North Africans in France fill similar low-status employment roles.

Historically, in the United States, there was a spatial association between the location of such employment opportunities—the inner-city central business district (CBD) and its margins—and the location of the oldest, most dilapidated, and least desirable housing. Proximity to job opportunity and the availability of cheap housing near the CBD, therefore, combined to concentrate the U.S. immigrant slum near the heart of the 19th-century central city. In the second half of the 20th century, the suburbanization of jobs, the rising skill levels required in the automated offices of the CBD, and the effective isolation of inner-city residents by the absence of public transportation or their inability to pay for private transport maintained the association of the least competitive minorities and the least desirable housing area. But now those locations lack the promise of the entry-level jobs that used to be close at hand.

That U.S. spatial pattern is not universal, however. In Latin American cities, the newest arrivals at the bottom of the economic and employment ladder are most apt to find housing in squatter or slum areas on the outskirts of the urban unit. In French urban agglomerations, the outer fringes frequently have a higher percentage of foreigners than the city itself.

Internal Controls

Although part of the American pattern of urban residential segregation may be explained by the external controls of host-culture resistance and discrimination, the clustering of specific groups into ethnically homogeneous neighborhoods is best understood

The Caribbean Map in London

Although the movement [to England] from the West Indies has been treated as if it were homogeneous, the island identity, particularly among those from the small islands, has remained strong. . . . [I]t is very evident to anyone working in the field that the process of chain migration produced a clustering of particular island or even village groups in their British destination. . . .

The island identities have manifested themselves on the map of London. The island groups can still be picked out in the clusters of settlements in different parts of the city.

There is an archipelago of Windward and Leeward islanders north of the Thames; Dominicans and St. Lucians have their core areas in Paddington and Notting Hill; Grenadians are found in the west in Hammersmith and Ealing; Montserratians are concentrated around Stoke Newington, Hackney, and Finsburry Park; Antiguans spill over to the east in Hackney, Waltham Forest, and Newham; south of the river is Jamaica.

That is not to say that Jamaicans are found only south of the river or that the only West Indians in Paddington are from St. Lucia. The mixture is much greater than that. The populations overlap and interdigitate: there are no sharp edges. . . . [Nevertheless, north of the river,] there is a west-east change with clusters of Grenadians in the west giving way to St. Lucians and Dominicans in the inner west, through to Vincentians and Montserratians in the inner north and east and thence to Antiguans in the east.

Source: Ceri Peach, "The Force of West Indian Island Identity in Britain," in Geography and Ethnic Pluralism, *eds. Colin Clarke, David Ley, and Ceri Peach (London: George Allen & Unwin, 1984).*

---

as the result of internal controls of group defensiveness and conservatism. The self-elected segregation of ethnic groups can be seen to serve four principal functions—defense, support, preservation, and group assertion.

First, it provides *defense,* reducing individual immigrant isolation and exposure by physical association within a limited area. The walled and gated Jewish quarters of medieval European cities have their present-day counterparts in the clearly marked and defined "turfs" of street gang members and the understood exclusive domains of the "black community," "Chinatown," and other ethnic or racial neighborhoods. In British cities, it has been observed that West Indians and Asians fill identical slots in the British economy and reside in the same sorts of areas, but they tend to avoid living in the *same* areas. West Indians avoid Asians; Sikhs isolate themselves from Muslims; Bengalis shun Punjabis. In London, patterns of residential isolation even extend to West Indians of separate island homelands, as the feature "The Caribbean Map in London" makes clear. Their own defined ethnic territory provides members of the group with security from the hostility of antagonistic social groups, a factor also underlying the white flight to "garrison" suburbs.

Second, the ethnic neighborhood provides *support* for its residents in a variety of ways. The area serves as a halfway station between the home country and the alien society. It provides supportive social and religious ethnic institutions, familiar businesses, job opportunities where language barriers are minimal, and friendship and kinship ties to ease the transition to a new society.

Third, the ethnic neighborhood may provide a *preservation* function, reflecting the ethnic group's positive intent to preserve and promote such essential elements of its cultural heritage as language and religion. The preservation function represents a fear of being totally absorbed into the charter society and a desire to maintain those customs and associations seen to be essential to the conservation of the group.

For example, Jewish dietary laws are more easily observed by and exposure to potential marriage partners within the faith is more certain in close-knit communities than when individuals are scattered.

Finally, ethnic spatial concentration can serve as a base for *group assertion,* a peaceful search for democratic political representation. Voter registration drives and political candidates drawn from ethnic neighborhoods represent concerted efforts to promote group interests at all governmental levels.

Shifting Ethnic Concentrations

Ethnic communities, once established, are not necessarily permanent. For Europeans who came in the 19th and early 20th centuries, and for more recent Hispanic and Asian immigrants, high concentrations were and are encountered in neighborhoods of first settlement. Second-generation neighborhoods usually become far more mixed. The older, dominant, urban ethnic groups in places given names like "Little Italy" are now often in the minority, as middle- and upper-middle-class members of the immigrant group move on. That mobility pattern appears to be repeating among Asian and Latino groups, but only, or most clearly, where those groups collectively account for a relatively small share of the total metropolitan area population. Black segregation and black communities, in contrast, appear more pronounced and permanent.

Ethnic clusters initially identified with particular central city areas are frequently or usually displaced by different newcomer groups (**Figure 6.20**). With recent diversified immigration, older homogeneous ethnic neighborhoods have become highly subdivided and polyethnic. In Los Angeles, for example, the great wave of immigrants from Mexico, Central America, and Asia has begun to push African Americans out of South Los Angeles and other well-established black communities, converting them from racially exclusive to multicultural areas. In New York, the borough of Queens, once the stronghold of European ethnic immigrants, has now become home to more than 110 different, mainly non-European nationalities. In Woodside in Queens, Latin Americans and Koreans are prominent among the many replacements of the formerly dominant German and Irish groups. Elsewhere within the city, West Indians now dominate the old Jewish neighborhoods of Flatbush; Poles, Dominicans, and other Central Americans have succeeded Germans and Jews in Washington Heights. Manhattan's Chinatown has expanded into old Little Italy, and a new Little Italy has emerged in Bensonhurst.

Further, the new ethnic neighborhoods are intermixed in a way that enclaves of the early 20th century never were. The restaurants, bakeries, groceries, specialty shops, and their customers and owners from a score of different countries (and even different continents) are now found within a two- or three-block radius. In the Kenmore Avenue area of East Los Angeles, for example, 1.3-square-kilometer (a half-square-mile) area of former Anglo neighborhood now houses more than 9,000 people representing Hispanics and Asians of widely varied origins, along with Pacific Islanders, Amerindians, African Americans, and a scattering of native-born whites. Students in the neighborhood school come from 43 countries and speak 23 languages, a localized ethnic intermixture unknown in the communities of single ethnicity so characteristic of earlier stages of immigration to the United States.

The changing ethnic spatial pattern is not yet clear or certain. Increasing ethnic diversity coupled with continuing immigration flow has, in some instances, expanded rather than reduced patterns of urban group segregation. The tendency for separate ethnic groups to cluster for security, economic, and social reasons cannot be effective if many different, relatively small ethnic groups find themselves in a single city setting. Intermixture is inevitable when individual groups do not achieve the critical mass necessary to establish a true identifiable separate community. But as continuing immigration and natural increases allow groups to expand in size, they are able to create more distinctive, self-selected ethnic clusters and communities.

Figure 6.20 The landscape offers evidence of shifting ethnic concentrations. As Jews left North Minneapolis for the suburbs, they were succeeded by African Americans. This former Orthodox synagogue is one of the few reminders of the once vibrant Jewish presence. The building is now used by a nondenominational Protestant Christian congregation and the altered facade mixes carved lions guarding Hebrew scrolls, Stars of David, crosses, and both Jewish and Christian messages.

©Mark Bjelland

Enclaves, Ghettos, and Ethnoburbs

When ethnic residential clusters endure, the clusters may be termed **colonies,** serving essentially as points of entry for members of the particular ethnic group. They persist only to the extent that new arrivals perpetuate the need for them. In American cities, many European ethnic colonies began to lose their vitality and purpose with the reduction of European immigration flows after the 1920s.

When an ethnic cluster does persist because its occupants choose to preserve it, their behavior reflects the internal cohesiveness of the group and its desire to maintain an enduring **ethnic enclave** or neighborhood. When the cluster is perpetuated by external constraints and discriminatory actions, it has come to be termed a **ghetto.** The term *ghetto* was first used in Venice, Italy, in the 16th century to refer to the area of the city where Jews were required to live. In reality, the colony, the enclave, and the ghetto are spatially similar. Growing ethnic groups that maintain voluntary spatial association frequently expand the area of their dominance by outward growth from the core of the city in a radial pattern. That process has long been recognized in Chicago (**Figure 6.21**) and has, in that and other cities, typically been extended beyond the central city boundaries into at least the inner fringe of the suburbs.

African Americans have traditionally found strong resistance to their territorial expansion from the white majority though white-black urban relations and patterns of black ghetto formation and expansion have differed in different sections of the country. A revealing typology of African American ghettos is outlined in **Figure 6.22**. In the South, the white majority, with total control of the housing market, was able to assign residential space to blacks in accordance with white, not black, self-interest. In the *early southern* ghetto of such pre–Civil War cities as Charleston and New Orleans, African Americans were assigned small dwellings in alleys and back streets within and bounding the white communities where they worked as (slave) house and garden servants. The *classic southern* ghetto for newly free blacks was composed of specially built, low-quality housing on undesirable land—swampy, perhaps, or near industry or railroads—and was sufficiently far from better-quality white housing to maintain full spatial and social segregation.

In the North, on the other hand, African Americans were open competitors with other claimants for space in a generalized housing market. The *early northern* ghetto represented a "toehold" location in high-density, aged, substandard housing on the margin of the CBD. The *classic northern* ghetto is a more recent expansion of that initial enclave to surround the CBD and to penetrate, through invasion and succession, contiguous zones as far as the numbers, rent-paying ability, and housing discrimination will allow. Finally, in new western and southwestern cities not tightly hemmed in by resistant white neighborhoods or suburbs, the black community may display a linear expansion from the CBD to the suburban fringe.

Increasingly, ethnic communities are found in the outer reaches of major metropolitan areas. In New York City, the outer

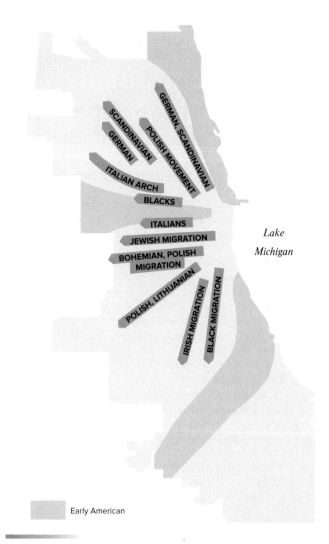

Figure 6.21 Chicago's many ethnic groups tended to expand their territory and migrate outward from the city center. "Often," Samuel Kincheloe observed in the 1930s, "[minority] groups first settle in a deteriorated area of a city somewhere near its center, then push outward along the main streets." More recently, many—particularly young, innovative, and entrepreneurial—immigrants have avoided traditional first locations in central cities and from their arrival, they have settled in metropolitan area suburbs and outlying cities, where economic opportunity and quality of life is perceived as superior to conditions in the primary inner city.

Source: The American City and its Church by Samuel Kincheloe. Copyright 1938 by Friendship Press, New York.

Queens neighborhood of Elmhurst houses immigrants from 114 different countries and is the city's most ethnically diverse community. In part due to rising affluence among immigrants, "Satellite Chinatowns" are found in Los Angeles's San Gabriel Valley, in San Francisco, and in Vancouver, Canada (**Figure 6.23**). This has given rise to the **ethnoburb**, a politically independent suburban community with a significant, though not exclusive, concentration of a single ethnic group (see the feature "Vancouver, Canada: Chinatown versus Ethnoburb"). Monterey Park, outside Los Angeles, and Richmond, British Columbia, outside Vancouver, are examples of Asian ethnoburbs. Ethnoburbs differ from traditional,

Vancouver, Canada: Chinatown versus Ethnoburb

Ethnoburbs differ from traditional ethnic enclaves (such as Chinatowns) due to changes in cities, the economy, and communications and travel. Whereas ethnic enclaves tend to reflect the lower status of recent immigrants, ethnoburbs are products of globalization, which has led to rising wealth in developing countries and increased flows of people and investment between distant places.

Vancouver's Chinatown dates to the 1880s, when the city was a frontier outpost. Chinese workers, mostly male, were recruited in large numbers to work in the construction of the transcontinental Canadian Pacific Railway. When the railway reached its western terminus at Vancouver in 1885, many Chinese workers settled in Vancouver's Chinatown. While Chinatown was a place where Chinese immigrants lived, worked, and socialized, it was shaped in large part by the racist attitudes of the host society. In 1885, the Canadian government imposed a "head tax" on incoming immigrants that applied to the Chinese, but not to European immigrants. Because Chinese men could not afford the head tax to bring over wives and other relatives, Chinatown was a struggling, mostly male community in its early years. Derogatory cartoons in newspapers, voting restrictions that forbade the Chinese from participating in elections, mob violence directed against the Chinese, and discriminatory policing were part of the experience of early Chinese immigrants in Vancouver (Figure 6A). Although not required by law to live in segregated areas, the Chinese clustered together for support and defense. Chinese social, cultural, and economic institutions built structures that lent a distinctive appearance to the Chinatown district.

After World War II, host society attitudes toward the Chinese softened, and Chinatown came to be viewed as an exotic destination for tourists. The relatively recent addition of a formal Chinese classical garden, distinctive red lampposts, and gateway arches lend a distinctive look to the Chinatown landscape.

Nonetheless, Chinatown is located adjacent to Vancouver's poorest, inner-city neighborhood and struggles to escape its marginalized image.

As a result of globalization and increased economic development in East Asia, new, wealthy, entrepreneurial Chinese immigrants began arriving in North American cities in the 1980s. Unlike previous immigrants, these people never formed inner-city ethnic enclaves but immediately settled in the suburbs of cities such as Los Angeles, New York, San Francisco, Toronto, and Vancouver. Asian immigration was made possible by the elimination of exclusionary immigration laws and encouraged by relaxed regulations for business immigrants. In the years leading up to the British government's transfer of Hong Kong to China in 1997, many Chinese from Hong Kong immigrated to Vancouver. More recently, wealthy Chinese immigrants have immigrated from Taiwan and Mainland China to Vancouver. The suburb of Richmond was one of several popular immigrant destinations and illustrates the differences between ethnoburbs and a traditional Chinatown.

Richmond is home to Vancouver's international airport and features many daily flights to Asia. Immigrants comprise almost 60 percent of the city's 200,000 residents. While the Chinese population is densely clustered in Chinatown, in the ethnoburb of Richmond, the Chinese are spread across a multiethnic suburb among South Asians, Filipinos, Japanese, Koreans, and whites. Along one street in Richmond, Buddhist temples and monasteries, Chinese Christian churches, a Sikh Gurdwara, Muslim mosque, and Hindu temple sit side by side. While the population at the center of Richmond is 80 percent Chinese and features a collection of Chinese-themed shopping malls and hotels, it is also home to many non-Chinese businesses, chain stores, and restaurants. While the traditional Chinatown features crowded, high-density housing, Richmond's housing includes many single-family houses alongside modern, mid-rise condominums. Compared to an enclave, the ethnoburb has less-defined boundaries. Compared to Chinatown, suburban Richmond's Chinese residents are younger, have higher levels of education, higher incomes, and are deeply connected to the global economy and its flows of people and investment. Unlike a traditional Chinatown, the ethnoburb is not the result of discrimination but a voluntary clustering to maximize ethnic social and business contacts in a familiar language and cultural environment.

Figure 6A Historic Vancouver newspaper cartoon reflecting racist attitudes toward the Chinese and criticizing the supposedly overcrowded and vice-ridden conditions in Chinatown.

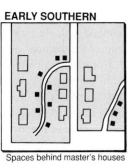
EARLY SOUTHERN

Spaces behind master's houses

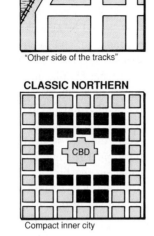
CLASSIC SOUTHERN

"Other side of the tracks"

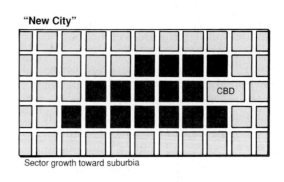
"New City"

Sector growth toward suburbia

EARLY NORTHERN

CBD

Foothold in inner city

CLASSIC NORTHERN

CBD

Compact inner city

Figure 6.22 A typology of black ghettos in the United States.

Source: David T. Herbert and Colin J. Thomas, Urban Geography, *London: David Fulton Publishers, 1987.*

(a)

(b)

Figure 6.23 (*a*) Vancouver's historic Chinatown developed when Chinese railroad workers settled there upon completion of the transcontinental railroad. It features distinctive architecture, ethnic specialty shops, and restaurants. Its population is dense, mostly Chinese, and relatively old. (*b*) Richmond, British Columbia, is an ethnoburb with a large, multiethnic immigrant population. Richmond is a prosperous suburb of Vancouver, and it is filled with expensive, single-family houses, condominiums, international hotels, Chinese-themed shopping malls, and an international airport. Richmond's Chinese make up almost half of the city's population, are young, well-educated, and maintain strong social and business ties with Hong Kong, Taiwan, and mainland China.

©*Mark Bjelland*

inner-city Chinatowns, which were the product of discrimination and the low socioeconomic status of the Chinese. Ethnoburbs, on the other hand, attract relatively wealthy, well-educated, highly mobile immigrants. Many of the immigrants in ethnoburbs display **transnationalism;** that is, they maintain strong ties with more than one country, often in the form of social and business connections with their homeland. Jet travel and the Internet make it quite possible to run a business on the other side of the Pacific Ocean. Whereas traditional Chinatowns are self-sustaining enclaves filled with small,

independent shops, ethnoburbs feature transnational banks, hotels, retail chains, and sellers of luxury goods.

Native-Born Dispersals

Immigration flows to the United States during the last third of the 20th century—unlike those of earlier mass-immigration periods—have begun to affect both the broad regional ethnic makeup of the United States and the internal migration pattern of native-born Americans. The consequence has been dubbed a "demographic balkanization," a spatial fragmentation of the population by race/ethnicity, economic status, and age across extended metropolitan areas and larger regions of the country.

Early 20th-century immigration streams resulted, as we have seen, in temporary ethnic segregation by urban neighborhoods and between central cities and suburbs. Immigration legislation of 1965 dropped the national-origin quotas that had formerly favored European immigrants, replacing that with a more inclusive formula emphasizing family reunification. That change, plus economic and political pressures in many countries of Asia and Latin America, has swelled the influx of poorer, less-skilled Asians and Hispanics. Highly dependent on family members and friends for integration into both the informal and formal American job market, the new arrivals are drawn to primary port-of-entry metropolitan areas by chain migration links. In those areas where immigrants account for most of the present and prospective population growth, the trend is toward increasingly multicultural, younger, and poorer residents, predominantly of Hispanic and Asian origin.

The high degree of areal concentration of recent immigrant groups initiated a selective native-born (particularly white) retreat, not only fleeing the cities for the suburbs but leaving entire metropolitan areas and states. California, with nearly one-quarter of its population foreign-born, saw a departure of one native-born white or black resident for nearly each foreign-born arrival. Individual urban areas echoed California's state experience. For domestic moves, top destinations were to cities and states away from coastal and southern border immigrant entry points. A visible spatial consequence, then, is an emerging pattern of increasing segregation and isolation by metropolitan areas and regions of the country. Immigrant assimilation may now be more difficult than in the past, and social and political divisions more pronounced and enduring.

6.5 Cultural Transfer

Immigrant groups arrive at their destinations with already existing sets of production techniques and skills. They bring established ideas of "appropriate" dress, foods, and building styles, and they have religious practices, marriage customs, and other cultural expressions in place and ingrained. That is, immigrants carry to their new homes a full complement of artifacts, sociofacts, and mentifacts. They may modify, abandon, or even pass these on to the host culture, depending on a number of interacting influences: (1) the background of the arriving group; (2) its social distance from the charter group; (3) the disparity between new home and origin-area environmental conditions; (4) the importance given by the migrants to the economic, political, or religious motivations that caused them to relocate; and (5) the kinds of encountered constraints that force personal, social, or technical adjustments on the new arrivals.

Immigrant groups rarely transferred intact all of their culture traits to North America. Invariably, there have been modifications as a result of the necessary adjustment to new circumstances or physical conditions. In general, if a transplanted ethnic trait was usable in the new locale, it was retained. Simple inertia suggested there was little reason to abandon the familiar and comfortable when no advantage accrued. If a trait or a cultural complex was essential to group identity and purpose—the religious convictions of the rural Amish, for example, or of urban Hasidic Jews—its retention was certain. But ill-suited habits or techniques would be abandoned if superior American practices were encountered, and totally inappropriate practices would be discarded. German settlers in Texas, for example, found that grape vines and familiar midlatitude fruits did not thrive there. Old-country agricultural traditions, they discovered, were not fully transferable and had to be altered.

Finally, even apparently essential cultural elements may be modified in the face of unalterable opposition from the majority population. The some 30,000 Hmong and Mien tribespeople who settled in the Fresno, California, area after fleeing Vietnam, Thailand, and Laos found that their traditional practices of medicinal use of opium, of "capturing" young brides, and of ritual slaughtering of animals brought them into conflict with American law and customs and with the more Americanized members of their own culture group.

The assimilation process is accelerated if the immigrant group is in many basic traits similar to the host society, if it is relatively well educated, relatively wealthy, and finds political or social advantages in being "Americanized." On the other hand, the immigrant group may seek physical separation by concentrating in specific geographic areas or raising barriers to assure separation from corrupting influences. Social isolation can be effective even in congested urban environments if it is buttressed by distinctive costume, beliefs, or practices (**Figure 6.24**). Group segregation may even result in the retention of customs, clothing, or dialects discarded in the original home area.

The assimilation process may be reversed by **culture rebound,** a belated adoption of group consciousness and reestablishment of identifying traits. These may reflect an attempt to reassert old values and to achieve at least a modicum of social separation. The wearing of dashikis, the popularity of Ghanaian-origin kente cloth, or the celebration of Kwanzaa by American blacks seeking identification with African roots are examples of culture rebound. Ethnic identity is fostered by the nuclear family and ties of kinship, particularly when reinforced by residential proximity. It is preserved by such group activities as distinctive feasts or celebrations and by marriage customs; by ethnically identified clubs, such as the Turnverein societies of German communities or the Sokol movement of athletic and cultural centers among the Czechs; and by ethnic churches (**Figure 6.25**).

6.6 The Ethnic Landscape

Landscape evidence of ethnicity may be as subtle as the greater number and size of barns in the German-settled areas of the Ozarks or the designs of churches or the names of villages. The evidence may be as striking as the buggies of the Amish communities,

Figure 6.24 Ultra-Orthodox Hasidic Jews, in their distinctive dress and beards, watch runners in the New York City marathon. Hasidic Jews seek social isolation to protect their way of life from the corrupting influences of modern urban life.

©New York Daily News Archive/New York Daily News/Getty Images

Figure 6.25 These young girls, dressed in traditional garb for a Los Angeles Greek Orthodox Church festival, show the close association of ethnicity and religion in the American mosaic.

©Tony Freeman/Photo Edit

the massive Dutch (really, German-origin) barns of southeastern Pennsylvania, or the adobe houses of Mexican American settlements in the Southwest. The ethnic landscape, however defined, may be a relic, reflecting old ways no longer pursued. It may contain evidence of artifacts or designs imported, found useful, and retained. In some instances, the physical or customary trappings of ethnicity may remain unique to one community or very few communities. In others, the diffusion of ideas or techniques may have spread introductions to areas beyond their initial impact. The landscapes and landscape evidences explored by cultural geographers are many and complex (and further explored in Chapter 7).

The distinctive landscape elements of ethnic communities come in different forms: farming practices, architecture, monuments, gardens, places of worship, specialty shops, ethnic institutions, and festivals that take over streets or city parks for a designated period of time. Although ethnic landscapes are created originally as expressions of cultural heritage, their continuation may be economically motivated. As cultural and economic forces work to homogenize places around the world, communities with a distinctive identity can attract tourist revenues. New Glarus, Wisconsin (America's "Little Switzerland"), Solvang, California ("Little Denmark"), Frankenmuth, Michigan ("Little Bavaria"), and Lindsborg, Kansas ("Little Sweden, USA") are good examples. Similarly, urban neighborhoods with identities such as Chinatown, Little Tokyo, Greektown, or Little Italy can attract shoppers and restaurant-goers who seek the novelty or imagined authenticity of an ethnic enclave.

New Glarus, Wisconsin illustrates the tension between assimilation and preservation. For tourists, it presents the image of an ethnic place supposedly untouched by assimilation and homogenization (**Figure 6.26**). The town was settled by Swiss immigrants who, over time, underwent assimilation while still keeping elements of their ethnic heritage. As the rural community struggled economically, it found that by playing up its ethnic heritage through adding chalet-style

Figure 6.26 New Glarus, Wisconsin, was settled by Swiss immigrants in the mid-1800s, but over time, it came to look like other small towns in the Midwest. More recently, the town has played up its Swiss heritage to attract tourists.

©Volkmar K. Wentzel/National Geographic/Getty Images

architecture, ethnic festivals, specialty shops, and museums, it could attract large numbers of tourists from nearby cities. But the town is not Disneyland, and the tourist image of the community can conflict with actually living and working in the town.

Ethnic Regionalism

Patterns of long-established ethnic regionalism are displayed in pronounced contrasts in the built landscape. In areas of intricate mixtures of ethnic homelands—eastern and southeastern Europe, for example—different house types, farmstead layouts, and even the use of color can distinguish for the knowledgeable observer the ethnicity of the local population. The one-story "smoking-room" house of the northern Slavs, with its covered entrance hall and stables all under one roof, marks their areas of settlement south of the Danube River. Blue-painted, one-story, straw-roofed houses indicate Croatian communities. In the Danube Basin, areas of Slovene settlement are distinguished by the Pannonian house of wood and straw-mud. In Spain, the courtyard farmstead marks areas of Moorish influence, just as white stucco houses trimmed with dark green or ochre paint on the shutters indicates Basque settlement.

It is difficult to delineate ethnic regions of the United States that correspond to the distinctive landscapes created by sharply contrasting cultural groups in Europe or other world areas. The reason lies in the mobility of Americans, the degree of acculturation and assimilation of immigrants and their offspring, and the significance of charter cultures and mass communications in shaping ideas, activities, institutions, and material artifacts. What can be attempted is the delimitation of areas in which particular immigrant-group influences have played a recognizable or determinant role in shaping tangible landscapes and intangible regional "character."

SUMMARY

Ethnic diversity is a reality in most countries of the world and is increasing in many of them. Immigration, refugee streams, guest workers, and job seekers all contribute to the mixing of peoples and cultures. The mixing is not complete, however. Ethnicity—affiliation in a group sharing common identifying cultural traits—is fostered by territorial separation or isolation. In much of the world, that separation identifies home territories within which the ethnic group is dominant and with which it is identified. In societies of immigrants—the United States and Canada, for example—such homelands are replaced by ethnic colonies, islands, enclaves, ghettos, and ethnoburbs of self-selected or imposed separation from the larger host society. Cluster migration helped establish such colonies in rural America; chain migration encouraged their development in cities.

The 19th- and early 20th-century American central city displayed pronounced areal segregation as immigrant groups established and clung to protective ethnic neighborhoods while they gradually adjusted to the host culture. A continual population restructuring of urban areas occurred as older groups underwent acculturation, amalgamation, or assimilation, and new groups entered the urban social mix. The durability of ethnic neighborhoods has depended on the degree of social distance separating the minority group from the host culture and on the significance the immigrant group places on long-term maintenance of their own cultural identity. That is, ethnic communities have been the product of both external and internal forces.

In other world regions, similar spatial separation of immigrant groups by racial, cultural, national, tribal, or village origin within the city is common. In Europe, because of the uncertain legal and employment status of many foreign populations and the restricted urban housing market they enter, ethnic enclaves have taken a different form, extent, and level of segregation than has been the case in the United States and Canada.

Ethnicity is one of the threads of diversity in the global cultural fabric. Throughout the world, ethnic groups have imprinted their presence on the landscapes they have developed or to which they have transported their culture. In house and farm building style, settlement patterns, and religious structures, the beliefs and practices of distinctive groups are reflected in the cultural landscape. Such interweaving of cultural identities and cultural landscapes and their interaction with the homogenizing force of popular culture are the subjects of Chapter 7.

KEY WORDS

acculturation
amalgamation theory
assimilation
chain migration
charter group
cluster migration
colony
culture rebound
ethnic cleansing
ethnic enclave

ethnic geography
ethnic group
ethnic island
ethnicity
ethnic province
ethnoburb
ethnocentrism
first effective settlement
ghetto
host society

natural selection
race
residential dissimilarity index
segregation
social distance
spatial assimilation
tipping point
transnationalism
xenophobia

FOR REVIEW

1. How does *ethnocentrism* contribute to preservation of group identity? In what ways might an ethnic group sustain and support new immigrants?

2. What is the difference between race and ethnicity? How are the concepts of *ethnicity* and *culture* related?

3. What have been some of the principal time patterns of immigration flows into the United States? Into Canada? How are those patterns important to an understanding of present-day conflicts over immigration in either country or both countries?

4. How may *segregation* be measured? Does ethnic segregation exist in the cities of world areas outside North America? If so, does it take different forms than in American cities?

5. What forces external to ethnic groups help create and perpetuate immigrant neighborhoods? What functions beneficial to immigrant groups do ethnic communities provide?

6. How is a ghetto different than an ethnic enclave? How is an ethnoburb different than a traditional ethnic enclave?

KEY CONCEPTS REVIEW

1. **What are the implications and bases of *ethnicity*, and how have historic immigration streams shaped multiethnicity in the United States and Canada?** Section 6.1–6.2.
   Ethnicity implies a "people" or "nation," a large group classified according to common religious, linguistic, or other aspects of cultural origin or background, or, often, to racial distinctions. In common with nearly all countries, the United States and Canada are multiethnic. Past and current immigration streams—earlier primarily European, more recently Asian and Latin American—have intricately mixed their populations.

2. **How were the dominant Anglo American culture norms established, and how completely, spatially and socially, are its ethnic minorities integrated?** Section 6.3.
   The first effective settlers of Anglo America created its English-rooted charter culture to which other, later immigrant groups were expected to conform. Assimilation or acculturation has not been complete, and areal expressions of ethnic differentiation persist in America in the form of ethnic islands, provinces, or regional concentrations. French Canadian, black, Amerindian, Hispanic, Asian American, and other, smaller groups display recognizable areal presences. Among immigrant groups, those concentrations may result from cluster and chain migration.

3. **What patterns of ethnic diversity and segregation exist in the world's urban areas, and how are they created or maintained?** Section 6.4–6.5.
   Ethnic communities, clusters, and neighborhoods are found in cities worldwide. They are a measure of the social distance that separates minority from majority or other minority groups. Segregation measures the degree to which culture groups are not uniformly distributed within the total population. Although different world regions show differing patterns, all urban segregation is based on external restrictions of isolation and discrimination or internal controls of defense, mutual support, and cultural preservation by the ethnic group. Ethnic colonies, enclaves, ghettos, and ethnoburbs are the spatial result.

4. **What have been some of the cultural landscape consequences of ethnic concentrations in Anglo America and elsewhere?** Section 6.6.
   Landscape evidence of ethnicity may be subtle or pronounced. Clustered and dispersed rural settlement customs; house and barn types and styles; distinctive, largely urban, "Chinatowns," "Little Havanas," and other cultural communities; and even house colors, murals, or lawn decorations are landscape imprints of multiethnicity in modern societies.

CULTURAL IDENTITIES AND CULTURAL LANDSCAPES:

Diversity and Uniformity

The Amish, one of the most distinctive folk culture groups in the United States and Canada, pursue traditional farming methods.
©Photo Spirit/Shutterstock

Key Concepts

Cultural Identities

7.1 Folk culture characteristics, culture hearths, and culture regions

7.2 Folk food, drink, and music

7.3 Popular culture characteristics and diffusion

7.4 Popular food, drink, and music

7.5 Reactions against globalized popular culture

7.6 Vernacular regions

Cultural Landscapes

7.7 Tools for reading landscapes

7.8 Cultural landscapes: survey systems, houses, landscapes of consumption, and heritage landscapes

*C*hina's urban landscape features replicas of world landmarks including Athens' Parthenon, Sydney's Opera House, Egypt's Great Sphinx of Giza, London's Tower Bridge, Paris' Arc de Triomphe, Venice's Grand Canal, the Roman Colosseum, the Moscow Kremlin, and the White House in Washington D.C. Some of these landmarks are functional, serving as highway bridges, concert venues, theme parks, or shopping malls. Others are purely ornamental such as the 108-meter (354-foot) tall version of Paris' Eiffel Tower that squats in the center of a gated neighborhood.

As China opened its economy to the world, its civic leaders and designers repeated a long-established tradition among American elites—the Grand Tour of European capital cities and landmarks. In addition to Europe, their tours included some of North America's most distinctive landscapes such as the sprawling suburbs of southern California and the neon-lit Las Vegas strip. The Chinese tourists brought back more than photographs. They returned with building measurements, furnishings, and designers, aiming to replicate what they'd seen abroad.

Some Chinese middle-class urbanites reside in neighborhoods that carry the name and capture the look of some of the world's most desirable places. New developments in Chinese cities have opened with names such as Manhattan Gardens, Orange County, Palm Springs, Thames Town, and Vancouver Forest. The Orange County development near Beijing mimics the white stucco and red-tile roofs of single-family houses in sunny southern California. Beijing's Vancouver Forest neighborhood imitates its namesake's housing styles and tree-lined streets. Thames Town features Tudor and Georgian architecture just like London, classic red British phone booths, and statues of Harry Potter and Winston Churchill. Like many other Chinese real estate investment schemes, the Thames Town development has failed to attract the planned number of residents. Instead, wedding parties and photographers are among the most common users of its quaint landscape.

These uncanny Chinese copies of foreign landmarks and landscapes reflect the rising economic power of China, its fascination with Western culture, and the growth of its own popular, consumer culture. Nowhere is the growth of Chinese consumer culture more apparent than in its immense shopping malls. The three largest shopping malls in the world are all located in China. The New South China Mall features twice as much leasable space as the famed Mall of America in Minnesota. China's malls brim with consumer goods from stores such as Gap and H&M, luxury brands such as Gucci and Louis Vuitton, and global chain restaurants such as KFC and McDonalds. Some critics complain that the extensive borrowing from elsewhere has led to the neglect of China's rich cultural fabric, its distinctive cultural landscapes, and its historic folk-building traditions. As such, these additions to the Chinese landscape demonstrate the importance and interweaving of cultural landscapes and cultural identities.

7.1 Cultural Identities

The kaleidoscope of culture presents an endlessly changing design, different for every society, world region, and national unit. Ever present in each of its varied patterns, however, are persistent fragments of diversity amidst the spreading color of uniformity. One distinctive element of diversity derives from *folk culture*—the material and nonmaterial aspects of daily life preserved by small, local groups partially or totally isolated from the mainstream currents of the larger society around them. Additional sources of diverse cultural identities are provided by language, religion, and ethnicity, as we have seen in Chapters 5 and 6. Gender, and norms related to it, adds an additional element of cultural diversity. Finally, given time, easy communication, and common interests, globalized *popular* culture may provide a unifying color to the kaleidoscope, reducing differences among formerly distinctive groups, though perhaps not totally eradicating them.

In this chapter, we will trace the tensions between diversity and uniformity, particularly in U.S. and Canadian contexts, where diversified immigration provided the ethnic mix, frontier and rural isolation encouraged folk differentiation, and modern technology produced the leveling of popular culture. Along the way, we shall see the close interconnections between folk and popular culture, as well as reactions against the unifying tendencies of globalized popular culture. For geographers, the world is composed of multiple cultural landscapes bearing the imprint of human culture: values, preferences, activities, social relations, and technologies. Landscape features as varied as land survey systems, houses, shopping malls, and heritage landscapes can all be interpreted to better understand diverse cultural identities and the tensions between diversity and uniformity (**Figure 7.1**).

Folk Culture

Folk connotes traditional and nonfaddish, the characteristic or product of a cohesive, largely self-sufficient group that is isolated from the larger society surrounding it. **Folk culture,** therefore, may be defined as the collective heritage of institutions, customs, skills, dress, stories, music, and way of life of a small, stable, close-knit, usually rural community. Tradition controls folk culture, and resistance to change is strong. The homemade and handmade dominate in tools, food, music, story, and ritual. Buildings are erected without

Figure 7.1 Disney films and merchandise are popular around the world. Part of their popularity may stem from the fact that many of the story lines were adapted from classic folk tales. Disney movies, merchandise, and theme parks symbolize the globalization of popular culture. Starting with Disneyland, which opened in southern California in 1955, Disney has built theme parks in Florida, Paris, Tokyo, Hong Kong, and Shanghai.

©Flame/Alamy Stock Photo

architect or blueprint using locally available building materials. Folk societies, because of their subsistence, self-reliant economies, and limited technologies, are particularly responsive to physical environmental circumstances. Thus, foodstuffs, herbs, and medicinal plants reflect what is naturally available or able to be grown locally, and buildings reflect local climate and available materials.

Folk culture is often understood in opposition to popular or mass culture. It is seen as the unchanging, rural way of life, largely relegated to nonmodern, "traditional" peoples untouched by outside influences of mass media and market economies. Where folk culture exists in developed countries like the United States, it is found only among socially or geographically isolated rural groups—for example, the Amish, some Native American communities, or the presumably reclusive mountain folk of Appalachia. The prevailing notion in most of America is that the artifacts, beliefs, and practices of folk culture are curious reminders of the past, to be relegated to museums or tourist destinations offering the quaint and exotic "other." Thus, the relationship between folk and popular culture tends to be portrayed as one of conflict with folk culture doomed to eventual extinction by the forces of modernization and globalization. As we shall see, reality is more complex, with complicated linkages between folk and popular culture.

Although folk cultures are conservative and tend to resist innovations, they are not static. They often demonstrate flexibility and creativity when they encounter new or changing environmental or social circumstances, and they adapt accordingly. Indeed, all cultures were subsistence rural folk cultures until people built cities and developed a class hierarchy in which social and economic stratification began to differentiate the elite from commoners. The popular or mass culture that now so totally dominates modern life is a product of the industrialization and urbanization trends that began in the late 18th and early 19th centuries in Europe and the United States. Recently, globalization of popular culture has affected nearly every inhabited corner of the world. People everywhere tend to discard or alter elements of their folk culture when confronted with the attractions of modernity. Despite the tourism trade's efforts to sell "pristine" traditional cultures as a commodity, few if any groups remain, even in the developing world, that are still totally immersed in folk culture.

Yet folk cultural elements persist in all advanced societies. In Japan, traditional culture is tenaciously preserved in a highly industrialized and urbanized society that enthusiastically embraces nearly every fad produced by the Western culture that they have adopted. And the popularity in the United States of folk-themed movies and music (such as Mumford and Sons) and the proliferation of folk music and folk life festivals are evidence that folk culture is far from irrelevant to modern society. Rather, it should be viewed as an underlying foundation for popular culture, intersecting and influencing popular culture.

Folk life in its unaltered form, however, is a cultural whole composed of both tangible and intangible elements. **Material culture** is made of physical, visible things: everything from musical instruments to furniture, tools, and buildings. **Nonmaterial culture,** in contrast, is the intangible part, the mentifacts and sociofacts expressed in oral tradition, folk song and folk story, and customary behavior. Ways of speech, patterns of worship, outlooks, and philosophies are parts of the nonmaterial component passed to following generations by teachings and examples.

Within the United States, true folk societies are rare; the impacts of industrialization, urbanization, and mass communication have been too pervasive for their full retention. The Old Order Amish, with their rejection of electricity, the internal combustion engine, and other "worldly" accoutrements in favor of buggies, hand tools, and traditional dress are one of the least altered—and few—folk societies of the United States (**Figure 7.2**). Yet the Amish are very adept at dealing with the

(a)

Figure 7.2 (a) An Amish schoolhouse surrounded by horse-drawn buggies is evidence of a surviving folk culture. Motivated by religious convictions favoring simplicity, the Old Order Amish, a small conservative branch of Protestant Christianity, shun modern luxuries and most connections with the wider secular society. Children take a horse and buggy, not a school bus, on their daily trip to this rural school in east central Illinois. (*b*) Distribution of Old Order Amish communities in the United States. Starting from Lancaster County in eastern Pennsylvania, the Amish have spread west in search of available agricultural land.

©*Jean Fellmann*

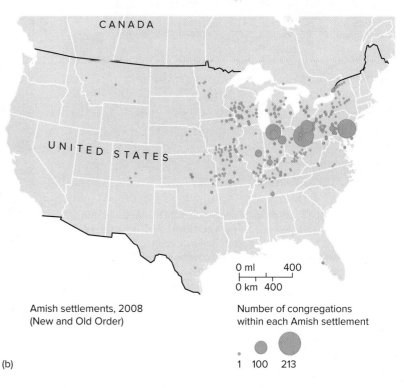

(b)

Amish settlements, 2008
(New and Old Order)

Number of congregations
within each Amish settlement

1   100   213

modern world, often with the assistance of their non-Amish (or *English*, in their vernacular) neighbors. For example, they own and operate successful cheese-making and furniture businesses; occasionally ride in, but do not drive, motor vehicles; use their *English* neighbors' telephones; and use propane gas to power their refrigerators instead of the forbidden electrical power grid.

Canada, on the other hand, has more clearly recognizable folk and decorative art traditions. One observer has noted that nearly all of the national folk art traditions of Europe can be found, in one form or another, well preserved and practiced somewhere in Canada. From the earliest arts and crafts of New France to the domestic art forms and folk artifacts of the Scandinavians, Germans, Ukrainians, and others who settled in western Canada in the late 19th and early 20th centuries, folk and ethnic traditions have been adapted to the Canadian context.

Indeed, in many respects, the geographies of ethnicity and folk culture are intertwined. The variously named *Swiss* or *Mennonite* or *Dutch* barn (**Figure 7.3**), introduced into Pennsylvania by German immigrants, has been cited as physical evidence of ethnicity; in some of its many modifications and migrations, it may also be seen as a folk culture artifact of Appalachia. The folk songs of, say, western Virginia can be examined either as

nonmaterial folk expressions of the Upland South or as evidence of the ethnic heritage derived from rural English forebears. In the New World, the debt of folk culture to ethnic origins is clear and pervasive. With the passage of time, of course, the dominance of origins recedes and new cultural patterns and roots emerge.

Culture Hearths of the United States and Canada

The United States and Canada are "lumpy stews" composed of groups of people who came with distinct ethnic identities and underwent partial or complete assimilation. They brought more than tools, kitchen items, and clothing. They brought ideas of what implements were proper to use, how to cook, dress properly, find a spouse, and practice their faith. They brought familiar songs to be sung and stories to be told, and ideas of how a house should look and a barn be built. They came, in short, with all the mentifacts and sociofacts to shape the artifacts of their way of life in their new home (**Figure 7.4**). (Mentifacts, sociofacts, and artifacts are discussed in Chapter 2.)

Their material and nonmaterial culture frequently underwent immediate modification in the New World. Climates and soils were often different from their homelands; new types of crops and livestock were found. Building materials differed. The settlers still retained the essence and the spirit of the old but made it simultaneously new and American. The first colonists, their descendants, and still later arrivals created not one but many cultural landscapes of America, defined by the structures they built, the settlements they created, and the customs they followed. The natural landscape of America became settled, and superimposed on the natural landscape as modified by its Amerindian occupants were the cultural traits and characteristics of the European immigrants.

Figure 7.3 The Pennsylvania Dutch barn, with its origins in southern Germany, has two levels. Livestock occupy the ground level; on the upper level, reached by a gentle ramp, are the threshing floor, haylofts, and grain and equipment storage. A distinctive projecting forebay provides shelter for ground-level stock doors and unmistakably identifies the Pennsylvania Dutch barn. The style, particularly in its primitive log form, diffused from its eastern origins, underwent modification, and became a basic form in the Upland (i.e., off the Coastal Plain) South, Ohio, Indiana, Illinois, and Missouri. It is an example of a distinctive ethnic imprint on the landscape that became part of the material folk culture of the United States.

Figure 7.4 The reconstructed Plimoth Plantation in Massachusetts offers visitors a 17th century village experience. The first settlers in the New World carried with them fully developed cultural identities. Even their earliest settlements reflected established ideas of house and village form. Later, they were to create a variety of distinctive cultural landscapes reminiscent of their homeland areas, though modified by American environmental conditions and material resources.

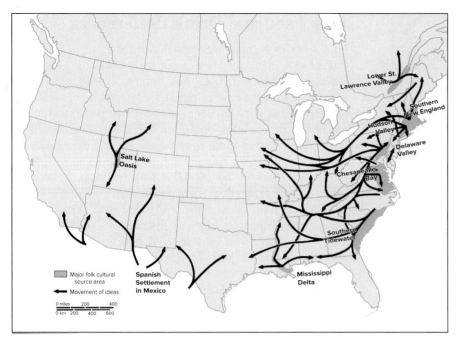

Figure 7.5 Folk culture hearths and diffusion paths for the United States and Canada.

Sources: Based on Henry Glassie, Pattern in the Material Folk Culture of the Eastern United States, *37–38, 1968, University of Pennsylvania Press; Michael P. Conzen, ed.,* The Making of the American Landscape *(Winchester, Mass.: Unwin Hyman, 1990), 373. Allen G. Noble, Wood, Brick, and Stone, Vol. 1 (Amherst: University of Massachusetts Press, 1984); and* Annals of the Association of American Geographers, *Richard Pillsbury, Vol. 60, 446, Association of American Geographers, 1970.*

The early arrivers established footholds along the East Coast. Their settlement areas became cultural hearths, nodes of introduction into the New World—through *relocation diffusion*—of concepts and artifacts brought from the Old. Each of the North American hearths had its own mix of peoples and, therefore, its own landscape distinctiveness. Locales of innovation, they were source regions from which relocation and *expansion diffusion* carried their cultural identities deeper into the continent (**Figure 7.5**). Later arrivals, as we have seen in Chapter 6, added to the cultural mix, and in some cases, they set up independent secondary hearths in advance of or outside the main paths of diffusion. French settlement in the lower St. Lawrence Valley re-created the long lots and rural house types of northwestern France. Upper Canada was English and Scottish, with strong infusions of New England folk housing carried by Loyalists who left that area during the Revolutionary War. Southern New England bore the imprint of settlers from rural southern England, while the Hudson Valley hearth showed the imprint of Dutch, Flemish, English, German, and French Huguenot settlers.

In the Middle Atlantic area, the Delaware River hearth was created by a complex of English, Scotch-Irish, Swedish, and German influences. The Delaware Valley below Philadelphia also received the Finnish Karelians, who introduced the distinctive backwoods subsistence lifestyles, self-sufficient economies, and log cabin building techniques of their forested homeland. It was their pioneering "midland" culture that was the catalyst for the rapid advance of the frontier and successful settlement of much of the interior of the continent and, later, of the Pacific Northwest.

Coastal Chesapeake Bay held English settlers and some Germans and Scotch-Irish. The large landholdings of the area led to dispersed settlement and prevented a tightly or clearly defined culture hearth from developing. However, distinctive house types later diffused from there. The Southern Tidewater hearth was dominantly English modified by West Indian, Huguenot, and African influences. The French again were part of the Mississippi Delta hearth, along with Spanish and Haitian elements.

Later in time and deeper in the continental interior, the Salt Lake hearth marks the penetration of the distant West by the Mormons, a group identified by their religious distinctiveness. Spanish American borderlands, the Upper Midwest Scandinavian colonies, English Canada, and the ethnic clusters of the Prairie provinces could be added to the North American map of distinctive immigrant culture hearths.

The ethnic hearths gradually lost their identification with immigrant groups and became source regions of American architecture and implements, ornaments and toys, cookery and music. The evidence of the homeland was there, but the products became purely indigenous. In the isolated, largely rural American hearth regions, the ethnic culture imported from the Old World was partially transmuted into the folk culture of the New.

Folk Culture Regions of the United States and Canada

When folk **customs**—repeated, characteristic acts, behavioral patterns, artistic traditions, and conventions regulating social life—are shared by a people living in a distinctive area, a folk culture region may be recognized. Frontier settlers carrying to new, interior locations the artifacts and traditions of those hearth areas created a small set of indistinctly bounded eastern folk cultural regions (**Figure 7.6**).

From the small *Mid-Atlantic* region, folk cultural items and influences were dispersed into the North, the Upland South, and the Midwest. Furniture styles, log construction, decorative arts, house and barn types, and distinctive "sweet" cookery were among the European imports converted in the Mid-Atlantic hearth to American folk expressions.

The folk culture of the *Lowland South,* by contrast, derived from English originals and African admixtures. French influences in the Mississippi Delta hearth combined with elements from the highland areas added to the amalgam. Dogtrot and I houses became common; English cuisine was adapted to include black-eyed peas, turnip greens, sweet potatoes, small-bird pies, and syrups from sugarcane and sorghum. African origins influenced the widespread use of the banjo in music.

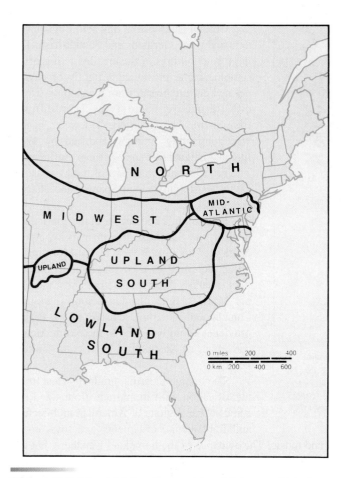

Figure 7.6 Material folk culture regions of the eastern United States.

Source: Redrawn from Henry Glassie, Pattern in the Material Folk Culture of the Eastern United States, *p. 39, 1968, University of Pennsylvania Press.*

Folk Food and Drink Preferences

Food is one of the most important distinguishing elements of cultural groups. Ethnic foods are among the central attractions of the innumerable fairs and "fests" held throughout the United States and Canada. In the case of ethnic foods, what is celebrated is the retention in a new environment of the food preferences, diets, and recipes that had their origin in a distant homeland. Folk food habits, on the other hand, are products of local circumstances; the diet consists of the natural foods derived from hunting, gathering, and fishing or the cultivated foods and domestic animals suited to the local environmental conditions.

Second, most areas of the world have been occupied by a complex mix of peoples migrating in search of food and carrying food habits and preferences with them in their migrations. In the Americas, Australia, New Zealand, and a few other regions of recent colonization, we are aware of these differing *ethnic* origins and the recipes and customs that they imply. In other world regions, ethnic and cultural intermixture is less immediately apparent. In Korea, for example, what outsiders see as a distinctive ethnic cookery best known, perhaps, for *kimch'i*—brined, pickled, and spiced vegetables in endless combinations and uniquely Korean—also incorporates Japanese and Chinese foodstuffs and dishes.

Third, food habits are not just matters of survival but are intimately connected with the totality of *culture*. People eat what is available and also what is, to them, edible. Sheep's brains and eyeballs, boiled insects, animal blood, and pig intestines, which are delicacies in some cultures, may be unclean or disgusting to others. Further, in most societies, food and eating are considered a social experience, not just a personal one, and a specially prepared meal is the true mark of hospitality.

The interconnections among folk, ethnic, and customary food habits are evident in the U.S. diet. Of course, the animals and plants, the basic recipes and flavorings, and the specialized festive dishes of American folk groups have ethnic origins. Many originated abroad and were carried to and preserved in remote New World areas. Many were derived from the diet of the Amerindians and adapted in widely different regional contexts. Turkey, squash, pumpkin, and cranberries are among the foods adapted from Amerindians, as is corn (maize), which appeared over time south of Pennsylvania as Southern grits, Southwestern tortillas, and cornbread. Such classic American dishes as Brunswick stew (a thick stew made with vegetables and two meats, such as squirrel and rabbit or chicken), the clambake, smoked salmon, cornflakes, and beef jerky were originally Indian fare. Gradually, the environmental influences and isolation, over time created culinary distinctions among populations recognized as American rather than as ethnic immigrants.

Cookbook categories of New England, Creole, Southern, Chesapeake, Southwestern, and other regional fare may be further refined into cookbooks containing Boston, Pennsylvania Dutch, Charleston, New Orleans, Southern Tidewater, and other more localized recipes. Their diversity is ample proof of the diffusion into national and international popular culture of formerly local folk cultural distinctions. Specific American dishes that have achieved fame and wide acceptance developed locally in

The *Upland South* showed a mixture of influences carried from the Southern Tidewater hearth and brought south from the Mid-Atlantic folk region along the Appalachian highlands by settlers of German and Scotch-Irish stock. The physical isolation of the Upland South and its Ozark Mountains outlier encouraged the retention of traditional folk culture long after it had been lost in more accessible locations. Log houses and farm structures, rail fences, traditional art and music, and home-crafted quilts and furniture make the Upland South region a prime repository of folk artifacts and customs in the United States.

The *North*—dominated by New England, but including New York State, English Canada, Michigan, and Wisconsin—showed a folk culture of decidedly English origin. The saltbox house and Boston baked beans in stoneware pots are characteristic elements. The New England–British domination is locally modified by French Canadian and central European influences.

The *Midwest*—a conglomerate of inputs from the Upland South, from the North, and, particularly, from the Mid-Atlantic region—is the least distinctive, most intermixed, and most Americanized of the cultural regions. Everywhere the interior contains evidences, both rural and urban, of artifacts carried by migrants from the eastern hearths and by newly arriving European immigrants.

response to food availability. New England seafood chowders and baked beans; southern pone, johnnycake, hush puppies, and other corn-based dishes; the wild rice of the Great Lakes states; Louisiana crayfish (crawfish); gumbo; and salmon and shellfish dishes of the Pacific coast are but a few of many examples of folk foods and recipes originally and still characteristic of specific cultural areas but now also—through cookbooks, television cooking shows, and supermarket products—made part of the national food experience.

In the United States, drink also represents a hybrid mixture of ethnic imports and folk adaptations. A colonial taste for rum was based on West Indian and Tidewater sugarcane and molasses. European rootstock was introduced, with mixed results, to develop vineyards in most seaboard settlements; the native scuppernong grape was tried for wine making in the South. Peach, cherry, apple, and other fruit brandies were distilled for home consumption. Whiskey was a barley-based import accompanying the Scots and the Scotch-Irish to America, particularly to the Appalachians. In the New World, the grain base became native corn, and whiskey making became a deeply rooted folk custom integral to the subsistence economy.

Whiskey also had significance in the cash economy. Small farmers of isolated areas far from markets converted part of their corn and rye crops into whiskey to produce a concentrated, low-volume, high-value commodity conveniently transportable by horseback over bad roads. Such farmers viewed a federal excise tax imposed in 1791 on the production of distilled spirits as an intolerable burden not shared by those who could sell their grain directly. The tax led first to a short-lived tax revolt, the Whiskey Rebellion of 1794 in western Pennsylvania, and subsequently to a tradition of moonshining—producing untaxed liquor in unlicensed stills. **Figure 7.7** suggests the close association between its isolated Appalachian upland environment and illicit whiskey production in East Tennessee in the 1950s.

Folk Music

Folk music in North America is not merely intertwined with popular culture; it is the foundation for American popular music. In turn, American popular music that was derived from folk sources exerts a global influence that since the late 1800s has fostered both popular and folk music genres throughout the rest of the world. And those folk sources continue to serve as inspirations and themes for the commercial music industry, films and film scores, musical theater, concert music, and television. Here again, folk and popular culture intermingle and influence each other.

Old World songs were carried by settlers to the New World. Each group of immigrants established an outpost of a European musical community, making the American folk song, in the words of Alan Lomax, "a museum of musical antiques from many lands." But the imported songs became Americanized, hybridization between musical traditions occurred, and

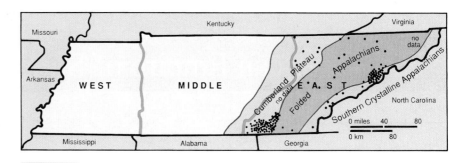

Figure 7.7 In the mid-1950s, official estimates put weekly moonshine production at 24,000 gallons in mountainous eastern Tennessee, at 6,000 gallons in partially hilly middle Tennessee, and at 2,000 gallons in flat western Tennessee. The map shows the approximate number of stills seized each month at that time in East Tennessee. Each dot indicates one still.

Source: Redrawn from Loyal Durand, "Mountain Moonshining in East Tennessee," Geographical Review *46 (New York: American Geographical Society, 1956), 171.*

the American experience added its own songs of frontier life, of farming, courting, and laboring (see the feature "The American Empire of Song"). Eventually, distinctive American styles of folk music and recognizable folk song cultural regions developed (**Figure 7.8**).

The *Northern* song area—including the Maritime provinces of Canada, New England, and the Middle Atlantic states—in general featured unaccompanied solo singing in clear, hard tones. Its ballads were close to English originals, and the British

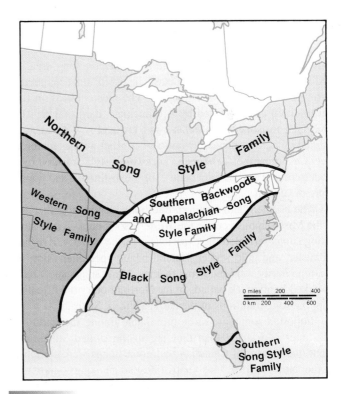

Figure 7.8 Folk song regions of eastern United States. Alan Lomax has indirectly outlined folk culture regions of the eastern United States by defining areas associated with different folk song traditions.

Source: Redrawn "Map depicting folk song regions of the Eastern U.S." by Rafael Palacios, from Folk Songs of North America *by Alan Lomax, 1960.*

The American Empire of Song

The map sings. The chanteys surge along the rocky Atlantic seaboard, across the Great Lakes and round the moon-curve of the Gulf of Mexico. The paddling songs of the French Canadians ring out along the Saint Lawrence and west past the Rockies. Beside them, from Newfoundland, Nova Scotia, and New England, the ballads march towards the West.

Inland from the Sea Islands, slave melodies sweep across the whole South, from the Carolinas to Texas. And out on the shadows of the Smoky and Blue Ridge mountains, the old ballads, lonesome love songs, and hoedowns echo through the upland South into the hills of Arkansas and Oklahoma. There, in the Ozarks, the Northern and Southern song families swap tunes and make a marriage.

The Texas cowboys roll the little "doughies" north to Montana, singing Northern ballads with a Southern accent. New roads and steel rails lace the Southern backwoods to the growl and thunder of Negro chants of labor—the axe songs, the hammer songs, and the railroad songs. These blend with the lonesome hollers of levee-camp mule-skinners to create the blues, and the blues, America's *cante hondo,* uncoils its subtle, sensual melancholy in the ear of all the states, then all the world.

The blues roll down the Mississippi to New Orleans, where the Creoles mix the musical gumbo of jazz—once a dirty word, but now a symbol of musical freedom for the West. The Creoles add Spanish pepper and French sauce and blue notes to the rowdy tantara of their reconstruction-happy brass bands, stir up the hot music of New Orleans and warm the weary heart of humanity. . . . These are the broad outlines of America's folk-song map.

Source: "Introduction" from *Folk Songs of North America* by Alan Lomax, 1960, Doubleday.

---

connection was continuously renewed by new immigrants, including Scots and Irish. The traditional ballads and popular songs brought by British immigrants provided the largest part of the Anglo Canadian folk song heritage. On both sides of the border, the fiddle was featured at dances, and in the United States, fife-and-drum bands became common in the early years of the Republic.

The *Southern Backwoods and Appalachian* song area, extending westward to East Texas, involved unaccompanied, high-pitched, and nasal solo singing. The music, based on English tradition and modified by Appalachian "hardscrabble" life, developed in isolation in upland and lowland settlement areas. Marked by moral, spiritual, emotional conflict with an undercurrent of haunting melancholy, the backwoods style emerged in the modern period as the major source for the distinctive and popular genre of "country" music.

The northern and southern traditions blended together west of the Mississippi to create the *Western* song area. There, storytelling songs reflected the experiences of the cowboy, riverman, sodbuster, and gold seeker. Natural beauty, personal valor, and feminine purity were recurring themes. Many songs appeared as reworked lumberjack ballads of the North or other modifications from the song traditions of the eastern United States.

Imported songs are more prominent among the traditional folk tunes of Canada than they are in the United States; only about one-quarter of Canadian traditional songs were composed in the New World. Most native Canadian songs—like their U.S. counterparts—reflected the daily lives of ordinary folk. In Newfoundland and along the Atlantic coast, those lives were bound up with the sea, and songs of Canadian origin dealt with fishing, sealing, and whaling. Remote Cape Breton Island, off the east coast of Canada, remains a refuge of traditional Scottish folk music styles that have mostly disappeared elsewhere.

Particularly in Ontario, it was the lumber camps that inspired and spread folk music. Anglo Canadian songs show a strong Irish character in pattern and tune and traditionally were sung solo and unaccompanied.

The *Black* folk song tradition, growing out of racial and economic oppression, reflects a union of American folk song, English country dancing, and West African musical patterns. The African American folk song of the rural South or the northern ghetto was basically choral and instrumental in character; hand clapping and foot tapping were used to establish rhythm. A strong beat, a leader-chorus style, and deep-pitched mellow voices were characteristic.

Rounding out the North American scene, there are the river and fur trader songs of the French Canadians and the strong Mexican American musical tradition in the Southwest.

Different folk music traditions metamorphosed and spread in the 20th century as distinctive styles of popular music. Jazz emerged in New Orleans in the later 19th century as a union of ragtime and the blues, a type of Southern black music based on work songs and spirituals. Urban blues—performed with a raw vocal delivery accompanied by electric guitars, harmonicas, and piano—was a Chicago creation, brought there largely by artists who had migrated from Mississippi. Country music spread from its Southern hearth region with the development of the radio and the phonograph in the 20th century. It became commercialized, electrified, and amplified but retained folk music elements at its core. Bluegrass style, a high-pitched derivative of Scottish bagpipe sound and church congregational singing, is performed unamplified, true to its folk origins. Bluegrass groups often take their name from local places or landscape features, emphasizing the ties of people, performers, and the land. As these examples of musical style and tradition show, the ethnic merges into the folk and the folk blends into the popular—in music and in many other elements of culture.

Popular Culture and National Uniformities

In the early 20th century, rural America was a mosaic of unique regional cultural landscapes. Socially, the cities of the eastern and midwestern parts of the country were a world apart from life on the farm. Brash and booming with the economic success of rampant industrialization, the cities were in constant flux. Building and rebuilding, adding and absorbing immigrants and rural in-migrants, increasingly interconnected by telegraph and by passenger and freight railroads, their culture and way of life were far removed from the surrounding agricultural areas.

It was in the countryside that regional cultural differentiation was most clearly seen. Although the flow of young people to the city, responding to the push of farm mechanization and the pull of urban jobs and excitement, was altering traditional social orders, the automobile, electrification, and the lively mass medium of radio had not yet obscured the distinction between urban and rural. The family farm, kinship and community ties, the traditions, ways of life, and artifacts of small town and farm existence still dominated rural life. But those ways and artifacts, and the folk cultural regions they defined, were all eroded and erased with the modernization of North American life and culture.

Regional character is transient. New immigrants, new economic challenges, and new technologies serve as catalysts of rapid change. By World War I and the Roaring Twenties, the automobile, radio, motion pictures, and a national press began to homogenize America. The slowing of the immigrant stream and assimilation of second-generation immigrants blurred some of the most regionally distinctive cultural identifications. Mechanization, mass production, and mass distribution through mail-order catalogs diminished self-sufficiency and household crafts. Popular culture began to replace traditional culture in everyday life for the majority of the population throughout the United States and Canada.

As early as the middle 19th century, women's magazines dictated taste in fashion and household furnishings, at least for urban elites. Mail order catalogs appearing in the late 19th century served the same purpose for more ordinary goods, garments, and classes of customers. Popular culture, based on fashions, standards, or fads developed in national centers of influence, diffused across wide areas and diverse social strata. Popular culture promises liberation through exposure to a much broader range of available opportunities—in clothing, foods, tools, recreations, and lifestyles—than the limited choices imposed by custom and isolation.

As we have seen, folk and popular culture are distinct but not necessarily opposites. Folk or ethnic culture is the domain of distinctive small groups, and above all, tradition. **Popular culture,** in contrast, refers to the general mass of people, mostly urban or suburban, constantly adopting, conforming to, and quickly abandoning ever-changing trends and fads promoted by the mass media and social media and sold in the market economy. Even so, folk culture often forms the inspiration and backdrop for new popular cultural forms. For example, universally enjoyed popular and spectator sports such as soccer, football, golf, and tennis originated as local and regional folk games, many of them hundreds of years old.

The popular mass culture of the latter part of the 20th century that locked millions of American viewers into sharing the "must see" offerings of variety shows, situation comedies, and evening newscasts on three national television networks had largely passed by the early 21st century. It was replaced by a culture of multiple entertainment and information niches. With hundreds of cable and satellite television channels to choose from, millions of Web sites, and proliferating social media, the mass-culture era has been transformed into one of fragmented subcultures. Individuals interact electronically with likeminded persons, become their own music and entertainment programmers, use their computer and cell phone as personal media, and forego at least some of the contact with the larger society that is implied by notions of mass popular culture.

Presumably, all mass-produced consumer goods should be equally available to all segments of a society. Our travel experiences, however, have taught us that tastes and styles differ from place to place. Similar *regionalism* can be found throughout the popular culture realm. For example, while most of the world is unfamiliar with the game of cricket, it is remarkably popular in Britain and the countries of its former empire, such as Australia, India, Pakistan, South Africa, and Sri Lanka.

Popular culture uniformity is frequently, though not exclusively, associated with national populations: the American or Canadian way of life distinguished from that of the English, the Japanese, or others. Even these distinctions are eroding as popular culture in many aspects of music, movies, sports (soccer, for example), and fashion becomes internationalized (**Figure 7.9**). Popular culture becomes dominant with the wide dissemination of common influences and with the mixing of cultures that force both ethnic and folk communities to become part of a larger homogeneous society.

Figure 7.9 Soccer is the most globalized of sports. It is extremely popular in Africa, Europe, and South America, as these Brazilian fans demonstrate. Its world championship, the World Cup, is watched by an estimated 600 million fans. The market for professional players is also highly globalized.

©kaisersosa67/iStock/Getty Images

Cultural Globalization

Popular culture exerts a leveling force, reducing but not eliminating locally distinctive lifestyles and material and nonmaterial cultures. Uniformity replaces diversity. Landscapes of popular culture also tend to acquire uniformity through the installation of standardized facilities. Within the United States, for example, national motel chains announced by identical signs, advertised by repetitious billboards, and featuring uniform facilities and services may comfort travelers with the familiar but also denies them one of the benefits of travel—local variety. Chain gas stations, discount stores, and other enterprises offer familiar standardized products and services wherever one resides or journeys.

Other material and nonmaterial items are subject to the same widespread uniformities. The latest movies are simultaneously released throughout the country; the same children's toys and adults' games are everywhere instantly available to satisfy the generated wants.

Wilbur Zelinsky reported on the speed of diffusion of a manufactured desire:

> In August, 1958, I drove from Santa Monica, California, to Detroit at an average rate of about 400 miles (650 km) per day; and display windows in almost every drugstore and variety store along the way were being hastily stocked with hula hoops just off the delivery trucks from Southern California. A national television program the week before had roused instant cravings. It was an eerie sensation, surfing along a pseudo-innovation wave.[2]

Many of these North American elements of popular culture are oriented toward the automobile, the ubiquitous means of local and interregional travel. Advertising and distinctive design assures instant recognition as chain outlets cluster along highway retail strips, guaranteeing that whatever regional character still remains, the commercial areas are everywhere the same—*placeless.* Critics perceive that the diffusion of popular culture promotes **placelessness,** the replacement of local identity and variety with a homogeneous and standardized landscape. Increasingly, news reports bring stories of communities protesting the arrival of a Wal-Mart "big box" store, the multiplication of uniform highway strip malls, and the like. For some people, that uniformity and loss of local control is unacceptable, and individuals and whole communities fight the pervasive influence of popular culture. For these protesters, globalized popular culture destroys valuable traditions, unique local identities, and a way of life that is perhaps more in tune with place and the environment. In its defense, popular culture brings a cultural uniformity that is vastly richer in choices than that which was lost.

The globalization of popular culture is seen in the rapid diffusion of brands and styles of clothing, food and drink, movies, television shows, and music. However, those uniformities are transitory. Whereas folk cultures have ingrained traditions that change only slowly and locally, popular culture tends to change rapidly and uniformly over wide expanses. That is, popular culture diffuses rapidly, even instantaneously, in our age of immediate global communication and sharing of ideas through television, radio, and the Internet. Those same media and means assure the widespread quick replacement of old fads with new. Chain stores and restaurants quickly go in and out of fashion. Now, news reports tell of communities mourning the loss of their chain stores, put out of business by online shopping in the latest wave of globalized consumption.

Imagine this scene: Wearing a Yankees baseball cap, a Hollister shirt, Abercrombie and Fitch jeans, and Nike shoes, a middle-class teenager in Shanghai, China, goes with his friends to see the latest Hollywood release. After the movie, he uses his smart phone to text-message his mother that they plan to eat at a nearby McDonald's. Meanwhile, his sister sits at home, listening to the latest American music on her iPhone, uploading pictures to Facebook while playing multiplayer video games with gamers from around the world. The activities of both young people are evidence of the globalization of popular culture that is Western and particularly—though certainly not invariably—American in origin. U.S. movies, television shows, software, music, food, brand names, and fashions are marketed worldwide. They influence the beliefs, tastes, and aspirations of people in virtually every country, though their effect is most pronounced on the young. They, rather than their elders, want to emulate the stars in movies and popular music. They are also the group most apt to use English words and slang in everyday conversation, though the use of English as the worldwide language of communication in economics, technology, and science is an even broader indication of current cultural merging.

Rapid introduction and quickly falling prices of high-tech communication and entertainment devices have had a profound effect on lifestyles wherever personal freedom and Westernized economies, incomes, and cultures prevail. These new communication technologies speed the diffusion of popular culture and expand opportunities for education, recreation, and information. Africa has lagged the rest of the world in telecommunications access because it lacked landline telephone networks in many areas. But with the advent of prepaid mobile phones, Africa has begun to catch up. It would be a mistake to believe that the United States leads in the adoption and development of new communication technologies. Europe leads the United States in the adoption of mobile telephones, and residential broadband Internet service is substantially faster in Japan and South Korea than in the United States.

Popular Food and Drink

Fast-food restaurants—franchised or corporate owned—use a standardized logo, building design, and menu across cultural and political borders (**Figure 7.10**). With large budgets for advertising and expansion, they are major carriers of the globalization of popular culture. For communities outside the mainstream of popular culture, their arrival is a status symbol. For the traveler, they provide the assurance of a known product, at the cost of insulating the palate from the regionally distinctive. Even food outlets identified with ethnic identities tend to

---

[2]*The Cultural Geography of the United States.* Rev. ed. (Englewood Cliffs, N.J.: Prentice-Hall, 1992), p. 80, fn 18.

Figure 7.10 Western fast-food chains, classics of standardized popular culture, have gone international—and bilingual—as this KFC outlet in Xian, China, reveals.

©Jon C. Malinowski/Human Landscape Studio

Figure 7.12 Anti-Starbucks graffito on a San Francisco sidewalk contesting the corporate homogenization of the urban landscape.

©Jerry Fellmann

lose their cultural character. Pizza has become American, not Italian (**Figure 7.11**), just as the franchised Mexican American taco and burrito have escaped their regional and ethnic confines and been carried nationwide and worldwide.

Starting in the 1990s, specialty coffee establishments serving cappuccinos, lattes, and other variants began displacing local beverages such as tea or soft drinks. Like other globalized products, specialty coffee has a distinct geography and place identity. Drinking coffee connects consumers to the world, offering the sensation of globe-trotting by purchasing beans from Sumatra, Colombia, Kenya, or Ethiopia. The American hearth region for the specialty coffee culture was Seattle, Washington, the original home of Starbucks, the most prominent corporate identity in the trade. Starbucks was established in 1971 in Seattle's Pike Place Market, an eclectic public market and distinctive local landmark, popular with both tourists and locals. Starbucks didn't invent specialty coffee but was inspired by the espresso bars of Milan, Italy. Of course, the Italians didn't invent coffee either—the crop was first domesticated in present-day Ethiopia. Today, Starbucks has 27,000 stores in 75 countries and continues to spread, an expansion diffusion process that is not universally appreciated (**Figure 7.12**).

The standardization brought about by globalized popular culture, of course, is not complete. Seemingly universal popular icons are always differentially adapted and modified for easy acceptance by different national societies. The term **glocalization** describes this adaptation of globalized products to fit local contexts. Regional food and drink preferences persist.

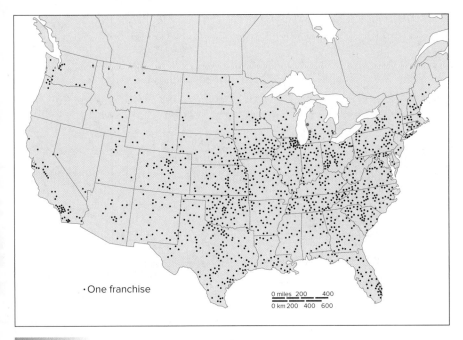

· One franchise

0 miles 200   400
0 km 200 400 600

Figure 7.11 The locations of pizza parlors of a single national chain.

Source: Floyd M. Henderson and J. Russel, unpublished drawing.

Domino's and Pizza Hut, for example, have a combined total of some 6,000 overseas outlets in more than 100 countries, but they do not serve a standard product worldwide. In India, customers likely will order their pizza with spicy chicken sausage or pickled ginger. In Japan, a best seller is pizza topped with potatoes, mayonnaise, and ham or bacon bits. Hong Kong customers prefer Cajun spice pizza flavoring; Thais favor hot spices mixed with lemon grass and lime; and in Australia, the number one topping for pizza is eggs. McDonald's sells the Chicken Maharaja Mac, McVeggie and McCurry Pan in India, kosher Big Macs in Israel, and beer in Germany. Outside the United States, McDonald's advertising emphasizes the use of products from local farms. The store name and logo may be universal, but the product varies to fit local tastes.

In contrast to the standardization of popular food and drink, there has been a movement termed **neolocalism** that emphasizes a return to local or regional food and drink products and a rejection of more homogeneous national and global products. Neolocalism cultivates a distinctive local place identity and can be seen in the popularity of microbrewed beer. Beer was once a local product before refrigeration, interstate highways, national advertising, and industry mergers concentrated production in the hands of two companies, Anheuser Busch and Miller, both located in interior cities with large German immigrant populations, St. Louis and Milwaukee. By contrast, the source or hearth for the microbrewery or craft beer movement in the late 1980s was the West, specifically the Colorado Front Range cities, the San Francisco area, and the coastal cities of Oregon and Washington. The spatial distribution of the approximately 6,300 craft breweries in the United States, for example, is not uniform across the country but exhibits regional concentrations (**Figure 7.13**). The names and labels used in craft beers, as well as the decor inside the brew pubs, are often based on local history, landmarks, or distinctive landscapes that create and strengthen the product's place identity (**Figure 7.14**). For fans, drinking a microbrewed beer is a way of demonstrating one's attachment and loyalty to the local region.

Popular Music and Dance

The music and dance styles of folk and popular culture and high-status and low-status groups intertwine in complex ways. For example, the waltz began as a folk dance in Austria and southern Germany, was refined into an elite style of dance in Vienna, diffused through the cities of Europe and the Americas, and finally sparked a popular culture craze among the rising middle and working classes.

In popular music and dance, geography still matters. Differences in regional tastes and hearth regions for innovation are evident. Music is symbolically expressive of the experiences and emotions of people with particular geographic and group identities. Country music lyrics, for example, contain themes that resonate with the experience of the rural working class. At the local scale, both rap and grunge originated as expressions of the alienation felt by particular segments of the youth population in the Bronx and in Seattle, respectively. Regional musical expressions of culture include Cajun music of south Louisiana, Tejano music of the TexMex borderlands, and the polka of the Upper Midwest. To enter the popular and mass cultural spheres, however, particular regional genres of music must diffuse at the national and global

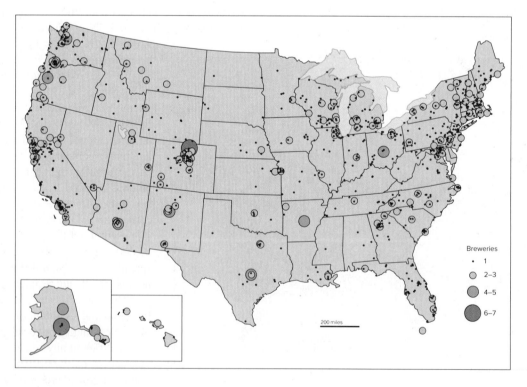

Figure 7.13 Microbreweries by zip code, 2002. Heavy concentrations of microbreweries are found in the source regions of the West.

Source: "Microbreweries as Tools of Local Identity," by Steven M. Schnell and Joseph F. Reese from Journal of Cultural Geography, 21, 1 (2003), Figure 1, 50.

Figure 7.14 Local imagery often appears in craft beer labels. Craft breweries work to cultivate a regional identity in order to compete with national and international brands. They draw upon their customer's sense of place, often using images of distinctive local landscapes and emphasizing the use of local ingredients.

©Bell's Brewery, Inc.

scales and, in turn, be modified to express the collective cultural identities of people occupying different places. Country music was originally associated with the Upland South. It has long since lost that regional exclusivity, and Nashville has become a product, not a place. By the late 1970s (**Figure 7.15**), no American with access to radio was denied exposure to slide guitar and melancholy lyrics.

Globalization of popular music with folk culture roots is clearly demonstrated by *world music*. It is usually described as music

strongly rooted in the folk and/or ethnic traditions of non-Western cultures but often blended with Western music to retain its sense of the exotic and yet be acceptable to Western tastes. Originally used to describe music from Africa and its diaspora in the New World, the term *world music* now more broadly includes the music of folk, ethnic, and minority groups in any culture. Much of world music is hybrid in nature, a fusion of various music genres from different global origins. In that blending process, local musical forms are

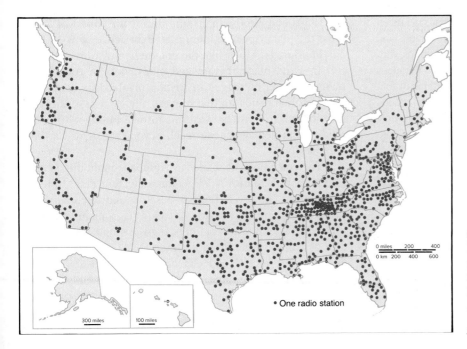

Figure 7.15 Country music radio stations. Although still most heavily concentrated in the Upland South, radio stations playing only country music had become commonplace nationally by the late 1970s.

Source: Redrawn from George O. Carney, "From Down Home to Uptown," Journal of Geography, 76 (Indiana, PA: National Council for Geographic Education, 1977), 107.

Hip-Hop Undergoes Globalization and Glocalization

Unlike the distant folk background of some world music, the beginnings of hip-hop culture are recent and easily mapped. Hip-hop culture emerged in the Bronx borough of New York City in the 1970s at a time when the South Bronx was undergoing a massive downward spiral. The disruptive Cross-Bronx expressway had been sliced through the neighborhood, the middle class was fleeing to the suburbs, manufacturing jobs were disappearing by the thousands, the city teetered on the edge of bankruptcy, and it had cut back essential services to the South Bronx. Many landlords responded by abandoning their tenement buildings, sometimes setting them afire to collect insurance money. Out of this urban wasteland arose hip-hop culture.

DJs at dance clubs and in street parties began stylized talking with a simple four-count beat—which became known as rapping—over a background of funk rock and disco music. Hip-hop culture, which involves rap music, DJs, graffiti, and break dancing, exhibits West African and Jamaican influences and is a contemporary expression of older forms of "talking music," such as spirituals, work songs, talkin' country blues, and the rhythmic sermons of black preachers. From these beginnings, rap developed as underground "protest music," with lyrics voicing the experiences of socially and economically disadvantaged black and Latino youth—alienation, police harassment, drug use, sexual conduct, race relations, and the like. Themes of place, home, and identity abound in rap music, with many songs depicting life in the "hood."

From its hearth in New York City, hip-hop diffused to a second center in Los Angeles (particularly the suburb of Compton) where "gangsta" or "reality" rap developed. Subsequently, hip-hop spread nationally to suburbs and corporate boardrooms, becoming the biggest-selling genre of popular music in America. Other centers of rap music have emerged, such as the South Coast region (including Atlanta, New Orleans, Miami, and Memphis) and a Midwest region (centered on Detroit).

Rap music has often been controversial due to lyrics and imagery that frequently celebrate violence, hopelessness, and the oppression of women. For example, among Tibetan immigrant youth, the adoption of hip-hop culture has generated objections because of negative stereotypes and its associations with the black underclass. Hip-hop has also diffused globally and undergone *glocalization*—adaptation to suit local cultures and experiences. In Cuba, popular rap lyrics protested racial discrimination and inequality that wasn't supposed to exist in Fidel Castro's Cuba. In Mexico, rapper Control Machete has written lyrics that denounce U.S. border controls, while Colombian rap group La Etnia deals with poverty, prostitution, homelessness, and violence in their lyrics. And from the Americas to South Africa, black youth have used hip-hop as a way to forge a black identity that spans the Atlantic.

Sources: *Arlene Tickner, "Aqui en el Ghetto: Hip Hop in Colombia, Cuba, and Mexico,"* Latin American Politics and Society, *50(3):121–146 (2008); Emily Yeh and Kunga Lama, "Hip-hop Gangsta or Most Deserving of Victims? Transnational Migrant Identities and the Paradox of Tibetan Racialization in the USA,"* Environment and Planning A, *2006, 38: 809–829; Marc Perry, "Global Black Self-Fashionings: Hip Hop as Diasporic Space,"* Identities: Global Studies in Culture and Power, *2008, 15: 635–664.*

"deterritorialized" and globalized, taking on transnational and cross-cultural characteristics not tied to a single culture or location. World music, then, not only establishes African, West Indian, or Asian music in Western popular culture but creates new combinations and syncretisms such as Afro-Celtic, a fusion of African rhythms with traditional Celtic folk music, or reggae. World music also broadens the awareness and modifies the purity of such Western ethnic or folk genres as Celtic, zydeco, klezmer, Rom, bluegrass, jazz, and others. Whatever its folk, cultural, or national origin, world music represents a process of transformation of the musically unique and "other" into forms accepted by globalized popular culture.

Although popularly perceived as an authentic Jamaican music form, reggae displayed the hybrid and globalized character of world music even before that separate music category was recognized. Reggae is a fusion of African rhythms, earlier Caribbean music forms such as ska and rock steady, and European melodies with pronounced influences from modern American jazz, rhythm blues, and soul. As employed by Rastafarians (members of an African-originated religion associated with the poorer black population of Jamaica), reggae lyrics often address issues of poverty, subordination, oppression, black pride, and pan-Africanism. The globalization of reggae music, however, depended on contracts with the internationalized entertainment industry to package and market the reggae product.

Reggae first diffused to England from Jamaica in the early 1970s, taken up by Jamaicans who migrated there in the 1950s and 1960s. In Britain, lyrics were modified to express the place-specific immigrant experience of West Indian neighborhoods. Much like the commercialization of rock and roll from its black rhythm and blues roots in America, the mainstreaming of reggae in Britain involved its adoption by white British bands. In that process, reggae as a musical platform of cultural protest was transformed into a cultural commodity that helped reggae reach beyond the Afro-Caribbean community.

The emergence of reggae as a global commodity is identified with Bob Marley's 1972 album *Catch a Fire*. In recording sessions, however, London-based Island Records thought the album's music was too "Jamaican," and an alternative rock album with a strong roots reggae sound was produced. The global was thus fused with the local. To stress the local, however, the record and CD covers purposefully portrayed "dreadlocked revolutionaries" to authenticate their exotic place origins to Western consumers. For a more recent example of the connections between the local and the global in popular music, see the feature "Hip-Hop Undergoes Globalization and Glocalization."

Birds of a Feather . . . or Lifestyle Segmentation

How does Starbucks or McDonald's decide where to open a new outlet? Will there be enough families with children to justify a play area? What sort of design theme and special menu items will work in this location? Why do some stores consistently outperform other locations? Geodemographic analysis, a marketing application of human geography, attempts to answer these and other related questions. Starting from the premise that "birds of a feather flock together," geodemographic market analysts argue that "you are where you live." Geodemographic marketing analysts point out that the populations cluster into lifestyle segments, so where a person lives is a useful predictor of the kind of car she will drive, her recreational pursuits, her household and personal purchases, the music she will listen to, and the magazines and newspapers she will read. Using a geographic information system (GIS) to map data from the census, and consumer spending records (now you know why store clerks ask for your phone number or postal code), geodemographic marketers have mapped and classified urban and rural neighborhoods. Claritas is one such company; it has classified U.S. ZIP codes into 66 different lifestyle segments. Some of the company's lifestyle segments include the following:

Shotguns and Pickups: White working-class couples with large families who live in small houses or trailers. This group has a moderate median household income. They are typically high school graduates and are likely to own Ford F-series pickup trucks, go hunting, shop at Sears Hardware, and read *North American Hunter* magazine.

Young Digerati: Affluent young families living in trendy urban neighborhoods filled with boutiques, fitness clubs, coffee shops, and microbreweries. The young digerati often have graduate degrees and are early adopters and leaders in the use of new technology. This group has higher than average median household incomes. They are likely to read the *Economist* magazine, go snowboarding, watch independent films, and drive hybrid cars or Audis.

Multi-Culti Mosaic: Lower middle-class, ethnically diverse families living in immigrant gateway neighborhoods. This group has lower than average median household incomes. They are likely to buy Spanish-language music, shop at chain pharmacies, and read *Seventeen*.

Blue Blood Estates: Wealthy suburban families with a significant percentage of Asian Americans. This group consists mostly of well-educated professionals and business executives and has the highest median household incomes. They are likely to drive expensive European automobiles, read architectural magazines, play tennis, and live in manicured suburbs.

The company realizes that the designations don't define the tastes and habits of everyone in a community, but it maintains that the clusters summarize typical behavior.

Reactions Against Globalized Popular Culture

As we have seen in the case of popular food, drink, and music, the globalization of popular culture does not erase all regional differences. Globalization of popular culture generates counterreactions. Just as speakers of minority languages have resisted the dominance of English, the globalization of popular culture is resented by many people, rejected by some, and officially opposed or controlled by certain governments. The Canadian government imposes minimum "Canadian content" requirements on television and radio broadcasters, for example, and Iran, Singapore, China, and other states attempt to restrict Western radio and television programming from reaching their people. Governments of many countries—Bahrain, China, Cuba, Ethiopia, Iran, Myanmar, North Korea, Pakistan, Saudi Arabia, Syria, Uzbekistan, and Vietnam among the most restrictive—impose pervasive Internet surveillance and censorship and demand that U.S.-based search engines filter content to conform with official restrictions and limitations. Many more countries impose partial censorship on the Internet.

In other instances, religious and cultural conservatives may decry what they see as the imposition of Western values, norms, and excesses through such mass culture industries as advertising, the media, and professional sports. Whether or not movies, music, television programming, or clothing fads accurately reflect the essence of Western culture, critics argue that they force on other societies alien values of materialism, violence, self-indulgence, sexual promiscuity, and defiance of authority and tradition. More basically, perhaps, globalization of popular culture is seen as a form of dominance made possible by Western control of the means of communication and by Western technical, educational, and economic superiority. What may be accepted or sought by the young and better educated in many societies may simultaneously be strongly resisted by those of the same societies more traditional in outlook and belief.

Culture Regions

Of course, not all expressions of popular culture are spatially or socially uniform. Areal variations do exist in the extent to which particular elements of popular culture are adopted. Spatial patterns in sports, for example, reveal that the games played, the migration paths of their fans and players, and the sports landscape constitutes part of the geographic diversity of the world. For example, Brazil is a hotbed for soccer, the Dominican Republic for baseball, and Canada still dominates the production of hockey players, despite the increasing globalization of these sports. Figure 7.16a, for example, shows that television interest in professional baseball is not universal despite the sport's reputation as "the national pastime." Studies and maps of many regional differences in food and drink preferences, leisure activities, and personal and political tastes such as **Figure 7.16** are important to marketers and demonstrate ongoing regional contrasts along with the commonalities of popular culture (see the feature Birds of a Feather . . . or Lifestyle Segmentation).

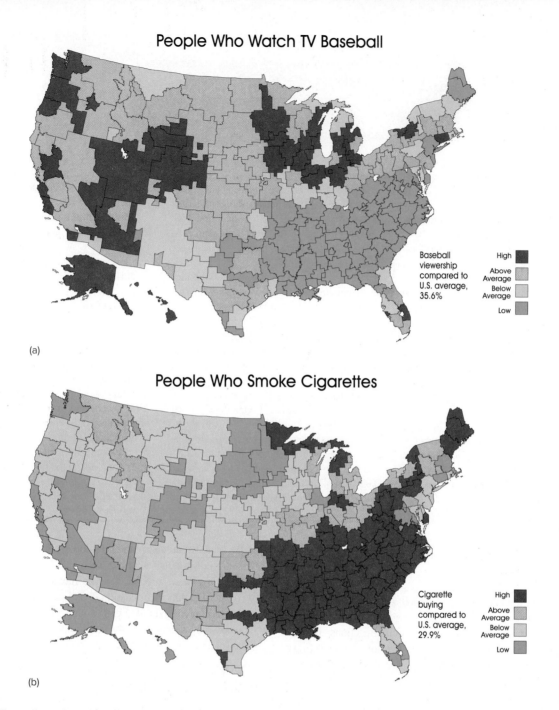

People Who Watch TV Baseball

Baseball viewership compared to U.S. average, 35.6%

High
Above Average
Below Average
Low

(a)

People Who Smoke Cigarettes

Cigarette buying compared to U.S. average, 29.9%

High
Above Average
Below Average
Low

(b)

Figure 7.16 Expressions of popular culture display regional variations. (*a*) Part of the regional variation in television viewing of baseball reflects the game's lack of appeal in the African American community and, therefore, its low viewership in the South and in metropolitan centers where, in addition, attendance at games is an alternative to watching TV. (*b*) Even bad habits regionalize. The country's cigarette belt includes many of the rural areas where tobacco is grown.

Source: From Michael J. Weiss, Latitudes and Attitudes: An Atlas of American Tastes, Trends, Politics and Passions, *Boston: Little, Brown and Company, 1994.*

Ordinary people have a clear view of space. They are aware of variations from place to place in the mix of phenomena, both physical and cultural. They use and recognize as meaningful such common regional names as Corn Belt, Sunbelt, and "the Coast." More important, people individually and collectively agree on where they live. They occupy regions that have reality in their minds and that are reflected in regional journals, in regional museums, and in regionally based names employed in businesses, by sports teams, or in advertising slogans. These are **vernacular** or **popular regions;** they have reality as part of folk culture or popular perceptions rather than as political impositions or scholarly constructs. Geographers are increasingly recognizing that vernacular regions are significant concepts affecting the way people view space, assign their loyalties, and interpret their world. One geographer has drawn the boundaries of the large popular regions of North America on the basis of place names and locational identities found in the white pages of central-city telephone directories (see Figure 1.19). The subnational vernacular regions recognized accord reasonably well with cultural regions defined by more rigorous methods employed by geographers (**Figure 7.17**).

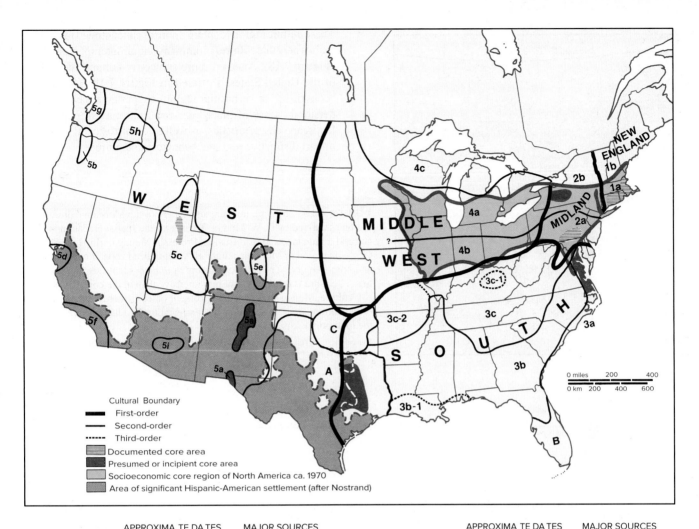

Cultural Boundary
- ▬▬ First-order
- ─── Second-order
- - - - Third-order
- ▤ Documented core area
- ▓ Presumed or incipient core area
- ▢ Socioeconomic core region of North America ca. 1970
- ▨ Area of significant Hispanic-American settlement (after Nostrand)

REGION	APPROXIMATE DATES OF SETTLEMENT AND FORMATION	MAJOR SOURCES OF CULTURE (listed in order of importance)
NEW ENGLAND		
1a. Nuclear New England	1620–1750	England
1b. Northern New England	1750–1830	Nuclear New England, England
THE MIDLAND		
2a. Pennsylvania Region	1682–1850	England and Wales, Rhineland, Ulster 19th-Century Europe
2b. New York Region or New England Extended	1624–1830	Great Britain, New England, 19th-Century Europe, Netherlands
THE SOUTH		
3a. Early British Colonial South	1607–1750	England, Africa, British West Indies
3b. Lowland or Deep South	1700–1850	Great Britain, Africa, Midland, Early British Colonial South, aborigines
3b-1. French Louisiana	1700–1760	France, Deep South, Africa, French West Indies
3c. Upland South	1700–1850	Midland, Lowland South, Great Britain
3c-1. The Bluegrass	1770–1800	Upland South, Lowland South
3c-2. The Ozarks	1820–1860	Upland South, Lowland South, Lower Middle West
THE MIDDLE WEST		
4a. Upper Middle West	1800–1880	New England Extended, New England,19th-Century Europe, British Canada
4b. Lower Middle West	1790–1870	Midland, Upland South, New England Extended, 19th-Century Europe
4c. Cutover Area	1850–1900	Upper Middle West, 19th-Century Europe

REGION	APPROXIMATE DATES OF SETTLEMENT AND FORMATION	MAJOR SOURCES OF CULTURE (listed in order of importance)
THE WEST		
5a. Upper Rio Grande Valley	1590–	Mexico, Anglo America, aborigines
5h. Willamette Valley	1830 1900	Northeast U.S.
5c. Mormon Region	1847–1890	Northeast U.S., 19th-Century Europe
5d. Central California	(1775–1848) 1840–	(Mexico) Eastern U.S., 19th Century Europe, Mexico, East Asia
5e. Colorado Piedmont	1860–	Eastern U.S., Mexico
5f. Southern California	(1760–1848) 1880–	(Mexico) Eastern U.S., 19th and 20th-Century Europe, Mormon Region, Mexico, East Asia
5g. Puget Sound	1870–	Eastern U.S., 19th and 20th-Century Europe, East Asia
5h. Inland Empire	1880–	Eastern U.S., 19th and 20th-Century Europe
5i. Central Arizona	1900–	Eastern U.S., Southern California, Mexico
REGIONS OF UNCERTAIN STATUS OR AFFILIATION		
A. Texas	(1690–1836) 1821–	(Mexico) Lowland South, Upland South, Mexico, 19th-Century Central Europe
B. Peninsular Florida	1880–	Northeast U.S., the South, 20th-Century Europe, Antilles
C. Oklahoma	1890–	Upland South, Lowland South, aborigines, Middle West

Figure 7.17 Culture areas of the United States based on multiple lines of evidence.

Source: From Wilbur Zelinsky, The Cultural Geography of the United States, *Rev. ed., 1992, 118 -119. Prentice-Hall, Inc., Upper Saddle River, NJ.*

These culture regions contain a core of concentrated intensity, a domain where that culture is dominant, and an outer sphere of influence (**Figure 7.18**). Another, more subjective cultural regionalization of the United States is offered in **Figure 7.19**. The generalized "consensus" or vernacular regions suggested are based on an understood "sense of place" derived from current population and landscape characteristics, as well as on historical differences that impart distinctive regional behaviors and attitudes.

Figure 7.18 The core, domain, and sphere of the Mormon culture region as defined by D. W. Meinig. To express the spatial gradation in Mormon cultural dominance and its diffusion, Professor Meinig defined the Salt Lake City *core* region of Mormon culture as "a centralized zone of concentration . . . and homogeneity." The broader concept of *domain* identifies "areas in which the . . . culture is dominant" but less intensive than in the core. The *sphere* of any culture, Meinig suggests, is the zone of outer influence, where only parts of the culture are represented or where the culture's adherents are a minority of the total population.

Source: Redrawn from Annals of the Association of American Geographers, D. W. Meinig, Vol. 55, 214, Association of American Geographers, 1965.

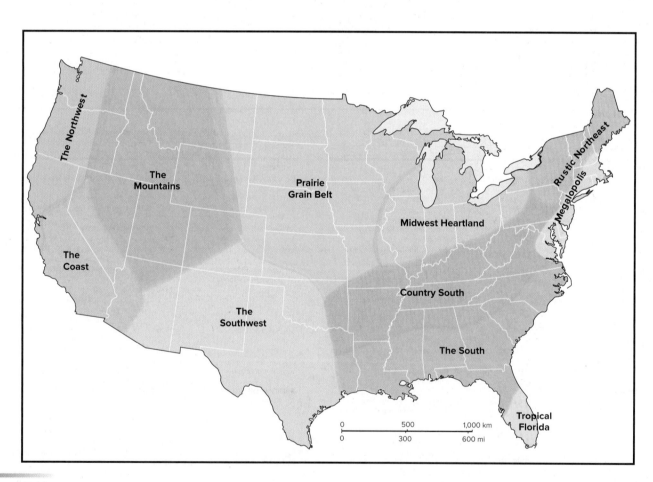

Figure 7.19 Generalized U.S. culture regions. "In spite of strong tendencies toward cultural homogenization and place obliteration, . . . regional identities persist," in the view of geographer Larry Ford, who suggests the 11 culture regions shown. Whatever the reasons for that persistence, "the different [culture] regions of the United States continue to have their own personalities and senses of place."

7.2 Cultural Landscapes

Folk and popular cultures involve more than food, drink, music, and folklore. Cultures make a distinct impression on the landscape, as they fashion the raw material of the natural landscape into a human habitat. Influential geographer Carl Sauer described it as follows:

> The cultural landscape is fashioned from a natural landscape by a culture group. Culture is the agent, the natural area is the medium, the cultural landscape is the result.[3]

Landscapes are more than pretty scenes. They are all around us, and they tell us a great deal about those who shaped them. The **cultural landscape** is a blended work of nature and culture and offers some of the best evidence of human-environment interactions. American geographer Peirce Lewis wrote,

> "[O]ur human landscape is our unwitting autobiography, reflecting our tastes, our values, our aspirations, and even our fears, in tangible visible form."[4]

In other words, the everyday landscape—farms, gardens, suburbs, cemeteries, trailer parks, shopping malls, and more—offers the savvy observer important insights into a culture's attitudes, priorities, and way of life. While a highway lined with billboards, neon signs, and fast-food restaurants is designed to communicate messages to all who pass by, many landscapes are just there. Nonetheless, all landscapes can be "read" and interpreted. For example, a cemetery tells us something about how a culture approaches death, and a derelict trailer park hidden on the "wrong side of the tracks" says something about a society's way of dealing with economic inequality. As students of geography, we develop a vocabulary and basic skills to begin a lifetime of reading the landscape, interpreting its messages, and attempting to understand the culture that produced it.

National and regional cultural identities remain embedded in urban and rural landscapes (**Figure 7.20a**). Certain landmarks and landscapes take on special significance in the identity, emotions, and politics of a country. Such landscapes exert great symbolic power and can be contested by different groups in society (see feature "Monuments and Memorials"). In the United States, politicians use the rhetoric of "Main Street" and "Wall Street," referring to places that really exist but are also deeply symbolic and associated with particular vices or virtues (**Figure 7.20b**). Landmarks come to symbolize a city in the mental imagery of residents and tourists alike. London's Big Ben, Paris's Eiffel Tower, New York's Statue of Liberty, and Sydney's Opera House are just a few examples of symbolic landmarks. Symbolic landscapes and landmarks have great emotional power and are often used in advertising

(a)

(b)

Figure 7.20 Cultural landscapes are the joint creations of humans and nature. (*a*) Cultural landscapes such as this scene from Italy are often intimately connected to regional or national identities. Often such national landscapes are the subject of countryside protection rules. (*b*) Ordinary landscapes can become symbolic landscapes. In the United States, the small-town Main Street has come to symbolize the values of face-to-face community, small businesses, honesty, and integrity. This scene of Main Street in Northfield, Minnesota, is typical of the 19th-century business districts in small towns across the United States and Canada. Local residents have defended it against the intrusion of big box stores.

(a) ©Rob Tilley/Blend Images; (b) ©Mark Bjelland

images and political speeches. Not surprisingly then, the targets of the September 11, 2001, terrorist attacks were the World Trade Center and the Pentagon, landscape symbols of economic globalization and U.S. military power. And again, not surprisingly, the 2011 Occupy movement began with the very symbolic occupation of Wall Street by protesters.

Time is an important element in the creation of cultural landscapes. Both urban and rural landscapes, especially those in Europe and Asia, bear the imprint of human reshaping over thousands of years of use. Today's largest cities are often located on the site of ancient settlements. The **sequential occupation** of the

---

[3]Carl Sauer, *The Morphology of Landscape*, University of California Publications in Geography Number 22, 1925, 19–53.
[4]Peirce Lewis, "Axioms for Reading the Landscape: Some Guides to the American Scene," in *The Interpretation of Ordinary Landscapes*, D. W. Meinig and F. B. Jackson, eds., Oxford: Oxford University Press 1979, 12.

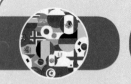

Monuments and Memorials

Monuments, memorials, civic squares, public buildings, and historic sites are part of a nation's symbolic landscape. They contribute to the shared cultural identity of a people. They express and transmit to successive generations an understanding of who they are as a people, who and what deserves honor, and how the past is to be remembered. To tour the National Mall in Washington D. C., making pilgrimage to its many monuments and memorials, is a way of experiencing one's identity as an American citizen. It is one thing to read history in a book, quite another to stand in the same location, touch the walls, walk the terrain, and experience firsthand the landscapes where world-changing events took place. It is because of the power of landscape that debates are so fierce over how public spaces, monuments, and historic sites in the United States represent the Civil War and the treatment of Native Americans and African American slaves.

Berlin, Germany was at the center of many of the most tumultuous and destructive episodes in 20th-century history and yet is also a busy capital region with 5 million residents. For decades, Berliners have worked to build landscapes of remembrance into the fabric of the city. Near a remnant section of the Berlin Wall, the Topography of Terror site documents the Gestapo prison that operated in the heart of the city. Artwork, installed throughout the city, tells the stories of the city's violent past. Walking to a suburban train station one passes a concrete retaining wall with rounded cut outs in vaguely human forms—a reminder of lives lost in the Holocaust. Or, on an overgrown train platform, metal plaques, one for each rail shipment, list the dates and number of Jews deported to concentration camps at that location.

During the Soviet era, communist leaders rebuilt city centers to reflect their ideals and show their power. They created broad avenues and wide open squares for civic events such as the annual May Day parade. They also made a practice of demolishing prominent Christian churches and replacing them with government buildings. Located on the Moskva River near the Kremlin in Moscow, the gold domes of

the Russian Orthodox Cathedral of Christ the Savior made it the tallest Orthodox church building in the world. The cathedral was detonated in 1931 on orders from Josef Stalin to make way for the highly symbolic Palace of the Soviets. After the collapse of communism, the Russian Orthodox Church reasserted its power in Russian society and immediately set about rebuilding the Cathedral of Christ the Savior. With donations from a million local residents, the soaring gold domes of the cathedral were rebuilt in an exact replica of the destroyed building.

Leipzig, Germany's Paulinerkirche was a marvel of 13th-century gothic architecture. It faced Augustusplatz, the city's largest civic square where it played a prominent role in German culture and history, hosting performances by musicians such as J. S. Bach and Felix Mendelsohn. It survived the World War II bombings intact but in 1968, East German communist leaders detonated the church to make space for university classrooms. With the collapse of communism in 1989, proposals and funders arose to reconstruct the gothic Paulinerkirche. Opponents of the rebuilding argued that a gothic church would be too expensive and lack authenticity. Further, declining church attendance meant that Leipzig had more than enough existing church space. City and university officials

argued that by replacing the destroyed church with an exact copy, the rebuilders would be sanitizing Germany history and hiding the story of the willful destructiveness of the communist era. In 2009, instead of a gothic replica, a university building with a small prayer chapel was erected on the site of the original church. The new university building mimics the form of the gothic original, including a steeple spire and a steeply pitched roof, but is finished in a modern, concrete and glass architectural style.

Questions to Consider

1. Who or what is honored in the memorials and civic spaces in your community? What do they communicate about your community's shared identity and values?

2. When important historic buildings or monuments have been lost or demolished, how do you think we should honor that past? Would you support the approach to rebuilding taken at the Cathedral of Christ the Savior in Moscow, the approach taken in Leipzig, or a different approach altogether?

3. Where tragic or horrific events have taken place, is it important that they be remembered tangibly in the landscape? If so, who should participate in creating memorials and monuments?

Figure 7A The Martin Luther King Jr. Memorial in Washington, D.C.
Source: NPS Photo

Figure 7.21 A landscape can be viewed as a palimpsest that often contains visible traces of multiple past cultures. In Rome, drivers travel on streets laid out in ancient times while skirting 2,000-year-old ruins.

©r.nagy/Shutterstock

Figure 7.23 Vernacular landscapes can help us understand the values and lifestyles of their creators. This highway strip in the United States reflects a fast-moving, automobile-based way of life. Placeless landscapes are created by the spread of chain restaurants, hotels, and gas stations.

©Mark Bjelland

landscape by different groups of people almost always leaves behind visible traces, some of which influence future human uses of the land. Ancient paths and roads have often become the busy streets of the contemporary world. The trace of ancient medieval walls remains visible as circular parkways in many European cities. The routes of abandoned railroad lines are often reused for trail networks. Thus, the landscape has been compared to a **palimpsest**—an ancient parchment or vellum document that has been written on over and over again, with the earlier writings scraped away. Just as historians attempt to read the faint traces of earlier writings on a palimpsest, geographers look to the landscape for visible traces of the past (see **Figure 7.21**).

Elite landscapes communicate the refined taste and exclusivity of society's most privileged and powerful members. Through imposing architecture, gardens, gates, and the generous use of space, elite landscapes communicate sophistication, power, and status (**Figure 7.22**), In doing so, elite landscapes tend to legitimate and reproduce the status of their occupants. Vernacular or ordinary landscapes of houses, trailer parks, roadside motels, and fast-food restaurants are equally important to the geographer seeking to understand a culture (**Figure 7.23**).

While learning to pay attention to and interpret cultural landscapes is among the pleasures of geography and a source of important insights, we must remember that much remains hidden. Geographer Don Mitchell reminds us of how the landscape works to normalize a particular set of social relationships—relationships that may be unjust or unhealthy. For example, the fertile fields of California's Central Valley communicate a message of prosperity emerging from the harmonious human cultivation of the Earth. Hidden from view, however, are the shacks of exploited migrant workers and the dangerous agricultural chemicals that make that productive landscape possible. Similarly, derelict landscapes composed of abandoned buildings, rusting factories, broken windows, and overgrown lots speak loudly of cultural change and a people's attitude toward progress and the past (**Figure 7.24**). In the remainder of this chapter, we explore two major elements of the cultural landscape: systems for surveying

Figure 7.22 Historic or elite architecture, exclusivity, and spaciousness convey the high status of those who occupy elite landscapes.

©Mark Bjelland

Figure 7.24 The direct landscape of the vacant Packard automobile manufacturing plant in Detroit, Michigan offers landscape evidence of the creative destruction of capitalism.

©Josh Cornish/Shutterstock

and dividing land and the houses built in different times and places. We then examine a sampling of different types of landscapes: places of consumption and heritage landscapes.

Land Survey Systems

Flying over North America, one sees a distinct difference between the checkerboard fields of the Midwest, the narrow fields of southern Quebec, and the irregular shaped fields of the Atlantic Seaboard. These differences stem from different land survey systems. In a capitalist system, the division of land into individual holdings is necessary for both agricultural and urban development. The charter group settling an area had to create a system for surveying, dividing, claiming, and allocating land and that system leaves a lasting legacy, shaping the look and feel of both rural and urban landscapes.

For the most part, the English established land-division policies in the Atlantic Seaboard colonies. The **metes-and-bounds** system, which had long been used in England, was used to describe property boundaries using landform or water features or such temporary landscape elements as prominent trees, unusual rocks, or cairns. Not surprisingly, the metes-and-bounds system led to boundary uncertainty and dispute and lengthy descriptions of property boundaries (**Figure 7.25a**). It also resulted in *topographic* road patterns, such as those found in Pennsylvania and derelict landscape of the other eastern states, where routes are often controlled by the contours of the land rather than the regularity of a geometric survey.

When independence was achieved, the federal government decided that the public domain should be systematically surveyed and subdivided before being opened for settlement. Inspired by Enlightenment ideals of rational order and the prospect of rapidly dividing and settling the vast continent, the United States adopted **a rectangular survey** system in the Land Ordinance of 1785. The resulting **Public Land Survey System (PLSS)** established township and range survey lines oriented in the cardinal directions and divided the land into townships (9.7 km) 6 miles square, which were further subdivided into sections that were (1.6 km) 1 mile on a side (**Figure 7.25b**). The sections were divided into quarter sections of 160 acres, the size of a typical grant under the Homestead Act of 1862. The resultant rectangular system of land subdivision and ownership was extended to all parts of the United States included within the public domain, creating the basic checkerboard pattern of minor civil divisions, the regular pattern of section-line and quarter-line country roads, and the block patterns of fields and farms.

Elsewhere in North America, the French and the Spanish constituted charter groups and established their own traditions of land description and allotment. The French imprint has been particularly enduring. The **long-lot system** was introduced into the St. Lawrence Valley and followed French settlers wherever they established colonies in the New World: the Mississippi Valley, Detroit, Green Bay, Louisiana, and elsewhere. The long-lot holding was typically about 10 times longer than wide, stretching far back from a narrow river

(a)   (b)

Figure 7.25 The original metes-and-bounds property survey of a portion of the Virginia Military District of western Ohio is here contrasted with the regularity of surveyor's townships, made up of 36 numbered sections, each 1 mile (1.6 km) on a side.

Source: Redrawn from Original Survey and Land Subdivision, Monograph Series No. 4, Norman J.W. Thrower, 46, Association of American Geographers, 1966.

frontage (**Figure 7.26**). The back of the lot was indicated by a roadway roughly parallel to the line of the river, marking the front of a second series (or *range*) of long lots. The system had the advantage of providing each settler with a fair access to fertile land along the floodplain, lower-quality river terrace land, and remote poorer-quality back areas on the valley slopes

serving as woodlots. Dwellings were built at the front of the holding, in a loose settlement alignment called a *côte,* where access was easy and the neighbors were close.

Although English Canada adopted a rectangular survey system, the long lot became the legal norm in French Quebec, where it controls land surveying even in areas where river access is insignificant. In the Rio Grande Valley of New Mexico and Texas, Spanish colonists introduced a long-lot system similar to the French.

Settlement Patterns

Settlement patterns reflect ways different cultures understand the relationship between the individual and the wider group. In some cultures, the extended family is the basis for the settlement system (**Figure 7.27**). In much of the world, farmers historically lived in small, agricultural villages creating a clustered rural

Figure 7.26 A portion of the Vincennes, Indiana–Illinois, topographic quadrangle (1944) showing evidence of the original French long-lot survey. Note the importance of the Wabash River in both long-lot and Vincennes street-system orientations. This U.S. Geological Survey map was originally published at the fractional scale of 1:62,500.

Source: U.S. Geological Survey map.

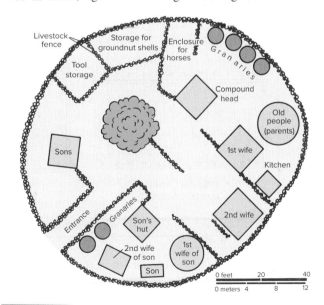

Figure 7.27 The extended family compound of the Bambara of Mali.

Source: Redrawn from Reuben K. Udo, The Human Geography of Tropical Africa *(Ibadan: Heinemann Education Books (Nigeria) Ltd., 1982), 50.*

population. The French and Hispanic long lots encouraged the alignment of closely spaced farmsteads along river or road frontage (**Figure 7.28a**). While the metes-and-bounds and rectangular survey systems look very different, unlike long-lot systems, they both produce the dispersed pattern of isolated farmsteads that typifies the rural United States and English Canada (**Figure 7.28b**). It is an arrangement conditioned by the block pattern of land survey, by the homesteading tradition of "proving up" claims through living on them, and the regular pattern of rural roads. The dispersed rural settlement pattern for much of the United States both reflects and reinforces the individualism of U.S. culture.

Other survey systems permitted different culturally rooted settlement choices. The New England village reflected the transplanting of an English tradition. The central village, with its meeting house and its commons area, was surrounded by larger fields subdivided into strips for allocation among the community members. The result was a distinctive pattern of nucleated agricultural villages and fragmented farms. Agricultural villages were found as well in Mormon settlement areas, in the Spanish American Southwest, and as part of the cultural landscapes established by utopian religious communities, such as the Oneida Community of New York; the Rappites's Harmony, Indiana; Fountain Grove, California; and others. On the prairies of Canada, the Mennonites were granted lands not as individuals, but as communities. Their agricultural villages with surrounding communal fields (**Figure 7.28c**) reflected their religious ideals and re-created in Manitoba the landscape of their European homelands.

The spatial pattern of streets, blocks, and lots exerts a major influence on the feel and social life of urban settlements. Urban street grids may be irregular, rectangular, or contain winding loops and lollipops. Irregular street grids are common in areas of metes-and-bounds surveying, and a rectangular street grid

can be easily created in areas of rectangular surveying. The best-known rectangular street grid is the 1811 New York City plan for Manhattan featuring 11 major avenues and 155 perpendicular crosstown streets. Frederick Law Olmsted, the designer of Central Park and many other famous urban parks, created elite, romantic suburbs in the 1800s using winding, picturesque streets and detached single-family houses on large lots. After World War II, middle-class housing developers copied the look of elite suburbs and created the winding street patterns that typify most North American suburbs. While often attacked by transportation planners, the winding, dead-end streets of contemporary suburbs reflect the cultural ideal of the house as a quiet, private retreat.

Houses

Houses are among the most important material expressions of a culture and most visible features of the cultural landscape. Nearly a century ago, French geographer Jean Brunhes called houses a "central fact" of human geography. Using Abraham Maslow's hierarchy of human needs as a framework, we can see that houses meet both basic and higher-level human needs. Houses meet physiological needs for shelter, safety needs for the protection of one's person and property, and needs for belonging by placing individuals within a household and a community. At the higher end of the needs hierarchy, houses meet needs for self-esteem by serving as a marker of achievement, and they meet self-actualization and self-expression needs through their design, color, and interior and exterior decor.

Vernacular house styles—those built in traditional form but without formal architectural plans or drawings—are of particular interest to cultural geographers. Throughout the world, folk societies established types of housing appropriate to their economic and family needs, available materials and technologies, and local

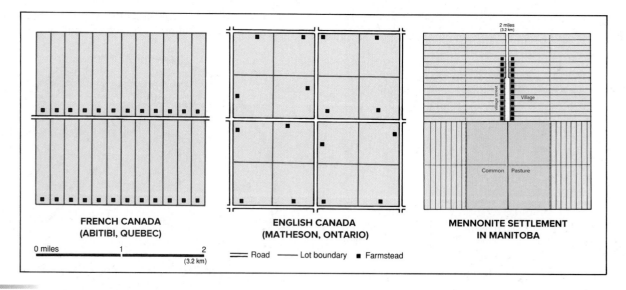

Figure 7.28 Contrasting land survey and settlement systems in Canada. Adjacent areas of Canada demonstrate the effects of different survey systems and cultural heritages on rural settlement patterns. The long-lot survey of Quebec in French Canada (*a*) creates a linear village. The regular plots of Ontario in English Canada (*b*) display the isolated farmsteads that are characteristic of much of the United States and Canada. The German-speaking Mennonites, a Protestant Christian group, settled in Manitoba in the 1870s. They created an agricultural village that reflected the group's European homeland and strong emphasis on community (*c*). Individual farmers were granted strip holdings in the separate fields to be farmed in common with the other villagers. The farmsteads themselves, with elongated rear lots, were aligned along both sides of a single village street in an Old World pattern.

Sources: Redrawn from Annals of the Association of American Geographers, George I. McDermott, *Vol. 51, 263, Association of American Geographers, 1961; and from Carl A. Dawson,* Group Settlement: Ethnic Communities in Western Canada, *Vol. 7, Canada Frontiers of Settlement (Toronto: Macmillan Company of Canada, 1936), 111.*

environmental conditions. Thus, folk housing is wonderfully diverse, reflecting ingenious adaptations to local conditions. Industrialization, transportation improvements, and the influences of popular culture have revolutionized styles of house construction so that today, there is more diversity in a given place but fewer differences between places. In some cases, folk housing styles are being revisited as contemporary homebuyers attempt to recover a local sense of place and/or reduce the amount of energy used to heat and cool their houses.

The Mongol or Turkic *yurt* or *ger,* a movable low, rounded shelter made of felt, skin, short poles, and rope, is a housing solution well adapted to the needs and materials of nomadic herdsmen of the Central Asian grasslands (**Figure 7.29a**). A similar solution with different materials is reached by the Maasai, another nomadic herding society living on the grasslands of eastern Africa. Their temporary home was traditionally the *manyatta,* an immovable low, rounded hut made of poles, mud, and cow dung that was abandoned as soon as local grazing and water supplies were consumed. As the structures in **Figure 7.29** suggest, folk housing solutions in design and materials provide a worldwide mosaic of nearly infinite diversity and ingenuity.

(a)
(b)
(c)
(d)
(e)
(f)

Figure 7.29 The common characteristics of preindustrial folk housing are an essential uniformity of design within a culture group and region, a lack of differentiation of interior space, a close adaptation to the conditions of the natural environment, and ingenious use of available materials in response to the dictates of climate or terrain. (*a*) A *yurt* in the Tian Shan mountains, Kyrgyzstan; (*b*) The traditional Zulu hut in South Africa. (*c*) stone house of Nepal; (*d*) Icelandic sod farm house; (*e*) reed dwelling of the Uros people on Lake Titicaca, Peru; (*f*) traditional thatched roof housing in Zimbabwe decorated with geometric paintings.

(a) ©MEP/Getty Images; (b) ©andyKRAKOVSKI/iStock/Getty Images; (c) ©Courtesy of Professor Colin E. Thorn; (d) ©windcoast/Shutterstock; (e) ©Photofrenetic/Alamy Stock Photo; (f) ©Ralph A. Clevenger/Getty Images;

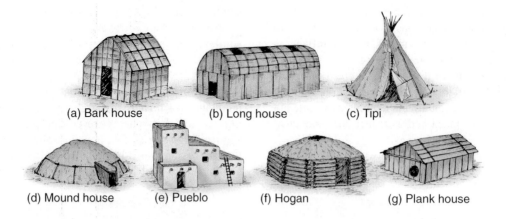

(a) Bark house   (b) Long house   (c) Tipi

(d) Mound house   (e) Pueblo   (f) Hogan   (g) Plank house

Figure 7.30 Native American housing types reflected varied physical environments and ways of life.

Source: From John Burchard and Albert Bush-Brown, The Architecture of America: A Social and Cultural History, *(Boston: Little, Brown and Company, 1961), 57.*

America, like every other world region, had its own indigenous vernacular architecture (**Figure 7.30**). This was the architecture of Amerindians—the bark houses of the Penobscots, the long houses of the Iroquois, the tipis of the Crow, the mounds of the Mandans, the pueblos of the Zuñi, the hogans of the Navajos, and the plank dwellings of Puget Sound. Despite their elegance and suitability for particular environments, these styles were largely swept away by European settlers. The primary exception is the thick-walled pueblos, which are much more comfortable in desert climates than the wood structures built by some European settlers.

Hearths and Diffusion Streams

Within the United States and Canada, the variety of ethnic and regional origins of immigrants and the differences in environmental conditions led to architectural contrasts among the settlement hearths of the Atlantic Seaboard. The landscapes of structures and settlements creating those contrasts speak to us of their creators' origins, travels, adaptations to new locales, and retention of the customs of other places. The folk cultural heritage is now passing; old farm structures are replaced or collapse with disuse as farming systems change. Old houses are torn down, remodeled, or abandoned, and trendy, contemporary styles replace the evidence of earlier occupants. Preservationists may succeed in retaining and refurbishing some structures but gradually the landscapes—the voices—of the past are lost. Many of those fading voices first took on their North American accents in the culture hearths suggested in Figure 7.5. They are still best heard in the house types associated with them. Each separately colonized area produced its own distinctive vernacular housing mix, and a few served eventually as hearth districts from which imported and developed house forms diffused. Folk house styles for much of the United States and Canada can be traced to four source regions on the Atlantic Coast, each feeding a separate diffusion stream: Northern, Middle Atlantic, Southern Coastal, and Mississippi Delta. By 1850, diffusion from these four eastern architectural hearths had produced a clearly defined folk housing geography in the eastern half of the United States and subsequently, by relocation and expansion diffusion, had influenced vernacular housing throughout the United States.

The Northern Hearths

In the North, cold, snowy winters posed different environmental challenges than did the milder, frequently wetter climates of northwestern Europe, and American timber was more accessible than in the homelands. In the north, the *St. Lawrence Valley* is a cultural landscape shaped by French settlement. There, in French Canada, beginning in the middle of the 17th century, a common house type was introduced based on styles still found in western France today (**Figure 7.31**); they were found as well

Figure 7.31 Vernacular house types in the Lower St. Lawrence hearth region of Quebec, Canada, included the *Norman cottage* with steeply pitched, hipped roofs and wide or upturned eaves, nearly identical to houses in the Normandy region of northern France.

©Professor John A. Jakle

(a)

Figure 7.32 Dutch colonial houses, found in New York's Hudson Valley, have a distinctive gambrel roof that gives more headroom in the top floor by using two slopes for the roof.

©*Jon C. Malinowski/Human Landscape Studio*

in other areas of French settlement in North America—Louisiana, the St. Genevieve area of Missouri, and northern Maine. The *Hudson Valley*'s complex mix of Dutch, French, Flemish, English, and German settlers produced a comparable mixture of common house forms. The Dutch were initially dominant, and their houses were characterized by a split "Dutch door" (whose separately opened upper half let air in and closed lower half kept children in and animals out) and a gambrel roof (the slope changes to create more headroom in the upper floor), often with flared eaves (**Figure 7.32**). The rural southern English colonists who settled in *southern New England* brought with them the heavily framed houses of their home counties: sturdy posts and stout horizontal beams sided by overlapping clapboards and distinguished by steep roofs and massive chimneys (**Figure 7.33**). While the house styles of the St. Lawrence Valley and Hudson Valley did not diffuse widely, the New England house styles did. As the New England styles diffused westward, it was the Upright and Wing (**Figure 7.33c**) in particular that settlers spread across New York, Ohio, Michigan, Indiana, and Illinois, and into Wisconsin, Iowa, and Minnesota (**Figure 7.34**).

(b)

The Middle Atlantic Hearths

The *Delaware Valley* and *Chesapeake Bay* were ethnically diverse sites of vernacular architecture more influential on North American housing styles than any other early settlement area. The log cabin, later carried into Appalachia and the trans-Appalachian interior, evolved there, as did the vernacular *four-over-four* house—so called in reference to its basic two-story floor plan with four rooms up and four down (**Figure 7.35**). This house type formed the basis for the rowhouses found in the largest cities of the Middle Atlantic region, such as Baltimore and Philadelphia. There, too, was introduced what would later be called the *I house*—a two-story structure one room deep, with two rooms on each floor.

(c)

Figure 7.33 New England house types included (*a*) the *saltbox* house with an asymmetrical gable roof covering a lean-to addition for extra room; (*b*) the Georgian-style *New England large house* with up to 10 rooms, a lobby entrance, and paired chimneys; and (*c*) the *upright-and-wing* house (the wing represented a one-story extension of a basic gable-front house plan) which diffused widely in both rural and urban areas from western New York to the Midwest.

(*a*) ©*Professor John A. Jakle;* (*b*) ©*WANGKUN JIA/123RF;* (*c*) ©*Professor John A. Jakle*

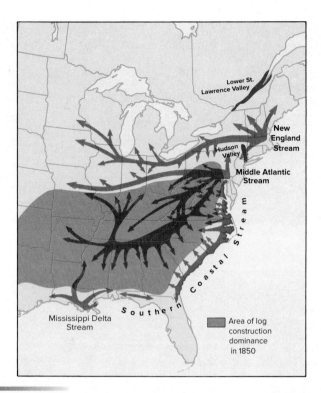

Figure 7.34 Architectural source areas and the diffusion of building styles from the eastern hearths. The variation in the width of paths suggests the strength of the influence of the various hearths on vernacular housing away from the coast. The Southern Coastal Stream was limited in its influence to the coastal plain. The Delaware Valley hearth not only exerted a strong impact on the Upland South but also became—along with other Middle Atlantic hearths—the dominant vernacular housing source for the lower Midwest and the continental interior. By 1850, and farther west, new expansion cores were emerging around Salt Lake City, in coastal California, and in the Willamette Valley area of Oregon—all bearing the imprint of housing designs that first emerged in eastern hearths.

Sources: F. Kniffen, Annals of the Association of American Geographers, *Vol. 55: 560, 1965; Fred Kniffen and Henry Glassie, "Building in Wood in the Eastern United States," in Geographical Review 56:60, 1966 The American Geographical Society; and Terry G. Jordan and Matti Kaups,* The American Backwoods Frontier, *8 -9, 1989 The Johns Hopkins University Press.*

Figure 7.35 House types of the Middle Atlantic hearths included the *four-over-four house.*

©*Professor John A. Jakle*

The I house resembles the four-over-four house with its gables at the sides of the house. The I house became prominent in the Upper South and southern Midwest in the 19th century. Its major diffusion directions were southward along the Appalachian Uplands, with offshoots in all directions, and westward across Pennsylvania. Multiple paths of movement from this hearth converged in the Ohio Valley Midwest, creating an interior "national hearth" of several intermingled streams (see Figure 7.5), and from there spread north, south, and west. In this respect, the narrow Middle Atlantic region played for vernacular architecture the same role that its Midland dialect did in shaping the linguistic geography of the United States, as discussed in Chapter 5.

The earliest diffusion from the Middle Atlantic hearth was the backwoods frontier culture that carried the Finnish-German log carpentry to all parts of the forested East. The identifying features of that building tradition were the dogtrot and saddle-bag house plans and double-crib barn designs. The basic unit of both house and barn was a rectangular "pen" ("crib" if for a barn) of four log walls that characteristically stood in tandem with an added second room that joined the first at the chimney end of the house. The resultant two-room central chimney design was called a *saddlebag house.* Another even more common expansion of the single-pen cabin was the *dogtrot* (**Figure 7.36**), a simple roofing-over of an open area left separating the two pens facing gable toward gable. Log construction techniques and traditions were carried across the intervening grasslands to the wooded areas of the northern Rockies and the Pacific Northwest during the 19th century.

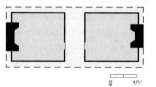

Figure 7.36 The "dogtrot" house was built of two log pens joined by a covered open area.

©*Denise McCullough*

The Southern Hearths

In the American South, a new ethnic and cultural mix and a hot and humid climate led to new styles of vernacular housing long before air conditioning or electric fans were available. In the *Southern Coastal* hearth, along the southeastern Atlantic coastal region of South Carolina and Georgia, the malaria, mosquitoes, and extreme heat plaguing their inland plantations during the summer caused the wealthy to prefer hot-season residence in coastal cities such as Charleston, where sea breezes provided relief. The I house was adapted as the characteristic *Charleston single house,* turned sideways with sun-blocking porches extending along the entire length of the house (**Figure 7.37**). The third source area, in the Lower Chesapeake, spread its remarkably uniform influence southward as the *Southern Coastal Stream,* diffusing its impact inland along numerous paths into the Upland South. In that area of complex population movements and topographically induced isolations, source-area architectural styles were transformed into truly indigenous local folk housing forms. The French established a second North American culture hearth in the *Mississippi Delta* area of New Orleans and along the lower Mississippi during the 18th century. There, French influences from Nova Scotia and the French Caribbean islands—Haiti, specifically—were mixed with Spanish and African cultural contributions. Again, heat and humidity were an environmental problem requiring distinctive housing solutions, one of which was the narrow *shotgun* house, which provided excellent cross ventilation (**Figure 7.38**). The shotgun house is a folk cultural contribution that traces its roots to Africa through free Haitian blacks who settled in the delta before the middle of the

19th century. Crossing cultural boundaries, the shotgun house became a common residence type after the Civil War for both poor blacks and whites migrating to cities in the South and southern Midwest in search of employment. The French and Caribbean influences of the *Delta Stream,* in contrast, were much more restricted and localized.

The Interior and Western Hearths

Various immigrant groups, some from the eastern states, and others from abroad responded to the extremes of a continental climate and different building materials by developing new house styles in the *Interior* and North American *West.* Settlers on the Great Plains initially built sod dugouts or sod or rammed earth houses in the absence of native timber stands. Later, after the middle of the century, "balloon frame" construction, utilizing newly available cheap wire nails and light lumber milled to standard dimensions, became the norm in the interior where builders adapted the simple front-gable or upright-and-wing forms. The strong, low-cost housing the new techniques and materials made possible owed less to the architectural traditions of eastern America than it did to the simplicity and proportional dimensions imposed by the standardized materials.

(a)

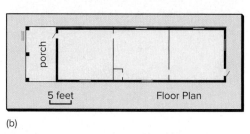

(b)

Figure 7.38 The Mississippi Delta region is the hearth for the shotgun house. (*a*) This simple, three-room *shotgun house* in New Orleans, Louisiana, shows the features typical of this design: narrow and deep, one room wide, multiple rooms deep, a front-facing gable and porch, off-center front and rear doors, and rooms connected by internal unaligned doors. (*b*) One variant of a shotgun cottage floor plan.

©Mark Bjelland

Figure 7.37 The Charleston single house of the Southern Tidewater hearth. The name refers to its single row of three or four rooms arranged from front to back and lined on the outside of each floor by a long veranda extending along one side of the structure.
It resembles an I house turned on its side and shaded with a veranda from the hot southern sun.

©Mark Bjelland

The thick-walled *Spanish adobe house,* with its small windows and flat or low-pitched roof supported by timbers, entered the United States through the Hispanic borderlands. However, most of its features owed more to indigenous Pueblo Indian design than to Spanish origins (**Figure 7.39**). In the Far West, Hispanic and Russian influences were locally felt, although housing concepts imported from the East predominated. In the Utah area, Mormon immigrants established the *central-hall house,* related to both the I house and the four-over-four house, as the dominant house type.

A variety of ethnic and architectural influences met and intermingled in the Pacific Northwest. By the 1870s, an architecturally distinctive Chinatown had emerged in Seattle, and similar enclaves were established in Portland, Tacoma, Vancouver, and Victoria. But most immigrants to Oregon and Washington came from Midwestern roots, representing a further westward migration of populations whose forebears (or who themselves) were part of the Middle Atlantic culture hearths. Some—the earliest—carried to the Oregon and Washington forested regions the "midland" American backwoods pioneer culture and log-cabin tradition first encountered in the Delaware Valley hearth; others brought the variety of housing styles already well represented in the continental interior.

National Housing Styles

Just as popular culture has homogenized tastes in clothes, food, drink, music, and dance, national house styles began to replace folk styles in the 1800s, at least for those who could afford it. National house styles diffused across the continent, adopted first by urban elites who wanted the latest architecture. Thus, the look of a house came to say less about where it was from and more about when it was built and the status of its owner. In the 1800s and early 1900s, popular styles included neoclassical houses modeled on Greek temples, Italianates modeled on Italian villas, the Second Empire style inspired by French designs, the elaborate Queen Anne (or Victorian) style (**Figure 7.40a**), and revivals of various historic styles, including colonial-era styles and English Tudor cottages and castles (**Figure 7.40b**).

Dramatic changes in house design accompanied the widespread adoption of automobiles. As cars become the normal and expected mode of transportation, the cost of distance dropped dramatically, and cities and houses began to spread outward. At the forefront of these changes was the popular ranch-style house.

(a)

(b)

Figure 7.40 Popular national house styles of the 1800s and early 1900s diffused across North America. (*a*) The *Queen Anne* (or Victorian) style was popular in the late 19th and early 20th centuries with its asymmetry, flamboyance, and elaborately decorated surfaces. This highly romantic and picturesque style appeared as the Industrial Revolution was sweeping Great Britain and North America. (*b*) Tudor Revival houses, which imitate the steeply pitched roofs and half-timbering of Medieval English houses, were popular during the early decades of the 20th century.

(a) ©Mark Bjelland; (b) ©Mark Bjelland

Figure 7.39 The pueblo house of the Southwest with thick-walled, adobe construction, retains a relatively constant temperature, staying warm in winter and cool in summer.

©mtcurado/Getty Images

Figure 7.41 The ranch house was built in large numbers in the burgeoning suburbs of the United States from the 1950s through the 1970s. The widespread adoption of the automobile after World War II allowed houses and neighborhoods to sprawl outward. The garage became part of the house in this unpretentious style that facilitated relaxed, informal lifestyles centered on the backyard patio or pool.
©Mark Bjelland

The ranch house was inspired by the ranch houses of the West and Southwest regions and by modern architecture's rejection of ornamentation. Sprawling, low-slung, unpretentious, and closely connected to the outdoors through sliding, glass patio doors, the ranch house came to symbolize casual lifestyles in southern California, the Sun Belt, and suburbs everywhere (**Figure 7.41**). The ranch became the most popular house built in the United States during the late 1950s and 1960s, when the annual rates of house construction boomed. Eventually, it too went out of style, but not before becoming the most common house type in the United States.

Gaining momentum since 1980 has been a return to building new houses in traditional styles (**Figure 7.42**). This return to regional vernacular styles enhances regional identities and erases some of the homogeneity imposed by national trends and national homebuilders. Because cities tend to add new housing at their outer fringe, the different popular house styles are arranged in concentric rings around the city center, just like a tree's growth rings. During boom times, cities add thick rings of new housing, and in poor times, a particular style might be absent.

Building Styles Around the World

Because most house builders are local or national rather than global, each country has its own traditions and styles of construction. The influences of various folk traditions contribute to ongoing differences in the look of housing in different countries. Even within Europe, there are substantial differences in house types between countries. Taking advantage of their abundant forests, the Scandinavian countries prefer wood houses, while brick rowhouses or duplexes dominate in Britain. Responding to China's newfound prosperity, middle-class home builders have sometimes borrowed the styles of other countries, creating entire suburban communities that imitate the look of Canadian ski towns, alpine villages, and the suburban housing of Orange County, California.

However, in the construction of large apartment, office, factory, or civic buildings, there is less variation from country to country. Here, the diffusion of Western styles, the influence of the modern movement in architecture, and engineering design standards have created a look termed *the International Style*. The International Style uses industrial construction materials of glass, steel, and concrete and is less likely to use local stone or wood. The International Style, by definition, is the

(a)

(b)

Figure 7.42 Architects and urban designers have sometimes revived regional housing styles. (*a*) Beginning in the 1980s, the Kentlands neighborhood in suburban Washington, D.C., was developed following traditional 18th- and early 19th-century Middle Atlantic designs. (*b*) An early 2000s suburb near St. Paul, Minnesota, featured revivals of common Upper Midwest vernacular house styles such as this upright-and-wing.
©Mark Bjelland

Figure 7.43 Many of the international style buildings in this image of the Nairobi, Kenya skyline could be from anywhere in the world. The diffusion of the modern style of architecture and adoption of universal engineering design standards lends a standardized look to high-rise office buildings and residences.

©Ralph A. Clevenger/Corbis/Getty Images

same everywhere and disregards tradition and local vernacular styles. Thus, an International Style office building in Nairobi, Kenya, can look remarkably similar to one in Houston, Bangkok, or Paris (**Figure 7.43**).

Landscapes of Consumption: The Shopping Mall

Traditionally, manufacturing was seen as the key to national strength. However, since World War II, consumption rather than production has been promoted as the dominant engine of the American economy. Shopping is viewed as an enjoyable pursuit, and sometimes even our patriotic duty, as reflected in common phrases such as "born to shop," "shop 'til you drop," or "he who dies with the most toys wins." The most prominent landscape expression of our commodity-driven popular culture, of course, is the shopping mall, sometimes dubbed the "palace of consumption."

Major regional malls have been created in every part of the world that boasts a population of middle-class consumers large enough to satisfy their carefully calculated purchasing-power requirements. With their mammoth parking lots and easy access from expressways or highways, most malls are part of the automobile culture that helped create them after World War II. Enclosed, temperature-controlled, without windows or other acknowledgment of a world outside, some assume monumental size, approximating the retail space contained in the central business districts of older medium-sized and large cities (**Figure 7.44**). The largest mall in the United States, the aptly named Mall of America in Bloomington, Minnesota, boasts

more than 500 stores, an indoor amusement park, and aquarium. But the world's three largest shopping malls are located in China, and the fourth-largest is in the Philippines.

On the one hand, there is a sameness and placelessness about shopping malls. Shopping center developers and owners have built nearly identical malls and filled them with a nearly identical mix of national and international chain stores and restaurants. A handful of large real estate companies own the majority of regional malls in the United States, as well as having overseas portfolios. They cater to a full range of homogenized consumer desires that have been molded by popular culture and advertising. On the other hand, many successful malls have been tailored to the income, culture, and tastes of the local area with fine-tuned **geodemographic analysis** of the local population, its age distribution, and social and economic characteristics. Inside the mall, the designs for individual stores reflect the fragmentation of society into various subcultures with storefronts, layouts, and music subtly conveying messages about the appropriate age, tastes, and income levels of the target consumer.

The ubiquity of malls and the uniformity of their offerings homogenize the books, movies, wall art, furnishings, and fashions that the consumer can choose among. Whatever fashion trend may be dictated nationally or internationally is instantly available locally, hurried to market by well-organized chains responding to well-orchestrated customer demand. A handful of centers of innovation for fashion—Milan, Paris, London, and New York—dictate what shall be worn, and a few designer names dominate the popularly acceptable range of choices.

Figure 7.44 The shopping mall is a palace of consumption. Massive, enclosed, and buffered from its surroundings, the modern metropolitan shopping mall is a prime carrier of popular culture. This mall is located in Dubai in the United Arab Emirates, but it could be in nearly any city with sufficient buying power to attract mall developers. The standardized offerings of the mall contrast with the individuality of regional and folk cultures.

©In Green/Shutterstock

Because popular culture is, above all, commercialized culture, a market success by one producer is instantly copied by others. Thus, even the great number of individual shops within the mall is only an assurance of variations on the same limited range of clothing (or other) themes, not necessarily diversity of choice. Yet, of course, such a range of choices would not be possible within constrained folk or ethnic groups.

While regional shopping malls have successfully outcompeted traditional shopping districts, leading to vacant downtown and Main Street stores in many towns and cities, their own long-term success is not guaranteed. The construction of large, enclosed malls has slowed to a virtual halt in the United States since 2000. A plethora of vacant "dead malls" exist wherever shopping trends, demographic changes, or new competitors have driven away consumers. Because the earlier, impersonal, massive enclosed malls could inspire feelings of alienation, developers have redesigned older malls and constructed new ones to replicate the landscape features of traditional urban places and evoke the sense of a distinctive community. Familiar urban landscape elements are recreated in the design of these malls, but in romanticized forms. The façades create the appearance of multiple buildings, and the central corridors recall the small-town Main Streets of the past, where shoppers jostled with friendly strangers. In the middle may be a large open space that functions as a simulated town square, often with benches and shrubbery imitating the central city park of a small-town center. The ever-present food court is often an idealized reproduction of past urban places: the sidewalk cafes of Paris, or the city squares (piazzas) of Italy, for example. It has been suggested that these new-style shopping centers, with their contrived sense of place, are idealized Disneyland versions of the places that they simulate. They are part of a broader trend inspired by Disneyland and Las Vegas of creating **themed landscapes** that simulate different times and places (**Figure 7.45**). The success of Disneyland's Main Street U.S.A. and other themed environments in the Magic Kingdom inspired the designers of chain stores, chain restaurants, shopping malls, and housing subdivisions to create their own themed environments. In the mall, one casual clothing store catering to teenagers elaborately recreates the look of a California surf shack, while in the food court, a chain restaurant creates a simulated Amazonian rainforest experience.

The redesign of shopping malls is exemplified in the recent popularity of lifestyle centers. The new centers differ from older enclosed malls by being smaller outdoor or open-air assemblages with storefronts facing a pseudo Main Street or plaza. Many lifestyle centers are influenced by the **New Urbanism**, a planning movement that promotes walkability and mixed-use buildings with offices and residences on the upper floors. (**Figure 7.46**). Customers in lifestyle centers are able to park near their destinations, shop, and leave

Figure 7.45 Themed landscapes are created by real estate developers to attract tourists and shoppers. Las Vegas has casino hotels "themed" as New York City, Greco-Roman temples, Egyptian pyramids, medieval castles, Venetian palaces, and a French hotel, complete with a replica of the Eiffel Tower.

©W. Buss/Stockbyte/Getty Images

Figure 7.46 Lifestyle centers represent a recent trend to create themed environments for shopping, dining, and entertainment. The Santana Row shopping center in San Jose, California, as shown here, replaced a failing 1950s single-story shopping mall. The designers of Santana Row were inspired by the New Urbanism movement and by the architecture and public spaces of southern European cities.

©MasaoTaira/iStock/Getty Images

without encountering the crowded interior space of traditional malls. Unlike the "placeless" environment associated with the massive enclosed mall, lifestyle centers attempt to create the uniqueness, ambience, and "localized" feel of small towns, although most of the upscale retailers and restaurants are national chains that would not be found in a typical small town. While lifestyle centers pretend to be public spaces, they are privately owned and controlled, unlike traditional shopping streets or city squares. Activities that would be allowed in a traditional city square, such as rallies, proselytizing, and begging, are usually not allowed in a mall or lifestyle center because they aren't conducive to a pleasant shopping atmosphere.

Heritage Landscapes

Just as many countries have decided to preserve their best examples of wild, natural landscapes, some believe that we should designate and preserve important cultural landscapes. With the exception of a few mountaintops, the entire European landscape has been altered by human activities stretching over millennia. The concept of preserving wilderness, as developed in the United States, does not readily fit the context of Europe and much of Asia. However, the many tourists that visit Europe each year are evidence that a settled **heritage landscape** can be particularly attractive, not in spite of the human presence, but because of a culture's harmonious interaction with the natural landscape or creation of beautiful, human-scale architecture. The United Nations Educational, Scientific, and Cultural Organization (UNESCO) operates a world heritage preservation program that has designated nearly 1,000 cultural sites, natural landscapes, and cultural landscapes from all over the world. The designated cultural landscapes are the joint creations of humans and nature and are selected for their unique scenery, archeology, architecture, history of human habitation, or symbolic significance. A number of the places listed on the World Heritage List are relatively wild landscapes, but have been included not for their ecological significance but for their cultural significance. These are wild landscapes that are viewed as sacred places by indigenous peoples. Unlike most national parks or wilderness preserves, the cultural landscapes on the World Heritage List are often inhabited (**Figure 7.47**). For example, the UNESCO world heritage list includes the city centers of a number of well preserved historic European cities, a classical Chinese garden, British mining and early industrial landscapes, several of Iraq's ancient cities, and the University of Virginia campus, designed by Thomas Jefferson.

Figure 7.47 The village of Halstatt, Austria has been designated by UNESCO as one of more than 700 world heritage cultural landscapes. The Halstatt area has been occupied by humans for three millennia and prospered in Medieval times as a salt mining center. In the 19th century the town's beauty attracted painters and poets. Today, it is immensely popular with tourists for its blend of natural scenery and intricate, human-scale architecture. It is so popular with East Asian tourists, in fact, that it has inspired Chinese developers to build an exact, life-size replica on a tropical, lakeshore in Guangdong Province, China.

©robertharding/Photodisc/Getty Images

Among the diversity of cultural identities and cultural landscapes, we can identity tensions between folk, local, or regional identities and the globalizing tendencies of popular culture. Ethnic culture and folk culture tend to create distinctions among peoples and to impart a special character to the areas where they are dominant. Popular culture diffuses widely in a world of increased spatial interaction and tends to homogenize culture and make places alike.

Early arriving ethnic groups were soon Americanized, and their imported cultures were converted from the distinctly ethnic traits of foreigners to the folk cultures of the New World. The foothold settlements of the first colonists became separate culture hearths in which imported architectural styles, food preferences, music, and other elements of material and nonmaterial culture were mixed, modified, abandoned, or disseminated along clearly traceable diffusion paths into the continental interior. Ethnic culture became folk culture when nurtured in isolated areas and made part of traditional America as it adapted to local circumstances. Those folk cultures contributed both to Eastern regional diversity and to the diffusion streams affecting Midwestern and Western cultural mixtures.

Popular culture fads, foods, music, dress, toys, games, sports, and other fashions tend to be adopted within a larger society, irrespective of the ethnic or folk distinctions of its parts. The more modernized and urbanized a country, the more that it is uniformly subjected to the mass media, the spread of popular culture, and the loss of folk customs. Popular culture and folk culture are intertwined, with popular culture depending upon folk customs for new ideas and folk cultures incorporating elements of popular culture. Glocalization, the adaptation of globalized products to local tastes, and neolocalism, the renewed emphasis on local identities for products, are reactions to the globalization of popular culture.

Over time, culture groups refashion the natural landscape into a cultural landscape. Even the most ordinary of landscapes are important to the geographer, for they tell a story of how humans have modified the Earth over time in accord with changing cultural values. Some cultural landscapes are deeply symbolic, others reflect the high status of their creators, and still others help us remember the past. Systems for dividing and settling land, styles of housing, and landscapes of consumption are major elements of the cultural landscape. As folk styles of building have given way to national and international styles, there has arisen a countervailing desire to recover traditional housing styles and preserve globally significant cultural landscapes. Behind the diffusion and globalization of popular culture that have made up a theme of this chapter is the reality of economic globalization. Therefore, in the next chapter we turn our attention to *economic geography,* the world of work and the spatial patterns of production and exchange.

KEY WORDS

cultural landscape	metes-and-bounds	popular region
custom folk culture	neolocalism	Public Land Survey System (PLSS)
geodemographic analysis	New Urbanism	rectangular survey
glocalization	nonmaterial culture	sequential occupation
heritage landscapes	palimpsest	themed landscapes
long-lot system	placelessness	vernacular house
material culture	popular culture	vernacular region

FOR REVIEW

1. What contrasts can you draw between *folk culture* and *popular culture?* What different sorts of material and nonmaterial elements identify them? How are folk culture and popular culture interconnected and interrelated?

2. How many of the early settlement cultural hearths of North America can you name? Did early immigrants create uniform *built environments* within them? If not, why not?

3. When and under what circumstances did popular culture begin to erode the folk and ethnic cultural differences between Americans? Thinking only of your own life and habits, what traces of folk culture do you carry? To what degree does popular culture affect your decisions on dress? On entertainment? On recreation?

4. What kinds of connections can you discern between the nature of the physical environment and the characteristics of different *vernacular house* styles in North America and in other parts of the world? Does that relationship with the physical environment continue with the national and international styles of building?

5. Describe how the metes-and- bounds, French long-lot system, and rectangular survey produce a different look to the land. How do different rural settlement patterns reflect different values and different types of social interaction?

6. Describe how folk house styles vary across the United States and Canada. How have house styles changed over time? What are the major house styles in your community, and in what different time periods were they produced?

7. Choose a specific cultural landscape, describe its elements, and explain how it has been fashioned over time and how it expresses the values and aspirations of a particular culture group.

KEY CONCEPTS REVIEW

1. **What is folk culture, what material and nonmaterial elements can be observed, and what folk culture hearths and regions are found in the United States and Canada?** Section 7.1

Folk culture, often based on ethnic backgrounds, tends to be localized by population groups and areas. It acts to distinguish groups within mixed-culture societies. In the United States and Canada, diversified immigrant groups settling different, particularly eastern, regions brought their own building traditions to *hearth* regions: those of the *North,* in the northeastern United States and southeastern Canada; the *Middle Atlantic;* and the *South.* The universal elements of nonmaterial folk culture include food and drink preferences and ingredients, music, recreation, and folklore traditions.

2. **What is popular culture, how does it diffuse, and how does it interact with folk cultures?** Section 7.1

Popular culture implies the tastes and habits of the general mass of a society and diffuses widely and rapidly through mass media, advertising, and transnational corporations. Popular culture is based on changing trends in clothing, food, houses, services, sports, and entertainment and embodies the ever-changing and yet dominant "way of life" of a society. Shopping malls, standardized national chains of restaurants, motels, and retail stores, many oriented toward automobile mobility, characterize the contemporary landscape and lifestyle in the United States and Canada. Although popular culture tends to obliterate folk cultures, their relationship can be more complex. The rise of neolocalism, the deliberate identification of a product or service with a particular place, shows the continuing importance of local cultural identities. Similarly, the rise of glocalization, the adaptation of globalized products for local or regional tastes, shows the persistence of regional distinctiveness.

3. **What are cultural landscapes and why are they of interest to geographers?** Section 7.2

Cultural landscapes are the product of a culture group refashioning the natural landscape over time. Cultural landscapes reflect a culture's relationships with the natural environment, and its social systems, values, and aspirations.

4. **How do systems of land survey and settlement, housing styles, and landscapes of consumption reflect cultural values?** Section 7.2

The long-lot system reflected the importance of river access, the rectangular survey reflected the desire for order, rationality, and ease of settlement, and the agricultural village and commons of the Mennonites showed the importance of shared community to that group. Folk housing styles reflected different ethnic influences, as well as the natural environment. Popular house styles have followed trends, revived historic styles, and been dramatically influenced by the dominant role of the automobile in contemporary society. Landscapes of consumption play a major role in the diffusion of popular culture and the erosion of regional distinctiveness. However, palaces of consumption have increasingly been designed as themed environments, attempting to create or simulate a distinctive place identity.

CHAPTER

8

ECONOMIC GEOGRAPHY:
Primary Activities

An abandoned farm house and barn among the vast wheat fields of the Great Plains.
©Design Pics/Kelly Redinger

Key Concepts

8.1 How productive activities and economies are classified

8.2 The types and prospects of subsistence agriculture; commercial agriculture: technological change, geographic patterns, and concerns

8.3 Natural resource-based activities: fishing, forestry, and mining

8.4 Trade in primary products

*T*he crop bloomed luxuriantly that summer of 1846. The disaster of the preceding year seemed over, and the potato, the sole sustenance of some 8 million Irish peasants, would again yield the bounty needed. Yet within a week, wrote Father Mathew, "I beheld one wide waste of putrefying vegetation. The wretched people were seated on the fences of their decaying gardens . . . bewailing bitterly the destruction that had left them foodless." Colonel Gore found that "every field was black," and Father O'Sullivan noted that "the fields . . . appeared blasted, withered, blackened, and . . . sprinkled with vitriol. . . ." The potato was irretrievably gone for a second year; famine and pestilence were inevitable.

Within five years, the settlement geography of the most densely populated country in Europe was forever altered. The United States, Canada, and Great Britain received 2 million Irish immigrants, who provided the cheap labor needed for the canals, railroads, and mines that they were creating in a rush to economic development. New patterns of commodity flows were initiated as American maize for the first time found an Anglo-Irish market—as part of Poor Relief—and then entered a wider European market, which had also suffered general crop failure in that bitter year. Within days, a microscopic organism, the cause of the potato blight, had altered the economic and human geography of two continents.

Although the Irish famine of the 1840s was a regional tragedy, it dramatically demonstrated the intricate interrelations among widely separated peoples and places. It demonstrated the fundamental importance of patterns of economic geography and subsistence. Chapters 8, 9, and 10 explore the dynamic economic innovations that are reworking the landscapes of human activities. Food and raw material production still dominate the economies in some parts of the world, but increasingly, people are engaged in activities that involve the processing of raw materials into finished products and the provision of personal, business, and professional services within an increasingly interconnected world economy. Changing patterns of subsistence, livelihood, exchange, and the pursuit of "development" are the focus of economic geography.

Simply stated, **economic geography** is the study of how people earn their living, how livelihood systems vary from place to place, and how economic activities are spatially interrelated and linked. It applies geography's general concern with spatial variation to the production, exchange, and consumption of goods and services. In reality, of course, we cannot comprehend the totality of the economic pursuits of 7 billion human beings. Instead, economic geographers seek consistencies. They develop generalizations to help understand complex patterns. Studying economic geography reveals the dynamic, interlocking diversity of human activities and the impact of economic activity on all other facets of human life and culture. It reveals the increasing interdependence of differing national and regional economic systems. The potato blight, although it struck one small island, ultimately affected the economies of continents. In like fashion, the exploitation of the natural resources of the United States, the *deindustrialization* of its economy, and the shift to postindustrial service and knowledge activities, are altering the relative wealth of countries, flows of international trade, domestic employment and income patterns, and more (**Figure 8.1**).

Figure 8.1 This oil tanker is part of a world of increasing economic interdependence. Oil is a global commodity, with its price set by global markets.
©Malcolm Fife/Getty Images

8.1 The Classification of Economic Activity and Economies

Understanding livelihood patterns is made more difficult by the complex environmental and cultural realities controlling the economic activities of humans. Many production patterns are rooted in the spatially variable circumstances of the *physical environment*. The staple crops of the humid tropics, for example, are not part of the agricultural systems of the midlatitudes; livestock types that thrive in American feedlots or on Western ranges are not adapted to the Arctic tundra or to the margins of the Saharan desert. The unequal distribution of useful petroleum and mineral deposits make some regions wealthy and others dependent. Forestry and fishing depend on still other natural resources that are unequal in occurrence, type, and value.

Within the bounds of what is environmentally possible, *cultural considerations* may shape economic decisions. For example, culturally based food preferences rather than environmental limitations may dictate the choice of crops or livestock. Corn (maize) is a preferred grain in Africa and the Americas; wheat in North America, Australia, Argentina, southern Europe, and Ukraine; and rice in much of Asia. Pigs are not raised in Muslim areas, where religious belief prohibits pork consumption.

Level of *technological development* of a culture will affect its recognition of resources or its ability to exploit them. **Technology** refers to the totality of tools and methods available to and used by a culture group in producing items essential to its subsistence and comfort. Preindustrial societies have no knowledge of or need for the iron ore, coal, petroleum, or uranium underlying their hunting, gathering, or gardening grounds. *Political decisions* may encourage or discourage—through subsidies, protective tariffs, or production restrictions—patterns of economic activity. And, ultimately, production is controlled by *economic factors* of demand, whether that demand is expressed through a free-market mechanism, government controls, or the consumption requirements of a single family producing for its own needs.

Categories of Activity

One approach to categorize the world's productive work is to view economic activity as ranged along a continuum of both increasing complexity of product or service and increasing distance from the natural environment. Seen from that perspective, four distinctive stages of economic activities may be distinguished: primary, secondary, tertiary, and quaternary (**Figure 8.2**).

Primary activities are those that harvest or extract something from the Earth. They are at the beginning of the production cycle, where humans are in closest contact with the resources of the environment. Such activities involve basic food and raw material production. Hunting and gathering, grazing, agriculture, fishing, forestry, and mining and quarrying are examples. **Secondary activities** are those that add value to materials by changing their form or combining them into more

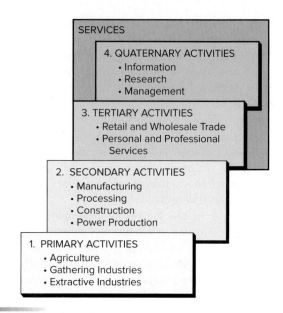

Figure 8.2 The categories of economic activity.

useful—therefore more valuable—commodities. That processing of raw materials into finished products may range from simple handicraft pottery to the assembly of electronic goods or space vehicles (**Figure 8.3**). Copper smelting, steel making, metalworking, automobile production, food processing, textile and chemical industries—indeed, the full array of *manufacturing and processing industries*—are included in this phase of the production process. Also included are the production of *energy* and the *construction* industry.

Tertiary activities provide *services* to the primary and secondary sectors and *goods* and *services* to businesses and to individuals. The service sector includes wholesale and retail trade, which constitute the vital link between producers and consumers. Business services include accounting, advertising, financial services, insurance, legal services, and real estate. Consumers may use some of these same services, although often from different providers. Examples of consumer service providers include health care, eating and drinking establishments, repair and maintenance providers, and personal service establishments such as hair salons. **Quaternary activities** are a specialized subset of service activities involving research, information, and administration. In advanced economies, competitiveness and productivity are closely tied to the gathering, analysis, and dissemination of information. Generally, economic development brings a dramatic shift in the distribution of economic activity across the categories of economic activity (Table 8.1). Industrialization leads to an increase in the secondary sector at the expense of agriculture. Further economic development tends to shift the economic structure toward services such that the world's most advanced economies are now largely **post-industrial information economies.** The United States demonstrates this shift, with just 1 percent of its gross domestic product (GDP) derived from agriculture and more than three-fourths from service activities. Still, primary activities are essential to all human life and are dominant globally on a land area basis.

Figure 8.3 These fattened cattle outside a meat packing plant in Colorado are products of *primary production*. The slaughter and processing of these cattle into various beef cuts, hamburger, and beef by-products such as dog food and fertilizer is a *secondary activity*.

©*Glow Images*

The term *industry*—in addition to its common meaning as a branch of manufacturing activity—is frequently employed as a substitute, identical in meaning to *activity,* as a designation of these categories of economic enterprise. That is, we can speak of the steel, or automobile, or textile "industry," with all the impressions of factories, mills, raw materials, and products each type of enterprise implies. But with equal logic, we can refer in a more generalized way to the "entertainment" or the "travel" industry or, in the present context, to "primary," "secondary," and "service" industries.

These categories of production and service activities help us see an underlying structure to the nearly infinite variety of things people do to earn a living and to sustain themselves. But by themselves, they tell us little about the organization of the larger economy of which the individual worker or establishment is a part. For that broader organizational understanding, we look to *systems* rather than *components* of economies.

Types of Economic Systems

Broadly viewed, economies fall into one of three major types of systems: *subsistence, commercial,* or *planned*. None of these economic systems is "pure." That is, none exists in isolation in an increasingly interdependent world. Each, however, displays certain underlying characteristics based on its distinctive forms of resource management and economic control.

In a **subsistence economy,** goods and services are created for the use of the producers and their kinship groups. Therefore, there is little exchange of goods and only limited need for markets. In the **market (commercial) economies** that have become dominant in nearly all parts of the world, producers or their agents, in theory, freely market their goods and services, the laws of supply and demand determine price and quantity, and market competition is the primary force that shapes production decisions and distributions. In the extreme form of **planned economies** associated with communist societies, producers or their agents disposed of goods and services through government agencies that controlled both supply and price. The quantities produced and the locational patterns of production were strictly programmed by central planning departments.

With a few exceptions—such as Cuba and North Korea—rigidly planned economies no longer exist in their classical form; they have been modified or dismantled in favor of free market structures or only partially retained in a lesser degree

Table 8.1

Stage of Economic Development and the Structure of Economic Activity

Country/Category	Value Added as Percentage of GDP, 2016[1]		
	Agriculture	Manufacturing	Services
Least Developed			
Central African Republic	43	7	41
Mali	42	5	40
Newly Industrialized			
Malaysia	9	22	53
Thailand	8	27	56
Industrial			
Czech Republic	3	27	60
South Korea	2	29	59
Postindustrial			
Australia	3	7	73
United States	1	12	79

Source: *The World Bank*, Open Database, *2018. Values do not add to 100 percent due to mining, construction, and utilities*
[1]Some values are from earlier years where 2016 data was missing.

of economic control associated with governmental supervision or ownership of selected sectors of increasingly market-oriented economies. Nevertheless, their landscape evidence lives on in formerly communist societies, such as the former Soviet Union and its satellite bloc countries. The physical structures, patterns of production, and regional interdependencies they imposed continue to influence the economic decisions of successor societies.

In actuality, few people are members of only one of these systems. A farmer in India may produce rice and vegetables privately for the family's consumption but also save some of the produce to sell. In addition, members of the family may market cloth or other handicrafts they make. With the money derived from those sales, the Indian peasant is able to buy, among other things, clothes for the family, tools, and fuel. Thus, that Indian farmer is a member of at least two systems: subsistence and commercial.

In the United States, government controls or subsidies on the production of various types of goods and services (such as growing corn, producing alcohol, constructing and operating nuclear power plants, and engaging in licensed personal and professional services) mean that the country does not have a purely market economy. To a limited extent, its citizens participate in a controlled and planned environment, as well as in a free-market environment. Many African, Asian, and Latin American market economies have been decisively shaped by government or international development and monetary agency policies that encourage the production of export commodities rather than food for domestic markets. Other developing countries promote through import restrictions and currency manipulation the development of their own industrial capacity. Example after example would show that "free" markets depend upon institutions and government controls.

No matter what economic system prevails, transportation is a key variable. No advanced economy can flourish without a well-connected transport network. Subsistence societies—or subsistence areas of developing countries—are characterized by their isolation from regional and world trade routes (**Figure 8.4**). That isolation restricts their progression to more advanced forms of economic structure.

Inevitably, spatial patterns of economic activities and systems are subject to change. For example, the commercial economies of Western European countries are being restructured by both increased free market competition and supranational regulation under the World Trade Organization (WTO) and the European Union (EU; see Chapter 12 for more on this topic). The countries of Latin America, Africa, Asia, and the Middle East that traditionally were dominated by subsistence economies are now benefiting from technology transfer and integration into expanding global production and exchange patterns. For example, the phenomenal growth of the Chinese economy is rewriting the map of economic activity and shifting the global balance of economic power. Economic globalization increases linkages among distant regions and spreads wealth more widely, but also undermines the stability of established production locations. In short, the creative destruction of capitalism produces results that vary widely from place to place.

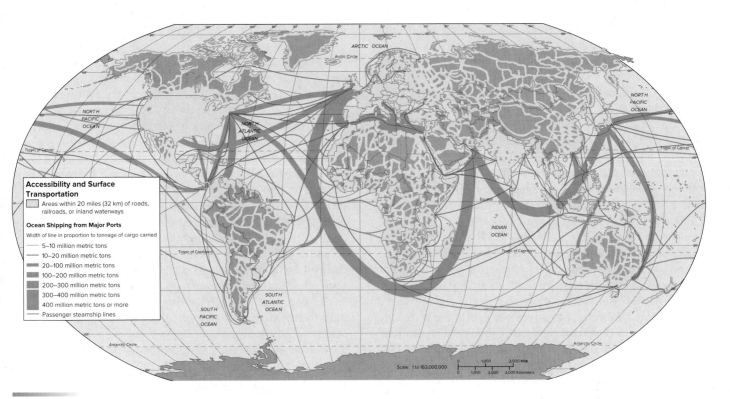

Figure 8.4 Patterns of surface transportation and accessibility. Accessibility is a key measure of economic development and of the degree to which a world region can participate in interconnected market activities. Isolated areas of countries with advanced economies suffer a price disadvantage because of high transportation costs. Lack of accessibility in subsistence economic areas slows their modernization and hinders their participation in the world market.

Source: Allen, Paul, Student Atlas of World Geography, *McGraw-Hill.*

In the remainder of this chapter, we will center our attention on the primary industries. In Chapter 9, we will consider the secondary and service sectors of the economy.

8.2 Primary Activities: Agriculture

Humankind's basic economic concern is producing or securing sufficient food resources to meet daily energy requirements and normal nutritional needs. Those supplies may be acquired directly, through hunting, gathering, farming, or fishing, or indirectly, through performance of other primary, secondary, or service sector endeavors that yield sufficient income to obtain needed daily sustenance. Statistics from the Food and Agriculture Organization (FAO) at the United Nations show that in 2010, 2.6 billion people or 38 percent of the world's population depended on agriculture, hunting, fishing, and forestry for their livelihoods.

Since the 1960s, neo-Malthusians (see Chapter 4) have revived Thomas Robert Malthus's fears that the world's steadily increasing population would exceed food supplies. Instead, although global population has tripled since 1950, the total number of undernourished people has dropped since 1990. The FAO has set the minimum daily requirement for caloric intake at 2,350 per person. By that measure, annual food supplies are more than sufficient to meet world needs. That is, if total food resources were evenly distributed, everyone would have access to amounts sufficient for adequate daily nourishment. In reality, however, about 800 million people or 11 percent of the world's population are inadequately supplied with food and nutrients. This stark contradiction between sufficient worldwide food supplies and widespread malnutrition reflects, among other reasons, inequalities in national and personal incomes; lack of access to fertile soils, credit, and education; local climatic conditions or catastrophes; and lack of transportation and storage facilities. By mid-century, the increasingly interconnected world population will expand to a projected 9.8 billion, and concerns with individual states' food supplies will inevitably remain a persistent international issue. World and regional issues of food security are explored in Chapter 10.

Before there was farming, *hunting* and *gathering* were the universal forms of primary production. These preagricultural pursuits are now practiced by at most a few thousand people worldwide, primarily in isolated and remote pockets within the low latitudes and among the sparse populations of very high latitudes. The interior of New Guinea, rugged areas of interior Southeast Asia, diminishing segments of the Amazon rain forest, and a few districts of tropical Africa and northern Australia still contain such preagricultural people. Much of the Arctic region, of course, is ill suited for any form of food crop production. Hunter-gatherer numbers are few and declining, and wherever they are brought into contact with technologically more advanced cultures, their way of life is eroded or lost.

Agriculture, defined as the growing of crops and the tending of livestock, has replaced hunting and gathering as the most significant of the primary activities. It is spatially the most widespread, found in all world regions where environmental circumstances—including adequate moisture, good growing season length, and productive soils—permit (**Figure 8.5**). The United Nations (UN) estimates that more than one-third of the world's land area (excluding Greenland and Antarctica) is in some form of agricultural use, including permanent pastureland. Crop farming alone covers some 15 million square kilometers (5.8 million square miles) worldwide, about 12 percent of the Earth's total land area. In many developing economies, at least two-thirds of the labor force is directly involved in farming and herding. In some, such as Burundi in Africa, the figure is more than 90 percent. Overall, however, employment in agriculture is steadily declining in developing economies, echoing but trailing the trend in commercial economies, in which direct employment in agriculture involves only a small fraction of the labor force (**Figure 8.6**). Globally, in 2017, 26 percent of the world's economically active population worked in agriculture. In the United States, just 2 percent of workers were in agriculture and in the United Kingdom, it was just 1 percent. Indeed, a declining number or proportion of farm workers, along with farm consolidation and increasing output, are typical in all highly developed commercial agricultural systems. On the other hand, agriculture remains a major component in the economies of many of the world's developing countries, producing for domestic markets and providing a major source of national income through exports (**Figure 8.7**).

It has been customary to classify agricultural societies on the twin bases of the importance of off-farm sales and the level of mechanization and technological advancement. *Subsistence, traditional* (or *intermediate*), and *advanced* are terms that recognize both aspects. These are recognized stages along a continuum. At one end lies production solely for family sustenance, using rudimentary tools and native plants. At the other is the specialized, highly capitalized, industrialized agriculture for off-farm sale that marks advanced economies. Between these extremes is the middle ground of traditional agriculture, where farm production is in part destined for home consumption and in part oriented toward off-farm sale. Further complicating the contrast between subsistence and advanced agriculture is the rise of organic agriculture and the local foods movement in advanced industrial countries. We will begin our survey of the geography of agriculture by examining the *subsistence* and *advanced* ends of the agricultural continuum.

Subsistence Agriculture

By definition, a *subsistence* economic system involves nearly total self-sufficiency on the part of its members. Production for exchange is minimal, and any exchange is noncommercial;

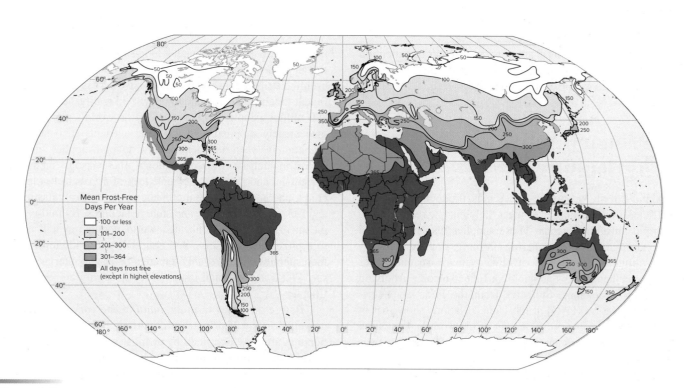

Figure 8.5 Average length of growing season. The number of frost-free days is an important environmental control on agriculture, as is the availability of precipitation sufficient in amount and reliability for crop production. Because agriculture is not usually practicable with less than a 90-day growing season, large parts of Russia and Canada have only limited cropping potential. Except where irrigation water is available, arid regions are similarly outside of the margins of regular crop production.

Source: Wayne M. Wendland.

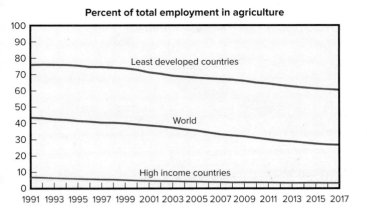

Percent of total employment in agriculture

Figure 8.6 In economies worldwide, the percentage of the labor force in agriculture has been steadily declining—and is projected to decrease to even lower levels. Still, large differences remain between the world's high income countries and the least developed countries.

Sources: World Bank, Open Database, 2018.

each family or close-knit social group relies on itself for its food and other most essential requirements. Farming for the immediate needs of the family is, even today, the predominant occupation of humankind. In most of Africa, South and East Asia, and much of Latin America, a large percentage of people are primarily concerned with feeding themselves from their own land and livestock.

Two chief types of subsistence agriculture may be recognized: *extensive* and *intensive*. The essential contrast between them is yield per unit of land area and, therefore, the population

that can be supported. **Extensive subsistence agriculture** involves large areas of land and minimal labor input per hectare. Both production per land unit and population densities are low. **Intensive subsistence agriculture** involves the cultivation of small landholdings through the expenditure of great amounts of labor per acre. Yields per unit area and population densities are both high (**Figure 8.8**).

Extensive Subsistence Agriculture

Of the several types of *extensive subsistence* agriculture, two are of particular interest: nomadic herding and shifting cultivation.

Nomadic herding, the wandering yet controlled movement of livestock solely dependent on natural forage, is the most extensive type of land-use system (Figure 8.8). That is, it requires the greatest amount of land area per person sustained. Over large portions of semiarid and desert areas of Asia, in certain highland zones, and on the fringes of and within the Sahara, a relatively small number of people graze animals for consumption by the herder group, not for market sale. Sheep, goats, and camels are most common, while cattle, horses, and yaks are locally important. The reindeer of Lapland were formerly part of the same system.

Whatever the animals involved, their common characteristics are hardiness, mobility, and an ability to subsist on sparse forage. The animals provide a variety of products: milk, cheese, blood, and meat for food; hair and wool for clothing; skins for clothing and shelter; and excrement for fuel. For the herder, the animals represent primary subsistence, savings, and insurance against an uncertain future. Nomadic movement is tied to sparse

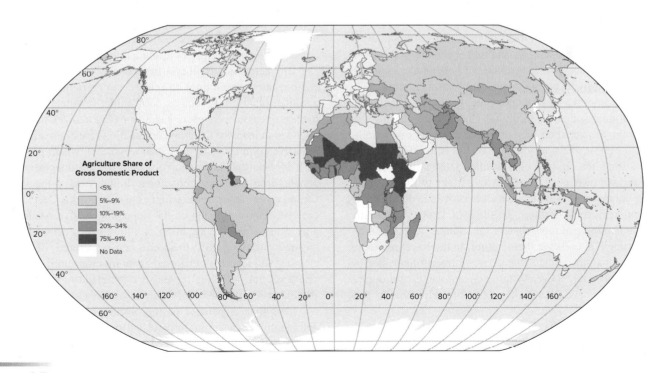

Figure 8.7 Share of agriculture in GDP, 2016. Agriculture makes the largest percentage contributions to GDP in the world's least developed countries.

Source: The World Bank, Open Database, 2018.

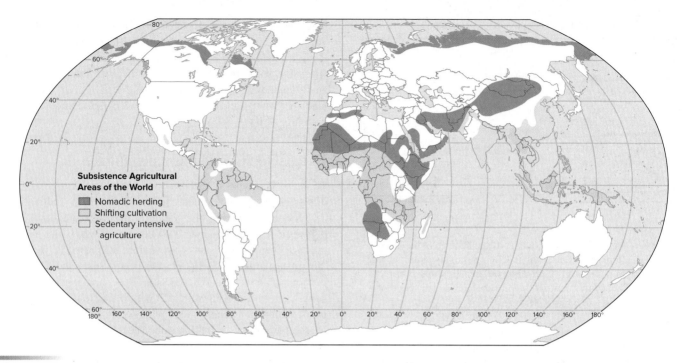

Figure 8.8 Nomadic herding, supporting relatively few people, was the age-old way of life in the dry or cold regions of the world. Shifting or swidden agriculture maintains soil fertility by tested traditional practices in tropical wet and wet-and-dry climates. Large parts of Asia support millions of people engaged in sedentary intensive cultivation, with rice and wheat the chief crops.

and seasonal rainfall or to cold temperatures and to the varying appearance and exhaustion of forage. Extended stays in a given location are neither desirable nor possible. *Transhumance* is a special form of seasonal movement of livestock to exploit specific locally varying pasture conditions. Used by permanently or seasonally sedentary pastoralists and pastoral farmers, transhumance involves either the regular vertical migration from mountain to valley pastures between summer and winter months or horizontal movement between established lowland grazing areas to reach pastures temporarily lush from monsoonal (seasonal) rains.

As an economic system, nomadic herding is declining. Many economic, social, and cultural changes are causing nomadic groups to alter their way of life or to disappear entirely. First and foremost, the nomadic way of life is at odds with the modern world's imperatives of controlled international borders and private land ownership. On the Arctic fringe of Russia, herders under communism were made members of state or collective herding enterprises. In northern Scandinavia, Lapps (Saami) are engaged in commercial more than in subsistence livestock farming and struggle to preserve their culture. In the Sahel region of Africa on the margins of the Sahara, oases formerly controlled by herders have been taken over by farmers, and the great droughts of recent decades have forever altered the formerly nomadic way of life of thousands.

A much differently based and distributed form of extensive subsistence agriculture is found in all the warm, moist, tropical areas of the world. There, many people engage in a kind of nomadic farming. Once put into agricultural use, the soils of those areas rapidly lose many of their nutrients (in hot, wet climates, organic matter rapidly decomposes and heavy rains and groundwater

dissolve and *leach* the nutrients from the soil). After several harvests, the soils are depleted and the farmers move on. In a sense, the farmers rotate fields rather than crops to maintain soil productivity. This type of **shifting cultivation** has a number of names, the most common of which are *swidden* (an English localism for "burned clearing") and *slash-and-burn*. Each region of its practice has its own name—for example, *milpa* in Middle and South America, *chitemene* in Africa, and *ladang* in Southeast Asia.

Characteristically, the farmers hack down the natural vegetation, burn the cuttings to release the nutrients into the soil, and then plant crops such as corn, millet (a cereal grain), rice, manioc or cassava, yams, or sugarcane (**Figure 8.9**). Increasingly included in many of the crop combinations are high-value, labor-intensive commercial crops (such as coffee), which provide the cash income that is evidence of the growing integration of all peoples into exchange economies. Initial yields—the first and second crops—may be very high, but they quickly decrease with each successive planting on the same plot. As that occurs, cropping ceases, native vegetation is allowed to reclaim the clearing, and gardening shifts to another newly prepared site. The first clearing will ideally not be used again for crops until, after many years, natural fallowing replenishes its fertility (see the feature "Swidden Agriculture").

Less than 3 percent of the world's people are still predominantly engaged in tropical shifting cultivation on about one-seventh of the world's land area (Figure 8.8). Because the essential characteristic of the system is the intermittent cultivation of the land, each family requires a total area equivalent to the garden plot in current use, plus all land left fallow for regeneration. Population densities are traditionally low, for much land

(a)

(b)

Figure 8.9 Swidden agriculture in Liberia, Africa. First, the vegetation is cut and burned (*a*). Then, the field is planted (*b*). Stumps and trees left in the clearing will remain after the burn.

(a, b) ©Albert Swingle

is needed to support few people. Here as elsewhere, however, population density must be considered a relative term. In actuality, although crude (arithmetic) density is low, people per unit area of *cultivated* land may be high.

Shifting cultivation is one of the oldest and most widely spread agricultural systems of the world. It is found on the islands of Borneo, New Guinea, and Sumatra but is now retained only in small parts of the uplands of Southeast Asia in Vietnam, Thailand, Myanmar, and the Philippines. Nearly the whole of inland Central and West Africa, Brazil's Amazon basin, and large portions of Central America were engaged in this type of extensive subsistence agriculture.

It may be argued that shifting cultivation is an ingenious, highly efficient cultural adaptation where land is abundant in relation to population. Shifting cultivation generally involves *polyculture,* the production of many different types of crops in a single field. Polyculture reduces vulnerability to pests and diseases and spreads the harvests through the year to provide food security. Polyculture keeps the soil covered by vegetation, reducing the potential for soil erosion. Traditional shifting cultivation has many advantages over commercial agriculture: no chemical fertilizers, herbicides, or pesticides are used and energy is provided by humans or draft animals rather than fossil fuels. Nonetheless, as population densities increase, the system becomes less viable. The basic change, as noted in Chapter 4, is that land is no longer abundant in relation to population in many of the less-developed wet, tropical countries. Their growing populations have cleared and settled the forestlands formerly only intermittently used in swidden cultivation. The **Boserup thesis,** proposed by the economist Ester Boserup, is based on the observation that population increases necessitate increased inputs of labor and technology to compensate for reductions in the natural yields of swidden farming. It holds that population growth forces an increased use of technology in farming and—in a reversal of the Malthusian idea that the supply of food is fixed or only slowly expandable—triggers the switch from extensive to intensive subsistence agriculture, which sharply increases food production.

Intensive Subsistence Systems

Intensive subsistence agriculture is particularly important in the densely populated areas of Asia as shown in Figure 8.8. As a descriptive term, *intensive subsistence* is no longer fully applicable to changing practices in which subsistence and commercial agriculture are increasingly combined. Although families may still be fed primarily with the produce of their individual plots, the exchange of farm commodities within the system is considerable. Production of food for sale in rapidly growing urban markets is increasingly vital for the rural economies of subsistence farming areas and for the sustenance of the growing proportion of national and regional populations no longer themselves engaged in farming. Nevertheless, hundreds of millions of Indians, Chinese, Pakistanis, Bangladeshis, and Indonesians plus further millions in other Asian, African, and Latin American countries remain small-plot, mainly subsistence producers of rice, wheat, corn, millet, or pulses (peas, beans, and other legumes). Most live in monsoon Asia, and we will devote our attention to that area.

Intensive subsistence farmers are concentrated in such major river valleys and deltas as the Ganges and the Chang Jiang (Yangtze) and in smaller valleys close to coasts—level areas with fertile alluvial soils. These warm, moist districts are well suited to the production of rice, a crop that under ideal conditions can provide large amounts of food per unit of land. Rice cultivation requires a great deal of time and attention, for planting rice shoots by hand in standing fresh water is a tedious art (**Figure 8.10**). In the cooler and drier portions of Asia, wheat is grown intensively, along with millet and, less commonly, upland rice.

Figure 8.10 Transplanting rice seedlings requires hard manual labor. The newly flooded diked fields, previously plowed and fertilized, will have their water level maintained until the grain is ripe. This photograph was taken in Vietnam. The scene is repeated wherever subsistence wet-rice agriculture is practiced.

©guenterguni/E+/Getty Images

Rice is known to have been cultivated in parts of China and India for more than 7,000 years. Today, wet, or lowland, rice is the mainstay of subsistence agriculture and diets of populations from Sri Lanka and India to Taiwan, Japan, and Korea. It is grown on more than 80 percent of the planted area in Bangladesh, Thailand, and Malaysia. Almost exclusively used as a human food, rice provides 25 percent to 80 percent of the calories in the daily diet of some 3 billion Asians, or half the world's population. Its successful cultivation depends on the controlled management of water, relatively easy in humid tropical river valleys with heavy, impermeable, water-retaining soils though more difficult in upland and seasonally dry districts. Throughout Asia, the necessary water management systems have left their distinctive marks on the landscape. Permanently diked fields to contain and control water, levees against unwanted water, and reservoirs, canals, and drainage channels to control its availability and flow are common sights. Terraces to extend level land to valley slopes are occasionally encountered as well (see Figure 4.23).

Intensive subsistence farming is characterized by large inputs of labor per unit of land, by small plots, by the intensive use of fertilizers, mostly animal manure, and by the promise of high yields in good years (see the feature "The Economy of a Chinese Village"). Vegetables and some livestock are part of the agricultural system, and fish may be reared in rice paddies and ponds. Food animals include swine, ducks, and chickens. Religion is a factor as well; because Muslims eat no pork, hogs are absent in their areas of settlement. Hindus generally eat little meat, mainly goat and lamb but not pork or beef. The large number of cattle in India is vital for labor, as a source of milk and cheese, and as producers—through excrement—of fertilizer and fuel.

Urban Subsistence Farming

Not all of the world's subsistence farming is based in rural areas. Urban agriculture is a rapidly growing activity, with some 800 million city farmers worldwide. Occurring in all regions of the world but most prevalent in Asia, urban agricultural activities range from small garden plots, to backyard livestock breeding, to fish raised in ponds and streams. Using the garbage dumps of Jakarta, the rooftops of Mexico City, and meager dirt strips along roadways in Kolkata (Calcutta) or Kinshasa, millions of people are feeding their own families and supplying local markets with vegetables, fruit, fish, and even meat—all produced within the cities themselves and all without the expense and spoilage of storage or long-distance transportation.

In all parts of the developing world, urban food production has reduced the incidence of adult and child malnutrition in rapidly expanding cities. City farming is, as well, a significant outlet for underemployed residents. In some cities in the developing world, one-fifth to two-thirds of all families are engaged in agriculture.

There are both positive and negative environmental consequences of urban agricultural activities. On the plus side, urban farming helps convert waste from a problem to a resource by reducing runoff and erosion from open dumps and by avoiding the costs of wastewater treatment and solid waste disposal. In Khartoum, Sudan, for example, about 25 percent of the city's garbage is consumed by farm animals; in Kolkata, India, city sewage is used to supply lagoons which, in turn, produce some 6,000 tons of fish annually. Nearly everywhere, human and animal wastes, vegetable debris, and table scraps are composted

Swidden Agriculture

The following account describes shifting cultivation among the Hanunóo people of the Philippines. Nearly identical procedures are followed in all swidden farming regions:

When a garden site of about one-half hectare (a little over one acre) has been selected, the swidden farmer begins to remove unwanted vegetation. The first phase of this process consists of slashing and cutting the undergrowth and smaller trees with bush knives. The principal aim is to cover the entire site with highly inflammable dead vegetation so that the later stage of burning will be most effective. Because of the threat of soil erosion the ground must not be exposed directly to the elements at any time during the cutting stage. During the first months of the agricultural year, activities connected with cutting take priority over all others. It is estimated that the time required ranges from 25 to 100 hours for the average-sized swidden plot.

Once most of the undergrowth has been slashed, chopped to hasten drying, and spread to protect the soil and assure an even burn, the larger trees must be felled or killed by girdling (cutting a complete ring of bark) so that unwanted shade will be removed. The successful felling of a real forest giant is a dangerous activity and requires great skill. Felling in second growth is usually less dangerous and less arduous. Some trees are merely trimmed but not killed or cut, both to reduce the amount of labor and to leave trees to reseed the swidden during the subsequent fallow period.

The crucial and most important single event in the agricultural cycle is swidden burning. The main firing of a swidden is the culmination of many weeks of preparation in spreading and leveling chopped vegetation, preparing firebreaks to prevent flames escaping into the jungle, and allowing time for the drying process. An ideal burn rapidly consumes every bit of litter; in no more than an hour or an hour and a half, only smoldering remains are left.

The Hanunóo, swidden farmers of the Philippines, note the following as the benefits of a good burn: (1) removal of unwanted vegetation, resulting in a cleared swidden; (2) extermination of many animal and some weed pests; (3) preparation of the soil for dibble (any small hand tool or stick to make a hole) planting by making it softer and more friable; (4) provision of an evenly distributed cover of wood ashes, good for young crop plants and protective of newly-planted grain seed. Within the first year of the swidden cycle, an average of between 40 and 50 different types of crop plants have been planted and harvested.

The most critical feature of swidden agriculture is the maintenance of soil fertility and structure. The solution is to pursue a system of rotation of 1 to 3 years in crop and 10 to 20 in woody or bush fallow regeneration. When population pressures mandate a reduction in the length of fallow period, productivity of the region tends to drop as soil fertility is lowered, marginal land is utilized, and environmental degradation occurs. The balance is delicate.

Source: *Based on Harold C. Conklin,* Hanunóo Agriculture, *FAO Forestry Development Paper No. 12.*

---

or applied to garden areas, and nearly everywhere, vegetable gardens and interspersed fruit trees, ornamental plants, and flowers enhance the often drab urban scene. Negative consequences include the use of untreated human waste as fertilizers, which exposes both producers and consumers to infectious diseases such as cholera and hepatitis.

Expanding Crop Production

Continuing population pressures on existing resources are a constant spur for ways to increase the available food supply. Two paths to promoting increased food production are apparent: (1) expand the land area under cultivation and (2) increase crop yields from existing farmlands.

The first approach—increasing cropland area—is not a promising strategy. Approximately 70 percent of the world's land area is agriculturally unsuitable, being too cold, too dry, too steep, or totally infertile. Of the remaining 30 percent, most of the area well suited for farming is already under cultivation, and of that area, millions of hectares annually are being lost through soil erosion, salinization, desertification, and the conversion of farmland to urban, industrial, and transportation uses. Only the rain forests of Africa and the Amazon Basin of South America retain sizable areas of potentially farmable land. The soils of those regions, however, are fragile, are low in nutrients, have poor water retention, and are easily eroded or destroyed following deforestation.

When population pressures dictate land conversion, serious environmental deterioration may result. Clearing of wet tropical forests in the Philippines, the Amazon Basin, and Indonesia has converted dense woodland to barren desolation within a very few years as soil erosion and nutrient loss have followed forest destruction. In Southeast Asia, some 10 million hectares (25 million acres) of former forestland are now wasteland, covered by useless sawgrass that supplies neither forage, food, nor fuel. By most measures, world food output cannot reasonably be increased by simple expansion of cultivated areas.

The Economy of a Chinese Village

The village of Nanching is in subtropical southern China on the Zhu River delta near Guangzhou (Canton). Its traditional subsistence agricultural system was described by a field investigator. The system is still followed in its essentials in other rice-oriented societies.

In this double-crop region, rice was planted in March and August and harvested in late June or July and again in November. March to November was the major farming season. Early in March, the earth was turned with an iron-tipped wooden plow pulled by a water buffalo. The very poor who could not afford a buffalo used a large iron-tipped wooden hoe for the same purpose.

The plowed soil was raked smooth, fertilizer was applied, and water was let into the field, which was then ready for the transplanting of rice seedlings. Seedlings were raised in a seedbed, a tiny patch fenced off on the side or corner of the field. Beginning from the middle of March, the transplanting of seedlings took place. The whole family was on the scene. Each took the seedlings by the bunch, 10 to 15 plants, and pushed them into the soft inundated soil. For the first 30 or 40 days, the emerald green crop demanded little attention except keeping the

water at a proper level. But after this period came the first weeding; the second weeding followed a month later. This was done by hand, and everyone old enough for such work participated. With the second weeding went the job of adding fertilizer. The grain was now allowed to stand to "draw starch" to fill the hull of the kernels. When the kernels had "drawn enough starch," water was let out of the field, and both the soil and the stalks were allowed to dry under the hot sun. Then came the harvest, when all the rice plants were cut off a few inches above the ground with a sickle. Threshing was done on a threshing board. Then the grain and the stalks and leaves were taken home with a carrying pole on the peasant's shoulder. The plant was used as fuel at home.

As soon as the exhausting harvest work was done, no time could be lost before starting the chores of plowing, fertilizing, pumping water into the fields, and transplanting seedlings for the second crop. The slack season of the rice crop was taken up by chores required for the vegetables that demanded continuous attention, since every peasant family devoted a part of the farm to vegetable gardening. In the hot and damp period of late spring and summer, eggplant and several

varieties of squash and beans were grown. The green-leafed vegetables thrived in the cooler and drier period of fall, winter, and early spring. Leeks grew year round.

When one crop of vegetables was harvested, the soil was turned and the clods broken up by a digging hoe and leveled with an iron rake. Fertilizer was applied, and seeds or seedlings of a new crop were planted. Hand weeding was a constant job; watering with the long-handled wooden dipper had to be done an average of three times a day, and in the very hot season when evaporation was rapid, as frequently as six times a day. The soil had to be cultivated with the hoe frequently as the heavy tropical rains packed the earth continuously. Instead of the two applications of fertilizer common with the rice crop, fertilizing was much more frequent for vegetables. Besides the heavy fertilizing of the soil at the beginning of a crop, usually with city garbage, additional fertilizer, usually diluted urine or a mixture of diluted urine and excreta, was given every 10 days or so to most vegetables.

Source: *Adapted from C. K. Yang,* A Chinese Village in Early Communist Transition *(Cambridge, MA: Massachusetts Institute of Technology, 1959).*

Intensification and the Green Revolution

Increased productivity of existing cropland rather than expansion of cultivated area has been the key to agricultural production over the past few decades. Between 1960 and 2009, world grain yields rose nearly 140 percent. Crop output, however, varies considerably from year to year, adversely or favorably affected by weather, insect damage, plant diseases, and other growing season conditions. Overall, despite dramatic population growth, grain production per capita today is higher than it was in the 1970s. The vast majority of that production growth was due to increases in yields rather than expansions in cropland. The largest increases were in Asia, primarily China and India, and South America. Unfortunately, grain yields have been nearly stagnant in sub-Saharan Africa.

Two interrelated approaches to those yield increases mark recent farming practices. First, throughout much of the developing world, production inputs such as water, fertilizer, pesticides, and labor have been increased to expand yields on a relatively constant supply of cultivable land. Irrigated area, for example, have doubled since 1960. Global consumption of fertilizers has dramatically increased since the 1950s, and inputs of pesticides and herbicides

have similarly grown. Traditional practices of leaving land fallow (uncultivated) to renew its fertility have been largely abandoned, and double and triple cropping of land where climate permits has increased in Asia and even in Africa, where marginal land is put to near-continuous use to meet growing food demands. Second, many of these intensification practices are linked to the **Green Revolution**—the shorthand reference to a complex of seed and management innovations adapted to the needs of intensive agriculture and designed to bring larger harvests from a given area of farmland.

Using conventional plant breeding techniques of cross-pollination, American researcher Norman Borlaug developed a dwarf, high-yielding wheat variety at a Mexican research center in the 1940s. Borlaug was later awarded the Nobel Peace Prize, and Mexico soon went from importing half its wheat to being a wheat exporter. Similarly, the International Rice Institute in the Philippines developed dwarf rice strains that yielded many more grains per plant. These high-yielding dwarf varieties respond dramatically to heavy applications of fertilizer, resist plant diseases, and can tolerate much shorter growing seasons than traditional native varieties can. Adopting the new varieties and applying the irrigation, mechanization, fertilization, and pesticide practices they require have created a new "high-input, high-yield" agriculture. Most poor farmers on marginal and rain-fed (nonirrigated) lands, however, have not benefited from the new plant varieties, which require abundant water and chemical inputs.

Expanded food production made possible through the Green Revolution has helped alleviate some of the shortages and famines predicted for subsistence agricultural regions since the early 1960s, saving an estimated one billion people from starvation. According to World Bank calculations, almost 90 percent of people in developing countries now have adequate diets, versus 55 percent in 1950. As **Figure 8.11** shows, however, not all world regions share those positive results equally. Although total food production has more than doubled in Africa since 1960, population growth has steadily reduced that continent's per capita food output. Although the *number* of undernourished people globally remains near the 800 million mark because of population growth, total world food supply has increased even faster than population, and the United Nations predicts that it will continue to do so through at least 2050.

A price has been paid for the successes of the Green Revolution. Irrigation, responsible for an important part of increased crop yields, has destroyed large tracts of land; excessive salinity of soils resulting from poor irrigation practices is estimated to have a serious effect on the productivity of 20 million to 30 million hectares (80,000–120,000 square miles) of land around the world, out of a world total of some 270 million hectares of irrigated land. And the huge amount of water required for Green Revolution irrigation has led to serious groundwater depletion, conflict between agricultural and growing urban and industrial water needs in developing countries—many of which are in dry climates—and to worries about scarcity and future wars over water.

And very serious genetic consequences are feared from the loss of traditional and subsistence agriculture. With it is lost the food security that distinctive locally adapted native crop varieties provided and the nutritional diversity and balance that multiple-crop intensive gardening assured. Unlike commercial monoculture farming, subsistence polyculture farming minimized the risk of major crop failures. Many different crops and many different varieties of each crop guaranteed some yield despite adverse weather, disease, or pest problems.

Commercial agriculture, however, aims at profit maximization, not food security. Poor farmers unable to afford the capital investment that the Green Revolution demands have been displaced by a commercial monoculture, one often oriented toward specialty and industrial crops designed for export rather than to food for domestic markets. Traditional rural society has been disrupted, and landless peasants have been added to the urbanizing populations of affected countries. The presumed benefits of the Green Revolution are not available to all subsistence agricultural areas or advantageous to everyone engaged in farming (see the feature "Women and the Green Revolution"). Africa is a case in point (see Figure 8.11). Green Revolution crop improvements have concentrated on wheat, rice, and corn. Of these, only corn is important in Africa, where principal food crops include millet, sorghum, cassava, manioc, yams, cowpeas, and peanuts. Belated research efforts directed to African crops, the continent's great range of growing conditions, and its abundance of yield-destroying pests and viruses have denied it the dramatic

Index of Per Capita Food Production, 1960–2009

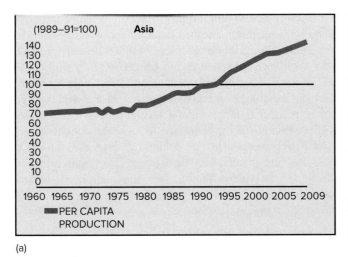

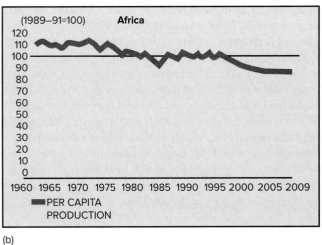

(a)

(b)

Figure 8.11 Trends in per capita food production, 1960–2009. Globally, per capita production of food increased steadily over the 50-year span shown. The intensification and expansion of farming in Asia resulted in food production outpacing population growth, in contrast to Malthus's predictions. It was a different picture in Africa, where total production of food grew steadily but didn't keep pace with population growth, resulting in less food per person. The food security problem in sub-Saharan Africa continues, with about 30 percent of the population suffering chronic malnutrition.

Source: Data from Food and Agriculture Organization.

regionwide increase in food production experienced elsewhere in the developing world. No crop research or genetic modifications can fully compensate for an underlying limitation of African agricultural productivity: 80 percent of the continent's sub-Saharan farmland is severely depleted of the basic nutrients needed to grow crops, a condition that has been steadily worsening since the 1990s.

Some successes have been reported. The most widely cultivated tuber and second most important food staple in sub-Saharan Africa, cassava has been transformed from a low-yielding subsistence hedge against famine to a high-yielding cash crop. Virus-resistant varieties of sweet potatoes and both white and yellow corn and faster-growing bananas are already available (though not yet widespread) and other food and fiber crops are receiving attention from African biotechnology scientists in Kenya, South Africa, and Egypt, with contributions from American and other Western investigators. And in some physically favored areas benefiting in part from foreign investment but particularly reflecting local small farmer enterprise, encouraging pockets of crop specialization and growth in farm productivity, agribusiness creation, and rural income have been emerging. Uganda, for example, enjoys two growing seasons, ample rainfall, rich volcanic soils, and millions of small farmers rapidly expanding production of cash crops, most aimed at export markets in Asia, Europe, and North America.

In many areas that had shown the greatest past successes, Green Revolution gains are falling off. The FAO now considers the productivity gains of Green Revolution technologies for Asian rice cultivation to have leveled off. Little prime land and even less water remain to expand farming in many developing countries. Climate change has created new challenges including higher temperatures, expanded pest ranges, and more frequent droughts.

Another means of increasing production is biotechnology—through the use of **genetically modified (GM) crops**. Crops are genetically modified by moving desired genes from one organism to another or from one species to another in ways that do not occur naturally. Despite resistance, the production of engineered crops is spreading rapidly. In 1996, the first year that GM crops were commercially available, about 1.7 million hectares (4.3 million acres) were placed in biotechnology cultivation. By 2016, the area planted to GM crops had increased to 457 million hectares (1,130 million acres). Initially the acreage devoted to GM crops grew faster in the developed countries. However, because of European concerns about the health of GM foods and the potential for genetically modified crops to cross-pollinate with wild plants, producing "superweeds," GM crop adoption is now proceeding faster in the developing countries. The bulk of GM crop use is in Argentina, Brazil, Canada, China, India, and the United States. Globally, the principal GM crops have been GM soybeans, GM corn (including white corn for food in South Africa), transgenic cotton, and GM canola. Herbicide resistance (Roundup Ready soybeans) and insect resistance (Bt corn and cotton) have been the most important of the genetic crop modifications introduced, as well as the ones responsible for the significant increase in productivity and reduction in costs of the crops involved.

Even in those world regions favorable for Green Revolution introductions, its advent has not always improved diets or reduced dependency on imported food. Often, the displacement of native agriculture involves a net loss of domestic food availability. In many instances, through governmental directive, foreign ownership or management, or domestic market realities, the new commercial agriculture is oriented toward food and industrial crops for the export market or toward specialty crop and livestock production for the expanding urban market rather than food production for the rural population.

The genetic diversity of our food supply is now a major concern. The monocultures of the Green Revolution have reduced plant genetic diversity, while globalization has created more uniform diets around the world. Only 150 crops are commercially grown and three grains: rice, wheat, and corn provide 60 percent of the world's calories. The genetic diversity of these crops has dropped dramatically. For example, in Sri Lanka, the number of rice varieties has dropped by 95 percent. "Seed banks" rather than native cultivation have been created to preserve genetic diversity for future plant breeding and as insurance against climate change or catastrophic pest or disease susceptibility of inbred varieties. Much of the world's crop genetic diversity is located in developing countries where crops were first domesticated and where subsistence farmers continue to use diverse varieties (**Figure 8.12**). However, conserving the world's crop genetic diversity will be a challenge because these places face strong population growth pressures and efforts to commercialize and intensify agriculture.

Commercial Agriculture

Few people or areas still retain the isolation and self-sufficiency that are characteristic of subsistence economies. Nearly all have been touched by a modern world of trade and exchange and have adjusted their traditional economies in response. Modifications of subsistence agricultural systems have inevitably made them more complex by some of the diversity and linkages that mark the advanced economic systems of the more developed world. Farmers in those systems produce not for their own subsistence, but primarily for a market off the farm itself. They are part of integrated exchange economies in which agriculture is but one element in a complex structure that includes mining, manufacturing, processing, and the service activities of the economy. In those economies, farm production responds to market demand, as expressed through price signals, and is related to the consumption requirements of the larger society rather than to the immediate needs of farmers themselves.

Production Controls

Modern agriculture is characterized by *specialization*—by enterprise (farm), by area, and even by country; by *off-farm sale* rather than subsistence production; and by *interdependence* of producers and buyers linked through markets. Farmers in a free market economy supposedly produce those crops that their estimates of market price and production cost indicate will yield the

Women and the Green Revolution

Traditional agricultural labor is often divided by gender, as seen in Figure 8.9 where men prepare the swidden plot and women, often with children on their back, plant the crops. Women farmers grow at least half of the world's food (and up to 80 percent in some African countries). They are responsible for an even larger share of food consumed by their own families: 80 percent in sub-Saharan Africa, 65 percent in Asia, and 45 percent in Latin America and the Caribbean. Further, women comprise between one-third and one-half of all agricultural laborers in developing countries. For example, African women perform about 90 percent of the work of processing food crops and 80 percent of the work of harvesting and marketing.

Women's agricultural dominance in developing states is increasing, as male family members continue to leave for cities in search of paid urban work. In Mozambique, for example, for every 100 men working in agriculture, there are 153 women. In nearly all other sub-Saharan countries, the female component runs between 120 and 150 per 100 men. The departure of men for near or distant cities means, in addition, that women must assume effective management of their families' total farm operations.

Despite their important role, however, women do not share equally with men in the rewards from agriculture, nor benefit from new agricultural technologies. First, most women farmers are involved in subsistence farming and food production for the local market, which yields little cash return. Second, they have far less access than men to credit at bank or government-subsidized rates that would make it possible for them to acquire the Green Revolution technology, such as hybrid seeds and fertilizers. Third, women cannot own land in some cultures and so are excluded from agricultural improvement programs and projects aimed at landowners. For example, many African agricultural development programs are based on the conversion of communal land, to which women have access, to private holdings, from which they are excluded. In Asia, inheritance laws favor male over female heirs, and female-inherited land is managed by husbands.

At the same time, the Green Revolution and its greater commercialization of crops has generally required an increase in labor per hectare, particularly in tasks typically reserved for women, such as weeding, harvesting, and postharvest work. If women are provided no relief from their other daily tasks, the Green Revolution for them may be more burden than blessing. But when mechanization is added to the new farming system, women tend to be losers. Frequently, such predominantly female tasks as harvesting or dehusking and polishing of grain—all traditionally done by hand—are given over to machinery, displacing rather than employing women. Even the application of chemical fertilizers (a "man's task") instead of cow dung ("women's work") has reduced the female role in agricultural development programs. The loss of those traditional female wage jobs means that already poor rural women and their families have insufficient income to improve their diets even in the light of substantial increases in food availability through Green Revolution improvements.

If women are to benefit from the Green Revolution, new cultural norms—or culturally acceptable accommodations within traditional household, gender, and customary legal relations—will be required. These must permit or recognize women's landowning and other legal rights not now clearly theirs, access to credit at favorable rates, and admission on equal footing with males to government assistance programs. The FAO is working to promote gender-based equity in agricultural development.

---

greatest return. Theoretically, farm products for which demand at a given price increases will command an increased market price. That, in turn, should induce increased production to meet the demand. In some developing countries, that market equilibrium is broken and the farm economy distorted when government policy requires uneconomically low food prices for urban workers. It may also suffer material distortion under governmental programs protecting local producers by inhibiting farm product imports or subsidizing production by guaranteeing prices for selected commodities.

Where free market conditions prevail, however, the crop or the mix of crops and livestock that individual commercial farmers produce is a result of an appraisal of profit possibilities. Farmers must assess and predict prices, evaluate the physical nature of farmland, and factor in the possible weather conditions. The costs of production (seed, fuel, fertilizer, capital equipment, labor, and so on) must be reckoned. A number of unpredictable conditions may thwart farmers' aspirations for profit. Among them are the uncertainties of growing season conditions that follow the original planting decision, the total yield that will be achieved (and therefore the unit cost of production), and the supply and price situation that will exist months or years in the future, when crops are ready for market.

Beginning in the 1950s in the United States, specialist farmers and corporate purchasers developed strategies for minimizing those uncertainties. Processors sought uniformity of product quality and timing of delivery. Vegetable canners—of tomatoes, sweet corn, and the like—required volume delivery of raw products of uniform size, color, and ingredient content on dates that accorded with cannery and labor schedules. And farmers wanted the support of a guaranteed market at an assured price to minimize the uncertainties of their specialization and stabilize the return on their investment.

The solution was contractual arrangements or vertical integration (where production, processing, and sales are all coordinated within one firm) that unite contracted farmer and purchaser-processor. Broiler chickens of specified age and weight, cattle fed to an exact weight, wheat with a minimum protein content, popping corn with prescribed characteristics, potatoes of the kind and quality demanded by particular fast-food

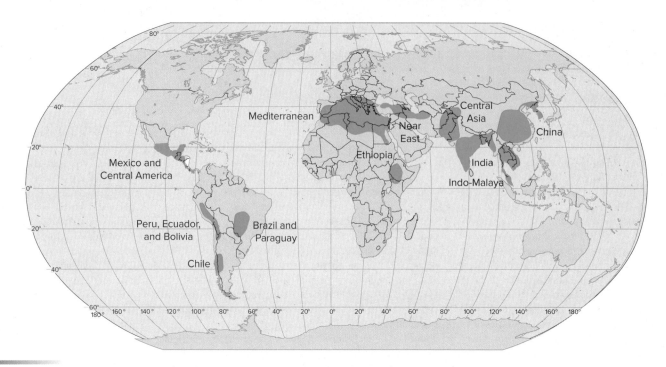

Figure 8.12 Areas with high current genetic diversity of crop varieties. Loss of crop varieties characterizes the commercial agriculture of much of the developed world. In place of the many thousands of species and subspecies (varieties) of food plants grown since the development of agriculture 15,000 or more years ago, fewer than 100 species now provide most of the world's food supply. Most of the diversity loss has occurred in the last 100 years or so. In the United States, for example, 96 percent of commercial vegetable varieties listed by the Department of Agriculture in 1903 are now extinct. Crop breeders, however, require genetic diversity to develop new varieties that are resistant to evolving plant pest and disease perils. That need requires the protection of plant stocks and environments in those temperate and subtropical zones where food plants were first domesticated and are home to the wild relatives of our current food crops. Comparable losses of species diversity are being felt in livestock as well. Half the livestock breeds that existed in Europe in 1900 are already extinct, and almost half the remainder are at risk or endangered.

Sources: J. G. Hawkes, The Diversity of Crop Plants *(Cambridge, MA: Harvard University Press, 1983); and Walter V. Reid and Kenton R. Miller,* Keeping Options Alive: The Scientific Basis for Conserving Biodiversity *(Washington, DC: World Resources Institute, 1989), Figure 5, p. 24.*

chains, and similar product specification became part of production contracts between farmer and buyer-processor. In the United States, the percentage of total farm output produced under contractual arrangements or by vertical integration has risen dramatically. For example, the vast majority of hogs are sold under some form of contract today while in 1980 only 5 percent were sold that way. The term *agribusiness* is applied to the growing merging of the older, farm-centered crop economy and newer patterns of more integrated production and marketing systems.

Contract farming is spreading as well to developing countries, though it is often criticized as another adverse expression of globalization subjecting small farmers to exploitation by powerful Western agribusiness. The FAO, however, argues that well-managed contract arrangements are effective in linking the small farmers of emerging economies with both foreign and local sources of advanced advice, seeds, fertilizers, machinery, and profitable markets at stable prices. The agency cites successful examples of contract farming in northern India, Sri Lanka, Nepal, Indonesia, Thailand, and the Philippines and sees in the arrangements a most promising approach to market-oriented production in areas still dominated by subsistence agriculture.

Even for family farmers not bound by contractual arrangements to suppliers and purchasers, the older assumption that

supply, demand, and the market price mechanism are the effective controls on agricultural production is not wholly valid. In reality, those theoretical controls are joined by a number of nonmarket governmental influences that may be as decisive as market forces in shaping farmers' options and spatial production patterns. If there is a glut of wheat on the market, for example, the price per ton will come down, and the area sown to it should diminish. It will also diminish regardless of supply if governments, responding to economic or political considerations, impose acreage controls.

Distortions of market control may also be introduced to favor certain crops or commodities through subsidies, price supports, market protections, and the like. The political power of farmers in the EU, for example, secured for them generous subsidies. European farm supports are seen as necessary to protect farmers from price volatility and low prices brought about by increasing global yields. In Japan, the home market for rice is largely protected and reserved for Japanese rice farmers, even though their production efficiencies are low and their selling price is high by world market standards. In the United States, programs of farm price supports, ethanol subsidies, acreage controls, financial assistance, and other governmental involvements in agriculture have been of recurring and equally distorting effect (**Figure 8.13**).

Figure 8.13 Ethanol plants have expanded throughout the Corn Belt. Federal subsidies for growers and mandates for blending ethanol into gasoline have bolstered the corn ethanol industry. Critics decry the heavy subsidies and the use of food crops for transportation fuel.
©JKendall/iStock/Getty Images

A Model of Agricultural Location

Early in the 19th century, before such governmental influences were the norm, Johann Heinrich von Thünen (1783–1850) observed that uniformly fertile areas of farmland were used differently. Around each major urban market center, he noted, there developed a set of concentric land-use rings of different farm products (**Figure 8.14**). The ring closest to the market featured intensive agriculture producing heavy, bulky, or perishable commodities that were both expensive to ship and in high demand. The high prices that they could command in the urban market made their production an appropriate use of high-value land near the city. Surrounding rings of farmlands farther away from the city were used for less perishable commodities with lower transport costs, reduced demand, and lower market prices. Less intensive farming such as grain farming replaced the market gardening of the inner ring. At the outer margins of profitable agriculture, farthest from the single central market, livestock ranching and similar extensive land uses were found. After all, transport costs were low for livestock in von Thünen's day because cattle could walk to market.

To explain this pattern of concentric rings of activity, von Thünen constructed a formal spatial model—the **von Thünen model**—perhaps the first one developed to analyze human activity patterns. He concluded that the uses to which parcels were put was a function of the differing "rent" values placed on seemingly identical lands. Those differences, he claimed, reflected the costs transporting farm products to the central market town ("A portion of each crop is eaten by the wheels," he observed). The greater the distance, the higher the cost to the farmer, because transport charges had to be added to other expenses. When a commodity's production costs plus its transport costs just equaled its value at

market, a farmer was at the economic margin of its cultivation. A simple trade-off emerged: the greater the transportation costs, the lower the rent that could be paid for land.

Because in the simplest form of the model, transport costs are the only variable, the relationship between land rent and distance from market can be easily calculated by reference to each competing crop's *transport gradient*. Perishable, bulky, or heavy commodities such as dairy products, fruits, vegetables, and bedding plant and tree nurseries would encounter high transport rates per unit of distance; other items, such as grain, would have lower rates. Land rent for any farm commodity decreases with increasing distance from the central market, and the rate of decline is determined by the transport gradient for that commodity. Crops that have both the highest market price and the highest transport costs will be grown nearest to the market. Less perishable crops with lower production and transport costs will be grown at greater distances from the market (**Figure 8.15**). Because in this model, transport costs are uniform in all directions away from the center, a concentric zonal pattern of land use called the *von Thünen rings* results.

The von Thünen model may be modified by introducing ideas of differential transport costs (**Figure 8.16**), variations in topography or soil fertility, or changes in commodity demand and market price. With or without such modifications, von Thünen's analysis helps explain the changing crop patterns and farm sizes evident on the landscape at increasing distance from major cities, particularly in regions dominantly agricultural in economy. Farmland close to markets takes on high value, is used *intensively* for high-value crops, and is subdivided into relatively small units. Land far from markets is used *extensively* and in larger units.

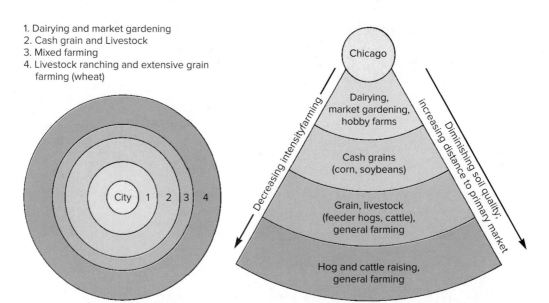

1. Dairying and market gardening
2. Cash grain and Livestock
3. Mixed farming
4. Livestock ranching and extensive grain farming (wheat)

(a)

(b)

Figure 8.14 (*a*) von Thünen's model. Recognizing that as distance from the market increases, the value of land decreases, von Thünen developed a descriptive model of intensity of land use that holds up reasonably well in practice. The most intensively produced crops are found on land close to the market; the less intensively produced commodities are located at more distant points. The zones of the diagram represent modern equivalents of the land-use sequence von Thünen suggested in the 1820s. As the metropolitan area at the center increases in size, the agricultural specialty areas are displaced outward, but the relative position of each is retained. Compare this diagram with Figure 8.17. (*b*) This schematic view applies the model to the agricultural zones in the sector south of Chicago. There, farmland quality decreases southward as the boundary of recent glaciation is passed and hill lands are encountered in southern Illinois. On the margins of the city near the market, dairying competes for space with livestock feeding and suburbanization. Southward into flat, fertile central Illinois, cash grains dominate. In southern Illinois, livestock rearing and fattening and general farming are the rule.

Source: (b) Modified from Bernd Andreae, Farming Development and Space: A World Agricultural Geography, *translated by Howard F. Gregor (Berlin; Hawthorne, N.Y.: Walter de Gruyter and Co., 1981).*

Historian and geographer William Cronon uses the von Thünen model to help explain the frontier stages during the settlement of the American West. Starting in Chicago, the center of the U.S. agricultural economy, a westward traveler would first encounter a zone of intensively farmed dairies, market gardens, and orchards before reaching the corn and wheat cash grain operations of the Iowa prairie. Farther west, the traveler would reach the open range cattle ranches of the Great Plains, then a zone of trapping, hunting, and Indian trade, and beyond that lay the wilderness. Thus, the settlement of the interior of the continent and the closing of the frontier was intimately connected to the growth of cities and the expansion of urban markets for

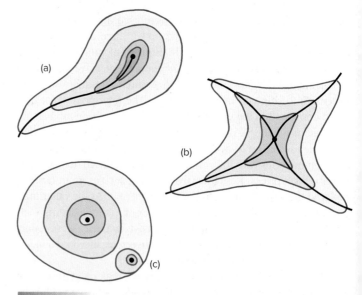

Figure 8.16 Ring modifications. Modifications of model assumptions will alter the details but not the underlying pattern of the *von Thünen rings.* For example, a growth in demand, and therefore market price of a commodity, would merely expand its ring of production. An increase in transport costs would shrink the production area, while reductions in freight rates would extend it. (*a*) If transport costs are reduced in one direction, the shape—but not the sequence—of the rings will be affected. (*b*) If several roads are constructed or improved, land-use sequences assume a star shape. (*c*) The addition of a smaller outlying market results in the emergence of a set of von Thünen rings that is subordinate to it.

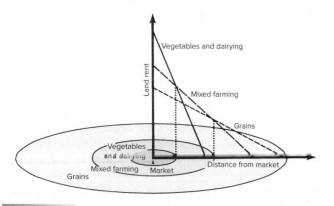

Figure 8.15 Transport gradients and agricultural zones.

produce, grain, and cattle. In industrial and postindustrial economies, the basic forces determining land use near cities are often those associated with urban expansion itself. For example, peripheral city growth, hobby farms, you-pick agritourism farms, and the withholding of land from farming in anticipation of future suburban development may locally alter the von Thünen rings. Nonetheless, we can affirm the validity of von Thünen's fundamental insights that the city and countryside are intimately connected and that the distance from the city shapes rural land values and land uses.

Intensive Commercial Agriculture

Following World War II, agriculture in the developed world's market economies turned increasingly to concentrated methods of large-scale production. Machinery, chemicals, irrigation, and dependence on a restricted range of carefully selected and bred plant varieties and animal breeds all were employed in a concerted effort to maximize efficiency and productivity on each unit of farmland.

The goal, of course, was to increase off-farm sales as American agriculture increasingly shifted from an objective of partial self-sufficiency to a total commitment to the commercial, exchange economy. Prior to 1950, most U.S. farms had a significant subsistence orientation; they were *general farms* growing a variety of crops, some for sale and some for feed for farmstead livestock—a milk cow or two, chickens for the pot and for household eggs, and a few hogs and steers, partly for farm slaughter and use. Their extensive kitchen gardens supplied vegetables and fruits for farm family seasonal consumption and home canning. In 1949, the average American farm sold only $4,100 worth of products. By 2012, however, most farms had a full commitment to the market, average off-farm sales rose to $187,000, and farm families—like other Americans—shopped at supermarkets for their food needs. With the increases in capital investment and the need for larger farms to maximize return on that investment, many inefficient small farms have been abandoned. Consolidation has reduced the number and enlarged the size of farms still in production. To stay in business, many operations have had to expand by factors of 10 or 100, whether measured in crop acreage of number of livestock. From a high of 6.8 million in 1934, the number of U.S. farms dropped to 5.7 million in 1949 and to 2.1 million by 2012, with many of the smallest units counting as "farms" only because of a generous Department of Agriculture definition.

The reorientation of farm production goals in the United States and in most other highly developed market economies has led to significant changes in regional farm production patterns. Reflecting the drive for enhanced, more specialized output and the investment of large amounts of capital (for machinery, fertilizers, and specialized buildings, for example), all modern agriculture is *intensive*. But the several types of farm specializations differ in how much capital is invested per hectare of farmed land (and, of course, in the specifics of those capital inputs). Those differences underlie generalized distinctions between traditional intensive and extensive commercial agriculture.

The term **intensive commercial agriculture** is now usually understood to refer specifically to the production of crops that give high yields and high market value per unit of land. These include fruits, vegetables, bedding plants for urban gardens, and dairy products, all of which are highly perishable, as well as some "factory farm" production of livestock. Dairy farms and **truck farms** (horticultural or "market garden" farms that produce a wide range of vegetables and fruits) are found near most medium-size and large cities. Because the product is perishable, transport costs increase because of the special handling that is needed, such as the use of refrigerated trucks and custom packaging. Note the distribution of truck and fruit farming in **Figure 8.17**, which also suggests the importance of climatic conditions in commercial fruit and vegetable growing.

Feed grain and livestock farming involves the growing of grain on a producing farm to be fed to livestock, which constitute the farm's cash product. In Western Europe, three-fourths of cropland is devoted to production for animal consumption; in Denmark, 90 percent of all grains are fed to livestock for conversion not only into meat but also into butter, cheese, and milk. Although livestock-grain farmers work their land intensively, the value of their product per unit of land is usually less than that of the truck farm. Consequently, in North America at least, feed grains and livestock farms are centered in the Midwest, farther from the main markets than are horticultural and dairy farms.

Normally, the profits for marketing livestock (chiefly hogs and beef cattle in the United States) are greater per pound than those for selling corn or other feed, such as alfalfa and clover. As a result, farmers convert their corn into meat on the farm by feeding it to the livestock, efficiently avoiding the cost of buying grain. They may also convert farm grain at local feed mills to the more balanced feed modern livestock rearing requires. Where land is too expensive to be used to grow feed, especially near cities, feed must be shipped to the farm. The grain-livestock belts of the world are close to the great coastal and industrial zone markets. The Corn Belt of the United States and the livestock region of Western Europe are two examples.

In the United States—and commonly in all developed countries—the traditional feed grains and livestock operations of small and family farms have been largely supplanted by very-large-scale concentrated animal feeding operations or *livestock factory farms* involving thousands or tens of thousands of closely quartered animals. From its inception in the 1920s, the intensive, industrialized rearing of livestock, particularly beef and dairy cattle, hogs, and poultry, has grown to dominate meat, dairy, and egg production. To achieve their objective of producing a marketable product in volume at the lowest possible unit cost, operators of livestock factory farms confine animals to pens or cages, treat them with antibiotics and vitamins to maintain health and speed growth, provide processed feeds that often contain the low-cost animal by-products or crop residue, and deliver them under contract to processors, packers, or their parent company (**Figure 8.18**). Although serious concerns have been voiced about animal waste management and groundwater, stream, and atmospheric pollution, contract-based concentrated

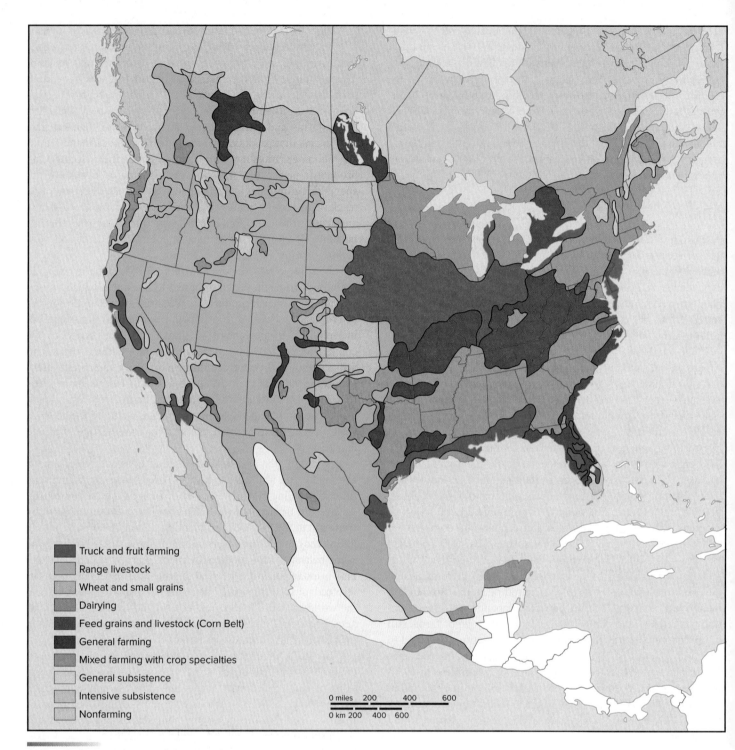

Figure 8.17 Generalized agricultural regions of North America.

Sources: U.S. Bureau of Agricultural Economics; Agriculture Canada; and Secretaría de Agricultura y Recursos Hidráulicos, Mexico.

Legend:
- Truck and fruit farming
- Range livestock
- Wheat and small grains
- Dairying
- Feed grains and livestock (Corn Belt)
- General farming
- Mixed farming with crop specialties
- General subsistence
- Intensive subsistence
- Nonfarming

0 miles 200 400 600

0 km 200 400 600

feeding operations now provide almost all supermarket meat and dairy products. The location of this form of intensive commercial farming, however, is often determined not by land value or proximity to market, but by land-use restrictions and environmental standards imposed by state and county governments.

Extensive Commercial Agriculture

Farther from the market, on less expensive land, there is less need to use the land intensively. Cheaper land and lower profits per unit of land leads to larger farm units. **Extensive commercial agriculture** is typified by large wheat farms and livestock ranching.

There are, of course, limits to the land-use explanations attributable to von Thünen's model. Although it is true that farmland values decline westward with increasing distance from the northeastern market of the United States, they show no corresponding increase with increasing proximity to the massive West Coast market region until the specialty agricultural areas of the coastal states themselves are reached. The western states are characterized by extensive agriculture, but as a consequence of environmental, not distance, considerations. Climatic conditions obviously affect the productivity and the potential agricultural use of an area, as do soils and topography. In North America, of course, increasing distance westward from eastern markets happens to be associated with increasing aridity and the beginning of mountainous terrain. In general, rough terrain and dry climates, rather than simple distance from market, underlie the widespread occurrence of extensive agriculture.

Large-scale wheat farming requires sizable capital inputs for planting and harvesting machinery—a large tractor might cost $300,000 and a combine (harvester) $500,000. However, the inputs per unit of land are low; and wheat farms are very large. Nearly half the farms in Saskatchewan, for example, are more than 400 hectares (1,000 acres). The average farm in Kansas is larger than 300 hectares (740 acres), and in North Dakota, more than 525 hectares (1,300 acres). In North America, the spring wheat (planted in spring, harvested in autumn) region includes the Dakotas, eastern Montana, and the southern parts of the Prairie provinces of Canada. The winter wheat (planted in fall, harvested in midsummer) belt centers on Kansas and includes adjacent sections of neighboring states (**Figure 8.19**). Argentina is the only South American country to have comparable large-scale wheat farming. In the Eastern Hemisphere, the system is fully developed only east of the Volga River in northern Kazakhstan and the southern part of western Siberia, and in southeastern

(a)

(c)

(b)

Figure 8.18 Industrial poultry and livestock farming is an example of intensive commercial agriculture. (*a*) Most of the chicken consumed in developed countries comes from factory-style poultry farms such as this. Thousands of broiler chickens are raised together in a single barn for about 45 days before slaughtering. (*b*) On this hog farm in Georgia, 900 animals are fed and raised indoors in large rectangular barns for the four or five months that it takes them to grow to market weight. Animal manure is collected in lagoons. (*c*) Large cattle feedlot operations supply much of the country's beef supply.

(a) ©Digital Vision./Photodisc/Getty Images; (b) Source: Jeff Vanuga, USDA Natural Resources Conservation Service; (c) ©Cathryn Dowd

Figure 8.19 Contract harvesters follow the ripening wheat northward through the plains of the United States and Canada.
©*Glow Images*

and western Australia. Because wheat is an important crop in many agricultural systems—today, wheat ranks first in total production among all the world's grains and accounts for more than 20 percent of the total calories consumed by humans collectively—it is a truly global commodity, bought and sold internationally (**Figure 8.20**).

Livestock ranching differs significantly from feed grains and livestock farming and, by its commercial orientation, from the nomadism it superficially resembles. A product of the 19th-century growth of urban markets for beef and wool in Western Europe and the northeastern United States, ranching has been primarily confined to areas of European settlement. It is found in the western

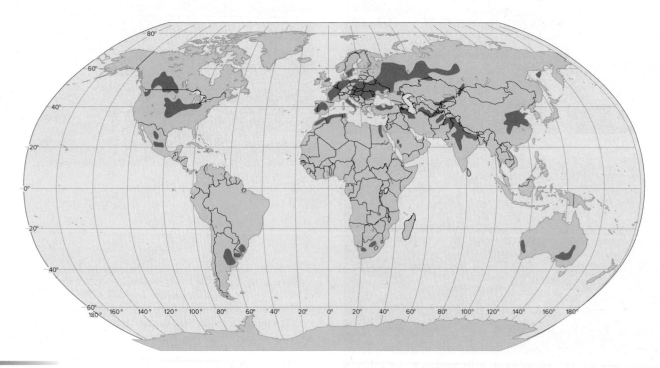

Figure 8.20 Principal wheat-growing areas. Only part of the world's wheat production comes from large-scale farming enterprises. In western and southern Europe, eastern and southern Asia, and North Africa, wheat growing is part of general or intensive subsistence farming. Recently, developing country successes with the Green Revolution and subsidized surpluses of the grain in Europe have altered traditional patterns of production and world trade in wheat.

United States and adjacent sections of Mexico and Canada (see the range livestock region on Figure 8.17); the grasslands of Argentina, Brazil, Uruguay, and Venezuela; the interior of Australia; the uplands of South Island, New Zealand; and the Karoo and adjacent areas of South Africa (**Figure 8.21**). All except New Zealand and the humid pampas of South America have semiarid climates. All these areas, even the most remote from markets, were a product of improvements in transportation by land and sea, refrigerated train cars and trucks, and meat-canning technology.

In all of the ranching regions, livestock range has been reduced as crop farming has encroached on their more humid margins, and as feedlots have supplemented traditional grazing. Recently, the growing global demand for beef has been blamed for expanded cattle ranching and extensive destruction of tropical rain forests in Central America and the Amazon Basin, although in recent years, Amazon Basin deforestation has reflected the expansion of soybean farming more than beef production.

In areas of livestock ranching, young cattle or sheep are allowed to graze over thousands of acres. In the United States, when the cattle have gained enough weight so that weight loss in shipping will not be a problem, they are sent to feed grain livestock farms or to feedlots near slaughterhouses for accelerated fattening. Because ranching can be an economic activity only where alternative land uses are nonexistent and land quality is low, ranching regions of the world characteristically have low population densities, low capital investment per land unit, and relatively low labor requirements.

Special Crops

Under special circumstances, usually related to unique physical geography, some places far from markets may become intensively developed agricultural areas. Two special cases are Mediterranean agriculture and plantation agriculture (Figure 8.21).

Most of the arable land in the Mediterranean basin itself is planted to grains, and much of the agricultural area is used for grazing. **Mediterranean agriculture** as a specialized farming economy, however, is known for grapes, olives, oranges, figs, vegetables, and similar commodities. These crops need warm temperatures year round and a great deal of sunshine in the summer. The Mediterranean agricultural lands indicated in Figure 8.21 are among the most productive in the world. Farmers benefit from a predictable climate with few storms or severe weather problems. Also, the precipitation pattern of Mediterranean climates—winter rains and dry summers—lends itself to the controlled use of water. Of course, much capital must be spent for the irrigation systems. This is another reason for the intensive use of the land for high-value crops that are, for the most part, destined for export to industrialized countries or areas outside the Mediterranean climatic zone and even, in the case of Southern Hemisphere locations, to markets north of the equator.

Climate is also considered the vital element in the production of what are commonly known as *plantation crops*. **Plantation agriculture** involves the introduction of foreign investment, management, and marketing into an indigenous culture and economy, often employing a nonnative labor force to produce an introduced

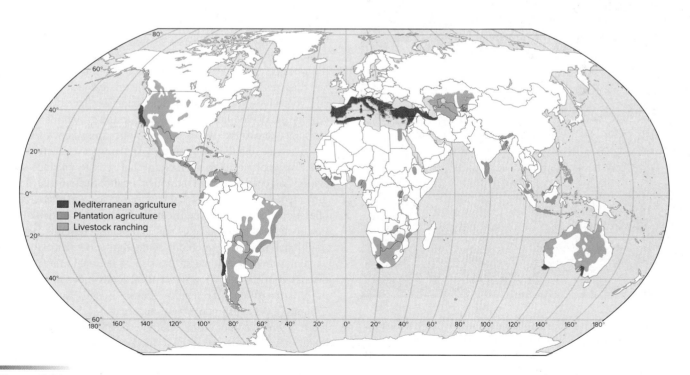

Figure 8.21 Livestock ranching and special crop agriculture. Livestock ranching is primarily a midlatitude enterprise catering to the urban markets of industrialized countries. Mediterranean and plantation agriculture are similarly oriented to the markets provided by advanced economies of Western Europe and North America. Areas of Mediterranean agriculture—all of roughly comparable climatic conditions—specialize in similar commodities, such as grapes, oranges, olives, peaches, and vegetables. The specialized crops of plantation agriculture are influenced by both physical geographic conditions and present or, particularly, former colonial control of the area.

crop for foreign markets. The plantation itself is an estate whose resident workers produce one or two specialized export crops. Those crops, although native to the tropics, were frequently foreign to the areas of plantation establishment: African coffee and Asian sugar in the Western Hemisphere and American cacao, tobacco, and rubber in Southeast Asia and Africa are examples (**Figure 8.22**). Plantation developers from Western countries such as Britain, France, the Netherlands, and the United States became interested in the tropics partly because the climate allowed them to produce agricultural commodities that could not be grown at home. Custom and convenience usually retain the term *plantation* even where native producers of local crops dominate, as they do in cola nut production in Guinea, spice growing in India or Sri Lanka, or sisal production in the Yucatán.

Figure 8.22 Coffee is a classic plantation crop, with operations typically established by foreign investors in locations offering a suitable physical environment (climate and soils) and producing an introduced crop for export to distant foreign market. While coffee was first domesticated in Africa, it is grown in many tropical countries, mostly for export to developed countries in the midlatitudes. Note that the female laborers on this Kenyan coffee plantation are still responsible for child care even as they work in the fields.

©Christopher Pillitz/The Image Bank/Getty Images

The major plantation crops and the areas where they are produced include tea (India and Sri Lanka), jute (India and Bangladesh), rubber (Malaysia and Indonesia), cacao (Ghana and Nigeria), cane sugar (Cuba and the Caribbean area, Brazil, Mexico, India, and the Philippines), coffee (Brazil and Colombia), and bananas (Central America). As Figure 8.21 suggests, for ease of access to shipping, most plantation crops are cultivated along or near coasts because production for export rather than for local consumption is the rule.

Sustainable Agriculture

The adoption of large-scale, highly industrialized commercial agriculture has increased overall agricultural productivity and benefited successful farmers and agribusinesses. But it has not come without significant costs. The negative effects of industrialized agriculture on the health of rural communities, ecosystems, and food systems have convinced many of the need for a more sustainable mode of agriculture. As farms have grown larger and replaced human labor with machinery, the population involved in farming has dramatically declined. This has led to depopulation of rural areas and struggles to maintain the institutions and basic services necessary for a high quality of life. Industrialized agriculture relies upon heavy inputs of fertilizers, pesticides, and herbicides, each of which has had a negative effect on rural populations, wildlife, surface waterways, and coastal systems that receive agricultural runoff (for more, see Chapter 13). Industrialized agriculture relies on large quantities of fossil fuels to fuel the machinery, manufacture petrochemical-based fertilizers, and distribute the food around the world. Heavy reliance on nonrenewable fossil fuels to produce food is obviously unsustainable over the longer term. Finally, there are serious concerns about the quality of the food supply and its relationship to human health, the obesity epidemic, and diet-related diseases such as cancer and diabetes. Concerns range from the relative lack of fresh fruits and vegetables in the diets of rich countries, the safety of foods with pesticide and hormone residues, to the growth of antibiotic resistance due to the overuse of antibiotics in livestock feed.

The sustainable agriculture movement is a collection of alternative approaches to agriculture that seek to enhance social, ecological, economic, and individual health. Sustainable agriculture advocates emphasize local knowledge, local markets, smaller operations, and farm diversification. Sustainable agriculture relies upon local knowledge of soil and plant conditions, and uses traditional agriculture methods adapted to particular places. Just as neolocalism is evident in the rise of regional craft beers (as discussed in Chapter 7), a wider local foods movement is growing in strength (see the feature "Eating Locally on the College Campus"). As consumers show concern about the health of their food, where it comes from, and how far it has traveled to their table, they've turned to organic foods, farmer's markets, urban gardening, and community-supported agriculture. Where industrial agriculture creates monocultures and specialized producers, sustainable agriculture advocates believe in farm diversification and biodiversity. They argue that a region with many smaller producers growing many different crops and raising different livestock is more resilient and creates a healthier local economy and

Eating Locally on the College Campus

Stereotypes associate college life with a diet of fast food hamburgers and greasy pizza. However, many college students are starting to pay close attention to their food choices. They are asking where their food comes from and how it affects human health, animal welfare, and the environment. Some are even getting involved in producing their own food in campus gardens.

Several Midwestern colleges and universities have made commitments to growing more of their own food. A couple of generations ago, many of their students came from farms and had experience with fieldwork, milking cows, gathering eggs, butchering chickens, gardening, and canning. Back then, farms were small, nonspecialized general farming operations. Today, cash grain farmers plant thousands of hectares of just two crops—corn and soybeans. Livestock farmers raise thousands of hogs, turkeys, or chickens and dairies may house thousands of milk cows. Although the Midwest is among the most productive of agricultural landscapes, very little is produced for direct local consumption. Meat and milk are shipped to distant markets, soybeans are refined for industrial processes, and corn goes to animal feed, to Asian or European markets, or to produce ethanol fuel. And no longer do Midwest colleges and universities get their students from farms or supply their cafeterias from local farms. Instead, the students come from cities and suburbs, and cafeteria food is delivered by large refrigerated trucks.

At Calvin University in Grand Rapids, Michigan, the dining services has altered menus to incorporate the abundant greens produced by the campus farm. Students also collect and process hundreds of bottles of maple syrup from campus trees as well as collecting honey from bee hives located in the campus farm.

At Gustavus Adolphus College in Saint Peter, Minnesota, the Big Hill Farm grew out of an undergraduate senior project and is an important piece in the college's commitment to environmental sustainability. The college supports the farm with land, tillage equipment, and paid summer student internships. A grant paid for a greenhouse, which extends the short Minnesota growing season. The farm and the student dining service have a mutually beneficial relationship. Food waste from the dining service is composted and used as fertilizer and soil amendment on the farm, while the farm produces lettuce, tomatoes, peppers, beans, onions, melons, pumpkins, berries, and much more for the cafeteria. Where possible, the farm uses rare heirloom seed varieties to promote crop diversity. Extra produce is sold at the local farmer's market. Students are now eating healthier, locally grown food, and the college is reducing its impact on the environment. Students who work on the farm learn lifetime skills and describe their labors as both exhausting and deeply rewarding.

Figure 8A Student workers at the Calvin University campus farm.
Courtesy of Alicia De Jong

environment. In many ways, sustainable agriculture is a return to methods of agriculture that were widely used prior to World War II, Thus, critics question whether sustainable agriculture will be able to maintain the productivity gains of industrial agriculture and feed the world's growing population.

8.3 Primary Activities: Resource Exploitation

In addition to agriculture, primary economic activities include fishing, forestry, and the mining and quarrying of minerals. These industries involve the direct exploitation of natural resources that are unequally distributed in the environment.

Fishing, forestry, and fur trapping are **gathering industries** based on harvesting the natural bounty of renewable resources that can easily be depleted through overexploitation. Livelihoods based on these resources are areally widespread and involve both subsistence and market-oriented components. Mining and quarrying are **extractive industries,** removing nonrenewable metallic and nonmetallic minerals, including the mineral fuels, from the Earth's crust. They are the initial raw material phase of modern industrial economies.

Resource Terminology

Resources or **natural resources** are the naturally occurring materials that a society perceives to be useful to its economic and material well-being. Their occurrence and spatial distribution are the result of physical processes over which people have little or no direct control. The fact that things exist, however, does not mean that they are resources. To be considered such, a given substance must be *understood* to be a resource—and this is a cultural, not purely a physical, circumstance. Native Americans may have viewed the resource base of Pennsylvania, West Virginia, or Kentucky as composed of forests for shelter and fuel and as the habitat of the game animals (another resource) on which they depended for food. European settlers viewed the forests as the unwanted covering of the resource that *they* perceived to be of value: soil for agriculture. Still later, industrialists appraised the underlying coal deposits, ignored or unrecognized as a resource by earlier occupants, as the item of value for exploitation (**Figure 8.23**).

Resources may be classified as *renewable* or *nonrenewable*. **Renewable resources** are materials or energy sources that are replenished by natural processes. The sun's energy, wind, water, food crops, soils, forests, fish, and animals are renewable resources. Even renewable resources can be exhausted if exploited to extinction or destruction. Soil can be eroded or its fertility destroyed, and an animal species may be driven to extinction. That is, some resources are renewable only if carefully managed. The **maximum sustainable yield** of a resource is the maximum rate of use that will not impair its ability to be renewed or to maintain the same future productivity. For fishing and forestry, for example, that level is marked by a catch or harvest equal to the net growth of the replacement stock. If that maximum exploitation level is exceeded, the renewable resource becomes a nonrenewable one—an outcome increasingly likely in the case of Atlantic cod and some other food fish species. **Nonrenewable resources** exist in finite amounts or are generated in nature so slowly that for all practical purposes, their supply is finite.

Figure 8.23 Resources are defined by a culture's values and perceptions. The indigenous population treated this forested West Virginia landscape as a prime hunting area. The original hardwood forest covering these hills were removed by European American settlers who saw greater resource value in the underlying soils. The soils, in turn, were selectively stripped away for access to the still more valuable coal deposits below. Future generations may value this landscape more for its beauty and recreational opportunities.

©Jon Bilous/Shutterstock

Fishing

Although fish and shellfish account for just 17 percent of all human consumption of animal protein, an estimated 1 billion people—primarily in developing countries of eastern and southeastern Asia, Africa, and parts of Latin America—depend on fish as their primary source of protein. Fish are also very important in the diets of most advanced states, both those with and those without major domestic fishing fleets. Globally, the average person consumes 20 kg (44 pounds) of fish per year. Although most of the world annual fish harvest is consumed by humans, up to one-fifth is processed into fish meal to be fed to livestock or used as fertilizer. Those two quite different markets have increased both the demand for and the annual harvest of fish. Indeed, so rapidly have demand pressures on the world's fish stocks expanded that evidence is unmistakable that at least locally, their *maximum sustainable yield* is being exceeded.

The annual fish supply comes from three sources:

- The *inland catch,* from ponds, lakes, and rivers
- *Fish farming (or aquaculture),* in which fish are produced in a controlled and contained environment
- The *marine catch,* all wild fish harvested in coastal waters or on the high seas

The inland catch supplies a modest fraction of the global fish catch, while fish farming continues to grow, producing about half of the world's fish harvest. While the world's ocean catch is still the most important source of fish, its annual harvest has been stable or declining since the late 1980s (**Figure 8.24**). Most of the marine catch is made in coastal wetlands, estuaries, and the relatively shallow coastal waters above the *continental shelf*—the gently sloping extension of submerged land bordering most coastlines and reaching seaward for varying distances up to 150 kilometers (about 100 miles) or more. Near shore, shallow embayments and marshes provide spawning grounds, and river waters supply nutrients to an environment highly productive of fish. Increasingly, these areas are also seriously affected by pollution from runoff and ocean dumping, an environmental assault so devastating in some areas that fish and shellfish stocks have been destroyed, with little hope of revival.

Commercial marine fishing is largely concentrated in northern waters, where warm and cold currents join and mix and where such familiar food species as herring, cod, mackerel, haddock, and flounder congregate or *school* on the broad continental shelves and *banks*—extensive elevated portions of the shelf where environmental conditions are most favorable for fish production (**Figure 8.25**). Two of the most heavily fished regions are the Northeast Pacific and Northwest Atlantic. Tropical fish species tend not to school and, because of their high oil content and unfamiliarity, are less acceptable in the commercial market. They are, however, of great importance for local consumption. Only a very small percentage of total marine catch comes from the open seas that make up more than 90 percent of the world's oceans.

Modern technology and more aggressive fishing fleets from more countries greatly increased annual marine capture in the years after 1950. That technology included use of sonar, radar, helicopters, and satellite communications to locate schools of fish;

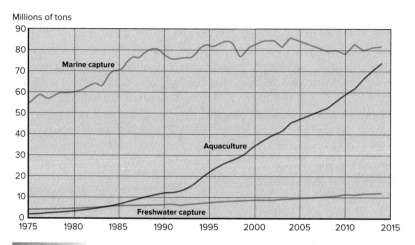

Figure 8.24 Officially recorded annual fish harvests, 1975–2014. The marine catch shows evidence of overfishing, while aquaculture continues its rapid growth. The FAO estimates that 20 to 40 million tons per year of unintended marine capture of juvenile or undersized fish and nontarget species are discarded each year.

Source: Food and Agriculture Organization (FAO).

more efficient nets and tackle; and factory trawlers to follow fishing fleets to prepare and freeze the catch. In addition, more nations granted ever-larger subsidies to expand and reward their marine trawler operations. The rapid rate of increase led to inflated projections of the continuing or growing productivity of fisheries and to optimism that the resources of the oceans were inexhaustible.

Quite the opposite has proved to be true, however. In fact, in recent years, the productivity of marine fisheries has been stagnant because *overfishing* (catches above reproduction rates) and

pollution of coastal waters have seriously endangered the supplies of traditional and desirable food species. The United Nations reports that about one-third of wild fish stocks are being overfished. The plundering of U.S. and Canadian coastal waters has imperiled a number of the most desirable fish species; in 1993, Canada shut down its cod industry to allow stocks to recover, and U.S. authorities report that 67 North American species are overfished and 61 harvested to capacity.

Overfishing is partly the result of the accepted view that the world's oceans are common property, a resource open to anyone's use with no one responsible for its maintenance, protection, or improvement. The result of this "open seas" principle is but one expression of the so-called **tragedy of the commons**[1]—the economic reality that when a resource is available to all, each user, in the absence of collective controls, thinks he or she is best served by exploiting the resource to the maximum even though this exploitation means its eventual depletion. In 1995, more than 100 countries adopted a treaty—that became legally binding in 2001—to regulate fishing on the open oceans outside territorial waters. Applying to such species as cod, pollock, and tuna—that is, to migratory and high-seas species—the treaty requires fishermen to report the size of their catches to regional organizations that would set quotas and subject vessels to boarding to check for violations. These and other fishing control

---

[1]The *commons* refers to undivided land available for the use of everyone; usually, it meant the open land of a village that all used as pasture. The Boston Common, for instance, originally had this function.

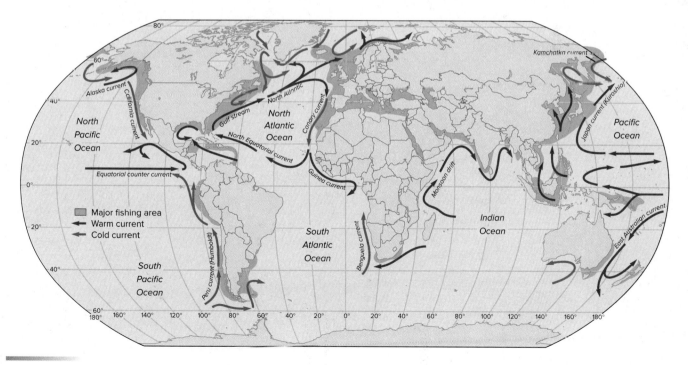

Figure 8.25 The major commercial marine fisheries of the world. The waters within 325 kilometers (200 miles) of the U.S. coastline account for almost one-fifth of the world's annual fish and shellfish harvests. Overfishing, urban development, hydroelectric dams that block access to spawning grounds, and the contamination of bays, estuaries, and wetlands have contributed to the depletion of the fish stocks in those coastal waters.

measures could provide the framework for future sustainability in the fishing industry, although they appear to be too late to save the cod fishery on the Atlantic Coast of Canada. There, the collapse of large fish stocks in the 1990s and the virtual disappearance of cod, haddock, flounder, and hake has induced ecological changes that may prevent the fishery from ever recovering. The collapse of the Atlantic cod fishery caused the loss of 40,000 jobs and immense economic hardship in Canada's Maritime provinces.

One approach to increasing the fish supply is through fish farming or **aquaculture,** the breeding of fish in freshwater ponds, lakes, and canals or in fenced-off coastal bays and estuaries or enclosures (**Figure 8.26**). Aquaculture production has provided nearly half of the total fish harvest in recent years; its contribution to the human food supply is even greater than raw production figures suggest. Whereas one-third of the conventional fish catch is used to make fish meal and fish oil, virtually all farmed fish are used as human food. Fish farming has long been practiced in Asia, where fish are a major source of protein, but now takes place on every continent. Marine aquaculture can create serious problems, including water pollution from fish wastes, transfer of disease to wild fish, and genetic damage to wild fish from escaped alien or genetically altered farmed fish. Despite concerns about its potential negative consequences, aquaculture is the fastest-growing sector of the world food economy.

Forestry

Before the rise of agriculture, the world's forests and woodlands probably covered some 45 percent of the Earth's land area, not counting Antarctica and Greenland. They were a sheltered and productive environment for earlier societies that subsisted on gathered fruits, nuts, berries, leaves, roots, and fibers collected

Figure 8.26 Production of fish, scallops, oysters and other seafood through aquaculture is one of the fastest-growing sectors in world food production. Floating aquaculture platforms, such as these in Japan, are often operated near coastlines where they can contribute to water pollution.

©GYRO PHOTOGRAPHY/amanaimagesRF/Getty Images

from trees and woody plants. Few such cultures still exist, although the gathering of forest products remain an important supplemental activity, particularly among subsistence agricultural societies.

Even after millennia of land clearance for agriculture and, more recently, commercial lumbering, cattle ranching, and fuel-wood gathering, forests still cover roughly 30 percent of the world's land area, not counting Greenland and Antarctica. As an industrial raw material source, however, forests are more restricted in area. Although forests of some type reach discontinuously from the equator northward to beyond the Arctic Circle and southward to the tips of the southern continents, *commercial forests* are restricted to two very large global belts. One, nearly continuous, is found in upper-middle latitudes of the Northern Hemisphere; the other is located in the equatorial zones of South and Central America, Central Africa, and Southeast Asia (**Figure 8.27**). These forest belts differ in the types of trees they contain and in the type of market and uses that they serve.

The northern coniferous, or softwood, forest is the largest and most continuous stand, extending around the globe from Scandinavia across Siberia to North America, then eastward to the Atlantic and southward along the Pacific Coast. The pine, spruce, fir, and other conifers are used for construction lumber and to produce pulp for paper, rayon, and other cellulose products. On the south side of the northern midlatitude forest region are the deciduous hardwoods: oak, hickory, maple, birch, and the like. These, as well as the trees of the mixed forest lying between the hardwood and softwood belts, have been greatly reduced in areal extent by centuries of agricultural and urban settlement and development. In both Europe and North America, however, their area has been held constant through conservation, protection,

and reforestation. They still are commercially important for hardwood applications: furniture, veneers, railroad ties, and the like. The tropical lowland hardwood forests are exploited primarily for fuelwood and charcoal, although an increasing quantity of special-quality woods are cut for export as lumber. In fact, developing countries in the tropics account for most of the world's hardwood log exports (**Figure 8.28**).

The uses of forests differ significantly between developed and developing countries. In developed countries, trees are primarily cut for various wood product industries, including paper, packaging, personal care products (toilet paper), construction, and furniture. The global leaders in producing forest products are the United States, Canada, China, and Russia. Chiefly because of their distance from major industrial wood markets, the developing countries other than China have seen only limited of industrial wood production. The logic of von Thünen's analysis of transportation costs and market accessibility helps explain the pattern. In developing countries, the primary uses of forests are fuelwood and charcoal. Growing populations that depend upon fuelwood and charcoal for energy place pressure on tropical forest stands. For decades, fuelwood gathering was blamed for tropical deforestation. However, recent evidence suggests that fuelwood gathering is not a major source of forest loss and the real problem is forest conversion to agriculture and forest fires, which may be increasing due to climate change. Rates of deforestation are highest in Africa, Asia, and South America, while Europe and North America have seen an increase in forested area. Deforestation in Brazil's Amazon basin has been well-publicized, but policies to protect the rainforest have dramatically reduced the rates of forest loss since 2000.

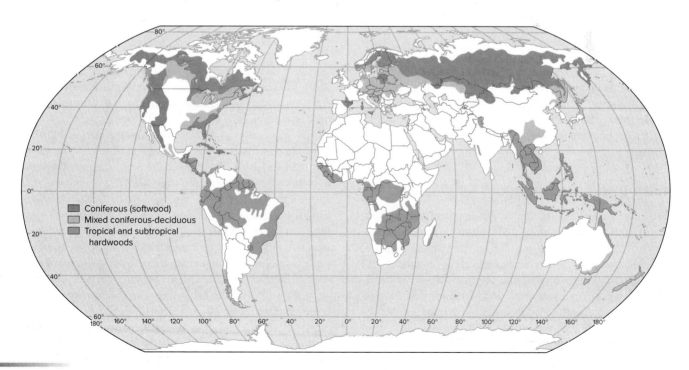

Figure 8.27 Major commercial forest regions. Much of the original forest, particularly in midlatitude regions, has been cut over. Many treed landscapes that remain do not contain commercial stands. Significant portions of the northern forest are not readily accessible and at current prices cannot be considered commercial. Deforestation of tropical hardwood stands involves more clearing for agriculture and firewood than for roundwood production.

Figure 8.28 Loading tropical hardwood lumber for export from Congo.
©Seth Lazar/Alamy Stock Photo

Forests have many noneconomic values, including wildlife habitat, soil conservation, preservation of biological diversity, and carbon storage, which helps slow climate change. Thus, forest loss has serious implications not only economically, but also ecologically.

Mining and Quarrying

Societies at all stages of economic development can and do engage in agriculture, fishing, and forestry. The extractive industries—mining and drilling for nonrenewable mineral wealth—emerged only when technological development and economic necessity turned coal, iron ore, and various minerals into valuable resources. Now those industries provide the raw material and energy for the world's advanced economies. Metals such as iron, copper, and rare earths are the basis for industrial products ranging from automobiles to smart phones. Nonmetallic minerals such as gravel and building stone are widely used in construction, and fossil fuels have provided the energy riches that have made possible the high standards of living in developing countries. The geographic distribution of mineral resources and fossil fuels is highly uneven, and thus, these raw materials are a major part of the international trade connecting the developed and developing countries of the world.

In physically workable and economically usable deposits, minerals constitute only a tiny fraction of the Earth's crust—far less than 1 percent. The Industrial Revolution took place in locations with rich and accessible deposits of the requisite materials such as the coal and iron ore deposits of South Wales and the English Midlands. Economies grew fat by skimming the cream. It has been suggested that should some catastrophe occur to return human cultural levels to a preagricultural state, it would be extremely unlikely that humankind ever again could move along the road of industrialization with the resources available for its use.

Our successes in exploiting mineral resources have been achieved, that is, at the expense of depleting the most easily extractable world reserves and with the penalty of increasing monetary costs as the highest-grade deposits are removed. Costs increase as more advanced and expensive technologies must be applied to extract the desired materials from ever-greater depths in the Earth's crust or from new deposits of lower mineral content. While the Earth's nonrenewable resources are indeed finite, the exact quantity available depends upon both available technology and prices. In fact, as a consequence of advances in exploration and extraction technologies, known reserves of all fossil fuels and of most commercially important metals are now larger than they were in the middle of the 20th century. **Usable reserves,** also known as proved reserves, are those deposits that can be recovered with reasonable certainty, assuming existing economic and operating conditions (**Figure 8.29**). Usable reserves are not the same as the ultimate crustal limit of a resource. For example, between 1987 and 2010, proved oil reserves increased from 0.91 trillion barrels of oil to 1.38 trillion barrels, despite heavy petroleum consumption—reflecting continued exploration, technological improvements, and price changes. That increasing abundance of at least nonfuel resources is reflected in the steady decrease in raw material prices since the 1950s that has so adversely affected some export-oriented developing world economies.

Metals

Because usable mineral deposits are the result of geological processes, it follows that the larger the country, the more probable it is that such past processes will have occurred within their national territory. And in fact, Russia, Canada, China, the United States, Brazil, and Australia possess abundant and diverse mineral resources. It is also true, however, that many smaller, developing countries are major sources of one or more critical raw materials and therefore become important participants in the growing international trade in minerals.

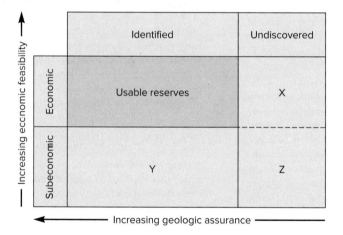

Figure 8.29 The variable definition of reserves. Assume that the large rectangle includes the total world stock of a particular resource. Some deposits of that resource have been discovered and are shown in the left column as "identified." Deposits not yet known are "undiscovered reserves." Deposits that are economically recoverable with current technology are at the top of the diagram. Those below, labeled "subeconomic" reserves, are not attractive for any of several reasons of mineral content, accessibility, cost of extraction, and so on. Only the pink area can be properly referred to as usable reserves. These are deposits that have been identified and can be recovered at current prices and with current technology. *X* denotes reserves that would be attractive economically but are not yet discovered; identified but not economically attractive reserves are labeled *Y;* and *Z* represents undiscovered deposits that would not be attractive now even if they were known.

Source: U.S. Geological Survey.

The production of most metallic minerals, such as copper, lead, and iron ore, is affected by a balance of three forces: the quantity available, the richness of the ore, and the distance to market. A fourth factor, land acquisition and royalty costs, may equal or exceed other considerations in mine development decisions (see the feature "Public Land, Private Profit"). Even if these conditions are favorable, mines may not be developed or even remain operating if supplies from competing sources are more cheaply available in the market. In the 1980s, more than 25 million tons of iron ore–producing capacity was permanently shut down in the United States and Canada. Similar declines occurred in North American copper, nickel, zinc, lead, and molybdenum mining as market prices fell below domestic production costs. Beginning in the early 1990s, as a result of both resource depletion and low cost imports, the United States became a net importer of nonfuel minerals for the first time. Of course, increases in mineral prices may be reflected in opening or reopening mines that, at lower returns, were deemed unprofitable. However, the developed industrial countries of market economies, whatever their former or even present mineral endowment, find themselves at a competitive disadvantage against producers in developing countries with lower-cost labor and state-owned mines with abundant, rich reserves.

When the ore is rich in metallic content (in the case of iron and aluminum ores), it is profitable to ship it directly to the market for refining. Of course, the highest-grade ores tend to be mined first. Consequently, the demand for low-grade ores has been increasing in recent years as richer deposits have been depleted (**Figure 8.30**). Low-grade ores are often upgraded by various types of separation treatments at the mine site to avoid the cost of transporting waste materials not wanted at the market. Concentration of copper is nearly always mine-oriented (**Figure 8.31**); refining takes place near areas of consumption. The large amount of waste in copper (98 percent to 99 percent or more of the ore) and in most other industrially significant ores should not be considered the mark of an unattractive deposit. Indeed, the opposite

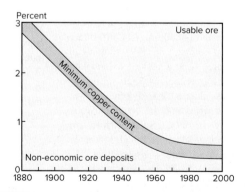

Figure 8.30 Minimum metal content of copper ore for profitable mining. In 1830, 3 percent copper ore in rock was needed to justify its mining; today, rock with 0.5 percent ore content is mined. As the supply of a metal decreases and its price increases, the concentration needed for economic recovery goes down. It also goes down as improved and more cost-effective technologies of rock mining and ore extraction come into play.

Source: Data from the U.S. Bureau of Mines.

Public Land, Private Profit

When U.S. president Ulysses S. Grant signed the Mining Act of 1872, the presidential and congressional goal was to encourage western settlement and development by allowing any "hard-rock" miners (including prospectors for silver, gold, copper, and other metals) to mine federally owned land without royalty payment. It further permitted mining companies to gain clear title to publicly owned land and all subsurface minerals for no more than $12 a hectare ($5 an acre). Under those liberal provisions, mining firms have bought 1.3 million hectares (3.2 million acres) of federal land since 1872; and each year, they remove some $1.2 billion worth of minerals from government property. In contrast to the royalty-free extraction privileges granted to metal miners, oil, gas, and coal companies pay royalties of as much as 12.5 percent of their gross revenues for exploiting federal lands.

Whatever the merits of the 1872 law in encouraging economic development of lands otherwise unattractive to homesteaders, modern-day mining companies throughout the western states have secured enormous actual and potential profits from the law's generous provisions. In Montana, a company claim to 810 hectares (2,000 acres) of land would cost less than $10,000 for an estimated $4 billion worth of platinum and palladium; in California, a gold mining company in 1994 sought title to 93 hectares (230 acres) of federal land containing a potential of $320 million of gold for less than $1,200. Foreign as well as domestic firms may be beneficiaries of the 1872 law. In 1994, a South African firm arranged to buy 411 hectares (1,016 acres) of Nevada land with a prospective $1.1 billion in gold from the government for $5,100. A Canadian firm in 1994 received title to 800 hectares (nearly 2,000 acres) near Elko, Nevada, that cover a likely $10 billion worth of gold—a transfer that

Interior Secretary Bruce Babbitt dubbed "the biggest gold heist since the days of Butch Cassidy." And in 1995, Babbitt conveyed about $1 billion worth of travertine (a mineral used in whitening paper) under 45 hectares (110 acres) of Idaho to a Danish-owned company for $275.

The "gold heist" characterization summarized a growing administration and congressional feeling that what was good in 1872 and today for metal mining companies was not necessarily beneficial to the American public that really owns the land. In part, that feeling results from the fact that mining companies commit environmental sins that require public funding to repair or public tolerance to accept. The mining firms may destroy whole mountains to gain access to low-grade ores and leave toxic mine tailings, surface water contamination, and open-pit scarring of the landscape as they move on or disappear. Projected public cleanup costs of more than 500,000 abandoned mine sites, thousands of miles of damaged or dead streams, and several billion tons of contaminated waste are estimated at a minimum of $35 billion.

A congressional proposal introduced in 1993 would have required mining companies to pay royalties of 8 percent on gross revenues for all hard-rock ores extracted, and prohibited them from outright purchase of federal land. The royalty provision alone would have yielded nearly $100 million annually at 1994 levels of company income. Mining firms claim that imposition of royalties might well destroy the U.S. mining industry. They stress both the high levels of investment that they must make to extract and process frequently low-grade ores and the large number of high-wage jobs they provide as their sufficient contribution to the nation. The Canadian company involved in the Elko site, for example, reports that since it acquired the claims in 1987 from their previous owner, it has expended more than $1 billion and also has made donations for town sewer lines and schools and created 1,700 jobs. The American Mining Congress estimates the proposed

8 percent royalty charge would cost 47,000 jobs out of the 140,000 that exist today, and even the U.S. Bureau of Mines assumes a loss of 1,100 jobs.

Public resistance to Western mining activities is taking its toll. State and federal regulatory procedures, many dragging on for a decade or more, have discouraged opening new mines; newly enacted environmental regulations restricting current mining operations (for example, banning the use of cyanide in gold and silver refining) reduce their economic viability. In consequence, both investment and employment in U.S. mining is in a steady decline, eroding the economic base of many Western communities.

Questions to Consider

1. Do you believe that the Mining Act of 1872 should be repealed or amended? If not, what are your reasons for arguing that the law should remain on the books as it is? If so, would you advocate the imposition of royalties on mining company revenues? At what levels, if any, should royalties be assessed? Should hard-rock and energy companies be treated equally for access to public land resources? Why or why not?

2. Would you propose to prohibit outright land sales to mining companies? If not, should sales prices be determined by surface value of the land, or by the estimated (but unrealized) value of mineral deposits that it contains?

3. Do you think that cleanup and other charges now borne by the public are acceptable, given the capital investments and jobs created by mining companies? Do you accept the industry's claim that imposing royalties would destroy American metal mining? Why or why not?

4. Do you favor continued state and federal restrictions on mining operations, even at the cost of jobs and community economies? Why or why not?

Figure 8.31 Molybdenum mine and concentrating mill at the open pit mine in Climax, Colorado. Concentrating mills crush the ore, separating molybdenum-bearing material from the rocky mass containing it. The great volume of waste removed ensures that most concentrating operations are found near the ore bodies. Because concentrating is a "weight-reducing" process, it saves transportation costs to locate it close to the mine.
©John A. Karachewski

may be true. Because of the cost of extraction or the smallness of the reserves, many higher-content ores are left unexploited in favor of the use of large deposits of even very-low-grade ore. The attraction of the latter is a size of reserve sufficient to justify the long-term commitment of development capital and, simultaneously, to ensure a long-term source of supply.

At one time, high-grade magnetite iron ore was mined and shipped from the Iron Range of northeastern Minnesota. Those deposits are now exhausted. However, immense amounts of capital have been invested in the processing of the virtually unlimited quantities of low-grade iron-bearing rock (taconite) into high-grade iron ore pellets. Naturally, such processing of low-grade deposits takes place at the mine site. Large investments do not guarantee the profitable exploitation of the resource. The metals market is highly volatile. Rapidly and widely fluctuating prices can change profitable mining and refining ventures to losing undertakings very quickly. Marginal gold and silver deposits are opened or closed in response to trends in precious metals prices. Taconite processing in the Lake Superior region nearly ceased as the U.S. steel industry declined, but it has restarted in recent years in response to overseas market demand. In capitalist economies, cost and market controls dominate economic decisions. In planned economies, cost may be a less important consideration than other concerns, such as national development and resource independence.

Nonmetallic Minerals

From the standpoint of volume and weight of material removed, the extraction of nonmetallic materials is the most important branch of the extractive industries. The minerals mined are usually classified by their end use. Of widest distribution, greatest use, and least long-distance movement are those used for *construction:* sand and gravel, building stone, and the gypsum and limestone that are the ingredients of cement. Transportation costs play a large role in determining where low-value minerals will be mined. Minerals such as gravel, limestone for cement, and aggregate are found in such abundance that they have value only when they are near the site where they are to be used. For example, gravel for road building has value if it is at or near the road-building project, not otherwise. Transporting gravel hundreds of miles is an unprofitable activity (**Figure 8.32**). For example, while the United States imports 100 percent of its rare earths, it produces 99 percent of its own crushed stone.

The mined *fertilizer* minerals include potash and phosphate, which move in international trade because of their unequal distribution and market value. *Precious* and *semiprecious* stones are also important in the economy of some countries, including the rich diamond deposits of several central African countries and Sri Lanka's gemstones.

Fossil Fuels

The advanced economies have gotten the way they are through their control and use of energy. By using external energy sources, humans can perform tasks beyond the wildest dreams of our human ancestors. This is largely because fossil fuels are incredibly energy-dense. One 42-gallon (166 liter) barrel of oil contains the energy equivalent of 50,000 person-hours of labor. Compare how far $10 of gasoline will take you in an automobile versus how far you could get paying your friends to push that automobile down energy consumption, the higher the gross national income per capita. Further, the application of energy can allow us to transform low-value raw materials into valuable commodities. High-quality iron ore may be depleted, but by massive applications of energy, the iron contained in rocks of very low iron content, such as taconite, can be concentrated for industrial uses.

Because of the close relationship between energy use and economic development, a basic disparity between societies is made clear. Countries that can afford high levels of energy consumption through production or purchase continue to expand their economies and to increase their levels of living. Those without access to energy or those unable to afford it are left behind.

Except for the brief and localized importance of waterpower at the outset of the Industrial Revolution, modern economic advancement has been heavily dependent on the *fossil fuels:* coal, oil, and natural gas. These nonrenewable energy sources represent the capture of the sun's energy by plants and animals in earlier geologic time and its storage in the form of hydrocarbon compounds in sedimentary rocks within the Earth's crust.

Coal was the earliest in importance and is still the most plentiful of the fossil fuels. As the first major industrial energy source, nearby coal deposits were essential to early manufacturing development, as we shall see in Chapter 9. Although coal is a nonrenewable resource, world supplies are so great—on the order of 1.1 trillion metric tons—that its resource life expectancy is measured in centuries, not in the much shorter time scales cited for oil and natural gas. Worldwide, the most extensive deposits are concentrated in the industrialized middle latitudes of the Northern Hemisphere (Table 8.2). China is the world's largest coal producer, responsible for 46 percent of the world production in 2016. The United States is a distant second, contributing 10 percent of world coal production in 2016.

Coal is not a resource of constant quality, varying in *rank* (a measure—from lignite to anthracite—of increasing carbon content and fuel quality) and *grade* (a measure of its waste material content, particularly ash and sulfur). The value of a coal deposit depends on these measures and on its accessibility, which is a function of the thickness, depth, and continuity of the coal seam. Much coal can be mined relatively cheaply by open-pit (surface) techniques,

Figure 8.32 The Thornton quarry south of Chicago, depicted here with the city skyline in the distance, is one of the largest aggregate quarries in the world. It produces crushed limestone rock for road base and concrete mixes. Proximity to the market is necessary for low-value nonmetallic minerals unable to bear high transportation costs.

©Henryk Sadura/Shutterstock

the road. While slavery made a few rich by harnessing the forced labor of many, fossil fuels can make many "rich" by harnessing fossil fuel energy—at least while prices stay low. Energy consumption goes hand in hand with industrial production and with increases in personal wealth. In general, the greater the level of

Table 8.2

Proved Petroleum, Natural Gas, and Coal Reserves, January 1, 2017

	Share of World Total Petroleum (%)	Share of World Total Natural Gas (%)	Share of World Total Coal (%)
North America[a]	13.3	6.0	22.8
Europe and Central Asia	9.5	30.4	28.3
Of which: Russian Federation	6.4	17.3	14.1
Central and South America	19.2	4.1	1.2
Africa	7.5	7.6	1.2
Middle East[b]	47.7	42.5	0.1
Australia	0.2	1.9	12.7
Japan	–.–	–.–	–.–
China	1.5	2.9	21.4
Other Asia Oceania	1.1	5.0	12.4
Total World	100.0	100.0	100.0
Of which OPEC[c]	71.5	NA	NA

[a]Includes Canada, Mexico, United States
[b]Middle East includes the Arabian Peninsula, Iran, Iraq, Israel, Jordan, Lebanon, and Syria.
[c]OPEC: Organization of Petroleum Exporting Countries. Member nations are, by world region:
   South America: Venezuela
   Middle East: Iran, Iraq, Kuwait, Qatar, Saudi Arabia, and United Arab Emirates (Abu Dhabi, Dubai, Ras-al-Khaimah, and Sharjah)
   North Africa: Algeria and Libya
   West Africa: Nigeria
   Asia Pacific: Indonesia
Source: *Data from the BP Amoco,* Statistical Review of World Energy, *2017.*

in which huge shovels strip off surface material and remove the exposed coal (see Figure 13.17). Much coal, however, is available only by expensive and more dangerous shaft mining, as in Appalachia, China, and most of Europe. In spite of their generally lower heating value, western U.S. coals are attractive because of their low sulfur content. They do, however, require expensive transportation to market or high-cost transmission lines if they are used to generate electricity for distant consumers (**Figure 8.33**).

Oil, first extracted commercially in the 1860s in both the United States and Azerbaijan, became a major power source, raw material for the chemical industry, and favorite transportation fuel. It is used as a raw material for a number of important industries, from plastics to fertilizers. However, its supplies are limited, and it is unlikely to retain its present position of importance in the world energy market.

Estimating how much oil (or natural gas) remains in the world and how long it will last is the subject of heated debate. Given the world economy's dependence on petroleum, the stakes are undeniably high. Some experts subscribe to *peak oil theory*, which is based on the work of American geologist M.

King Hubbert. Hubbert began by showing that the rate of oil pumped from a single well over time followed a symmetrical bell curve, rising, peaking, and finally declining. He also noted that the same bell curve applied to the total production from an entire wellfield or an entire country. By knowing the shape of the rising portion of the curve, one can predict the timing of the peak and the inevitable decline that will follow. Peak oil theorists warn that once the world reaches peak oil production, the inevitable decline in production will cause shortages and skyrocketing prices. Estimates of when the world economy would hit peak oil undergo regular revision. Optimists are encouraged by new horizontal drilling techniques and deep sea drilling that have opened up large new oilfields in North Dakota and the Gulf of Mexico. They also point to a steady drop in the price of solar and other alternative energy sources, which they believe will reduce oil demand long before supply becomes an issue. In a different form of assessment, British Petroleum reported that the world had 1.7 trillion barrels of proven oil reserves at the end of 2016—enough to last for 51 years at the 2016 rate of extraction.

Figure 8.33 Long-distance transportation to eastern markets adds significantly to the cost of the low-sulfur western coal useful in meeting federal environmental protection standards. To minimize these costs, unit trains carrying only coal engage in a continuous shuttle movement between western strip mines and eastern utility companies.

©Medioimages/Superstock

Petroleum is among the most unevenly distributed of the major resources (Table 8.2). A total of 80 percent of proved oil reserves are concentrated in just eight countries: Saudi Arabia, Venezuela, Canada, Iran, Iraq, Kuwait, United Arab Emirates, and the Russian Federation. The underlying role of petroleum in geopolitics and military conflicts over the past 30 years is undeniable. The distribution of petroleum supplies differs markedly from that of the coal deposits on which the urban-industrial markets developed, but the substitution of petroleum for coal did little to alter earlier patterns of manufacturing and population concentration. Because oil is easier and cheaper to transport than coal, it was moved in enormous amounts to the existing centers of consumption via intricate and extensive national and international systems of transportation, a textbook example of spatial interaction, complementarity, and transferability (see Chapter 3, particularly Figure 3.2).

Natural gas has been called the nearly perfect energy resource. It is a highly efficient, versatile fuel that requires little processing, and its emissions do not contribute to urban air pollution or acid precipitation, although they do contain carbon dioxide. Geologists estimate that world recoverable gas reserves are sufficient to last another 50 years at current levels of consumption. New discoveries in the United States, such as the Marcellus Shale formation deep under New York, Ohio, Pennsylvania, and West Virginia, are drawing public attention to the rich possibilities of natural gas extraction, as well as the potential negative effects. *Ultimately recoverable reserves,* those that may be found and recovered at much higher prices, might last another 200 years.

As we saw for coal and petroleum, reserves of natural gas are very unevenly distributed (Table 8.2). In the case of gas, however, inequalities of supply are not so readily accommodated by massive international movements. Like oil, natural gas flows easily and cheaply by pipeline, but unlike petroleum, it does not move freely in international trade by sea. Transoceanic shipment involves costly terminal equipment for liquefaction and special vessels to contain the liquid under appropriate temperature conditions.

Where the fuel can be moved, even internationally, by pipeline, its consumption has increased dramatically. For the world as a whole, gas consumption has risen to about one-quarter of global energy consumption.

8.4 Trade in Primary Products

International trade has expanded rapidly since the end of World War II, increasing more than eightfold since 1980. Primary commodities—agricultural goods and fuels—contribute significantly to the total dollar value of those international flows. During much of the first half of the 20th century, the world distribution of supply and demand for those items in general resulted in a colonial pattern of commodity flow: from raw-material producers located within less-developed countries to processors, manufacturers, and consumers of the more developed ones (**Figure 8.34**). The reverse flow carried manufactured goods from the industrialized states for sale to the developing countries. That two-way trade benefited the developed states by providing access to a continuing supply of industrial raw materials and foods not available domestically, as well as markets for their manufactured goods. While the two-way exchange gave less-developed countries some capital to invest in their own development and to purchase imports, they lagged behind in industrialization.

Today, however, world trade flows and export patterns of the emerging economies have changed. Raw materials have greatly decreased, and manufactured goods correspondingly increased in the export flows from developing states. Even with that overall decline in raw material exports, however, trade in unprocessed goods remains dominant in the economic well-being of many of the world's poorest economies. Increasingly, the terms of the traditional trade flows on which they depend have been criticized as unequal and damaging to commodity-exporting countries.

Commodity prices are volatile; they may rise sharply in periods of product shortage or international economic growth. During much of the 1980s and 1990s, however, commodity price movements were downward, to the great detriment of material-exporting economies. Prices for agricultural raw materials, for example, dropped by 30 percent between 1975 and 2000, and those for metals and minerals decreased by almost 40 percent. Such price declines cut deeply into the export earnings of many emerging economies. Of the 141 developing countries, 91 rely on commodities for more than 60 percent of their export earnings and thus are vulnerable to commodity price volatility. Sub-Saharan African countries are particularly dependent on export earnings from a small number of mineral or agricultural commodities. For example, Burundi earned almost half of its export income through tea and coffee exports in 2015.

Whatever the current world prices of raw materials may be, raw material exporting states have long expressed resentment as a group at what they perceive as commodity price manipulation by rich countries and corporations to ensure low-cost supplies. Although collusive price-fixing has not been demonstrated, other disadvantages of being a commodity supply region are evident. Technology, for example, has provided industries in advanced countries with a vast array of materials that now substitute for the ores and metals produced by developing states. Glass fibers replace copper wire in telecommunication applications; synthetic rubber replaces natural rubber; carbon fibers are superior in performance and strength to the metals that they replace; and a vast and enlarging array of plastics are the accepted raw materials for commodities and uses for which natural rivals are not even considered. That is, even as the world industrial economy expands, demands and prices for traditional raw materials remain relatively low.

While prices paid for developing country commodities tend to be low, prices charged for the manufactured goods offered in exchange by the developed countries tend to be high. To capture processing and manufacturing profits for themselves, some developing states have placed restrictions on the export of unprocessed commodities. Malaysia, the Philippines, and Cameroon, for example, have limited the export of logs in favor of increased domestic processing of sawlogs and exports of lumber. Some developing countries have also encouraged domestic manufacturing to reduce imports and to diversify their exports. Frequently, however, such exports meet with tariffs and quotas designed to protect the home markets of the industrialized states.

In 1964, in reaction to the whole range of perceived trade inequities, developing states promoted the establishment of the United Nations Conference on Trade and Development (UNCTAD). Its central constituency—the "Group of 77," which later expanded to 130 developing states—continues to press for a new world economic order based in part on an increase in the prices and values of exports from developing countries, a system of import preferences for their manufactured goods, and a restructuring of international cooperation to stress trade promotion and recognition of the special needs of poor countries. The WTO, established in 1995 (and discussed in detail in Chapter 12) was designed in part to reduce trade barriers and inequities. It has, however, been judged by its detractors as ineffective on issues of importance to developing countries. Chief among the complaints is the continuing failure of the high-income countries to eliminate generous protections for their own agricultural and mineral industries.

In 2001, members of the WTO met in Doha, Qatar, to begin negotiations on opening world markets. Low-income, developing countries argued for the elimination of agricultural subsidies and protectionist policies in the European Union and United States. In turn, the rich economies insisted on significant concessions from poorer countries on trade in both manufactured goods and, particularly, services. The "Doha Round" of trade negotiations continued through 2008, when trade negotiations broke down. Agriculture was the primary roadblock in global trade talks. The goals of greater fairness in world trade, while maintaining special consideration for the economic development needs of poorer countries, have been elusive. The 2015 Nairobi Package included agreements to eliminate export subsidies for agricultural products, a step toward greater openness in world trade.

Figure 8.34 Sacks of cocoa beans are loaded at Tema, Ghana for overseas shipment. Much of the developing world depends on exports of primary products to the developed economies for the major portion of its income. Fluctuations in market demand and price of some of those commodities can have serious and unexpected consequences.

©Julian Nieman/Alamy Stock Photo

SUMMARY

How different peoples and cultures earn their living and use the diversified resources of the Earth is a fundamental concern in human geography. The economic activities that support us and our society influence our well-being, everyday life, and our perceptions of the world. At the same time, the totality of our culture—technology, religion, customs, and so on—and the circumstances of our natural environment influence the economic choices that we face and the livelihood decisions that we make.

Looking for patterns in the nearly infinite diversity of human economic activity, it is useful to distinguish three types of economic systems: subsistence, commercial, and planned. The first is concerned with production for the immediate consumption of individual producers and family members. In the second, economic decisions ideally respond to impersonal market forces and the profit motive. In the third, at least some nonmonetary social goals, rather than personal ones, influence production decisions. The three system forms are not mutually exclusive; all societies contain some mixture of features of at least two of the three types, and some economies have elements of all three. Recognition of each type's respective features and controls, however, helps us to understand the forces shaping economic decisions and patterns in different cultural and regional settings.

Our search for regularities is advanced by classifying economic activities according to the stages of production and the degree of specialization they represent. We can divide our continuum into primary activities (food and raw material production), secondary production (processing and manufacturing), tertiary activities (wholesaling, retail trade, personal service, and business services) and quaternary activities (administration, information, and research).

Agriculture is the most extensively practiced of the primary industries. In subsistence economies—whether it takes the form of extensive or intensive, shifting or sedentary production—this industry is responsive to the immediate consumption needs of the producer group and reflective of local environmental conditions. Agriculture in advanced economies involves the application of capital and technology with the goal of increasing productivity and efficiency. Commercialized agriculture increasingly involves large-scale, highly specialized agribusiness operations. Arising in reaction to industrialized agriculture, the sustainable agriculture movement seeks to use local knowledge to improve the health of communities, the food supply, and the environment. Agriculture, fishing, forestry, trapping, and the extractive (mining) industries are closely tied to the uneven distribution of Earth's resources. Their spatial patterns reflect those resource potentials, but they also are influenced by the integration of all societies and economies through international trade and mutual dependence. The flows of primary products and of manufactured goods suggest the hierarchy of production, marketing, and service activities, which will be the subject of Chapter 9.

KEY WORDS

agriculture	intensive subsistence agriculture	renewable resource
aquaculture	market economy	resource
Boserup thesis	maximum sustainable yield	secondary activity
commercial economy	Mediterranean agriculture	service activity
economic geography	natural resource	shifting cultivation
extensive commercial agriculture	nomadic herding	subsistence economy
extensive subsistence agriculture	nonrenewable resource	technology
extractive industry	planned economy	tertiary activity
gathering industry	plantation agriculture	tragedy of the commons
genetically modified (GM) crops	post–industrial	truck farm
Green Revolution	primary activity	usable reserves
intensive commercial agriculture	quaternary activity	von Thünen model

FOR REVIEW

1. What are the distinguishing characteristics of the economic systems labeled *subsistence, commercial,* and *planned?* Are they mutually exclusive, or can they coexist within a single political unit?

2. What are the ecological consequences of the different forms of *extensive subsistence* land use? In what world regions are such systems found? What, in your opinion, are the prospects for these land uses and for the way of life they embody?

3. How is *intensive subsistence* agriculture distinguished from *extensive subsistence* cropping? Why, in your opinion, have such different land-use forms developed in separate areas of the warm, moist tropics?

4. Briefly summarize the assumptions and dictates of von Thünen's agricultural model. How might the land-use patterns predicted by the model be altered by an increase in the market price of a single crop? A decrease in the transportation costs of one crop, but not of all crops?

5. What is the basic distinction between a *renewable* and a *nonrenewable* resource? Under what circumstances might the distinction between the two be blurred or obliterated?

6. What economic and ecological problems can you cite that do or might affect the viability and productivity of the *gathering industries* of forestry and fishing? What is meant by the *tragedy of the commons?* How is that concept related to the problems that you discerned?

7. Why have fossil fuels been so important in economic development? What are the mineral fuels, and what are the prospects for their continued availability? Which fossil fuels does your country have in abundance and which does it lack? What economic and social consequences might you anticipate if the price of fossil fuels should double?

KEY CONCEPTS REVIEW

1. **How are economic activities and national economies classified?** Section 8.1

   The innumerable economically productive activities of humans are influenced by regionally varying environmental, cultural, technological, political, and market conditions. Understanding the world's work is simplified by thinking of economic activity as arranged along a continuum of increasing complexity of product or service and increasing distance from nature. Primary industries (activities) harvest or extract something from the Earth. Secondary industries change the form of those harvested items. Tertiary activities render services to consumers or businesses. Quaternary activities focus on information and research. Those activity stages are carried out within national economies grouped as subsistence, commercial, or planned.

2. **What are the types and prospects of subsistence agriculture?** Section 8.2

   Subsistence farming—food production primarily or exclusively for the producers' family needs—still remains the predominant occupation of humans on a worldwide basis. Nomadic herding and shifting (*swidden*) cultivation are extensive subsistence systems. Intensive subsistence farming involves large inputs of labor and fertilizer on small plots of land. Both rural and urban subsistence efforts are increasingly marked by some production for market; they have also benefited from Green Revolution crop improvements.

3. **What characterizes commercial agriculture, and what are its controls and special forms?** Section 8.2

   The modern integrated world of exchange and trade increasingly implies farming efforts that reflect broader market requirements, not purely local or family needs. Commercial agriculture is characterized by specialization, off-farm sale, large-scale operations, and interdependence of farmers and buyers linked through complex markets.

   The von Thünen model of agricultural location suggests that intensive forms of commercial farming—fruits, vegetables, and dairy products—should be located close to markets. More extensive commercial agriculture businesses, including large-scale wheat farms and livestock ranches, are found at more distant locations. Special crops have unique requirements and spatial patterns as in the case of Mediterranean and plantation agriculture.

4. **What are the special characteristics and problems of nonagricultural primary industries?** Section 8.3

   The *gathering* industries of fishing, forestry, and trapping and the *extractive* industries of mining and quarrying involve the direct exploitation of spatially variable natural resources. Resources are natural materials that humans perceive as necessary and useful. They may be renewable—replenished—by natural processes or nonrenewable once extracted and used. Overexploitation can exceed the maximum sustainable yield of fisheries and forests and eventually destroy the resource. Such nonrenewable minerals and fuels are vital to industrial economies and yet proven reserves are finite.

5. **What is the status and nature of world trade in primary products?** pp. 276–277

   The primary commodities of agricultural goods, fish, forest products, furs, minerals, and fuels are a significant component of international trade. Traditional exchange flows of raw materials from developing states that then imported manufactured goods from advanced economies have changed in recent years. Increasingly, the share of manufactured goods in developing world exports is growing. However, material-exporting states argue that current international trade agreements are unfavorable to exporters of agricultural products and other primary commodities.

ECONOMIC GEOGRAPHY:
Manufacturing and Services

The financial district of Shanghai, China, along the Huangpu River.
©Jeremy Woodhouse/Getty Images

Key Concepts

*R*oute 837 connects the four U.S. Steel plants stretched out along the Monongahela River south of Pittsburgh. In the late 1960s, 50,000 workers labored in those mills, and Route 837 was choked with the traffic of their cars and of steel haulers' trucks. By 1979, fires were going out in the furnaces of the aging mills as steel imports from Asia and Europe flowed unchecked into domestic markets long controlled by American producers. By the mid-1980s, with employment in the steel plants of the "Mon" Valley well below 5,000, the highway was only lightly traveled, and only occasionally did anyone turn at the traffic lights into the closed and deserted mills. From Massachusetts to Wisconsin, competition with Japanese imports and the opening of new assembly plants in lower-wage countries such as Mexico led to manufacturing job losses and empty factories. Derelict factories with pollution in the soils and groundwater scared away potential investors and thus were left to rot.

At the same time, traffic was building along many highways in the northeastern part of the country. The four-lane Route 1 was clogged with traffic along the 42 kilometers (26 miles) of the "Princeton Corridor" in central New Jersey as that stretch of road in the 1980s had more office space, research laboratories, hotels, conference centers, and residential subdivisions under construction than anywhere else between Washington, D.C., and Boston. Farther south, in the Virginia and Maryland suburbs of Washington, D.C., traffic grew heavy along the Capital Beltway and Dulles Toll Road, where vast office building complexes, defense-related industries, and commercial centers were converting rural land to urban uses. And east of New York City, traffic jams were monumental around Stamford, Connecticut, in Fairfield County, as it became a leading corporate headquarters town with 150,000 daily in-commuters.

By the early 1990s, traffic in Fairfield County had thinned as corporate takeovers, leveraged buyouts, and "downsizing" reduced the number and size of companies and their need for both employees and office space. Vacancies exceeded 25 percent among the office buildings and research parks so enthusiastically built during the 1970s and 1980s, and vacant "corporate campuses" lined stretches of formerly clogged highways. But soon traffic was building elsewhere in the country as millions of Americans during the 1990s and the early 21st century gained technology-related jobs in California's "Silicon Valley," and a whole series of other emerging "high-tech" hot spots clustered around such industries as computers, lasers, software, medical devices, and biotechnology. Starting in the late 1990s and continuing unabated through this writing, the explosive growth of China's manufacturing exports led to factory closures in a wide swath of the world including textile and garment factories in the Carolinas. Starting in the mid-1990s, all sections of the United States again experienced the congested traffic and breakneck housing and commercial development that economic prosperity induces, only once more to endure job losses, office vacancies, economic reversals, and altered traffic flows following the 2007 housing market and financial industry crisis.

These contrasting and fluctuating patterns of traffic flow symbolize the ever-changing structure of the North American economy. The smokestack industries of the 19th and early 20th centuries have declined, replaced by research park industries, shopping

Figure 9.1 The economic changes on Minneapolis's riverfront typify the changes occurring in many postindustrial economies. Minneapolis's riverfront was once lined with flour mills and sawmills taking advantage of waterpower from the only waterfalls on the Mississippi River. Now, former manufacturing buildings have been converted to service-sector offices and expensive residences. The large blue structure is the Guthrie Theater, which completes the transformation of a heavy industrial landscape into a service sector landscape of entertainment and consumption.

©Mark Bjelland

centers, and office building complexes that in their turn experience cyclical prosperity and adversity. The continent's economic landscape and employment structure are continually changing (**Figure 9.1**). And North America is not alone. Change is the ever-present condition of contemporary economies, whether of the already industrialized, advanced countries or of those newly developing in an integrated world marketplace. Resources are exploited and exhausted, markets grow and decline, and patterns of economic advantage, of labor skills, and of industrial investment and productive capacity undergo alteration as countries and regions differentially develop, prosper, or experience reversals and decline. Such changes have a profound impact on the spatial structure and processes of economic activity.

9.1 Components of the Space Economy

All human activity creates observable spatial patterns. In the economic sphere, we recognize regions of industrial concentration, areas of employment and functional specialization, and specific factory sites, store locations, and tourist destinations. As geographers, we seek to understand and explain the underlying logic behind those spatial patterns of economic activity.

Primary industries are tied to the location of natural resources. Location is therefore predetermined by the distribution of minerals, fuels, forests, fisheries, or natural conditions suitable for agriculture and herding. The secondary, tertiary, and quaternary stages of economic activity, however, are increasingly divorced from the conditions of the physical environment. Processing, distribution, communication, and management work can be located in response to cultural and economic considerations rather than physical influences. They are movable, rather than spatially fixed activities. Locational decisions and economic patterns differ with the type of economic activity in question. Secondary industries involved in material processing and goods production have different spatial constraints than the retailing activities, tourist attractions, research parks, or office complexes of the service sector. Global competition and new distance-shrinking technologies regularly upset established economic patterns, creating new centers of activity and a new international division of the world's work.

Basic Economic Concepts

Principles of human spatial behavior apply to economic behavior as well. We already explored some of those principles in Chapter 3. We noted, for example, that the intensity of spatial interaction decreases with increasing separation of places—distance decay. We observed the importance of complementarity and transferability in the assessment of resource value and trade potential. Johann Heinrich von Thünen's model of agricultural land use, you will recall, was rooted in the relationship between transportation costs and land values. Conventional economic thinking is based on a set of simplifying assumptions about the motivations guiding human economic behavior. Economists assume, for example, that people are *economically rational;* that is, given all of the information relevant to a particular economic decision, they make locational, production, or purchasing decisions in light of their perception of what is most cost-effective and advantageous. From the standpoint of producers or sellers of goods or services, it is assumed each is intent on *maximizing profit* (from the standpoint of consumers, it is assumed each is intent on *maximizing value*). To reach that objective, each may consider a host of production and marketing costs and political, competitive, and other limiting factors—and, perhaps, respond to individual behavioral quirks—but the ultimate goal of profit-seeking remains clear. Finally, most economists assume that in commercial economies the best

measure of the efficiency of economic decisions is afforded by the *market mechanism.*

At root, that market control mechanism is measured by *price*—the price of land (rent), of labor (wages), of a college course (tuition), or of goods at the store. In turn, price is seen as a function of *supply* and *demand.* In large, complex economies where there are many producers, sellers, and buyers, and many alternative products competing in the marketplace, price is the neutral measure of comparative value and profitability. If demand for a good or service exceeds its available supply, scarcity will drive up the price that it can command in the marketplace. That increased price will enhance the profitability of the sale, which will encourage existing producers to increase output or induce new producers or sellers to enter the market (**Figure 9.2a**). That is, *the higher the price of a commodity, the more of it will be offered in the market.* Of course, this does not imply that more expensive commodities will be offered in greater quantities than less expensive commodities, only that more of a given commodity will be offered if more can be charged for it.

When the price is very high, however, relatively few people are inclined to buy. To dispose of their increased output, old and new producers of the commodity are forced to reduce prices to enlarge the market by making the commodity affordable to a larger number of potential customers. That is, *at lower prices, more of a commodity will be purchased* (**Figure 9.2b**). If the price falls too low, production or sale becomes unprofitable and inefficient suppliers are forced out of business, reducing supply. **Market equilibrium** is marked by the price at which supply equals demand, satisfying the needs of consumers and the profit motivation of suppliers (**Figure 9.2c**).

These basic economic models and assumptions treat supply, demand, and price as if all production, buying, and selling occurred at a single point. But as geographers, we know that human activities have specific locational settings and that neither people, nor resources, nor opportunities are uniformly distributed over the Earth. We appreciate that the place or places of production may differ from the locations of demand. We understand

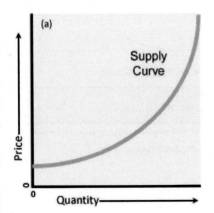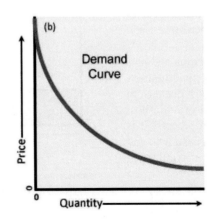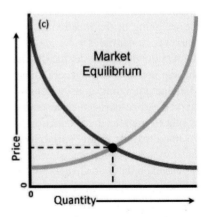

Figure 9.2 The regulating mechanism of the market may be visualized graphically: supply, demand, and market equilibrium. (*a*) The *supply curve* tells us that as the price of a commodity increases, more of it will be made available for sale. Countering any tendency for prices to rise to infinity is the market reality that the higher the price, the smaller the demand, as potential customers find other purchases or products more cost-effective. (*b*) The *demand curve* shows how the market will expand as prices are lowered and commodities are made more affordable and attractive to more customers. (*c*) *Market equilibrium* is marked by the point of intersection of the supply and demand curves and determines the price of commodities, the total demand, and the quantity bought and sold.

that there are spatial relations and interactions based on supply, demand, and equilibrium price. We realize there is a geography of *supply,* a geography of *demand,* and a geography of *cost.*

Other economic geographers question the economist's assumptions of economic rationality. They point to a wide range of human motivations and behaviors that aren't the result of purely rational economic calculations—some examples might include impulsiveness; envy; altruism; attachments to people, places, and things; nostalgia for the past; optimism for the future; or a willingness to settle for less than the optimum result. Despite being an oversimplification, the assumption of economic rationality is important to economic thinking. Fortunately, the assumption of economic rationality is most applicable to the decisions made by companies, the focus of this chapter.

9.2 Secondary Activities: Manufacturing

Secondary activities involve transforming raw materials into usable products, from pouring iron and steel to stamping out plastic toys, assembling computer components, or sewing jeans. In every case, the common characteristics are the application of power and specialized labor to the production of finished products in factory settings: in short, industrialization.

Unlike the gathering or extraction of primary commodities, manufacturing involves assembling and processing multiple inputs and distributing the output to markets in diverse locations. It therefore presents the question of where the processing should take place. If we assume free markets, rational producers, and informed consumers, then the decision where to locate a manufacturing facility should be based on costs and opportunities that vary from place to place. In the case of primary industries—those tied to the environment—possible locations are fixed by the locations of natural resources. The decision is only whether or not to exploit known resources. In the instance of secondary-quaternary economic activity, however, there are many possible locations and the locational decision is more complex. It involves the weighing of the locational "pulls" of a number of cost and market considerations.

On the *demand* side, the distribution of population and purchasing power defines general areas of marketing opportunities. Manufacturers must consider costs of raw materials, distance to markets, wage costs, fuel costs, capital availability, and a host of other inputs to the production and distribution process. It is assumed that the spatial variability of those costs is known, and that rational location decisions leading to profit maximization are based on that knowledge.

Locational Decisions in Manufacturing

Locational decisions for manufacturing may require multiple spatial scales of analysis. The first scale is international. The second scale is regional and examines the attractiveness of different sections of a country. Later decision stages become more focused, localized, and specific to an individual enterprise. They involve assessment of the special production and marketing requirements of particular industries and the degree to which those requirements can or will be met at different subregional scales—at the state (in the United States), community, and individual site levels. That is, we can ask at one level why the northeastern United States–southeastern Canada exerted an earlier pull on industry in general and, at other decision stages, why specific sites along the Monongahela Valley to the south of Pittsburgh in Pennsylvania were chosen by the U.S. Steel Corporation for its mills.

For a great many searches, two or several alternate locations would be equally satisfactory. In very practical terms, locational decisions at the state, community, and site levels may ultimately be based on the value of incentives offered by rival areas and agencies competing to lure the new or relocated manufacturing plant (see the feature "Contests and Bribery,"). In both practice and theory, locational factors are complex, interrelated, change over time in their relative significance, and differ between industries and regions. But all of them are tied to *principles of location* that are assumed to operate under all economic systems.

Principles of Location

The principles of industrial location are simply stated. Certain input costs of manufacturing are **spatially fixed costs;** that is, they are relatively unaffected no matter where the industry is located within a regional or national setting. Wage rates set by national or areawide labor contracts are an example. Fixed costs do not give any location an advantage over others. Other input costs of manufacturing are **spatially variable costs;** that is, they show significant differences from place to place (**Figure 9.3**). These will influence locational choices.

The ultimate aim of the economic activity is *profit maximization.* In an economic environment of full and perfect competition, the profit objective is most likely to be achieved if the manufacturing

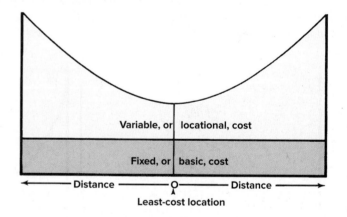

Figure 9.3 The spatial implications of fixed and variable costs. *Spatially fixed* costs represent the minimum price that must be paid at any location for the necessary inputs of production of a given item. Here, for simplicity, a single raw material is assumed and priced at its cheapest source. *Spatially variable (locational)* costs are the additional costs incurred at alternate locations in overcoming distance, attracting labor, purchasing the plant site, and so forth. In the example, only the transportation cost of the single material away from its cheapest (source) location is diagrammed to determine *O,* the optimal or least-cost location.

enterprise is situated at the *least total cost* location. Under conditions of imperfect competition, considerations of sales and market may be more important than production costs in fixing "best" locations. As we see later in this chapter, the location of markets becomes the predominant factor in locating tertiary services such as retail.

Spatially fixed costs are not of major importance in determining optimum, or least-cost, locations. Rather, the locational determinant is apt to be the variable cost that is an important component of total costs and shows the greatest spatial variation.

Transportation charges—the costs of bringing together the inputs and distributing products—are highly variable costs. As such, they may become the locational determinant, imparting an unmistakable *orientation*—a term describing locational tendencies—to the plant siting decision.

Individual establishments rarely stand alone; they are part of integrated manufacturing sequences and environments, in which *interdependence* increases as the complexity of industrial processes increases. Spatial interdependence may be a decisive locational determinant for some industries. *Linkages* among firms may localize manufacturing in areas of industrial clusters (agglomerations) where common resources—such as skilled labor—or multiple suppliers of product inputs—such as automobile component manufacturers—are found.

These principles are generalized statements about the locational tendencies of industries. Their relative weight, of course, varies among industries and firms. Their significance also varies depending on the extent to which purely economic considerations—as opposed, say, to political or environmental constraints—dictate locational decisions.

Raw Materials

All manufactured goods have their origins in the processing of raw materials, but only a few industries at the early stages of the production cycle use raw materials directly from farms or mines. Most manufacturing is based on the further processing and shaping of materials already processed by an earlier stage of manufacturing. In general, the more advanced the industrial economy of a nation, the smaller the role played by truly *raw* materials in its economic structure.

For those industries in which unprocessed commodities are a primary input, however, the source and characteristics of those raw materials is important. The quality, amount, or ease of mining or gathering of a resource may be a locational determinant if cost of raw material is the major variable cost and multiple sources of the primary material are available. Raw materials may attract the industries that process them when they are bulky, undergo great weight loss in the processing, or are highly perishable. Copper smelting and iron ore beneficiation are examples of weight- (impurity-) reducing industries that almost always are located next to ore supplies. Pulp, paper, and sawmills, which reduce bulky logs into neat, stackable outputs, are found in timber harvesting areas. Fruit and vegetable canning in California, meat packing in Iowa, and Florida orange juice concentration and freezing are comparable examples of raw **material orientation.** The reason is simple: It is cheaper and easier to transport a refined or stabilized product than one that is filled with waste material or subject to spoilage and loss.

Multiple raw materials might dictate an intermediate plant location. Least cost may be determined not by a single raw material input, but by the spatially variable costs of several inputs. Steel mills at Gary, Indiana, or Cleveland, Ohio, for example, were not based on local raw material sources but on minimizing the total cost of assembling the necessary iron ore from northern Minnesota, coking coal from Appalachia, and fluxing material inputs (**Figure 9.4**). Steel mills along the U.S. East Coast—at Sparrows Point, Maryland, or the Fairless Works near Philadelphia, for example—were located where imported ores were unloaded from ocean carriers, avoiding expensive transshipment costs. In this respect, both the Great Lakes and the coastal locations are similar.

Power Supply

For some industries, power supplies with low transferability may serve to attract energy-intensive activities. Such was the case early in the **Industrial Revolution**, when water power sites

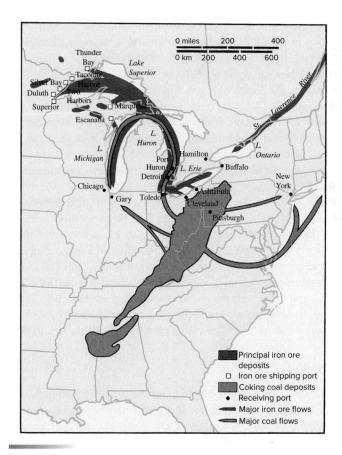

Figure 9.4 Material flows in the steel industry. When an industrial process requires the combination of several heavy or bulky ingredients, an intermediate point of assembly of materials is often a least-cost location. In the early 20th century, the iron and steel industry of the eastern United States showed this kind of localization—not at the source of any single input but where coking coal, iron ore, and limestone could be brought together at the lowest price. In fact, the city of Gary, Indiana, at the southern tip of Lake Michigan, was founded by the U.S. Steel Corporation in the early 1900s for the sole purpose of making steel at the lowest-cost location.

attracted textile mills and fuel (initially charcoal, later coking coal) drew the iron and steel industry. Metallurgical industries became concentrated in such coal-rich regions as the Midlands of England, the Ruhr district of Germany, and the Donets Basin of Ukraine.

Massive amounts of electricity are required to extract aluminum from its processed raw material, *alumina* (aluminum oxide). Electrical power accounts for between 30 percent and 40 percent of the cost of producing aluminum and is the major variable cost influencing plant location. The Kitimat plant on the northwest coast of Canada and the Bratsk plant near Lake Baikal in eastern Siberia are examples of industry placed far from raw material sources or market but close to vast supplies of cheap power—in these instances, hydroelectricity.

Labor

Labor costs are highly variable across space, increasingly affecting location decisions and industrial development. Traditionally, three different considerations—price, skill, and amount—of labor were considered important. For many manufacturers today, an increasingly important consideration is *labor flexibility,* implying more highly educated workers able to apply themselves to a wide variety of tasks and functions. For some activities, a cheap labor supply is a necessity. For others, labor skills may constitute the locational attraction and regional advantage. Machine tools in Sweden, precision instruments in Switzerland, and optical and electronic goods in Japan are examples of industries that have created and depend on localized labor skills. In an increasingly high-tech world of automation, electronics, and industrial robots, labor skills—even at high unit costs—are often more in demand than an unskilled, uneducated workforce. Manufacturing of lower-cost clothing is an example of an industry that requires a large, low-cost labor supply to be competitive.

In some world areas, of course, labor can be in too short a supply to satisfy the developmental objectives of government planners or private entrepreneurs. In the former Soviet Union, for example, longstanding economic plans called for the fuller exploitation of the vast resources of sparsely populated Siberia but failed because the area lacked sufficient workers and was generally unattractive to workers from other regions of the country.

Market

Goods are produced to supply a market demand. Therefore, the size, nature, and distribution of markets may be as important in industrial location decisions as raw material, energy, labor, or other inputs. When the transportation charges for sending finished goods to market are a relatively high proportion of the total value of the good, then the attraction of location near to the consumer is obvious and **market orientation** results.

The consumer may be another firm or the general public. When a factory is but one stage in a larger manufacturing process—firms making wheels, tires, windshields, bumpers, and the like in the assembly of automobiles, for example—location near the next stage of production is an obvious advantage. This advantage is increased if that final stage of production is also near the ultimate consumer market. Thus, automobile part plants have been scattered throughout the North American realm in response to the existence of large regional markets and the cost of distribution of the finished automobile. This market orientation is further reflected by the location in North America of auto manufacturing or assembly plants of Asian and European motor vehicle companies, although both foreign and domestic firms again appear to be reconcentrating the industry in the southeastern part of the United States.

People themselves, of course, are the ultimate consumers. Large urban concentrations represent markets, and major cities have always attracted producers of goods consumed by city dwellers. Admittedly, it is impossible to distinguish clearly between cities as markets and cities as labor force. In either case, many manufacturing activities are drawn to major population centers. For example, California, the most populous state in the United States, is also the leader in the manufacture of mattresses and soft drinks, products that gain significant bulk during manufacture. Certain producers are, in fact, inseparable from the immediate markets that they serve and are so widely distributed that they are known as **ubiquitous industries.** Newspaper publishing, bakeries, and dairies, all of which produce a highly perishable commodity designed for immediate consumption, are examples.

Transportation Modes

Transportation is such an essential factor of industrial location that it is difficult to isolate its separate role. Earlier observations about manufacturing plant orientations can be restated in purely transportation cost terms. For example, copper smelting or iron ore beneficiation—already described as examples of raw material orientation—may also be seen as industries engaged in *weight reduction* designed to minimize transportation costs. Some market orientation is of the opposite nature, reflecting *weight-gaining* production. Soft drink bottlers, for example, add large amounts of water to small amounts of concentrated syrup to produce a bulky product of relatively low value. All transport costs are reduced if only the concentrate is shipped to local bottlers, who add the water that is available everywhere and distribute only to local dealers. The frequency of this practice suggests the inclusion of soft drink bottlers among the ubiquitous industries.

The location of industry is inseparably tied to transportation systems. The Industrial Revolution involved a transportation revolution as successive improvements in the technology of movement of peoples and commodities enlarged the effective areas of spatial interaction and made integrated economic development and areal specialization possible. All advanced economies are well served by a diversity of transport modes (see Figure 8.4); without them, all that is possible is local subsistence activity. All major industrial agglomerations are important nodes for different transportation modes, each with its own characteristic advantages and limitations.

Water transportation is the cheapest means of long-distance freight movement (**Figure 9.5**). Inland waterway improvement and canal construction marked the first phase of the Industrial

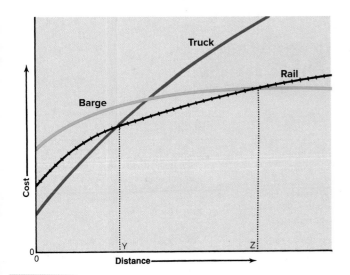

Figure 9.5 Different transport modes have cost advantages over differing distances. Generally, trucks are most efficient and least expensive over short hauls of up to about 500 kilometers (about 300 miles), railroads have the cost advantage for intermediate hauls of 500 to 3,200 kilometers (about 300 to 2,000 miles), and water (ship or barge) movement over longer distances (and, often, over shorter distances where speed of delivery of nonperishable commodities is not a consideration). Railroad and water shipments require much less energy and generate fewer greenhouse gas emissions per unit weight of freight hauled.

Revolution in Europe and was the first stage of modern transport development in the United States. Even today, river ports and seaports have locational attractiveness for industry unmatched by alternative centers not served by water carriers. Where water routes are in place, as in northwestern Europe or the Great Lakes–Mississippi systems of the United States, they are vital elements in regional industrial economies.

Railroads efficiently move large volumes of freight over long distances at low fuel and labor costs. They are, however, inflexible in route, slow to respond to changing industrial locational patterns, and expensive to construct and maintain. When traffic declines below minimum revenue levels, rail service becomes uneconomic and the lines are abandoned—a response of American railroads, which abandoned more than 125,000 miles of line between 1915 and 2005.

Trucks operating on modern roads and expressways have altered the competitive picture to favor highways over railways for many intercity movements. Road systems provide great flexibility of service and are more quickly responsive than railroads or waterways to new traffic demands and changing origin and destination points. Intervening opportunities are more easily created and regional integration more cheaply achieved by highway than by railroad (or waterway systems).

Major cost and time savings are achieved by the use of freight containers that can link trucking, railroads, and oceangoing vessels. Such *multimodal freight* movements seek the advantages of the most efficient carrier for each stage of the journey from cargo origin point to final destination through the use of internationally standardized shipping containers, which are tracked

by computer. The containers with undisturbed content may be transferred to ships for international ocean carriage, to railroads for long-haul land movement, and to truck trailers for shorter-haul distances and pickup and delivery. Their use is increasingly common on long "trailer-on-flat-car" trains and in the growing volume of international ocean trade (**Figure 9.6**).

Pipelines provide fast, efficient, and dependable transportation for a variety of liquids and gases. Pipeline corridors are laid out to serve the industries that use the transported commodity—particularly fertilizer plants, oil refineries, and petrochemical plants—which in turn encourages new plants using those commodities to choose nearby locations.

Air transport is vital for the movement of skilled workers, consultants, and decision makers and increases the attractiveness of airport sites for high-tech and other industries shipping or receiving high-value, low-bulk commodities. It is not, however, an effective competitor for most freight flow (see the feature "A Comparison of Transport Media").

Figure 9.6 Cargo containers in docks in Shanghai, China. Standardized cargo containers have revolutionized shipping, sharply reducing shipping times and making possible the increased economic interdependence in the world economy. In 2010, China exported $1.6 trillion worth of goods, with the largest categories being electrical machinery and equipment, power generation equipment, and clothing.

©Kevin Phillips/Getty Images

A Comparison of Transport Media

Mode	Uses	Advantages	Disadvantages
Railroad	Intercity medium- to long-haul bulk Fast, reliable service on separate and general cargo transport.	Rights-of-way; essentially nonpolluting; most energy-efficient; adapted to steady flow of single commodities between two points; routes and nodes provide intervening development opportunities.	High construction and operating costs; inflexible routes; underutilized lines cause economic drain.
Highway trucking	Local and intercity movement of general cargo and merchandise; pickup and delivery services; feeder to other carriers.	Highly flexible routes, origins, and destinations; individualized service; maximum accessibility; unlimited intervening opportunity; high speed and low terminal costs.	Low energy efficiency; major contributor to air pollution and greenhouse gas emissions; adds congestion to public roads; high maintenance costs; inefficient for large-volume freight.
Inland waterway	Low-speed haulage of bulk, nonperishable commodities.	High energy efficiency; low per mile costs; large cargo capacity.	High terminal costs; low route flexibility; not suited for short hauling; possible delays from ice or low water levels.
Pipelines	Continuous flows of liquids, gases, or suspended solids where volumes are high and continuity is required.	Fast, efficient, dependable; low per mile costs over long distances; maximum safety.	Highly inflexible in route and cargo type; high development cost.
Airways	Medium- and long-haul of high-value, low-bulk cargo where delivery speed is important.	High speed and efficiency; adapted to goods that are perishable, packaged, of a size and quantity unsuited to other modes; high route flexibility; access to areas otherwise inaccessible.	Very expensive; high greenhouse gas emissions; high mileage costs; inconvenient terminal locations; no intervening opportunities between airports.
Intermodal containerization	Employs standardized closed containers to move a shipment by any combination of water, rail, and truck without unpacking between origin and final destination.	Speed and efficiency of transit and lower shipping costs when multiple carriers are needed; reduced labor charges and theft losses.	Requires special terminals and handling machinery to load, off-load, and transfer containers.

Transportation and Location

Figure 9.7 shows the general pattern of industrial orientation based on transportation costs. Those costs are more than a simple function of the distance that goods are carried. Rather, they represent the application of differing **freight rates,** charges made for loading, transporting, and unloading of goods. In general, manufactured goods have higher value and greater fragility, require more special handling, and can bear higher freight charges than unprocessed bulk commodities. The higher transport cost for finished goods is one reason for locating high-value manufacturing near the market for the finished products.

Freight rates are composed of **terminal costs,** the charges for paperwork, loading, packing, and unloading of a shipment; and **line-haul** or *over-the-road costs,* the expenses for the actual movement of commodities once they have been loaded. Total transport costs represent the sum of all pertinent charges and are

curvilinear rather than linear functions of distance. That is, carrier costs tend to decline as the length of haul increases because scale economies for long-haul movement permit the averaging of total costs over a greater distance. The result is the **tapering principle** diagrammed in **Figure 9.8**.

One consequence of the necessary assignment of fixed and terminal costs to *every* shipment regardless of distance moved is that factory locations intermediate between sources of materials and final markets are less attractive than locations at either end of a single long haul. That is, two short hauls cost more than a single continuous haul over the same distance (**Figure 9.9**).

Two exceptions to this locational generalization are of practical interest. **Break-of-bulk points** are sites where goods have to be transferred or transshipped from one carrier to another—at ports, for example, where barge or ocean vessel must be unloaded and cargo reloaded to railcar or truck, or between railroad and truck line. When such transfer occurs, an additional fixed

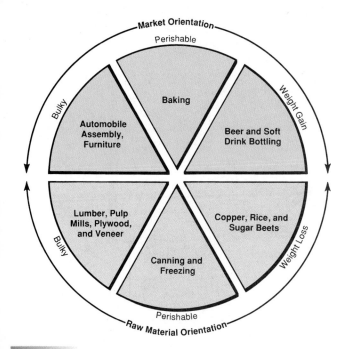

Figure 9.7 Spatial orientation tendencies. *Raw material orientation* occurs when there are few alternative material sources, when the material is perishable, or when—in its natural state—it contains a large proportion of impurities or nonmarketable components. *Market orientation* represents the least-cost solution when manufacturing uses commonly available materials that add weight to the finished product, when the manufacturing process produces a commodity bulkier or more expensive to ship than its separate components, or when the perishable nature of the product demands processing close to the market.

Source: Adapted from Interpreting the City: An Urban Geography, *Truman A. Hartshorn. 1980 John Wiley & Sons, Inc.*

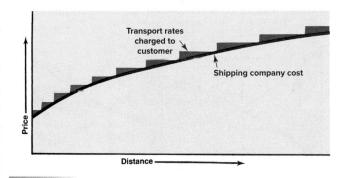

Figure 9.8 The tapering principle. The actual costs of transport, including terminal charges and line costs, increase at a decreasing rate as fixed costs are spread over longer hauls. The "tapering" of total cost for the shipping company varies among modes because their mixes of fixed and variable costs are different, as Figure 9.5 diagrams. Note that the actual transport rates charged to customers move in stepwise increments within certain distance ranges.

or terminal cost is levied against the shipment, perhaps significantly increasing its total transport costs (use of cargo containers reduces, but does not eliminate, those handling charges). There is a tendency for manufacturing to concentrate at such points to avoid the additional charges. Many of the world's important industrial cities developed at break-of-bulk locations.

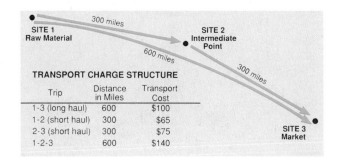

Figure 9.9 The short-haul penalty. Plant locations intermediate between material and market are generally avoided because of the realities of transportation pricing shown here. Two short hauls cost more than a single long haul simply because two sets of fixed costs must be assigned to the interrupted movement.

Industrial Location Theory

In practice, industrial locational decisions are based not on a single factor, but on the interplay of a number of considerations. Implicit in our review has been the understanding that each type or branch of industry has its own specific set of significant plant siting conditions. Classic industrial location theory was developed in the early 1900s, when the world economy was dominated by railroads, based on heavy industry and goals of national industrial self-sufficiency. Obviously, much has changed in a globalized economy shaped by institutions such as the World Trade Organization (WTO), transnational corporations, environmental protection agencies, and the like. Nevertheless, classic location theory concepts and their spatial implications remain relevant in understanding past and present-day industrial locational decisions.

Least-Cost Theory

The classical model of industrial location theory, the **least-cost theory,** is based on the work of Alfred Weber (1868–1958) and sometimes called **Weberian analysis.** It explains the optimum location of a manufacturing establishment based on minimizing three basic expenses: transport costs, labor costs, and agglomeration costs. **Agglomeration** refers to the clustering of productive activities and people for mutual advantage. Such clustering can produce "agglomeration economies" through shared facilities and services. Diseconomies such as higher rents or wage levels resulting from competition for these resources also may occur.

Weber concluded that transport costs are the major consideration determining location. That is, the optimum location will be found where the costs of transporting raw materials to the factory and finished goods to the market are at their lowest. He noted, however, if variations in labor or agglomeration costs are sufficiently great, a location determined solely on the basis of transportation costs may not in fact be the optimum one.

Weber made five simplifying assumptions:

1. An area is completely uniform physically, politically, culturally, and technologically. This is known as the **uniform,** or **isotropic, plain** assumption.

2. Manufacturing involves a single product to be shipped to a single market in a known location.

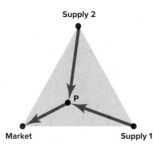

Figure 9.10 Weber's locational triangle. With one market for the finished product and two raw material sources, the optimum production point, *P*, lies within the triangle. The heavy arrows represent the weight or "pull" of the raw materials and the finished product. When the weight of the finished product exceeds that of the raw materials, the optimum location is pulled closer to the market. If the raw materials are heavier, the optimum location is closer to the raw material supply locations.

3. Inputs involve raw materials from more than one known source location.

4. Labor is infinitely available but immobile.

5. Transportation routes are not fixed but connect origin and destination by the shortest path; and transport costs directly reflect the weight of items shipped and the distance moved.

Given these assumptions, Weber derived the least transport cost location by means of the *locational triangle* (**Figure 9.10**). It diagrams the cost consequences of fixed locations of materials and market and movement of raw material supplies and finished goods. Except in the unlikely scenario that the weight of each raw material input and the finished product were all equal, the least transport cost location will be an intermediate point somewhere within the locational triangle. Its exact position will depend on distances, the respective weights of the raw material inputs, and the final weight of the finished product. Each input and the finished product exerts a pull proportional to its weight. Material orientation (that is, when the least cost location is close to raw material supplies) reflects a sizable weight loss during the production process. Market orientation (that is, when the least-cost location is near the market) indicates weight gain during the production process. The optimum placement of *P* can be found by mathematical or geographic information system (GIS) methods, but the easiest to visualize is by way of a mechanical model of weights and strings (**Figure 9.11**).

Modifications to Least-Cost Theory

For many theorists, the assumptions and simplicities of least-cost theory are unrealistically restrictive. They agree that the correct location of a production facility is where the net profit is greatest. However, they propose employing a **substitution principle** that recognizes that in many industrial processes it is possible to replace a declining amount of one input (e.g., labor) with an increase in another (e.g., capital for automated equipment) or to increase transportation costs while simultaneously reducing land rent. With substitution, a number of different points may be appropriate manufacturing locations. Further, they suggest a

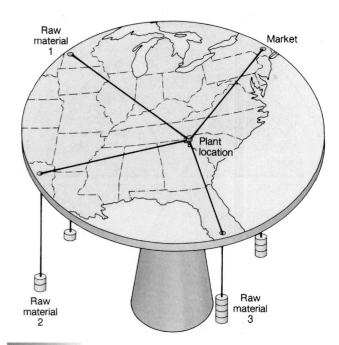

Figure 9.11 Plane table solution to a plant location problem. This mechanical model, suggested by Alfred Weber, uses weights to demonstrate the least transport cost point where there are several sources of raw materials. When a weight is allowed to represent the "pull" of raw material and market locations, an equilibrium point is found on the plane table. That point is the location at which all forces balance one another and represents the least-cost plant location.

whole series of points may exist where total revenue of an enterprise just equals its total cost of producing a given output. These points, connected, mark the **spatial margin of profitability** and define the larger area within which profitable operation is possible (**Figure 9.12**). Location anywhere within the margin assures

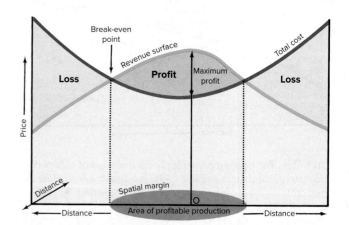

Figure 9.12 The spatial margin of profitability. In the diagram, *O* is the single optimal profit-maximizing location, but locations anywhere within the brown-shaded area (defined by the intersection of the total cost and total revenue surfaces) will be profitable. Some industries will have wide margins; others will be more spatially constricted. Skilled entrepreneurs may be able to expand the margins farther than less able industrialists. Importantly, a *satisficing* location may be selected by reasonable estimate, even in the absence of the perfect information required for an *optimal* decision.

Geography and Citizenship

Contests and Bribery

In 1985, it cost Kentucky more than $140 million in incentives—some $47,000 a job—to induce Toyota to locate an automobile assembly plant in Georgetown, Kentucky. That was cheap. In 1993, Alabama spent $169,000 per job to lure Mercedes-Benz to that state; Mississippi agreed to $400 million in spending and tax rebates to Nissan in 2001; and in 2002, Georgia gave DaimlerChrysler $320 million in incentives in successful competition with South Carolina to secure the company's proposed new factory. Earlier, Kentucky bid $350,000 per job in tax credits to bring a Canadian steel mill there.

The spirited bidding for jobs is not confined to manufacturing. A University of Minnesota economist calculates that his state will have spent $500,000 for each of the 1,500 or more permanent jobs created by Northwest Airlines at two new maintenance facilities. Illinois gave $240 million in incentives ($44,000 per job) to keep 5,400 Sears, Roebuck employees within the state, and New York City awarded $184 million to the New York Mercantile Exchange and more than $30 million each to financial firms Morgan Stanley and Kidder, Peabody to induce them to stay in the city. For some, the bidding between states and locales to attract new employers and employment gets too fierce. Kentucky withdrew from competition for a United Airlines maintenance facility, letting Indianapolis have it when Indiana's offered package exceeded $450 million. By 2004, following a slowdown in air travel, United walked away from a fully completed operational facility, leaving the city and state with $320 million of bonded debt and a complex of empty hangars and office buildings.

Inducements to lure companies are not just in cash and loans—though both figure in some offers. For manufacturers, incentives may include workforce training, property tax abatement, subsidized or free land and buildings, and below-market financing of bonds. Similar offers are regularly made by states, counties, and cities to wholesalers, retailers, and major office worker– and other service activity–oriented employers. The total annual loss of city and state tax revenue through abatements, subsidies, grants, and the like to benefit retained or attracted firms has been estimated at $30 billion to $40 billion. The objective, of course, is not just to secure the new jobs represented by the attracted firm, but to benefit from the general economic stimulus and employment growth that those jobs—and their companies—generate. Auto parts manufacturers are presumably attracted to new assembly plant locations; cities grow and service industries of all kinds—doctors, department stores, restaurants, food stores, and so on—prosper from the investments made to attract new employment.

Not everyone is convinced that those investments are wise. A majority of Minnesotans opposed the generous loans made by the state to keep Northwest Airlines headquarters and maintenance facilities in the state. The naysayers may have had a point: Before the loans were repaid, Northwest Airlines merged with Delta Airlines and moved their headquarters to Atlanta. In the late 1980s, the governor of Indiana, a candidate for Kentucky's governorship, and the mayor of Flat Rock, Michigan, were all defeated by challengers who charged that too much had been spent in luring the Suburu-Isuzu, Toyota, and Mazda plants, respectively. Established businesses resent what often seems neglect of their interests in favor of spending

their tax money on favors to newcomers. The Council for Urban Economic Development has studied the issue and actively lobbies against incentives, and many academic observers note that industrial incentives are a zero-sum game: Unless the attracted newcomer is a foreign firm, whatever one state achieves in attracting an expanding U.S. company comes at the expense of another state.

Some doubt that inducements matter much, anyway. Although, sensibly, companies seeking new locations will shop around and solicit the lowest-cost, best deal possible, their site choices are apt to be determined by more realistic business considerations: access to labor, suppliers, and markets; transportation and utility costs; weather; the nature of the workforce; and overall costs of living. Only when two or more similarly attractive locations have essentially equal cost structures might such special inducements as tax reductions or abatements be determinants in a locational decision.

Questions to Consider

1. As citizen and taxpayer, do you think it is appropriate to spend public money to attract new employment to your state or community? If not, why not? If yes, what kinds of inducements and what total amount offered per job seem appropriate to you? What reasons support your opinion?

2. If you believe that "best locations" for the economy as a whole are those determined by pure location theory, what arguments would you propose to discourage locales and states from making financial offers designed to circumvent decisions clearly justified on abstract theoretical grounds?

some profit and tolerates both imperfect knowledge and personal (rather than purely rational economic) considerations. Such suboptimal, but still acceptable, sites are considered **satisficing locations.**

For some firms, spatial margins may be very broad because transport costs are a negligible factor in production and marketing. Such firms are said to be **footloose**—that is, neither resource- nor market-oriented. For example, both the raw materials and the finished product in the manufacture of computers are so valuable,

light, and compact that transportation costs have little bearing on where production takes place.

Contemporary Industrial Location Considerations

Weber's classic industrial location theory was based on a highly simplified version of the world. Through assumptions such as the isotropic plan, it tried to explain the behavior of individual firms seeking production sites under competitive market conditions. But such theory no longer fully explains world or regional patterns of industrial localization or specialization. Moreover, it does not account for locational behavior that is uncontrolled by objective "factors," directed by national or regional economic development planning goals, or influenced by new production technologies and corporate structures.

Political Considerations

Location theories dictate that in a pure, competitive economy, the costs of material, transportation, labor, and plant should control locational decisions. However, just as the world is not an isotropic plain, a pure market economy does not exist, even in the United States. Political factors and constraints also affect, perhaps decisively, the location decision process. Least-cost locations rely upon governments to build the highways and regulate the interstate commerce that connects the raw materials, production facilities, and markets, Land use and zoning controls, as well as environmental regulations, also influence where industries locate.

Many governments actively seek to attract industry to underdeveloped regions or encourage the full utilization of a region's resources by creating quasi-governmental corporations, building industrial parks, or investing in large development projects. In the United States, the Bureau of Reclamation, Tennessee Valley Authority, and Appalachian Regional Commission are examples of federal government organizations that promote economic development. The Bureau of Reclamation's mission is to promote the development of the western United States, and it has built many large hydroelectric dams, including the Hoover Dam and Grand Coulee Dam, to provide water and electricity for cities, industry, and agriculture. State and local governments promote economic development by offering incentives and subsidies to industries who agree to build plants in specific places. These incentives are a way for communities and regions to lure footloose industries to specific places, but they often ending up pitting communities and regions against each other to the benefit of the private company (see the feature "Contests and Bribery").

Agglomeration Economies

Geographical concentration of economic activities is the norm. We take it for granted that certain places are associated with certain products. Hollywood makes films; Silicon Valley makes computer software; Detroit makes automobiles; and Pittsburgh used to make steel. Weber's least-cost theory made provision for *agglomeration,* the spatial concentration of people and activities for mutual benefit. That is, clustering of industrial activities may produce benefits for individual firms that they could not experience in isolation. Those **agglomeration economies** are a form of **external economies;** that is, benefits that firms enjoy due to factors outside the firm. The benefits of agglomeration economies come from linkages among firms and savings from shared transport facilities, worker training programs, social services, public utilities, communication facilities, and forms of industrial **infrastructure.**

Geographic clusters are centers of innovation as knowledge and new ideas are shared among related firms. Clustering of similar firms creates pools of skilled and ordinary labor, of capital, suppliers, ancillary business services, and, of course, a built-in market of other industries and urban populations. New firms, particularly, may find significant advantages in locating near other firms engaged in the same activity because specialized workers and support services specific to that activity are already in place. Thus, for example, it is not surprising that Facebook, which started at Harvard University in Massachusetts, moved its headquarters to Silicon Valley, where it could find the skilled software professionals the young company needed. A concentration of capital, labor, management skills, customer base, and infrastructure will tend to attract still more industries to the cluster. In Weber's terms, agglomeration economies alter locational decisions that otherwise would be based solely on transportation and labor costs, and once in existence, agglomerations will tend to grow (**Figure 9.13**). Through a **multiplier effect,** each new firm added to the agglomeration will lead to the further development of infrastructure and linkages. As we shall see in Chapter 11, the "multiplier effect" also implies total (urban) population growth, and thus the expansion of the labor pool and the local market that are part of agglomeration economies.

Agglomeration—concentration or clustering—of like industries in small areas dates from the early industrial age and continues with many of the newest industries. Familiar examples include the town of Dalton, Georgia, home to all but one of the top 20 U.S. carpet makers, and Akron, Ohio, which, before 1930, held almost the entire 100 or so tire manufacturers in the country. Silicon Valley dating from the 1960s and other more recent high-tech clusters simply continue the tradition.

On the other hand, agglomeration may have disadvantages. Overconcentration can result in diseconomies of congestion, high land values, pollution, and rising labor costs. When the costs of aggregation exceed the benefits, a firm will actually profit by relocating to a more isolated position, a process called **deglomeration.** It is a process seen in the suburbanization of industry within metropolitan areas or the relocation of firms to nonmetropolitan locations.

Just-in-Time and Flexible Production

Traditional theories sought to explain location decisions for plants engaged in mass production for mass markets where transportation lines were fixed and transport costs relatively high. Both conditions began to change significantly during the late 20th century. Assembly-line work that breaks the production process into many repetitive, low-skill tasks in order to produce large quantities of identical commodities for mass markets efficiently is known as "**Fordism,**" in honor of Henry Ford's pioneering role in implementing this idea. Increasingly, Fordist production processes have been moved to low-wage countries, and in advanced economies, Fordism was replaced by post-Fordist *flexible manufacturing* processes based on smaller production runs of a greater variety of goods aimed at smaller niche markets. Agglomeration economies are encouraged by newer manufacturing approaches practiced by both older, established

Figure 9.13 On a small scale, the planned industrial park furnishes its tenants some of the agglomeration economies offered by large urban concentrations. An industrial park provides a subdivided tract of land developed according to a comprehensive plan for the use of firms. Because the park developers, whether private companies or public agencies, supply the basic infrastructure of streets, water, sewage, power, transport facilities, and perhaps private police and fire protection, tenants are spared the additional cost of providing these services themselves. In some instances, factory buildings are available for rent, still further reducing start-up costs. Counterparts of industrial parks for manufacturers are the office parks, research parks, and science parks for "high-tech" firms and for service enterprises.

©geogphotos/Alamy Stock Photo

industries and by newer, post-Fordist plants. Traditional Fordist industries required the on-site storage of large lots of materials and supplies ordered and delivered well in advance of their actual need in production. That practice permitted cost savings through infrequent ordering and reduced transportation charges and made allowances for delayed deliveries and for inspection of received goods and components. The assurance of supplies on hand for long production runs of standardized outputs was achieved at high inventory and storage costs.

Just-in-time (*JIT*) manufacturing, in contrast, seeks to reduce inventories for the production process by purchasing inputs for arrival just in time to use and producing output just in time to sell. Rather than costly accumulation and storage of supplies, JIT requires frequent ordering of small lots of goods for precisely timed arrival and immediate deployment to the factory floor. JIT manufacturing is often associated with Toyota, whose engineers developed the concept from American management consultants and watching American companies. Such *lean manufacturing* based on frequent purchasing of immediately needed goods demands rapid delivery by suppliers and encourages them to locate near the buyer. The trend toward JIT manufacturing, thus, reinforces the spatial agglomeration tendencies evident in the older industrial landscape.

JIT is one expression of a transition from mass-production Fordism to *flexible production systems.* That flexibility is designed to allow producers to shift quickly and easily between different levels of output and, importantly, to move from one factory process or product to another as market demand dictates. Flexibility of that type is made possible by new technologies of easily reprogrammed computerized machine tools and by computer- aided design and computer-aided manufacturing systems. These technologies permit small-batch, JIT production and distribution responsive to current market demand, as monitored by computer-based information systems.

Flexible production to a large extent requires significant acquisition of components and services from outside suppliers rather than from in-house production. For example, modular assembly, where many subsystems of a complex final product enter the plant already assembled, reduces factory space and worker requirements. The premium that flexibility places on proximity to component suppliers adds still another dimension to industrial agglomeration tendencies. *Flexible production regions* have emerged in response to the new flexible production strategies and dependencies on outside suppliers. Those regions have a different set of labor-owner relations and are usually located some distance—spatially or socially—from established concentrations of entrenched Fordist industrialization.

Comparative Advantage, Offshoring, and the New International Division of Labor

The principle of **comparative advantage** extends the capitalist division of labor from individual workers to the economies of entire regions and countries. The principle of comparative advantage asserts that areas and countries can best improve their economies and living standards through specialization and trade. Each area or country should concentrate on the production of those items for which it has the greatest relative advantage over other areas and imports all other goods. This principle is one of the most important justifications for free trade between countries.

The logic of comparative advantage was recognized by economists in the 19th century when specialization and exchange involved shipments of grain, coal, or manufactured goods whose relative costs of production in different areas were clearly evident. Today, when other countries' comparative advantages may reflect lower costs for labor, land, and capital, the application of the principle is questioned by some critics. They observe that manufacturing activities may relocate from higher-cost developed country locations to lower-cost foreign production sites, taking jobs and income away from the developed country. The temptation is obvious when looking at the wide variation in hourly compensation costs (wages plus benefits) for manufacturing work. According to the U.S. Bureau of Labor Statistics, costs vary from $58 per hour in Norway and $35 per hour in the United States to just $6 per hour in Mexico and $1.90 per hour in the Philippines. Defenders of outsourcing, however, argue that the increased efficiencies due to such voluntary **outsourcing** increases overall prosperity.

Outsourcing has also come to mean subcontracting production and service sector work to outside (often nonunion) domestic companies. In manufacturing, outsourcing has become an important element in JIT acquisition of preassembled components for snap-together fabrication of finished products, often built only to fill orders actually received from customers. Reducing parts inventories and introducing build-to-order production demands a high level of flexible freight movement increasingly supplied by *logistics* firms that themselves may become involved in packaging, labeling, and even manufacturing products for client companies.

A clear example of the impact of outsourcing is evident in the changing nature of automobile manufacturing. Formerly, motor vehicle companies were largely self-contained production entities. At one time, Ford Motor Company even owned its own rubber plantations. Ford's River Rouge Complex near Detroit made its own steel and glass, turning raw materials into automobiles. Since the early 1990s, that self-containment has been abandoned because car companies have divested themselves of raw material production facilities and have in large part sold off their in-house parts production departments. Increasingly, they purchase parts from independent, often distant, suppliers. In fact, some observers of the changing vehicle production scene predict that established automobile companies will eventually convert themselves into "vehicle brand owners," retaining for themselves only such essential tasks as vehicle design, engineering, and marketing. All else, including final product assembly, is projected to be done through outsourcing to parts suppliers. Similar trends are already evident in consumer electronics, where a significant portion of production is outsourced.

A distinctive example of outsourcing is found along the northern border of Mexico. In the 1960s, Mexico enacted legislation permitting foreign (specifically, U.S.) companies to establish "sister" plants, called **maquiladoras,** within 20 kilometers (12 miles) of the U.S. border for the duty-free assembly of products destined for re-export. By the early 20th century, more than 3,000 such assembly and manufacturing plants had been established to produce a diversity of goods, including electronic products, textiles, furniture, leather goods, toys, and automotive parts. The plants generated direct and indirect employment for more than a million Mexican workers (**Figure 9.14**) and for large numbers of U.S. citizens, employees of growing numbers of American-side *maquila* suppliers and of diverse service-oriented businesses spawned by the "multiplier effect." The North American Free Trade Agreement (NAFTA), which created a single Canadian–U.S.–Mexican production and marketing community, simplifies outsourcing in the North American context. It has led

Figure 9.14 U.S. manufacturers, seeking lower labor costs, began in the 1960s to establish light manufacturing, component production, and assembly operations along the international border in Mexico. Outsourcing to such maquiladoras as this factory in Ciudad Juarez has moved a large proportion of U.S. electronics, small appliance, toy, and garment industries to offshore subsidiaries or contractors in Asia and Latin America. In the last few years, Mexican maquiladoras have been losing jobs to competition from lower-cost, more-efficient Chinese and other Asian producers. Comparative advantage is not a permanent condition.
©Joe Raedle/Getty Images

to significant shifts in the location of production activities, some of which have hurt local economies (**Figure 9.15**).

On the broader world scene, outsourcing often involves production of manufactured goods by developing countries that have benefited from the transfer of technology and capital from industrialized states. For example, electrical and electronic goods from China and Southeast Asia compete with and replace in the market similar goods formerly produced by Western firms. Such outsourcing has resulted in new global patterns of industrial regions and specializations. They have also strikingly changed the developing world's share of gross global output (see Chapter 10 for more on that subject). Outsourcing not only involves manufacturing activity and blue-collar jobs but also, as we shall see later in this chapter, may be used by companies to reduce their service worker costs When that reduction involves janitorial and similar services spatially tied to the home establishment, no job losses are felt. When, however, lower-paid foreign workers can satisfactorily replace technical, professional, and white-collar workers, the outsourcing action is known as service *offshoring* and has the immediate effect of exporting the jobs of highly paid skilled workers.

Offshoring is the practice of either hiring foreign workers or, commonly, contracting with a foreign third-party service provider to take over and run particular business processes or operations, such as call centers or accounting, billing, and similar nonproduction "back-office" aspects of manufacturing. Offshoring has become an increasingly standard cost-containment strategy, due to the steep decline in communication costs, faster

Figure 9.15 Economic change in postindustrial economies. Brantford, Ontario, was once Canada's major manufacturer of agricultural machinery and equipment. The closure of this Massey-Ferguson tractor factory was a consequence of the economic changes that followed the North American Free Trade Act. Production was shifted to lower-cost locations, increasing overall economic efficiency, but harming workers and factory towns. The plant closure typifies the painful structural changes that accompany deindustrialization. The abandoned plant is a **brownfield site** facing issues of blight and pollution. Some cities have been successful in attracting service jobs and redeveloping brownfield sites for commercial, housing, or recreational uses. In Brantford, however, this site and others like it remain derelict.
©Mark Bjelland

Internet bandwidth, and the growing technical skills of foreign workers. With an ever-increasing portion of the developing world acquiring the education and experience to provide skilled professional services of almost every kind at a level comparable to that formerly available only in advanced countries, traditional notions of comparative advantage are disappearing in the face of a new era of *hypercompetition,* at least in business and professional services. India in particular has emerged as the dominant competitor and beneficiary of services offshoring, echoing China's position as the preferred destination of production outsourcing.

The exploitation of comparative advantage and utilization of outsourcing and offshoring, by transferring technology from economically advanced to underdeveloped economies is transforming the world economy by introducing a **new international division of labor (NIDL).** In the 19th century and the first half of the 20th century, the division of labor involved exports of manufactured goods from the "industrial" countries and raw materials from the "colonial" or "undeveloped" economies. Roles have now altered (see the feature "Where Do Your Clothes Come From?"). Manufacturing no longer is the mainstay of the economy of developed countries, and the world pattern of industrial production is shifting to reflect the growing dominance of *newly industrializing countries* that were formerly peasant societies. In recognition of that shift, the NIDL builds on the current trend toward the increased subdivision of manufacturing processes

into smaller steps. That subdivision permits multiple outsourcing and offshoring opportunities based on differential land, labor, and capital costs and skill levels available in the globalized world economy, opportunities effectively exploited by transnational corporations.

Transnational Corporations (TNCs)

Outsourcing is just one small expression of the growing international structure of today's manufacturing and service enterprises. Businesses are increasingly stateless and economies borderless as giant **transnational corporations (TNCs)** become ever more important in the globalizing world economy. TNCs (also known as *multinational companies*) are private firms that have established branch operations in foreign nations. The total annual revenue of the world's largest TNCs rivals the gross domestic product (GDP) of entire countries. For example, Wal-Mart Stores in 2010 had $408 billion in revenues which, if it were a country, would have placed it 24th in the world, just behind Norway and ahead of Venezuela. The largest TNCs, with the exception of Wal-Mart, are engaged in petroleum exploration, refining and distribution, automobiles, or electrical and electronic equipment. TNCs are increasingly international in origin and administrative home, based primarily in a growing number of both economically advanced and newly industrializing countries. In 2008,

Where Do Your Clothes Come From?

One of the distinguishing characteristics of humans is that we (almost always) clothe our bodies. The clothing we wear expresses our culture, values, social status, and self-identity. In the United States, at least, clothes are required to carry a label indicating the country of origin. A quick check through your closet is likely to reveal the international nature of the clothing industry. Clothing is second only to agriculture as the leading product in international trade. Textiles and clothing production were at the heart of the original Industrial Revolution and are one of the leading ways for developing countries to begin industrialization. The clothing industry was the leader in globalizing production and creating a new international division of labor. The technological requirements are fairly simple, and the production process is labor-intensive, offering a comparative advantage to low-wage countries. After World War II,

Japan used clothing production to jump-start its manufacturing sector. The newly industrializing countries, particularly China, are following that pattern. China is now the world's leading clothing manufacturer and is gaining share at the expense of most other countries. The geography of clothing production is changing rapidly. In a vintage clothing store, many of the clothes will have been manufactured in the United States.

Broader trends in manufacturing are also evident in the clothing industry. Export-processing zones are common for garment manufacturing. *Maquiladoras* along the Mexico-U.S. border assumed an important role after the passage of the North American Free Trade Act. JIT manufacturing and lean, flexible manufacturing has become increasingly important, along with more rapid turnover of styles with the advent of fast-fashion as pioneered by global

brands such as the Swedish retailer H&M and the Spanish brand Zara. While most mass-market production has moved to developing countries, most major brands and customers are based in the developed countries. High-end fashion production must be closely connected to the designers and shows in the major fashion centers of New York, London, Paris, and Milan. Thus, most production of higher-priced, specialty fashion remains near those cities. Lower-cost fashion, especially for discount retailers, has moved relentlessly to the lowest-wage countries.

Working conditions in clothing factories (or *sweatshops* as they are sometimes known) remain a major concern of human rights watch groups. The workers, who are mostly female, may be subjected to long hours and unsafe working conditions, with little recourse to file complaints.

---

91 of the world's 100 largest nonfinancial TNCs had home offices in Europe, the United States, or Japan. However, cash-rich multinationals of such developing world states as China, Korea, Mexico, Malaysia, Taiwan, India, and Brazil were moving up the list. Through their own surging growth and through mergers and acquisitions, formerly developing world regional players have emerged as major global forces.

The direct impact of TNCs is limited to relatively few countries and regions. **Foreign direct investment (FDI)**—the purchase or construction of factories and other fixed assets by TNCs—has been a significant engine of globalization. Although more than half of FDI goes from one developed country to another developed country, a growing proportion is invested in less-developed economies, potentially stimulating their economic growth. The three main sources for outward FDI are the countries or regions that are home to the largest TNCs—the United States, Europe, Hong Kong, and Japan. Within Europe, the United Kingdom, Germany, and France are the leaders in FDI. The leading destinations for inward FDI are Hong Kong, China, Singapore, Mexico, Brazil, and India. Distance and proximity influence where FDI flows go. For example, FDI from the United States is more likely to go to Latin America, and Asian countries are more likely to invest in other Asian countries.

The portion of FDI going to the 50 least developed countries as a group—including nearly all African states—remains less than 5 percent. Despite poor countries' hopes for foreign investment to spur their economic growth, critics argue that it is

counterproductive. Economic control is lost to a foreign firm and may undermine political sovereignty as the TNC demands subsidies and tax breaks. TNCs may rely on foreign suppliers instead of local firms, bankrupt local competitors who lack the capital to compete, and then return their profits to the home country rather than reinvesting them in the host country.

Investment outflows from companies based in India, Brazil, South Africa, Malaysia, and China (among others) have swelled, with an increasing share going to other developing countries. Because more than 80 percent of the world's 7 billion consumers live in the expanding less-developed nations, TNCs based in newly industrializing countries have the strength of familiarity with those markets and have an advantage in supplying them with goods and services that are usually cheaper and more effectively distributed than those of many Western TNCs.

The advanced-country destination of more than half of FDI capital flows is understandable: TNCs are actively engaged in merging with or purchasing competitive established firms in already developed foreign market areas, and cross-border mergers and acquisitions have been the main stimulus behind FDI. Because most transnational corporations operate in only a few industries—computers, electronics, petroleum and mining, motor vehicles, chemicals, and pharmaceuticals—the worldwide impact of their consolidations is significant. Some dominate the marketing and distribution of basic and specialized commodities. In raw materials, a few TNCs account for 85 percent or more of world trade in wheat, corn, coffee, cotton, iron ore, and timber,

Figure 9.16 The number of transnational corporations has grown rapidly since the 1970s. Of the top 100 TNCs, 91 are headquartered in the Triad—Europe, the United States, and Japan. Their recognition and impact of TNCs, however, are global, as suggested by this scene from the developing world, a Nestlé factory (headquartered in Switzerland) in Tianjin, China.

(a) ©Zhang Peng/Getty Images

for example. Because they are international in operation with multiple markets, plants, and raw material sources, TNCs actively exploit the principle of comparative advantage and seize opportunities for outsourcing and offshoring. In manufacturing, they have internationalized the plant-siting decision process and multiplied the number of locationally separate operations that must be assessed. TNCs produce in that country or region where costs of materials, labor, or other production inputs are minimized, or where existing efficient company-owned factories can be easily expanded to produce for a global, rather than simply a national, market. At the same time, they can maintain operational control and pay taxes where the economic climate is most favorable. Research and development, accounting, and other corporate activities are placed wherever economical and convenient.

TNCs have become global entities because global communications make it possible (**Figure 9.16**). Many have lost their original national identities and are no longer closely associated with or controlled by the cultures, societies, and legal systems of a nominal home country. At the same time, their multiplication of economic activities has reduced any earlier identification with single products or processes and given rise to "transnational integral conglomerates" that span a large spectrum of both service and industrial sectors.

9.3 High-Technology Manufacturing

Classic location theories are less effective in explaining the location of high-technology (or *high-tech*) research, development, and manufacturing activities. For these firms, new and different patterns of locational orientation have emerged based more on human talent than the traditional factors of raw materials and transportation costs.

High technology is more a concept than a precise definition. It probably is best understood as the application of intensive scientific and engineering research and development to the creation and manufacture of new, technologically advanced products. Professional—"white collar"—workers make up a large share of the total workforce. They include research scientists, engineers, and skilled technicians. When these highly skilled specialists are added to administrative, supervisory, marketing, and other professional staffs, they may greatly outnumber the actual production workers in a firm.

Although only a few types of industrial activity are generally reckoned as exclusively high-tech—electronics, communication, computers, software, pharmaceuticals, biotechnology, and aerospace—advanced technology is increasingly a part of the structure and processes of all types of industry. Robotics on the assembly line, computer-aided design and manufacturing, electronic controls of smelting and refining processes, and the constant development of new products of the chemical industries are cases in point.

The impact of high-tech industries on patterns of economic geography is expressed in at least three ways. First, high-tech activities are major factors in employment growth, manufacturing output, and the total gross value added (GVA)[1] for many individual countries. Relatively high wages in high-tech occupations reflect the level of training and specialization they require.

Second, high-tech industries have tended to become regionally concentrated in centers of innovation, frequently forming self-sustaining, highly specialized agglomerations (**Figure 9.17**). Third, the offshoring of less-skilled production and assembly tasks has spurred the economic development of newly industrializing countries.

Concentrations of high-tech employment include California, the Pacific Northwest (including British Columbia), New England, New Jersey, Texas, and Colorado. And within these and other states or regions of high-tech concentration, specific locales have achieved prominence: "Silicon Valley" of Santa Clara County near San Francisco; Irvine and Orange County south of Los Angeles; the "Silicon Forest" near Seattle; North Carolina's Research Triangle; Utah's "Software Valley"; Routes 128 and 495 around Boston; "Silicon Swamp" of the Washington, D.C., area; "Silicon Alley" in Manhattan; Ottawa, Canada's "Silicon Valley North"; and the Canadian Technology Triangle, west of Toronto (**Figure 9.18**).

Within such concentrations, specialization is often the rule: biomedical technologies in Minneapolis and Philadelphia; biotechnology around San Antonio; computers and semiconductors in the "Silicon Hills" of Austin, Texas; biotechnology and telecommunications in New Jersey's Princeton Corridor; and telecommunications and Internet industries

---

[1]GVA is linked to GDP; the link can be defined as GVA + taxes on products – subsidies on products = GDP.

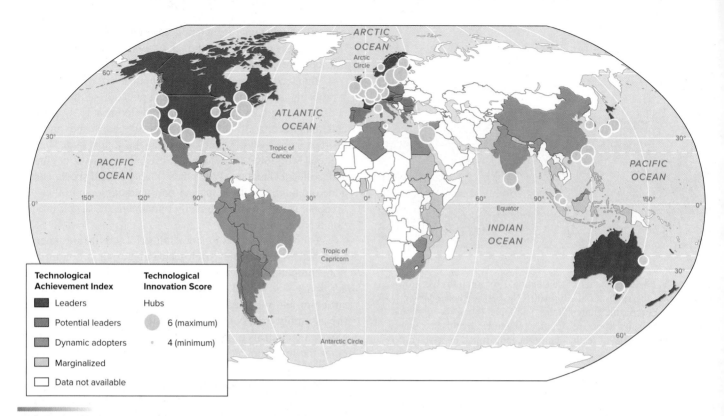

Figure 9.17 Global hubs of technological innovation. The technology innovation hubs shown with circles were identified by *Wired* magazine in 2000 based on the presence of research universities, research laboratories, established technology companies, venture capital, and entrepreneurial activity. The highest scoring regions were Silicon Valley (California), Boston (Massachusetts), Stockholm (Sweden), Israel, Research Triangle (North Carolina), and London (U.K.). The technological achievement index was generated by the United Nations.

Source: United Nations Human Development Report, *2001. Adapted from Bradshaw, White, Dymond, and Chacko, Contemporary World Regional Geography, 2009.*

Figure 9.18 Silicon Valley, the area around San Jose, California, just south of San Francisco, is the world's leading hub for high-tech innovation. The valley was once known for its orchards, but is now home to world-leading technology companies such as Apple, Cisco, eBay, Google, Hewlett-Packard, Intel, Oracle, and Yahoo. It was named for the silicon chips used in semiconductors in the computer and electronics industries. Silicon Valley illustrates agglomeration economies where in certain places, knowledge of how to make a particular product is "in the air." The region offers a high quality of life, institutions such as Stanford University and government research labs, and an incredible concentration of skilled workers who can share ideas in formal or informal social settings.

©David McNew/Getty Images

near Washington, D.C. Elsewhere, Scotland's Silicon Glen, England's Sunrise Strip and Silicon Fen, Wireless Valley in Stockholm, China's Zhong Guancum in suburban Beijing and the High-Tech Industries Zone in Xian, and Hitec City at Hyderabad, Pune, and Bangalore, India, are other examples of industrial landscapes characterized by low, modern, dispersed office-plant-laboratory buildings rather than by massive factories, mills, and railyards.

The map of high-tech industries shows that they respond to different factors than heavy manufacturing industries. At least five locational tendencies have been recognized: (1) Proximity to major research universities or government research laboratories that create a large pool of scientific and technical labor skills; (2) avoidance of areas with strong labor unionization, where rigid contracts slow innovation and workforce flexibility; (3) locally available venture capital and entrepreneurial skills; (4) a reputation for a good "quality of life"—climate, scenery, recreation, cultural activities, good schools and neighborhoods, and job opportunities for professionally trained spouses; and (5) availability of first-rate communication and transportation facilities to unite research, development, and manufacturing operations and to connect the firm with suppliers, markets, finances, and government agencies. Most major high-tech agglomerations have developed on the suburban edges of metropolitan areas, far from inner-city problems and disadvantages.

Agglomeration economies are extremely important to high-tech industries. The formation of new firms is frequent and rapid in industries where discoveries are constant and innovation is continuous. Because many are "spin-off" firms founded by employees leaving established local companies, areas of existing high-tech concentration tend to spawn new entrants and to provide necessary labor skills. In essence, talent is the essential raw material for high-tech firms and talent clusters in specific locations.

Not all phases of high-tech production, however, must be concentrated. The professional, scientific, and knowledge-intensive aspects of the high-tech economy are often located far from the component manufacturing and assembly operations. Highly automated or low-skill assembly tasks are footloose; they require highly mobile capital and technology investments, but they may be performed at a lower cost in lower-wage countries such as China, Taiwan, Singapore, Malaysia, or Mexico. Major high-tech companies such as Apple, Sony, and Microsoft outsource assembly of electronic devices to independent contract manufacturers. Most often the same factory produces similar or identical products under a number of different brand names (**Figure 9.19**). Unfortunately, extreme pressures to meet production deadlines have led to reports of abusive working conditions at a number of contract electronics manufacturers.

High-tech products often have complex, highly international **commodity chains**—steps in the production and distribution process. The iPhone was designed by U.S. engineers, but its manufacture uses rare metals from Asia and Africa and specialized components manufactured in Germany, Korea, Taiwan, and Japan, all of which come together in an assembly plant in China.

Through such outsourcing and technology transfers, high-tech activities are spread to newly industrializing countries—from the center to the periphery. This globalization through geographic transfer and diffusion represents an important impact of high-tech activities on world economic geographic patterns. For example, by 2005, China had surpassed the United States in exporting information-technology goods, such as laptop computers, mobile phones, and digital cameras. With rising education levels, countries such as China, India, Singapore, and South Korea are producing large numbers of highly trained scientists and engineers capable of doing much more than assembly work. Thus, computer software companies have begun taking advantage of India's strengths in engineering and computer science, making Bangalore and Hyderabad major world players in software development.

9.4 World Manufacturing Patterns and Trends

Growth and change have produced a distinctive world pattern of manufacturing. While **Figure 9.20** suggests a large number of industrial concentrations, in fact four regions are commonly recognized as most significant: *Eastern North America, Western and Central Europe, Eastern Europe,* and *Eastern Asia.* Together, the industrial plants within these established regional clusters account for an estimated three-fifths of the world's manufacturing output by volume and value.

Their continuing dominance is by no means assured. The first three—those of North America and Europe—were the beneficiaries of an earlier phase in the development and spread of manufacturing following the Industrial Revolution of the 18th and 19th centuries and lasting until after World War II. The countries within them now are increasingly postindustrial, and traditional manufacturing and processing are declining in relative importance.

The fourth—the East Asian district—is part of the wider, newer pattern of world industrialization that has emerged in recent years, the result of massive international *cultural convergence* and technology transfers in the latter half of the 20th century and early in the 21st. The older, rigid economic split between the developed and developing worlds has rapidly weakened as the full range of industrial activities from primary metal processing (e.g., the iron and steel industries) through advanced electronic assembly has been established within an ever-expanding list of countries.

Such states as Mexico, Brazil, China, and others of the developing world have created industrial regions of international significance, and the contribution to world manufacturing activity of the smaller newly industrializing countries (NICs) has been growing significantly. The list of NICs includes the four Asian tigers: South Korea, Hong Kong, Singapore, and Taiwan. They each developed by combining a well-educated workforce,

Figure 9.19 Contract electronics manufacturing. Assembly work that can be broken down into multiple, repetitive steps is often outsourced to overseas contractors, many based in Asia. The Taiwanese company FoxConn Technology Group operates large assembly plants in China, Brazil, Mexico, and the Czech Republic, employing more than 900,000 workers in 2010. It manufactures many of the most popular computer, consumer electronics, and communications devices, including the iPad, iPhone, PlayStation 4, Xbox One, and Amazon Kindle. Its largest plant in Shenzhen, China, nicknamed FoxConn City, is a walled, self-contained campus, complete with factories, dormitories, and all the services needed by its approximately 250,000 workers.

©STR/Getty Images

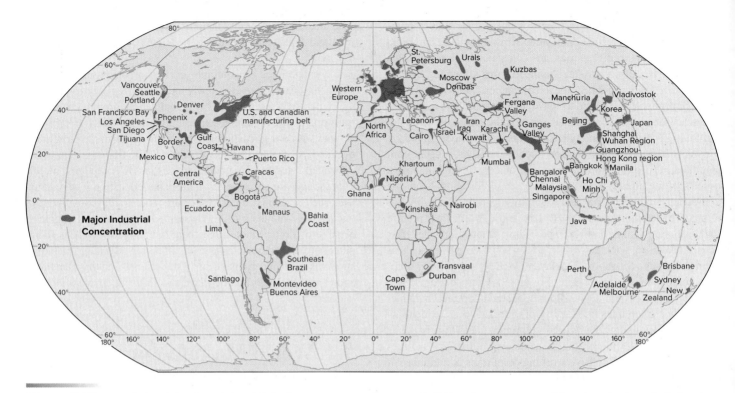

Figure 9.20 World industrial regions. Industrial districts are not as continuous or "solid" as the map suggests. Manufacturing is a relatively minor user of land even in the areas of greatest concentration.

infrastructure investments, and policies supporting export-oriented industrialization. Following them came the Asian dragons: Malaysia, Indonesia, and Thailand. The largest of all NICs is China, the world's most populous country, which has seen phenomenal economic growth since the 1970s, passing Japan in 2010 to become the world's second-largest economy after the United States. Latin American NICs include Mexico, Chile, and Brazil.

The spreading use of efficient and secure containerized shipment of high-value goods has been a major contributor to the competitive success of NICs. Even economies that until recently were dominated by the primary sector have become important players in world manufacturing. Foreign branch plant investment in low-wage Asian, African, and Latin American states has not only created an industrial infrastructure but also increased their gross national products (GNPs) and per capita incomes sufficiently to permit expanded production for growing domestic—not just export—markets.

Much of that new plant investment and expanded developing country industrial production has concentrated within the great number of *export processing zones (EPZs)* recently created within those countries. An EPZ may be either a delimited geographical area or, frequently, an export-oriented manufacturing enterprise located anywhere within a host country that benefits from special investment incentives. These incentives usually include exemptions from customs duties, preferential treatment from various regulatory and financial regulations, and the provision of high-quality infrastructure—airports, highways,

telecommunications, and electric and water facilities—usually provided by the local or national governments. Enterprises operating within or as an EPZ usually enjoy preferential conditions under which they can import equipment, components, and raw materials duty free to produce goods mainly for export. And exports from those zones generally are afforded tariff reductions or duty-free entry into receiving European and North American markets. Because of their obvious production-site advantages, therefore, EPZs are both favored locations for transnational corporation outsourcing and an economic development tool for developing countries competing for TNC investments.

The importance of manufacturing in the United States and Canada has been steadily declining. In 1960, the 28 percent of the labor force that engaged in manufacturing generated nearly one-third of the region's wealth. By 2010, manufacturing employment had dropped to just over 10 percent of a much larger labor force.

Deindustrialization—the declining relative share of manufacturing in a nation's economy—has picked up pace in the past two decades. Outsourcing and the new international division have shifted the spatial patterns of industrial production. While the map of production is dynamic, communities are fixed in space and can be devastated by the closure of large manufacturing plants. Cities that lose major employers can enter a downward cycle of falling incomes, declining tax revenues, and higher social services costs. High unemployment, closed factories, closed stores, abandoned houses, and less money for roads and schools become the norm. Between 1998 and 2008, the United States lost about one-fourth of its manufacturing jobs, many of them high-wage jobs. The decline in manufacturing employment was due to a combination of replacing labor with capital (equipment) and overseas competition. Particularly hard hit were the industrial cities of the manufacturing belt and the Southeast, where unemployment rates have been among the highest in the country.

The Industrial Revolution that began in England in the late 1700s and spread to the continent during the 19th century established Western and Central Europe as the world's premier manufacturing regions and the source areas for the diffusion of industrialization across the globe. Europe accounted for 80 percent of the world's industrial output by 1900, although, of course, its relative position has since eroded, particularly after World War II.

Water-powered textile mills in England began the Industrial Revolution, but it was coal that fueled the full industrialization of Europe. Consequently, coal fields were the sites of new manufacturing districts in England, northern France, Belgium, central Germany, the northern Czech Republic, southern Poland, and eastward to southern Ukraine.

9.5 Tertiary Activities

Primary activities are connected directly to the Earth through gathering, extracting, or growing raw materials. *Secondary* industries, we have seen in this chapter, turn the raw materials of primary industry into useful products through manufacturing or processing. A major and growing segment of both domestic and international economic activity, however, involves *services* rather than the production of commodities. These **tertiary activities** consist of business and labor specialties that provide services to the primary and secondary sectors, to the general community, and to individuals.[2] They provide intangible products ranging from education to haircuts, rather than tangible commodities such as finished goods.

As we have seen, regional and national economies undergo fundamental changes in emphasis in the course of their development. Subsistence societies exclusively dependent on primary industries may progress to secondary stage processing and manufacturing activities. In that progression, the importance of agriculture as an employer of labor or a contributor to national income declines as manufacturing expands. Many parts of the formerly underdeveloped world have made or are making that developmental transition, as we shall review in Chapter 10.

In contrast, many of the economically advanced countries that originally dominated the world manufacturing scene experienced deindustrialization in the late 20th and early 21st centuries. Rising labor costs in advanced economies, space-shrinking technologies for communications and transportation, the growth of transnational corporations, technology transfer

---

[2]*Quaternary activities* are a subset of tertiary services that consist of information and administrative services, including media, education, research, and information technology.

Table 9.1			
Contribution of the Service Sector to GDP			
	Percentage of GDP		
Country Group	**1960**	**1980**	**2010**
Low-income	32	30	50
Middle-income	47	46	56
High-income	54	59	73
United States	58	63	77
World		55	70

Source: *Data from World Bank,* World Development Indicators, 2011.

to developing countries, and outsourcing of processing or assembly work have produced a new international division of labor. The earlier competitive manufacturing advantages of the developed countries could no longer be maintained and were replaced by a new focus on service activities. Based on the contribution of each sector to their GDPs, it is the advanced economies that have most completely made that transition and are often referred to as *postindustrial* (Table 9.1).

Perhaps more than any other major country, the United States has reached postindustrial status. Its primary sector component fell from 66 percent of the labor force in 1850 to 1 percent in 2010, and the service sector rose from 18 percent to 86 percent (Figure 9.21). Virtually all job growth in the past two decades occurred in services. Comparable changes are found in other countries. Today, between 65 percent and 80 percent of jobs in such economies as Japan, Australia, Canada, Israel, and all major Western European countries are also in the services sector.

The significance of tertiary activities to national economies and the contrast between more developed and less-developed states are made clear not just by employment but also by the differential contribution of services to the GDPs of states. The

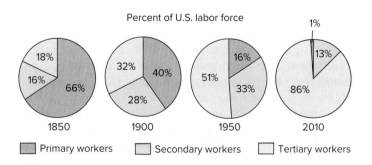

Figure 9.21 Changing sectoral allocation. The changing sectoral allocation of the U.S. labor force demonstrates the transition from a largely agricultural to postindustrial status.

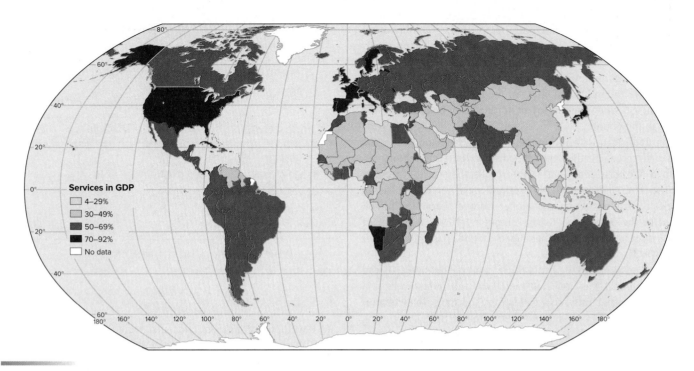

Services in GDP
- 4–29%
- 30–49%
- 50–69%
- 70–92%
- No data

Figure 9.22 Services accounted for 70 percent of global GDP in 2010, up sharply from 55 percent 30 years earlier. As the map documents, the contribution of services to individual national economies varied greatly; Table 9.1 indicates all national income categories shared to some degree in the expansion of service activities.

Source: World Bank, World Development Indicators.

relative importance of services displayed in **Figure 9.22** shows a marked contrast between advanced and subsistence societies. The greater the service share of an economy, the greater the average incomes and economic complexity of that society. That share has grown over time among most regions, and all national income categories as all economies have shared to some degree in economic growth and integration into the world economy. Indeed, the expansion of the tertiary sector in modernizing East Asia, South Asia, and the Pacific has exceeded the world average in recent decades as these regions catch up. In Latin America and the Caribbean, for example, services accounted for 64 percent of total output in 2010.

Types of Service Activities

Tertiary and *service* are broad, imprecise terms that cover a range of activities, from neighborhood barber to college professor to World Bank president. The category includes both traditional low-order consumer and retail activities and higher-order, knowledge-based professional services performed primarily for other businesses, not for individual consumption. Based on who purchases the services, we distinguish between **consumer services** and **producer services.** Consumer services are performed for individuals and include entertainment, tourism, restaurants, hotels, bars, maintenance services, education, health care, and the vast array of personal services. Producer services are performed for corporations and include finance, insurance, real estate, legal services, accounting, architecture, and engineering consulting services. Wholesale and retail trade are

categories of services that link producers and consumers. Transportation and communication services also serve both producers and consumers. In addition, government and nonprofit service providers are important components of the service economy.

Growth in the tertiary sector has numerous explanations. It reflects the development of ever more complex social, economic, and administrative structures, the effects of rising personal incomes and changes in family structure and individual lifestyles. For example, in subsistence economies, families care for their own children, produce and prepare their own food, and build and repair their own houses. In postindustrial societies, people hire childcare workers to care for the children, send their children to formal schools and universities, purchase prepared meals in restaurants, and hire contractors to build and/or repair their houses. Similar needs are met, but with very different employment structures.

As personal incomes rise, a greater proportion of income is spent on services rather than primary products or durable goods. If a person gets a raise, he or she might take a cruise vacation or dine out at restaurants more often, but probably will not add a second washing machine. Growth in the health care industry is driven by both rising incomes and the aging of society that inevitably occurs when a society completes the demographic transition. Growing complexity in the economy translates into the need for higher levels of education and training, as well as more government employees to collect taxes, control borders, alleviate poverty, ensure public safety, plan community development, monitor commerce, protect the environment, and maintain safe workplaces.

Part of the growth in the tertiary component is statistical, rather than functional. We saw in our discussion of manufacturing that *outsourcing* was increasingly used to reduce costs and improve efficiency. In the same way, outsourcing of services formerly provided in-house is also characteristic of current business practice. Cleaning and maintenance of factories, shops, and offices—formerly done by the company itself as part of internal operations—now are subcontracted to specialized service providers. The jobs are still done, perhaps even by the same personnel, but worker status has changed from *secondary* (as employees of a manufacturing plant, for example) to *tertiary* (as employees of a service company).

Locational Interdependence Theory for Services

The locational controls for tertiary enterprises are simpler than those for the manufacturing sector. Service activities are by definition market-oriented. Those dealing with transportation and communication are concerned with the location of people and commodities to be connected or moved; their locational determinants are, therefore, the patterns of population distribution and the spatial structure of production and consumption. Just as Weber offered a classic location theory for manufacturing enterprises, economist Harold Hotelling (1895–1973) used simplifying assumptions to create the **locational interdependence** model for retail services. In the locational interdependence model, the location decisions of firms are influenced by those of its competitors. Firms choose locations that give them a measure of *spatial monopoly* so that they maximize revenues, rather than minimizing costs as in Weber's model.

Imagine the location decisions of two firms in competition with each other, each selling identical goods to customers evenly spaced along a linear market. The usual example cited is of two ice cream vendors, each selling the same brand at the same price along a stretch of beach with a uniform distribution of people. Beachgoers will purchase the same amount of ice cream no matter where the store is located (that is, demand is *inelastic*—is not very sensitive to a change in the price or the effort required to obtain a commodity) and will patronize the store closest to them. **Figure 9.23** suggests that the two sellers would eventually cluster at the midpoint of the linear market (the beach) so that each vendor could supply customers to each side of the market area without yielding locational advantage to the other competitor.

This is a spatial solution that maximizes revenues for sellers but does not minimize costs for customers. The lowest total cost location would be for each vendor to locate at the midpoint of his or her half of the beach, as shown at the top of Figure 9.23, where the total effort expended by customers walking to the ice cream stands (or cost by sellers delivering the product) is least. To maximize market share, however, one seller might decide to relocate immediately next to the competitor (Figure 9.23b), capturing three-fourths of the market. The logical retaliation would be for the second vendor to jump back over the first to recapture market share. Ultimately, side-by-side location at

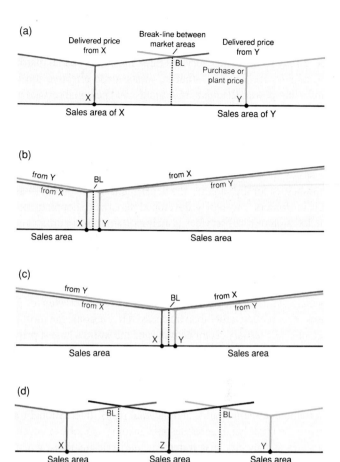

Figure 9.23 Locational interdependence in retail location. The Hotelling model assumes customers evenly spread in a linear market. An example would be vacationers along a beach. (*a*) The initial *socially optimal* locations that minimize total delivery costs (b) will be vacated in the search for market advantage (*c*), eventually resulting in *competitive equilibrium* at the center of the market. This competitive equilibrium poorly serves customers at the periphery. (*d*). Spatial dispersion may occur if another competitor enters the market or the sellers subdivide the market by agreement.

the center line of the beach is inevitable, and a stable placement is achieved because neither seller can gain any further advantage from moving. But now the average customer has to walk farther to satisfy his or her desire for ice cream than initially; that is, the total cost or delivered price (ice cream purchase plus effort expended) has increased. This situation will be a problem for vendors if demand for ice cream is actually *elastic,* which it probably is in most situations. If a third vendor enters the market, the optimal locations for each vendor change to a more dispersed pattern (Figure 9.23d).

The locational interdependence model offers some simple lessons. First, the locational controls for services depend on the locations of both customers and competitors and, under one set of conditions, may produce a clustered pattern and under another set, a dispersed pattern. Second, the Hotelling model suggests that a location solution that optimizes revenue for sellers may not be optimal from the point of view of the customers.

Consumer Services

The supply of consumer services must match the spatial distribution of *effective demand*—that is, wants made effective through purchasing power. Retailers, restaurants, and personal service providers are savvy about locating close to their customers and the most successful chains use *geodemographic* analysis to find optimal locations within cities. Prior to the 1960s, shopping for clothes, furniture, or housewares meant a trip downtown, where nearly all the stores were clustered. However, as middle-class residents left central cities for the suburbs, the department stores quickly chased their customers into newly developed suburban shopping malls. The location of retail services is an important topic in urban geography and receives more attention in Chapter 11.

Tourism

Special note should be made of *tourism*—travel undertaken for purposes of recreation rather than business. It has become the most important single tertiary sector activity, and the world's largest private industry in jobs and total value generated. On a worldwide basis, travel and tourism in 2016 accounted for almost 300 million jobs and about 10 percent of the world's GDP. Domestic tourism leads to spending on transportation, roadside services, lodging, meals, entertainment, theme parks, and national parks. International tourism, on the other hand, generates new income and jobs in developing states as they are "discovered" as tourist destinations, whether for their climate, unspoiled character, or unique culture and cultural landscapes. For half of the world's 50 poorest countries, tourism has become the leading service export sector.

The growth of tourism is part of a broader shift in emphasis from production to consumption that accompanies rising standards of living. Like manufacturing, the tourism industry has experienced a post-Fordist transition away from one-size-fits-all, mass-produced tourist destinations to numerous fragmented consumer niches. Consumers with enough money can choose between cruise vacations, beach vacations, African safaris, ecotourism to "unspoiled" wilderness areas, adventure tourism such as helicopter skiing in the Canadian Rockies or kayaking in Alaska, cultural tourism to exotic places such as Bali or Guatemala, heritage tourism in historic villages and cities, sex tourism, bicycle tours through Europe, wine tourism, gambling, and more (**Figure 9.24**).

The geographic pattern of tourist destinations is highly uneven. Just imagine the different challenges facing a tourism and convention organization working to promote Iowa versus one working on behalf of Hawaii. The geographic features of the destination matter, but so does the level of tourism infrastructure and proximity to potential customers. Modest hills that would hardly qualify as hills in the mountainous western United States have been turned into downhill ski resorts in the Midwest. The important factors are proximity to major cities, such as Chicago or Detroit, and the developer's willingness to add the necessary infrastructure such as snowmaking and lifts. Tourism is an important tool in economic development, but geographers have raised a number of critical questions about the industry. Many of the jobs in the tourism industry are low-skill, low-wage positions such as hotel maids, and the profits often return to developed countries that are the home of the transnational corporations who own and operate the resorts. Tourism can be exploitative, particularly

Figure 9.24 Tourism is the world's largest private industry and comes in many forms. Ecotourism seeks to empower local communities and preserve wildlife habitat and environmental quality.
©DavorLovincic/Getty Images

sex tourism. Tourist visits are often highly seasonal and this creates stresses on the destination, both during the on-season and off-season. Tourist destinations, like other expressions of popular culture, can go in and out of style, destabilizing the local economy. Tourism transforms places, often dramatically, and in some cases undermines the original tourist attraction. For example, in the United States, the areas just outside designated wilderness areas have become magnets for new housing, hotels, amusement parks, and other development. Cultural tourism inevitably changes the culture and cultural landscapes on which it is based. In response, ecotourism has emerged as an ethical form of tourism that focuses on education, minimizing environmental impacts, and using locally owned service providers.

Gambling

Gambling is a fast-growing industry that draws large numbers of tourists and in the process remakes places and local economies. In the United States, the gambling industry attracts almost 15 percent of all entertainment or recreation spending and generates more revenue than professional sports, museums, performing arts, fitness centers, golf courses, or amusement parks. The geography of gambling is determined by legal structures, political boundaries, and proximity to consumers. Gambling was once concentrated in a few select locations where it was permitted: Las Vegas, Nevada; Atlantic City, New Jersey; cruise ships (some of which never left shore); Monte Carlo, Monaco; and Macau, the only Chinese territory where gambling is permitted. The dominance of those gambling centers is being challenged by the rise of lotteries and Internet gambling. The Indian Gaming Regulatory Act of 1988 allowed states to permit casino gambling on Indian reservations, and today there are more than 450 Indian casinos in the United States (**Figure 9.25**). Indian reservations located near major population centers or interstate highways are major beneficiaries, and have often funneled their

Figure 9.25 Indian casinos have proliferated across the United States since the 1988 Indian Gaming Regulatory Act, which recognized the right of tribes to operate casinos as a way of addressing high unemployment and poverty on reservations. More than 450 Indian casinos bring in revenues, which are often invested in education, social services, infrastructure improvements, and natural resource conservation on the reservations. As in all service industries, location is essential, and reservations in or near major metropolitan areas have profited the most.
©Mark Bjelland

substantial profits into improving conditions on the reservation. Reservations in Florida, California, and Connecticut are among the most profitable, although many of the jobs go to outsiders. Unfortunately, reservations in remote locations have usually not benefited from casinos.

Producer Services

Producer services are specialized activities performed for other businesses. They allow producers to realize cost savings by outsourcing specialized tasks when they are needed, without the expense of adding to their own labor force.

One difference between consumer services and producer services is that knowledge and skill-based producer service establishments can be spatially divorced from their clients; they are not tied to resources, affected by the environment, or necessarily localized by market. Of course, when high-level personal, face-to-face contacts are required, service firms will often locate close to their clients, the primary, secondary, tertiary, or quaternary industries they serve. But the transportability of producer services also means that many of them can be spatially isolated from their client base.

As with other industries, the trade-off between costs and proximity is one of the central tensions when a producer services firm chooses office space. The clients for producer services firms are the major companies, many of which are headquartered in the largest cities, where real estate and labor costs are highest.

Producer services are quaternary *knowledge* activities that are highly dependent on communication. The spatial dispersion of some kinds of tasks has been facilitated by innovations in information and communication technologies. Satellite and fiber-optic cables, wireless communications, and the Internet permit the spatial separation of office work into *front-office* and *back-office* tasks. Front-office tasks involve face-to-face interactions with clients where projecting the correct corporate image is imperative. Front-office work requires and can bear the high costs of the most prestigious commercial real estate—in high-quality office buildings with prestigious addresses (Park Avenue, Wall Street) or well-known signature office buildings (Transamerica Tower, Seagram Building). The back office was once literally in the back of the same office building, but now it may be spatially distant from the headquarters of either the service or client firms. Insurance claims, credit card billings, mutual fund and stock market transactions, and consumer help requests are more cost effectively handled in low-rent, low-labor-cost locations—often in suburbs or small towns in rural states—than in the financial districts of major cities. Many suburban back-office operations employ part-time female workers who want to stay close to home because they have primary responsibility for childcare—leading to the label "pink-collar" work. While New York remains the center of the financial sector, the relatively small city of Sioux Falls, South Dakota, has several thousand employees engaged in back-office work for major banks and credit card companies.

Different types of service sector professionals have different locational needs and preferences. Political lobbyists and companies that do consulting work for federal government agencies need to be in Washington, D.C., and often cluster along the famed Capital Beltway. Investment and law firms prefer the most prestigious downtown addresses. Scientific and engineering firms prefer suburban office parks or research campuses. Advertising, architecture, and other design professions often try to project a more relaxed, creative image, frequently choosing old, brick factories or warehouse buildings that have been converted to office space.

The list of services employment is long. Its diversity and familiarity remind us of the complexity of modern life and of how far removed we are from subsistence economies. As societies advance economically, the share of employment and national income generated by the primary, secondary, tertiary, and quaternary sectors continually changes; the spatial patterns of human activity reflect those changes. The shift is steadily away from production and processing, and toward the trade, personal, and professional services of the tertiary and quaternary sectors. That transition is the essence of the now-familiar term *postindustrial*.

9.6 Services in World Trade

Just as service activities have been major engines of national economic growth, so too have they become an increasing factor in international trade flows and economic interdependence. Between 1980 and 2010, services increased from 15 percent of total world trade to 20 percent. Rapid advances and reduced costs in information and communications technology have been central elements in the internationalization of services, as wired and wireless communication and data transmission costs have dropped to negligible levels. Many services considered nontradable, even late in the 1990s, are now actively exchanged at long distance, as the growth of services offshoring clearly shows.

Developing countries have been particular beneficiaries of the new technologies. The increasing tradability of services has expanded the comparative advantage of developing states in labor-intensive service activities, such as data processing. At the same time, they have benefited from increased access to efficient, state-of-the-art equipment and techniques transferred from advanced economies.

That global integration has shifted to higher level economic and professional services. There are clear cost advantages to outsourcing skilled functions such as paralegal and legal services, accountancy, medical analysis and technical services, architectural and engineering design, and research and development. Wired and wireless transmission of data, documents, medical and technical records, charts, and X-rays make distant consumer and producer services immediately and efficiently accessible. Further, many higher-level services are easily subdivided and performable either in sequence or simultaneously in multiple locations. The well-known "follow the sun" practices of software developers who finish a day's tasks only to pass the work to colleagues elsewhere in the world, who then pass it back to them when their workday is over, are now increasingly used by professionals in many other fields. As transnational corporations use computers around the clock for data processing, they can exploit or eliminate time zone differences between home office countries and host countries of their affiliates. Such cross-border intra-firm service transactions are not usually recorded in trade statistics but are part of the growing volume of international services flows. When the practice involves highly educated and talented specialists receiving developing world compensation levels, the cost attractions for developed country companies are irresistible. Increasing volumes of back-office work for Western insurance, finance, accounting, and legal services firms are being performed overseas.

With its large population of well-educated English speakers, India has been particularly successful in attracting outsourced service sector jobs. Customer interaction services ("call centers") formerly based in the United States are now increasingly relocated to India, employing workers trained to use an American nickname and speak in perfect American English (Figure 9.26). Claims processing for life and health insurance firms formerly were concentrated in English-speaking Caribbean states though increasingly such business process outsourcing (BPO) has shifted to India, Eastern Europe, and China. In all such cases, the result is accelerated technology transfer in such key areas as information and telecommunications services.

Despite the increasing share of global services trade held by developing countries, world trade—imports plus exports—in

Figure 9.26 Call center in India. With its large number of well-educated, English speakers, India has benefited from the globalization of back-office service sector work. These New Delhi workers are recent university graduates who work night shifts to cater to European and North American customers.

©Bloomberg/Getty Images

services is still overwhelmingly dominated by a very few of the most advanced states (Table 9.2). The country and category contrasts are great, as a comparison of the "high-income" and "low-income" group documents. At a different level, the single small island state of Singapore has a larger share of world services trade than all of sub-Saharan Africa.

The same cost and skill advantages that enhance the growth and territorial expansion of domestic service sector firms also operate internationally. Principal banks of all advanced countries have established foreign branches, and the world's leading banks have become major presences in the primary financial capitals.

Table 9.2

Shares of World Trade in Commercial Services (Exports, 2016)

Country or Category	% of World
United States	15.1
United Kingdom	6.7
China (with Hong Kong)	6.3
Germany	5.7
France	4.8
Japan	4.2
India	3.3
Singapore	3.1
Ireland	3.0
Netherlands	3.0
High-income states	78.7
Low-income states	0.5
Sub-Saharan Africa	1.2

Source: *Data from World Bank,* World Development Indicators, 2016.

In turn, a relatively few world cities have emerged as international business and financial centers whose operations and influence are continuous and borderless (**Figure 9.27**). The world's key cities for banking, securities firms, and stock exchanges are spread across the globe, allowing almost continuous 24-hour per day trading. Meanwhile, a host of offshore banking havens have emerged to exploit gaps in regulatory controls and tax laws (**Figure 9.28**).

Accounting firms, advertising agencies, management consulting companies, and similar establishments of primarily North American or European origin have increasingly established their international presence, with main branches located in principal business centers worldwide. Those advanced and specialized service components help swell the dominating role of the United States and European Union in the structure of world trade in services.

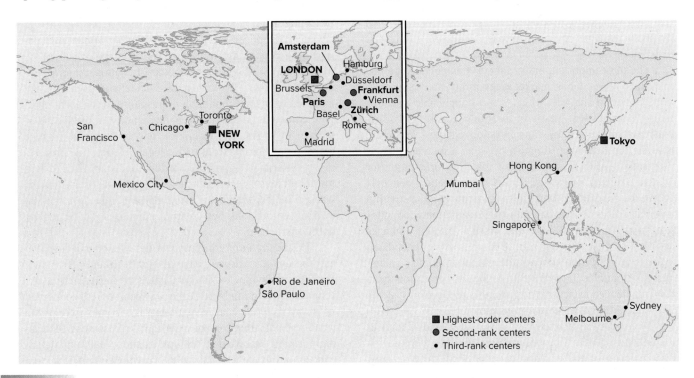

Figure 9.27 The hierarchy of international financial centers, topped by New York, London, and Tokyo, indicates the tendency of highest-order producer services to concentrate in a few world and national centers.

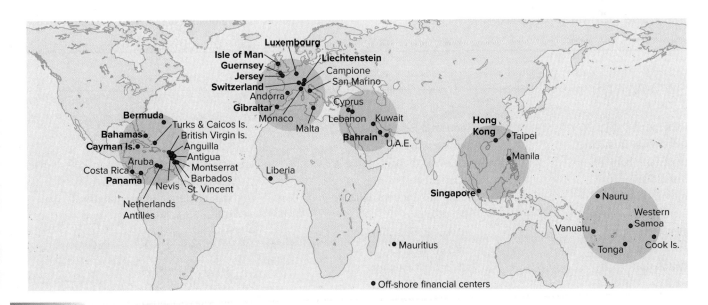

Figure 9.28 Offshore banking. Offshore financial centers, mostly in small island countries and micro-states, allow "furtive money" to avoid taxation and regulatory scrutiny. These financial havens have low tax rates and relaxed financial regulations. They are spread around the world to offer proximity to international financial centers and 24-hour trading. International pressure has led most of the tax havens to agree to greater openness and less protective secrecy.

Source: Peter Dicken, Global Shift, *6th ed. New York: Guilford Press, 2011, Figure 12.9.*

The spatial patterns of the world's manufacturing activities are dynamic and represent the landscape expression of past and recent industrial location decisions. Location theories help us understand how economic behavior motivated by profit responds to variable costs of manufacturing and distribution.

The most important industrial cost components are raw materials, power, labor, market accessibility, and transportation. Weberian analysis, which is most applicable to heavy industry, argues that least-cost locations are strongly or exclusively influenced by transportation charges. Weberian analysis leads us to distinguish between a market orientation for weight increasing and raw materials orientation for weight reducing processes. Profit maximization concepts accept the possibility of multiple satisficing locations within a spatial margin of profitability. Agglomeration economies lead to the growth of industrial clusters through linkages among firms, supporting infrastructure, and a specialized labor force. Location concepts developed to explain industrial distributions under Fordist production constraints have been challenged as new JIT and flexible production systems introduce different locational considerations.

Major industrial districts of the U.S. and Canadian manufacturing belt, Europe, and Eastern Asia contain the vast majority of the world's secondary industrial activity. The most advanced countries, however, are undergoing deindustrialization as newly industrializing countries with lower costs of production compete for markets. The patterns of high-tech industry are highly influenced by agglomeration economies and the location of workers with specialized skills rather than by transportation costs.

In the advanced economies, service activities have become more important as secondary-sector employment and share of gross national income decline. Service activities range from consumer services, such as hair salons, to advanced producer services, such as management and engineering consulting. Markets are a primary consideration in the location of service businesses. Locational interdependence theory suggests that firms situate themselves to assure a degree of market monopoly in response to the location of their competitors. Face-to-face interactions for producer service firms in finance, insurance, law, and accounting take place in prestigious, front-office locations. Routine processing of claims, calls, and bills is done in distant back offices located in lower-cost small cities or overseas. Tourism is the world's largest employer, bringing economic development to diverse locations but creating numerous challenges for host communities.

The nearly empty highway of the Monongahela Valley and the crowded expressways of high-tech and office park corridors are symbols of those changes in North America. As economic activity in North America becomes less concerned with raw materials and freight rates, it becomes freer of the locational constraints of an older industrial society. Increasingly, skills, knowledge, communication, and population concentrations are what attract the growing economic sectors in the most advanced economies. At the same time, much of the less-developed world is striving for the transfer of manufacturing technology from developed economies and view industrial growth as the path to future prosperity. Those aspirations for economic development and the geographically uneven patterns of development are topics in Chapter 10.

KEY WORDS

agglomeration
agglomeration economies
break-of-bulk point
brownfield site
commodity chains
comparative advantage
consumer services
deindustrialization
deglomeration
external economies
footloose firm
Fordism
foreign direct investment (FDI)
freight rates

Industrial Revolution
infrastructure
least-cost theory
line-haul costs
locational interdependence
maquiladoras
market equilibrium
market orientation
material orientation
multiplier effect
new international division of labor (NIDL)
offshoring
outsourcing
producer services

satisficing locations
secondary activities
spatially fixed costs
spatially variable costs
spatial margin of profitability
substitution principle
tapering principle
terminal costs
tertiary activities
transnational corporations (TNC)
ubiquitous industry
uniform (isotropic) plain
Weberian analysis

FOR REVIEW

1. What are the primary *principles of location* outlined in this chapter? Briefly explain each and note its contribution to a firm's locational search.

2. What is the difference between *fixed* and *variable* costs? Which of the two is of interest in the plant locational decision? What kinds of variable costs are generally reckoned as most important in locational theory?

3. *Raw materials, power, labor, market,* and *transportation* are the "factors of location" usually considered most important in industrial placement decisions. Summarize the role of each, and cite examples of where each could be decisive in a firm's location.

4. What were Weber's controlling assumptions in his theory of plant location? What "distortions" did he recognize that might alter the locational decision?

5. What is deindustrialization, where is it occurring, and what are its effects on communities and workers?

6. How have the concepts or practices of *comparative advantage* and *outsourcing* affected the industrial structure of advanced and developing countries? Which of the "factors of location" is most important in the choice to move production from North America to Mexico or Asia?

7. How are the locational considerations for *high-tech* industries significantly different from those of heavy manufacturing activities? What are the primary factors in the location of high-tech industries in Silicon Valley and other high-tech clusters?

8. What is the difference between *front-office* and *back-office* tasks, and where is each likely to be located?

9. What sorts of challenges does the tourism industry bring to host communities?

10. What are the major centers for finance and stock exchanges? Do you anticipate change in the location of leading financial centers?

KEY CONCEPTS REVIEW

1. **What are the principal elements of locational theory, and how do different classical theories employ them?** Section 9.1–9.2

   Costs of raw materials, power, labor, market access, and transportation are the assumed controls governing industrial location decisions. They receive different emphases and imply different conclusions in the theories considered here. *Least-cost* (Weber) analysis concludes transport costs are the fundamental consideration; *profit maximization* maintains a firm should locate where profit is maximized by using the substitution principle.

2. **How do agglomeration, JIT, comparative advantage, and TNCs affect location outcomes?** Section 9.2

   By sharing infrastructure, agglomerating companies may reduce their individual costs, while JIT supply flows reduce their inventory capital and storage charges. Comparative advantage recognizes that different regions or nations have different industrial cost structures. Companies use *outsourcing* to exploit those differences. Transnational corporations distribute their operations based on comparative advantage: manufacturing in countries where production costs are lowest; performing research, accounting, and other service components where economical or convenient; and maintaining headquarters in locations that minimize taxes. Outsourcing and TNC practices get away from the concept of a single optimal location that was the goal of classical location theories.

3. **What influences high-tech activity location, and what is the impact of high-tech growth on established world manufacturing regions?** Section 9.3–9.4

   Long-established industrial regions in the U.S. and Canadian manufacturing belt and Western, Central, and Eastern Europe developed over time in line with classical location analysis. Eastern Asia, the most recently developed major industrial region, has been influenced by both classical locational pulls, economic development goals of newly industrializing countries and outsourcing in search of lower-cost, reliable labor sources. High-tech industries tend to create regionally specialized agglomerations reflecting proximity to scientific research centers, technically skilled labor pools, venture capital availability, quality of life, and superior transport and communication facilities. Their emergence has altered traditional industrial emphases and distributional patterns.

4. **What are the different types and locational characteristics of service activities and how are they reflected in world trade patterns?** Section 9.5–9.6

   Tertiary industries include all non-goods production activities and provide services to goods producers, the general community, and individuals. Subdivided into consumer services and producer services, the tertiary sector is broad and employs most workers in advanced economies. The location of customers is an important consideration for both consumer and producer services. Front-office and back-office tasks are often spatially distant. The growing world trade in services, made possible by plummeting costs of information transmission, has altered international economic relations and encouraged cultural and functional integration.

CHAPTER
10
ECONOMIC DEVELOPMENT AND CHANGE

Women learning computer skills in a rural village in India.
©Visuals Stock/Alamy Stock Photo

Key Concepts

*I*n January 2010, the world's attention turned for a few days to Haiti, where a 7.0-magnitude earthquake struck near the capital city of Port-au-Prince. The powerful quake shook an estimated 3.5 million of Haiti's 9 million people. The devastation was astounding. As bodies were pulled from the rubble, the estimated death toll climbed to more than 220,000, while the true number will probably never be known. After the quake, inadequate food supplies, untreated injuries, and disease outbreaks raised the death toll even higher. With most of the country's infrastructure concentrated in the primate city of Port-au-Prince, the damage to the country's government and systems of transportation and utilities was a major setback. Key structures destroyed or seriously damaged including the presidential palace, national assembly building, national penitentiary, main cathedral, the airport control tower, port docks and cranes, luxury hotels, hospitals, and schools. Approximately 1.5 million Haitians were left homeless as their houses collapsed in the quake.

Six years later, long after the world's media attention moved on, an estimated 60,000 displaced Haitians still lived in overcrowded tent camps in parks, golf courses, and vacant lots, suffering from poor sanitation, crime, and inadequate food and water supplies (*Figure 10.1*). Hurricane Matthew in 2016 added to the loss of life and housing. With little to show for much of the promised international assistance, Haitians returned to rebuilding their country.

Figure 10.1 One of many tent camps for displaced persons after the January 2010 earthquake in Haiti. This camp was located in the capital city of Port-au-Prince, on Haiti's only golf course. About 1.5 million of Haiti's 9 million people were made homeless by the earthquake. Two years later, hundreds of thousands were still living in tent camps
©Anna Versluis

10.1 An Uneven World

News reports often refer to catastrophes such as the Haitian earthquake as "natural disasters." But the staggering death toll in Haiti would suggest that something more was going on. In recent years, earthquakes of similar strength have hit California, New Zealand, Chile, Indonesia, Japan, and Italy, with far less damage and loss of life. The devastating human consequences of the Haitian earthquake were not due to nature alone. Rather, poverty and the lack of private and public resources, public services, institutions, and social safety nets that we associate with "development" made this natural event much more devastating and the recovery much slower. Haiti is the poorest country in the Western Hemisphere, and like other developing countries, it has undergone a massive migration from rural areas to the crowded city of Port-au-Prince. Haiti's government is burdened by large international indebtedness for loans taken out by previous, corrupt governments, but in the years prior to the earthquake, the country had made substantial progress in stabilizing institutions and renegotiating its international debts. Haitians were aware of their country's earthquake dangers. But in a poor, indebted country where most houses are makeshift slum dwellings and day-to-day survival is a challenge for much of the population, updating and enforcing building codes, creating disaster preparation plans, and earthquake-proofing even key public buildings was not a priority. Partly to blame for the many deaths were poor construction materials and techniques that could not withstand the forces of earthquakes. Limited financial resources meant little or no government inspection, and builders sometimes succumbed to the temptation to skimp on construction quality. Extreme poverty meant that there was no insurance claim or bank account to fall back upon for many Haitians who lost homes, jobs, and family members.

Any view of the contemporary world quickly shows great—almost unbelievable—contrasts from place to place in levels of economic development and people's material well-being. Such differences are evident in the *artifacts*—energy sources and technologies differing societies use—the kinds of economic activities in which they engage, and in their social organizations—*sociofacts*. The ready distinction we make between the "Gold Coast" and the "slum" indicates that different groups have different access to the wealth, tools, resources, and decision-making power of the global and national societies of which they are a part.

At an international scale, we distinguish between "advanced" or "rich" nations, such as Canada or Switzerland, and "less developed" or "poor" countries, like Bangladesh or Burkina Faso, although neither country may be comfortable with those adjectives. Vast economic differences exist within countries, too. The poverty of rural South Africa stands in sharp contrast to the prosperous, industrialized, urbanized modernity of Johannesburg (**Figure 10.2**). Similarly, in the United States, farmers or coal miners in Appalachia live in a different economic and cultural reality than urban professionals.

(a)

(b)

Figure 10.2 Economic contrasts can be dramatic within a single country. *(a)* The economic power and global connectedness displayed in the city center of Johannesburg, South Africa stands apart from *(b)* the poverty and isolation of much of the rest of the country.

(a) ©Ariadna22822/Shutterstock; (b) ©HenkBadenhorst/Getty Images

10.2 Dividing the Continuum: Defining Development

Countries display different levels of development. **Development,** in that comparative sense, simply means the extent to which the human and natural resources of an area or country have been brought into full productive use. In common usage, it also suggests urbanization, modernization, and improvement in levels of material production and consumption. For some, it also suggests changes in traditional social, cultural, and political structures to resemble more nearly those displayed in countries and economies deemed *advanced*. For others, the concepts of *development* and *underdevelopment* were post–World War II inventions of Western culture. Countries were classified by the degree to which they conformed to Western standards of wealth, well-being, and achievement. Once visualized in this manner, the perceived conditions of underdevelopment could be addressed by international institutions such as the World Bank, International Monetary Fund Bank, and Inter-American Development Bank. Although the economic differences among countries are undeniable, some critics challenge development theory, criticizing it as a conscious means of exerting Western influence and control over postcolonial societies.

Whatever the philosophical merits of the two viewpoints, many of the attributes of development under its usual economic definition can be quantified by referring to statistics of national production, per capita income, energy consumption, nutritional levels, labor force characteristics, and the like. Taken together, such variables might comprise a scale of achievement against which the level of development of a single country may be compared. Such a scale would reveal that countries lie along a continuum from the least advanced in technology or industrialization to the most developed. Geographers (and others) attempt to classify and group countries along the continuum in ways that are useful and informative.

We must be careful in choosing the terms that we use to describe these differences, for we are speaking of vast inequalities in economic and political power, access to resources, and chances of survival. In the usage of many governments and nongovernmental organizations, the term *developed* stands in easy contrast to *underdeveloped, less developed,* or the *developing* world. The term *developing* itself was introduced by President Harry S. Truman in 1949 as a replacement for *backward,* the unsatisfactory and unflattering reference then in use. **Underdevelopment,** from a strictly economic point of view, suggests the possibility or desirability of applying additional capital, labor, or technology to the resource base of an area to permit the present population to improve its material well-being.

The catch-all category of *underdeveloped,* however, does not tell us where countries are on the continuum. With time, therefore, more refined subdivisions of development have been introduced, including such relative terms as *moderately, less,* or *least developed* countries.[1] Because development is commonly understood to imply industrialization and to be reflected in improvements in national and personal income, the additional terms *newly industrializing countries (NICs)* (which was explored in Chapter 9) and *middle-income countries* have been employed. More recently, *emerging economy* has become a common designation, providing a more positive image than "underdeveloped." The United Nations groups the formerly communist Russia, former USSR republics, and southeast European countries into a category termed *transition economies*. In a corruption of its original meaning, the term **Third World** is still occasionally applied to developing countries as a group, though when first used, that designation was a purely

---

[1]In 1971, the General Assembly of the United Nations listed 24 "least developed" countries identified by per capita gross domestic product (GDP), share of manufacturing in GDP, and adult literacy. In later years, the criteria were changed to reflect low national income (per capita GNI under $1,025); weak human assets (a composite index based on health and educational measures), and high economic vulnerability (based on vulnerability to natural disasters, agricultural instability, and inadequate diversification of a small economy). In addition, population of a "least developed country" had to be below 75 million. The list of those countries—also recognized as "poorest" countries—has grown over the years. There were 47 countries on the list in 2018: 32 in Africa, 9 in Asia, 5 small island states in Oceania, and Haiti in the Caribbean. Only five countries, Botswana, Cabo Verde, Equatorial Guinea, Maldives, and Samoa have "graduated" from the list. See Figure 10.3.

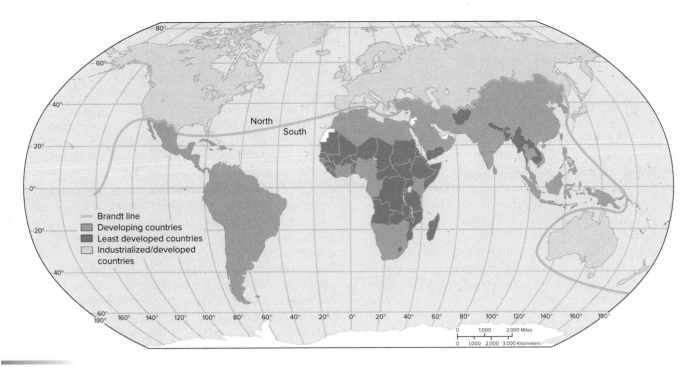

Figure 10.3 The "North-South" line of the 1980 *Brandt Report* suggested a simplified world contrast of development and underdevelopment based largely on degree of industrialization and per capita wealth. Following the dissolution of the USSR in 1991, the former Central Asian and trans-Caucasian Soviet republics were added to the "South." Based on measures of income, health, education, and economic vulnerability, 47 countries are classified by the United Nations as the "least developed countries." The broad category of "developing countries" ignores recent significant economic and social gains in many Asian and Latin American states, raising them now to "industrialized/developed" status. Some "least developed" states are small island countries not shown at this map scale.

Sources: UNCTAD and United Nations Development Programme.

political reference to nations not formally aligned with a *First World* of industrialized free market (capitalist) nations or a *Second World* of centrally controlled (communist-bloc) economies. In addition, the name *Fourth World* has sometimes been attached to indigenous peoples or other marginalized ethnic groups. Further, some development workers have altered the term *Third World* to *Two-Thirds World* or *Majority World* as a reminder that most of the world's population resides in developing countries.

Because all these terms clearly suggest the possibility of a country progressing from a lower to a higher developmental status, one would expect agreement on which category is applicable to a specific country and when a state has advanced from one category to another. Unfortunately, there is no agreed-upon classification. In common practice, the United Nations considers the United States, Canada, Japan, Australia, New Zealand, and most of Europe to be "developed" and the rest of Asia, the Americas, Caribbean, and Oceania to be "developing." While most of Europe is classified by the United Nations as "developed," the exceptions are five formerly communist states in southeastern Europe that are grouped with Russia, Georgia, and the Commonwealth of Independent States as "economies in transition." Various international agencies reach different conclusions: for the United Nations, Singapore and South Korea are both "developing" and "high-income" economies; for the International Monetary Fund (IMF), they are "advanced economies." And to equate "advanced" with "industrialized" economies, which is commonly done, neglects the fact that these economies are increasingly postindustrial service economies.

In 1980, the contrasting terms *North* and *South* were introduced (by the Independent Commission on International Development Issues, commonly called the Brandt Report[2]) as a broad generalization emphasizing the distinctions between the rich, advanced, developed countries of the Northern Hemisphere (to which Australia and New Zealand are added)—the *North*—and, roughly, the rest of the world—the *South* (**Figure 10.3**). This split agreed with the UN classification that placed all of Europe and North America, plus Australia, Japan, New Zealand, and the former USSR in a *more developed country* category, with all other states classed as *less developed countries.*

The variety of terms devised—not all of them accurately descriptive or acceptable to the countries that they designate—represent sincere efforts to categorize countries along a continuum of economic and social characteristics. In the remainder of this chapter, broad developmental contrasts between countries or regions will conform to the "North-South" and the United Nation's "developed-developing" categorizations. Although not ideal, we use the terms *developed* and *developing* because of their common usage in reports and statistical databases issued by the United Nations and other organizations. Our primary

[2]*North-South: A Programme for Survival.* The commission was established in 1977 at the suggestion of the chairman of the World Bank. Under its charge, "global issues arising from economic and social disparities of the world community" were to be studied and "ways of promoting adequate solutions to the problems involved in development" were to be proposed. The former Soviet Union was included within the North at that time; since the breakup of the Soviet Union in 1991, Georgia, Uzbekistan, and other former Soviet republics in Asia have been classified as "less developed" by the United Nations.

attention in maps and text, however, will be given to the developing countries of the Third World or "global South."

The terminology of development is usually applied to country units, but it is equally meaningful at the regional and local levels within them, for few countries are uniformly highly developed or totally undeveloped. Many emerging economies contain pockets—frequently the major urban centers—of productivity, wealth, and modernity not shared by the rest of the state. For example, Mexico is a leading NIC, but its manufacturing activity is concentrated in northern states and in metropolitan Mexico City. Many other parts of the country, and particularly its amerindian population in the south, remain untouched by industrial development. Similarly, Mumbai (Bombay) agglomeration, with less than 2 percent of India's population, dominates India's commercial, film, finance, and manufacturing sectors and boasts per capita incomes three times the national average. Even within the most advanced societies, some areas and populations remain outside the mainstream of progress and prosperity enjoyed by the majority.

And finally, we should remember that *development* is a culturally relative term. It is usually interpreted in Western, democratic, market economy terms that presumably can be generalized to apply to all societies. Others insist that it must be seen in light of the cultural and economic aspirations of different peoples, many of whom specifically reject Western cultural and economic standards.

10.3 Measures of Development

Discussions of development tend to begin with economic measures because income and national wealth strongly affect the degree to which societies can invest in food supplies, education, sanitation, health services, and other components of individual

and group well-being. Indeed, there is a close relationship between economic and social measures of development. Although there are exceptions because societies prioritize things differently, in general, the higher the per capita income, the higher the national ranking in such measures as access to safe drinking water, prevalence of sanitation, availability of physicians, and education and literacy levels.

In contrast, the relationship between social-economic and demographic variables is usually inverse. Higher educational or income levels are usually associated with lower infant mortality rates, birth and death rates, and rates of natural increase. However it is measured, the gap between the most and least developed countries in noneconomic characteristics is at least as great as it is in their economies and technologies. Closing that gap by reducing extreme poverty was the focus of the UN's Millennium Development Goals (MDGs) which coincided with the period 1990–2015. While much work remains, considerable progress was made under the MDG development framework (Table 10.1). Most notably, the number of persons living in extreme poverty was reduced by more than one billion. Building upon the success of the MDGs, the United Nations has launched a more ambitious development framework for the period 2015–2030 (See the feature "Sustainable Development Goals").

Gross National Income (GNI) and Purchasing Power Parity (PPP) per Capita

Two common measures are used to gauge economic activity—gross domestic product (GDP), which was introduced in Chapters 8 and 9, and **gross national income (GNI).** GDP is the total market value of all final goods and services produced annually within the borders of a country. GNI adds to GDP the

Table 10.1

Achievements of the Millennium Development Goals, 1990 to 2015

	Sub-Saharan Africa		Southern Asia		Southeastern Asia		Developing Regions	
	1990	2015	1990	2015	1990	2015	1990	2015
People living on less than $1.25/day, %	57	41	52	17	47	7	47	14
Undernourished people, %	33	23	24	16	31	10	23	13
Under age 5 mortality rate (per 1,000 live births)	179	86	126	50	71	27	100	47
Maternal mortality rate (deaths/100,000 live births)	990	510	530	190	320	140	430	230
Child deliveries attended by skilled personnel, %	43	52	32	52	49	82	57	70
Access to an improved drinking water source, %	48	68	73	93	72	90	70	89
Children in primary education, %	52	80	75	95	93	94	80	91

Sources: United Nations, Millennium Development Goals Report, *2015*

Sustainable Development Goals

In September 2015, the member states of the United Nations adopted a new sustainable development framework with specific goals to "end poverty, protect the planet, and ensure prosperity for all." Compared to the MDGs, this new framework placed greater emphasis on environmental concerns. The 17 interlinked goals that they identified—the Sustainable Development Goals (SDGs)—built upon the successes and shortcomings of the Millennium Development Goals. Convinced that development programs are most successful when targeted to specific, measurable goals, the SDGs create a focus for individual countries, donors, and international agencies.

The SDGs are each accompanied by measurable targets for ending poverty and unsustainable practices; they aim to do the following:

1. End poverty with a focus on those earning less than $1.90/day
2. End hunger and improve nutrition and food security
3. Ensure healthy lives by reducing maternal and child mortality and eradicating diseases
4. Ensure inclusive and quality education with a focus on increasing high school enrollment for girls
5. Achieve gender equality with a focus on reducing discrimination against women and providing equal access to economic and political life
6. Ensure access to clean water and sanitation for all
7. Ensure access to electricity for all through sustainable sources
8. Provide decent employment for all
9. Build resilient infrastructure such as transportation, communications, energy, and irrigation systems
10. Reduce inequalities within and among countries by focusing on the bottom 40 percent of the population
11. Make cities safe and sustainable by upgrading slums and making housing safe and affordable
12. Ensure sustainable consumption and production patterns through reducing waste and efficiently using natural resources
13. Take urgent action to both combat and adapt to climate change
14. Conserve and sustainably use the oceans and marine resources
15. Sustainably manage forests and combat desertification and biodiversity loss
16. Promote peace, justice, and strong institutions by reducing violence, promoting the rule of law, and reducing government corruption
17. Revitalize partnerships for development by ensuring that developed countries follow through on their commitments of financial resources for development assistance

---

total foreign income earned by its citizens (GNI was formerly known as **gross national product,** or **GNP**). To make values comparable, we make three adjustments to GNI. First, we convert each country's currency into a single measure, typically U.S. dollars. Second, we divide by the number of people in a country to get GNI per capita. Third, we note that market currency exchange rates fluctuate with business cycles and trading patterns and do not necessarily correspond to prices in a particular country. For example, you may have noticed that prices for most goods are lower in developing countries, and if you have traveled to Scandinavia, you probably noticed that prices are much higher there than in other developed countries. Thus, a **purchasing power parity (PPP)** correction is often applied to GNI to account for price differences. Let's say that a Big Mac costs $4 in the United States and 40 pesos in Mexico. Then, the purchasing power of $1 equals 10 pesos. Of course, the actual PPP calculation uses a bundle of goods rather than just Big Macs.

Even after applying the PPP correction, it still remains difficult to compare personal income figures across national borders. Personal incomes in Sweden are taxed at much higher rates than in the United States. But social welfare programs, higher education, and medicine receive greater central governmental funding in Sweden; the U.S. family must set aside a larger portion of its income for such services. Further, identical incomes will be spent on different amounts and types of goods and services in different countries. Residents of higher latitudes must buy fuel and heavy clothing not necessary in tropical climates.

The World Bank divides the world's countries into low-income, lower-middle, upper-middle, and high-income categories. As expected, the countries with the highest GNI per capita are those in northwestern Europe, where the Industrial Revolution began, their former midlatitude colonies—United States, Canada, Australia, and New Zealand—and Singapore, a small island city-state that was once a British colony and is now home to prosperous manufacturing and international trading industries (**Figure 10.4**). In the middle position are found many of the countries of Latin America and of southern and eastern Europe. With the exception of Haiti, the low-income countries are in Africa and Asia. Because the money value of the nontraded goods and services that subsistence farmers provide for themselves goes unrecorded in the GNI, the differences between countries are less extreme when the PPP correction is used (**Figure 10.5**).

Like any single development statistic, GNI tells only part of a complex story. Although it is still the primary measure used by the World Bank to gauge economic performance, it is under increasing attack as a poor measure of progress and human well-being. GNI simply measures the flow of money through the economy, without regard to purpose, and without

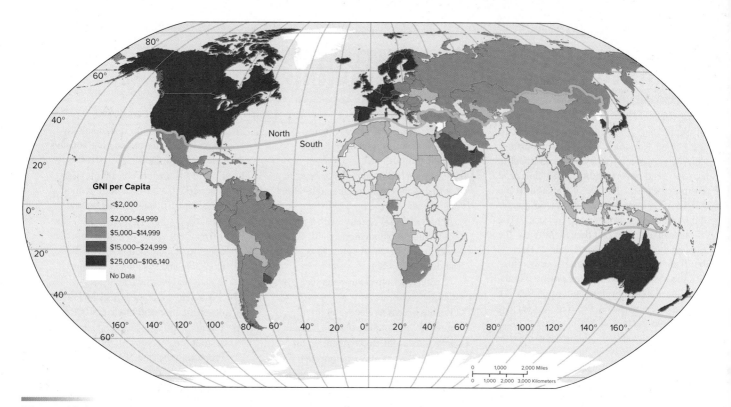

Figure 10.4 GNI per capita, 2016. GNI per capita is a frequently employed measure of economic advancement, although high incomes in sparsely populated, oil-rich countries may not have the same meaning in developmental terms as do comparable per capita values in advanced industrial states.

Sources: Data from World Bank, World Development Indicators, *2018.*

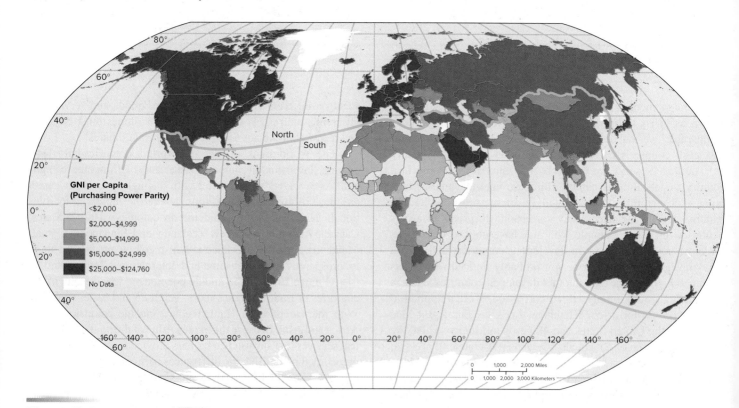

Figure 10.5 GNI per capita adjusted for purchasing power parity, 2016. When local currency measures of GNI or GDP are converted into PPPs, the wide gap between the high-income and least developed countries is reduced somewhat. Many low- and middle-income countries have lower costs of living and thus do better in the PPP analysis. Compare this map to Figure 10.4 to see how PPP changes our impressions of some countries' economic status.

Sources: Data from World Bank, World Development Indicators, *2018.*

regard for the condition of capital stocks that are necessary for productivity. Natural disasters, accidents, and illnesses trigger additional spending and increase the GNI, without increasing well-being. Similarly, increased spending on security or the military increases GNI, but it is at best a necessary evil and a rather poor indicator of societal health. It is quite possible to increase GNI in the short term while destroying stocks of capital. Like other forms of capital, the condition of the natural environment (natural capital) is not measured in GNI. For example, overfishing the oceans will increase GNI for a few years, but in the long term, it will cause ecological collapse that undermines future productivity in that industry.

Of course, GNI per capita says nothing about the distribution of income within a country; it could be evenly distributed across the population or concentrated in the hands of a few. Nor is per capita GNI an accurate summary of developmental status. It overemphasizes the purely monetary circumstances of countries and misses the activities of the **informal economy** that are especially important in developing countries. The informal economy is composed of activities, whether self-employment, work in a family business, or for an employer, that are unlicensed, lack formal contracts, and generate earnings that go unreported. Informal activities include traditional subsistence strategies such as raising one's own food, bartering, and unpaid household labor. Other informal activities include home sewing businesses. waste picking, domestic work, sex work, shoeshining, and some forms of street vending (**Figure 10.6**). Illegal activity is one part of the informal economy. Informal economic activities tend to have low entry requirements for capital, education, and technology and are the major job creators in the poorest parts of the world. The informal economy is a growing proportion of the total labor force and is vital to the livelihoods and economies in Africa, Asia, Latin America, and the Caribbean (Table 10.2).

Table 10.2

Size of the Informal Economy

	Non-Agricultural Jobs	Non-Agricultural Value Added
Sub-Saharan Africa	33–82%	36–62%
Latin America	40–74%	16–32%
Middle East and North Africa	31–59%	17–34%
South and East Asia	33–84%	46%
Eastern Europe and Commonwealth of Independent States	6–20%	9–28%

Source: *International Labour Organization,* Women and Men in the Informal Economy: A Statistical Picture, 2nd ed., 2013.

Energy Consumption per Capita

Per capita energy consumption has been used as another common measure of technological advancement because it loosely correlates with per capita income, degree of industrialization, and use of advanced technology. On average, the industrialized countries use about 10 times more energy on a per capita basis than developing economies do. Energy consumption is an imperfect measure, however, because some advanced countries such as Sweden have prioritized environmental sustainability and reduced their energy usage while maintaining a high standard of living. The consumption rather than the production of energy is the concern. Many of the highly developed countries consume large amounts of energy but produce relatively little of it. Japan, for example, must import energy supplies from abroad because its domestic resources are very limited. In contrast, many less-developed countries have very high per capita or total energy production figures, but primarily export the resource (petroleum). Libya, Nigeria, and Brunei are cases in point. Most of the less- and least-developed countries depend on animate energy (human and animal labor) to do work and firewood, crop residues, dung, and peat for cooking fuels (see the feature "The Energy Crisis in Less-Developed Countries"). The SDGs emphasize the importance of bringing basic electricity to all communities. Advances in practical and affordable technologies are improving that picture through the diffusion of high-efficiency stoves, solar stoves, solar lights, waste matter converters, and solar panels (**Figure 10.7**).

The advanced countries developed their economic strength through the use of cheap, energy-dense fossil fuels and their application to industrial processes. But energy is cheap only if immense capital investment is made to produce it at a low cost per unit. The less-developed nations have been unable to make those necessary investments, or they lack domestic energy resources, widening the gulf between the technological subsystems of the

Figure 10.6 Informal economic activities include this fruit vendor who sells from a bicycle on the streets of Kathmandu, Nepal.

©Erica Simone Leeds

The Energy Crisis in Less-Developed Countries

For the world's poor, the pressing energy issues are not rising prices for gasoline or high natural gas or electric bills during extreme weather. The crisis of the less-developed societies involves cooking food and basic lighting.

The World Bank reported in 2016 that 950 million people lacked electricity. Thus, many people in developing countries live in nighttime darkness. Globally, 3 billion people rely on wood, coal, charcoal, or animal dung for cooking and heating. The highest usage of fuel wood is in the poorest countries, such as Ethiopia and Nepal.

Traditional cooking fuels are typically used in unvented, inefficient stoves. Pollution from indoor cooking on unvented stoves is a major source of disease and premature death, especially among women, who do most of the cooking. Widespread use of traditional fuels has led to deforestation and fuel shortages in the drier areas of Africa, the mountainous districts of Asia, and in the Andean uplands of Latin America. As a result of shortages and deforestation in such widely scattered areas as Nepal and Haiti, families have been forced to change their diets to less nutritious foods that need no cooking. Depletion of forests near villages leads to longer wood gathering trips, taking time away from food- or income-producing activities. Growing populations ensure that the problem of fuel shortages will continue to plague developing countries.

Simple high-efficiency stoves, solar reflector ovens, and backyard fermentation tanks to convert human and animal waste into methane cooking gas (biodigesters) have improved the quality of life for the rural poor. With more efficient stoves, there is less pollution and less need for distant firewood gathering. Increasingly popular in Africa are small rooftop solar panels that can power a cell-phone charger and several indoor lights. In areas far from electrical or telephone lines, such a system allows children to do homework after dark and allows parents to receive telephone money transfers from family members working in distant cities or countries. In Nepal, miniature hydroelectric systems generate power for basic electric service to remote villages. One challenge of these small-scale innovations is that development agencies are comfortable making large loans for the construction of large power plants or hydroelectric dams costing hundreds of millions of dollars, but they do not have systems in place to make millions of hundred-dollar loans.

Figure 10A Solar lighting helps school-children in Africa.
©robertharding/Alamy Stock Photo

Figure 10.7 A woman adds waste to a biogas generator in Nepal. Human, animal, and vegetable wastes are significant energy sources in developing economies such as Pakistan, India, Thailand, and China, where such wastes are fermented to produce methane gas (*biogas*) as a fuel for cooking, lighting, and heating. The simple technology involves only a fermentation tank (foreground) fed with wastes—straw and other crop residues, manure, human waste, and kitchen scraps. These are left to decompose and ferment; the emitted methane gas passes into a large collection chamber and later is drawn through a tube into the farm kitchen. The remaining sludge is used for fertilizer in the fields.

©Christian Ender/Getty Images

rich and the poor countries of the world. Regardless, the desirability of imitating the developing countries' exorbitant energy use is increasingly suspect. If the developing countries were to do so, it would lead to intensified competition for limited, non-renewable energy supplies and rising greenhouse gas emissions (see Chapter 13 for more about this topic).

Percentage of the Workforce Engaged in Agriculture

A high percentage of employment in agriculture (**Figure 10.8**) is almost invariably associated with subsistence agriculture, low per capita gross national income, and low energy consumption—that is, with underdevelopment. Thus, many development programs have focused on commercializing the agriculture sector. The argument is that economic development creates a wider range of occupational choices than those available in a subsistence agricultural society. Mechanization of agriculture increases the productivity of a shrinking farm labor force; surplus rural workers are made available for urban industrial and service employment, and if jobs are found, national and personal prosperity increases. When a labor force is primarily engaged in subsistence agriculture, on the other hand, there is limited

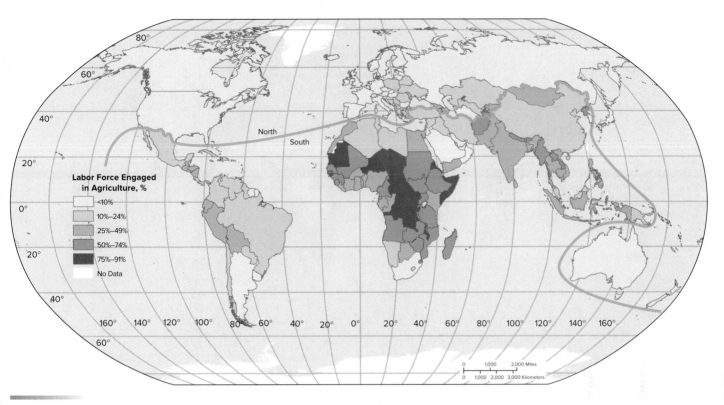

Figure 10.8 Percentage of labor force engaged in agriculture, 2016. For the world as a whole, agricultural workers make up about one-quarter of the total labor force. Highly developed economies usually have relatively low proportions of their labor force in the agricultural sector and have a highly commercialized and industrialized agricultural sector. Rapid population growth in developing countries has resulted in increased rural landlessness and poverty, causing many to migrate to cities.

Source: World Bank, World Development Indicators, *2018.*

capital accumulation or national economic growth. On the other hand, defenders of traditional subsistence agriculture point out that it is often less damaging to the environment, preserves traditional ways of life, and meets the vital food needs of vulnerable poor households.

Food Security and Nutrition

Although tragic famines, such as those suffered by Nigeria, South Sudan, and Yemen in 2017, generate headlines, long-term chronic undernourishment is a frequent outcome of poverty. Undernourishment has a crippling effect on individual well-being and also creates a major obstacle to community development. Hunger kills more people each year than AIDS, malaria, and tuberculosis combined. Availability of urban employment or access to arable land is far more important in determining national levels of undernourishment than a country's total per capita food production. During the Bangladesh famine of 1974, for example, total food availability per capita was at a long-term peak; starvation, according to World Bank reports, was the result of declines in real wages and employment in the rural sector and short-term speculative increases in the price of rice. In India in 2002, huge stockpiles of government-owned wheat rotted in storage while held for sale at prices beyond the

reach of malnourished or starving but impoverished citizens. The 2017 famine in Africa is mostly caused by military conflict and terrorism.

Food, as the essential universal necessity, is the ultimate indicator of economic well-being. Thus, a primary goal of development should be **food security**—the provision of sufficient quantities of safe and nutritious food. Calorie requirements to maintain moderate activity vary according to a person's type of occupation, age, sex, and size, as well as to climate conditions. The Food and Agriculture Organization (FAO) of the United Nations specifies 2,350 calories as the minimum necessary daily consumption level, but that figure has doubtful universal applicability. By way of a benchmark, per capita daily calorie availability in the United States is nearly 3,700. Despite the limitations of the FAO standards, **Figure 10.9** uses them to assess the degree of undernourishment of countries' populations.

Like other national indicators, caloric intake figures must be viewed with suspicion; the dietary levels reported by some states may reflect self-serving estimates or fervent hopes rather than actual food availability. National averages may seriously obscure the food deprivation of large segments of a population. But the data in Figure 10.9 support FAO's 2017 estimate that 815 million people (or roughly one in nine worldwide) were undernourished. The general trend since 2000 has been a significant decline in the

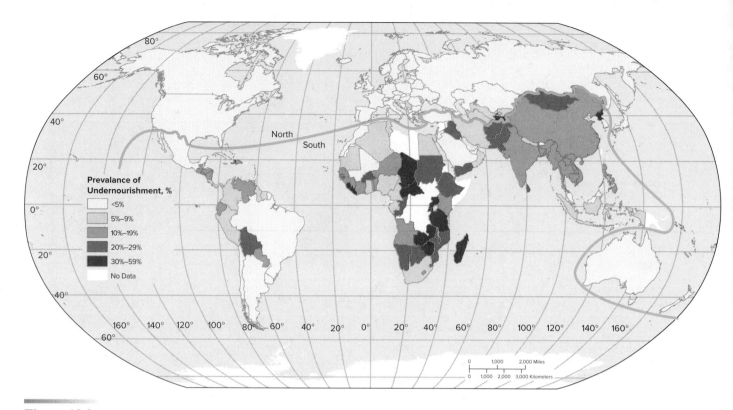

Figure 10.9 Prevalence of undernourishment among the population, 2016. Countries in Africa that are missing data have undergone civil strife and therefore have high rates of undernourishment. In 2016, according to the FAO, there were about 815 million undernourished people worldwide facing chronic hunger or starvation, undernutrition, and deficiencies of essential iron, iodine, Vitamin A, and other micronutrients. In contrast, the FAO indicates that all industrialized countries have average daily per capita caloric intake well above physiological requirements and many face an obesity epidemic.

Sources: Data from World Bank. World Development Indicators, *2018.*

percentage of the world population suffering undernourishment. The absolute number of undernourished persons, however, has declined more modestly because of population growth. Parts of Western Asia and sub-Saharan Africa, particularly South Sudan, Nigeria, Yemen, and Syria, have seen rising undernourishment due to terrorism, civil war, and climate shocks (**Figure 10.10**). Despite the sobering world statistics, a number of developing countries have succeeded in reducing hunger levels. Most of the improvements have been in East Asia, Southeast Asia, and Latin America.

Low caloric intake is usually coupled with lack of dietary balance, reflecting an inadequate supply of the carbohydrates, proteins, fats, vitamins, and minerals needed for optimum physical and mental development and maintenance of health. The WHO estimates that more than 2 billion people worldwide suffer from some form of micronutrient malnutrition that leads to high infant and child mortality, impaired physical and mental development, and weakened immune responses. Dietary insufficiencies in the first years of life retard later physical and cognitive development and increase vulnerability to infections, creating a cycle of malnutrition and poverty. In 2016, the United Nations estimated that 155 million children under age 5 suffered stunted growth. The problem is most critical in rapidly growing countries that have large proportions of their populations in the young age groups (see Chapter 4). South Asia, Southeast Asia, and sub-Saharan Africa show the highest incidence of childhood

stunting and wasting as measured by standardized weight-for-age and weight-for-height measures. At the same time, 13 percent of the world's population is considered obese, posing a different set of health concerns. In upper-middle and high-income countries

Figure 10.10 Malnourished Sudanese children receive assistance at an aid center. The uncertain supplies of food dispensed by foreign aid programs and private charities are not sufficient to assure them of life, health, vigor, or normal development.

©BRENNAN LINSLEY/AP Images

food in security and obesity often co-exist. Poverty and food insecurity are often accompanied by low access to nutritious food and to energy-dense diets that lead to obesity.

Education

A literate, educated labor force is essential to take advantage of advanced technology and to compete for skilled jobs in the global economy. Yet in the poorest societies, half or more of adults are illiterate; for the richest, the figure is 1 percent or less (**Figure 10.11**). The problem stems in part from a national poverty that denies funds sufficient for teachers, school buildings, books, and other educational necessities. In part, it reflects the lack of a trained pool of teachers and the inability to expand their number rapidly enough to keep up with the ever-increasing size of school-age populations. In African countries worst hit by the AIDS epidemic, deaths among existing teachers exceeded the supply of new teachers entering the profession in the late 1990s. For the same number of potential pupils, the richest countries may have 10 times as many teachers as do the poorest countries. Poverty and war-stricken countries such as the Central African Republic, Eritrea, and Somalia have the lowest number of teachers per capita.

Family poverty makes tuition fees prohibitive and keeps millions of school-age children in full-time work. When Burundi abolished primary school tuition fees in 1999, enrollment increased threefold. The largest numbers of school-age children not enrolled in school are found in sub-Saharan Africa and southern Asia. However, during the period of the MDGs, the world's poorest countries made tremendous improvements in education. Despite recording the most rapid population growth, sub-Saharan Africa increased the percentage in school from 52 percent to 80 percent. The number of children not in school was cut in half. The gender gap in education has been closing in recent decades, but it still is apparent in Africa. South Asia has gone from having a female:male ratio in primary schools of 64:100 to 103:100 today. Closing the gender gap has important development consequences, as seen in the correlations between levels of female education and birth rates, family size preferences, family nutrition practices, health maintenance, and life expectancies. The urban-rural education gap persists, however, with higher percentages of children out of school in rural areas compared to urban areas.

Safe Drinking Water and Sanitation

Development implies more than industrial expansion or agricultural improvement. Safe drinking water and toilets, while taken for granted in the North, make profound contributions to human health and quality of life (see Figure 4.18 in Chapter 4). Safe water supplies and sanitation go together because fecal contamination causes many water-borne diarrheal diseases such as cholera, dysentery, and typhoid fever. The WHO estimates that 525,000 children die each year due to diarrheal diseases, making it the second leading cause of death in children under 5 years old. Diarrheal diseases have a disproportionate effect on the very young and contribute to malnutrition and stunted growth. Lack of a nearby improved water supply often forces people, usually

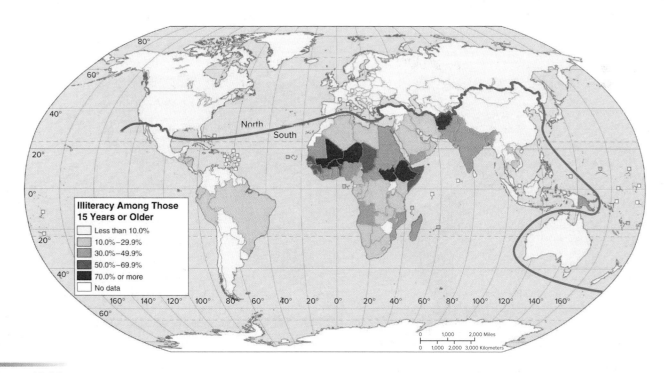

Figure 10.11 Adult illiteracy rate, 2012, as a percentage of the adult population (over 15 years of age). Illiteracy is defined as not being able to read and write short, simple statements relating to everyday life. With almost no exceptions, adult literacy is close to 100 percent in countries of the North. With only a few exceptions, literacy rates in all countries of the South improved dramatically during the period of the Millennium Development Goals. For the least developed countries, the improvement was from 45 percent literate in 1990 to 65 percent literate in 2016.

Source: Sutton, Chris. Student Atlas of World Geography, *8th ed., Map 40. Data from World Bank,* World Development Indicators, *2012.*

women, in developing countries to spend hours every day carrying water for their families.

Worldwide, about 660 million people lacked a dependable sanitary supply of water even though 2.6 billion have gained access to a private or shared piped water supply since 1990. A total of 2.4 billion people lack basic sanitation (**Figure 10.12**). While nearly a billion people worldwide must rely on open defecation, that is a reduction of almost half compared to 1990. The problems of water and sanitation are most pronounced in urban slums and in rural areas (**Figure 10.13**). In fact, in all world regions rural areas lag behind urban areas by substantial margins. However, as Table 10.1 notes, disparities in access to safe water are being steadily reduced. Improvements to sanitation lag behind safe water improvements, presumably because donors and aid agencies find water projects more attractive than sewage projects.

Health

Access to medical facilities and personnel is another spatial variable with profound implications for the health and well-being of populations. Within the less-developed world, vast numbers of people are effectively denied the services of physicians. The number of persons served by each physician varies widely between countries, from a low of 130 people per physician in Cuba to 40,000 per physician in Ethiopia. The shortage of doctors is a crisis in many sub-Saharan African, Central American, and South Asian countries. In the developing world, there are simply too few trained health professionals to serve the needs of expanding populations. Those few who are in practice tend to congregate in urban areas, or they leave for better pay in developed countries. Rural clinics are few in number and the distance to them so great that many rural populations lack access to even the most rudimentary medical treatment.

The **brain drain** of well-educated medical professionals leaving poor countries to work in developed countries is a major barrier to improving global heath. Developed countries such as the United States simply do not produce enough nurses and general-practice physicians for their needs. Instead, the United States imports about a fourth of its practicing physicians from abroad, mostly from developing countries. A recent study showed more Ethiopian doctors practicing in the Chicago metropolitan area than there are doctors in the entire country of Ethiopia, despite its population of more than 100 million people.

Health-related contrasts between advanced and developing countries have become matters of international concern and attention. Indeed, the SDGs contain targets that deal directly with child mortality, maternal mortality, and eradication of disease. We saw the importance in Chapter 4 of the transfer of advanced medical and public health technologies: insecticides, antibiotics, contraception, and immunization, for example. Most recently, childhood diseases and deaths in developing countries have come under coordinated attack by the WHO. Gains have been impressive, yet stark contrasts remain between most developed and least developed societies. The child mortality rate in developing countries dropped by more than half during the period of the MDGs. Still, 9 percent of sub-Saharan African children and 5 percent of south Asian children do not live to their fifth birthday.

Taken at their extremes, advanced and developing countries occupy two distinct worlds of disease and health. One is affluent; its death rates are low, and the chief killers of its mature

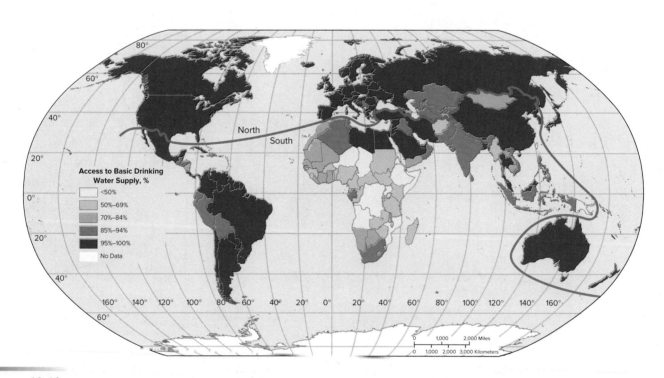

Figure 10.12 Percentage of population with access to a basic drinking water source, 2016. This map shows substantial improvements in the availability of improved drinking water supplies compared to rates in the 1990s. Still, rural areas lag behind urban areas.

Source: World Bank, World Development Indicators, *2018*

Figure 10.13 Because they have no access to safe drinking water or sanitary waste disposal, impoverished populations of a developing country's unserved rural districts and urban slums—like this one in Capetown, South Africa—are subject to water-borne and sanitation-related diseases.
©Louise Gubb/The Image Works

populations are cancer, heart disease, chronic respiratory disease, strokes, diabetes, and suicide. The other world is impoverished and prone to disease. The deadly dangers to its youthful populations are infectious, respiratory, and parasitic diseases made more serious by malnutrition.

As mentioned above, diarrheal diseases are concentrated in areas lacking safe drinking water supplies and sanitation (**Figure 10.14a**). Unfortunately, resurgence of old diseases may disrupt or reverse the hoped-for transition to better health in many world areas (See the feature "Our Delicate State of Health," Chapter 4). Almost 10 percent of the world's population now suffers from one or more tropical diseases such as malaria, many of which were formerly thought to be eradicable but now are spreading in drug-resistant forms. Malaria-carrying mosquitoes are mostly found in the tropics, especially sub-Saharan Africa, where that disease is a major barrier to human health and development (**Figure 10.14b**). Globally there are about 250 million malaria cases and one million malaria deaths each year. According to the WHO, an average African child suffers malarial fever two to five times each year. Solutions combine insecticide-treated mosquito nets, insecticide spraying, and anti-malarial drugs. Another such scourge, tuberculosis, is appearing as a major concern, especially among poorer populations outside tropical regions.

Low-income countries are also hard hit by the spread of AIDS. In 2016, more than two-thirds of persons living with HIV/AIDS were in sub-Saharan Africa (**Figure 10.14c**). Medicines to control the advance of HIV/AIDS were for a long time beyond the reach of most Africans infected with the disease. But agreements between the United Nations, pharmaceutical companies, the United States, and African governments brought down the price of antiretroviral drugs that treat HIV and reduce its spread. In 2015, the United Nations concluded that spread of HIV had been halted and reversed. The number of HIV infections and AIDS-related deaths were much lower than in 2000, averting 30 million new infections and 8 million deaths over a 15-year period. The SDGs call for ending the AIDS epidemic by 2030. The emergence of new threats such as severe acute respiratory syndrome (SARS), Ebola, and Zika virus will continue to challenge public health authorities (**Figure 10.15**).

Technology

Technology refers to the totality of tools and methods used by a culture group to produce items for subsistence and comfort. We saw in Chapter 2 how in antiquity, there emerged *culture hearths*—centers of technological innovation, new ideas, and techniques that diffused from the core region. The ancient

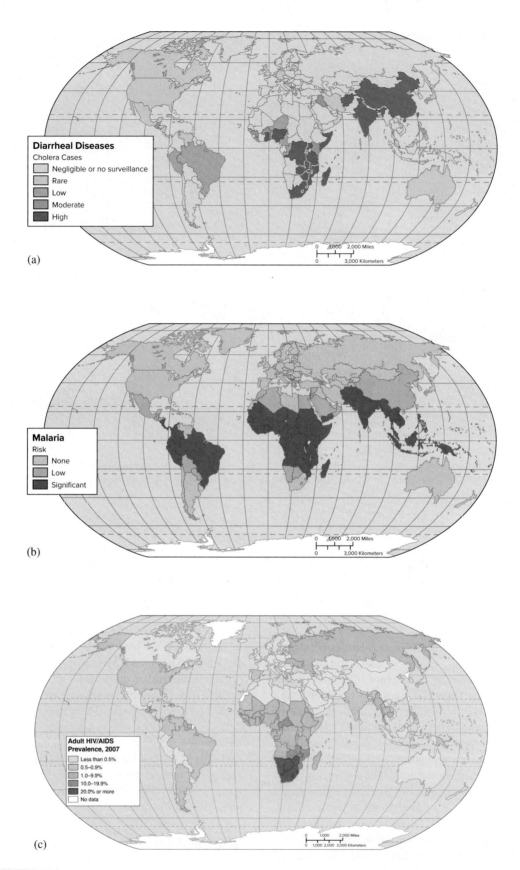

Figure 10.14 *(a)* Diarrheal diseases such as cholera are spread by bacterial contamination of drinking water supplies. They can be prevented by proper sanitation and improved drinking water systems. *(b)* Malaria is caused by parasites carried by a type of mosquito that thrives in the warm, moist conditions found in tropical locations. It killed 655,000 people in 2010, mostly African children. Malaria has had a measurable negative effect on the GNI of the most affected African countries. Prevention and control efforts are decreasing rates of infection. *(c)* In 2010, an estimated 34 million people were living with HIV/AIDS, 23 million in Africa where the disease first emerged.

Figure 10.15 A Guinean infection control supervisor trains health care workers in proper hand-washing technique during the 2014 Ebola outbreak.

Source: Centers for Disease Control and Prevention/Conne Ward-Cameron

hearths (see Figure 2.14) were locales of ongoing invention and innovation. Their modern counterparts are the highly urbanized, industrialized advanced nations whose creativity is recorded by patent registrations and new product and process introductions.

A **technology gap** has always existed between hearths of innovation at the core and the periphery. That gap widened with the Industrial Revolution and has continued to grow with innovations in railroads, steelmaking, electrical engineering, chemistry, automobiles, petrochemicals, computers, and information and communications technologies. In the modern world, as we saw in Chapter 2, there is a widespread sharing of technologies, organizational forms, and cultural traits. But not all countries are equally able to draw on advanced technology.

The technology gap matters. Understandably, all countries aspire to expand their resource base, increase its support levels through application of improved technologies, or enter more fully into income-producing exchange relationships with other world regions through economic development. One objective of development is **technology transfer,** the deliberate introduction of technologies and processes that mark the more advanced countries. Of course, not all technology is equally transferable. Computers, information management techniques, and cell phones easily make the move between advanced and emerging economies (**Figure 10.16**). Although Africa lags in landline telephones, it is the world's second-largest market for cell phones. Other technologies, particularly in the life science, materials innovation, and energy sectors, are more specific to the markets, capital resources, and needs of the rich countries and not adapted to those of the less-developed states. Even where transfer is feasible, imported innovations may require domestic markets sufficient to justify their costs, markets that poor countries will not possess at their current national income levels. And the purchase of technology presumes recipient country export earnings sufficient to pay for it, again a condition not always met by the poorest states.

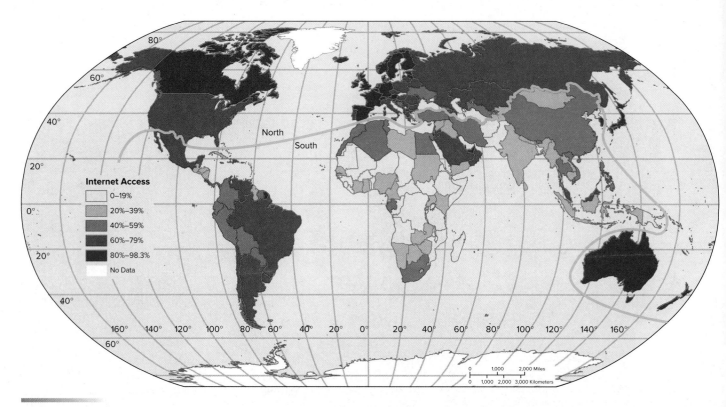

Figure 10.16 Internet use has diffused to all corners of the Earth, yet there are still countries where less than 10 percent of the population has used the Internet. Even in developed countries, certain low density rural areas do not have access to broadband Internet service.

Source: World Bank, World Development Indicators, *2018.*

10.4 Explanations of Development and Underdevelopment

It is one thing to map differences in income or health and to assign countries to categories such as "least developed" or "advanced"; it is quite another to explain the underlying causes and spatial pattern. The widespread disparities in the basic demographic and economic characteristics of countries raises fundamental questions such as why one group of countries is "developed" and another group is "less developed." We observe that the developed countries have high per capita incomes, a small proportion of workers engaged in agriculture, and most of their population living in cities where they enjoy abundant food and high-quality housing, health care, education, and public services. We note that in the least developed countries, incomes are very low, many of their workers are engaged in subsistence agriculture, and the cities are plagued by poor-quality services and slums. We can observe that the developed countries have completed the demographic transition, while the developing countries are in the middle stages of the transition, with higher-than-replacement fertility rates and shorter life expectancies. The statistical differences themselves, however, do not offer a full explanation of development and underdevelopment, nor do they show a clear pathway for countries to follow. Our explanations reflect our theories about how economies and politics work. Explanations are important because they suggest where we should fix blame

and what solutions are most likely to work. Some explanations look to physical geography, others to history, and still others to economic or sociological theories.

Physical Geography

The Brandt Report hints at one frequent but simplistic geographic explanation: Development is a characteristic of the rich "North"— the midlatitudes; poverty and underdevelopment are tropical conditions. Proponents of the latitudinal explanation support their conviction not only by reference to such thematic maps as Figures 10.4 and 10.9, but by noting that rich countries are mostly in temperate climate zones; the world's poorest states are mostly located in tropical latitudes.

Supporters of simple geographic explanations for inequality point out, for example, that Brazilians of the southeastern temperate highlands have average incomes several times higher than their compatriots of tropical Amazonia. Annual average incomes of Mexicans of the temperate north far exceed those of the southern Yucatán. Australians of the tropical north are poorer than Australians of the temperate south.

Influential economist Jeffrey Sachs argues for the contributing factors of latitude and climate in global economic inequality. He notes that in 1820, tropical regions had a per capita GNP 70 percent of that of the temperate regions. Economic growth in the temperate regions outpaced the tropics and by 1992, per capita

GNP in the tropics was just 25 percent that of the temperate regions. Tropical regions face the major ecological handicaps of low agricultural productivity, challenging soil conditions, and substantially higher incidence of plant, animal, and human disease. Confounding the search for a simple explanation, however, many of the poorer nations of the "South" lie partially or wholly within the midlatitudes or at temperate elevations—Afghanistan, North Korea, and Mongolia are examples—while equatorial Singapore and Malaysia prosper. Physical geography matters, but it is not destiny.

Other simple geographic explanations have some merit, yet are ultimately inconclusive:

1. Resource poverty is cited as a limit to developmental possibilities. Although some developing countries are deficient in raw materials, others are major world suppliers of both industrial minerals and agricultural goods—bauxite, cacao, and coffee, for example. Admittedly, a developing world complaint is that their commodities are underpriced in world markets or are restricted by tariffs and subsidized competitors. Those, however, are matters of marketing, politics, and economics, not of resources. Further, economists have long held that reliance on natural resource wealth and exports by less-developed countries undermines their prospects for growth by interfering with their development of manufacturing industries.

2. Overpopulation and overcrowding are frequently discussed as causes of poverty and underdevelopment, but Singapore prospers, with some 7,900 people per square kilometer (20,500 per square mile), while impoverished Mali is nearly empty, with 15 per square kilometer (39 per square mile) (**Figure 10.17**).

3. Landlocked countries have reduced access to global markets and greater costs for transport of goods. For theorists who believe that global trade is the key to development, reduced access to foreign markets is seen as a major impediment. Kazakhstan, Afghanistan, Chad, Niger, Zambia, and Zimbabwe are more than 2,000 kilometers (1,200 miles) from the nearest seacoast. Counterexamples such as Switzerland, which is both poor in resources and landlocked, suggest that these challenges can be overcome with investments in transportation infrastructure.

The Slave Trade and Colonialism

The 500-year history of colonialism played a vital role in shaping the political and economic geography of the contemporary world. Some scholars believe it laid the foundation for the present-day economic differences among countries. It allowed European countries to gain an initial economic advantage by gaining control of territory, natural resources, labor, and markets. Former colonial status is often blamed for underdevelopment because nearly all of the developing countries in the Global South were once colonies. In cases where the colonists largely replaced the original inhabitants—as in Australia, New Zealand, Canada, or the United States—the association of colonial past with present underdevelopment does not hold.

Figure 10.17 Landlocked and subject to severe droughts, Mali is one of the poorest of the least developed countries. Low densities of population are not necessarily related to prosperity, or high densities to poverty. Mali has only 15 people per square kilometer (39 per square mile). In this parched land, Dogon women often spend hours carrying water to their homes. Even in more humid South Africa, rural women on average spend more than three hours each day fetching water, according to a government survey.
©Lissa Harrison

The accusation is arguably valid for countries where—as in sub-Saharan Africa and southern Asia—colonizers left largely intact the indigenous populations but created political structures and physical infrastructures designed for exploitation for mother country profit rather than for balanced economic, social, and political development for the long-term benefit of the colony itself.

The first phase of colonialism dates from Christopher Columbus's first voyage to the New World and peaked just before the American Revolution. Spain and Portugal dominated this phase, which was focused on the Americas and Caribbean islands. The British, Dutch, and French followed in developing their own colonies. The African slave trade played a vital role in this phase of colonialism through the "triangular trade," which sent manufactured goods, liquor, and guns from Europe to Africa, slaves from Africa to the New World, and sugar and tobacco from the New World to Europe. In the early 1800s, the Spanish and Portuguese lost much of their colonial empires to Latin American independence movements. A second phase of colonialism in the late 1800s and early 1900s was dominated by the British and was focused on Africa, Asia, and Oceania. During this period, nearly all the lands in the tropics and subtropics were brought under the control of European powers or the United States (see Figure 12.6). Independence movements in the 20th century, culminating in the independence of Namibia in 1990, brought the colonial era to an end.

One legacy of colonialism is visible in country borders that are often poorly suited to nation-building. African borders were established without regard for the boundaries of ethnic groups.

For example, Gambia straddles the Gambia River and is about 10 kilometers (6 miles) wide on each bank of the river and extends 330 kilometers (205 miles) upstream (**Figure 10.18**). Typically, colonial infrastructure systems were designed to move resources from the interior to coastal ports without developing interconnections or networks to facilitate trade among colonies or develop the interior, Other negative legacies included dependence on the mother country for manufactured goods and authoritarian political institutions that did not transition well after independence.

Modernization Theory

Modernization theory is the most widely accepted understanding of the development process. It dates back to the optimism of the post–World War II era, when the United States began working with European countries and Japan to reconstruct both their war-damaged economies and the global economic order. European countries agreed to begin dismantling their colonial empires and cooperate with the United States, the United Nations, and other non-communist countries in bringing the benefits of science, technology, and industrial progress to the less-developed countries of the world. Modernization theory drew upon sociological and economic theories and shaped the original thinking about development in Western countries. Modernization theory begins by arranging all societies on a continuum with *traditional* on one end and *modern* on the other. Development, according to modernization theory, is the logical progression from traditional to modern as societies adopt the characteristics of advanced societies—advanced technology, urbanization, high per capita incomes, high quality of life, a completed demographic transition, individualism, democracy, and capitalism.

Economic historian Walt Rostow (1916–2003) generalized on the "sweep of modern history," theorizing that all developing economies pass through six stages of growth and advancement:

- *Traditional societies* of subsistence agriculture, simple technology, and poorly developed commercial economies are relatively static and limited to low productivity levels.
- *Preconditions for takeoff* are established when an "external intrusion" initiates political and economic change. The nation-state, rather than kinship units, becomes the primary organizing structure for society. Entrepreneurial elites invest in transportation systems and other productive and supportive infrastructure. Economic growth is accepted as a necessity.
- *Takeoff* to sustained growth is the critical developmental stage, lasting perhaps 20 to 30 years, during which rates of investment increase, farmers adopt mechanization and agriculture is commercialized, new industries are established, resources are exploited, and growth becomes the expected norm.
- *The drive to maturity* sees economic output growing faster than population. The country engages in specialization and international trade. Advanced technology is incorporated into all phases of economic activity; diversification carries the economy beyond the industrial emphases first triggering growth, and the economy becomes increasingly self-sufficient.

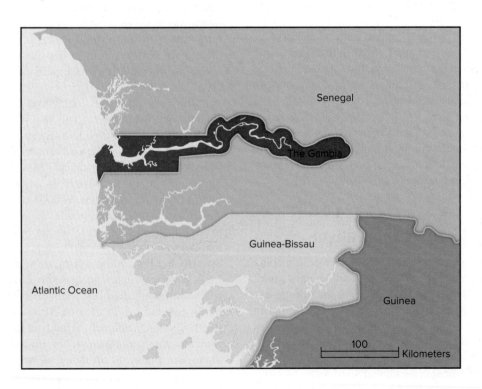

Figure 10.18 The Gambia's borders illustrate a common legacy of colonialism—boundaries designed for the exploitation of natural resources for export markets rather than nation-building. The narrow shape of the country, straddling the Gambia River, inhibits trade and communication between the northern and southern parts of the country.

Source: Redrawn based on E. Sheppard, P. Porter, D. Faust, and R. Nagar, A World of Difference: Encountering and Contesting Development, 2nd ed., Guilford, 2009.

- *The age of mass consumption* sees consumer goods and services begin to rival heavy industry as leading economic sectors and most of the population achieves consumption levels far above basic needs.
- *The postindustrial stage is marked by the rise of services that replace secondary activities* as the principal sector of the economy. Professional and technical skills assume preeminence in the labor force, and information replaces energy as the key productive resource.

Rostow's stages of economic growth turned the experience of western countries into a blueprint for developing countries in Latin America, Asia, and Africa. The 1960s, 1970s, and 1980s were all proclaimed by UN resolutions as "Development Decades," with a belief that foreign assistance, technology transfer from advanced economies, and infrastructure investments would trigger the upward, progressive march of development.

The Core-Periphery Model

In contrast to the optimism of modernization theory, critics point out repeated patterns of **uneven spatial development.** They observed that within developing countries, undergoing modernization there often includes a modern **core area** of capitalist production integrated with the global economy alongside a traditional *periphery* with subsistence wages. The **core-periphery model** helps us understand the dualism that exists within many spatial systems such as the urban-rural contrasts commonly seen in infrastructure, health, and education.

The core and periphery are linked parts of a wider system. If for any reason (perhaps a new industrial process or product) one section of a country experiences accelerated economic development, that section becomes increasingly attractive for investors, entrepreneurs, and migrants. Assuming that investment capital is limited, growth in the developing core must come at the expense of the peripheries of the country. Economic growth sets in motion a process of **circular and cumulative causation.** Through the multiplier effect and agglomeration economies (see Chapters 9 and 11), new businesses spur the growth of other businesses. Businesses pay wages to workers, who purchase goods and services such as groceries and housing, supporting the local retail and construction industries. Firms and employees pay taxes, funding public services such as schools, parks, and roads. High-quality public facilities and infrastructure improve business productivity and the quality of life, making it an attractive place for additional businesses to locate, starting the process all over again. Circular and cumulative causation works to polarize development and, according to economist Gunnar Myrdal, leads to a permanent division between prosperous (and dominating) cores and depressed (and exploited) peripheral districts that are milked of surplus labor, raw materials, and profits.

A more optimistic version of this model suggests that within market economies, income disparities tend to be reduced as developmental levels increase. Eventually, it is argued, income convergence will occur as **trickle-down effects,** or **spread effects,** work to diffuse benefits outward from the center in the

form of higher prices paid for needed materials or through the dispersion of technology to branch plants or contract suppliers to lower-cost regions of production. The experience of developed countries offers only mixed support for this belief. Nearly all residents of developed countries have access to education, safe drinking water, and other indicators of development. Still, in most developed countries, there remain strong contrasts between wealthy, fast-growing urban cores and prosperous "high-tech" concentrations on the one hand, and depressed, depopulated rural peripheries or declining, deindustrialized "rust belts" on the other.

Dependency Theory

Dependency theory emerged from dissatisfaction with modernization theory and development programs as they were applied to regions such as Latin America. Drawing upon Karl Marx's ideas, dependency theorists extended the core-periphery model to the international scene. They claimed that the development of the advanced core nations depended upon the underdevelopment of the peripheral nations. Unlike modernization theory, which sees poverty as the natural state of affairs, dependency theory argues that development creates underdevelopment. In other words, developing countries were made poor by their interactions with advanced countries, starting with colonialism and the slave trade and continuing in new forms to the present. Dependency theory sees the developing world as effectively held captive by the leading industrial nations. It is drained of wealth and deprived of growth by remaining largely a food and raw material exporter and an importer of manufactured commodities. Developing countries suffer because prices for the primary goods that they export remain low, while the prices for imported manufactured goods continue to rise. Development aid, where proffered, involves a forced economic reliance on donor countries that continues an imposed cycle in which, in a sense, selective industrialization leads not to independent growth, but to further dependent underdevelopment. A condition of **neocolonialism** is said to exist even after legal independence in which economic and even political control is exercised by developed states over the economies and societies of independent countries of the underdeveloped world. This control is said to be exercised through unequal terms of exchange and the power exerted by international bodies such as the WTO and IMF.

Support for dependency theory comes from the growing gap between the world's poorest and richest countries. Dependency theory holds that these differentials are not accidental but the logical result of the ability of developed countries and power elites to exploit other populations and regions to secure for themselves a continuous source of new profits. Today, transnational corporations, the theory contends, tend to dominate through their investments in key areas of developing economies. They introduce technologies and production facilities to further their own corporate goals, not to further the balanced development of the recipient economies. Dependency theory has been criticized for treating developing countries as passive victims and for viewing economic development as a zero-sum game. Critics of dependency theory point out that technological innovations increase economic productivity, creating a larger pie to be shared by all. Another failing of dependency theory is its inability to account for the rising prosperity in newly industrializing countries such as China.

World Systems Theory

World systems theory extends the core-periphery model to the entire capitalist global world economy. On the international scene, core-periphery contrasts are discerned between, particularly, Western Europe, Japan, and the United States as prosperous cores and the least developed countries on the periphery of the global economy. At all spatial scales, core-periphery models assume that in part, the growth and prosperity of core regions come at the expense of exploited peripheral zones. The core controls key high-level functions in the global economy, such as headquarters for transnational corporations, financial centers, and stock exchanges. In contrast with dependency theory, world systems theory acknowledges that countries can shift from a peripheral position to a core position. A **semi-periphery** of newly industrializing countries, such as South Korea and Brazil, occupies an intermediate position between core countries, such as the United States, and peripheral countries, such as Liberia.

10.5 Strategies for Development

Strategies for development have been strongly influenced by modernization theory and Rostow's stages of economic development. From the 1950s through 1970s, the leading model for economic development was the "Big Push" of massive coordinated investments in infrastructure and industry. The stimulus provided by this big push would cause the economy to "take off" through productivity increases, expansion of the consumer base, and the creation of backward- and forward-linked industries. Countries were to focus on industries where they had a *comparative advantage* in order to develop efficient industrial specializations, agglomeration economies, and expanding trade. Because the needed push was too large for the private sector, governments of developing countries were to take the initiative by borrowing large sums of money to finance the necessary investments.

The two major international lending agencies, the IMF and the World Bank, were the outcome of a conference held at Bretton Woods, New Hampshire, near the end of World War II. The IMF is mostly concerned with lending to governments to help stabilize their currencies or to pay for their imports or international debts when their reserves fall short. The World Bank has focused on lending for development, beginning with the restoration of war-devastated countries of Europe, and then shifting its attention to the world's developing countries. For example, from 1960–1975, the World Bank made 135 loans for power plants and 196 loans for roads or railroads in Africa, Asia, Latin America, the Caribbean, and the Middle East.

New Directions in Development

During the 1980s, dissatisfaction with inefficiencies and corruption in government-led development projects led to a different strategy called the *Washington Consensus*. The new approach, sometimes called **neoliberal globalization,** was supported by the IMF, World Bank, and U.S. Treasury. It was called *neoliberal* because it revived the liberal faith in the market mechanism and the private sector. Countries were to eliminate tariffs and quotas to promote global free trade, and governments of developing countries were asked to reduce regulations, privatize state-run industries, and remove barriers to foreign firms and foreign investors. Instead of excessive spending on social needs, governments were to balance their budgets carefully and keep taxes low. Neoliberal thinking is reflected in a UN report concluding that "good government," including protection of property rights under a stable political and legal system, should be the top priority in development.

Another approach to development that emerged in the 1980s was a focus on *human capital,* a composite of skills, habits, schooling, and knowledge that—more than labor force numbers or capital availability—contributes to successful economic development and sustained growth. Technological progress in recent decades, it is pointed out, has been notably dependent on more educated workforces equipped with high levels of capital investment. The current deep global imbalance in literate and technically trained people has been called the most potent force of divergence in well-being between the rich world and the poor. When developing countries offer incentives to attract foreign direct investment and technology transfer, the imported ideas and technology help create *human capital*—labor and intellectual skills. This creates the potential to develop industrial specializations, export-led development, and rising levels of living, as presumably they did for Taiwan, Singapore, and the other surging Asian economies.

A different sort of foreign assistance comes in the form of **remittances,** flows of money sent home by workers who have left their homes in developing countries to take jobs in developed countries. Geographic differences in population growth rates and economic opportunities are behind much of the global flow of international migrants. For example, relatively high wages and declining labor forces in Europe versus low wages and rapid population growth in Africa creates strong migration pressures.

Remittances are among the most important flows of money from rich countries to poor countries. In many poor countries, remittances rival foreign direct investment as their largest source of foreign capital. While many transfers go unrecorded, officially recorded remittances were $613 billion in 2017, according to the World Bank. In most cases, remittances are used to meet basic family needs, such as food, medicine, and school fees, or to allow children to stay in school longer before entering the workforce. They are sometimes used to fund construction of houses for family members or to purchase desired consumer goods that demonstrate the success of the migrant. While not always reaching their potential, remittances offer an important economic development opportunity for poor countries by transferring capital and providing startup funds for businesses ventures.

Development Prospects

Rostow's expectations of an inevitable progression of development have proved illusory. Many developing countries remain locked in one of the first two stages of his model, unable to achieve the takeoff to self-sustained growth despite importing technology and attracting foreign aid investment funds from the more developed world (see the feature "Does Foreign Aid Help?"). Indeed, it has become apparent to many observers that despite the efforts of the world community, the development gap between the most and the least advanced countries may widen rather than narrow over time. A case in point is sub-Saharan Africa; between 1975 and 2000, per capita income declined by almost 1 percent a year, leaving all but a tiny elite significantly poorer at the end of the period. Over the same years, income per head in the industrial market economies grew at a 1.8 percent annual rate. More recently, the picture brightened for the developing world as a whole. On the bright side, newly industrializing countries such as China lead the world in economic growth. Even sub-Saharan Africa posted growth in the 2000–2010 period, growing at an average annual rate of 2.3 percent. For many, faith in the likelihood of growth—even if not in definable "stages of development"—was renewed.

Challenges and Opportunities Facing Developing Countries

In addition to managing population growth and promoting economic development, developing countries often face special challenges. Major challenges facing developing countries include natural hazards, foreign debt, land ownership, and gender inequality. The case of Haiti illustrates several of these challenges. As noted in the chapter's opening vignette, natural disasters are often much more devastating when they strike developing countries. The reconstruction efforts following the 2010 Haitian earthquake demonstrate some of the challenges of providing effective development assistance. After the quake, an outpouring of international sympathy led to promises of $2.5 billion in emergency relief work and $4.5 billion in reconstruction assistance. While debating the aid packages, it was revealed that Haiti was already burdened by immense foreign debts, most of which were forgiven as part of the assistance. Delivery of the promised aid was delayed, and most of the relief and reconstruction work was performed by organizations and companies based in donor countries such as the United States, rather than by Haitian agencies, nongovernmental organizations, or businesses. This method of delivery aid only exacerbated Haiti's position of dependency and missed an opportunity to build the capacities of Haitian agencies and firms. The actual rebuilding of homes and businesses was complicated by issues of land ownership because only about 5 percent of Haiti's land holdings were properly titled and recorded prior to the earthquake. Finally, human rights advocates criticized the reconstruction planning for ignoring gender issues, such as the fact that women living in tent camps for years after the earthquake where highly vulnerable to sexual violence.

Foreign Debt

In the pursuit of development, many developing countries borrowed heavily in the 1960s and 1970s. Money was spent on hydroelectric dams, power plants, ports, and other large, government-directed development projects. Unfortunately, many expensive World Bank–financed development projects were disappointing failures and did not generate sufficient returns to pay back the loans (**Figure 10.19**). Other borrowed money was spent on weapons or was lost to corruption. In the 1980s, rising interest rates and economic stagnation combined to push many borrowers into crisis. Mexico defaulted on its debts in 1982, and other countries followed. The IMF intervened, refinancing development debts and requiring structural adjustment programs that forced governments to eliminate tariffs and, in many cases, reduce spending on education, health care, and social services. Thus, the burden of structural adjustment fell most directly upon the poor.

Loans were refinanced during structural adjustment but not forgiven. The neoliberal reforms put in place did not stimulate economic growth sufficient to pay down debts. External debt owed by developing countries grew from $567 billion in 1980 to $1.6 trillion in 1990 and $3.7 trillion in 2008. In some countries, foreign debts approached or even exceeded their

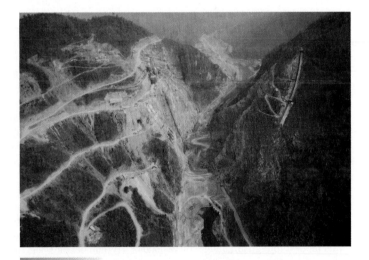

Figure 10.19 The El Cajón Dam in Honduras is an example of a technology transfer development project funded by the World Bank. Honduras lacks fossil fuel supplies but was felt to have a comparative advantage in hydroelectric power generation thanks to its rugged topography and powerful rivers. The 226-meter (741-foot)–high dam cost almost $800 million when it was built in the 1980s. The amount of debt required for the dam was staggering for a country with just 5 million residents and a per capita income of $590 per year in 1990. The dam failed to stimulate enough economic growth to pay back the loans and contributed to Honduras being classified by the IMF and World Bank as a HIPC. The electricity was delivered to the cities, where it was used by new, export-oriented textile industries. Meanwhile, many rural areas of the country still lack electricity, and peasant farmers living near the dam lost their livelihoods when their lands were flooded by the reservoir.

©David A. Harvey/Getty Images

Does Foreign Aid Help?

A 1998 World Bank report on "Assessing Aid" concluded that the raw correlation between rich-country aid and developing-country growth is near zero. Simply put, more aid does not mean more growth, certainly not for countries with "bad" economic policies (high inflation, large budget deficits, corrupt bureaucracy, and so on); for them, the report claims, aid actually retards growth and does nothing to reduce poverty. Similarly, other studies have found no clear link between aid and faster economic development. The $1 trillion that rich countries and international agencies gave and loaned to poor ones between 1950 and 2002 did not have the hoped-for result of eliminating poverty and reducing economic and social disparities between the rich and poor countries of the world.

In part, that was because economic growth was not necessarily a donor country's first priority. During the Cold War, billions flowed from both the Soviet Union and the United States to prop up countries whose leaders favored the donor state agendas. Even today, strategic considerations may outweigh charitable or developmental aims. Israel gets a major share of U.S. aid for cultural and historical reasons; Egypt, Lebanon, Pakistan, and Colombia get sizable portions for political and strategic reasons; and Afghanistan and Iraq have received billions for rehabilitation and restructuring. Up until 2005, in fact, the United States spent only 40 percent of its modest foreign aid budget on assistance to poorer states.

About one-quarter of all aid from whatever national source has been tied to purchases that must be made in the donor country, and additional large shares flow to former colonies of donor countries, regardless of need or merit. Even worse, a World Bank report admits, aid failures often reflect the fact that the bank and its sister agencies have wasted billions on ill-conceived projects.

In cases where the aid project involves advanced technology, much of the money is spent on hiring experts from the developed countries, rather than building the expertise and capabilities in the recipient country. The reconstruction aid following the 2010 Haitian earthquake illustrates some of the problem. According to the Center for Economic and Policy Research, only 6 percent of the reconstruction aid was disbursed through Haitian nonprofit organizations, and less than 1 percent through the Haitian government. Of the reconstruction contracts funded by the U.S. government, 40 percent of the funds were awarded to firms from the Washington, D.C., area, while only 2.4 percent went to Haitian companies. In other words, aid following the Haitian earthquake came in a top-down fashion, strengthening U.S. firms and institutions, rather than building the capabilities of the Haitian government and local companies and nongovernmental organizations.

More optimistic conclusions are drawn by other observers who note that: (a) foreign aid tends to reduce poverty in countries with market-based economic policies, but is ineffective where those policies do not exist; (b) aid is most effective in lowering poverty if it is given to poor, rather than less poor, countries; and (c) aid targeted to specific objectives—eradication of a disease or Green Revolution crop improvement, for example—is often remarkably successful, though spending on food aid or on aid tied to purchases from donor countries is of little use.

Although some countries—Botswana, the Republic of Korea, China, and different Southeast Asian states—made great progress thanks to development assistance, a large number of others saw their prospects worsen and their economies decline. Slow growth and rising populations lowered their per capita incomes, and poor use of aid and loans failed to improve their infrastructures and social service levels. Most critical for the economic and social development prospects of those countries was that the financing offered to them over the years in the hopes of stimulating new growth became an unmanageable debt burden.

So great and intractable did their debt problem become that the international community recognized a whole class of countries distinguished by their high-debt condition: Heavily Indebted Poor Countries (HIPCs) that were so far in debt that many of them were paying more in interest and loan payments to industrialized countries and international agencies than they were receiving in exports to or aid from those sources. Gradually, the rich world accepted that debt relief, not lectures on capitalism, is the better approach to helping the world's poor countries and people. In 1996 the World Bank, the IMF, and other agencies launched the first HIPC initiative, identifying 41 very poor countries and acknowledging that their total debt burden (including the share owed to international institutions) must be reduced to sustainable levels. In the following years, differing definitions of *sustainable* and criteria for debt relief were adopted but remained rooted in the requirement that benefiting countries must face an unsustainable debt burden, maintain good economic policies, and prepare a blueprint laying out how the country will fight poverty and promote health and educational programs. After 2000, both debt relief and continuing flows of aid were also tied to the United Nation's MDGs. Part of the philosophy behind this debt forgiveness was to clean the slate and shift future aid away from loans to outright grants.

The expressed hope of the international community now is that the answer to the question "Does foreign aid help?" will finally be "Yes." In a reconsideration of its former pessimism, the World Bank now concludes that, indeed, the answer is affirmative. It feels that foreign aid has been instrumental in increasing life expectancy at birth in developing countries by 20 years since 1960, cutting adult illiteracy in half since 1970, reducing the number of people in abject poverty by 200 million since 1980 even as world population increased by 2 billion, and more than doubling the per capita income in developing countries since 1965. The expectation now is that massive debt forgiveness will be reflected in accelerating social and economic improvement in emerging economies and further reduce the disparities between their conditions and those of more affluent developed states.

Questions to Consider

1. How should foreign aid programs be redesigned so that they strengthen existing institutions and firms within the recipient state rather than enriching agencies and firms in the donor country?

2. Do you think donor countries such as the United States should completely ignore all self-interest (including, for example, extra generosity toward friendly or politically compatible states) in making aid decisions? Why or why not?

3. Do you think international programs of debt forgiveness are fair to lending countries and their citizens? Why or why not?

4. One widely held opinion is that money now spent on direct and indirect foreign aid more properly should be spent on domestic programs dealing with poverty, unemployment, and homelessness. An equally strongly held contrary view is that foreign aid should take priority, for it is needed to address world and regional problems of overpopulation, hunger, disease, destruction of the environment, and the civil and ethnic strife that those conditions foster. Assuming that you had to choose one of the two polar positions, which view would you support, and why?

---

annual GNI. By 2000, African countries as a whole had external debt equivalent to more than 60 percent of GNI and made annual debt service payments equivalent to 27 percent of exports. Outrage at suffering on the part of the world's poor in order to finance debts owed to the world's richest nations sparked debt relief movements such as the Jubilee 2000 Movement. Rescheduled loans and debt relief has been offered to many of the world's poorest countriesthrough the IMF's Multilateral Debt Relief Initiative and the joint IMF-World Bank Heavily Indebted Poor Countries Initiative. These two programs provided $99 billion in debt relief to 36 countries. Debt relief, it was argued, was essential to meeting the MDGs. Freed from large public debts owed to the high income countries, African, Asian, Latin American, and Caribbean governments were able to direct public funds into health, sanitation, and education improvements. Still, debt troubles remain in a number of countries where debt payments exceed the growth rate of the economy.

Land Ownership

In both urban and rural settings, resolving issues of land ownership is critical to improving the lives of the poorest residents. Urban squatter settlements built without registered land ownership are vulnerable to slum clearance and difficult to improve. Even if a squatter settlement is improved with better infrastructure and higher-quality buildings, its residents will not actually capture the increased value unless they have legal title to the property.

In rural agricultural societies, land is the most essential resource. In the most densely settled portions of the developing world, population growth increasingly leads to landlessness. The problem is most acute in southern Asia, particularly on the Indian subcontinent, where the landless rural population is estimated to number some 300 million—nearly as large as the total population of the United States. Additional millions have access to parcels too small to feed the average household adequately (**Figure 10.20**). A landless agricultural labor force is also of increasing concern in Africa and Latin America.

Figure 10.20 Terraced fields in Nepal are a response to land scarcity. A high population density and land ownership concentrated in the hands of large estates means that poor farmers must cultivate small plots on steep slopes. Landlessness is a major cause of poverty in Nepal. About 10 percent of the rural population is landless, and more than half of the rural population own farms smaller than 0.5 hectares, which cannot produce enough to meet subsistence requirements. Nearly half of the low-caste Dalit households are landless.

©Bilderbuch/Design Pics

Landlessness is in part a function of an imbalance between the size of the agricultural labor force and the arable land resource. It is also frequently due to concentration of ownership in the hands of a few and consequent landlessness for many. Concentrated ownership of large tracts of rural land appears not just to affect the economic fortunes of the agricultural labor force itself, but also to depress national economic growth through inefficient use of a valuable but limited resource. Large estates are often farmed carelessly, are devoted to the production of export crops with little benefit for low-paid farm workers, or are even left idle. In some societies, governments concerned about undue concentration of ownership have imposed restrictions on total farm size—though not always effectively.

In Latin America, where huge farms are a legacy of colonial land grants and many peasants are landless, land reform—that is, redistribution of arable land to farm workers—has had limited effect. The Mexican revolution early in the 20th century resulted in the redistribution of nearly half the country's agricultural land over the succeeding 60 years, but the rural discord in Chiapas beginning in the 1990s reflects the persistence of underused large estates and peasant landlessness. The Bolivian revolution of 1952 was followed by a redistribution of 83 percent of the land. Some 40 percent of Peru's farming area was redistributed by the government during the 1970s. In other Latin American countries, however, land reform movements have been less successful. In Guatemala, for example, 85 percent of rural households are landless or nearly so, and the top 1 percent of landowners control 34 percent of arable land; in Brazil, the top 5 percent of farms by size control 70 percent of the arable land, leaving only 2 percent of the land for the bottom 50 percent of farmers.

In India, where two-thirds of rural families either have no land at all or own less than 2 hectares (5 acres), a government regulation limits ownership of "good" land to 7 hectares (18 acres). That limitation has been effectively circumvented by owners distributing title to the excess land to their relatives. Population growth has reduced the amount of land available to the average farmer on the Indonesian island of Java to only 0.3 hectare (0.75 acre), and the central government reports that more than half of Java's farmers now work plots too small to support them.

The rural landless are the most disadvantaged segment in the poorest countries of the world. They have far higher levels of malnutrition and incidence of disease and lower life expectancies than other segments of their societies. In Bangladesh, for example, the rural landless consume only 80 percent of the daily caloric intake of their landholding neighbors. To survive, many there, and in other countries where landlessness is a growing rural problem, migrate to urban areas, swelling the number of residents in squatter settlements.

10.6 Gender Inequality

The most common measure of development, GNI per capita, takes no account of the sex and age structures of the societies examined. Yet among the most prominent strands in the fabric of culture are the social structures, roles, and relationships (sociofacts) assigned to males and females. Gender relationships and roles vary across societies, introducing an important element of spatial variation. **Gender** in the cultural sense refers to socially constructed—not biologically based—distinctions between femininity and masculinity that are shaped by custom, religion, and other ideological forces. Because gender is such an important variable, development agencies increasingly track gender parity in education, literacy, and employment .

Gender roles and stage of development are related by a common belief that greater equality accompanies development. However, it appears that at least in the earlier phases of technological change and development, women generally lose rather than gain in status and rewards. Only recently, and only in the most developed countries, have gender-related contrasts been reduced within and between societies. Hunting and gathering cultures observed a general egalitarianism (see Figure 2.11 in Chapter 2). Gender roles are affected by innovations in agricultural societies (see the feature "Women and the Green Revolution," in Chapter 8). The Agricultural Revolution—a major change in the technological subsystem—altered the earlier structure of gender-related responsibilities. In the hoe agriculture found in much of sub-Saharan Africa and in South and Southeast Asia, women became responsible for most of the actual field work while retaining their traditional duties in child rearing, food preparation, and the like.

Plow agriculture, on the other hand, tended to subordinate the role of women and diminish their level of equality. Women may have hoed, but men plowed, and female participation in farm work was drastically reduced. Women are often more visibly productive in the market than in the field (**Figure 10.21**). As women's agricultural productive role declined, they were afforded less domestic authority, less control over their own lives, and few if any property rights independent of male family members.

Western industrial—*developed*—society emerged directly from the agricultural tradition, which subordinated females. Only within the later 20th century, and then largely only in the more developed countries, has that subordinate role pattern changed. The rate and extent of women's participation in the labor force has expanded everywhere, most dramatically in Latin America. Women's increased participation in the workforce reflects several changing conditions. Women have gained greater control over their fertility, thus increasing their opportunities for education and employment. Further, attitudes toward employed women have changed, and public policies on, for example, child care, maternity benefits, and the like are more favorable. Economic growth, including the expansion of service sector jobs open to women, was also important in many regions. Permissive attitudes and policies with regard to micro and small enterprises, including financing and credit programs, have in some areas played a major role in encouraging

Figure 10.21 Women dominate the once-a-week *periodic* markets in nearly all developing countries. Here, a Nepali woman sells produce in a local market. More than half of the economically active women in sub-Saharan Africa and southern Asia, and about one-third in northern Africa and the rest of Asia, are self-employed, working primarily in the informal sector.

©*Bartosz Hadyniak/Getty Images*

Empowering Women Financially

In 1976, a Bangladeshi economist, Muhammad Yunus, wandered into a poor village and got an idea that has captured international interest and changed accepted beliefs and practices of banking in developing countries. The concept behind the Grameen Bank that he established is simple: If individual borrowers are given access to credit, they will be able to identify and engage in viable income-producing activities, such as pottery making, weaving, sewing, buying and marketing simple consumer goods, and providing transportation and other basic services.

Declaring that "Access to credit should be a human right," Yunus was a pioneer in extending "microcredit" for "microenterprises," with women emerging as the primary borrowers and beneficiaries of Grameen Bank's practice of lending money without collateral and at low rates of interest. Under the original Grameen concept, to be eligible for the average loan of about US$160, women without assets must join or form a "cell" of five unrelated women, of whom only two can borrow at first though all five are responsible for repayment. When the first two begin to repay, two more can borrow, and so on. As a condition of the loan, clients also must agree to increase their savings, observe sound nutritional practices, and educate their children.

By 2011, the bank had made more than 8 million loans in 40,000 villages in Bangladesh alone. A reported 97 percent of the borrowers are women, and repayment rates reach above 95 percent. The average household income of Grameen Bank members has risen much faster than that of nonmembers in the same villages, with the landless benefiting most and marginal landowner families following closely. Because of enterprise incomes resulting from the lending program, there has been a sharp reduction in the number of Grameen Bank members living below the poverty line. There has also been a marked shift from low-status agricultural labor to self-employment in simple manufacturing and trading. That shift has encouraged a borrower and lender recognition that larger loans are needed to enable increasingly entrepreneurial women to build small businesses, hire employees, acquire office and manufacturing equipment, and the like. In consequence, some lenders now approve loans of several thousand dollars, although such larger loans are still much in the minority.

Muhammed Yunus was awarded the 2006 Nobel Peace Prize for developing his model of microcredit. His microcredit ideas have spread from their Bangladesh origins to other countries in Asia, Latin America, Eastern Europe, and Africa. Of those poorest clients, the vast majority are women. But the female recipients still represent only a fraction of the women worldwide who have virtually no access to credit—or to the economic, social, educational, and nutritional benefits that come from its availability.

---

women entrepreneurs (see the feature "Empowering Women Financially").

Considering all work—paid and unpaid economic activity and unpaid housework—women spend more hours per day working than do men in developed regions. In developing countries, the United Nations estimates, when unpaid agricultural work and housework are considered along with wage labor, women's work hours exceed men's by 30 percent and may involve at least as arduous—or even heavier—physical labor. The FAO reports that "rural women in the developing world carry 80 tons or more of fuel, water, and farm produce for a distance of 1 km during the course of a year. Men carry much less. . . ." Women are paid less than men for comparable employment everywhere, but in most world regions the percentage of economically active women holding wage or salaried positions is about equal to the rate for men. Exceptions are Latin America, where a higher proportion of active women than men are wage earners, and Africa, where wage-earning opportunities for women are few.

The present world pattern of gender-related institutional and economic assignments is varied. It is influenced by a country's level of economic development, by the persistence of customary restrictions that some religions and cultures impose on women, and by the specific nature of its economic—particularly agricultural—base. The differential impact of these and other conditions is evident in **Figure 10.22**. The pattern shows a distinct gender-specific regionalization among the countries of the developing world. Among the Arab or Arab-influenced Muslim areas of western Asia and North Africa, the recorded proportion of the female population that is economically active is low. Religious tradition restricts women's acceptance in economic activities outside the home. The same cultural limitations do not apply in the different rural economic conditions of Muslims in southern and southeastern Asia, where labor force participation by women in Indonesia and Bangladesh, for example, is much higher than it is among Western Muslims.

In Latin America, women have been overcoming cultural restrictions on their employment outside the home, and their active economic participation has been increasing. That participation is occurring almost entirely outside of the agricultural realm, where the high degree of farm labor tenancy as well as custom limits the role of females. Sub-Saharan Africa, highly diverse culturally and economically, is highly dependent on female farm labor and market income. The historical role of strongly independent, property-owning females formerly encountered under traditional agricultural and village systems, however, has increasingly been replaced by the subordination of women with the modernization of agricultural techniques and introduction of formal, male-dominated commercialized agriculture.

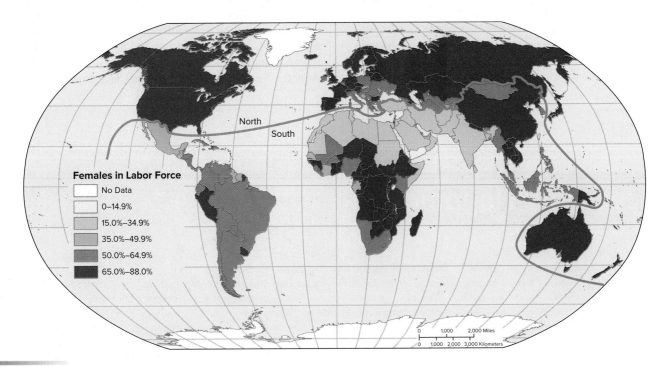

Figure 10.22 Participation of women (ages 15 and above) in the labor force, 2016. Worldwide, 52 percent of women were in the labor force in 2008. Many of the countries with the highest rates of female participation in the labor force were in sub-Saharan Africa. Cultural factors lie behind low rates of female labor force participation in South Asia and the Middle East–North Africa zone.

Source: World Bank, World Development Indicators, *2018.*

10.7 Alternative Measures of Development and Well-Being

Dissatisfaction with purely economic measures of development has spurred development agencies to seek composite measures that better measure human well-being. Development, it is maintained, goes beyond the purely economic and physical, and personal development may have little or nothing to do with objective statistical measures. The achievement of development must also be seen in terms of individual and collective well-being: a safe environment, freedom from want, opportunity for personal growth and enrichment, and access to goods and services beyond the absolute minimum to sustain life (see the feature "Measuring Happiness"). Health, safety, educational and cultural development, security in old age, political freedom, and similar noneconomic criteria are among the evidence of comparative developmental level that is sought in composite statistics. Also sought is a measure of development that is neither ethnocentric nor colored by political agendas. The values of one culture—for example, in housing space per person, in educational levels, or in distribution of national income—are not necessarily universally applicable or prioritized, and comparative statistics should not imply that they are.

One comparative ranking that has gained increasing recognition is employed by the UNDP. Its Human Development Index (HDI) combines per capita GNI (corrected for purchasing power), life expectancy, and education (years of schooling) (**Figure 10.23**).

The HDI reflects the program's conviction that the important human aspirations are leading a long and healthy life, receiving adequate education, and having access to economic resources sufficient for a high quality of life. The weighting of the three input variables lends a further element of subjectivity to the rankings.

The UNDP has also developed a measure of poverty in its Multidimensional Poverty Index (MPI). While the HDI measures average *achievement,* the MPI measures *deprivation* in the same three dimensions of development underlying the HDI. Deprivation within a household is measured by children not attending school, childhood deaths, malnourished children, and poor living standards at home. Household living standards are evaluated by the availability of electricity, an improved drinking water source, sanitation, finished flooring, and basic household assets. The dimensions of poverty are discussed in more detail in the feature "What Is Poverty?"

The UN's Gender Development Index (GDI) simply computes the HDI for women and men separately and then compares them. Countries with high HDI scores tend to have high GDI scores. The global GDI is 0.938, indicating that men tend to score higher than women on the HDI. Among world regions, South Asia has the lowest GDI at 0.822. A number of countries have GDI scores above 1.0 (largely because of higher levels of education and longer life expectancies for women). A "gender inequality index" (GII) devised by the UNDP, emphasizes female reproductive health, educational attainment, and participation in political, management, professional, and technical positions. The GII rankings are heavily biased toward measures that technologically advanced, career-oriented, Western cultures consider indicative of gender equality and progress; they do not consider alternate cultural standards or values. As calculated, the GII shows that gender equality in political, economic, and professional activities is not necessarily related to the level of national wealth or development. A number of Latin American, Caribbean, and sub-Saharan African countries have a higher share of seats in parliament held by women than the high income countries. European countries dominate the top of the GII rankings along with Australia, Singapore, Canada, and the United States.

Measuring Happiness

Is there a spatial pattern to happiness? And is it related to spatial patterns of development, underdevelopment, high incomes, or poverty? Psychologists have long used surveys of reported well-being to study happiness, and some economists have explored the relationships between economic variables and measures of life satisfaction. The results are fascinating, complicated, and sometimes surprising. Within a country, those with higher incomes tend to report greater happiness, but as standards of living rise, happiness does not necessarily increase. Generally, higher-income countries have higher reported levels of happiness, but the many exceptions suggest that other cultural factors are important influences. Ruut Veenhoven at Erasmus University in the Netherlands has compiled "The World Database of Happiness," containing the results of surveys of life enjoyment drawn from 155 countries. The following is a sample of top-ranked countries, middle-range countries, and bottom-ranked countries on surveys administered between 2005 and 2014 that asked people to rate their "satisfaction with life as a whole" on a 0 to 10 scale. While the United States ranked fairly high on the life satisfaction survey, with an average rating of 7.3, Canadians and Mexicans were even happier, despite lower average incomes. Costa Rica took first place for life satisfaction and all five of the Nordic European states (Denmark, Finland, Iceland, Norway, Sweden) ranked in the top 11. The countries with the lowest happiness ratings are in Africa.

Country	Happiness Score (0–10)	GNI/Capita (US$), 2017
Happiest Countries		
Costa Rica	8.5	$15,750
Denmark	8.4	$51,040
Mexico	8.3	$17,740
Iceland	8.1	$52,490
Canada	8.0	$43,420
Norway	8.0	$62,510
Switzerland	8.0	$63,660
Middle-Range Countries		
Croatia	6.0	$22,880
Pakistan	6.0	$5,580
Romania	6.0	$22,950
Turkey	6.0	$23,990
Least Happy Countries		
Sierra Leone	3.5	$1,320
Benin	3.0	$2,170
Burundi	2.9	$770
Togo	2.6	$1,370
Tanzania	2.5	$2,740

Sources: R. Veenhoven, *Average Happiness in 158 Nations 2005–2014, World Database of Happiness*, and Carol Graham, 2005. *"The Economics of Happiness,"* World Economics, 6(3): 41–55.

What Is Poverty?

According to the UNDP, of the world's 7.5 billion people (2017), 770 million lived on less than $1.90 per day. At this level, people struggle to meet their basic needs and are vulnerable to a precipitous decline in well-being if they were to experience an illness, natural disaster, crop failure, or economic shock. Although the dollar definition of poverty is applied as if it were a worldwide constant, in reality, poor people define the wealth of people quite differently. One researcher wrote:

> In some ways the "poor" cultures of the Third World are rich psychologically and spiritually, enjoying a contentment and sense of tradition sorely lacking in hectic, ulcer-ridden, depersonalised industrial societies. To many Buddhists, for example, inner peace is more valuable than a high Gross National Product. The highest divorce and suicide rates occur in the First and Second Worlds. If personal happiness were our criterion, the Third World might rank first.[a]

Research on how the poor view poverty suggests that poverty has multiple dimensions, some of which are noneconomic, such as a sense of shame, powerlessness, or insecurity. The poor define poverty not just in monetary terms, but as a lack of education, health, proper housing, physical safety, or decent employment. As a consequence, a multidimensional measure of poverty has been developed by the Oxford Poverty and Human Development Initiative at Oxford University. It measures poverty using a short survey to assess quality of employment, sense of empowerment, feelings of physical safety, shame, and psychological well-being. Survey questions ask, for example, about hazards that respondents might be exposed to while at work, whether they believe that they have control over personal and household decisions, whether they feel respected by others and society, and their degree of satisfaction with life and overall happiness. An overall measure of poverty is created by counting the number of different dimensions in which a poor person feels he or she is deprived.

One key to improving both the economic and social lot of the "poorest of the poor," the World Bank and United Nations argue, is to target public spending on their special needs of education and health care and to pursue patterns of investment and economic growth that can productively employ the underused and growing labor force that is so abundant in the least developed countries. These identified socioeconomic needs are important elements in the United Nation's interlocked SDGs.

---

[a]Merriam, A. H., "What Does the 'Third World' Mean?" in J. Norwine and A. Gonzalez, eds., *The Third World: States of Mind and Being*. Boston: Unwin and Hyman, 1988, 15–22.

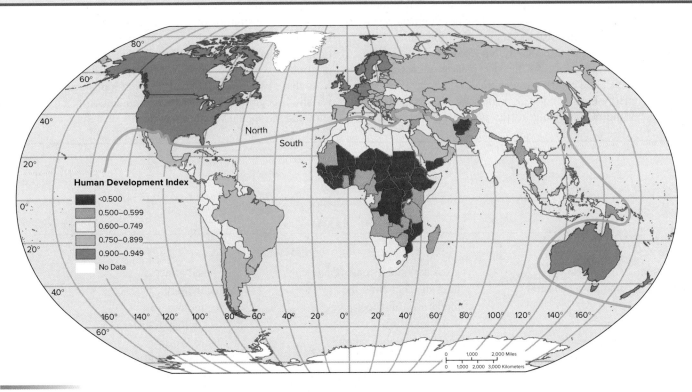

Figure 10.23 Country rankings according to the HDI, 2015. Because this index is intended to measure the absence of deprivation, it discounts incomes higher than needed to achieve an acceptable level of living and therefore does not distinguish between the richest countries. The three measures that are used by the UNDP—life expectancy school enrollment, and real (PPP) income—are highly correlated with one another. For that reason, it has been noted, the rankings derived by the HDI differ only slightly from income rankings adjusted for purchasing power parity; the Indian minister for human resources in 2002 objected that the HDI ignored "spiritual happiness" and "intellectual advances." The countries at the bottom of the HDI closely match the "least developed" countries recognized by the United Nations and shown on Figure 10.3.

Source: United Nations, Human Development Report, *2016.*

Development is generally understood as a process of change and improvement. It involves the fuller and more productive use of the resources of an area through the application of advanced levels of technology. The result is presumed to be improved conditions of life and human well-being through a fuller integration into and more equal share of the global economy. While this definition of development is widely accepted and has shaped development policies and programs, some critics point out that it reflects a particularly Western conception of progress.

A variety of comparative economic and noneconomic data are available to help identify the relative level of development for individual countries. *Gross national income (GNI)* and *purchasing power parity (PPP) per capita* document the basic core-periphery pattern while making clear the income diversity among the developing countries. A high percentage of a country's *workforce in agriculture* is associated with less-developed subsistence economies, with low labor productivity and low levels of national wealth. *Per capita consumption of commercial energy* reveals the immense size of the technology gap between most and least developed states, for energy use may be loosely equated with modern industrial and transportation facilities. While these economic measures are widely used by development agencies, they have serious shortcomings. GNI measures flows of money through the economy but does not measure the condition of capital stocks, the distribution of income, or whether spending contributes to human well-being. Further, using energy consumption as a measure of development neglects the damage done to the natural environment by using "modern" energy sources. Other development measures focus on factors contributing to human development: food security and nutrition, safe drinking water, and sanitation, education, and health.

Modernization theory has dominated thinking about development. Its optimistic forecasts of inevitable progression from traditional to modern society have undergone revision. Many countries appear unable to accumulate the capital, develop the skills, or achieve the technology transfer necessary to carry them along the path to full economic development and prosperity. Without that development, countries score poorly on noneconomic measures such as literacy, safe water, and conditions of health. With it, they can—as the experience of newly industrializing countries demonstrates—experience growing cultural and technological convergence with the most advanced states. The core-periphery model helps us understand how developed and less-developed regions can coexist in the same country. Dependency theory argues that development is always spatially uneven, producing underdevelopment in other regions. World systems theory describes the global economy as divided into core, semi-peripheral, and peripheral countries.

Development implies pervasive changes in the organizational and institutional structuring of peoples and space. Urbanization has invariably accompanied economic development, as has a more complete and rigorous political organization of space. We turn our attention in Chapters 11 and 12 to these two important expressions of human geographic variation, beginning first with an examination of city systems and the spatial variations inside urban areas.

KEY WORDS

brain drain	gross national income (GNI)	spread effect
circular and cumulative causation	informal economy	technology
core area	modernization theory	technology gap
core-periphery model	neocolonialism	technology transfer
dependency theory	neoliberal globalization	Third World
development	periphery	trickle-down effect
food security	purchasing power parity (PPP)	underdevelopment
gender	remittances	uneven spatial development
gross domestic product (GDP)	semi-periphery	

FOR REVIEW

1. What different ways and measures do we have to indicate degrees of development of particular countries or regions? Do you think these measures can be used to place countries or regions into uniform *stages of development?*

2. What kinds of material and nonmaterial economic and noneconomic contrasts can you cite that differentiate more developed from less-developed societies?

3. How does the core-periphery model help us understand observed contrasts between developed and developing regions of countries? In what way is *circular and cumulative causation* linked either to the perpetuation or the reduction of those contrasts? How does the concept of *trickle-down effects,* or *spread effects,* explain the equalization of development and incomes on a regional or international scale?

4. What factors of physical geography and history might help explain why some countries are *developed* and others are *underdeveloped?* What are the shortcomings of these explanations?

5. What is modernization theory? What are some of the criticisms of modernization theory made by alternative theories of development?

6. Imagine that you are assigned the task of devising a composite index of national development and well-being. What *kinds* of characteristics would you like to include in your composite index? Why? What specific *measures* of those characteristics would you like to cite?

7. Why did many developing countries experience debt crises, and what significance does it have for their future development?

8. Have both males and females shared equally in the benefits of economic development in its early stages? What regional contrasts within the developing world are evident in the economic roles assigned to women?

KEY CONCEPTS REVIEW

1. **How do we define development?** Section 10.1–10.2
   Development implies improvement in economic and quality-of-life aspects of a society.

2. **What economic measures mark a country's stage of development?** Section 10.3
   GNI and PPP per capita, per capita commercial energy consumption, percentage of labor force in agriculture, and average daily caloric intake are common, accepted measures of development.

3. **What are noneconomic aspects of development, and how are they related to measures of economic growth?** Section 10.3
   Education, literacy, safe drinking water, sanitation, and health services are among the many noneconomic indices of development. These noneconomic measures relate directly to human well-being and quality of life. They are correlated with income because higher-income countries can afford to spend more on improvement of quality-of-life conditions.

4. **What different explanations are offered for development and underdevelopment?** Section 10.4
   Modernization theory dominates conventional development policies and programs. It sees all societies positioned along a continuum between traditional and modern with modern societies characterized by advanced technology, high standards of living, democratic institutions, and capitalist production. The core-periphery model attempts to explain patterns of uneven spatial development. Dependency theory argues that development in some regions by necessity creates underdevelopment elsewhere. World systems theory posits a system of core, semi-periphery, and periphery countries within the global economy but allows for growth and change in that hierarchy. Rising productivity growth of newly industrializing countries offers some support for modernization theory.

5. **What special challenges face developing countries?** Section 10.5–10.6
   Natural hazards often have more devastating consequences when they strike poor countries. Foreign indebtedness has crippled the ability of many poor countries to invest in social needs, education and other measures that improve well-being. Debt relief has helped a number of countries. Landlessness is a major cause of poverty in rural areas. Gender inequality hampers the full participation of women in economic and political life. Education and empowerment for women contributes to many positive development outcomes.

6. **What are some alternative ways of measuring development?** Section 10.7
   Life expectancy, education, and living standards are the three components of the United Nation's HDI, an improvement over using single measures of development. Gender empowerment indices measure political and economic status of women. International studies of "happiness" suggest that income matters, but as only one of many factors in life satisfaction.

CHAPTER
11
URBAN SYSTEMS AND URBAN STRUCTURES

Hong Kong, China, the world's most densely populated city.

©George Hammerstein/Corbis

Key Concepts

11.1–11.3 An urbanizing world: the nature of cities, definitions, and favored locations

11.4–11.5 Systems of cities: the economic base, functions, hierarchies, and networks

11.6–11.8 Inside the city: land uses, social areas, and patterns of change

11.9 World urban diversity: European and non-Western cities

Cairo was a world-class city in the 14th century. Situated at the crossroads of Africa, Asia, and Europe, it dominated trade on the Mediterranean Sea. By the early 1300s, it had a population of half a million or more, with 10- to 14-story buildings crowding the city center. A Cairo chronicler of the period recorded the construction of a huge building with shops on the first floors and apartments housing 4,000 people above. One Italian visitor estimated that more people lived on a single Cairo street than in all of Florence. Travelers from all over Europe and Asia made their way to Cairo, and the shipping at its port of Bulaq outdistanced that of Venice and Genoa combined. The city contained more than 12,000 shops, some specializing in luxury goods from all over the world—Siberian sable, chain mail, musical instruments, luxurious cloth, and exotic songbirds. Travelers marveled at the size, density, and variety of Cairo, comparing it favorably with Venice, Paris, and Baghdad.

Today, Cairo (also known as Al-Qahirah) is a vast, sprawling metropolis, plagued by many of the problems common to rapidly urbanizing developing countries where population growth has outpaced economic development. The population of Egypt grew from 35 million in 1970 to more than 93 million by 2017, thanks to improved health care in general, a dramatic drop in infant mortality, high total fertility, and longer life expectancies. An estimated 19 million people reside in the Cairo greater metropolitan area. Cairo continues to grow, spreading onto valued farmland and decreasing food production for the country's increasing population. The United Nations projects that Cairo will reach 24.5 million people by 2030.

A steady stream of migrants arrives daily in Cairo, where, they hope, opportunities will be available for a better and brighter life. The city is the symbol of modern Egypt, a place where young people are willing to undergo deprivation for the chance to "make it." But real opportunities continue to be scarce. The poor, of whom there are millions, crowd into row after row of apartment houses, many of them poorly constructed. Tens of thousands more live in rooftop sheds or small boats on the Nile; a half million find shelter living between the tombs in the Northern and Southern Cemeteries—known as the Cities of the Dead—on Cairo's eastern edge. On occasion, buildings collapse; the earthquake of October 12, 1992, measuring 5.9 on the Richter scale, did enormous damage, leveling thousands of structures.

One's first impression when arriving in central Cairo is of opulence, a stark contrast to what lies outside the city center. High-rise apartments, regional headquarters buildings of transnational corporations, and modern hotels stand amid clogged streets, symbols of the new Egypt (*Figure 11.1*). New suburban developments and exclusive residential communities create enclaves for the wealthy, whose posh apartments are but a short distance from the slums housing a largely unemployed 20 percent of Cairo's population. Like cities nearly everywhere in the developing world, Cairo has experienced explosive growth that finds an increasing proportion of the country's population housed in an urban area without the economy or facilities to support them all. Street congestion and idling traffic generate air pollution comparable to Mexico City, Bangkok, and other highly polluted megacities. Both the Nile River and the city's treated drinking water show dangerous levels of lead and cadmium, the unwanted by-products of the local lead smelter.

Cairo is a classic case of the urban explosion in which more than half of the world's population lives in cities. This chapter introduces **urban geography,** which is divided into two broad categories of approach. The first looks at systems of cities, examining how cities support themselves economically, the functions that they perform in regional, national, and global economies, and how they exist in regular spatial patterns, networks, and hierarchies. Among their many purposes, cities serve as concentrations of people and activities to facilitate social interaction and the efficient exchange of information, goods, and services. Manufacturing, trade, and the exchange of ideas often require concentrations of workers, managers, merchants, and supporting institutions. Cities exist as functional nodes within a broader, hierarchical system of cities. The second approach to urban geography looks inside cities at their internal arrangements. Cities are unique places, with complex patterns of land use and social groups. In this chapter, we begin by examining the nature and evolution of cities, then we look at systems of cities, and finally, we turn our attention to life inside cities in different parts of the world.

Figure 11.1 Metropolitan Cairo, Egypt's remarkable population growth—from some 3 million in 1970 to an estimated 19 million today—has been mirrored in most developing countries. The rapid expansion of urban areas and populations brings housing shortages, inadequate transportation and other infrastructure development, unemployment, poverty, and environmental deterioration.

©Photov.com/AGE Fotostock

11.1 An Urbanizing World

In 2007, the world reached a major turning point. As seen on **Figure 11.2,** after astounding urban growth in the 20th century, the world's urban population surpassed the rural population. Some 436 metropolitan areas had in excess of 1 million people ("million cities") by 2015; in 1900, there were only 12. Expectations are for 662 "million cities" in 2030. A total of 29 metropolises had populations of 10 million or more people in 2015; the United Nations calls them *megacities* (**Figure 11.3**). In 1900, no city was of that size, and in 1950, there were just 2 (Table 11.1). It follows, of course, that because the world's total population has greatly increased over the centuries, so too would its urban component—from 3 percent in 1800 to more than half today. The urban share of the total has grown everywhere as urbanization has spread to all parts of the globe. Virtually all of the world's population growth in the first half of the 21st century will take place in cities—specifically, the cities of the developing world (**Figure 11.4**). Thus, the location of the world's largest cities will shift from Europe and North America to Asia, Latin America, and Africa.

The degree of **urbanization** differs from continent to continent and from region to region (**Figure 11.5**), but in nearly all countries, the proportion of their population living in cities is rising. The United Nations projects that urban majorities will exist in all regions of the world by 2030 (Table 11.2). While Africa and Asia are the least urbanized continents, cities are growing particularly fast there. While some cities will grow into megacities, cities with less than 1 million residents will grow faster than the very largest cities.

Industrialization spurred the rapid urbanization in the highly developed regions of Western Europe and North America. In many of the still-developing countries, however, urban expansion is only partly the result of the transition from agricultural to industrial economies. Rather, in many of those areas, people flee impoverished rural districts; by their numbers and high fertility rates, they accelerate city expansion. Even the high-income, highly developed states—with low or negative rates of natural population increase—will experience growing cultural diversity as international migrants seek opportunities in their cities. As Ernst Ravenstein's studies of migration suggested (see Chapter 3), international migrants—whatever their destination country—tend to settle in large cities. The result everywhere is growing urban cultural diversity, with attendant challenges of social fragmentation, segregation, isolation, and poverty.

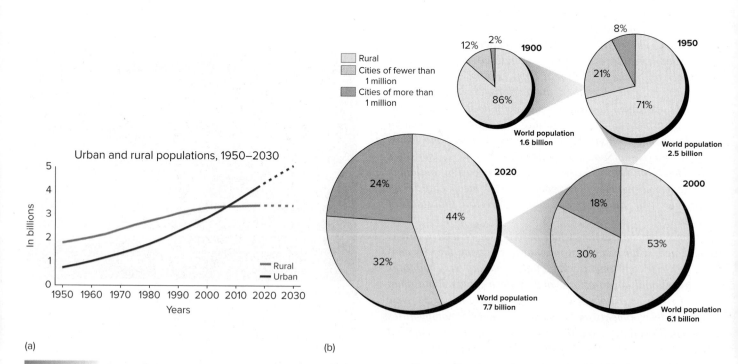

(a)   (b)

Figure 11.2 Trends of world urbanization document the steady decline in rural population proportions throughout the 20th century. *(a)* Since 1950, the growth rate of the rural component has slackened compared to the urban rate; by 2007, world urban numbers overtook the rural.

Source: United Nations, World Urbanization Prospects: The 2014, 2009 Revision and 2003 Revision, *Population Reference Bureau, and other sources.*

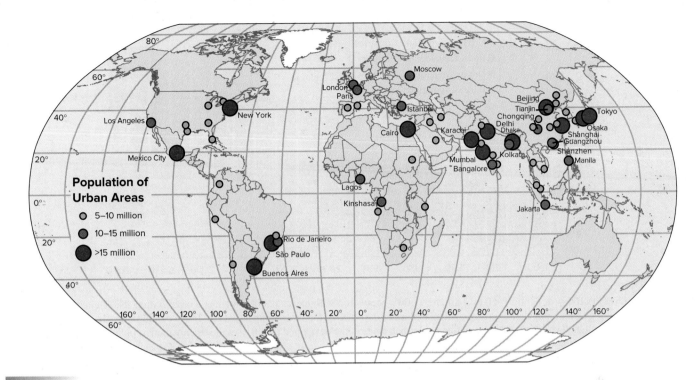

Figure 11.3 Metropolitan areas of 3 million or more in 2006. Only metropolitan areas with a population of 5 million or more are named. Massive urbanized districts are no longer limited to the industrialized, developed countries. They are now found on every continent, in all latitudes, as part of most economics and societies. Not all cities in congested areas are shown.

Source: Data from United Nations Population Division.

Table 11.1

World's Largest Urban Areas, 1900–2030 (Population in Millions)

1900		1950		2015		2030	
London	6.5	New York	12.3	Tokyo	38.0	Tokyo	37.2
New York	4.2	Tokyo	11.3	Delhi	25.7	Delhi	36.1
Paris	3.3	London	8.4	Shanghai	23.7	Shanghai	30.8
Berlin	2.7	Paris	6.5	São Paulo	21.1	Mumbai (Bombay)	27.8
Chicago	1.7	Moscow	5.4	Mumbai (Bombay)	21.0	Beijing	27.7
Vienna	1.7	Buenos Aires	5.1	Mexico City	21.0	Dhaka	27.4
Tokyo	1.5	Chicago	5.0	Beijing	20.4	Karachi	24.8
St. Petersburg	1.4	Kolkata (Calcutta)	4.5	Osaka-Kobe	20.2	Cairo	24.5
Manchester	1.4	Shanghai	4.3	Cairo	18.8	Lagos	24.2
Philadelphia	1.4	Osaka-Kobe	4.1	New York	18.6	Mexico City	23.9

Sources: *United Nations,* World Urbanization Prospects: The 2014 Revision, and Four Thousand Years of Urban Growth: An Historical Census, *Tertius Chandler. 1987, St. David's University Press.*

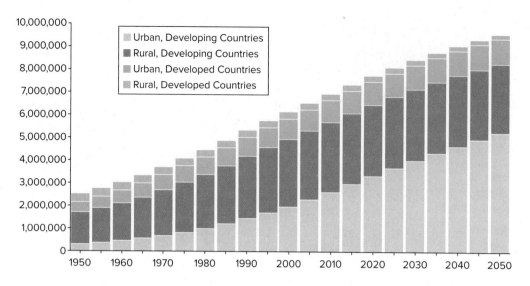

Figure 11.4 Over the 100-year period 1950 to 2050, the world's population will undergo both growth and urbanization. The United Nations estimates that virtually all future population growth will take place in the cities of the developing countries. In the coming decades, rural areas in both the developing and developed world will lose population, and urban areas in the developed countries will see only modest population growth.

Source: United Nations, World Urbanization Prospects: The 2014 Revision.

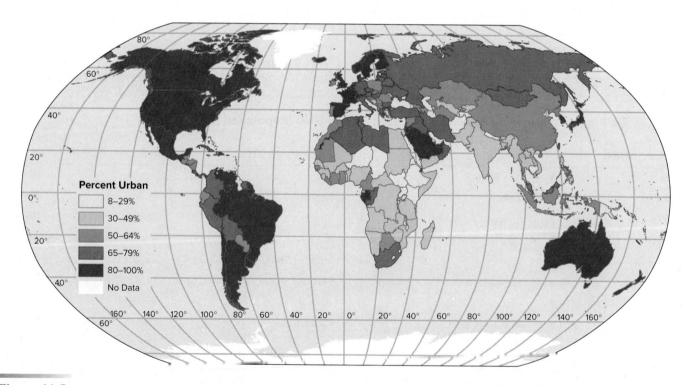

Figure 11.5 Percentage of population classified as urban, 2017. While Africa and South Asia lag behind other world regions in urbanization, they have the highest rates of urban growth.

Source: Population Reference Bureau, 2017.

Table 11.2

Estimated Urban Share of Total Population, Selected Areas: 1950, 2015, and 2050

Region or Country	1950	2015	2050
North America	64	82	87
Latin America and Caribbean	41	80	86
Europe	52	74	82
Oceania	62	71	74
Asia	18	48	64
Africa	14	40	56
More developed	55	78	85
Less developed	18	49	63
World	30	54	66

Sources: United Nations, World Urbanization Prospects: The 2014 Revision.

Merging Urban Regions

When separate major urban complexes expand along the superior transportation facilities connecting them, they may eventually meet, bind together at their outer margins, and create the extensive urban regions or **conurbations** suggested in Figure 11.3. No longer is there a single city with a single downtown area, set off by open countryside from any other urban unit in its vicinity. Rather, we must now recognize extensive regions of continuous urbanization made up of multiple centers that have come together at their edges.

A major North American example, *Megalopolis* was the term used by geographer Jean Gottmann to describe the nearly continuous urban string that stretches from north of Boston (southern Maine) to south of Washington, D.C. (southern Virginia). Other North American conurbations shown on **Figure 11.6** include the southern Great Lakes region, stretching from north of Milwaukee through Chicago and eastward to Detroit, Cleveland, and Pittsburgh; the Coastal California zone of San Francisco–Los Angeles–San Diego–Tijuana, Mexico; the Canadian "core region" conurbation from Montreal to Windsor, opposite Detroit, Michigan, where it connects with the southern Great Lakes region; the Vancouver–Willamette strip ("Cascadia") in the West; and the Piedmont, Gulf Coast and the Coastal Florida zones in the Southeast. Outside North America, examples of conurbations are numerous and growing, still primarily in the most industrialized European and East Asian (Japanese) districts, but forming as well in the other world regions where urban clusters and megacities emerged in developing countries that still were primarily rural in residential pattern.

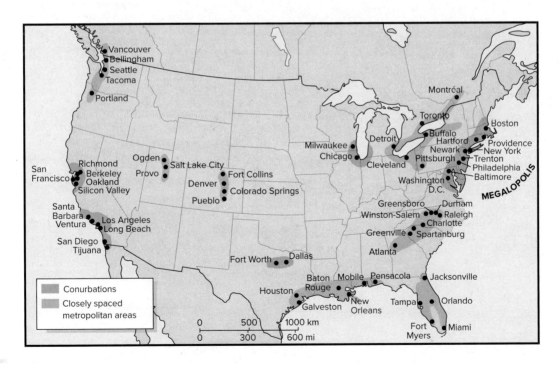

Figure 11.6 Megalopolis and other North American conurbations. The northeast U.S. Boston-to-Norfolk urban corridor comprises the original and largest *Megalopolis* and contains the economic, political, and administrative core of the United States. A Canadian counterpart core region anchored by Montreal and Toronto connects with U.S. contributions through Buffalo, New York, and Detroit, Michigan. For some of their extent, conurbations fulfill their classic definition of continuous built-up urban areas. In other portions, the urban areas are interspersed with land uses that are primarily rural.

11.2 Settlement Roots

The major cities of today had humble origins in the simple cluster of dwellings that was the starting point for human settlements everywhere. People are gregarious and cooperative. Even Stone Age hunters and gatherers lived and worked in groups, not as lone individuals or isolated families. All cultures are communal for protection, cooperative effort, sharing of tasks by age and sex, and for more subtle psychological and social reasons. Communal dwelling became the near-universal rule with the advent of sedentary agriculture wherever it developed, and the village became the norm of human society.

In most of the world, rural people still live in nucleated settlements (that is, in villages or hamlets), rather than in dispersed dwellings or isolated farmsteads. Only in North America, parts of northern and western Europe, and in Australia and New Zealand do rural dwellers tend to live apart, with houses and farm buildings located on land that is individually worked. Elsewhere in the world, villages and hamlets were and are the settlement norm, though size and form has varied by region and culture (**Figures 11.7 and 11.8**).

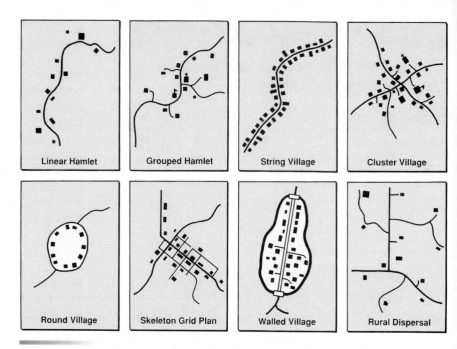

Figure 11.7 Basic settlement forms. The smallest organized rural clusters of houses and nonresidential structures are commonly called *hamlets,* and may contain only 10–15 buildings. *Villages* are larger agglomerations, although not as sizable or functionally complex as urban *towns*. The distinction between village and town is usually a statistical definition that varies by country. The rural dispersal settlement pattern is relatively rare, being found mostly in northern and Western Europe, North America, Australia, and New Zealand.

Source: Redrawn from Introducing Cultural Geography, *2nd ed., by J. E. Spencer and W. L. Thomas. 1978 John Wiley & Sons, Inc.*

Figure 11.8 Shirakawa village, Japan. Most of the world's rural population lives in small, agricultural villages rather than in dispersed farmsteads.
©Akira Kaede/Getty Images

Rural settlements in developing countries are often expressions of subsistence economic systems in which farming and fishing cultures produce no more than their individual families can consume. When trade does develop between two or more rural settlements, they begin to take on new physical characteristics as their inhabitants engage in additional types of occupations. The villages lose the purely social and residential character of subsistence agricultural settlements and assume urban features. The beginnings of urbanization are seen in the types of buildings that are erected and in the heightened importance of the main streets and of the roads leading to other settlements. No longer are the settlements self-contained; they become part of a system of towns and cities engaged in urban activities and exchange.

11.3 Origins and Evolution of Cities

Cities and civilization are inseparable; indeed, the words *city* and *civilization* have the same Latin root, *civis*. Dating from at least 8,000 years ago, cities originated in—or diffused from—the early culture hearths that first developed sedentary agriculture (see Figure 2.15 in Chapter 2 and **Figure 11.9**). As centers of cultural, economic, religious, and political life, they are among humanity's greatest achievements. Cities abound with contradictions, simultaneously displaying the extremes of luxury and misery, beauty and squalor. In cities, diverse peoples come together to exchange goods, services, and ideas and yet also stratify by race, ethnicity, and social class.

The earliest cities depended on the creation of agricultural surpluses. Many early cities included farms within their walls, but the main distinction between the city and the countryside stemmed from the nonagricultural work of most urban dwellers. This meant that food had to be provided to the urban population by the **hinterland**—the productive area surrounding a population center. Those in the newly emerging town who were not farmers were free to specialize in other activities—metal working, pottery making, cloth weaving, among others—producing goods for other urbanites and for the farm population on which they depended. Still others became scribes, merchants, priests, and soldiers, providing the services and refining the power structure on which the organized urban and rural society depended.

Social organization and a defined power structure, as reflected in a religious hierarchy and civil administration, were essential in urban development. Ancient cities centered on a temple or palace district housing the priests, the ruler, public storehouses and granaries, public baths, perhaps schools, and certainly a central marketplace. Cities became the seats of local and regional power, exercising control over the rural hinterland and extracting agricultural surplus from it for redistribution in the city. If possible, ancient cities were located in spots easy to defend—often on hilltops—but they were nearly always enclosed within protective

Figure 11.9 Aerial view of Erbil, Iraq. The site of modern Erbil—the ancient Assyrian city of Arabilu—has been continuously inhabited for about 8,000 years. The debris of millennia of urban settlement gradually raised the level of the land surface, producing a mound on which the city sits. The city—one of the oldest in the world—literally was constantly rebuilt at higher elevations on the accumulation of refuse from earlier occupants.
©Reza/Getty Images News/Getty Images

walls (**Figure 11.10**). The massive protective walls of early cities, however, could also limit the expansion of prospering, growing communities. Some cities, like Rome, went through multiple rounds of wall construction, with each new outer wall extending the urban area within which functions could be located and workers housed.

Figure 11.10 By Europe's Middle Ages, the ancient need for city protection remained, but fortifications and defensive structures had assumed elaborate and massive forms that were unknown and unneeded before siege weapons and siege warfare put all cities in jeopardy. The walls of Ávila, Spain, shown here, were built in the 12th century, extending 2,500 meters (8,200 feet) and encircling the entire city at that time; the modern part of Ávila lies outside.
©Pixtal/AGE Fotostock

Among those functions and workers were those engaged in long-distance trade, exchanging local goods and materials for raw materials and special products not obtainable locally. Merchants, wholesalers, clerks, scribes, carters, river men and sailors, and those who produced the vessels and supplied the necessary trade support services that came to characterize and dominate the functional base of the city. The importance of city location on navigable waterways, always a key to urban economic success, became ever more important.

In Europe and Asia, from about the 11th to the 18th century, local and distant trade, production of consumer goods by craftsmen organized into protective guilds, and increasing use of water-powered mills for grinding grain, making cloth, and sawing timber moved cities into intricate involvement in nearly modern forms of interregional and international economy. Massive trade fairs, international banking houses, and cooperative leagues of cities were precursors of today's global marketing, stock exchanges, and regional trade alliances (**Figure 11.11**).

With the Industrial Revolution, another shift in cities took place. Industrialization accelerated the rate of urban growth, initially in Europe and then elsewhere, where European control or influence was extended. Powered by water or steam, the new factories—operated by paid laborers, not by independent guild members—introduced mass production of standardized goods. Industrialization fundamentally changed cities, creating staggering economic and social stratification and, in many early industrial cities, dreadful conditions for the working classes. Cities, once centered on the temple, palace, marketplace, or waterfront wharves, and once surrounded by walls, were changed utterly; they had become places of industrial production centered on factories, canals, and railroads.

As industrialization diffuses around the globe, cities in the newly industrializing countries of Asia or Latin America have witnessed some of the same explosive growth and social polarization that historically accompanied industrialization. Meanwhile, in the more developed countries, the transition to a service economy has caused cities to take on a postindustrial character.

Figure 11.11 The densely built historic medieval section of Florence, Italy, is dominated by the Cathedral Santa Maria del Fiore. Florence prospered in the late Middle Ages as a center for textiles, artisanal craft industries, trade, and banking. It reached a population of 95,000 in 1300 before the plague decimated its population. In the 15th century, it was the center for the rediscovery of classical culture and was home to artists such as Sandro Botticelli, Leonardo da Vinci, and Michelangelo.

©Ken Welsh/AGE Fotostock

Smokestacks have disappeared and former factories, railroad yards, and industrial waterfronts have been redeveloped for parks, housing, and commercial uses. Consumption and service sector activities, rather than heavy industry, dominate the postindustrial city.

The Nature of Cities

Whether ancient or modern, all cities must perform functions—have an economic base—in order to generate the income necessary to support themselves. Second, no city exists in a vacuum; each is part of a larger society and economy with which it has essential relationships. That is, each is a unit in a system of cities and a focus for a surrounding rural area. Third, each urban unit has a more or less orderly internal arrangement of land uses, social groups, and economic functions. Because all urban functions and people cannot be located at a single point, cities themselves must take up space and organize the uses of that space. These arrangements may be partially or completely planned by central authorities or determined by individual decisions and market forces. Finally, all cities, large or small, ancient or modern, have experienced problems of land use, social conflict, and environmental quality. Yet cities, though flawed, remain the capstone of our cultures, the driving force in contemporary societies and economies, and the magnet for people everywhere.

All urban settlements exist for the efficient performance of functions required by the society that creates them. They reflect the saving of time, energy, and money that the agglomeration of people and activities provides. The more accessible the producer to the consumer, the worker to the workplace, the citizen to the town hall, the worshiper to the place of worship, or the lawyer or doctor to the client, the more efficiently they can perform their separate activities. Because interconnection is essential, the transportation system will have an enormous bearing on the total number of services that can be performed and the efficiency with which they can be carried out. The totality of people and urban functions constitutes a distinctive cultural landscape whose similarities and differences from place to place are the subjects for urban geographic analysis.

The Location of Urban Settlements

Urban centers are functionally connected to other cities and to rural areas. In fact, cities exist not only to provide services for themselves, but for others outside of it. The urban center is a consumer of food, a processor of materials, and an accumulator and dispenser of goods and services. But it must depend on outside areas for its essential supplies and as a market for its products and activities.

In order to adequately perform the tasks that support it and to add new functions as demanded by the larger economy, the city must be efficiently located. That efficiency may be marked by centrality to the area that it serves. It may derive from the physical characteristics of its site, or placement may be related to the

resources, productive regions, and transportation network of the country, so it can effectively perform a wide array of activities.

In discussing urban settlement locations, geographers usually distinguish between site and situation, concepts already introduced in Chapter 1 (see Figures 1.6 and 1.7). You will recall that *site* refers to the exact terrain features associated with the city, as well as—less usefully—to its absolute (globe grid) location. Classifications of cities according to site characteristics have been proposed, recognizing special locations. These include *break-of-bulk* locations, such as river crossing points where cargoes and people must interrupt a journey; *head-of-navigation* or *bay head* locations, where the limits of water transportation are reached; and *railhead* locations, where a railroad ended. For ancient and medieval cities, security and defense—islands or elevated sites—were considerations in choosing a location for a city. Waterpower sites and later coalfields were the prime city-building locations during the Industrial Revolution, as noted in Chapter 9.

If *site* suggests absolute location, *situation* indicates relative location in relation to the physical and cultural characteristics of surrounding areas. It is important to know what kinds of possibilities and activities exist in the area near a settlement, such as the distribution of raw materials, market areas, agricultural regions, mountains, and the places to which it is connected through rivers, oceans, and transportation systems. The functions and growth potentials of cities are more determined by their situation than their site.

The site or situation that originally gave rise to a city may not remain the essential ingredient for its continued growth and development. Agglomerations originally successful for whatever reason may by their success attract people and activities totally unrelated to the initial localizing forces. By what has been called a process of "circular and cumulative causation" (see Chapter 10), a successful urban unit may acquire new people and functions attracted by the existing markets, labor force, and urban facilities. In the same way, a site that originally favored the success of the new urban unit—on a navigable river or coal field, perhaps—may with the passage of time no longer be important in supporting any or all of its current economic activities.

Transportation Epochs

Break-of-bulk and head-of-navigation sites demonstrate the importance of transportation to the location of urban settlements. Whenever a new transportation system emerges, it changes the optimal locations for urban growth. Geographer John Borchert identified four epochs of inter-city transportation that shaped the location and growth rates of U.S. cities: (1) sail and wagon, 1790–1830; (2) iron railroads, 1830–1870; (3) steel railroads, 1870–1920; and (4) automobile and airplane travel, 1920–present. During the sail and wagon epoch, the major U.S. cities were all Atlantic ports such as New York City and Boston. These port cities served relatively small rural hinterlands. Canals helped expand the size of the hinterland. During the iron railroad era, inland waterway ports such as Chicago emerged as hubs of regional railroads that collected and distributed resources from the vast interior of the continent. During the steel railroad era, transcontinental railroads allowed westward expansion and the growth of Pacific port cities such as San Francisco and Seattle. In the era of automobile and airplane travel, urbanization dispersed to new areas, especially those with natural amenities. The Sun Belt regions in the southern and western United States have grown particularly fast in this epoch.

The Economic Base

We saw that from their ancient beginnings, cities depended on close relationships with their hinterlands. They provided the market where rural produce could be exchanged for the goods produced and the defense or religious functions performed by the city. Such rural service functions remain important. However, not all of the activities carried on within a city are intended to connect that city with the outside world. Some are necessary simply to support the city itself. Understanding the growth or decline of cities hinges on grasping the relationship between the two sectors.

Economic base theory was developed by noticing that the economic well-being of small, remote, resource-dependent towns was proportional to the value of goods they sold to outside markets. Part of the employed population of a city is engaged either in the production of goods or the performance of services for areas and people outside the city itself. They are workers engaged in *export* activities, whose efforts result in money flowing into the community. Collectively, they constitute the **basic sector** of the city's total economic structure. According to economic base theory, the basic sector makes up the **economic base** of the community and is essential to the health of the local economy.

Other workers support themselves by producing goods or services for residents of the city itself. Their efforts, necessary to the well-being and the successful operation of the settlement, do not generate new money but comprise a **nonbasic sector** of its economy. These activities circulate money within the community and are responsible for the internal functioning of the city. They are crucial to the continued operation of its stores, professional offices, city government, local transit, and school systems.

The total economic structure of an urban area equals the sum of its basic and nonbasic activities. In actuality, it is difficult to classify work as belonging exclusively to one sector or the other. It is often assumed that most manufacturing work is basic, although services today are increasingly traded. Some part of most jobs involves financial interaction with residents of other areas. Doctors, for example, may have mainly local patients and thus are members of the nonbasic sector, but the moment they provide a service to someone from outside the community, they bring new money into the city and become part of the basic sector.

Most centers perform many export functions, and the larger the urban unit, the more functions it performs. Nonetheless, even in cities with a diversified economic base, one export activity or a very small number of activities tends to dominate the economic specialization within a system of cities (**Figure 11.12**).

Assuming that it was possible to divide the employed population of a city into totally separate basic and service (nonbasic) components, a ratio between the two employment groups could be established. This *basic/nonbasic ratio,* shown in **Figure 11.13,** indicates that as a settlement increases in size, the number of nonbasic personnel grows faster than the number of new basic workers. The graph suggests that service sector jobs, most of which are nonbasic, will be more common in larger cities. In cities with a population of 1 million, the ratio is about two nonbasic workers for every basic worker. This means that adding 10 new basic employees expands the labor force by 30 (10 basic, 20 nonbasic). The resultant increase in total population is equal to the added workers plus their dependents. Thus, a **multiplier effect** exists, in which every new basic sector job creates additional nonbasic jobs. When news media report that a new manufacturing plant will create a certain number of new jobs in addition to those at the plant, they are referring to the multiplier effect. The size of the multiplier effect is determined by the community's basic/nonbasic ratio.

The changing numerical relationships shown in Figure 11.13 are understandable when we consider how settlements add functions and grow in population. A new industry selling services to other communities requires new workers who increase the basic workforce. These new employees in turn demand certain goods and services, such as clothing, food, and medical assistance, which may be provided locally. Those who perform such services must themselves have services available to them. For example, a grocery clerk must also buy groceries. The more nonbasic workers a city has, the more nonbasic workers are needed to support them. The reason that the size of the city influences the basic/nonbasic ratio is because money circulates more efficiently in large cities. In a town too small to support a clothing store, grocery store, or hospital, worker paychecks will be spent purchasing those goods and services from other communities, with no increase in the small town's nonbasic employment. On the other hand, large cities are more self-sufficient and can meet more of their needs internally, so that each new basic job generates additional nonbasic jobs.

The growth of cities may be self-generating—circular and cumulative as industries that specialize in the production of material objects for export, like automobiles and paper products, bring money into a community and set off chain reactions of additional economic activity, much of it involving *service* activities. In recent years, service industries have developed to the point where new service activities serve older ones. In addition to the spending by firms and their workers, the property taxes that they pay support public services such as schools, parks, and transportation systems. These public investments make the city an attractive place to live and do business, in turn attracting more new firms and residents.

In much the same way as settlements grow in size and complexity, so do they decline. When the demand for the goods and services produced by a city falls, less money comes into the community, and both the basic and nonbasic components are affected. Cities that experience deindustrialization often undergo a spiral of decline, losing additional nonbasic jobs and having less money to pay for local public services.

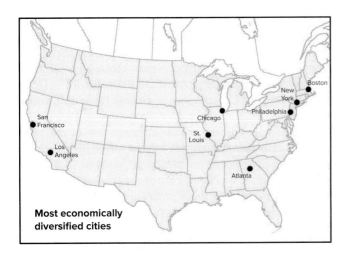

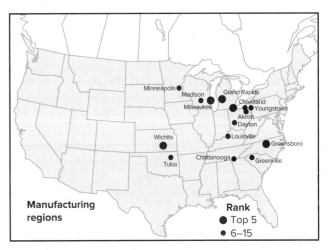

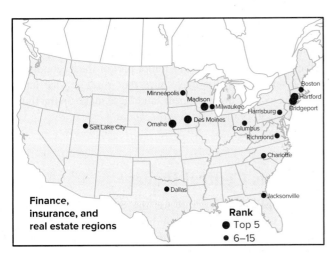

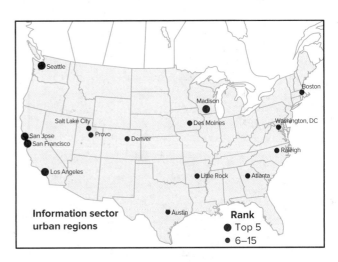

Figure 11.12 Functional specialization of selected U.S. metropolitan regions. Three categories of employment were chosen to show patterns of specialization for some U.S. metropolitan areas. In addition, the category "Most Economically Diversified" includes representative examples of cities with a generally balanced employment distribution. Note that the most diversified urban areas tend to be the largest.

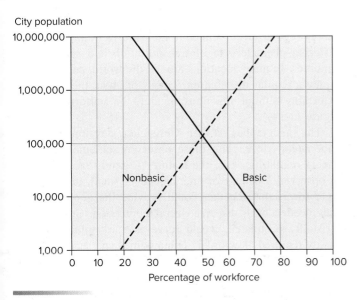

Figure 11.13 Generalized relationship between city size and the proportion of the workforce in basic and nonbasic activities. As settlements become larger, a greater proportion of the workforce is employed in nonbasic activities. Larger centers are, therefore, more self-contained.

There is, however, a resistance to decline that delays its impact. That is, settlements can grow rapidly as migrants respond quickly to the need for more workers, but under conditions of decline, those that have developed roots in the community are hesitant to leave or may be financially unable to move to another locale.

11.4 The Functions of Cities

Urban-based economic activities account for more than 50 percent of the gross national product (GNP) in all countries and up to 80 percent or more in the more urbanized states. Modern cities take on multiple functions. These include manufacturing, retailing, wholesaling, transportation, public administration, housing cultural and educational institutions, and, of course, the housing of their own citizens. Most cities, however, specialize in, or are dominated by, only one or a very few of the full range of economic activities. Only a relative few very large members of a national system of cities are importantly multifunctional and truly diversified.

No matter what their size, cities exist for the efficient performance of necessary functions. Those functions reflect cities' roles as transportation nodes and central places. The spatial pattern of *transportation centers* is that of alignment along seacoasts, rivers, canals, or railways. Transportation routes form the orienting axes along which cities developed and on which at least their initial functional success depended (**Figure 11.14**).

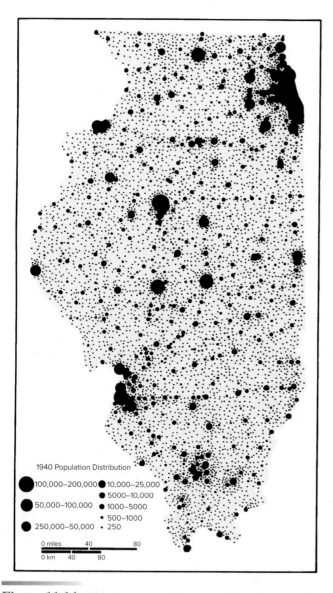

1940 Population Distribution

● 100,000–200,000   ● 10,000–25,000
● 50,000–100,000   ● 5000–10,000
                                   ● 1000–5000
                                   · 500–1000
● 250,000–50,000   · 250

0 miles   40   80
0 km   40   80

Figure 11.14 Urban alignments in Illinois. Railroads preceded settlement in much of the U.S. and Canadian continental interior, and urban centers were developed—frequently by the railroad companies themselves—as collecting and distributing points expected to grow as the farm populations increased. Located at constant 8- to 10-kilometer (5- to 6-mile) intervals in Illinois, the rail towns were the focal points of an expanding commercial agriculture. The linearity of the town pattern in 1940, at the peak of railroad influence, unmistakably marks the rail routes. Also evident are such special-function clusterings as the Chicago and St. Louis metropolitan districts and the mining towns of southern Illinois. In addition to the linear and cluster patterns, the smallest towns show the uniform distribution characteristic of central places.

Special-function cities are those engaged in mining or manufacturing, and they must be located where the raw materials occur, or agglomeration economies are present. Special-function cities show a pattern of urban clustering—as the mining and manufacturing cities of the Ruhr district of Germany, the Midlands of England, or the Donets Basin in Ukraine, for example.

Beginning with the Industrial Revolution of the 18th century, manufacturing industries were a major impetus for urban growth. The handicraft production of goods was, of course, always a part of the functional base of even the earliest urban units. Only with the rise of mass production, however, did industry become a primary basic urban function, producing wealth and, through the *multiplier effect,* expanding the numbers of basic and nonbasic workers through the export of manufactured goods throughout the larger economy.

All settlements serve as a central location for the provision of goods and services for a surrounding tributary area. For many, including mining or major manufacturing centers, service to tributary areas is only a very minor part of their economic base. Some settlements, however, have that rural service and trade function as their dominant role, and these make up the third simplified category of cities: *central places.*

Cities as Central Places

Central places are nodes for the distribution of economic goods and services to surrounding nonurban populations. For as long as cities have existed, they have served as marketplaces, not only for their own residents, but also for the population beyond the city limits. Small cities provide a range of goods and services that suffice for most everyday needs. But specialized, "higher-order" expensive or unique commodities and skilled specialized services can be found only in the largest cities. To serve the rural populations, central places show size and spacing regularities unrelated to the patterns of alignment and clustering characteristic of transportation and special-function cities. In many locations around the world, central places display a regular distribution. Towns of about the same size, performing similar functions, tend to be uniformly spaced (Figure 11.14).

The geographer Walter Christaller developed **central place theory** to explain those observed settlement size and spacing regularities (see the feature "Central Place Theory"). He observed a pattern of interdependent small, medium, and larger towns that together could provide the goods and services needed by a dispersed rural population. Small towns, Christaller postulated, would serve as marketplaces for frequently required "low-order" commodities and services, while expensive "high-order" luxury goods would be available only in larger communities that were central to a number of surrounding small towns. That is, people would have to travel only short distances for low order items, such as gasoline, convenience groceries, or haircuts, and longer distances for expensive and infrequently demanded goods and services, such as art museums, professional sports, luxury automobiles, or specialized medical treatments.

Christaller's explanation and description of the urban size and spacing regularities he observed have been shown to be generally applicable in widely differing regions of the world. When varying incomes, cultures, physical landscapes, and transportation systems are taken into consideration, his theory holds up rather well. It is particularly applicable, of course, to agricultural areas with a uniform distribution of consumers and purchasing power. If we combine a Christaller-type approach with the ideas of industrial location that help us understand the cluster patterns of special-function cities (see Chapter 9) and the alignments of transportation-based cities, we have a fairly good understanding of the distribution of most towns and cities. Central place theory is less relevant, of course, to arid regions such as the southwestern United States, where the lack of water prevented a uniform rural settlement and urban growth is much more concentrated.

The interdependence of small, medium, and large cities can also be seen in their influence on one another. A small city may influence a local region of some 1,000+ square kilometers (400 square miles) if, for example, its newspaper is delivered to that district. Beyond that area, another city may be the dominant influence. **Urban influence zones** are the areas outside a city that are still affected by it. As the distance away from a community increases, its influence on the surrounding countryside decreases (recall the idea of distance decay that was discussed in Chapter 3). The sphere of influence of an urban unit is usually proportional to its size.

For example, a large city located 100 kilometers (62 miles) away from a small city may influence that and other smaller communities through its banking services, TV stations, professional sports teams, and large shopping malls. There is an overlapping hierarchical arrangement, and the influence of the largest cities is felt over the widest areas, a *market area* dominance that is basic to central place theory.

Intricate relationships and hierarchies are common. Consider Grand Forks, North Dakota, which for local market purposes dominates the rural area immediately surrounding it. However, Grand Forks is influenced by political decisions made in the state capital, Bismarck. For a variety of cultural, commercial, and banking activities, Grand Forks is influenced by Minneapolis. As a center of wheat production, Grand Forks and Minneapolis are subordinate to the grain market in Chicago. Of course, the pervasive agricultural and other political controls exerted from Washington, D.C., on Grand Forks, Minneapolis, and Chicago indicate the size and complexity of urban zones of influence.

11.5 Systems of Cities

The systems of cities approach to urban geography considers the functions that cities perform in regional, national, or international economies, their relationships with the surrounding rural land and other cities, and how they are arranged in spatial patterns, networks, and hierarchies.

The Urban Hierarchy

Perhaps the most effective way to recognize how systems of cities are organized is to consider the **urban hierarchy,** a ranking of cities based on their size and functional complexity. One can measure the numbers and kinds of functions that each city or metropolitan area provides. The hierarchy is, then, like a pyramid; a few large and complex cities are at the top and many smaller, simpler ones are at the bottom. There are always more smaller cities than larger ones.

When a spatial dimension is added to the hierarchy as in **Figure 11.15,** it becomes clear that an areal system of metropolitan centers, large cities, small cities, and towns exists. Goods, services, communications, and people flow up and down the hierarchy. The few high-level metropolitan areas provide specialized functions for large regions, while the smaller cities serve smaller districts. The separate centers interact with the areas around them, but because cities of the same level provide roughly the same services, those of the same size tend not to serve each other unless they provide some very specialized activity, such as housing the political capital of a region or a major hospital or university. Thus, the settlements of a given level in the hierarchy

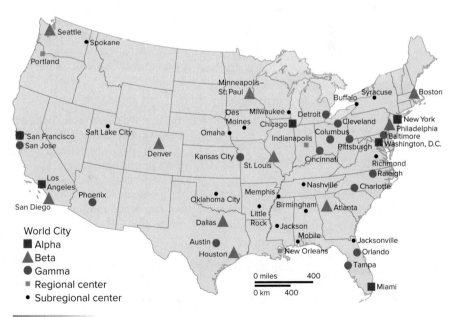

Figure 11.15 A functional hierarchy of U.S. metropolitan areas. With globalization, many former regional centers have become world cities. The hierarchy includes smaller urban centers (not shown) that depend on or serve the larger centers.

Source: Redrawn from P. J. Taylor, et al. The World According to GaWC 2016. Globalization and World Cities Research Network and P. L. Knox, ed., The United States: A Contemporary Human Geography. Harlow, United Kingdom. Longman, 1988, Fig. 5.5, p. 144.

Central Place Theory

In 1933, the German geographer **Walter Christaller** attempted to explain the size and spacing regularities he observed for towns in Southern Germany. In doing so, he developed a framework called *central place theory,* which provided the descriptive understandings that he sought. Christaller did recognize that his explanatory theory would best describe an idealized and somewhat artificial situation with the following characteristics:

1. Towns that provide the surrounding rural agricultural population with such fundamental goods as groceries and clothing would develop on an **isotropic plain,** that is, one with no topographic barriers, channelization of traffic, or variations in farm productivity.

2. The rural population would be dispersed in an even pattern across that uniform plain.

3. The characteristics of the people would be uniform; that is, they would possess similar tastes, demands, and incomes.

4. Each kind of product or service available to the dispersed population would have its own **threshold,** or minimum number of consumers needed to support its supply. Because such goods as sports cars or fur coats are either expensive or not in great demand, they would have a high threshold, while a fewer number of customers within smaller tributary areas would be sufficient to support a small grocery store.

5. Consumers would purchase goods and services from the nearest opportunity (store or supplier).

When all of Christaller's assumptions are considered simultaneously, they yield the following results:

1. Because each customer patronizes the nearest center offering the needed goods, the agricultural plain is automatically divided into noncompeting market areas—*complementary regions*—where each individual town (and its merchants) has a sales monopoly.

2. Those market areas will take the form of a series of hexagons that cover the entire plain, as shown in the diagram. There will be a central place at the center of each of the hexagonal market areas.

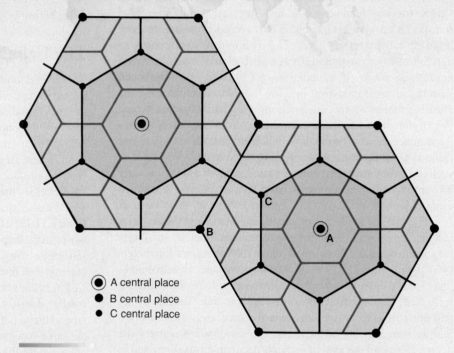

● A central place
● B central place
● C central place

Figure 11A Complementary regions and the pattern of central places. The two (A) central places are the largest on this diagram of one of Christaller's models. The (B) central places offer fewer goods and services for sale and serve only the areas of the intermediate-sized hexagons. The many (C) central places, which are considerably smaller and more closely spaced, serve still smaller market areas. The goods offered in the (C) places are also offered in the (B) and (A) places, but the latter offer considerably more and more specialized goods. Notice that places of the same size are equally spaced.

Source: Arthur Getis and Judith Getis, "Christaller's Central Place Theory." Journal of Geography, 1966.

3. The size of the market area of a central place will be proportional to the number of goods and services offered from that place and the largest central places (with the largest market areas) will supply all the goods and services the consumers in that area demand and can afford.

4. As the diagram indicates, the central place pattern shows a "nesting" of complementary regions in which part or all of multiple lower-order service areas are contained within the market area of a higher-order center.

In addition, Christaller reached two important conclusions. First, towns at the same size (functional level) in the central place system will be evenly spaced, and larger towns (higher-order places) will be farther apart than smaller ones. This means that many more small than large towns will exist.

In the diagram, the ratio of the number of small towns to towns of the next larger size is 3 to 1. This distinct, steplike series of towns in discrete classes differentiated by both size and function is called a **hierarchy of central places**.

Second, the system of towns is interdependent. If one central place were eliminated, the entire system would have to readjust. Consumers need a variety of products and services, each of which has a different minimum number of customers required to support it. The towns containing many goods and services become regional retailing centers, while the smaller central places serve just the people in their immediate vicinity. The higher the threshold of a desired product, the farther, on average, the consumer must travel to purchase it.

are not independent but interrelated with communities of other levels in that hierarchy. Together, all centers at all levels in the hierarchy constitute an urban system.

World Cities

Standing at the top of national systems of cities are a relatively few places that may be called **world cities** (or global cities). These large urban centers are command and control points for the global economy. When manufacturing dominated the economy, much of what an individual company did—production, management, sales, accounting, etc.—took place in a single city, often in the same building. Now, the globalized economy and transnational corporations (TNCs) have scattered those functions and jobs across the world. But all those activities must be coordinated somewhere, and that place is the world city.

London and New York, the world's two largest cities in 1950, are generally recognized as the two most dominant world cities today. They are no longer the world's two most populous cities, but they contain the highest number of producer services offices and TNC headquarters, and they dominate commerce in their respective parts of the world. Each is directly linked to a number of other primary- and secondary-level world cities. All are bound together in complex networks that control the organization and management of the global system of finance, production, and trade. **Figure 11.16** shows the links between the dominant centers and some of the major and secondary world cities, which include Paris, Dubai, Hong Kong, Singapore, Beijing, Shanghai, and Tokyo. These cities are all interconnected by advanced communication systems between governments, major corporations, stock and futures exchanges, securities and commodity markets, major banks, and international organizations.

World cities are home to society's most powerful and elite members, and thus they are centers for arts, culture, and the consumption of luxury goods. Some critics suggest that the forces of economic globalization that create world cities also increase inequality. Certainly, the incredible wealth generated in world cities leads to high costs for housing, creating affordability problems for lower-wage service workers.

Rank-Size and Primacy

In addition to considering city systems on a global scale, urban geographers also inquire about the organization of city systems within regions or countries. The observation that there are many more small than large cities within an urban system ("the larger the fewer") is itself a statement about expected city hierarchies. In some countries, especially those with complex economies and a long urban history, the city size hierarchy is summarized by the **rank-size rule.** It tells us that the nth largest city of a national system of cities will be $1/n$ the size of the largest city. That is, the second-largest settlement will be 1/2 the size of the largest, the 10th biggest will be 1/10 the size of the first-ranked city, and so on. Although no national city system exactly meets the predictions of the rank-size rule, that of Russia, Canada, and the United States closely approximate it.

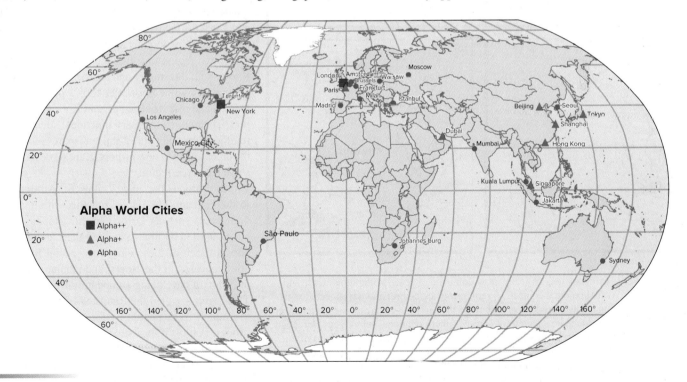

Figure 11.16 This classification of world (or global) cities is based on international business services of advertising, accounting, banking, and law. Compare this map to Figure 11.3. The location of the alpha world cities shows the greater economic power of the developed countries and the rising importance of newly industrializing Asian cities. These cities are bound together in complex networks, all interconnected by the flow of financial and economic information. A classification based on a different set of economic activities would yield a slightly different list but a similar hierarchy. Beta and gamma world cities are not shown.

Source: Adapted from P. J. Taylor, et al. The World According to GaWC 2016. *Globalization and World Cities Research Network.*

Rank-size ordering is less applicable to countries with developing economies or where the city system is dominated by a **primate city,** one that is far more than twice the size of the second-ranked city. In fact, there may be no obvious "second city" at all, for a characteristic of a primate city hierarchy is one very large city, few or no intermediate-sized cities, and many subordinate smaller settlements. For example, Seoul (at 9.8 million in 2010) contains about 20 percent of the total population of South Korea, and Luanda, Angola, and Bangkok, Thailand, are each home to about 40 percent of their country's urban population.

The capital cities of many developing countries display that kind of overwhelming primacy. In part, their primate city pattern is a heritage of their colonial past, when economic development, colonial administration, and transportation and trade activities were concentrated at a single point (**Figure 11.17**); Nairobi (Kenya), Dakar (Senegal), and many other African capital cities are examples.

In other instances—Egypt (Cairo) or Mexico (Mexico City), for example—development and population growth have tended to concentrate disproportionately in a capital city whose very size attracts further development and growth. Many European countries, such as Austria, the United Kingdom, and France, also show a primate structure due to the historic concentration of economic and political power in a capital city that was the administrative and trade center of a larger colonial empire.

Network Cities

The history of urban growth includes episodes of intense competition between cities, often over dominance of transportation networks. In recent years, a new kind of urban spatial pattern, the network city, has begun to appear as nearby cities work together. A **network city** evolves when two or more previously independent cities with potentially complementary functions develop high-speed transportation corridors and communications infrastructure to facilitate cooperation.

For example, since the reunion of Hong Kong and China proper in 1997, an infrastructure of highway and rail lines and communications improvements has been developed to help integrate Hong Kong with Guangzhou, the huge, rapidly growing industrial and economic hub on the mainland. In Japan, three distinctive, nearby cities—Kyoto, Osaka, and Kobe—are joining together to compete with the Tokyo region as a major center of commerce. Kyoto, with its temples and artistic treasures, is the cultural capital of Japan; Osaka is a primary commercial and industrial center; and Kobe is a leading port. Their complementary functional strengths are reinforced by high-speed rail transport connecting the cities and by a new airport (Kansai) designed to serve the entire region.

In Europe, the major cities of Amsterdam, Rotterdam, and The Hague, together with intermediate cities such as Delft, Utrecht, and Zaanstad, are connected by high-speed rail lines and a major airport. Each of these cities has special functions not duplicated in the others, and planners have no intention of developing competition between them. This region—called the Randstad—is second only to London in its popularity for international head offices, putting it in a strong position to compete for dominant world-city status.

No similar network city has yet developed in the United States. The New York–Philadelphia, Chicago-Milwaukee, San Francisco–Oakland, or Los Angeles–San Diego city pairings do not yet qualify for network city status because there has been no concerted effort to bring their competing interests together into a single structure of complementary activities.

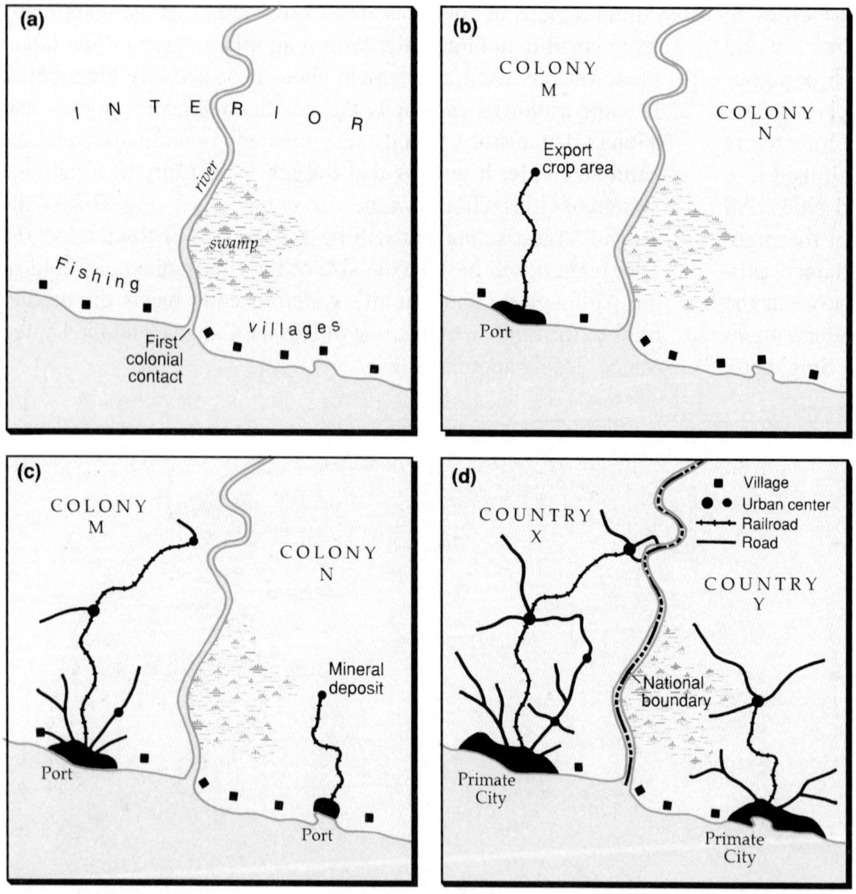

Figure 11.17 Primate cities typically grow dominant as centers of colonial trade and administration. At first colonial contact (*a*) settlements are coastal and unconnected with one another. Joining a newly productive hinterland by European-built railroads to a new colonial port (*b*) begins to create a pattern of core-periphery relations and to focus European administration, trade, and settlement at the port. Mineral discoveries and another rail line in a neighboring colony across the river (*c*) mark the beginnings of a new set of core-periphery relationships and of a new multifunctional colonial capital nearby but unconnected by land with its neighbor. With the passage of time and further transport and economic development, two newly independent nations (*d*) display *primate city* structures in which further economic and population growth flows to the single dominating centers of countries lacking balanced regional transport networks, resource development, and urban structures. Both populations and new functions continue to seek locations in the primate city where their prospects for success are greatest.

Source: Adapted from E. S. Simpson, The Developing World: An Introduction *(Harlow, Essex, England: Longman Group UK Limited, 1987).*

11.6 Inside the City

The location, structure, patterns, and spatial interactions of systems of cities make up only half of the story of urban settlements. The other half involves the distinctive cultural landscapes of cities themselves. An understanding of the nature of cities is incomplete without knowledge of their internal characteristics. So far, we have explored the origins and functions of cities within hierarchical urban systems. Now we look into the city itself to better understand how its land uses are distributed, how social areas are formed, and how institutional controls such as zoning regulations affect its structure. We will begin on familiar ground and focus our discussion primarily on U.S. cities. Later in this chapter, we will review urban land-use patterns and social geographies in different world settings. First, however, it is important to understand the common terms that we will use throughout this section.

Defining the City Today

Urban settlements come in different sizes, shapes, and types. Their common characteristic is that they are nucleated, nonagricultural settlements. At one end of the size scale, urban areas are hamlets or small towns with at most a single short main street of shops; at the opposite end, they are complex multifunctional metropolitan areas or megacities (**Figure 11.18**). The word *urban* is often used to describe such places as a town, city, suburb, and metropolitan area, but it is a general term, not used to specify a particular type or size of settlement. Although the terms designating the different types of urban settlement, like *city,* are employed in common speech, not everyone uses them in the same way. What is recognized as a city by a resident of rural Vermont or West Virginia might not be by an inhabitant of California or New Jersey. One should keep in mind, as well, that the same term may be understood or defined differently in different parts of the world. In the United States, the Census Bureau describes an *urban* place as having 2,500 or more inhabitants. In Greece, *urban* refers to municipalities in which the largest population center has 10,000 or more inhabitants, and Nicaragua uses the term to denote administrative centers with streets, lights, and at least 1,000 inhabitants. It is useful in this chapter to agree on the meanings of common terms with different usages.

The words **city** and *town* denote nucleated settlements with multiple functions, including a central business district (CBD) and both residential and nonresidential land uses. **Towns** are smaller in size and have less functional complexity than cities, but they still have a nuclear business concentration. **Suburb** implies a subsidiary area, a functionally specialized segment of a larger urban complex. It may be mostly residential, industrial, or commercial, but by the specialization of its land uses and functions, a suburb is not self-sufficient. A suburb, however, can be an independent political entity with its own local government. A **central city** is the principal core of a larger urban area, separately incorporated and ringed by its dependent suburbs.

Some or all of these urban types may be grouped into larger composite units. An **urbanized area** is a continuously built-up landscape defined by building and population densities, with no reference to political boundaries. It may be viewed as the *physical city* and may contain a central city and many contiguous cities, towns, and suburbs. A **metropolitan area,** on the other hand, is a large-scale *functional* entity, perhaps containing several urbanized areas, discontinuously built up but nonetheless operating as an integrated economic whole. The edge of the urbanized area is visible as the boundary where urban development meets the open countryside, By contrast, the boundary of the metropolitan area is often just a line on a map that is not apparent on the ground. **Figure 11.19** shows these areas in a hypothetical American metropolitan area.

(a)

(b)

Figure 11.18 The differences in size, density, and land-use complexity are immediately apparent between *(a)* a city (New York City) and *(b)* a town (Shepherdstown, West Virginia). Clearly, one is a city and one is a town, but both are *urban* areas.

(a) ©TongRo Images/Alamy Stock Photo; (b) ©Mark Bjelland

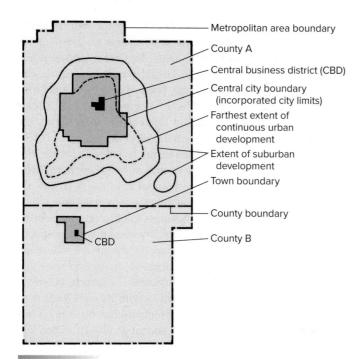

Figure 11.19 A hypothetical spatial arrangement of urban units within a metropolitan area. Sometimes official limits of the central city are very extensive and contain areas commonly considered suburban or even rural. On the other hand, older eastern and midwestern U.S. cities (and others, such as San Francisco in the west) more often have restricted limits and contain only part of the high-density land uses and populations of their metropolitan or urbanized areas as shown in this diagram. In this diagram, County B is part of the same metropolitan area as County A because of strong commuting and socialization ties between the counties.

The Bureau of the Census has redefined the concept of *metropolitan* from time to time to summarize the realities of the changing population, physical size, and functions of urban regions. The current *metropolitan statistical areas* are comprised of a central county or counties with at least one urbanized area of at least 50,000 residents, plus adjacent outlying counties with a high degree of social and economic integration with the central county as measured by commuting volumes. A list of the largest U.S. metropolitan statistical areas in 2017 is given in Table 11.3. Using similar criteria, the U.S. Census Bureau also defines micropolitan areas, where the urban core has between 10,000 and 50,000 residents.

Classic Patterns of Urban Land Use

Recurring patterns of land use and population density exist within urban areas. There are regularities in the way cities are internally organized, especially within one particular culture region, such as North America or Western Europe. Accessibility, a competitive market in land, and the innumerable individual residential, commercial, and industrial locational decisions made over time have shaped unplanned internal urban land-use patterns. Giving rise to three sharply different urban land-use layouts were the dominant transportation modes—first, walking, then mass transit

Table 11.3

The 30 Largest U.S. Metropolitan Statistical Areas, July 2017

Rank	Metropolitan Areas Identified by Their Principal Cities	Population
1	New York	20,321,000
2	Los Angeles	13,354,000
3	Chicago	9,533,000
4	Dallas–Fort Worth	7,400,000
5	Houston	6,892,000
6	Washington, D. C.	6,217,000
7	Miami	6,159,000
8	Philadelphia	6,096,000
9	Atlanta	5,885,000
10	Boston	4,837,000
11	Phoenix	4,737,000
12	San Francisco	4,727,000
13	Riverside–San Bernardino	4,581,000
14	Detroit	4,313,000
15	Seattle	3,867,000
16	Minneapolis–St. Paul	3,601,000
17	San Diego	3,338,000
18	Tampa–St. Petersburg	3,091,000
19	Denver	2,888,000
20	Baltimore	2,808,000
21	St. Louis	2,807,000
22	Charlotte	2,525,000
23	Orlando	2,510,000
24	San Antonio	2,474,000
25	Portland	2,453,000
26	Pittsburgh	2,333,000
27	Sacramento	2,325,000
28	Las Vegas	2,204,000
29	Cincinnati	2,179,000
30	Kansas City	2,129,000

Source: U.S. Bureau of the Census.

systems, and later the automobile—available during successive periods of urban growth.

The pedestrian and pack animal movement of people and goods within the small, compact pre-industrial walking city could no longer serve the increasing number of people and functions seeking accommodation within the expanding industrial city of the late 19th and early 20th centuries. Mass transit lines—horse car, cable car, electric streetcar lines, and eventually elevated and subway rail systems—were successively installed and extended; they controlled the development and layout of cities in, particularly, the northeastern United States, southeastern Canada, and older cities of the interior and west. Radiating outward from the town center, the transit systems immediately gave differential accessibility to the different areas of the growing city. Properties along and near the lines were usable and valuable because reachable; land beyond easy walking distance of the radial transit lines was unusable and left vacant. Transit lines generally converged at a hub in the CBD, making the central area the most valuable in the entire region. The result was a compact, high-density city with a single dominant center and sharp break at the boundary between urban and nonurban uses (**Figure 11.20**).

The Central Business District

Within the older central city, the radiating mass transit lines focused on the original city center (downtown), giving that area the highest accessibility within the growing urban complex. The center, therefore, held the greatest attraction for those functions profiting most from accessibility to the whole region. Building lots within the emerging **central business district (CBD)** could command the highest rental and purchase prices. The intersection where the major mass transit lines converged was called the **peak land value intersection.**

In a market system, the value of urban land was determined by competitive bidding among potential users. Public uses—parks, municipal buildings, schools—were allocated land according to criteria other than ability to pay. In the

private market, however, uses with the greatest need and demand for accessibility bid most for, and occupied, the most central parcels within the CBD. Those uses were typically the department stores and other retail outlets catering to the shopping needs of the majority of urban residents. The urban core, that is, became the highest-order central place, offering the full range of low-order and high-order goods. Parcels a short distance from the peak land value intersection generally became sites for tall office buildings (skyscrapers), the principal hotels, and similar land uses that helped produce the distinctive skyline of the older, high-order commercial city.

Outside the CBD

Just outside the core area of the city, industry controlled land next to essential cargo routes: rail lines, waterfronts, rivers, or canals. Lower-order commercial centers developed at the outlying intersections—transfer points—of the mass transit network. Strings of stores, light industries, and high-density apartment structures could afford and benefit from location along high-volume transit routes. The least accessible locations within the city were left for the least-competitive bidders: low-density residences. A diagrammatic summary of this repetitive allocation of space among competitors for urban sites is shown in **Figure 11.21.**

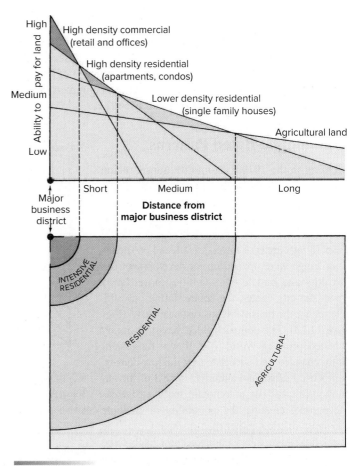

Figure 11.21 Generalized urban land-use pattern. The model depicts the location of various land uses in an idealized city where the highest bidder gets the most accessible land.

Figure 11.20 Townhouses, such as these in Boston's Back Bay area, as well as apartment buildings and duplexes, were a characteristic response to the price and scarcity of developable urban land in the era before automobiles became widely available. Where detached single-family dwellings were built, they were typically on smaller lots than became the norm once widespread automobile use allowed cities to spread outward in the second half of the 20th century.

©Mark Bjelland

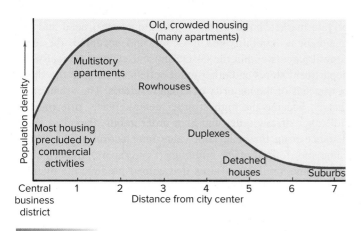

Figure 11.22 A summary population density curve. As distance from the area of multistory apartment buildings increases, the population density declines.

The competitive bidding for land should yield—in theory, at least—two separate but related distance-decay patterns: both land values and population densities decrease as distance from the CBD increases. Land values decline in a distinct pattern: within the CBD, there is a sharp drop in values a short distance from the peak land value intersection, the most accessible and costly parcel of the CBD; then the values decline less steeply to the margins of the built-up urban area. With the exception of a tendency to form a *hollow at the center,* the CBD, the population density pattern of the central city showed a comparable distance-decay arrangement, as suggested by **Figure 11.22.** The low population density at the city center, of course, reflected the superior rent-paying abilities of commercial and industrial users, displacing residential uses.

Automobile-Based Patterns

Starting in the 1940s, automotive transportation became dominant in the movement of people and goods and streetcar systems lost riders and were often converted to bus systems. As highway systems were extended outward after World War II, vast areas of lower-priced land on the urban fringe were opened up for development. As wealthy and middle class families moved away from the city center, the zones shifted outward, flattening the density versus distance curve (**Figure 11.23**). The compact older mass transit city created prior to World War II was fundamentally changed and succeeded by the low-density, unfocused urban and suburban sprawl of the automobile city. The automobile made vast areas accessible, creating the possibility of multiple business districts rather than a single CBD. The peak land value intersection was now likely to be the intersection of a major radial highway with a circumferential (beltway) highway or even an entire highway corridor. Still, the concepts of accessibility and competitive bidding for land shown in

Figure 11.21 apply. In newer automobile-based development, major commercial uses occupy the most accessible and most expensive land along major highway corridors. Higher-density housing, such as apartments, townhouses, and condominiums, often border these commercial districts, and lower-density single-family housing is found in more secluded, less accessible locations. In most communities, these patterns are not the product of pure free market bidding but are dictated by land use and zoning plans that try to anticipate the results of competitive bidding for each piece of land.

Regional Differences

The timing of an urban region's growth determines the relative mix of walking city, mass-transit city, and automobile city. Only the oldest parts of eastern cities such as Old Quebec and Boston's Beacon Hill still display remnants of the walking city. Cities in the East and Midwest, such as Philadelphia and Chicago, have large areas that developed when mass transit was dominant. The density and design of the newer cities of the West and Southwest, as well as the suburban growth areas of older centers, have been influenced primarily or exclusively by the automobile and motor truck, not by mass transit and railroads. The land use contrast between regions is not absolute, of course, as older cities have adapted to the automobile and rapidly growing cities in the West and Southwest have added light-rail transit systems. Even so, the different patterns have not been totally erased because cities, like other cultural landscapes, are built up over time, layer upon layer. Thus, the ever-changing 21st-century American city shows the intermingling of influences from different eras of city building. What the future holds for

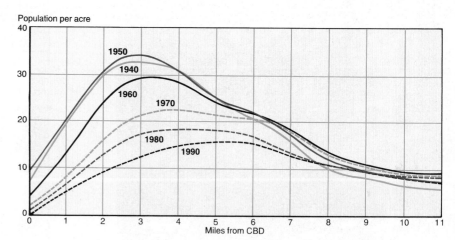

Figure 11.23 Population density gradients for Cleveland, Ohio, 1940–1990. The progressive depopulation of the central core and flattening of the density gradient over time to the city margin is clearly seen as Cleveland passed from mass transit to automobile domination. The Cleveland pattern is consistent with other cities where widespread adoption of the automobile caused density gradients to flatten over time. Some cities, such as Chicago, Toronto, Seattle, and Vancouver, have partially reversed this trend by increasing the amount of downtown housing.

Source: Anupa Mukhopadhyay and Ashok K. Dutt, "Population Density Gradient Changes of a Postindustrial City—Cleveland, Ohio 1940–1990," GeoJournal 34(4):517, 1994.

our cities is hard to say, but many urban geographers and planners are arguing for a return to the transit-oriented pattern of urban growth for reasons of energy conservation and environmental sustainability.

Models of Urban Form

We all have mental maps that help us summarize and make sense of the diverse places we've experienced in large cities. The meanings we associate with terms such as *inner-city* or *West End* reflect the content of those mental maps. Simple, graphic models of urban growth and land-use patterns began to appear during the 1920s and 1930s. Those models generalized the varied urban universe and helped explain some regularities in city growth and structure. More recently, urban geographers have begun to offer models that address the newer patterns of the decentralized automobile city.

The common starting point of the early models is the distinctive CBD found in every older central city. The core of the CBD displays the intensive land-use development already discussed: the major shopping concentration, tall office buildings, and streets crowded by pedestrians. Framing the core is a fringe area of warehousing, transportation terminals, and light industries. Just beyond the fringe, residential land uses begin.

The **concentric zone model (Figure 11.24a)** was developed by University of Chicago sociologists to explain the structuring of U.S. cities, specifically ethnically diverse, mass transit–based cities like Chicago in the 1920s. It describes the urban community as a set of nested rings of mostly residential uses at increasing distances in all directions from the CBD fringe. The first, a zone of transition, is characterized by change and deterioration and contains warehouses and factories mixed in with high-density, low-income slums, rooming houses, and perhaps ethnic ghettoes. Moving outward, the next ring is a zone of workers' homes, usually smaller, older homes on small lots. The third zone houses better residences, single-family homes or higher-rent apartments for those able to exercise choice in housing location and afford the longer journey to CBD employment. Finally, just beginning to emerge when this model was proposed, was an outer zone of low-density suburban development.

The concentric zone model is dynamic. Each type of land use and each residential group tends to move outward into the next outer zone as the city matures and expands. That movement was seen as part of a ceaseless process of invasion and succession that yielded a restructured land-use pattern and population segregation by income level. The least attractive housing is in the inner-city zone of transition where smelly factories are interspersed with aging, crowded apartments. As one travels outward, the housing is progressively newer and more spacious, and the social and economic status of the residents rises accordingly.

The **sector model** (Figure 11.24b) was devised in the 1930s by the land economist Homer Hoyt, who mapped housing values in major U.S. cities. The sector model posits that high-rent

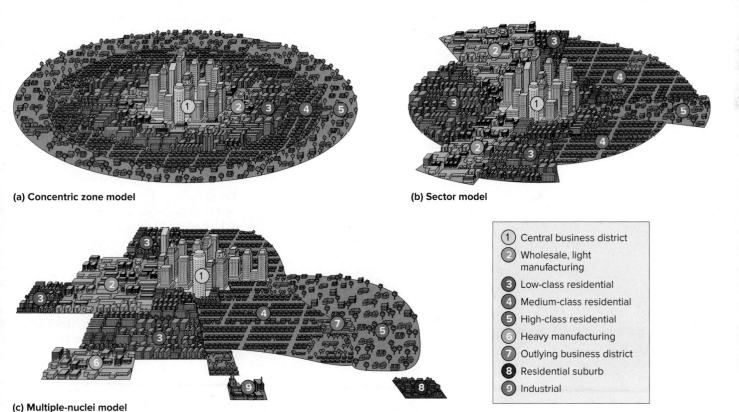

(a) Concentric zone model

(b) Sector model

(c) Multiple-nuclei model

1. Central business district
2. Wholesale, light manufacturing
3. Low-class residential
4. Medium-class residential
5. High-class residential
6. Heavy manufacturing
7. Outlying business district
8. Residential suburb
9. Industrial

Figure 11.24 Three classic models of the internal structure of cities.

Source: Redrawn from "The Nature of Cities" by C.D. Harris and E.L. Ullman, in Vol. 242 of The Annals of the American Academy of Political and Social Science. *1945 The American Academy of Political and Social Science, Philadelphia.*

residential areas are dominant in city expansion. The high-rent sector is typically established in areas of natural amenities such as lakeshore, large parks, or prominent ridges or hills. The high-rent sector grows outward from the city center along major transportation routes such as streetcar and elevated railroad lines or suburban commuter routes. Low-income populations occupy districts adjacent to the areas of industry and associated heavy transportation corridors, such as freight railroad lines. Middle-income housing fills in between the low-income and high-income districts.

The sector model is also dynamic, marked by a *filtering-down* process as older areas are left behind by the outward movement of their original higher-income inhabitants, with the lower-income populations moving into the recently vacated areas. The expansion of the city is radial, shaped by radial transportation systems. The social status of inner-ring neighborhoods extends outward into the suburbs. The accordance of the sector model with the actual pattern observed in Dallas–Fort Worth, Texas is suggested in **Figure 11.25**.

The basic assumption of the concentric circle and sector models—that urban growth and development proceeded outward from a single central core—was countered by the **multiple-nuclei model** (Figure 11.24c) proposed by geographers Chauncy Harris and Edward Ullman. In their view, large cities developed outward from several nodes of growth, not just one. Certain activities have specific locational requirements: the retail district needs accessibility; a port function needs a waterfront site; heavy industry requires level land adjacent to railroads. Peripheral expansion of the separate centers eventually leads to coalescence and the meeting of incompatible land uses along the lines of juncture. The urban land-use pattern, therefore, is not regularly structured from a single center in a sequence of circles or a series of sectors but based on separately expanding clusters of contrasting activities.

Although there have been many social, economic, and technological changes since these three models were developed, the patterns that they explained remain as vestiges and controls on the current landscape of older central cities. North American cities prior to 1945 resembled the concentric zone or sector models with a clearly defined and dominant CBD, but both new and expanding older cities grew more sprawling and complex in the automobile era following World War II. The multiple-nuclei model gives a better insight into the urban structure of the more recent past, but should be supplemented by newer visualizations of contemporary metropolitan complexes or *galactic cities*.

The **peripheral model** (also known as the **galactic city model**) as shown in **Figure 11.26** takes into account the major changes in urban form that have taken place since World War II, especially the suburbanization of what were once central city functions. The peripheral model focuses on the peripheral belt that lies within the metropolitan area, but outside the central city itself.

In these models, circumferential highways and expressways outside the central city make large tracts of land available for development in the low-density sprawl characteristic of individual rather than mass-transit movement of people. Residences are segregated by price level into relatively homogenous suburban clusters, and individual nodes in the peripheral belt are centers for employment or services: shopping malls, industrial parks, distribution and warehouse concentrations, office parks, airport-associated clusters containing hotels, meeting facilities, car rental agencies, and the like.

Much of the life of the residents of the periphery takes place outside the central city, as they shop for food, clothing, and services in the shopping malls, seek recreation in country clubs and entertainment complexes, and find employment in outlying industrial or office parks. While residents of the periphery may feel no need to travel to the old CBD with its problems of congestion, expensive parking, and homeless people, the periphery, however, remains a functional part of the metropolitan complex. Job markets are regional and the office parks and shopping malls of the periphery rely on low-wage service sector workers who often travel from the urban core where they find affordable housing. The urban core also retains an important cultural role, housing key institutions such as art museums, performance halls, universities, and stadiums.

The models of urban form just discussed aid our understanding of urban structure and development, but it must be stressed that a model is not a map, and that many cities contain elements and characteristics of more than a single model.

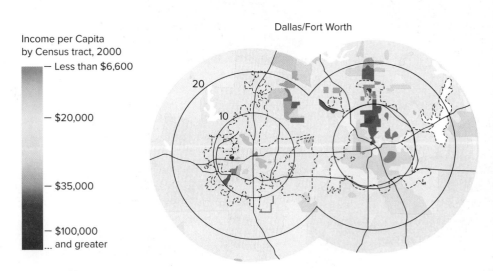

Income per Capita
by Census tract, 2000
— Less than $6,600

— $20,000

— $35,000

— $100,000
... and greater

Dallas/Fort Worth

20

10

Figure 11.25 Incomes in Dallas–Fort Worth, Texas, 2000. The high-income sector extending more than 20 miles north from downtown Dallas (east side of map) illustrates the applicability of the sector model. The inner rings of low-income residents are consistent with the concentric ring model.

Source: Cartography by Bill Rankin, data from the U.S. Census Bureau.

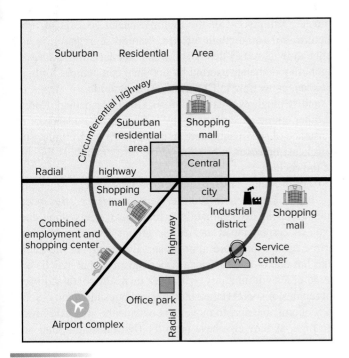

Figure 11.26 Metropolitan peripheral model. The galactic city's multiple downtowns and special function nodes and corridors are linked by the metropolitan expressway systems in this conceptualization proposed by Chauncy Harris.

Source: From Urban Geography, *Vol. 18, No. 1, pp. 15–35. Bellwether Publishing, Ltd.*

11.7 Social Areas of Cities

Vestiges of the ring and sectoral features depicted in the early models of U.S. cities are evident in the observed social segregation within urban areas. The larger and more economically and socially complex the city, the stronger the tendency for residents to sort themselves into groups based on *social status, family status,* and *ethnicity.* In a large metropolitan region with a diversified population, this territorial behavior may be a defense against the unknown or the unwanted, a desire to be among similar kinds of people, a response to income constraints, or a result of social and institutional barriers. Most people feel more at ease when they are near those with whom they can easily identify. In traditional societies, these groups are the families and tribes. In modern society, people tend to group according to income or occupation (social status), stages in the life cycle (family status), and language or race (ethnic characteristics); see the feature "Birds of a Feather" in Chapter 7.

Many of these social area groupings are fostered by the size and the value of available housing. Land developers, especially in cities, produce homes of similar quality and type in specific areas. Of course, as time elapses, there is a change in the condition and quality of that housing. Land uses may change and new groups may replace previous tenants, leading to the evolution of new neighborhoods with different social characteristics.

Social Status

The social status of an individual or a family is determined by income, education, occupation, and home value, although it may be measured differently in different cultures. In the United States, high income, a college education, a professional or managerial position, and high-value housing confer high status. High-value housing can mean an expensive rental apartment, a spacious loft in a former warehouse, or a large suburban house with extensive grounds. A good housing indicator of social status is persons per room or floor area per person. A low number of persons per room tends to indicate high status. Low status characterizes people with low income and lower levels of education, living in low-value housing.

Patterns of social status agree with the sector model. In most cities, people of similar status are grouped in sectors that fan out from the innermost urban residential areas (**Figure 11.27**). If the number of people within a given social group increases, they tend to move away from the central city along an arterial connecting them with the old neighborhood. Major transport routes leading to the city center are the usual migration routes out from the core. Chicago's elite Gold Coast along Lake Michigan and its low-income Southside neighborhoods display the extremes of social status (**Figure 11.28**).

Today, social status divisions are often perpetuated by political boundaries between separate municipalities or school districts. Communities on either side of the divide may differ greatly in income. Many residential developments are also

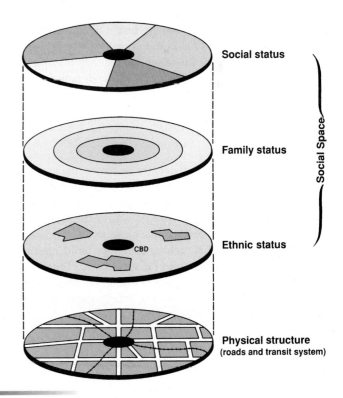

Figure 11.27 The social geography of American and Canadian cities.

Source: Redrawn from Robert A. Murdie, Factorial Ecology of Metropolitan Toronto. *Research Paper 116. Department of Geography Research Series, University of Chicago, 1969.*

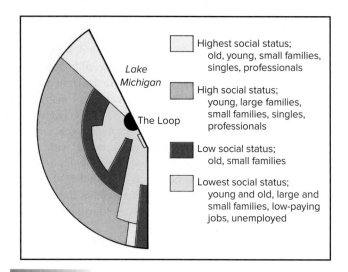

Figure 11.28 A diagrammatic representation of the major social areas of the Chicago region. The CBD of Chicago is known as the "Loop."

Source: Redrawn from Phillip Rees, "The Factorial Ecology of Metropolitan Chicago," M. A. thesis, University of Chicago, 1968.

income-segregated because their houses are of similar value. To preserve the upscale nature of a development and protect land values, self-governing community associations may be formed to enact and enforce land use restrictions (see the feature "The Gated Community"). Pervasive and detailed, these restrictions may specify such things as the size, construction, and color of exterior walls and fences, the size and permitted uses of rear and side yards, and the design of outside lights and mailboxes. Some go so far as to tell residents what ages they must be, what pets they may keep, where they may park their boats or recreational vehicles, and what landscaping and lawn ornaments are allowed.

Family Status

As the distance from the city center increases, the average age of the adults declines, the size of their family increases, or both. Within a particular sector, singles, young professionals without children, and older people whose children have left home (empty-nesters), tend to live close to the city center. In contrast, young families seeking space for child rearing and good schools live farther from the city center. Those without children at home may covet the accessibility of the cultural and business life of the urban core. The arrangement that emerges is a set of concentric circles divided according to family status, as Figure 11.27 suggests.

Ethnicity

For some groups, ethnicity is a more important factor in residential location than social or family status. Areas of a single ethnic group appear in the social geography of cities as separate clusters

or nuclei, reminiscent of the multiple-nuclei concept of urban structure. For some ethnic groups, cultural segregation is both sought and vigorously defended, even in the face of pressures for neighborhood change exerted by potential competitors for housing space, as we saw in Chapter 6. The durability of "Little Italys" and "Chinatowns" and of Polish, Greek, Armenian, Korean, and other ethnic enclaves in many U.S. cities is evidence of the persistence of self-maintained segregation. As each ethnic group assimilates, however, a growing share of its members live outside the enclave.

Certain ethnic or racial groups, especially African Americans, have had segregation forced on them. Sometimes this occurs through housing discrimination or real estate agents who "steer" people of certain racial and ethnic groups into neighborhoods that the agents think are appropriate. Every city in the United States has one or more African American areas that may be considered cities within a city, with their own self-contained social geographies of social status, income, and housing quality. Social and economic barriers to movement outside the area have always been high, as they also have been for Hispanics and other non-English-speaking minorities.

As whites and Asians increase their household incomes, they tend to move to neighborhoods that match their economic standing. Due to persistent residential segregation, as Census data document, blacks with similar income growth are less able to move to integrated neighborhood settings. Although segregation has moderated somewhat, at the start of the 21st century, the average African American city resident lived in a census tract that was more than 75 percent minority and three-fifths black. Figure 6.19 in Chapter 6 illustrates the concentration of whites, blacks, Hispanics, and other ethnic groups in Chicago. Elsewhere, black segregation varies by region. Black-white separation is highest in metropolitan areas in the Northeast and Midwest in cities such as Milwaukee; greatest integration is found in the metropolitan south and west and, notably, in military towns like Norfolk, Virginia, and San Diego.

All three factors in the social geography of cities have undergone widespread change in recent years. The diversity of household types has proliferated. Two-parent families with children living at home make up less than one-fourth of all U.S. households. Today, the suburbs house large numbers of singles and childless couples. Areas near the CBD have become popular for young professionals, some of whom have no plans to have children. Lesbian and gay couples and families often choose to live in urban centers, but increasing numbers are choosing suburbs as well. With more women in the workforce than ever before, and as a result of multiple-earner families, residential site selection has become a more complex undertaking. The heavy losses of manufacturing jobs and the rise of the service sector with its extremes of high-paying jobs (finance, insurance, and law, for example) and low-paying jobs have led to greater extremes of wealth and poverty and fewer middle-income neighborhoods. Immigration continues to diversify cities, but many immigrant groups now head directly to the suburbs. The city structure is constantly changing, reflecting changes in family and employment makeup.

The Gated Community

Approximately one in six Americans—some 50 million people—lives in a master-planned community. Particularly characteristic of the fastest-growing parts of the country, most of these communities are in the south and west, but they are increasingly common everywhere. In many regions, more than half of all new houses are being built in private developments. Master-planned communities in the United States trace their modern start back to the 1960s, when Irvine, California, and Sun City, Arizona, were built, but their roots can be found much earlier. Tuxedo Park, New York, for example, was planned and built in 1886 as a fully protected, socially exclusive community, and in the 1920s, Kansas City's Country Club District was established as a restricted residential development with land use controlled by planning and deed restrictions and a self-governing homeowners association providing a variety of governmental, cultural, and recreational services.

A subset of the master-planned community is the **gated community,** a fenced or walled residential area with checkpoints staffed by security guards and access limited to designated individuals and identified guests. More than 10 million Americans live in these middle- and high-income gated communities within communities. With private security forces, surveillance systems monitoring common recreational areas such as community swimming pools, tennis courts, and health clubs and—often—with individual home security systems, the walled enclaves provide a sense of refuge from high crime rates, drug abuse, and other social problems of urban America.

Gated and sheltered communities are not just an American phenomenon; they are increasingly found in all parts of the world. More and more guarded residential enclaves have been built in such stable Western European states as Spain, Portugal, and France. Developers in Indian cities have also used gated communities to attract wealthy residents. Trying to appeal to Indians returning to that country after years in areas like the Boston high-tech corridor and Silicon Valley, developers have built enclaves with names like Regent Place and Golden Enclave that boast American-style two-story houses and barbecues in the backyards.

Elsewhere, as in Argentina or Venezuela in South America, Lebanon in the Near East, or Ghana in Africa—with little urban planning, unstable city administration, and inadequate police protection—not only rich but also middle-class citizens are opting for protected residential districts. In China and Russia, the sudden boom in private and guarded settlements reflects a new form of post-communist social class distinction, while in South Africa, gated communities serve as effective racial barriers.

Figure 11B This gated community near Orlando is one of many in Florida.
©Ilene MacDonald/Alamy Stock Photo

Institutional Controls

Over the past century, and particularly since World War II, institutional controls have strongly influenced the land-use arrangements and growth patterns of most U.S. cities. Indeed, the governments—local and national—of most Western urbanized societies have instituted myriad laws to control all aspects of urban life, with particular emphasis on the ways in which individual property can be developed and used. In the United States, emphasis has been on land-use planning, subdivision control and zoning ordinances, and building, health, and safety codes. All have been designed to assure an orderly pattern of urban development, and all are based on broad applications of the police powers of municipalities to ensure public health, safety, and well-being, even when private property rights are infringed.

These nonmarket controls on land use are designed to minimize incompatibilities (residences adjacent to heavy industry, for example), set aside appropriate locations for public uses (the transportation system, waste disposal facilities, government buildings, parks, and so on), and private uses (colleges, shopping centers, housing, and so on) needed for a balanced, orderly community. In theory, such careful planning should prevent the emergence of slums, so often the result of undesirable adjacent uses, and should stabilize neighborhoods by reducing market-induced pressures for land use change.

However, zoning ordinances and land-use planning have frequently been criticized as being unresponsive to contemporary needs or unduly restrictive. To keep factories out of neighborhoods, zoning rules often strictly separate different kinds of land uses. In practice, this strict separation of different land uses can lead to sprawling developments where walking, biking, and transit use are quite difficult.

Zoning and subdivision control regulations that specify large lot sizes for residential buildings or forbid apartments have been criticized as devices to exclude the poor from upper-income areas. Some zoning laws have been criticized as discriminating against particular forms of residences: apartments, special housing for the aged, halfway houses, homeless shelters, and so forth. As a consequence, housing for society's least fortunate often ends up highly concentrated in less desirable areas (**Figure 11.29**). Bitter court battles have been waged, with mixed results, over "exclusionary" zoning practices that allegedly discriminate against the poor.

In most of Asia there is no zoning, and it is quite common to have small-scale industrial activities operating in residential areas. In both Europe and Japan, neighborhoods have been built and rebuilt gradually over time to contain a wide variety of building types from several eras intermixed on the same street. In North America, such mixing is rare.

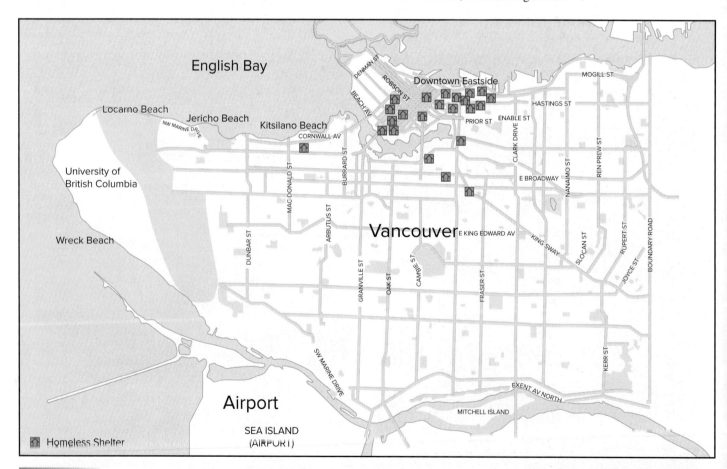

Figure 11.29 Concentration of homeless shelters in Vancouver, Canada. Zoning regulations and resistance by established middle-class neighborhoods leads to a concentration of shelters in the least desirable or least powerful parts of cities. The west side of Vancouver has always been the more desirable side of the city. The west side waterfront is a continuous line of public beaches stretching from Kitsilano Beach to Wreck Beach. By contrast, the east side waterfront is home to heavy industry, the container ship port, and grain terminals. The concentration of shelters shown here is on the Downtown Eastside, the poorest neighborhood in Vancouver, Canada.

11.8 Changes in Urban Form

The 20th century started with mass transit dominating the physical and social structure of the U.S. city. It ended with the automobile controlling the movement of people everywhere and determining the pattern and fate of cities and metropolitan areas. In the course of that century, new technological and institutional structures fundamentally changed the frameworks within which metropolitan areas developed.

First, the improvement of the automobile increased its reliability, range, and convenience, freeing its user from dependence on fixed-route public transit for access to work, home, or shopping. The new transport flexibility opened up vast new acreages of nonurban land to urban development. That flexibility of movement of people and, through semitrailer and pickup trucks, of heavy and light freight, was augmented by the substantial completion during the 1970s of the interstate highway system and its supplements, the major metropolitan expressways. The improved routeways made 30- to 45-kilometer (20- to 30-mile) commutes acceptable.

Second, during the 1930s and after World War II (1939–1945), both the Federal Housing Administration (FHA) and the Veterans Administration (VA), by easing the terms of home mortgage requirements, vastly increased the number of persons eligible to own their home rather than rent. Those agencies stimulated a housing boom by offering much more generous mortgage loan terms. Previously, buyers had to provide large down payments (sometimes 50 percent or more) and repay their high-interest loans within a short time, often 10 years or less. The VA program permitted veterans to purchase homes with virtually no down payment, and both the VA and FHA lengthened low-interest repayment periods to 15 to 25 years or more. Further, the acceptance of a maximum 40-hour workweek in 1938 guaranteed millions of Americans the time for a commuting journey not possible when 6-day workweeks, and 10-hour workdays were common. Tax deductibility of home mortgage interest and tax exclusion of capital gains on profits from the sale of a home were further inducements for Americans to purchase their own houses.

These structural and economic changes altered the prevailing patterns of accessibility and behavior and significantly modified the land value curve and population density gradient established in the mass-transit city. Over the past half century or more, U.S. metropolitan areas have experienced massive decentralization of people and activities as residents, businesses, and industries moved outward into suburbs. The end of the 20th century and early years of the 21st, however, have witnessed a modest reversal of those trends, with some population and economic rebound in the core areas of many cities.

Suburbanization and Edge Cities

Demand for housing, pent up by years of economic depression and wartime restrictions, was loosed in a flood after 1945, and a massive suburbanization altered the existing land-use and functional patterns of urban America. In the second half of the 20th century, the two most prominent patterns of change were *metropolitan growth* and, within metropolitan areas, *suburbanization* (**Figure 11.30**).

Figure 11.30 Satellite images of urban growth. Las Vegas, Nevada is among the fastest growing metropolitan regions in the United States. The region was home to 273,000 people in 1972 when the first image was captured. By 2017, the city had sprawled outwards into the desert and was home to 2.2 million residents.

Source: U.S. Geological Survey/Earthshots: Satellite Images of Environmental Change, 2017.

Suburban expansion reached its maximum pace during the decade of the 1970s when developers were converting open land to urban uses at the rate of 80 hectares (200 acres) an hour. Residential land uses led the initial rush to the suburbs. Typically, uniform but spatially discontinuous housing developments were built beyond the boundaries of older central cities. The new design was an unfocused sprawl, not tied to mass transit lines. It also represented a massive relocation of purchasing power to which retail merchants were quick to respond. The planned major regional shopping center became the suburban counterpart of the higher-order central places of the central city. Smaller shopping malls and strip shopping centers gradually completed the retailing hierarchy.

Faced with a newly suburbanized labor force, industry followed the outward move, attracted as well by the economies derived from modern single-story plants with plenty of parking space for employees. Industries no longer needed to locate near railway facilities; freeways presented new opportunities for lower-cost, more flexible truck transportation. Service industries were also attracted by the purchasing power and large, well-educated labor force now present in the suburbs, and complexes of office buildings developed, like the shopping malls, at freeway intersections and along freeway frontage roads and major connecting highways.

In time, in the United States, new metropolitan land-use and functional patterns emerged that could no longer be satisfactorily explained by the classic ring, sector, or multiple-nuclei models. Yet traces of the older-generation concepts remained applicable. Multiple nuclei of specialized land uses appeared, expanded, and coalesced. Sectors of high-income residential use continued their outward extension beyond the central city limits, usurping the most scenic and most desirable suburban areas and segregating them by price and zoning restrictions. Middle-, lower-middle-, and lower-income groups found their own income-segregated portions of the fringe. Ethnic and racial minorities are increasingly locating in suburbs. The share of minorities in suburbs of major cities is the same as their share in the overall U.S. population. By 2010, more than half the African American and Hispanic population in large metropolitan areas lived in the suburbs. Asians are even more likely than African Americans or Hispanics to live in suburbs, often in affluent *ethnoburbs.*

By the 1990s, a new urban feature had emerged on the perimeter of most major metropolitan areas—the **edge city.** Edge cities are defined by their large nodes of office and commercial buildings and characterized by having more jobs than residents within their boundaries. No longer dependent on the central city, select suburbs were reborn as vast collectively self-sufficient outer cities, marked by landscapes of industrial parks, high-rise office clusters, massive retailing complexes, and a proliferation of apartment and condominium districts and gated communities.

The new suburbia began to rival older CBDs in complexity and the amount of office and retail space. Collectively, the new centers surpassed the central cities as generators of employment and income. Together with the older CBDs, the suburbs perform the many advanced producer services that mark the postindustrial metropolis. During the 1980s, more office space was created in the suburbs than in the central cities of the United States. Tysons Corner, Virginia (between Arlington and Reston), for

example, became the ninth-largest business district in the United States. Regional and national headquarters of leading corporations, banking, professional services of all kinds, major hotel complexes and recreational centers—all formerly considered immovable keystones of CBDs—became part of the new outer cities.

Edge cities now exist in all regions of the urbanized United States. The South Coast Metro Center in Orange County, California; the City Post Oak-Galleria center on Houston's west side; Bellevue and Redmond, east of Seattle; King of Prussia and the Route 202 corridor northwest of Philadelphia; the Meadowlands, New Jersey, west of New York City; and Schaumburg, Illinois, in the western Chicago suburbs are only a few examples of this new urban form. Location factors for edge cities include proximity to major highway corridors, international airports, and areas of high social status. Often, edge city development takes place in the more affluent sector of the metropolitan region because corporate headquarters often relocate in the direction of its executive's home.

The metropolis has become polynucleated and urban regions are increasingly *galactic*—that is, galaxies of economic activity nodes organized primarily around the freeway systems, as suggested in Figure 11.26. Commuting across the galaxy is far more common than journeys to work between suburbs and central cities. In recent years, suburban outliers and edge cities have been coalescing, creating continuous metropolitan belts in the pattern shown in Figure 11.6.

On the leading edges of that pattern are the outer suburbs or *exurbs,* vast sprawling areas of centerless growth beyond the pull of central cities or edge cities. That unfocused low-density development continues and increases population segregation by income and further disperses places of employment and the intermittent commercial developments that always follow purchasing power. While minority groups are rapidly suburbanizing, the exurbs are overwhelmingly white. The aging inner suburbs that were developed in the first decades after World War II are now beginning to suffer the transfer of wealth and erosion of functions that earlier afflicted the center cities themselves.

Geographers studying Los Angeles have proclaimed the obsolescence of the older models of urban structure, most of which were based on Chicago with its dominant CBD and concentric rings of growth. Instead, they describe what they call postmodern urbanism: an urban region with no center, no edge, and no coherent pattern. It is a metropolitan area marked by radical fragmentation into a collage of theme parks, gated communities, corporate citadels, ethnoburbs, street warfare zones, consumption opportunities, spectacle sites, and edge cities—with only a communications network and highway system to hold it together (**Figure 11.31**) .

Central City Decline

The superior accessibility that determined the success and internal structure of the mass-transit city faded with the advent of the cheap and reliable automobile and motor truck and development of interstate highways, metropolitan expressways, and air transportation. The dominance of the CBD was based on its being the focus of urban mass-transit (streetcar, subway, elevated) systems

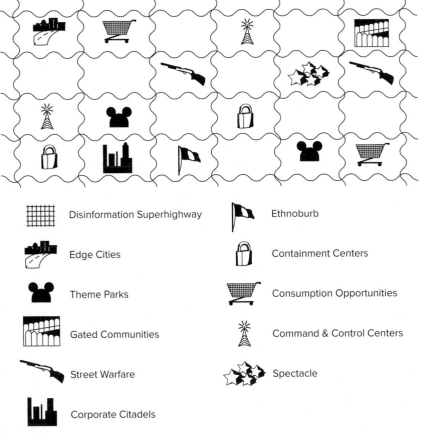

Icon	Label	Icon	Label
	Disinformation Superhighway		Ethnoburb
	Edge Cities		Containment Centers
	Theme Parks		Consumption Opportunities
	Gated Communities		Command & Control Centers
	Street Warfare		Spectacle
	Corporate Citadels		

Figure 11.31 The postmodern city. This model created by M. Dear and S. Flusty was inspired by Los Angeles and depicts an urban expanse without a center, edge, or coherent pattern. Instead, the city is fragmented into various independent zones, held together by highways and telecommunications.

Source: Michael Dear and Steven Flusty, 1998. "Postmodern Urbanism," Annals of the Association of American Geographers 88 (1): 66.

and intercity rail lines. When its accessibility eroded with the decline or abandonment of those carrier networks, central cities lost their primary situational advantage and the foundation of their internal land-use patterns. The dynamic that provided functional superiority to central cities increasingly worked to their detriment. Populations moved out, functions and jobs dispersed to the fringes following the relocating labor force and its purchasing power, and the central city was increasingly viewed as aging, congested, and inefficient. Once vibrant industrial districts were left behind as blighted, polluted sites—**brownfields.**

The redistribution of population caused by suburbanization resulted in both spatial and political segregation of social groups. The upwardly mobile—younger, whiter, wealthier, and better educated—took advantage of the automobile and the freeway to leave the central city. The poorer, older, least-advantaged urbanites were left behind in declining neighborhoods (**Figure 11.32**). The central cities and the suburbs became increasingly differentiated. Large areas within those cities now contain only the poor and minority groups, including women (see the feature "Women in the City"), a population little able to pay the rising costs of the social services that their numbers, neighborhoods, and condition require.

The services needed to support the poor include welfare payments, social workers, extra police and fire protection, health delivery systems, homeless shelters, and subsidized housing (see the feature "The Homeless"). Central cities, by themselves, are unable

Figure 11.32 Abandoned housing is common in Gary, Indiana as a result of white flight and deindustrialization. Some areas of cities have witnessed significant disinvestment and population decline. Properties in such neighborhoods have such weak resale and redevelopment potential that owners abandon them. Buildings left to decay become a source of danger and blight for those that remain. Since its peak in 1960, Gary has lost more than half its population.
©Mark Bjelland

Women in the City

Cities are not viewed or experienced in the same way by men and women; fear of sexual harassment or rape, for example, may restrict women's mobility in certain places or at certain times of day, denying them the access to public space enjoyed by men. Maurice Yeates has noted that women's needs, problems, and patterns with respect to urban social space are quite different from men's:

> In the first place, women are more numerous in large central cities than are men. Washington, D.C. is one of the most female-dominant (numerically) of any city in North America, with a "sex ratio" of 112 females for every 100 males. In New York City the ratio is 111 females per 100 males. The preponderance of women in central cities is related to an above-average number of household units headed by women, and to the larger numbers of women among the elderly.

A second characteristic is that female-headed households with children constitute the bulk of the poor. This feminization of poverty among all races is a consequence of the high costs of child care combined with the low wages and part-time work in many "women's jobs."

A third characteristic of women in urban areas is that they have shorter journeys to work and rely more heavily upon public transportation than do men, a reflection of the lower incomes received by women, the differences in location of "female jobs," and the greater involvement of women in childrearing. Women on the whole simply cannot afford to spend as much on travel costs as men and make greater use of public transportation, which in the United States is usually inferior and often dangerous. The concentration of employment of women in clerical, sales, service jobs, and nursing also influences travel distances because these "women's jobs" are spread around the metropolitan area more than "men's jobs," which tend to be concentrated. It might well be argued that the more widespread location of "women's jobs" helps maintain the relative inaccessibility of many higher-paid "men's jobs" to a large number of women.

Given the allocation of roles, the resulting inequities, and the persistence of these inequities, there are spatial issues that impinge directly upon women. One is that many women find that their spatial range of employment opportunities is limited as a result of the inadequate availability of child-care facilities within urban areas. A second spatial issue relates to the structure of North American metropolitan areas that reflects a particular set of assumptions about family life and male and female roles. Suburbs, in particular, reflect a male-paid work and female-home/children ethos. The suburban structure confines women to places with few meaningful choices. It has been argued that suburban women really desire a greater level of accessibility to a variety of conveniences and services, more efficient housing units, and a range of public and private transportation that will assure higher levels of mobility. These requirements imply higher-density urban areas.

Source: *Text excerpt from* The North American City, *5th ed., by Maurice Yeates. 1997 Pearson Education, Inc., Upper Saddle River, NJ.*

---

to support such an array of social services because they have lost the tax base represented by suburbanized commerce, industry, and upper-income residential uses. Lost, too, were the job opportunities that were formerly a part of the central city structure. Increasingly, the poor and minorities are trapped in a central city without the possibility of nearby employment and are isolated by distance, immobility, and unawareness—by *spatial mismatch*—from the few remaining low-skill jobs, which are now largely in the suburbs.

In an effort to help struggling central cities, the federal government, particularly after the landmark Housing Act of 1949, initiated urban renewal programs that remade inner city areas in the 1950s and 1960s. Under a wide array of programs, slum areas were cleared, public housing was built (**Figure 11.33**), cultural complexes and industrial parks were created, and city centers were reconstructed. Critics bemoan the federal bulldozer's destruction of heritage architecture and tightly knit working-class communities during urban renewal. Sadly, the modernist public housing projects that were constructed during urban renewal often became places of concentrated poverty and high crime. Many have since been torn down.

With the continuing erosion of the urban economic base and the loss of residents, the battle to maintain or revive the central city is frequently judged to be a losing one, at least in cities with a declining industrial base and concentrated poverty. Detroit, Michigan, is a classic example of decline, having dropped from a peak population of 1,800,000 in 1950 to just 670,000 in 2017 (meanwhile, the broader metropolitan area population of 4.3 million residents has been relatively stable since 1970). The experience in the western United States has been rather different. The fastest-growing U.S. metropolitan areas are concentrated in the west and south (**Figure 11.34**).

For the most part, these newer "automobile" metropolises placed few restrictions on physical expansion. That unrestricted growth has often resulted in the coalescence of separate cities into ever-larger metropolitan complexes. Unlike cities in the east and midwest, cities in the west were usually allowed by their state legislatures to expand their borders so that central cities were able to capture new growth taking place at the urban periphery. This allowed western cities to grow into ever-larger metropolitan complexes, but it also meant that central cities had a mixture of both new and older housing and poor and middle-class residents within their borders.

The speed and volume of growth in the west means city governments face the economic, social, and environmental consequences of unrestricted outward expansion. Scottsdale, Arizona, for example, covered a single square mile (2.6 square kilometers) in 1950; by the end of the 1990s, it had grown to nearly 200 square miles (500 square kilometers), four times the physical size of San Francisco. Phoenix, with which Scottsdale has now coalesced, is 70% larger in area than New York City, which has five times as many people.

Figure 11.33 Many elaborate—and massive—public housing projects have been failures. The Pruitt-Igoe complex in St. Louis was built in the early 1950s to replace crowded, deteriorating tenement buildings. The project was designed in the modernist international style by Minoru Yamasaki, who went on to design the World Trade Center towers in New York City. The complex of 33 nearly identical 11-story buildings was praised by *Architectural Forum* magazine as the "best high apartment" of the year and called "vertical neighborhoods for poor people." It quickly became a dangerous, crime-ridden complex and was demolished in the 1970s. The growing awareness that it was a mistake to segregate poor people in high-rise developments in neighborhoods lacking economic opportunities led to the U.S. Department of Housing and Urban Development's Hope VI program. That program funds the demolition of severely distressed public housing, replacing it with mixed-income housing. The Hope VI program has led to the demolition of more than 100,000 public housing units in cities around the country, including many well-known projects such as the Robert Taylor Homes in Chicago and the Desire Projects in New Orleans.

(a, b) ©Bettmann/Getty Images

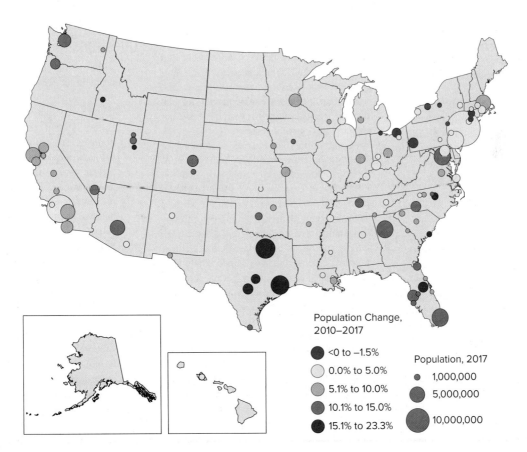

Figure 11.34 The pattern of metropolitan growth and decline in the United States, 2010–2017. Shown are the 100 largest metropolitan areas in 2010. The cities of the southern and southwestern Sun Belt showed the greatest relative growth. Modest growth or stability marked most metropolitan regions of the Northeast and Midwest. Meanwhile, deindustrializing regions in the Manufacturing Belt witnessed actual population decline.

Source: Data from U.S. Bureau of the Census.

The Homeless

The United States has a large homelessness problem. Every large city is apt to have hundreds, or even thousands, of people who lack homes of their own. One sees them pushing shopping carts containing their worldly goods, lining up at soup kitchens or rescue missions, and sleeping in parks or doorways. Reliable estimates of their numbers simply do not exist; official counts place the numbers of homeless Americans at more than 550,000 on any given night. The rates of homelessness are highest in urban areas with high housing costs such as Washington D. C., and Honolulu.

Their existence and persistence raise a multitude of questions; the answers, however, are yet to be agreed upon by public officials and private Americans. Who are the homeless, and why have their numbers increased? Who should be responsible for coping with the problems they present? Are there ways to eliminate homelessness?

Some people believe the homeless are primarily the impoverished victims of a rich and uncaring society. They view them as ordinary people, but ones who have had a bad break and been forced from their homes by job loss, divorce, domestic violence, or incapacitating illness. They point to the increasing numbers of families, women, and children among the homeless, less visible than the "loners" (primarily men) because they tend to live in cars, emergency shelters, or doubled up in substandard buildings. Advocates of the homeless argue that federal government cuts in the 1980s and 1990s in budgets for building low-income and subsidized housing reduced the potential supply of affordable housing. During the same period, city governments demolished low-income housing, especially single-room-occupancy hotels in the name of urban revitalization. In addition, federal regulations and reduced state funding for mental hospitals cast institutionalized patients onto the streets to join people displaced by gentrification, job loss, or rising rents.

A contrary view is presented by those who see the homeless as responsible for their own plight. In the words of one

Figure 11C A homeless encampment in Los Angeles, CA.
©Mark Bjelland

commentator, the homeless are "deranged, pathological predators who spoil neighborhoods, terrorize passersby, and threaten the commonweal." They point to studies showing that nationally between 66 percent and 85 percent of all homeless suffer from alcoholism, drug abuse, or mental illness, and argue that people are responsible for the alcohol and drugs they ingest; they are not helpless victims of disease.

Communities have tried a number of strategies to cope with their homeless populations. Some set up temporary shelters, especially in cold weather; some subsidize permanent housing and/or group homes. They encourage private, nonprofit groups to establish soup kitchens and food banks. Others attempt to drive the homeless out of town, or at least to parts of town where they will be less visible. They forbid loitering in city parks or on beaches after midnight, install sleep-proof seats on park benches and bus stations, and outlaw aggressive panhandling.

Neither point of view appeals to those who believe that homelessness is more than simply a lack of shelter, that it is a matter of people who need help getting off the streets and into mental health or substance abuse treatment. What the homeless need, they say, is a "continuum of care"—an entire range of services that includes education, treatment for drug and alcohol abuse and mental illness, and job training.

Questions to Consider

1. What is the nature of the homeless problem in your community? Is there an area where the homeless are concentrated? Why there?

2. Where should responsibility for the homeless lie: at the federal, state, or local governmental level? Is it best left to private groups, such as religious groups and charities? Or is it ultimately best recognized as a personal matter to be handled by homeless individuals themselves? Why?

3. Some people argue for a "housing first" strategy to help the homeless. Once a homeless person has a stable place to sleep, they can begin to address employment, mental health, and addiction issues. What are the advantages and disadvantages of this "housing first" strategy?

4. Periodically, cities will crack down on their homeless population and push them out of highly visible areas. Should the homeless have a right to be in the city? Do others have a right to enjoy the city without encountering homeless persons?

Figure 11.35 Urban sprawl characterizes growth in the Las Vegas, Nevada, metropolitan area. As in many desert cities, urban growth in Las Vegas has strained the region's limited water resources.

Source: Photo by Lynn Betts, USDA Natural Resources Conservation Service

The phenomenal growth of Las Vegas, Nevada, has similarly converted vast areas of desert landscape to low-density urban use, straining water resources and the environment (**Figure 11.35**). Seeking an alternative to endless outward sprawl, some metropolitan areas seek to restrain rather than encourage physical growth. Portland, Oregon, drew a growth boundary line around itself in the late 1970s, prohibiting conversion of surrounding forests, farmlands, and open space into suburbs (see the feature "Sustainable Cities").

Central City Renewal and Gentrification

Central cities hit their low point in the 1970s when New York City went bankrupt, the Bronx was burning, and crime was at an all-time high. More recently, pundits proclaimed the end of cities as the latest digital communications technologies (fax machines, the Internet, wireless devices, and so on) would eliminate the need for face-to-face interaction. Instead of replacing face-to-face interactions, digital communications have become centralizing forces by facilitating the growth of knowledge- and creativity-based industries and activities such as finance, law, design, advertising, and corporate consulting. These industries seem to prefer geographically centralized locations. Cities—particularly large metropolitan cores—provide the first-rate telecommunications and fiber optics infrastructures and the access to skilled workers, customers, investors, research, educational institutions, and cultural institutions needed by the modern, postindustrial economy. As a reflection of cities' renewed attractions, employment and gross domestic product (GDP) in the country's 50 largest urban centers began to grow in the 1990s, reversing a pattern of stagnation and decline in the preceding decades. Demand for downtown office space was met by extensive new high-rise and skyscraper construction and urban renewal.

Urban centers became attractive places of consumption, promoted by popular television shows and movies. Some of the new office workers chose to live in central city neighborhoods that offer well-built, character housing stock in highly accessible locations, spurring a central city residential revival called **gentrification.** Gentrification is the rehabilitation of housing in older, deteriorated inner-city areas by middle- and high-income groups (**Figure 11.36a**). Welcomed by many as a positive, privately financed force in the renewal of depressed urban neighborhoods, gentrification also has serious negative social and housing impacts on the low-income, frequently minority families that are displaced. Gentrification is another expression of the continuous remaking of urban land-use and social patterns in accordance with the rent-paying abilities of alternate potential occupants. Yet the rehabilitation and replacement of housing leads to inflated rents and prices that push out established residents, disrupting the social networks they have created.

The districts usually targeted for gentrification are those close to downtown jobs, with easy access to transit, low housing costs, and interesting older architecture (Figure 11.36b). Gentrification is often led by artists who lack financial resources but exert a great deal of cultural influence. The artists are in turn replaced by an influx of younger, wealthier professionals who have helped revitalize and repopulate inner city zones. Nearly all large North American cities have witnessed a significant increase in the residential population in the neighborhoods adjacent to the central business district. Individual home buyers and rehabbers opened the way; commercial developers followed—but often only after local, state, or federal governments made the first investments in slum clearance, brownfields cleanup, and construction of new infrastructure, parks, and cultural facilities.

The reason for that growth lies in both changing tastes and demographics. Young professionals are marrying and having children later or, often, are divorced, never-married, or same-sex couples. For them, suburban life and shopping malls hold few attractions, while central city residence offers high-tech and executive jobs within walking or biking distance and cultural, entertainment, and boutique shopping opportunities. The younger group has been joined by empty-nesters, couples who no longer have children living at home, who find big houses on suburban lots no longer desirable. By their interests and efforts, these two groups have largely or completely remade and upgraded such old city neighborhoods

(a)

(b)

Figure 11.36 *(a)* Gentrified historic housing in Old Town in Alexandria, Virginia, near Washington, D.C. *(b)* Former mills along the Mississippi River in Minneapolis, Minnesota, have undergone gentrification through conversion to trendy loft-style condominiums. The City of Minneapolis initiated the conversion of derelict mills by cleaning pollution and building riverfront parks and trails.

(a, b) ©Mark Bjelland

as Williambsurg in Brooklyn, Denver's LoDo, Minneapolis' Mill District and North Loop, Portland's Pearl District, Seattle's Belltown, Vancouver's Yaletown, and virtually all of San Francisco.

Another important part of the renewed vigor of central cities comes from new immigrants who spread beyond the usual coastal *gateway* cities. Immigrants have become deeply rooted in their new communities by buying and renovating homes in inner-city areas, spending money in neighborhood stores, and most importantly establishing their own businesses (**Figure 11.37**). They also are important additions to the general urban labor force, providing the skilled and unskilled workers needed in expanding office-work, service, and manufacturing sectors.

Figure 11.37 Immigrants from all over the world, but especially Latin America and Asia, have established their own businesses, fixing, making, or selling things, adding to the vitality of central cities. By purchasing and renovating houses in struggling neighborhoods, immigrants have helped revitalize many inner-city neighborhoods.

©David Grossman/Alamy Stock Photo

11.9 World Urban Diversity

The city, Figure 11.3 reminds us, is a global phenomenon. It also varies among regions, reflecting diverse cultural heritages and economies. The categories and models that we have used in this chapter to study the functions, land-use arrangements, suburbanization trends, and other aspects of the U.S. city do not always apply to cities in other parts of the world. Those cities have been created under different historical, cultural, and technological circumstances. They have developed different functional and structural patterns, some so radically different from our U.S. model that we would find them unfamiliar and uncharted landscapes indeed. Even Canadian cities differ significantly from their counterparts in the United States (see the feature "The Canadian City"). The city is universal; its characteristics are cultural and regional.

The West European City

Although each is unique historically and culturally, Western European cities share certain common features. They have a much more compact form and occupy less total area than U.S. cities of comparable population, and most of their residents are apartment dwellers (except the United Kingdom, where most live in rowhouses). Residential streets of the older sections tend to be narrow, and front, side, or rear yards or gardens are rare.

European cities also enjoy a long historical tradition. Medieval origins, Renaissance restructurings, and industrial growth have given the cities of Western Europe distinctive features. Despite wartime destruction and postwar redevelopment, many still bear the impress of past occupants and technologies, even back to Roman times in some cases. An irregular system of narrow streets may be retained from the random street pattern developed in medieval times of pedestrian and pack-animal movement. Main streets radiating from the city center and cut by circumferential "ring roads" tell us the location of primary roads leading

Sustainable Cities

Most people associate environmentalism with wilderness, not cities. But what if living in large, dense cities was actually better for the environment? Some scholars and city planners think so. They point to the fact that residents of New York City use the least energy per capita and have the lowest per capita greenhouse gas emissions in the United States. The explanation is simple: the high density of New York City makes walking, bicycling, or transit use the most effective way to get around the city. Rowhouses and apartments have fewer sides exposed to the outside, where they can gain or lose heat. As a result, they are much more efficient to heat and air-condition than detached houses. In contrast to the sprawling, low-density residential pattern that became the norm after World War II, Manhattan has a population density of 27,000 people per square kilometer (69,000 per square mile). However, to suggest that city living is good for the environment goes against the grain for most people, who still associate big cities with smog, noise, and garbage barges.

Cities can be made more sustainable by designing buildings and landscapes with the environment in mind. Green roofs, solar panels, and pavements that allow water to infiltrate are just a few of the many techniques to reduce a city's impact on the environment. Cities can also become more sustainable by developing neighborhoods at higher densities, adding sidewalks, and making sure basic services such as grocery stores and schools are within walking distance. The New Urbanism movement draws together city planners, architects, designers, and developers who favor dense, mixed-use neighborhoods. Transportation is one of the most important sources of pollution, but many trips can be replaced by bicycling if safe routes are available. In parts of the Netherlands and Denmark, bicycling is the dominant mode of transportation. Bicycling is also increasingly popular in North American cities such as New York, Minneapolis, and Portland. Cities can also become more

Figure 11D Greenwich Millennium Village is located near the Docklands in East London. This neighborhood was sponsored by the U.K. government as a model of environmentally sustainable urban development. The former heavy industrial site was cleaned of pollution and converted to parks, trails and a high-density community scaled to walking and bicycling.
©*Mark Bjelland*

sustainable by transforming their polluted industrial waterfronts into attractive postindustrial spaces. In cities across Europe and North America, derelict, contaminated brownfield sites have been cleaned up and redeveloped for parks, housing, and retail (see Figure 11.36b and the feature "The Canadian City").

In London, the heavily polluted former site of a plant that manufactured gas from coal was turned into a model community for sustainable urban development. Called "Greenwich Millennium Village," the project was built by the British government to demonstrate the highest standards for energy efficient and environmentally sustainable development. According to the British government, building high-density housing on former brownfield sites is a

necessity to preserve the much-loved countryside that surrounds British cities. At Greenwich Millennium Village, wood waste from tree trimming is used to generate electricity and heat the buildings. Housing for different incomes is mixed together. The streets in the core of the village are limited to pedestrians and bicyclists. A dedicated bus lane bisects the development, connecting it to the London Underground (subway). More than three-quarters of the residents use public transit to commute to work, about twice the London average and five times the national average. A continuous belt of landscaped parks line the River Thames waterfront and a pond and ecology park occupy what was once one of the most polluted pieces of land in the United Kingdom.

into town through the gates in city walls now gone and replaced by circular boulevards. Broad thoroughfares, public parks, and plazas mark Renaissance ideals of city beautification and the esthetic need felt for processional avenues and promenades.

European cities were developed for pedestrians and still retain the compactness and character appropriate to walking. The sprawl of U.S. suburbs is generally absent. At the same time, compactness

and high density do not necessarily mean skyscraper skylines. Much of urban Europe predates the steel frame building and the elevator. City skylines tend to be low, three to five stories in height, sometimes (as in central Paris) held down by building ordinance (**Figure 11.38**), or by prohibitions on private structures exceeding the height of a major public building, often the central cathedral. Where those older restrictions have been relaxed, however, taller

The Canadian City

Even within the seemingly homogeneous culture realm of the United States and Canada, cities in the two countries show subtle but significant differences. Although the urban expression is similar in the two countries, it is not identical. The Canadian city, for example, is more compact than its U.S. counterpart of equal population size, with a higher density of buildings and people and less suburbanization of populations and functions.

Space-saving and multiple-family housing units are more the rule in Canada, so a similar population size is housed on a smaller land area, with much higher densities, on average, within the central area of cities. The Canadian city is better served by and more dependent on public transportation than is the U.S. city. Because Canadian metropolitan areas have only one-quarter as many miles of expressway lanes per capita as U.S. metropolises—and at least as much resistance to constructing more—suburbanization is less extensive north of the border than south.

The differences are cultural as well. Cities in both countries are ethnically diverse (Canadian communities, in fact, have the higher proportion of foreign-born residents), but U.S. central cities exhibit far greater internal distinctions in race, income, and social status, and more pronounced contrasts between central city and suburban residents. That is, there has been much less "flight to the suburbs" by middle-income Canadians. As a result, the Canadian city shows greater social stability, higher per capita average income, more retention of shopping facilities, and more employment opportunities and urban amenities than its U.S. central city counterpart. In particular, it does not have the rivalry from well-defined competitive edge cities of suburbia that so spread and fragment United States metropolitan complexes.

Figure 11E Vancouver, British Columbia has embraced high-density living and a strong CBD as solutions to the traffic, land-use, and environmental problems of urban sprawl. The commercial and civic buildings of Vancouver's CBD are nearly obscured by the high-rise condominiums and apartments that have been built on former brownfield sites along the waterfront. Vancouver's CBD is well served by public transit and walking and bicycle trails. On average, Canadian metropolitan areas are almost twice as densely populated as those of the United States. Further, on a per capita basis, Canadian urbanites are two-and-a-half times more likely to use public transportation than American city dwellers.

©joe daniel price/Getty Images

Figure 11.38 European cities often have a low profile in their central areas, as shown in this scene of Paris. Although taller buildings—20, 30, even 50 or more stories in height—have become more common in major European cities since World War II, they are not the universal mark of CBDs that they are in North America, South America, and Asia, nor are they necessarily welcomed symbols of city progress and pride.

©Stockbyte/Getty Images

office buildings have been erected—such as in the financial districts of London and Frankfurt, Germany.

Compactness, high densities, and apartment dwelling encouraged the development and continued importance of public transportation, including well-developed subway systems. The private automobile has become much more common of late, though most central city areas have not yet been significantly restructured with wider streets and parking facilities to accommodate it. The automobile is not the universal need in Europe that it has become in U.S. cities. Home and work are generally more closely spaced in Europe—often within walking or bicycling distance—while most sections of towns have first-floor retail and business establishments (below upper-story apartments), bringing both places of employment and retail shops within convenient distance of residences.

In many cities, the historic core is now increasingly gentrified and residential units for the middle class, the self-employed, and the older generation of skilled artisans share limited space with preserved historic buildings, monuments, and tourist attractions. At the same time, many are affected by the processes of decentralization; some of their residents now choose to live in suburban locations as car ownership and use becomes more commonplace.

The West European city is not characterized by inner-city deterioration and out-migration. Its core areas tend to be stable in population and attract, rather than repel, the successful middle class and upwardly mobile. Nor does it always feature the ethnic neighborhoods of U.S. cities although some, like London, do (see The Caribbean Map in London, Section 6.4). Non-European immigrant communities, where present in a city, tend to be clustered in older, working class districts or in peripheral public housing apartment blocks. Segregation of new immigrants into remote suburban apartments has been a particular problem in France as it leads to social isolation and a lack of opportunities for youth.

Eastern European Cities

Cities of Eastern Europe, including Russia and the former European republics of the Soviet Union, once part of the communist world, make up a separate urban class. These post-communist cities share many of the traditions and practices of West European cities, but differ from them in the centrally administered planning principles that were, in the communist period (1945–1990), designed to shape and control both new and older settlements. For reasons both ideological and practical, the particular concerns were as follows: first, to limit the size of cities to avoid supercity growth and metropolitan sprawl; second, to ensure an internal structure of neighborhood equality and self-sufficiency; and third, to segregate land uses. The planned Eastern European city fully achieved none of these objectives, but by attempting them, it emerged as a distinctive urban form.

The planned city of the communist era is compact, with relatively high building and population densities reflecting the nearly universal apartment dwelling, and with a sharp break between urban and rural land uses on its margins. It depends heavily on public transportation. During the communist period, governments dictated that the central area of cities should be reserved for public use, assemblies, parades, and celebrations. In the Russian prototype, neither a CBD nor major outlying business districts were required or provided.

Residential areas were expected to be largely self-contained in the provision of at least low-order goods and services, minimizing the need for a journey to centralized shopping locations. They were made up of *microdistricts,* assemblages of uniform apartment blocks housing perhaps 10,000 to 15,000 persons, surrounded by broad boulevards, and containing centrally sited nursery and grade schools, grocery and department stores, theaters, clinics, and similar neighborhood necessities and amenities placed often at the outskirts of the city (**Figure 11.39**).

Figure 11.39 This scene from Poprad, Slovakia, shows the typical housing estates built throughout Eastern Europe during the socialist era. Superblocks of identical, mass-produced apartment houses formed self-contained districts, complete with their own shopping, schools, and other facilities.

©PHB.cz (Richard Semik)/Shutterstock

These characteristic patterns are changing as market principles of land allocation are adopted. Historic apartments and townhouses that were badly neglected during the communist era have been restored and are the most fashionable places to live. Newfound prosperity has expressed itself in the construction of Western-style shopping malls, and spacious privately owned apartments and single-family houses for the newly rich and middle-class. Meanwhile, population decline due to low birth rates **(see Appendix B)** and out-migration in pursuit of better-paying jobs in Western Europe has led to shrinking cities in Eastern Europe and problems of high vacancy rates in the drab, mass-produced apartment tower blocks of the communist era.

Rapidly Growing Cities of the Developing World

The fastest-growing cities and the fastest-growing urban populations are found in the developing world (Figure 11.4). Industrialization has come to most of them only recently. Modern technologies in transportation and public facilities are sometimes lacking, and the structures of cities and the everyday lifeworld of their inhabitants are far different from the urban world familiar to North Americans. The developing world is vast in extent and diverse in its physical and cultural content; generalizations about it or its urban landscapes lack certainty and universality.

The backgrounds, histories, and current economies and administrations of developing world cities vary greatly. Some are ancient, having been established many centuries before the more developed cities of Europe. Some are still pre-industrial, with only a modest central commercial core; they lack industrial districts, public transportation, or any meaningful degree of land-use separation. Others, though increasingly Western in form, are only beginning to industrialize. And some have taken on industrial, commercial, and administrative functions on the Western model and, at least in their central areas, assumed as well the appearance of fully modern urban centers **(Figure 11.40)**.

Despite the variety of urban forms found in such diverse regions as Latin America, Africa, the Middle East, and South and Southeast Asia, we can identify some features common to most of them. First, most of what are currently categorized as developing countries have a colonial legacy, and several major cities were established principally to serve the needs of the colonizing country. The second aspect is that of underdevelopment of urban facilities. The tremendous growth that these cities are experiencing as their societies industrialize has left many of them with inadequate physical infrastructure and public utilities and no way to keep up with population growth. Third, most cities in developing countries are now characterized by neighborhoods hastily built by new migrants, away from city services, and often occupying land illegally. Such squatter settlements are a large and growing component of these cities and reflect both the city's attractiveness as a destination and lack of opportunities for all. Finally, in many cases, governments have responded with drastic remedies, sometimes going so far as to move the national capital away from the overcrowded primate city to a new location or to create entirely new cities to house planned industrial or transportation functions.

Figure 11.40 Dubai, in the United Arab Emirates, grew rich on oil revenues, but it is now a diversified center for finance, tourism, and business. It has grown rapidly from 50,000 people in 1965 to 2.4 million in 2015. It features many modern high-rise commercial buildings, the world's tallest skyscraper, and the world's largest enclosed shopping mall.
©*Alasdair Drysdale*

Influences of the Past

Cities in developing countries originated for varied reasons and continue to serve several functions based on their position as market, production, government, or religious centers. Their legacy and purpose influence their urban form.

Many are the product of colonialism, established as ports or outposts of administration and exploitation, built by Europeans on a Western model current at the time of their development. For example, the British built Kolkata (Calcutta), New Delhi, and Mumbai (Bombay) in India and Nairobi in Kenya and Harare (formerly Salisbury) in Zimbabwe. The French developed Hanoi and Ho Chi Minh City (Saigon) in Vietnam, Dakar in Senegal, and Bangui in the Central African Republic. The Dutch planned Jakarta (formerly Batavia) in Indonesia as their main outpost, Belgium placed Kinshasa (formerly Leopoldville) in what is now the Democratic Republic of the Congo, and the Portuguese founded a number of cities in Angola and Mozambique.

Urban structure is a product not just of the time when a city was founded, or who the founders were, but also of the role it plays in its own cultural setting. Land-use patterns in capital cities reflect the centralization of government functions and the concentration of wealth and power in a single city of a country **(Figure 11.41a)**. The physical layout of a religious center is conditioned by the religion it serves, whether Hinduism, Buddhism, Islam, Christianity, or other faith. Typically, a monumental structure—a temple, mosque, or cathedral—and associated buildings rather than government or commercial offices occupy the city center. Multifunctional centers display a greater diversity of land uses and structures (Figure 11.41b). Traditional market centers for a wide area (Timbuktu in Mali and Lahore in Pakistan), or cultural capitals (Addis Ababa in Ethiopia and Cuzco in Peru), have land-use patterns that reflect their special functions. Similarly, port cities such as Dubai (United Arab Emirates), Haifa (Israel), or Shanghai (China) have a land-use structure different from that of an industrial or mining center such as Johannesburg (South Africa). Adding to the complexity is the fact that cities with a long history reflect the changes wrought by successive rulers and/or colonial powers, and recent rapid growth.

Yet, by observation and consensus, some common features of developing-world cities are recognizable. For example, wherever automobiles or modern transport systems are an integral part of the modernizing city, the metropolis begins to take on Western characteristics. Also, all of the large cities have modern centers of commerce, not unlike their North American counterparts (see Figure 11.40).

All, too, wherever located, have experienced massive in-migrations from rural areas. Many, particularly in sub-Saharan Africa, have absorbed large numbers of foreign immigrants seeking asylum or economic opportunity. Most have had even faster rates of natural increase than of immigration. The predicted consequences, according to the United Nations, will be that nearly all of the global population increase in the coming decades will take place in the urban areas of the world's developing countries. Many of those populations are and will continue to be drawn into a globalizing world economy searching for locations able to offer the cheapest labor for the *footloose* operations of transnational corporations. UN-Habitat has termed the economic consequences a "race to the bottom," as different cities compete for low-skill, low-wage manufacturing jobs. Increased urban poverty and greater social and economic inequality and segregation are the foreseen consequences for much of the urbanizing developing countries. In all their cities, large numbers of people support themselves in the *informal* sector—as food vendors, peddlers of cigarettes or trinkets, streetside barbers or tailors, sex workers, errand runners or package carriers—outside the usual forms of wage labor (see Table 10.2 and Figure 10.6). Although informal sector work is vital to the survival of many or even most city residents in developing countries, it doesn't pay the taxes that governments need to provide the basic services that will help the country emerge from poverty.

(a)

(b)

Figure 11.41 Developing-world cities vary greatly in structure and appearance, reflecting their differing culture regions, histories, and functions. (*a*) Monumental government buildings mark single-function Brasilia, the capital of Brazil. Brazil's capital was moved to Brasilia in the 1950s to promote the development of the country's remote interior. (*b*) The central area of multifunctional Guanajuato, Mexico, is dominated by religious structures, government buildings, a central plaza, and homes of the city's elite. The architecture in the central area displays the city's Spanish colonial heritage.

(a) ©Julia Waterlow/Getty Images; (b) ©Jose Fuste Raga/Getty Images

Urban Primacy and Rapid Growth

The population of many developing countries is disproportionately concentrated in their national and regional capitals. Few developing countries have mature, functionally complex systems of cities with multiple small and medium-size centers. Instead, one primate city dominates their urban systems (see Figure 11.17). One-fifth of all Nicaraguans live in metropolitan Managua, and Libreville contains a third of the populace of Gabon. Vast numbers of the rural poor are attracted to these developed seats of wealth and power. Poverty and rapid population growth in rural areas is the push factor and the bright lights and promise of opportunities in the big city provide the pull. All too often, though, the reality of life in the primate city doesn't live up to its promise.

With their agglomeration economies, cities are engines of economic growth. Thus, the economies of developing countries are even more highly concentrated in their largest cities. Buenos Aires, with a third of Argentina's population, generated almost two-thirds of the country's GDP. Examples of primate cities exerting a disproportionate economic influence are repeated over and over in the developing world.

Many large cities in developing countries with rapidly growing economies have a vibrant and modern city center and elite residential sector (**Figure 11.42**). Such districts contain amenities that could be found in major Western centers and are the places where the wealthiest members of society work and often live. This is also the part of the city that businesspeople, officials, tourists, and other visitors are most likely to see. Some, particularly Asian, cities have made great investments in these city centers, often as much for prestige as for practical purposes. In fact, the booming cities of Asia now are leaders in skyscraper construction. The world's tallest building is in Dubai, United Arab Emirates, and 9 of the world's 10 tallest buildings in 2015 were located in East or Southeast Asia, with just 1 in the United States.

Yet the presence of gleaming downtowns cannot disguise the fact that most of these cities simply cannot keep pace with the massive growth they are experiencing (**Figure 11.43**). The pace of urbanization promises unceasing pressure on governments to provide adequately for the housing, employment, and public service needs of that burgeoning population. In many individual cities, growth rates create daunting challenges; Lagos, Nigeria, for example, had 325,000 residents in 1950 but had grown to an estimated 13.1 million by 2015. It continues to add about 1,700 new residents each day and the United Nations estimates that it will reach 24 million by 2030. The challenge facing Lagos is equivalent to crowding in an additional Baltimore, Maryland, or Abilene, Texas, each year. The massive rural-to-urban movement contributing to such growth rates and population increases is augmented by the additional births produced by the youthful immigrants.

Figure 11.42 These high rise buildings along Copacabana Beach are part of the city's elite residential sector. This spine of higher-income housing extends outward from the CBD and is well-served with sewers and other services. It stands in stark contrast with the favelas elsewhere in the city (Figure 11.44).
©iStockphoto.com/Sfmthd

Squatter Settlements

Most developing-world cities are ringed by vast, high-density squatter settlements (also known as informal settlements) lacking in public facilities and services. With regional variations (Table 11.4),

Table 11.4

Percentage of Urban Population Living in Slums

Region	2000	2014
Sub-Saharan Africa	65	55
South Asia	46	31
Southeast Asia	40	27
East Asia	37	25
Western Asia	21	25
Oceania	24	24
Latin America and Caribbean	29	20
North Africa	20	11

Source: United Nations, The Sustainable Development Goals Overview, *2016.*

Figure 11.43 The dualism of prosperity and poverty is apparent in Kuala Lumpur, Malaysia. In the background, the Petronas Towers—for a time the world's tallest—and the Radisson Hotel rise in sharp contrast to the downtown shanties in their shadow.

©Espen Helland/robertharding/Alamy Stock Photo

slum dwellers accounted for about 30 percent of the urban population in developing regions in 2014, a substantial improvement over 1990 when 46 percent lived in slums. Although progress has been substantial, gains are partially offset by population growth. More than 880 million people were living in slum conditions in 2014. In sub-Saharan Africa, the rates are highest and the problem is most challenging. Thus, one of the United Nations' Sustainable Development Goals is to improve conditions in cities and reduce both the percentage and absolute number of people living in slums.

A substantial proportion of the population of most developing-world cities is crowded into informal settlements built by their inhabitants, often without legal title to the land. These informal communities—*favela* in Brazil, *barrio* in Mexico, *kampung* in Indonesia, *gecekondu* in Turkey, or *katchiabadi* in Pakistan—usually have little or no access to publicly provided services such as water supply, sewerage and drainage, paved roads, and garbage removal. In such megacities as Rio de Janeiro (**Figure 11.44**), São Paulo, Mexico City, Bangkok, Chennai (Madras), Cairo, or

Lagos, millions find refuge in the shacks and slums of the *informal housing sector*.

Only a fraction of the new housing in Third World cities is produced by the formal housing sector; the rest develops informally, ignoring building codes, zoning restrictions, property rights, and infrastructure standards. Squatter settlements often emerge on peripheral land that is undeveloped or on land that is too steep or flood-prone for conventional development. Overcrowding often transforms peripheral squatter settlements into vast zones of disease and squalor subject to constant danger from landslides, fire, and flooding. The informality (and often illegality) of the squatter housing solution means that those who improvise and build their own shelters lack registration and recognized ownership of their houses or the land on which they stand. Without such legal documentation, no capital accumulation based on housing assets is possible and no collateral for home improvement loans or other purposes is created.

As many as 3 million residents in Nairobi, Kenya, live in slums, most without electricity, running water, or sewers; in that

Figure 11.44 The *favelas* in Rio de Janeiro, Brazil house a substantial proportion of the population. These slums were originally built by new migrants to the city who could not afford existing housing. Urban slums such as these are often overcrowded and lacking in urban services. However, over time residents often form associations to help secure urban services and upgrade their community.

©dndavis/Getty Images

With some of the fastest urban growth rates, sub-Saharan African cities have the highest percentage of their urban populations living in slums, despite the gleaming modern skyscrapers of their capital city cores. Almost two-thirds of city residents are slum dwellers; they are, in addition, afflicted with low life expectancies, high levels of infant and child mortality, HIV/AIDS prevalence, and illiteracy, particularly among women and girls. The prevalence of peripheral slum concentrations in all developing world regions reflects an *inverse concentric zone* pattern in which the elite and upper class reside in central areas and social status declines with increasing distance from the center.

In most world regions, there has been substantial recent improvement in the percentage of urban dwellers living in slums. Conflict-torn regions such as Syria in West Asia are areas where conditions have worsened. Sometimes residents of squatter settlements have successfully lobbied governments for water, sewers, roads, and other infrastructure, and over time, they have become more established neighborhoods. One of the major steps in upgrading slums is to give residents some form of secure right to the land on which their dwelling is located. As incomes stabilize over time, shacks can be upgraded to regular houses and slums can become stable neighborhoods. Unless the land is unsafe or unstable, slum upgrading is preferable to demolition and relocation, which displace people and break apart dense social networks.

Latin American City Model

While Latin American cities have their own unique characteristics, many of the traits common to cities in the developing world can be observed in the **Latin American city model** (**Figure 11.45**). At the center lie the traditional market area, key government and religious buildings, and a modern CBD. Extending outward from the center is a commercial spine that features high-status establishments and terminates at a suburban mall. The spine features amenities such as tree-lined boulevards, is well supplied with sewers and urban services, and is surrounded by an elite residential sector. Residential zones generally decrease in status with distance from

city's sprawling slum district of Mathare Valley, some 190,000 people are squeezed into 15 square kilometers (6 square miles), and the population is increasing by 10,000 inhabitants each year.

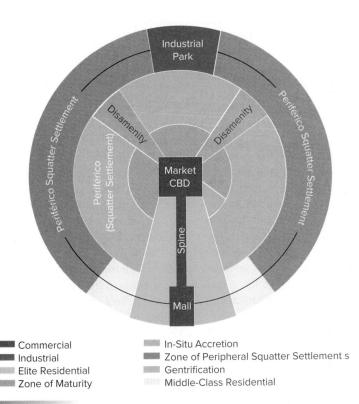

Legend:
- Commercial
- Industrial
- Elite Residential
- Zone of Maturity
- In-Situ Accretion
- Zone of Peripheral Squatter Settlements
- Gentrification
- Middle-Class Residential

Figure 11.45 The Latin American city model shows the wealthy living in the inner city and along a commercial spine that extends outward in one sector. Income generally declines with distance from the CBD.

Source: Larry R. Ford, "A New and Improved Model of Latin American City Structure," Geographical Review 86 (1996): 438.

the center because the inner city has a positive connotation. The zone of maturity has the better-quality residences, while the zone of in-situ accretion is mixed in status but undergoing improvement. Squatter settlements are found at the urban periphery and in disamenity zones, such as near dumps, in flood-prone areas, or on steep slopes. An industrial corridor terminates at a peripheral industrial park and a circumferential roadway (*perférico*) connects the industrial park and suburban mall.

Planned Cities

Some national capitals have been removed from their earlier primate city sites and relocated outside the core regions of their countries. The objective of these "forward capitals" has been to achieve a more central location or to encourage more uniform national development; examples include Islamabad (Pakistan), Brasilia (Brazil), Abuja (Nigeria), and Putrajay (Malaysia). Other relocations have been planned or announced, including a 2004 decision to relocate South Korea's primary government administrative agencies 150 kilometers (93 miles) to the southeast of Seoul.

A number of developing countries have also created or are currently building some new cities intended to draw population away from overgrown metropolises; others are designed to house transportation facilities or industrial agglomerations. For example, China's government is building dozens of new cities with the goal of relieving pressure on the capital city of Beijing.

Thailand opened Suvarnabhumi, a major airport an hour outside of Bangkok, intended to become the air transport hub for Southeast Asia with a new city nearby planned as a major new industrial center for the nation. China, India, Malaysia, and other Asian industrializing states are also planning—or have constructed—high-tech manufacturing and service centers, catering to the outsourcing needs and market opportunities of a globalizing economy (**Figure 11.46**).

Figure 11.46 The Cyber Gateway Building in Hyderabad's Hitec City houses firms like the multinational software companies Microsoft, IBM, and Toshiba, as well as Indian companies like Wipro, which provides information technology services and product design. Hitec City also houses professional schools in business and information technology.

©Idealink Photography/Alamy Stock Photo

SUMMARY

The city is the essential focus of activity for every society that advances beyond the subsistence level. Although they are among the oldest marks of civilization, only in the past decade have cities become the home of the majority of the world's people. Virtually all population growth worldwide in the first half of the 21st century will be captured by cities in the developing world.

All settlements growing beyond their village origins take on functions uniting them to the countryside and to a larger system of settlements. As they grow, they become functionally complex. Their economic structure, composed of both *basic* and *nonbasic* activities, may become diverse. Basic activities represent the functions performed for the larger economy and urban system; nonbasic activities satisfy the needs of the urban residents themselves. Functional classifications distinguish the economic roles of urban centers, while simple classification of them as transportation and special-function cities or as central places helps define and explain their role within the broader system of cities.

Systems of cities are reflected in the urban hierarchy and in part described by the rank-size rule, primacy, and central place theory. When a city is far larger than all others in its country, it is termed a *primate city*. Many countries display this dominating city pattern, but there are only a few *world cities* dominating the global economy.

Repetitive physical and social patterns are found inside North American cities. At the core, the central business district (CBD), localized originally by mass-transit line convergence, has the highest land values and accessibility. Outside the CBD, lower-order commercial centers are also oriented to transport routes. Low-density residential uses occupy less valuable and less accessible land. These patterns inspired geographers to summarize early 20th-century urban form by the concentric zone, sector, and multiple-nuclei models. Automobile-based expansion brought more recent recognition of metropolitan peripheral models and the postmodern city model. The period following World War II brought massive changes in urban organization, with the decline and regeneration of the central city accompanied by the rise and expansion of suburbs. Urban social patterns have been influenced by the tendency for urban dwellers to sort themselves spatially by family status, social status, and ethnicity. In Western countries, these patterns are also influenced by governmental controls that help determine land uses.

Urbanization is a global phenomenon, and North American models of city systems, land use, and social area patterns differ substantially from cities in the rest of the world, reflecting diverse heritages and economic structures. Western European cities differ from those in Eastern Europe, where land uses reflect earlier communist principles of city structure. Explosive growth in developing-world cities has challenged their ability to provide all their residents with employment, housing, safe water, sanitation, and other essential services and facilities.

KEY WORDS

basic sector
brownfields
central business district (CBD)
central city
central place
central place theory
Christaller, Walter
city
concentric zone model
conurbation
economic base
edge city
galactic city model

gated community
gentrification
hierarchy of central places
hinterland
isotropic plain
Latin American city model
metropolitan area
multiple-nuclei model
multiplier effect
network city
nonbasic (service) sector
peak land value intersection
peripheral model

primate city
rank-size rule
sector model
suburb
threshold
town
urban geography
urban hierarchy
urban influence zone
urbanization
urbanized area
world city

FOR REVIEW

1. Consider the city or town in which you live, attend school, or with which you are most familiar. In a brief paragraph, discuss that community's *site* and *situation.* Point out the connection, if any, between its site and situation and the basic functions that it originally performed or now performs.

2. Describe the *multiplier effect* as it relates to the population growth of urban units.

3. Is there a hierarchy of retailing activities in the community with which you are most familiar? Of how many and of what kinds of levels is that hierarchy composed? What localizing forces affect the distributional pattern of retailing within that community?

4. Briefly describe the urban land-use patterns predicted by the *concentric circle,* the *sector,* and the *multiple-nuclei* models of urban development. Which one, if any, best corresponds to the growth and land-use pattern of the community most familiar to you?

5. In what ways do *social status, family status,* and *ethnicity* affect the residential choices of households? What distributional patterns are associated with each factor? Does the social geography of your community conform to the predicted pattern?

6. Where has gentrification appeared in the city with which you are most familiar? Who or what has been displaced?

7. In what ways does the Canadian city differ from the pattern of its U.S. counterpart?

8. Why are metropolitan areas in developing countries expected to grow larger than many Western metropolises by 2030?

9. What are *primate cities?* Why are primate cities so prevalent in the developing world? What might be some disadvantages of a primate city distribution?

10. How does the Latin American city model differ from models of urban structure developed in the United States?

KEY CONCEPTS REVIEW

1. **What common features define the origin, nature, and locations of cities?** Section 11.1–11.3

   Cities arose 4,000–6,000 years ago as distinctive evidence of the growing cultural and economic complexity of early civilizations. Distinct from the farm villages of subsistence economies, true cities provided an increasing range of functions—religious, military, trade, production, etc.—for their developing societies. Their functions and importance were affected by the sites and situations chosen for them. The massive recent increase in number and size of cities worldwide reflects the universality of economic development and total population growth in the latter 20th century.

2. **How are cities structured economically and how are systems of cities organized?** Section 11.3–11.5

   The economic base of a city—the functions it performs—is divided between basic and nonbasic (or service) activities. Through a multiplier effect, adding basic workers increases both the number of service workers and the total population of a city. The amount of growth reflects the base ratio characteristic of the city. Cities may be hierarchically ranked by their size and functional complexity. Rank-size, primate, and central place hierarchies are commonly cited but distinctly different.

3. **How are cities structured internally and how do people distribute themselves within them?** Section 11.6–11.8

   Cities are themselves distinctive land use and cultural area landscapes. In the United States, older cities show repetitive land-use patterns that are largely determined by land value and accessibility considerations. Classical land-use models include the concentric circle, sector, and multiple nuclei patterns. Distinct social area arrangements have been equated with those land-use models. Newer cities and growing metropolitan areas have created different land-use and social area structures with suburbs, edge cities, and galactic metropolises as recognized urban landscape features.

4. **Are there world regional and cultural differences in the land-use and population patterns of major cities?** Section 11.9

   Cities are regional and cultural variables; their internal land-use and social area patterns reflect the differing historical, technological, political, and cultural conditions under which they developed. We are most familiar with the U.S. city, but we can easily recognize differences between it and Western European, Eastern European, and developing-world cities, themselves of great regional, physical, and cultural complexity.

CHAPTER 12

THE POLITICAL ORDERING OF SPACE

The Palace of Westminster (which contains the Houses of Parliament), located on the north bank of the River Thames in London, is the seat of the government of the United Kingdom of Great Britain and Northern Ireland.

©Ingram Publishing/SuperStock

Key Concepts

12.1 National political units: geographic characteristics and boundary concerns

12.2 Nationalism, unifying centripetal, and destabilizing centrifugal forces

12.3 International political systems: the United Nations, maritime law, and regional alliances

12.4 Local and regional political forms: representation and fragmentation

*T*hey met together in the cabin of the little ship on the day of the landfall. The journey from England had been long and stormy. Provisions ran out, a man had died, a boy had been born. Although they were grateful to have been delivered to the calm waters off Cape Cod that November day of 1620, their gathering in the cramped cabin was not to offer prayers of thanksgiving but to create a political structure to govern the settlement they were now to establish (**Figure 12.1**). The Mayflower Compact was an agreement among themselves to "covenant and combine our selves together into a civill Body Politick . . . to enact, constitute, and frame such just and equall Lawes, Ordinances, Acts, Constitutions, and Offices . . . convenient for ye Generall good of ye Colonie. . . ." They elected one of their company governor, and only after those political acts did they launch a boat and put a party ashore.

The land they sought to colonize had for more than 100 years been claimed by the England they had left. The New World voyage of John Cabot in 1497 had invested their sovereign with title to all of the land of North America and a recognized legal right to govern his subjects dwelling there. That right was delegated by royal patent to colonizers and their sponsors, conferring upon them title to a defined tract and the authority to govern it. Although the Mayflower settlers were originally without a charter or patent, they recognized themselves as part of an established political system. They chose their governor and his executive department annually by vote of the General Court, a legislature composed of all freemen of the settlement.

As the population grew, new towns were established too distant for their voters to attend the General Court. By 1636, the larger towns were sending representatives to cooperate with the executive branch in making laws. Each town became a legal entity, with election of local officials and enactment of local ordinances the prime purpose of the town meetings that are still common in New England today.

The Mayflower Compact, signed by 41 freemen as their initial act in a New World, was the first step in a continuing journey of political development for the settlement and for the larger territory of which it became a part. From company patent to crown colony to rebellious commonwealth under the Continental Congress to state in a new country, Massachusetts (and Plimoth Plantation) were part of a continuing process of the political organization of space.

That process is as old as human history. From clans to kingdoms, human groups have laid claim to territory and have organized themselves and administered their affairs within it. Indeed, the political organizations of society are as fundamental an expression of culture and cultural differences as are forms of economy or religious beliefs. Geographers are interested in that structuring because it is both an expression of the human organization of space and closely related to other spatial evidences of culture, such as religion, language, and ethnicity.

Political geography is the study of the organization and distribution of political phenomena, including their impact on other spatial components of society and culture. It includes the study of **geopolitics,** how spatial relations among regions influence their current and past political activities and political relations. Geopolitics was first propounded by the English geographer Halford Mackinder in his *Heartland Theory* from 1904. This theory suggested that the area of central Asia and Eastern Europe, given its central location in Europe-Asia-Africa, was key to the control of surrounding territories in these largest continents and, by extension, the rest of the inhabited world. Nationality is a basic element in cultural variation among people, and political geography traditionally has had a primary interest in country units, or *states* (**Figure 12.2**). Of particular concern have been spatial patterns that reflect the exercise of central governmental control, such as questions of boundary delimitation. Increasingly, however, attention has shifted both upward and downward on the political scale. On the world scene, international alliances, regional compacts, and producer cartels—some requiring the surrender of at least a portion of national sovereignty—have increased in prominence since World War II, representing new forms of spatial interaction. At the local level, voting patterns, constituency boundaries and districting rules, and political fragmentation have directed public attention to the significance of area in the domestic political process.

In this chapter, we discuss some of the characteristics of political entities, examine the problems involved in defining jurisdictions, seek the elements that lend cohesion to a political entity, explore the implications of partial surrender of sovereignty, and consider the significance of the fragmentation of political power. We begin with states (countries) and end with local political systems.

Emphasis in this chapter on political entities should not make us lose sight of the reality that states are rooted in the operations of the economy and society they represent, that social

Figure 12.1 Signing the Mayflower Compact, probably the first written plan for self-government in America. A total of 41 adult males signed the Compact aboard the *Mayflower* before going ashore. (*Courtesy of the Pilgrim Society, Plymouth, MA*)

Source: Library of Congress Prints & Photographs Division [LC-DIG-ppmsca-07842]

Figure 12.2 These flags, symbols of separate member states, grace the front of the United Nations (UN) building in Geneva, Switzerland. Although central to political geographic interest, states are only one level of the political organization of space.

©*Arthur Getis.*

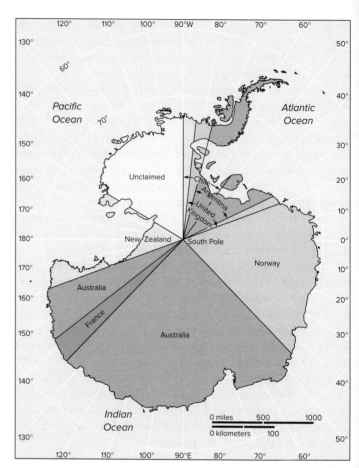

Figure 12.3 Territorial claims in Antarctica. Seven countries claim sovereignty over portions of Antarctica, and those of Argentina, Chile, and the United Kingdom overlap. The Antarctic Treaty of 1959 froze those claims for 30 years, banned further land claims, and made scientific research the primary use of the continent. The treaty was extended for 50 years in 1991. Antarctica is neither a sovereign state—it has no permanent inhabitants or local government—nor a part of one.

and economic disputes are as significant as border confrontations, and that in some ways, transnational corporations (TNCs) and other nongovernmental agencies may exert more influence in international affairs than do the separate states in which they are housed or operate. Some of those expanded political considerations are alluded to in the discussions that follow; others were described more fully in Chapter 9.

12.1 National Political Systems

One of the most significant elements in human geography is the nearly complete division of the Earth's land surface into separate country units, as shown on the "Countries of the World" map inside this book's cover. Even Antarctica is subject to the rival territorial claims of seven countries, although these claims have not been pressed because of the Antarctic Treaty of 1959 (**Figure 12.3**). This division into country units is relatively recent. Although countries and empires have existed since the days of ancient Egypt and Mesopotamia, only in the last century has the world been almost completely divided into independent governing entities. Now, people everywhere accept the idea of the state and its claim to sovereignty within its borders as normal.

States, Nations, and Nation-States

Before we begin our consideration of political systems, we need to clarify some terminology. Geographers use the words *state* and *nation* somewhat differently than the way they are used in everyday speech; sometimes confusion arises because each word has more than one meaning. A state can be defined as either (1) any of the political units forming a federal government (e.g., one of the United States) or as (2) an independent political entity

holding sovereignty over a territory (e.g., the United States). In this latter sense, *state* is synonymous with *country* or *nation*. That is, a nation can also be defined as (1) an independent political unit holding sovereignty over a territory (e.g., a member of the United Nations). But that term can also be used to describe (2) a community of people with a common culture and territory (e.g., the Kurdish nation). The second definition is *not* synonymous with state or country.

To avoid confusion, we shall define a **state** on the international level as an independent political unit occupying a defined, permanently populated territory and having full sovereign control over its internal and foreign affairs. We will use *country* as a synonym for the territorial and political concept of *state*. Not all recognized territorial entities are states. Antarctica, for example, has neither established government nor permanent population, and it is, therefore, not a state. Nor are *colonies* or *protectorates* recognized as states. Although they have defined extent, permanent inhabitants, and some degree of separate governmental structure, they lack full control over all of their internal and external affairs. More important, they lack recognition as states by the international community, a decisive consideration in the proper use of the term *state*.

We use *nation* in its second sense, as a reference to people rather than to political structure. A **nation** is a group of people with a common culture occupying a particular territory, bound together by a strong sense of unity arising from shared beliefs and customs. Language and religion may be unifying elements, but even more important are an emotional conviction of cultural distinctiveness and a sense of ethnocentrism; members of a nation self identify as such. The Cree nation exists because of its cultural uniqueness, not by virtue of territorial sovereignty.

The composite term **nation-state** properly refers to a state whose territorial extent coincides with that occupied by a distinct nation or people or, at least, whose population shares a general sense of cohesion and adherence to a set of common values (**Figure 12.4a**). That is, a nation-state is an entity whose members feel a natural connection with one another by virtue of sharing language, religion, or some other cultural characteristic that is strong enough both to bind them together and to give them a sense of distinction from others outside the community. Although all countries strive for consensus values and loyalty to the state, few can claim to be true nation-states because few are or have ever been wholly uniform ethnically. Iceland, Slovenia, Poland, Japan, and the two Koreas are often cited as examples.

A *binational* or *multinational state* is one that contains more than one nation (**Figure 12.4b**). Often, no single ethnic group dominates the population. In the constitutional structure of the former Soviet Union before 1988, one division of the legislative branch of the government was termed the Soviet of Nationalities. It was composed of representatives from civil divisions of the Soviet Union populated by groups of officially recognized "nations": Ukrainians, Kazakhs, Tatars, Estonians, and others. In this instance, the concept of nationality was territorially less than the extent of the state.

Alternatively, a single nation may be dispersed across and be predominant in two or more states. This is the case with the *part-nation-state* (**Figure 12.4c**). Here, a people's sense of nationality exceeds the areal limits of a single country. An example is the Arab nation, which dominates 17 states.

Finally, there is the special case of the *stateless nation,* a people without a state. The Kurds, for example, are usually regarded as the largest stateless nation, including some 20 million people divided among six states and dominant in none (**Figure 12.4d**). Kurdish nationalism has survived over the centuries, and many Kurds nurture a vision of an independent Kurdistan. Other stateless nations include the Roma (Gypsies), Basques, and several Amerindian groups.

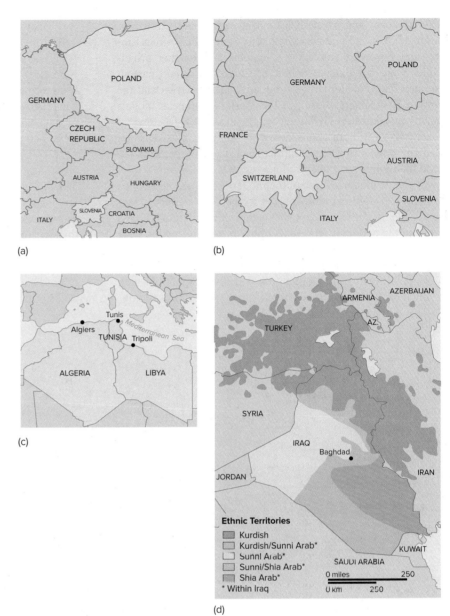

(a)

(b)

(c)

(d)

Figure 12.4 Types of relationships between states and nations. (*a*) **Nation-states.** Poland and Slovenia are examples of states occupied by a distinct nation or people. (*b*) **A multinational state.** Switzerland shows that a common ethnicity, language, or religion is not necessary for a sense of nationalism. (*c*) **A part-nation-state.** The Arab nation extends across and dominates many states in northern Africa and the Middle East. (*d*) **A stateless nation.** An ancient group with a distinctive language, the Kurds are concentrated in Turkey, Iran, and Iraq. Smaller numbers of these people live in Syria, Armenia, and Azerbaijan.

Ethnic Territories
- Kurdish
- Kurdish/Sunni Arab*
- Sunni Arab*
- Sunni/Shia Arab*
- Shia Arab*
- * Within Iraq

0 miles 250
0 km 250

The Evolution of the Modern State

The concept and practice of the political organization of space and people arose independently in many parts of the world. Certainly, one of the distinguishing characteristics of very early culture hearths—including those shown in Figure 2.15 in Chapter 2—was the political organization of their peoples and areas. The larger and more complex the economic structures they developed, the more sophisticated became their mechanisms of political control and territorial administration.

Our Western orientations and biases may incline us to trace ideas of spatial political organization through their Near Eastern,

Mediterranean, and Western European expressions. However, Mesopotamian and classical Greek city states, the Roman Empire, and European colonizing kingdoms and warring principalities were not unique. Southern, southeastern, and eastern Asia had their counterparts, as did sub-Saharan Africa and the Western Hemisphere. Although the Western European models and colonization strongly influenced the forms and structures of modern states in both hemispheres, the cultural roots of statehood run deeper and reach further back in many parts of the world than European example alone suggests.

The now universal idea of the modern state was developed by European political philosophers in the 18th century. Their views advanced the concept that people owe allegiance to a state and the people it represents rather than to its leader, such as a king or feudal lord. The new concept coincided in France with the French Revolution and spread over Western Europe, to England, Spain, and Germany. It was also compatible with the birth of the United States on the other side of the Atlantic.

Many states are the result of European expansion during the 17th, 18th, and 19th centuries, when much of Africa, Asia, and the Americas was divided into colonies. Usually these colonial claims were given fixed and described boundaries where none had earlier been formally defined. Of course, precolonial native populations had relatively fixed home areas of control, within which there was recognized dominance and border defense and from which there were, perhaps, raids of plunder or conquest of neighboring "foreign" districts. Beyond understood tribal territories, great empires arose, again with recognized outer limits of influence or control: Mogul and Chinese, Benin and Zulu, Incan and Aztec. Upon the less formally organized spatial patterns of effective tribal control, European colonizers imposed their arbitrary new administrative divisions of the land. In fact, groups that had little in common were often joined in the same colony (**Figure 12.5**). The new divisions, therefore, were not usually based on meaningful cultural or physical lines. Instead, the boundaries simply represented the limits of the colonizing empire's power.

Figure 12.5 The discrepancies between ethnic groups and country boundaries in Africa. Cultural boundaries were ignored by European colonial powers. The result was significant ethnic diversity in nearly all African countries and conflict countries over borders.

Source: Redrawn from World Regional Geography: A Question of Place *by Paul Ward English, with James Andrew Miller. 1977 Harper & Row.*

As these former colonies have gained political independence, they have retained the idea of the state. They have generally accepted—in the case of Africa, by a conscious decision to avoid precolonial territorial or ethnic claims that could lead to war—the borders established by their former European rulers (**Figure 12.6**). The problem that many of the new countries face is *nation-state building*—developing feelings of loyalty to the state among their arbitrarily associated citizens. For example, the Democratic Republic of the Congo, the former Belgian Congo (and Zaire before that, among other names), contains some 270 frequently antagonistic ethnic groups. Julius Nyerere, then president of Tanzania, noted in 1971, "These new countries are artificial units, geographic expressions carved on the map by European imperialists. These are the units we have tried to turn into nations."

The idea of separate statehood grew slowly at first and, more recently, has accelerated rapidly. At the time of the Declaration of Independence of the United States in 1776, there were only some 35 empires, kingdoms, and countries in the entire world. By the beginning of World War II in 1939, their number had only doubled to about 70. Following that war, the end of the colonial era brought a rapid increase in the number of sovereign states. From the former British Empire and Commonwealth, there have come the independent countries of India, Pakistan, Bangladesh, Malaysia, Myanmar (Burma), and Singapore in Asia, and Ghana, Nigeria, Kenya, Uganda, Tanzania, Malawi, Botswana, Zimbabwe, and Zambia in Africa. Even this extensive list is not complete. A similar process has occurred in most of the former overseas possessions of the Netherlands, Spain, Portugal, and France. By 1990, independent states totaled some 180, and their number increased again following—among other political geographic developments—the disintegration during the 1990s of the USSR, Czechoslovakia, and Yugoslavia, which created more than 20 countries where only three had existed before (**Figure 12.7**).

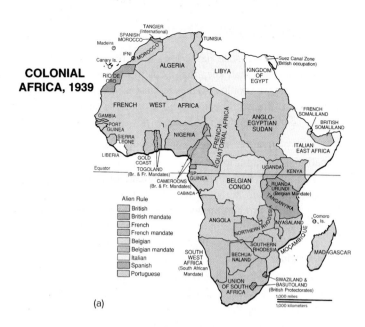

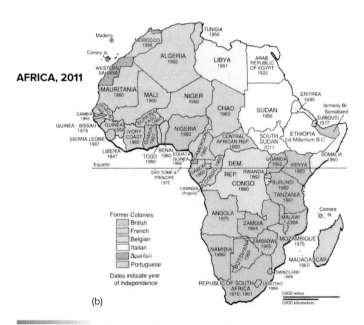

Figure 12.7 After the former USSR dissolved in 1991, 15 newly independent countries emerged from the 15 republics that once comprised it.

Figure 12.6 Africa—from colonies to states. (*a*) **Africa in 1939** was a patchwork of foreign claims and alien rule, some dating from the 19th century, others of more recent vintage. For example, Germany lost its claim to South-West Africa, Tanganyika, Togoland, and the Cameroons after World War I, and Italy asserted control over Ethiopia during the 1930s. (*b*) **Africa in 2011** was a mosaic of separate states. Their dates of independence are indicated on the map. French West Africa and French Equatorial Africa have been extensively subdivided, and Ethiopia and Somaliland emerged from Italian control. Most of the current countries retain the boundaries of their former colonial existence, though the continent's structure of political influence and regional power has changed through civil wars and neighboring state interventions. These marked the decline of earlier African principles of inviolability of borders and noninterference in the internal affairs of other states.

Challenges to the State

The state and nation-state hav e long been the focus of political geography, and we shall keep that focus in much of the following discussion. But we should also realize that the validity of that state-centric view of the world is increasingly under assaults from multiple new agents of economic and social power. Among them are:

- The globalization of economies and the emergence of transnational corporations whose economic and production decisions are unrelated to the interests of any single state, including their home office state. Those decisions—outsourcing of production and services, for example—may be detrimental to the employment structure, tax base, and national security of any single state and limit the applicability of national economic planning and control.

- The proliferation of international and supranational institutions initially concerned with financial or security matters but all representing the voluntary surrender of some traditional state autonomy. The World Trade Organization (WTO), regional trade blocs like the North American Free Trade Association, the European Union (EU), and a host of other international conventions and agreements all limit the independence of action of each of their members and thus diminish absolute state primacy in economic and social matters.

- The emergence and multiplication of nongovernmental organizations (NGOs), whose specific interests and collective actions cut across national boundaries and unite people sharing common concerns about, for example, globalization, the environment, economic and social injustice, AIDS efforts, and the like. The well-publicized protests and pressures exerted by NGOs channel social pressures to influence or limit governmental actions.

- The massive international migration flows that tend to undermine the state as a cultural community with assured and expected common values and loyalties. The Internet, cheap telephone calls, and easy international travel permit immigrant retention of primary ties with their home culture and state, discouraging their full assimilation into their new social environment or the transfer of loyalties to their adopted country.

- The increase in nationalist and separatist movements in culturally composite states, weakening through demands for independence or regional autonomy the former unquestioned primacy of the established state.

Some of these agents and developments have been touched on in earlier chapters; others will be reviewed in this chapter, particularly in the "Centrifugal Forces: Challenges to State Authority" section. All represent recent and strengthening forces that, in some assessments of political geographic reality, weaken the validity of a worldview in which national governments and institutionalized politics are all-powerful.

Spatial Characteristics of States

Every state has certain spatial characteristics by which it can be described and that set it apart from all other states. A look at the world political map inside the cover of this book confirms that every state is unique. The size, shape, and location of any one state combine to distinguish it from all others. These characteristics are of more than academic interest because they also affect the power and stability of states.

Size

The area that a state occupies may be large, as is China, or small, as is Liechtenstein. The world's largest country, Russia, occupies more than 17 million square kilometers (6.5 million square miles), some 11 percent of the Earth's land surface. It is nearly 1 million times as large as Nauru, one of the *ministates* or *micro-states* found in all parts of the world (see the feature "The Ministates").

An easy assumption would be that the larger a state's area, the greater is the chance that it will include the ores, energy supplies, and fertile soils from which it can benefit. In general, that assumption is valid, but much depends on accidents of location. Mineral resources are unevenly distributed, and size alone does not guarantee their presence within a state. Australia, Canada, and Russia, though large in territory, have relatively small areas capable of supporting productive agriculture. Great size, in fact, may be a disadvantage. A very large country may have vast areas that are remote, sparsely populated, and hard to integrate into the mainstream of economy and society. Small states are more apt than large ones to have a culturally homogeneous population. They find it easier to develop transportation and communication systems to link the sections of the country, and, of course, they have shorter boundaries to defend against invasion. Size alone, then, is not critical in determining a country's stability and strength, but it is a contributing factor.

Shape

Like size, a country's shape can affect its well-being as a state by fostering or hindering effective organization. Assuming no major topographical barriers, the most efficient form would be a circle with the capital located in the center. In such a country, all places could be reached from the center in a minimal amount of time and with the least expenditure for roads, railway lines, and so on. It would also have the shortest possible borders to defend. Uruguay, Zimbabwe, and Poland have roughly circular shapes, forming **compact states** (**Figure 12.8**).

Prorupt states are nearly compact but possess one or sometimes two narrow extensions of territory. Proruption may simply reflect peninsular elongations of land area, as in the case of Myanmar and Thailand. In other instances, the extensions have an economic or strategic significance, recording international negotiation to secure access to resources or water routes or to establish a buffer zone between states that would otherwise adjoin. Whatever their origin, proruptions tend to isolate a portion of a state.

The least efficient shape administratively is represented by countries like Norway, Vietnam, or Chile, which are long

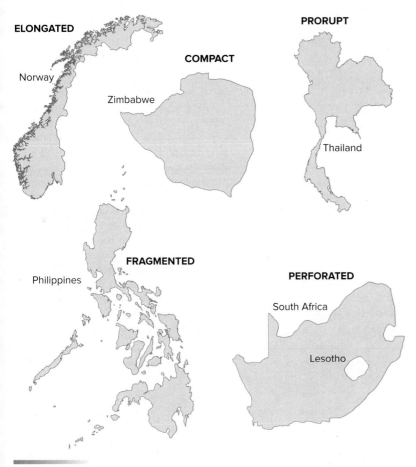

ELONGATED
Norway

COMPACT
Zimbabwe

PRORUPT
Thailand

FRAGMENTED
Philippines

PERFORATED
South Africa
Lesotho

Figure 12.8 Shapes of states. The sizes of the countries should not be compared. Each is drawn on a different scale.

and narrow. In such **elongated states,** the parts of the country far from the capital are likely to be isolated because great expenditures are required to link them to the core. These countries are also likely to encompass more diversity of climate, resources, and peoples than compact states, perhaps to the detriment of national cohesion or, perhaps, to the promotion of economic strength.

A fourth class of **fragmented states** includes countries composed entirely of islands (e.g., the Philippines and Indonesia), countries that are partly on islands and partly on the mainland (Italy and Malaysia), and those that are chiefly on the mainland but whose territory is separated by another state (the United States). Fragmentation makes it harder for the state to impose centralized control over its territory, particularly when the parts of the state are far from one another. This is a problem in the Philippines and Indonesia, the latter made up of more than 13,000 islands stretched out along a 5,100-kilometer (3,200-mile) arc. Fragmentation helped lead to the disintegration of Pakistan. It was created in 1947 as a spatially divided state with East and West Pakistan separated by 1,610 kilometers (1,000 miles). That distance exacerbated economic and cultural differences between the two, and when the eastern part of the country seceded in 1971 and declared itself the independent state of Bangladesh, West Pakistan was unable to impose its control.

A special case of fragmentation occurs when a territorial outlier of one state, an **exclave,** is located within another country. Before German unification, West Berlin was an outlier of West Germany within the eastern German Democratic Republic. Europe has many such exclaves. Kleinwalsertal, for example, is a patch of Austria accessible only from Germany. Point Roberts, Washington is a 13-square-kilometers (5-square-miles) quasi-exclave of the United States. From Lake of the Woods in Minnesota to the Pacific Ocean, the United States–Canada boundary follows the 49th parallel. Point Roberts is located on the southern tip of the Tsawwassen Peninsula, which connects to the Canadian mainland but extends 3.2 kilometers (2 miles) south of the 49th parallel. Point Roberts borders Canada to its north and water on the three remaining sides. After second grade, Point Roberts schoolchildren have to ride a bus through two international border crossings and 37 kilometers (23 miles) of Canada to reach their school in the United States.

The counterpart of an exclave, an **enclave,** helps to define the fifth class of country shapes, the **perforated state.** A perforated state completely surrounds a territory that it does not rule as, for example, the Republic of South Africa surrounds Lesotho. The enclave, the surrounded territory, may be independent or may be part of another country. Two of Europe's smallest independent states, San Marino and Vatican City, are enclaves that perforate Italy. As an *exclave* of former West Germany, West Berlin perforated the national territory of former East Germany and was an *enclave* in it. The stability of the perforated state can be weakened if the enclave is occupied by people whose value systems differ from those of the surrounding country.

Location

The significance of size and shape as factors in national well-being can be modified by a state's location, both absolute and relative. Although both Canada and Russia are extremely large, their *absolute location* in upper middle latitudes reduces their size advantages when agricultural potential is considered (they have short growing seasons and harsh winters). To take another example, Iceland has a reasonably compact shape, but its location in the North Atlantic Ocean, just south of the Arctic Circle, means that most of the country is barren, with settlement confined to the rim of the island.

A state's *relative location,* its position compared to that of other countries, is as important as its absolute location. **Landlocked** states, those lacking ocean frontage and surrounded by other states, are at a commercial and strategic disadvantage (**Figure 12.9**). They lack easy access to both maritime (seaborne) trade and the resources found in coastal waters and submerged lands. Typically, a landlocked country arranges to use facilities at a foreign port and to have the right to travel to that port. Bolivia,

The Ministates

Totally or partially autonomous political units that are small in area and population pose some intriguing questions. Should size be a criterion for statehood? What is the potential of ministates to cause friction among the major powers? Under what conditions are they entitled to representation in international assemblies like the United Nations?

Of the world's growing number of small countries, more than 40 have less than 1 million people, the population size adopted by the United Nations as the upper limit defining *small states,* though not too small to be members of that organization. Nauru has about 12,000 inhabitants on its 21 square kilometers (8.2 square miles). Other areally small states like Singapore (580 sq km; 224 square miles) have populations (5.9 million) well above the UN criterion. Many are island countries located in the Caribbean, the Pacific or Indian Ocean (such as Grenada, Tuvalu, and Maldives), but Europe (Vatican City and Andorra), Asia (Bahrain, Kuwait, and Brunei), Africa (Djibouti and Equatorial Guinea), and South and Central America (Suriname, Belize) have their share as well.

Many ministates are vestiges of colonial systems that no longer exist. Some of the small countries of West Africa and the Arabian peninsula fall into this category. Others, such as Mauritius, served primarily as refueling stops on transoceanic shipping lanes. However, some occupy strategic locations (such as Bahrain, Malta, and Singapore), and others contain valuable minerals (Kuwait and Trinidad). The possibility of claiming 370-kilometer-wide (200 nautical mile) zones of adjacent seas (see the feature "Specks and Spoils," Section 12.2) adds to the attraction of yet others.

Their strategic or economic value can expose small islands and territories to unwanted attention from larger neighbors. The 1982 war between Britain and Argentina over the Falkland Islands (claimed as the Islas Malvinas by Argentina) and the Iraqi invasion of Kuwait in 1990 demonstrate the ability of such areas to bring major powers into conflict and to receive world attention that is out of proportion to their size and population.

The proliferation of tiny countries raises the question of their representation and their voting weight in international assemblies. Should there be a minimum size necessary for participation in such bodies? Should countries receive a vote proportional to their population? New members accepted into the United Nations in 1999 and 2000 included four small Pacific island countries, all with populations at the time of about 100,000 or less: Nauru, Tonga, Kiribati, and Tuvalu. Within the United Nations, the Alliance of Small Island States (AOSIS) has emerged as a significant power bloc, controlling more than one-fifth of General Assembly votes.

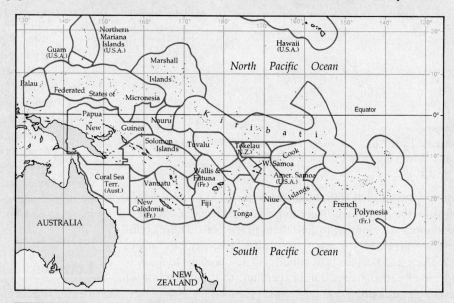

Figure 12A Pacific Ocean island mini-states, with nautical boundaries of territorial claims.

for example, has secured access to the Chilean port of Arica, the Peruvian port of Ilo, and the Argentine city of Rosario on the Paraná River (**Figure 12.10**). The number of landlocked states—more than 40—increased greatly with the dissolution of the Soviet Union and the creation of new, smaller nation-states out of such former multinational countries as Yugoslavia and Czechoslovakia.

In a few instances, a favorable relative location constitutes the primary resource of a state. Singapore, a state of only 580 square kilometers (224 square miles), is located at a crossroads of world shipping and commerce. Based on its port and commercial activities and buttressed by its more recent industrial development, Singapore has become a notable Southeast Asian

economic success. In general, history has shown that countries benefit from a location on major trade routes, not only from the economic advantages such a location carries, but also because they are exposed to the diffusion of new ideas and technologies.

Cores and Capitals

Many states have come to assume their present shape, and thus the location they occupy, as a result of growth over centuries. They grew outward from a central region, gradually expanding into surrounding territory. The original nucleus, or **core area,** of a state usually contains its most developed economic base,

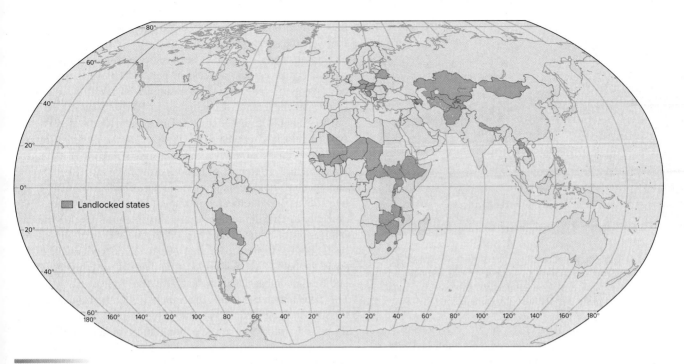

Figure 12.9 Landlocked states.

densest population, and largest cities, as well as the most highly developed transportation systems. All of these elements become less intense away from the national core. Transportation networks thin, urbanization ratios and city sizes decline, and economic development is less concentrated on the periphery than in the core. The outlying resource base may be rich but generally is of more recent exploitation with product and benefit tending to flow to the established heartlands. The developed cores of states, then, can be contrasted to their subordinate peripheries just as we saw the *core-periphery* idea applicable in an international developmental context in Chapter 10.

Easily recognized and unmistakably dominant national cores include the Paris Basin of France, London and southeastern England, Moscow and the major cities of European Russia, northeastern United States and southeastern Canada, and the Buenos Aires megalopolis in Argentina. Not all countries have such clearly defined cores—Chad, Mongolia, or Saudi Arabia, for example—and some may have two or more rival core areas. Ecuador, Nigeria, Democratic Republic of the Congo, and Vietnam are examples of multicore states.

The capital city of a state is usually within its core region and frequently is the very focus of it, dominant not only because it is the seat of central authority but because of the concentration of population and economic functions as well. That is, in many countries the capital city is also the largest city and sometimes a *primate* city, dominating the structure of the entire country. Paris in France, London in the United Kingdom, and Mexico City are examples of that kind of political, cultural, and economic primacy (Figure 11.17).

This association of capital with core is common in what have been called the **unitary states,** countries with highly centralized governments, relatively few internal cultural contrasts, a strong sense of national identity, and borders that are clearly cultural as

well as political boundaries. Most European cores and capitals are of this type. It is also found in many newly independent countries whose former colonial occupiers established a primary center of exploitation and administration and developed a functioning core in a region that lacked an urban structure or organized government.

Figure 12.10 Like many other landlocked countries, Bolivia has gained access to the sea through arrangements with neighboring states. Unlike most landlocked countries, however, Bolivia can access ports on two oceans.

Figure 12.11 Canberra, the planned capital of Australia, was deliberately sited away from the country's two largest cities, Sydney and Melbourne. Planned capitals are often architectural showcases, providing a focus for national pride. Eucalyptus hills rising behind the city.
©Taras Vyshnya/Shutterstock.

With independence, the new states retained the established infrastructure, added new functions to the capital, and, through lavish expenditures on governmental, public, and commercial buildings, sought to create prestigious symbols of nationhood.

In **federal states,** associations of more or less equal provinces with strong regional governmental responsibilities, the national capital city may have been newly created or selected to serve as the administrative center. Although part of a generalized core region of the country, the designated capital was not its largest city and acquired few of the additional functions to make it so. Ottawa, Canada; Washington, D.C.; and Canberra, Australia, are examples (**Figure 12.11**).

A new form of state organization, *regional* government or *asymmetric federalism,* is emerging in Europe as formerly strong unitary states acknowledge the autonomy aspirations of their several subdivisions and grant to them varying degrees of local administrative control while retaining in central hands authority over matters of nationwide concern, such as monetary policy, defense, foreign relations, and the like. That new form of federalism involves recognition of regional capitals, legislative assemblies, administrative bureaucracies, and the like. The asymmetric federalism of the United Kingdom, for example, now involves separate status for Scotland, Wales, and Northern Ireland, with their own capitals at Edinburgh, Cardiff, and Belfast, respectively. That of Spain recognizes Catalonia and the Basque country with capitals in Barcelona and Vitoria, respectively. In Canada, the establishment of the self-governing Inuit arctic territory of Nunavut (see Chapter 6, footnote 2) in 1999 has been followed by other recognized Indian claims of home territory land use control: that of the Haida in British Columbia, the Dogrib (Tlicho) in the Northwest Territories, and the Cree in northern Quebec, for example.

All other things being equal, a capital located in the center of the country provides equal access to the government, facilitates communication to and from the political hub, and enables the government to exert its authority relatively easily. Many capital cities, such as Washington, D.C., were centrally located when they were designated as seats of government but lost their centrality as the state expanded.

Some capital cities have been relocated outside of peripheral national core regions, at least in part to achieve the presumed advantages of centrality. Two examples of such relocation are from Karachi inland to Islamabad in Pakistan and from Istanbul to Ankara, in the center of Turkey's territory. A particular type of relocated capital is the *forward-thrust capital city,* one that has been deliberately sited in a state's interior to signal the government's awareness of regions away from an off-center core and its interest in encouraging more uniform development. In the late 1950s, Brazil relocated its capital from Rio de Janeiro to the new city of Brasilia to demonstrate its intent to develop the vast interior of the country. The West African country of Nigeria has been building the new capital of Abuja near its geographic center since the late 1970s, with relocation there of government offices and foreign embassies in the early 1990s.

The British colonial government relocated Canada's capital six times between 1841 and 1865, in part seeking centrality to the mid-19th-century population pattern and in part seeking a location that bridged that colony's cultural divide (**Figure 12.12**). A Japanese law of 1997 calling for the relocation of the parliament building, Supreme Court, and main ministries out of Tokyo by 2010 was more related to earthquake fears and a search for seismic safety than to enhanced convenience or governmental efficiency. Putrajaya, the new administrative seat of Malaysia 40 kilometers (25 miles) south of the present capital, Kuala Lumpur; Astana, the new national capital of Kazakhstan located on a desolate stretch of Siberian steppe; and Naypyidaw, a desolate, rocky site 322 kilometers (200 miles) north of Myanmar's present capital of Yangon (Rangoon) are other examples of recent new national capital creations.

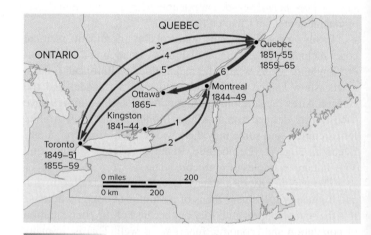

Figure 12.12 Canada's migratory capital. Kingston was chosen as the first capital of the united Province of Canada in preference to either Quebec, capital of Lower Canada, or Toronto, that of Upper Canada. In 1844, governmental functions were relocated to Montreal, where they remained until 1849, after which they shifted back and forth—as the map indicates—between Toronto and Quebec. An 1865 session of the provincial legislature was held in Ottawa, the city that became the capital of the Confederation of Canada in 1867.

Source: Redrawn from David B. Knight, A Capital for Canada (Chicago: University of Chicago, Department of Geography, Research Paper No. 182, 1977), Figure 1. p. vii.

Boundaries: The Limits of the State

We noted earlier that no portion of the Earth's land surface is outside the claimed control of a national unit, that even uninhabited Antarctica has had territorial claims imposed upon it (see Figure 12.3). Each of the world's states is separated from its neighbors by *international boundaries,* or lines that establish the limit of each state's jurisdiction and authority. Boundaries indicate where the sovereignty of one state ends and that of another begins (**Figure 12.13**).

Within its own bounded territory, a state administers laws, collects taxes, provides for defense, and performs other such governmental functions. Thus, the location of the boundary determines the kind of money people in a given area use, the legal code to which they are subject, the army they may be called upon to join, and the language and perhaps the religion children are taught in school. These examples suggest how boundaries serve as powerful reinforcers of cultural variation over the Earth's surface.

(a)

(b)

Figure 12.13 International boundaries. *(a)* The U.S.–Canada border on the Rainbow bridge over the Niagara River. *(b)* The U.S.–Mexico border fence, at Nogales, Arizona/Mexico.

Territorial claims of sovereignty, it should be noted, are three-dimensional. International boundaries mark not only the outer limits of a state's claim to land (or water) surface, but are also projected downward to the center of the Earth in accordance with international consensus allocating rights to subsurface resources. States also project their sovereignty upward, but with less certainty because of a lack of agreement on the upper limits of territorial airspace. Properly viewed, then, an international boundary is a line without breadth; it is a vertical interface between adjacent state sovereignties. In practice, the width of administrative boundaries like those of international states depends on the efforts of diplomats, surveyors, and administrative officials; they are as thin as legal decisions and the spatial precision of boundary determination allow.

Before boundaries were delimited, nations or empires were likely to be separated by *frontier zones,* ill-defined and fluctuating areas marking the effective end of a state's authority. They were essentially the fuzzy boundaries of functional regions rather than the potentially crisp boundaries of administrative regions. Such zones were often uninhabited or only sparsely populated and were liable to change with shifting settlement patterns. Many present-day international boundaries lie in former frontier zones, and in that sense, the boundary line has replaced the broader frontier as a marker of a state's authority.

Natural and Geometric Boundaries

Geographers have traditionally distinguished between *natural* and *geometric* boundaries. **Natural** (or **physical**) **boundaries** are those based on recognizable physiographic features, such as mountains, rivers, and lakes. Although they might seem to be attractive as borders because they actually exist in the landscape and are visible dividing elements, many natural boundaries have proved to be unsatisfactory. That is, they do not effectively separate states, either because they are difficult to measure precisely, do not maintain a fixed location over time, or otherwise prove unsatisfactory to one or the other state on either side.

Many international boundaries lie along mountain ranges, for example in the Alps, Himalayas, and Andes, but while some have proved to be stable, others have not. Mountains are rarely total barriers to interaction. Although they do not invite movement, they are crossed by passes, roads, and tunnels. High pastures may be used for seasonal grazing, and the mountain region may be the source of water for irrigation or hydroelectric power. Nor is the definition of a boundary along a mountain range a simple matter. Should it follow the crests of the mountains or the *watershed divide* (the line dividing two drainage areas)? The two are not always the same. Border disputes between China and India are in part the result of the failure of mountain crests and headwaters of major streams to coincide (Figure 12.13).

Rivers can be even less satisfactory as boundaries. In contrast to mountains, rivers foster interaction. River valleys are likely to be agriculturally or industrially productive and to be densely populated. For example, for hundreds of kilometers, the Rhine River serves as an international boundary in Western Europe. It is also a primary traffic route lined by chemical plants, factories, and power stations and dotted by the castles and cathedrals that make it one of

Europe's major tourist attractions. It is more a common intensively used resource than a barrier in the lives of the nations that it borders. And if that weren't problem enough, the locations of rivers are surprisingly dynamic over time. A flood can change the exact course of a river in a matter of days, and slower processes of natural erosion and deposition change the locations of rivers over time periods of only decades or centuries.

The alternative to natural boundaries is **geometric** (or **artificial**) **boundaries.** Frequently delimited as segments of parallels of latitude or meridians of longitude, they are found chiefly in Africa, Asia, and the Americas. The western portion of the U.S.-Canada border, which mostly follows the 49th parallel, is an example of a geometric boundary. Many such boundaries were established when the areas in question were colonies, the land was only sparsely settled, and detailed geographic knowledge of the frontier region was lacking.

Boundaries Classified by Settlement

Boundaries can also be classified according to whether they were laid out before or after the principal features of the cultural landscape were developed. An **antecedent boundary** is one drawn across an area before it is well populated (that is, before most of the cultural landscape features were put in place). To continue our earlier example, the western portion of the U.S.-Canada boundary is such an antecedent line, established by a treaty between the United States and Great Britain in 1846.

Boundaries drawn after the development of the cultural landscape are termed **subsequent.** One type of subsequent boundary is **consequent** (also called *ethnographic*), a border drawn to accommodate existing religious, linguistic, ethnic, or economic differences among countries. An example is the boundary drawn between Northern Ireland and Eire (Ireland). Subsequent **superimposed boundaries** may also be forced on existing cultural landscapes, a country, or a people by a conquering or colonizing power that is unconcerned about preexisting cultural patterns. The colonial powers in 19th-century Africa superimposed boundaries upon established African cultures without regard to the tradition, language, religion, or tribal affiliation of those whom they divided (see Figure 12.5).

When Great Britain prepared to leave the Indian subcontinent after World War II, it was decided that two independent states would be established in the region: India and Pakistan. The boundary between the two countries, defined in the partition settlement of 1947, was thus both a *subsequent* and a *superimposed* line. As millions of Hindus migrated from the northwestern portion of the subcontinent to seek homes in India, millions of Muslims left what would become India for Pakistan. In a sense, they were attempting to ensure that the boundary would be *consequent;* that is, that it would coincide with a division based on religion.

If a former boundary line that no longer functions as such is still marked by some landscape features or differences on the two sides, it is termed a **relic boundary (Figure 12.14).** The abandoned castles dotting the former frontier zone between Wales and England are examples of a relic boundary. They are also evidence of the disputes that sometimes attend the process of boundary making.

Figure 12.14 Like Hadrian's Wall in the north of England or the Great Wall of China, the Berlin Wall was a demarcated boundary. Unlike them, it cut across a large city and disrupted established cultural patterns. The Berlin Wall, therefore, was a subsequent superimposed boundary. The dismantling of the wall in 1990 marked the reunification of Germany; what remains standing as a historic monument is a relic boundary. These remnant pieces of the Wall in Portland, Maine, are essentially sculptures. Figure 13.12a (needs new first line of caption as shown below, with part *(b)* the same).
©Dave Moyer.

Boundary Disputes

Boundaries create many possibilities and provocations for conflict. Since World War II, almost half of the world's sovereign states have been involved in border disputes with neighboring countries. Just like households, states are far more likely to have disputes with their neighbors than with more distant parties. It follows that the more neighbors a state has, the greater its likelihood of conflict. Although the causes of boundary disputes and open conflict are many and varied, they can reasonably be placed into four categories.

Positional disputes occur when states disagree about the interpretation of documents that define a boundary and/or the way the boundary was delimited. Such disputes typically arise when the boundary is antecedent, preceding effective human settlement in the border region. Once the area becomes populated and gains value, the exact location of the boundary becomes important. The boundary between Argentina and Chile, originally defined during Spanish colonial rule and formalized by treaty in 1881, was to follow "the most elevated crests of the Andean Cordillera dividing the waters" between east- and west-flowing rivers. Because the southern Andes had not been adequately explored and mapped, it was not apparent that the crest lines (highest peaks) and the watershed divides do not always coincide.

Figure 12.15 Territory once disputed by Argentina and Chile in the Southern Andes Mountains. The treaty establishing the boundary between the two countries preceded adequate exploration and mapping of the area, leaving the precise location of the mountain crests and watershed divides in doubt. Friction over the disputed territory nearly led to war before an accord was reached in 1902. Subsequent disputes over details were resolved in an accord signed in 1998.

In some places, the water divide is many kilometers east of the highest peaks, leaving a long, narrow area of some 52,000 square kilometers (20,000 square miles) in dispute (**Figure 12.15**). In Latin America as a whole, the 21st century began with at least 10 unresolved border disputes, some dating back to colonial times.

Territorial disputes over the ownership of a region often, though not always, arise when a boundary that has been superimposed on the landscape divides an ethnically homogeneous population. Each of the two states then has some justification for claiming the entire territory inhabited by the ethnic group in question. We noted previously that a single nation may be dispersed across several states (see Figure 12.4d). Conflicts can arise if the people of one state want to annex a territory whose population is ethnically related to that of the state but now subject to a foreign government. This type of expansionism is called **irredentism** and is often a consequence of superimposed boundaries. In the 1930s, Adolf Hitler used the existence of German minorities in Czechoslovakia and Poland to justify German invasion and occupation of those countries. More recently, Somalia has had many border clashes with Ethiopia over the rights of

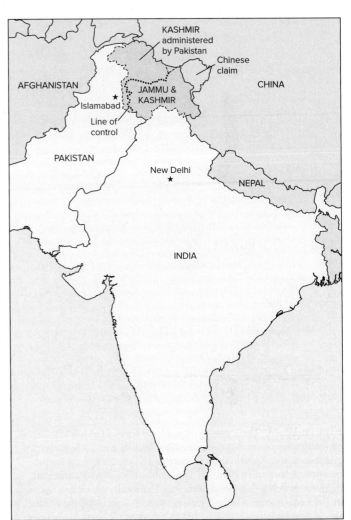

Figure 12.16 Kashmir, a disputed area, was left unresolved by the British when they dismantled their empire and partitioned India and Pakistan in 1947. At the time of partition the leader of the princely state of Kashmir chose to join India despite the Muslim majority in the Kashmir—giving both countries a claim to the territory. The dispute has sparked two wars between India and Pakistan and the territory has a *de facto* partition along a Line of Control that has been patrolled by United Nations peacekeepers since 1949.

Somalis living in that country, and the area of Kashmir has been a cause of dispute and open conflict between India and Pakistan since the creation of the two countries (**Figure 12.16**).

Closely related to territorial conflicts are **resource disputes.** Neighboring states are likely to covet the resources—whether they be valuable mineral deposits, fertile farmland, or rich fishing grounds—lying in border areas and to disagree over their use. In recent years, the United States has been involved in disputes with both its immediate neighbors: with Mexico over the shared resources of the Colorado River and Gulf of Mexico, and with Canada over the Georges Bank fishing grounds in the Atlantic Ocean. One of the causes of the 1990–1991 war in the Persian Gulf was the huge oil reservoir known as the Rumaila field, lying mainly under Iraq with a small extension into Kuwait (**Figure 12.17a**). Because the two countries were unable to agree on percentages of ownership of the rich reserve, or a formula for

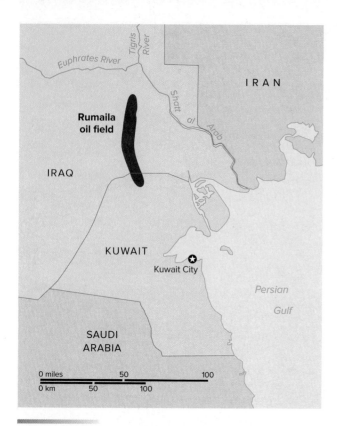

Figure 12.17 The Rumaila oil field. (*a*) One of the world's largest petroleum reservoirs, Rumaila straddles the Iraq-Kuwait border. Iraqi grievances over Kuwaiti drilling were partly responsible for Iraq's invasion of Kuwait in 1990. (*b*) To stem the flow of undocumented migrants entering California from Baja California, the United States in 1993 constructed a fence 3 meters (10 feet) high along the border. While American construction crews worked on the north side of the wall, protesters who opposed the wall painted graffiti on its south side. As of early 2009, approximately 1,080 kilometers (670 miles) of fence had been constructed in high priority sections along the approximately 3,220-kilometer (2,000-mile) U.S.-Mexico border. (See Box "Geography and Citizenship: Porous Borders" in Chapter 3)

sharing production costs and revenues, Kuwait pumped oil from Rumaila without any international agreement. Iraq helped justify its invasion of Kuwait by contending that the latter had been stealing Iraqi oil in what amounted to economic warfare.

Finally, **functional disputes** arise when neighboring states disagree over policies to be applied along a boundary. Such policies may concern immigration, the movement of traditionally nomadic groups, customs regulations, or land use. U.S. relations with Mexico, for example, have been affected by the movement of illegal immigrants and the flow of drugs entering the United States from Mexico (**Figure 12.17b**).

Centripetal Forces: Promoting State Cohesion

At any moment in time, a state is characterized by forces that promote unity and national stability and by others that disrupt them. Political geographers refer to the former as **centripetal forces** (from Latin for "seeking the center"). These are factors that bind together the people of a state, that enable it to function and give it strength. **Centrifugal forces,** on the other hand, destabilize and weaken a state. If centrifugal forces are stronger than those promoting unity, the very existence of the state will be threatened. In the sections that follow, we examine four centripetal (uniting) forces—nationalism, unifying institutions, effective organization and administration of government, and systems of transportation and communication—to see how they can promote cohesion.

Nationalism

One of the most powerful of the centripetal forces is **nationalism,** an identification with the state and the acceptance of national goals. Nationalism is based on the concept of allegiance to a single country and the ideals and the way of life it represents; it is an emotion that provides a sense of identity and loyalty and of collective distinction from all other peoples and lands.

States purposely try to instill feelings of allegiance in their citizens, for such feelings give the political system strength. People who have such allegiance are likely to accept common rules of action and behavior, and to participate in the decision-making process establishing those rules. In the extreme, citizens who nationalistically identify with their state might go as far as volunteering for the armed forces and engaging in war in order to defend their state. In light of the divisive forces present in most societies, not everyone, of course, will feel the same degree of commitment or loyalty. The important consideration is that the majority of a state's population accepts its ideologies, adheres to its laws, and participates in its effective operation. For many countries, such acceptance and adherence has come only recently and partially; in some, it is frail and endangered.

Recall that true nation-states are rare; in only a few countries do the territory occupied by the people of a particular nation and the territorial limits of the state coincide. Most countries have more than one culture group that considers itself separate in some important way from other citizens. In a multicultural society, nationalism helps integrate different groups into a unified population. This kind of consensus nationalism has emerged in countries such as the United States, where different culture groups have joined together to create political entities commanding the loyalties of all their citizens.

States promote nationalism in a number of ways. *Iconography* is the study of the symbols that help unite people. National anthems and other patriotic songs; flags, national sports teams, and officially designated or easily identified flowers and animals; and rituals and holidays are all developed by states to promote nationalism and attract allegiance (**Figure 12.18**). By ensuring that all citizens, no matter how diverse the population may be, will have at least these symbols in common, they impart a sense of belonging to a political entity called, for example, Japan or Canada. In some countries, certain documents, such as the Magna Carta in the United Kingdom or the Declaration of Independence in the United States, serve the same purpose. Royalty may fill the need; in Sweden, Japan, and the United Kingdom, the monarchy functions as the symbolic focus of allegiance. Such symbols are significant, for symbols and beliefs are

Figure 12.18 The ritual of the pledge of allegiance is just one way in which schools in the United States seek to instill a sense of national identity in students.

©McGraw-Hill Education/Jill Braaten, photographer.

major components of the ideological subsystem of every culture (see Section 2.6). When a society is very heterogeneous, composed of people with different customs, religions, and languages, belief in the national unit can help weld them together.

Unifying Institutions

Institutions as well as symbols help to develop the sense of commitment and cohesiveness essential to the state. Schools, particularly elementary schools, are among the most important of these. Children learn the history of their own country and relatively little about other countries. Schools are expected to instill the society's goals, values, and traditions, to teach the common language that conveys them, and to guide youngsters to identify with their country.

Other institutions that advance nationalism are the armed forces and, sometimes, a state church. The armed forces are of necessity taught to identify with the state. They see themselves as protecting the state's welfare from what are perceived to be its enemies. In about one-quarter of the world's countries, the faith of the majority of the people has by law been designated the state religion. In such cases, the religion may become a force for cohesion, helping unify the population. This is true of Buddhism in Thailand, Hinduism in Nepal, Islam in Pakistan, and Judaism in Israel. In countries like these, the religion and the church are so identified with the state that belief in one is transferred to allegiance to the other.

The schools, the armed forces, and the church are just three of the institutions that teach people what it is like to be members of a state. As institutions, they operate primarily on the level of the sociological subsystem of culture, helping to structure the outlooks and behaviors of the society. But by themselves, they

are not enough to give cohesion, and thus strength, to a state. Indeed, each of the institutions we have discussed can also be a destabilizing centrifugal force.

Organization and Administration

A further bonding force is public confidence in the effective organization of the state. Can it provide security from external aggression and internal conflict? Are its resources distributed and allocated in such a way as to be perceived to promote the economic welfare of all its citizens? Are all citizens afforded equal opportunity to participate in governmental affairs (see the feature "Legislative Women")? Do institutions that encourage consultation and the peaceful settlement of disputes exist? How firmly established are the rule of law and the power of the courts? Is the system of decision making responsive to the people's needs?

The answers to such questions, and the relative importance of the answers, will vary from country to country. But these and similar questions are implicit in the expectation that the state will, in the words of the Constitution of the United States, "establish justice, insure domestic tranquility, provide for the common defence, [and] promote the general welfare. . . ." If those expectations are not fulfilled, the loyalties promoted by national symbols and unifying institutions may be weakened or lost.

Transportation and Communication

A state's transportation network fosters political integration by promoting interaction among areas and by joining them economically and socially. The role of a transportation network in uniting a country has been recognized since ancient times. The saying that all roads lead to Rome had its origin in the impressive system of roads that linked Rome to the rest of its empire. Centuries later, a similar network was built in France, joining Paris to the various departments of the country. Often the capital city is better connected to other cities than the outlying cities are to one another. In France, for example, it can take less time to travel from one city to another by way of Paris than by direct route.

Roads and railroads have played a historically significant role in promoting political integration. In the United States and Canada, they not only opened up new areas for settlement but increased interaction between rural and urban districts. Because transportation systems play a major role in a state's economic development, it follows that the more economically advanced a country is, the more extensive its transport network is likely to be (see Figure 8.4 in Chapter 8). At the same time, the higher the level of development, the more resources there are to be invested in building transport routes. The two reinforce each other.

Transportation and communication, while encouraged within a state, are frequently curtailed or at least controlled between states as a conscious device for promoting state cohesion through limiting external spatial interaction (**Figure 12.19**). The mechanisms of control include restrictions on trade through tariffs or embargoes, legal barriers to immigration and emigration, and limitations on travel through passports and visa requirements.

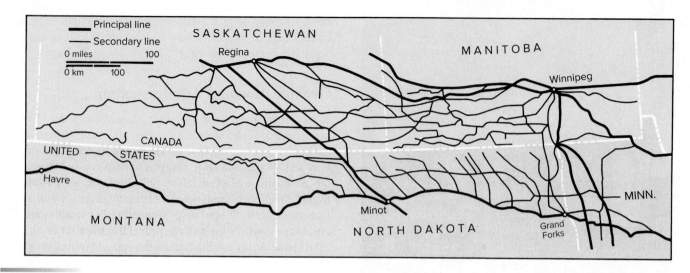

Figure 12.19 Canadian-U.S. railroad discontinuity. Canada and the United States developed independent railway systems connecting their respective prairie regions with their separate national cores. Despite extensive rail construction during the 19th and early 20th centuries, the pattern that emerged even before recent track abandonment was one of discontinuity at the border. Note how the political boundary restricted the ease of spatial interaction between adjacent territories. Many branch lines approached the border, but only eight crossed it. In fact, for more than 480 kilometers (300 miles), no railway bridged the boundary line. The international border—and the cultural separation it represents—inhibits other expected degrees of interaction. Telephone calls between Canadian and U.S. cities, for example, are far less frequent than would be expected if distance alone were the controlling factor.

Centrifugal Forces: Challenges to State Authority

State cohesion is not easily achieved or, once gained, invariably retained. Destabilizing *centrifugal forces* are ever-present, sowing internal discord and challenges to the state's authority (see the feature "Terrorism and Political Geography"). Transportation and communication may be hindered by a country's shape or great size, leaving some parts of the state not well integrated with the rest. A country that is not well organized or administered stands to lose the loyalty of its citizens. Institutions that in some states promote unity can be a divisive force in others.

Organized religion, for example, can be a potent centrifugal force. It may compete with the state for people's allegiance—one reason that the former USSR and other communist governments suppressed religion and promoted atheism. Conflict between majority and minority faiths within a country—as between Catholics and Protestants in Northern Ireland or Hindus and Muslims in Kashmir and Gujarat state in India—can destabilize social order. Opposing sectarian views within a single, dominant faith can also promote civil conflict. Recent years have seen particularly Muslim militant groups attempt to overturn official or constitutional policies of secularism or replace a government deemed insufficiently ardent in its imposition of religious laws and regulations. Islamic fundamentalism led to the 1979 overthrow of the shah of Iran; more recently, Islamic militancy has been a destabilizing force in, among other countries, Afghanistan, Algeria, Iraq, Tunisia, Egypt, Syria, and Saudi Arabia.

Nationalism, in contrast to its role as a powerful centripetal agency, is also a potentially disruptive centrifugal force. The idea of the nation-state (see Section 12.1) is that states are formed around and coincide with nations. It is a small step from that to the notion that every nation has the right to its own state or territory. Centrifugal forces are particularly strong in countries containing multiple nationalities and unassimilated minorities, racial or ethnic conflict, contrasting cultures, and a multiplicity of languages or religions. Such states are susceptible to nationalist challenges from within their borders. A country whose population is not bound by a shared sense of nationalism is split by several local primary allegiances and suffers from **subnationalism.** That is, many people give their primary allegiance to traditional groups or nations that are smaller than the population of the entire state.

Any country that contains one or more important national minorities is susceptible to challenges from within its borders if the minority group has an explicit territorial identification and believes that its right to *self-determination*—the right of a group to govern itself in its own state or territory—has not been satisfied. A dissident minority that has total or partial secession from the state as its primary goal is said to be guided by **separatism** or **autonomous nationalism.** In recent years, such nationalism has created currents of unrest within many countries, even long-established ones. In some countries, the central government attempts to deal with separatism by ceding a measure of additional self-governance to so-called *autonomous regions.*

Canada, for example, houses a powerful secessionist movement in French-speaking Quebec, the country's largest province. In October 1995, a referendum to secede from Canada and become a sovereign country failed in Quebec by a razor-thin margin. Quebec's nationalism is fueled by strong feelings of collective identity and distinctiveness, by a desire to protect its language and culture, and by the conviction that the province's ample resources and advanced economy would permit it to manage successfully as a separate country.

Legislative Women

Women, a majority of the world's population, fare poorly in general in the allocation of such resources as primary and higher education, employment opportunities and income, and health care. That their lot is improving is encouraging. In nearly every developing country, women have been closing the gender gap in literacy, school enrollment, and acceptance in the job market.

But in the political arena—where power ultimately lies—women's share of influence is increasing only slowly and selectively. In 2017, only 15 countries out of a world total of about 200 had women as heads of government: presidents or prime ministers. Nor did they fare much better as members of parliaments. Women in 2017 held just 24 percent of all the seats in the world's legislatures.

Only in about 50 countries did women in 2017 occupy one-quarter or more of the seats in the lower or single legislative house. Of these, Rwanda and Bolivia were the leaders, with 61 percent and 53 percent of their members female, respectively. A number of countries had no female representatives at all. In contrast, in nearly 40 percent of countries, women make up less than 10 percent of the legislature.

Although in the Parliament of the European Union, women comprised 37 percent of members in 2017, they held only 18 percent of the seats in the Greek parliament and 22 percent of those in Ireland,

Japan made an even poorer showing, with a 9 percent female membership. Nor did the United States show a very significant number of women members. At the start of the 115th Congress (2017–2019), only 21 women served in the Senate (21 percent female) and 83 in the House of Representatives (19 percent female). Those aren't very impressive numbers, but at this time, both numbers are actually at their highest-ever levels. American women have made slightly greater gains in state legislatures than at the national level in recent years. At the state level, women's legislative membership has increased from 4 percent in 1969 to 25 percent in 2017, although wide disparities exist among the states.

In the later 1990s, women's legislative representation began to expand materially in many developed and developing democracies, and their "fair share" of political power began to be formally recognized or enforced. In Western countries, particularly, improvement in female parliamentary participation has become a matter of plan and pride for political parties and, occasionally, for governments themselves.

Political parties from Mexico to China have tried to correct female underrepresentation, usually by setting quotas for women candidates, and a few governments—including Belgium and Italy—have tried to require their political parties to improve their balance. France went further than any other

country in acknowledging the right of women to equal access to elective office when it passed a constitutional amendment in 1999 requiring *parité*—parity, or equality. A year later, the National Assembly enacted legislation requiring the country's political parties to fill 50 percent of the candidacies in all elections in the country (municipal, regional, and European Parliament) with women, or lose a corresponding share of their state-provided campaign funding. India similarly proposed to reserve a third of the seats in parliament for women.

Quotas are controversial, however, and often are viewed with disfavor even by avowed feminists. Some argue that quotas are demeaning because they imply women cannot match men on merit alone. Others fear that other groups—for example, religious groups or ethnic minorities—also would seek quotas to guarantee their proportionate legislative presence.

A significant presence of women in legislative bodies makes a difference in the kinds of bills that get passed and the kinds of programs that receive governmental emphasis. Regardless of party affiliation, women are more apt than their male counterparts to sponsor bills and vote for measures affecting child care, elderly care, women's health care, medical insurance, and bills affecting women's rights and family law.

In Western Europe, five countries (the United Kingdom, France, Belgium, Italy, and Spain) house separatist political movements whose members reject total control by the existing sovereign state and who claim to be the core of a separate national entity (**Figure 12.20**). Their basic demand is for *regional autonomy,* usually in the form of self-government or "home rule" rather than complete independence. Accommodation of those demands has resulted in some degrees of **devolution**—the transfer of some central powers to regional or local governments—and in the forms of asymmetric federalism discussed earlier with the United Kingdom and Spain as examples.

Separatist movements affect many states outside Western Europe. Many countries containing disparate groups that are more motivated by enmity than affinity have powerful centrifugal tendencies. The Basques of Spain and the Bretons of France have their counterparts in the Palestinians in Israel, the Sikhs

in India, the Moros in the Philippines, the Tamils in Sri Lanka, and many other groups. Separatist movements are expressions of **regionalism,** minority group self-awareness and identification with a region rather than with the state.

The countries of Eastern Europe and the republics of the former Soviet Union have seen many instances of regionally rooted nationalist feelings. Now that the forces of ethnicity, religion, language, and culture are no longer suppressed by communism, ancient rivalries are more evident than at any time since World War II. The end of the Cold War aroused hopes of decades of peace. Instead, the collapse of communism and the demise of the USSR spawned dozens of smaller wars. Numerous ethnic groups large and small are asserting their identities and what they perceive to be their right to determine their own political status.

The national independence claimed in the early 1990s by the 15 former Soviet constituent republics did not ensure the

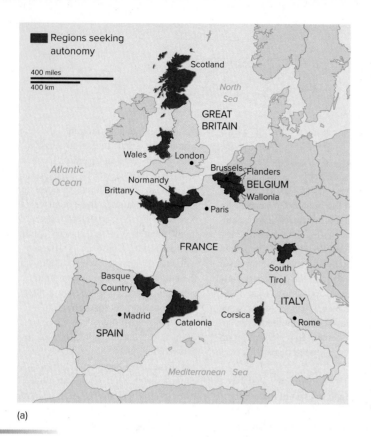

(a)

(b)

Figure 12.20 Regions in Western Europe seeking autonomy. Despite longstanding state attempts to assimilate these historic nations culturally, each contains a political movement that has sought or is seeking a degree of self-rule recognizing its separate identity. Separatists on the island of Corsica, for example, want to secede from France, and separatists in Catalonia demand independence from Spain. The desires of nationalist parties in both Wales and Scotland were partially accommodated by the creation in 1999 of their own parliaments and a degree of regional autonomy, an outcome labeled "separation but not divorce" from the United Kingdom. (b) Demonstrators carry a giant Basque flag during a march to call for independence for the Basque region.

satisfaction of all separatist movements within them. Many of the new individual countries are themselves subject to strong destabilizing forces that threaten their territorial integrity and survival. The Russian Federation itself, the largest and most powerful remnant of the former USSR, has 89 components, including 21 *ethnic republics* and a number of other nationality regions. Many are rich in natural resources, have non-Russian majorities, and seek greater autonomy within the federation. Some, indeed, want total independence. One, the predominantly Muslim republic of Chechnya, in 1994 claimed the right of self-determination and attempted to secede from the federation, provoking a bloody civil war that escalated again in 1996 and 1999; the Chechyan insurgency continued for another decade.

As the USSR declined and disbanded, it lost control of its communist satellites in Eastern Europe. That loss and resurgent nationalism led to a dramatic reordering of the region's political map. East Germany was reunited with West Germany in 1990, and three years later, the people of Czechoslovakia agreed to split their country into two separate, ethnically based states: the Czech Republic and Slovakia. More violently, Yugoslavia shattered into five pieces in 1991–1992, but with the exception of Slovenia, the boundaries of the five new republics did not match the territories occupied by nationalities, a situation that plunged the region into war as nations fought to redefine the boundaries

of their countries. One tactic used to transform a multinational area into one containing only one nation is **ethnic cleansing,** the killing or forcible relocation of less powerful minorities. It occurred in Croatia, Bosnia-Herzegovina, and the Kosovo province of southern Serbia.

More peacefully, several European governments have recently moved in the direction of regional recognition and devolution. In France, 22 regional governments were established in 1986; Spain has a program of devolution for its 17 *autonomous communities*, a program that Portugal is beginning to emulate. Italy, Germany, and the Nordic countries have, or are developing, similar recognitions of regional communities with granted powers of local administration and relaxation of central controls.

The two preconditions common to all separatist movements are *territory* and *nationality*. First, the group must be concentrated in a core region that it claims as a national homeland, seeking to regain control of land and power that it believes were unjustly taken from it. Second, certain cultural characteristics must provide a basis for the group's perception of separateness, identity, and unity. These might be language, religion, or distinctive group customs and institutions that promote feelings of group identity at the same time that they foster exclusivity. Normally, these cultural differences have persisted over several generations and have survived despite strong pressures toward assimilation.

Terrorism and Political Geography

"Where were you when the world stopped turning?" asks Alan Jackson in his song about the September 11, 2001, terrorist attacks on the United States. Your parents probably know the answer to his question, and they probably always will. Of course, the world didn't really stop turning, but that's how it felt to millions of Americans with no previous exposure to terrorism.

What is terrorism? How does it relate to political geography? Do all countries experience terrorism? Is terrorism new? Is there a way to prevent it? Attempting to answer these questions, difficult as they are, may help us understand the phenomenon.

Terrorism is the calculated use of violent acts against civilians and symbolic targets to publicize a cause, intimidate, or coerce a civilian population, or affect the conduct of a government. *International terrorism,* such as the September 11 attacks, include acts that transcend national boundaries. International terrorism is intended to intimidate people in other countries. *Domestic terrorism* consists of acts by individuals or groups against the citizens or government of their own country. *State terrorism* is committed by the agents of a government. *Subnational terrorism* is committed by nongovernmental groups. Whatever its agency or level, terrorism is a weapon designed to intimidate populations and, often, to influence government actions or policies.

State terrorism is probably as old as the concept of a state itself. As early as 146 BCE, for example, Roman forces sacked and completely destroyed the city of Carthage, burning it to the ground, slaughtering its population, and sowing salt on the fields so that no crops could grow (although many historians believe the latter is an historical "rumor" that did not actually take place). Governments have used systematic policies of violence and intimidation to dominate and control their own populations further. Nazi Germany, the Pol Pot regime in Cambodia, and Stalinist Russia are 20th-century examples of state terrorism. Heads of state ordered the murder, imprisonment, or exile of enemies of the state—politicians, intellectuals,

dissidents, or anyone who dared to criticize the government. In Rwanda, the former Yugoslavia, and Saddam Hussein's Iraq, state terrorism aimed against ethnic and religious minorities provided the government with a method of consolidating power; in each case, genocide or mass murder of ethnic minority groups was the result.

Subnational terrorism is of more recent vintage, coinciding with the rise of the nation-state. Subnational terrorism can be perpetrated by those who feel wronged by their own or another government. For example, ethnic groups in a minority who feel that the central government has taken their territory and absorbed them into a larger political entity, such as the Basques in Spain, have used terrorist activities to promote their cause and resist the government. Ethnic and religious groups that have been split by national boundaries imposed by others, such as Palestinian Arabs in the Middle East, have used terrorism to make governance impossible. Political, ethnic, or religious groups or individuals who feel oppressed by their own government, such as the Oklahoma City bombers in the United States, have committed acts of domestic terrorism.

Nearly every country has experienced some form of terrorism at some point since the mid-19th century. These acts have been as various as the anarchist assassinations of political leaders in Europe during the 1840s and in the United States in the late 19th century, the abduction of Canadian government officials by the Front Libération du Québec (FLQ) in 1970, and the release of sarin gas in the Tokyo subways in 1995 by the group Aum Shinrikyo.

The political and religious aims of these attackers, however, can cause confusion on the world stage. In 2001, the Reuters news agency told its reporters to stop using the word *terrorism* because "one person's terrorist is another's freedom fighter." The definition of *terrorism* rests on the ability to define motives and the outlook of the observer.

Although it may be difficult to distinguish among types of terrorism, it is even more difficult to prevent it. Generally speaking,

governments and international bodies respond to terrorist acts in one of four ways:

- Reducing or addressing the causes of terrorism. In some cases, political change can reduce a terrorist threat. For example, the 1998 Good Friday Agreement in Northern Ireland led to a reduction of terrorist acts; the Spanish government's granting of some regional autonomy to the Basques helped lessen the violent actions of the ETA (Basque separatists) and reduced the support of many Basque people for such acts.

- Increasing international cooperation in the surveillance of subnational groups. Spurred by terrorist crimes in Bahrain and Saudi Arabia, the Arab States of the Persian Gulf agreed in 1998 to exchange intelligence regarding terrorist groups, to share information regarding anticipated terrorist acts, and to assist one another in investigating terrorist crimes.

- Increasing security measures in a country. In the United States, following September 11, 2001, the government created a Department of Homeland Security, federalized air traffic screening, and increased efforts to reduce financial support for foreign terrorist organizations. In concert, the European Union froze the assets of any group on its list of terrorist organizations.

- Using military means, either unilaterally or multilaterally, against terrorists or governments that sponsor terrorists. Following the World Trade Center and Pentagon assaults on September 11, the United States led a coalition of countries in attacking the government of Afghanistan, which had harbored Osama bin Laden's al-Qaeda terrorist organization.

Each response to terrorism is expensive, politically difficult, and potentially harmful to the life and liberty of citizens. Governments must decide which response or combination of responses is likely to have the most beneficial effect.

Other characteristics common to many separatist movements are a *peripheral location* and *social* and *economic inequality*. Troubled regions tend to be peripheral, often isolated in rural pockets, and their location away from the seat of central government engenders feelings of alienation and exclusion. Further, the dominant culture group is often seen as an exploiting class that has suppressed the local language, controlled access to the civil service, and taken more than its share of wealth and power. Poorer regions complain that they have lower incomes and greater unemployment than prevail in the rest of the state and that "outsiders" control key resources and industry. Separatists in relatively rich regions believe that they could exploit their resources for themselves and do better economically without the constraints imposed by the central state.

12.2 Cooperation Among States

The modern state is fragile and, as we have seen, its primacy may be less assured in recent years. In many ways, countries are now weaker than ever before. Many are economically frail, others are politically unstable, and some are both. Strategically, no country is safe from military attack, for technology now enables countries to shoot weapons halfway around the world. Some people believe that no national security is possible in the nuclear age.

Recognizing that a country cannot by itself guarantee either its prosperity or its own security, many states have opted to cooperate with others. These cooperative ventures are proliferating quickly, and they involve countries everywhere. They are also adding a new dimension to the concept of political boundaries because the associations of states themselves have limits that are marked by borders of a higher spatial order than those among individual states. Such boundaries as the current division between the North Atlantic Treaty Organization (NATO) and non-NATO states, or between the EU area and other European countries represent a different scale of the political ordering of space.

Supranationalism

Associations among states represent a new dimension in the ordering of national power and national independence. Recent trends in economic globalization and international cooperation suggest to some that the sovereign state's traditional responsibilities and authorities are being diluted by a combination of forces and partly delegated to higher-order political and economic organizations. Corporations, and even economic and communication agencies, often operate in controlling ways outside of nation-state jurisdiction.

The rise of transnational corporations dominant in global markets, for example, limits the economic influence of individual countries. Cyberspace and the Internet are controlled by no one and are largely immune to the state restrictions on the flow of information exerted by many governments. Those information flows help create and maintain the growing number of international NGOs, estimated at more than 20,000 in number and including such well-known groups as Greenpeace, Amnesty International, and Doctors Without Borders. NGOs,

through petitions, demonstrations, court actions, and educational efforts, have become effective influences on national and international political and economic actions. And increasingly, individual citizens of any country have their lives and actions shaped by decisions not only of local and national authorities, but by those of regional economic associations (e.g., the North American Free Trade Agreement/NAFTA), multinational military alliances (e.g., NATO), and global political agencies (the United Nations).

The roots of such multistate cooperative systems are ancient—for example, the leagues of city-states in the ancient Greek world or the Hanseatic League of free German cities in Europe's medieval period. The creation of new ones has been particularly active since 1945. They represent a world trend toward a **supranationalism** comprised of associations of three or more states created for mutual benefit and to achieve shared objectives. Although many individuals and organizations decry the loss of national independence that supranationalism entails, the many supranational associations in existence early in the 21st century are evidence of their attraction and pervasiveness. Almost all countries, in fact, are members of at least one—and most are members of many—supranational groupings. Virtually all are at least members of the United Nations (Kosovo and Vatican City are not members, although the United States considers them to be independent states).

The United Nations and Its Agencies

The United Nations is the only organization that tries to be universal, and even it is not all-inclusive. With its membership expanded from 51 countries in 1945 to 193 by 2011, when South Sudan was admitted, the United Nations is the most ambitious attempt ever undertaken to bring together the world's nations in international assembly and to promote world peace. Stronger and more representative than its predecessor, the League of Nations, it provides a forum where countries may discuss international problems and regional concerns and a mechanism, admittedly weak but still significant, for forestalling disputes or, when necessary, for ending wars (**Figure 12.21**). The United Nations also sponsors 40 programs and agencies aimed at fostering international cooperation with respect to specific goals. Among these are the World Health Organization (WHO), the Food and Agriculture Organization (FAO), and the United Nations Educational, Scientific, and Cultural Organization (UNESCO). Many other UN agencies and much of the UN budget are committed to assisting member states with matters of economic growth and development.

Member states have not surrendered sovereignty to the United Nations, and the world body is legally and effectively unable to make or enforce a world law. Nor is there a world police force. Although there is recognized international law adjudicated by the International Court of Justice, rulings by this body are sought only by countries agreeing beforehand to abide by its arbitration. The United Nations has no authority over the military forces of individual countries.

A pronounced change both in the relatively passive role of the United Nations and in traditional ideas of international

Figure 12.21 UN peacekeeping forces on duty in East Timor (Timor Leste). Under the auspices of the United Nations, soldiers from many different countries staff peacekeeping forces and military observer groups in many world regions in an effort to halt or mitigate conflicts. Demand for peacekeeping and observer operations is indicated by the recent deployment of UN forces in Bosnia, Croatia, Cyprus, Eritrea, Haiti, Kosovo, Lebanon, Pakistan/India, Sierra Leone, Somalia, and elsewhere.

©ANTONIO DASIPARU/Getty Images.

relations has begun to emerge. Long-established rules of total national sovereignty that allowed governments to act internally as they saw fit, free of outside interference, are fading as the United Nations increasingly applies a concept of *interventionism*. The Persian Gulf War of 1991 was UN-authorized under the old rules prohibiting one state (Iraq) from violating the sovereignty of another (Kuwait) by attacking it. After the war, the new interventionism sanctioned UN operations within Iraq to protect Kurds in that country. Later, the United Nations intervened with troops and relief agencies in Somalia, Bosnia, and elsewhere, invoking an "international jurisdiction over inalienable human rights" that prevails without regard to state frontiers or sovereignty considerations.

Whatever the long-term prospects for interventionism replacing absolute sovereignty, for the short term, the United Nations remains the only institution where the vast majority of the world's countries can collectively discuss matters of international political and economic concerns and attempt peacefully to resolve their differences. It has been particularly influential in formulating a law of the sea.

Maritime Boundaries

Boundaries define political jurisdictions and areas of resource control. But claims of national authority are not restricted to land areas alone. Water covers more than two-thirds of the Earth's surface, and increasingly, countries have been projecting their sovereignty seaward to claim adjacent maritime areas and resources. A basic question involves the right of states to control water and the resources that it contains. The inland waters of a country, such as rivers and lakes, have traditionally been considered within the sovereignty of that country. Oceans, however, are not within any country's borders. Are they, then, to be open to all states to use, or may a single country claim sovereignty and limit access and use by other states?

For most of human history, the oceans remained effectively outside individual national control or international jurisdiction. The seas were a common highway for those daring enough to venture on them, an inexhaustible larder for fishermen, and a vast refuse pit for the muck of civilization. By the end of the 19th century, however, most coastal countries claimed sovereignty over a continuous belt 3 or 4 nautical miles wide (a *nautical mile* equals 1.15 statute miles, or 1.85 kilometers). At the time, the 3-nautical-mile limit represented the farthest range of artillery and thus the effective limit of control by the coastal state. Though recognizing the rights of others to innocent passage, such sovereignty permitted the enforcement of quarantine and customs regulations, allowed national protection of coastal fisheries, and made claims of neutrality effective during other people's wars. The primary concern was with security and unrestricted commerce. No separately codified laws of the sea existed, however, and none seemed to be needed until after World War I.

A League of Nations Conference for the Codification of International Law, convened in 1930, inconclusively discussed maritime legal matters and served to identify areas of concern that were to become increasingly pressing after World War II. Important among these was an emerging shift from interest in commerce and national security to a preoccupation with the resources of the seas, an interest fanned by the Truman Proclamation of 1945. Motivated by a desire to exploit offshore oil deposits, the federal government under this doctrine laid claim to all resources on the continental shelf contiguous to its coasts. Other states, many claiming even broader areas of control, hurried to annex their own adjacent marine resources. Within a few years, a quarter of the Earth's surface was appropriated by individual coastal countries.

An International Law of the Sea

Unrestricted extensions of jurisdiction and disputes over conflicting claims to maritime space and resources led to a series of UN conferences on the Law of the Sea. Meeting over a period of years, delegates from more than 150 countries attempted to achieve consensus on a treaty that would establish an internationally agreed-upon "convention dealing with all matters relating to the Law of the Sea." The meetings culminated in a draft treaty in 1982 called the **United Nations Convention on the Law of the Sea (UNCLOS)**.

The convention delimits territorial boundaries and rights by defining four zones of diminishing control (**Figure 12.22**):

- A *territorial sea* of up to 12 nautical miles (19 kilometers) in breadth over which coastal states have sovereignty, including exclusive fishing rights. Vessels of all types normally have the right of innocent passage through the territorial sea, though under certain circumstances, noncommercial (primarily military and research) vessels can be challenged.

- A *contiguous zone* to 24 nautical miles (38 kilometers). Although a coastal state does not have complete sovereignty in this zone, it can enforce its customs, immigration, and sanitation laws and has the right of hot pursuit out of its territorial waters.

- An **exclusive economic zone (EEZ)** of up to 200 nautical miles (370 kilometers) in which the state has recognized rights to explore, exploit, conserve, and manage the natural resources, both living and nonliving, of the seabed and waters (see **Figure 12.23** and the feature "Specks and Spoils"). Countries have exclusive rights to the resources lying within the continental shelf when this extends farther, up to 350 nautical miles (560 kilometers), beyond their coasts. The traditional freedoms of the high seas are to be maintained in this zone.

- The *high seas* beyond the EEZ. Outside any national jurisdiction, they are open to all states, whether coastal or landlocked. Freedom of the high seas includes the right to sail ships, fish, fly over, lay submarine cables and pipelines, and pursue scientific research. Mineral resources in the international deep seabed area beyond national jurisdiction are declared the common heritage of humankind, to be managed for the benefit of all the peoples of the Earth.

By the end of the 1980s, most coastal countries, including the United States, had used the UNCLOS provisions to proclaim and reciprocally recognize jurisdiction over 12-nautical-mile territorial seas and 200-nautical-mile economic zones. Despite reservations held by the United States and a few other industrial countries about the deep seabed mining provisions, the convention received the necessary ratification by 60 states and became international law in 1994.

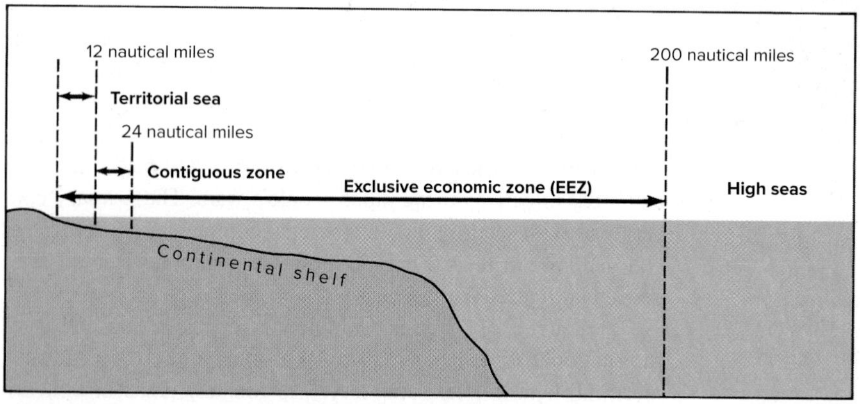

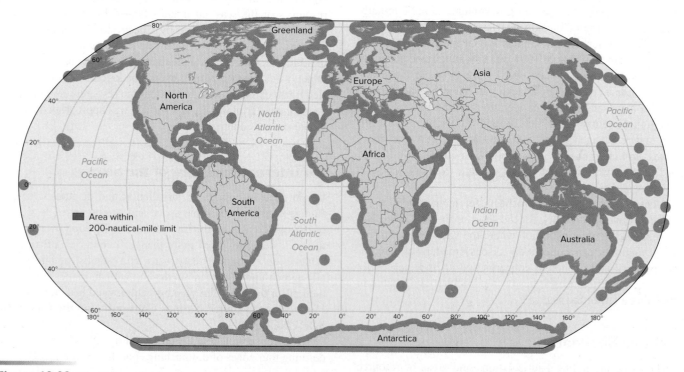

Figure 12.22 Territorial claims permitted by UNCLOS.

Figure 12.23 The 200-nautical-mile EEZ claims of coastal states. The provisions of the UNCLOS have in effect changed the maritime map of the world. Three important consequences flow from the 200-nautical-mile EEZ concept: (1) islands have gained a new significance, (2) countries have a host of new neighbors, and (3) the EEZ lines result in overlapping claims. EEZ lines are drawn around a country's possessions, as well as around the country itself. Every island, no matter how small, has its own 200-nautical-mile EEZ. This means that although the United States shares continental borders only with Canada and Mexico, it has maritime boundaries with countries in Asia, South America, and Europe. All told, the United States may have to negotiate some 30 maritime boundaries, which is likely to take decades. Other countries, particularly those with many possessions, will have to engage in similar lengthy negotiations.

Specks and Spoils

The UNCLOS gives to owners of islands claims over immense areas of the surrounding sea and, of course, to the fisheries and mineral resources in and under them. Tiny specks of land formerly too insignificant in size or distant in location to arouse the emotions of any nation now are avidly sought and fervently claimed. Remote Rockall, a British islet far west of Scotland, was used by Britain in 1976 to justify extending its fishing rights claim farther into the North Atlantic than otherwise was possible. Argentina nearly went to war with Chile in 1978 over three islands at the tip of South America at the Atlantic end of the Beagle Channel. Chile had lodged its claim of ownership hoping to gain access to known South Atlantic fish resources and hoped-for petroleum deposits. In 1982, Argentina seized the Falkland Islands from Britain, ostensibly to reclaim the Islas Malvinas (their Spanish name) as national territory, but with an underlying economic motive as well. British forces retook the islands and subsequently used sovereignty over them to claim a sea area three times as large as Britain. Japan has encased a disappearing islet in concrete to maintain territorial claims endangered through erosion of the speck of land supporting them.

The Paracel and Spratly Islands, straddling trade routes in the South China Sea, have attracted more attention and claimants than most island groups, thanks to presumed large reserves of oil and gas in their vicinities. The Japanese seized the Paracels from China during World War II and at the end of that war surrendered them to Nationalist Chinese forces that soon retreated to Taiwan. South Vietnam took them over until 1974, when they were forcibly ejected by the mainland Chinese. In 1979, a united Vietnam reasserted its claims, basing them on 17th- and 18th-century maps. China countered with reference to 3rd-century explorations by its geographers and maintained its control.

The location of the Paracels to the north, near China in the South China Sea, places them in a different and less difficult status than that of the Spratlys, whose nearest neighbors are the Philippines and Malaysia. Mere dots in the sea, the largest of the Spratlys is about 100 acres—no more than one-eighth the size of New York's Central Park. But under the UNCLOS, possession of the island group would confer rights to the resources (oil, it is hoped) that could be found beneath about 400,000 square kilometers (150,000 square miles) of sea. That lure has made rivals of six governments and posed the possibility of conflict. Until early in 1988, Vietnam, the Philippines, Malaysia, Taiwan, and tiny Brunei had all maintained in peaceful coexistence garrisons on separate islets in the Spratly group. Then China landed troops on islands near the Vietnamese holdings, sank two Vietnamese naval ships, and accused Vietnam of seizing "Chinese" territory on the pretext of searching for their missing sailors. Although China agreed in 1992 that ownership disputes in the Spratlys should be resolved without violence, it also passed a law the following year repeating its claims to all the islands and its determination to defend them. In early 1995, China occupied *Mischief Reef,* close to—and already claimed by—the Philippines, but in late 2002, it agreed with other Southeast Asian countries to avoid future disputes over the islands.

Assertions of past discovery, previous or present occupation, proximity, and simple wishful thinking have all served as the basis for the proliferating claims to seas and seabeds. The world's oceans, once open and freely accessible, are increasingly being closed by the lure of specks of land and the spoils of wealth that they command.

Figure 12B The Paracel Islands and Spratly Islands, contested island territories in the South China Sea.

UN Affiliates

Other fully or essentially global supranational organizations with influences on the economic, social, and cultural affairs of states and individuals have been created. Most are specialized international agencies, autonomous and with their own differing memberships but with affiliated relationships with the United Nations and operating under its auspices. Among them are the FAO, the International Bank for Reconstruction and Development (World Bank), the International Labor Organization (ILO), UNICEF, the WHO, and—of growing economic importance—the WTO.

The WTO, which came into existence at the start of 1995, has become one of the most significant of the global expressions of supranational economic control. It is charged with enforcing the global trade accords that grew out of years of international negotiations under the terms of the General Agreement on Tariffs and Trade (GATT). The basic principle behind the WTO is that the member countries (164 as of July 2016, with additional states preparing for membership or seeking admission) should work to cut tariffs, dismantle nontariff barriers to trade, liberalize trade in services, and treat all other countries uniformly in matters of trade. Any preference granted to one should be available to all.

Increasingly, however, regional rather than global trade agreements are being struck, and free trade areas are proliferating. Only a few WTO members are not already part of some other regional trade association. Such areal associations, some argue, make world trade less free by scrapping tariffs on trade among member states but retaining them separately or as a group on exchanges with nonmembers.

Regional Alliances

In addition to their membership in such international agencies, countries have shown themselves willing to relinquish some of their independence to participate in smaller multinational systems. These groupings can be economic, military, or political, and many have been formed since 1945. Cooperation in the economic sphere seems to come more easily to states than does military and political collaboration.

Economic Alliances

Among the oldest, most powerful, and far-reaching of the regional economic alliances are those that have evolved in Europe, particularly the European Union and its several forerunners. Shortly after the end of World War II, the Benelux countries (Belgium, the Netherlands, and Luxembourg) formed an economic union to create a common set of tariffs and to eliminate import licenses and quotas. Formed at about the same time were the Organization for European Cooperation (1948), which coordinated the distribution and use of Marshall Plan funds, and the European Coal and Steel Community (1952), which integrated the development of that industry in the member countries. A few years later, in 1957, the *European Economic Community (EEC),* or *Common Market,* was created, composed at first of only six states: France, Italy, West Germany, and the Benelux countries.

To counteract these Inner Six, as they were called, other countries in 1960 formed the European Free Trade Association (EFTA). Known as the Outer Seven, they were the United Kingdom, Norway, Denmark, Sweden, Switzerland, Austria, and Portugal (**Figure 12.24**). Between 1973 and 1986, three members (the United Kingdom, Denmark, and Portugal) left EFTA for membership in the Common Market and were replaced by Iceland and Finland.

The **European Union (EU)** grew out of the Common Market. It added new members slowly at first, as Greece, Spain, and Portugal joined during the 1980s; Austria, Finland, and Sweden joined in 1995. As it gathered momentum, more countries were admitted to the European Union during the early 2000s, including the island states of Malta and Cyprus and 10 former Soviet bloc nations from Estonia in the north to Bulgaria and Slovenia in the South. The most recent new addition to the EU was Croatia, who joined on July 1 of 2013 (**Figure 12.25**). These additions brought the number of member nations to 28, increased the total population of the European Union to some half-billion people, and expanded its economy to rival that of the United States. The

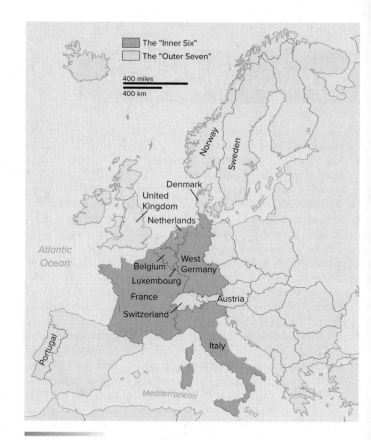

Figure 12.24 The original Inner Six and Outer Seven of Europe.

European Union is now the world's largest and richest bloc of nation-states.

Over the years, members of the European Union have taken many steps to integrate their economies and coordinate their policies in such areas as transportation, agriculture, and fisheries. A council of ministers, a commission, a European parliament, and a court of justice give the European Union supranational institutions with effective ability to make and enforce laws. By January 1, 1993, the European Union had abolished most remnant barriers to free trade and the free movement of capital and people among its members, creating a single European market. In another step toward economic and monetary union, the European Union's single currency, the *euro,* replaced separate national currencies in 1999.[1] And all applicant members in 2002 added 80,000 pages of EU law to their own legal systems.

We have traced this European development history, not because the full history of the EU is important to remember, but simply to illustrate the fluid process by which regional alliances are made. Countries come together in an association, some drop out, and others join. New treaties are made, and new coalitions emerge. Indeed, a number of such regional economic and trade

---

[1]In fact, no fewer than 10 EU members do not use the euro but retain their own currency. At the same time, six small European countries who are not EU members actually use the euro as their currency!

Figure 12.25 The 28 members of the European Union. The most recent admission was Croatia on July 1, 2013. The European Union earlier stipulates that in order to join, a country must have stable institutions guaranteeing democracy, the rule of law, human rights, and protection of minorities; a functioning market economy; and the ability to accept the obligations of membership, including the aims of political, economic, and monetary union. Although no country as of 2017 has left the EU or been suspended, the United Kingdom held a public referendum in June of 2016 in which nearly 52 percent of citizens voted to leave the EU. Prime Minister Theresa May started the formal withdrawal process in 2017, but it is understandably complex and not certain to come to fruition.

associations have been added to the world supranational map. None are as encompassing in power and purpose as the European Union, but all represent a concession of national independence to achieve broader regional goals.

NAFTA, launched in 1994, links Canada, Mexico, and the United States in an economic community aimed at lowering or removing trade and movement restrictions among the countries. It is perhaps the best known to North American students. The Americas as a whole, however, have other similar associations with comparable trade enhancement objectives, though frequently they—in common with other world regional alliances—have social, political, and cultural interests also in mind. The Caribbean Community and Common Market (CARICOM), for example, was established in 1974 to further cooperation among its members in economic, health, cultural, and foreign policy arenas. Mercosur, the Southern Cone Community Market, which unites Brazil, Argentina, Uruguay, and Paraguay in the proposed creation of a customs union to eliminate levies on goods moving among them, is a South American example.

A similar interest in promoting economic, social, and cultural cooperation and development among its members underpins the Association of Southeast Asian Nations (ASEAN), formed in 1967. A similar, but much less wealthy African example is the Economic Community of West African States (ECOWAS). The Asia Pacific Economic Cooperation (APEC) forum includes China, Japan, Australia, Canada, and the United States among its 18 members and has a grand plan for "free trade in the Pacific" by 2020. More restricted bilateral and regional preferential trade arrangements have also proliferated, numbering more than 450 by 2017, up from only 50 in 1990, with more under negotiation. They create a maze of rules, tariffs, and commodity agreements that result in trade restrictions and preferences contrary to the free trade intent of the WTO.

Three further points about regional international alliances are worth noting. The first is that the formation of a coalition in one area often stimulates the creation of another alliance by countries left out of the first. Thus, the union of the Inner Six gave rise to the treaty among the Outer Seven. Similarly, a counterpart of the Common Market was the Council of Mutual Economic Assistance (CMEA), also known as Comecon, which linked the former communist countries of Eastern Europe and the USSR through trade agreements.

Second, the new supranational unions tend to be composed of contiguous states (**Figure 12.26**). This was not the case with the recently dissolved empires, which included far-flung territories. Contiguity facilitates the movement of people and goods. Communication and transportation are simpler and more effective among adjoining countries than among those far removed from one another, and common cultural, linguistic, historical, and political traits and interests are more to be expected in spatially proximate countries.

Finally, it does not seem to matter whether countries are alike or distinctly different in their economies, as far as joining economic unions is concerned. There are examples of both. If the countries are dissimilar, they may complement one another. This was one basis for the European Common Market. Dairy products and furniture from Denmark are sold in France, freeing that country to specialize in the production of machinery and clothing. On the other hand, countries that produce the same raw materials hope that by joining together in an economic alliance, they might be able to enhance their control of markets and prices for their products. The Organization of Petroleum Exporting Countries (OPEC), mentioned in Chapter 8, is a case in point. Other attempts to form commodity cartels and price agreements between producing and consuming nations include the International Tin Agreement, the International Coffee Agreement, and others.

(a)

Economic Unions

☐ NAFTA
▨ Central American
  Common Market
■ CARICOM
■ CARICOM
  associates
▨ Andean Community of Nations
▨ Andean Community
  associates
▨ MERCOSUR
■ MERCOSUR
  associates
☐ CAFTA

(b)

Figure 12.26 (*a*) **NAFTA** is intended to unite Canada, the United States, and Mexico in a regional free trade zone. Under the terms of the treaty, tariffs on all agricultural products and thousands of other goods were to be eliminated by the end of 1999. In addition, all three countries are to ease restrictions on the movement of business executives and professionals. If fully implemented, the treaty will create one of the world's richest and largest trading blocs. (*b*) **Western Hemisphere economic unions** in 2009. In addition to these subregional alliances, President George H. W. Bush in 1990 proposed a "free trade area of the Americas" to stretch from Alaska to Cape Horn.

Military and Political Alliances

Countries form alliances for other than economic reasons. Strategic, political, and cultural considerations may also foster cooperation. The League of Arab States (now the Arab League), for example, was established in 1945 primarily to promote social, political, military, and foreign policy cooperation among its 22 members. In the Western Hemisphere, the Organization of American States (OAS), founded in 1948, concerns itself largely with social, cultural, human rights, and security matters affecting the hemisphere. A similar concern with peace and security underlay the Organization of African Unity (OAU), formed in 1963 by 32 African countries and, by 2017, expanded to 55 members and renamed the African Union.

Military alliances are based on the principle that unity assures strength. Such pacts usually provide for mutual assistance in the case of aggression. Once again, action breeds reaction when such an association is created. The formation of NATO, a defensive alliance of many European countries and the United States, was countered by the establishment of the Warsaw Treaty Organization, which joined the USSR and its satellite countries of Eastern Europe. Both pacts allowed the member states to base

armed forces in one another's territories, a relinquishment of a certain degree of sovereignty uncommon in the past.

Military alliances depend on the perceived common interests and political goodwill of the countries involved. As political realities change, so do the strategic alliances. NATO was created to defend Western Europe and North America against the Soviet military threat. When the dissolution of the USSR and the Warsaw Pact removed that threat, the purpose of the NATO alliance became less clear. Since the 1990s, however, the organization has added more than 10 members and has taken on a greater role in peacekeeping activities (**Figure 12.27**).

All international alliances recognize communities of interest. In economic and military associations, common objectives are clearly seen and described, and joint actions are agreed on with respect to the achievement of those objectives. More generalized mutual concerns or appeals to historical interest may be the basis for primarily *political alliances*. Such associations tend to be rather loose, not requiring their members to yield much power to the union. Examples are the Commonwealth of Nations (formerly the British Commonwealth), composed of many former British colonies and dominions, and the OAS, both of which offer economic as well as political benefits.

Figure 12.27 The NATO military alliance in 2017 had 29 members. Countries that have recently joined the alliance include Estonia, Latvia, Lithuania, Slovakia, Romania, Bulgaria, and Slovenia in 2004, Albania and Croatia in 2009, and Montenegro in 2017. Proponents of expansion argue that NATO is necessary in order to create a zone of stability and security throughout Europe. Opponents contend that enlargement is a divisive move that will cast a shadow over the future of relations with Russia, which is opposed to expansion so close to its borders.

There are many examples of abortive political unions that have foundered precisely because the individual countries could not agree on questions of policy and were unwilling to subordinate individual interests to make the union succeed. The United Arab Republic, the Central African Federation, the Federation of Malaysia and Singapore, and the Federation of the West Indies fall within this category.

Although many such political associations have failed, observers of the world scene speculate about the possibility that "superstates" will emerge from one or more of the economic or political alliances that now exist. For example, will a "United States of Europe," under a single government, be the logical

outcome of the successes of the European Union? No one knows, but so long as the individual state continues to be regarded as the highest form of political and social organization, and as the body in which sovereignty rests, such total unification is unlikely.

12.3 Local and Regional Political Organization

The most profound contrasts in cultures tend to occur between, rather than within, states, one reason political geographers traditionally have been primarily interested in country units. The

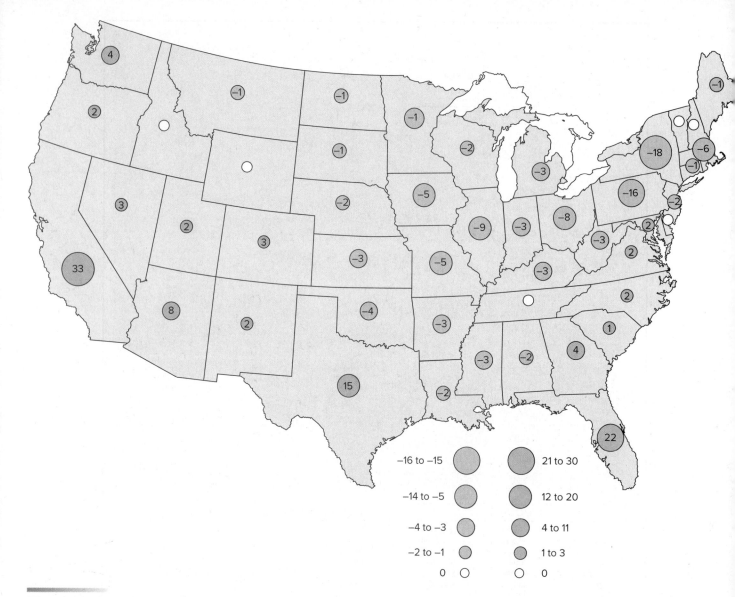

Figure 12.28 Geographic shifts in congressional apportionment between 1930 and 2010 illustrate dramatic population movements to the South and West. Seats in the U.S. House of Representatives are reapportioned after each census with the goal of achieving an equitable distribution based on population while maintaining at least one member for each state. Since 1930, New York has lost 18 seats and Pennsylvania 16, while California has gained 33, Florida gained 22, and Texas gained 15. After the 2010 census, 12 seats were shifted. Alaska and Hawaii are not shown because they were not states in 1930.

Source: Office of the Clerk, U.S. House of Representatives.

emphasis on the state, however, should not obscure the fact that it is at the local level that most of us find our most intimate and immediate contact with government and its influence on the administration of our affairs. In the United States, for example, an individual is subject to the decisions and regulations made by the school board, the municipality, the county, the state, and, perhaps, a host of special-purpose districts (water, fire, sewer, etc.)—all in addition to the laws and regulations issued by the federal government and its agencies. Among other things, local political entities determine where children go to school, the minimum-size lot on which a person can build a house, and where one may legally park a car. Adjacent states of the United States may be characterized by sharply differing personal and business tax rates; differing controls on the sale

of firearms, alcohol, and tobacco; variant administrative systems for public services; and different levels of expenditures for them (**Figure 12.28**).

All of these governmental entities are *spatial systems*. Because they operate within defined geographic areas and because they make behavior-governing decisions, they are topics of interest to political geographers. In the concluding sections of this chapter, we examine two aspects of political organization at the local and regional level. Our emphasis will be on the U.S. and Canadian scene simply because their local political geography is familiar to most of us. We should remember, however, North American structures of municipal governments, minor civil divisions, and special-purpose districts have counterparts in other regions of the world.

The Geography of Representation: The Districting Problem

There are more than 85,000 local governmental units in the United States. Slightly more than half of these administrative regions are municipalities, townships, and counties. The remainder are school districts, water-control districts, airport authorities, sanitary districts, and other special-purpose bodies. Around each of these districts, boundaries have been drawn. Although the number of districts does not change greatly from year to year, many boundary lines are redrawn in any single year. When the size or shape of a district is based on population numbers or distribution, such **redistricting** is made necessary by shifts in population, as areas gain or lose people. Alternatively, or in addition, the number of representatives allotted to each district may be modified, a process known as **reapportionment.**

For example, every 10 years following the U.S. Census, updated figures are used to reapportion the 435 seats in the House of Representatives among the 50 states. Redrawing the congressional districts to reflect population changes is required by the Constitution, the intention being to make sure that each legislator in the House represents roughly the same number of people. Since 1964, Canadian provinces and territories have entrusted redistricting for federal offices to independent electoral boundaries commissions. Although a few states in the United States also have independent, nonpartisan boards or commissions draw district boundaries, most rely on state legislatures for the task (with the consequent control of redistricting by whichever party is currently in power in that state). Across the United States, decennial census data are also used to redraw the boundaries of legislative districts within each state, as well as those for local offices, such as city councils and county boards.

Analyzing how the shape and location of voting district boundaries influences election outcomes is one aspect of **electoral geography,** which also addresses the spatial patterns yielded by election results and their relationship to the socioeconomic characteristics of voters. In a democracy, it might be assumed that electoral districts should contain roughly equal numbers of voters, that they should be reasonably compact, and that the final proportion of elected representatives of a particular political party or ideology should roughly correspond to the proportion of voters who vote for candidates of that party or ideology. But in fact, these proportions can be quite discrepant because the way in which district boundary lines are drawn can maximize, minimize, or effectively nullify the representational power of a group of voters.

Gerrymandering is the practice of drawing the boundaries of electoral districts so as to give particular candidates or classes of candidates an electoral advantage beyond the share of the electorate that supports them (**Figure 12.29**). This practice may advantage particular classes of candidates in a way that most people would find undemocratic; examples include *partisan, incumbent,* and *discriminatory racial* gerrymandering (see the feature "Gerrymandering in the United

The Gerry-Mander. (Boston, 1811.)

Figure 12.29 The original gerrymander. The term *gerrymander* originated in 1811 from the shape of an electoral district formed in Massachusetts while Elbridge Gerry was governor. When an artist added certain animal features, the district resembled a salamander and quickly came to be called a gerrymander.

States"). Alternatively, it may advantage classes of candidates in a way that attempts to right historical injustices and achieve greater fairness in the electoral system; the classic example of this is *affirmative racial* gerrymandering. Whether in the service of ends that are just or unjust, several characteristic strategies have been employed over the years to achieve gerrymandered districts. Gov. Elbridge Gerry (Figure 12.29) designed districts to concentrate his Federalist opponents in just a few districts, diluting their power outside those few districts, a technique known as *packing.* Electoral districts in Mississippi were once drawn intentionally to make sure black voters were never in the majority in any district, even though a large proportion of the voters in the state were black. Distributing and diluting classes of voters in this way is known as *cracking.*

Let's look at a concrete, if hypothetical, example (**Figure 12.30**). Assume that X and O represent two groups with an equal number of voters, but different policy or candidate preferences. Although there are equal numbers of Xs and Os, the way that electoral districts are drawn can dramatically affect voting results. In Figure 12.30a, the Xs are concentrated in one district but diluted in the other three, and will probably elect only one representative of the four. The power of the Xs

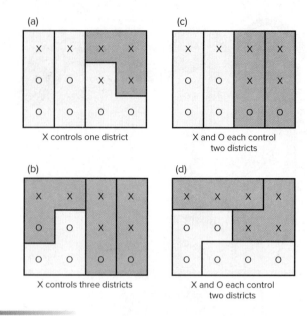

(a)

X controls one district

(b)

X controls three districts

(c)

X and O each control
two districts

(d)

X and O each control
two districts

Figure 12.30 Alternative districting strategies. *X*s and *O*s might represent Republicans and Democrats, urban and rural voters, blacks and whites, or any other distinctive groups.

is maximized in Figure 12.30b, where they are the majority in three of the four districts but completely absent from the fourth. The voters are evenly divided in Figure 12.30c, where the *X*s are the majority in two districts and the minority in the other two. Finally, the *X*s completely control two districts and the *O*s completely control the other two in Figure 12.30d. This pattern might result from both groups agreeing to design electoral districts so that incumbent candidates have "safe seats." Such incumbent gerrymandering offers little chance for electoral change.

Figure 12.30 depicts a hypothetical district, compact in shape with an even population distribution and only two groups competing for representation. In actuality, voting districts are often oddly shaped because of such factors as the city limits, current population distribution, and transportation routes—as well as past gerrymandering. Further, in any large area, many groups vie for power. Each electoral interest group promotes its version of fairness in the way boundaries are delimited. Minority interests, for example, seek representation in proportion to their numbers so that they will be able to elect representatives who are concerned about and responsive to their needs (see the feature "Gerrymandering in the United States").

In practice, gerrymandering is not always and automatically successful. First, a districting arrangement that appears to be unfair may be appealed to the courts. Further, voters are not unthinking party loyalists; key issues may cut across party lines, scandal may erode support for a candidate, or personal charm increase votes unexpectedly; and the amount of candidate financing or number of campaign workers may determine election outcomes if compelling issues are absent.

The Fragmentation of Political Power

Boundary drawing at any electoral level is never easy, particularly when political groups want to maximize their representation and minimize that of opposition groups. Furthermore, the boundaries that we may want for one set of districts may *not* be those that we want for another. For example, sewage districts must take natural drainage features into account, whereas police districts may be based on the distribution of the population or the number of kilometers of street to be patrolled, and school attendance zones must consider the numbers of school-aged children and the capacities of individual schools.

As these examples suggest, the United States is subdivided into great numbers of political administrative units whose areas of control are spatially limited. The 50 states are partitioned into more than 3,000 counties (*parishes* in Louisiana), most of which are further subdivided into townships, each with a still lower level of governing power. This political fragmentation is further increased by the existence of nearly 88,000 special-purpose districts whose boundaries rarely coincide with the standard major and minor civil divisions of the country or even with each other (**Figure 12.31**). Each district represents a form of political allocation of territory to achieve a specific aim of local need or legislative intent.

Canada, a federation of 10 provinces and 3 territories, has a similar pattern of political subdivision. Each of the provinces contains minor civil divisions—municipalities—under provincial control, and all (cities, towns, villages, and rural municipalities) are governed by elected councils. Ontario and Quebec also have counties that group smaller municipal units for certain purposes. In general, municipalities are responsible for police and fire protection, local jails, roads and hospitals, water supply and sanitation, and schools, duties that are discharged either by elected agencies or appointed commissions.

The existence of such a great number of districts in urban areas may cause inefficiency in public services and hinder the orderly use of space. *Zoning ordinances,* for example, control the uses to which land may be put. They are determined by each municipality and are a clear example of the effect of political decisions on the division and development of space. Unfortunately, in large urban areas, the efforts of one community may be hindered by the practices of neighboring communities. Thus, land zoned for an industrial park or shopping mall in one city may abut land zoned for single-family residences in an adjoining municipality. Each community pursues its own interests, which may not coincide with those of its neighbors or the larger region.

Inefficiency and duplication of effort characterize not just zoning, but many of the services provided by local governments. The efforts of one community to reduce or stop air and water pollution may be, and often are, counteracted by the rules and practices of other towns in the region, although state and national environmental protection standards are now reducing such potential conflicts. Social as well as physical problems spread beyond city boundaries. Thus, nearby suburban communities are

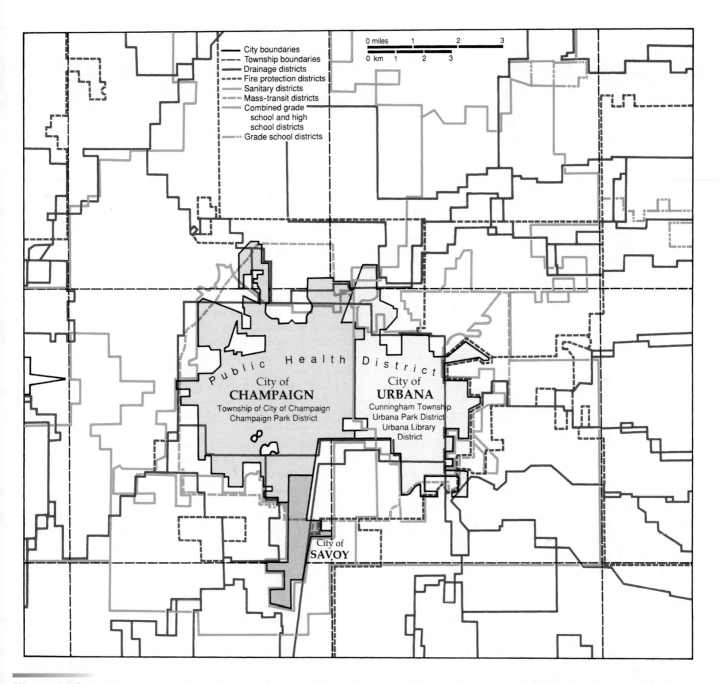

Figure 12.31 Political fragmentation in Champaign County, Illinois. The map shows a few of the independent administrative agencies with separate jurisdictions, responsibilities, and taxing powers in a portion of a single Illinois county. Among other such agencies forming the fragmented political landscape are Champaign County itself, a forest preserve district, a public health district, a mental health district, the county housing authority, and a community college district. If Champaign isn't complex enough for you, consider that nearby metropolitan Chicago has more than 1,500 units of government!

affected when a central city lacks the resources to maintain high-quality schools or to attack social ills. The provision of health care facilities, electricity and water, transportation, and recreational space affects the whole region and, many professionals think, should be under the control of a single consolidated metropolitan government.

The growth in the number and size of metropolitan areas has increased awareness of their administrative and jurisdictional problems. Too much governmental fragmentation and too little local control are both seen as metropolitan problems demanding attention and solution. The one concern is that multiple jurisdictions prevent the pooling of resources to address metropolitan-wide needs. The other is that local community needs and interests are subordinated to addressing the social and economic problems of a core city for which outlying communities feel little affinity or concern.

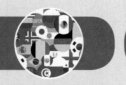

Gerrymandering in the United States

Gerrymandering has a long and checkered history in the United States, and it continues to be an issue of great relevance in American politics. For example, as we have discussed, there are histories of both discriminatory and affirmative racial gerrymandering in the United States. The first is designed to prevent racial/ethnic groups (especially African Americans) from gaining political power in legislative bodies like the U.S. Congress, while the second is intended to ensure racial/ethnic groups (especially African Americans, but also Hispanic Americans) get proportional representation in legislative bodies like the U.S. Congress. Irregularly shaped Congressional voting districts were created by several state legislatures after the 1990 census to make minority representation in Congress more closely resemble minority presence in the state's voting-age population. States that intentionally created majority-minority districts after the 1990 census included Florida, Georgia, Illinois, Louisiana, North Carolina, Texas, and Virginia. All represented a deliberate attempt to balance voting rights and race; all were specifically intended to comply with the federal Voting Rights Act of 1965, which provides that members of racial minorities shall not have "less opportunity than other members of the electorate . . . to elect representatives of their choice."

In North Carolina, for example, although 24 percent of the 1990 population of that state was black, past districting had divided black voters among a number of districts (cracking), with the result that not a single black congressional representative had been elected from that state in the 20th century. In 1991, the Justice Department ordered North Carolina to redistrict so that at least two districts would contain black majorities. Because of the way the black population was distributed, the only way to form black-majority districts was to string together cities, towns, and rural areas in very elongated, sinuous belts. The two newly created districts had slim (53 percent) black majorities.

The redistricting in North Carolina and other states had immediate effects. Black membership in the House of Representatives increased from 26 in 1990 to 39 in 1992; blacks constituted nearly 9 percent of the House, as opposed to 12 percent in the total population. But within a year, those electoral gains were threatened as lawsuits challenging the redistricting were filed in a number of states. The chief contention of the plaintiffs was that the irregular shapes of the districts violated compactness and amounted to reverse discrimination against whites.

Because at least some of the newly created districts had very contorted boundaries, as argued by opponents, they were ruled unconstitutional by the Supreme Court and had to be redrawn. The state legislatures' attempts at fairness and adherence to congressional mandate contained in the Voting Rights Act were held not to meet the need for reasonably compact shapes, a criterion not in the U.S. constitution but in many state constitutions (similarly, contiguity—the need to create districts that are continuously interconnected—is to be found in many state constitutions). *Compactness* is the property of a district's shape wherein its boundary is nearly the same distance from its center in all directions; the most compact two-dimensional shape possible is a circle. There are a variety of mathematical approaches to measuring compactness, such as calculating the area of a district divided by the length of its perimeter, a quantity which is generally larger in more compact districts. A very noncompact district, such as an octopus shape with several arms protruding out from the center body, has a very long perimeter for its modest area, so its area/perimeter has a small value.

In June 1993, a sharply divided Supreme Court ruled in *Shaw v. Reno* that North Carolina's 12th Congressional District might violate the constitutional rights of white voters and ordered a district court to review the case. The 5–4 ruling gave evidence that the country had not yet reached agreement on how to comply with the Voting Rights Act. It raised a central question: Should a state maximize the rights of racial minorities or not take racial status into consideration? A divided Court provided answers in 1995, 1996, and 1997 rulings that rejected congressional redistricting maps for Georgia, Texas, and North Carolina on the grounds that "race cannot be the predominant factor"

in drawing election district boundaries. It has become evident that the various criteria deemed desirable in creating electoral districts cannot all be satisfied, as they typically contradict one another to some extent. In *Easley v. Cromartie* (2001), the Court approved both a redrawn 12th District and the use of race as a redistricting consideration, so long as it was not the "dominant and controlling" one.

But just as the term *gerrymandering* was coined in reference to Elbridge Gerry's attempts to reduce the electoral power of his political opponents, probably the most significant issue circling around gerrymandering in the United States today revolves around attempts by one of the two major political parties (Democratic, Republican) to reduce access to legislative power by voters of the opposing party (and its largely opposing right- and left-leaning ideologies). Numerous states currently have, or recently had, legislative districts that apparently reflected partisan gerrymandering by the party in power. California's Congressional District 23 reflected gerrymandering by Democrats before an independent "citizen's commission" redrew it in 2011 (critics mockingly called it the "ribbon of shame"). Texas is widely seen to have electoral districts gerrymandered by Republicans; in 2012, for instance, Republicans won 60 of 99 Assembly seats while winning fewer than half of the votes statewide.

Currently, 37 states give the job of redistricting to their state legislatures, so that whichever party controls that legislature effectively controls redistricting; the two states that give the job to political commissions (made up of elected officials) similarly tend to give control to the party in power. In just four states, independent commissions handle the job, and the seven states with only one congressional district each do not carry out redistricting at this level. The exact makeup of independent commissions varies across states that use them, but in general, they do not allow elected officials or their employees or advocates to participate in the process. Likewise, redistricting for state legislative districts (whether upper- or lower-houses) is the responsibility of the legislatures in 37 states, politician commissions in 7 states, and independent commissions in 6 states. In

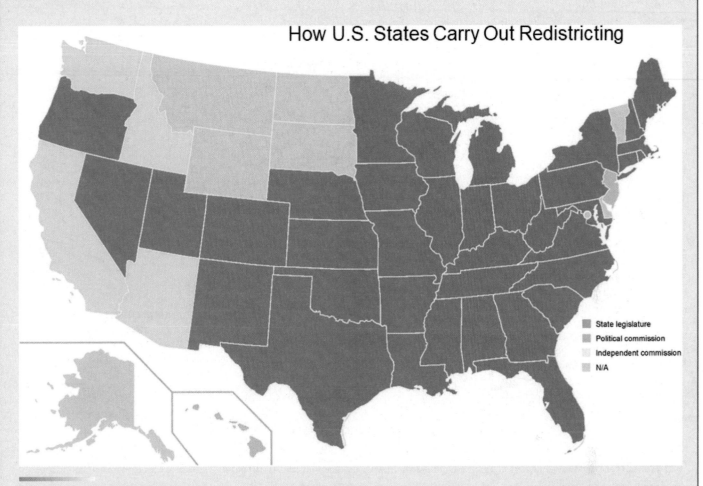

How U.S. States Carry Out Redistricting

State legislature
Political commission
Independent commission
N/A

Figure 12C U.S. states redistrict after each decennial Census; this is carried out either by their state legislature (effectively controlled by the party in power), a political commission (also partisan), or an independent commission (nonpartisan). Six states have only a single congressional district.

almost all states, the governor can veto the redistricting plan, which would require a new plan to be devised unless the legislature overrides the veto with their vote. Courts typically get involved when this process cannot produce a satisfactory districting map.

The U.S. Supreme Court has heard (and continues to hear) cases charging inappropriate racial gerrymandering but has been reluctant to hear cases charging partisan gerrymandering. The Court has wanted to avoid interfering with explicitly partisan political decisions. But within the past decade or so, the Court has begun to hear cases concerning partisan gerrymandering. Cases in 2004 and 2006 got stuck on the thorny (but fascinating) problem of how to use spatial and mathematical evidence to consistently decide when redistricting

is too partisan, whether for state offices or the U.S. Congress. The Court heard arguments in Fall of 2017 in the case of *Gill v. Whitford*, wherein Wisconsin Democratic party officials have claimed that partisan gerrymandering by the Republican-controlled legislature and governor has violated the freedom of speech clause of the First Amendment and the equal protection clause of the Fourteenth Amendment. That case has still not been decided, in part because other related cases are being, or about to be, heard. These include a charge by Republicans that Democrats have carried out partisan gerrymandering in Maryland, and charges of illegal Republican gerrymandering in North Carolina. The state court of Pennsylvania struck down congressional districts drawn by Republicans, but the U.S.

Supreme Court declined to review this decision. In Texas, the Court recently began hearing arguments that could radically alter the political map of the state. The plaintiffs in *Abbott v. Perez* claim that the Republican-created district map actually constitutes discriminatory racial gerrymandering against Latino and African Americans. Because these two groups in Texas tend to vote Democratic, this claim of racial gerrymandering is a way for Democrats to attack what is claimed to be underlying partisan gerrymandering. It is hard to overstate the important implications for political representation in the United States of these cases concerning possible partisan gerrymandering. For Texas, a charge of voting rights violations could also return the state to federal supervision of its redistricting process,

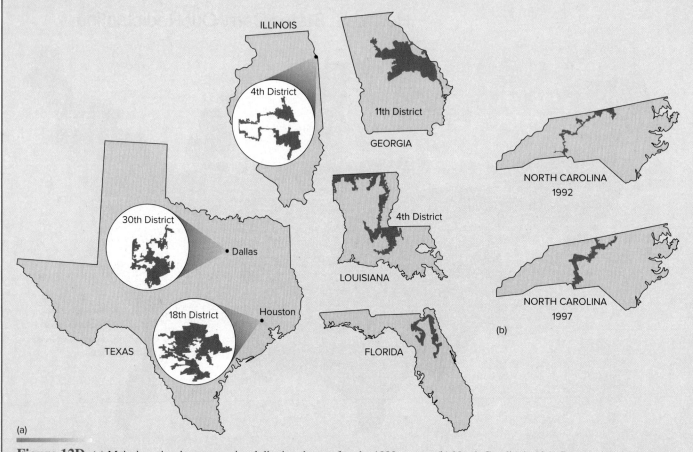

Figure 12D (a) Majority-minority congressional districts drawn after the 1990 census. (b) North Carolina's 12th Congressional District before and after court challenges.

a burden it had been placed under since the Voting Rights Act of 1965 and only came out from under in 2013.

Polls indicate that a clear majority of Americans, whether Democratic, Republican, independent, or otherwise, believe that partisan gerrymandering is bad for the democratic process. The use of geographic information systems (GIS) in redistricting has increased the speed and power with which officials can evaluate alternative districting schemes. Coupled with increasingly sophisticated data about citizens' partisan voting preferences, the danger of even more powerful and impactful partisan gerrymandering is only increasing. There are probably no other issues where geography is so central to important and fundamental ethical issues regarding democracy and electoral fairness.

Questions to Consider

1. What is your opinion of compactness as a criterion for redistricting? Is it a legitimate and appropriate criterion? Why or why not?

2. Do you believe that race should be a consideration in the electoral process? Why or why not? If so, should voting districts be drawn to increase the likelihood that representatives of racial or ethnic minorities will win elections? If not, how can we make sure that the voting power of minorities will not be unacceptably diluted?

3. Critics of affirmative racial gerrymandering contend that blacks have been and can continue to be elected in white majority districts and, further, that white politicians can and do adequately represent the needs of all voters in their districts, including blacks. Do you agree? Why or why not?

4. Do you think the United States should implement an analogue to affirmative racial gerrymandering that might be called *affirmative partisan gerrymandering*? Its intent would be to ensure that the partisan makeup of legislative bodies would be close to the proportions of voter sentiment across the partisan spectrum. How might this be achieved?

5. Just four states give control of congressional redistricting to independent commissions designed to balance the influence of voices across the partisan spectrum and remove the influence of elected officials already in power; only six do so at the level of state legislatures. What do you think of independent commissions as an approach to electoral redistricting? Should they be outlawed or, in contrast, should they be made mandatory for all states?

SUMMARY

The sovereign state is the dominant entity in the political subdivision of the world. It constitutes an expression of cultural separation and identity as pervasive as that inherent in language, religion, or ethnicity. A product of 18th-century political philosophy, the idea of the state was diffused globally by colonizing European powers. In most instances, the colonial boundaries they established have been retained by newly independent countries as their international boundaries.

The greatly varying spatial characteristics of states contribute to national strength and stability. Size, shape, and relative location influence countries' economies and international roles, while national cores and capitals are the heartlands of states. Boundaries, the legally defined borders that determine a state's size and shape, also determine the limits of its sovereignty. They may or may not reflect preexisting cultural landscapes, and in any given case, may or may not prove to be viable. Whatever their nature, boundary conflicts are at the root of many international disputes. Maritime boundary claims, particularly as reflected in the UNCLOS, add a new dimension to traditional claims of territorial sovereignty.

State cohesiveness is promoted by a number of centripetal forces. Among these are national symbols, a variety of institutions, and confidence in the aims, organization, and administration of government. Also helping to foster political and economic integration are transportation and communication connections. Destabilizing centrifugal forces, particularly ethnically based separatist movements, threaten the cohesion and stability of many states.

Although the state remains central to the partitioning of the world, a broadening array of political entities affects people individually and collectively. Recent decades have seen a significant increase in supranationalism in the form of a number and variety of global and regional alliances to which states have surrendered some sovereign powers. At the other end of the spectrum, expanding North American urban areas and governmental responsibilities raise questions of fairness in districting procedures and of effectiveness when political power is fragmented.

KEY WORDS

antecedent boundary
artificial boundary
autonomous nationalism
centrifugal force
centripetal force
compact state
consequent (ethnographic) boundary
core area
devolution
electoral geography
elongated state
enclave
ethnic cleansing
European Union (EU)
exclave
exclusive economic zone (EEZ)
federal state

fragmented state
functional dispute
geometric boundary
geopolitics
gerrymandering
irredentism
landlocked
nation
nationalism
nation-state
natural boundary
perforated state
physical boundary
political geography
positional dispute
prorupt state
reapportionment

redistricting
regionalism
relic boundary
resource dispute
separatism
state
subnationalism
subsequent boundary
superimposed boundary
supranationalism
territorial dispute
terrorism
unitary state
United Nations Convention on the
  Law of the Sea (UNCLOS)

FOR REVIEW

1. What are the differences between a *state*, a *nation*, and a *nation-state?* Why is a colony not a state? How can one account for the rapid increase in the number of states since World War II?

2. What attributes differentiate states from one another? How do a country's size and shape affect its power and stability? How can a piece of land be both an *enclave* and an *exclave?*

3. How can boundaries be classified? How do they create opportunities for conflict? Describe and give examples of three types of border disputes.

4. How does the *United Nations Convention on the Law of the Sea (UNCLOS)* define zones of diminishing national control? What are the consequences of the concept of the 200-nautical-mile *exclusive economic zone?*

5. Distinguish between *centripetal* and *centrifugal* political forces. What are some of the ways that national cohesion and identity are achieved?

6. What characteristics are common to all or most regional autonomist movements? Where are some of these movements active? Why do they tend to be on the periphery rather than at the national core?

7. What types of international organizations and alliances can you name? What were the purposes of their establishment? What generalizations can you make regarding economic alliances?

8. Why does it matter how boundaries are drawn around electoral districts? Theoretically, is it always possible to delimit boundaries "fairly"? Support your answer.

9. What reasons can you suggest for the great political fragmentation of the United States? What problems stem from such fragmentation?

KEY CONCEPTS REVIEW

1. **What are the types and geographic characteristics of countries and the nature of their boundaries?** Section 12.1

   States are internationally recognized independent political entities. When culturally uniform, they may be termed nation-states. Their varying physical characteristics of size, shape, and location have implications for national power and cohesion. Boundaries define the limits of states' authority and underlie many international disputes.

2. **How do states maintain cohesiveness and instill nationalism?** Section 12.1

   Cohesiveness is fostered through unifying institutions, education, and efficient transport and communication systems. It may be eroded by separatist wishes and tendencies of minority groups.

3. **Why are international alliances proliferating, and what objectives do they espouse and serve?** Section 12.2

   In an economically and technologically changing world, alliances are presumed to increase the security and prosperity of states. The United Nations claims to represent and promote worldwide cooperation; its Law of the Sea regulates use and claims of the world's oceans. Regional alliances involving some reduction of national independence promote economic, military, or political objectives of groups of states related spatially or ideologically. They are expressions of the growing trend toward *supranationalism* in international affairs.

4. **What problems are evident in defining local political divisions in North America, and what solutions have been proposed or instituted?** Section 12.3

   The great political fragmentation within, particularly, the United States reflects the creation of special-purpose units to satisfy a local or administrative need. States, counties, townships, cities, and innumerable special-purpose districts all have defined and often overlapping boundaries and functions. Voting rights, reapportionment, and local political boundary adjustments represent areas of continuing political concern and dispute. In the United States, racial gerrymandering is a current legal issue in the definition of voting districts.

HUMAN IMPACTS ON THE ENVIRONMENT

High-rise buildings, fishing boats, and air pollution form the background of a garbage-littered beach in Mumbai, India.
©RAJESH NIRGUDE/AP Images

Key Concepts

13.1 Earth's interlinked environmental systems: atmosphere, lithosphere, hydrosphere, and biosphere; Impacts on the atmosphere: air pollution, acid rain, ozone layer, and climate change.

13.2 Impacts on the land: land cover change, deforestation, desertification, and soil erosion.

13.3 Impacts on water resources: water supply and water quality.

13.4 Disposal of wastes, waste exports, and environmental justice.

13.5 Future prospects.

When the daily tides come in, a surge of water high as a person's head moves up the rivers and creeks of the world's largest delta, formed where the Ganges and Brahmaputra rivers meet the Bay of Bengal in the South Asian country of Bangladesh. Within that Wisconsin-sized country that is one-fifth water, millions of people live on thousands of alluvial islands known as chars. These form from the silt of the rivers and are washed away by their currents and by the force of cyclones that roar upstream from the bay during the annual cyclone period. As the chars are swept away so, too, are thousands and tens of thousands of their land-hungry occupants who fiercely battled each other with knives and clubs to claim and cultivate them.

Late in April 1991, an atmospheric low-pressure area moved across the Malay Peninsula of Southeast Asia and gained strength in the Bay of Bengal, generating winds of nearly 240 kilometers (150 miles) per hour. As it moved northward, the storm sucked up and pushed a wall of water 6 meters (20 feet) high. With a full moon and highest tides, the cyclone and its battering ram of water slammed across the chars and the deltaic mainland. When it had passed, some of the richest rice fields in Asia were gray with the salt that ruined them, islands totally covered with paddies were left as giant sand dunes, others—densely populated—simply disappeared beneath the swirling waters. An estimated 138,000 lives were lost to the storm and hundreds of thousands more to subsequent starvation, disease, and exposure.

Each year, lesser variants of the tragedy are repeated; each year, survivors return to rebuild their lives on old land or new still left after the storms or created as the floods ease and some of the annual 2.5 billion tons of river-borne silt is deposited to form new chars. Deforestation in the Himalayan headwaters of the rivers increases erosion there and swells the volume of silt flowing into Bangladesh. Dams on the Ganges River in India alter normal flow patterns, releasing more water during floods and increasing silt deposits during seasonal droughts. Population growth adds to the number of desperate people seeking homes and fields on lands more safely left as the realm of river and sea. And global climate change already underway brings rising sea levels and stronger storm systems, increasing the vulnerability of places like the delta of the Ganges and Brahmaputra rivers.

13.1 Physical Environments and Human Impacts

The people of the chars live with an immediate environmental contact that is outside the experience of most of us in the highly developed, highly urbanized countries of the world. In fact, much of the content of the preceding chapters has detailed ways that humans isolate themselves from the physical environment and superimpose cultural landscapes on it to accommodate the growing needs of their growing numbers.

Many cultural landscape changes are minor in themselves. The forest clearing for swidden agriculture and the terracing of hillsides for subsistence farming are modest alterations of nature. Plowing and farming the prairies, harnessing major river systems by dams and reservoirs, building cities and their connecting highways, or opening vast open-pit mines are much more substantial modifications. In some cases, the new landscapes are apparently completely divorced from the natural ones that preceded them, as in enclosed, air-conditioned shopping malls and office towers. The original minor modifications have cumulatively become totally new cultural creations.

But suppression of the physical landscape does not mean eradication of human-environmental interactions. They continue, though in altered form, as humans increasingly become the dominant agents of environmental change. More often than not, the changes that we have set in motion create unplanned cultural landscapes and unwanted environmental conditions. We have altered our climates, polluted our air, water, and soil, destroyed natural vegetation and land contours while stripping ores and fuels from the Earth.

Environment is an overworked word that means the totality of things that in any way affect an organism. Humans exist within a natural environment—the sum of the physical world—that they have modified by their individual and collective actions. Human impacts on the environment can be summarized through the simple **IPAT equation:**

$$I = PAT$$

where:

   I = Impact on the environment
   P = Population
   A = Affluence (often measured by per capita income)
   T = Technology

As indicated by the IPAT equation, population growth and rising standards of living both lead to greater use of natural resources and greater waste production. However, the technology factor in the equation accounts for the very different impacts associated, for example, with various sources of energy such as coal, nuclear, or wind energy. Further, in many instances, increased standards of living have led to improvements in local air and water quality because rising prosperity has allowed societies to invest in pollution controls and cleaner technologies. Unfortunately, some changes aimed at improving the quality of the local environment, such as shifting to taller smokestacks or shipping toxic wastes overseas, come at the expense of the environment elsewhere or at the expense of the global environment (**Figure 13.1**). Each of the factors in the IPAT equation relate to topics in human geography—population geography, economic geography, and the various technologies that societies use to house, feed, and transport themselves. Thus, adverse consequences of human impact on the environment are the unforeseen creations of the cultural landscapes that we have been examining and analyzing, and their study highlights the unity of physical and human geography.

Awareness of the severity of human impacts on the environment has led to the concept of sustainable development. **Sustainable development** meets the needs of today without jeopardizing the ability of future generations to meet their needs. Sustainable development must work within the limits of natural systems. For example, sustainable groundwater use is limited by the amount of water replenished by rainfall and snowmelt, and sustainable

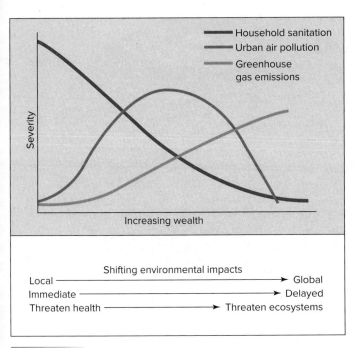

Figure 13.1 The geographic scale of environmental impacts shifts as incomes rise. For the world's poor, the primary environmental problems are disposing of human wastes and ensuring clean water supplies. As standards of living increase, communities can afford water and sewage treatment systems but also increase their consumption of raw materials, synthetic chemicals, and fossil fuel energy. This tends to shift environmental problems from local, immediate threats to human health to longer-term, delayed global impacts on ecosystems such as destruction of the ozone layer, acid precipitation, and global climate change.

Source: Graph from World Energy Assessment, UNDP, 2000, Figure 3.10, p. 95.

fisheries limit the harvest of fish to a rate that maintains the long-term resilience, quantity, and diversity of the stock. Sustainable forestry harvests trees at a rate below that of natural forest regeneration. Sustainable agriculture loses soil at a rate below that of soil formation. Thus, a starting point for considering sustainable development is with a basic picture of the Earth's environmental systems.

Earth's Environmental Systems

The **ecosphere** is the thin zone of air, water, earth, and living matter that extends from the mountaintops to the bottom of the ocean, within which life is found. The ecosphere is composed of four overlapping, interrelated parts: the atmosphere, hydrosphere, lithosphere, and biosphere. The **atmosphere** is a thin blanket of air enveloping the Earth, with more than half of its mass within 6.5 kilometers (4 miles) of the surface and 98 percent within 26 kilometers (16 miles). The **hydrosphere** consists of the perpetually moving surface and subsurface waters in oceans, rivers, lakes, glaciers, groundwater, water vapor, and clouds. The **lithosphere,** the upper reaches of the Earth's crust, contains the soils that support plant life, the minerals that plants and animals require, and the fossil fuels and ores that humans exploit. The **biosphere** consists of the living matter of plants and animals. Cycles of energy in the atmosphere, water movement in the hydrologic cycle, the rock cycle, and the cycling of energy and chemical

elements through the biosphere are all intricately linked. The ecosphere contains all that is needed for life, all that is available for life to use, and, presumably, all that ever will be available.

Geographic differences in climate, water availability, landforms, and biological communities generate a vast mosaic of environments on Earth. The most important components of climate are temperature and precipitation. Temperature is a measure of the air's energy, all of which comes from the sun. On an annual basis, tropical (equatorial) locations, where the sun is more directly overhead, receive the most energy from the sun. This uneven heating of the Earth's surface results in warm air in the tropics and cold air near the poles (**Figure 13.2**), The midlatitudes experience more seasonal variation, with less fluctuation near oceans and greater fluctuation in the middle of large landmasses. Summers become cooler and shorter farther toward the poles until finally, permanent ice cap conditions prevail. Temperature differences between the tropics and poles produce pressure differences that drive winds, storms, and ocean currents that redistribute the Earth's unevenly distributed energy.

The intense solar energy received in the tropics causes air to expand, rise, and eventually form clouds and precipitation. Some distance away from the equator, that air sinks, warms, and dries out. The rising and sinking air motions caused by uneven heating lead to a relatively consistent global pattern of atmospheric circulation cells (**Figure 13.3**). These circulation patterns create alternating bands of high precipitation and low precipitation areas on Earth (**Figure 13.4**). The Hadley cell, shown in Figure 13.3, is responsible for the wet tropical climates near the equator and dry, desert climates at about 30 degrees latitude north and south of the equator (**Figure 13.5**). In addition to the vertical air motions, the circulation cells interact with a spinning planet to produce the familiar surface winds such as the trade winds in the tropics and the westerlies in the midlatitudes.

Just as the rising and sinking air motions shown in Figure 13.3 create both rain forests and deserts, air that is forced up and over mountains creates wet climates on one side and persistent dryness on the other. A good example is the western United States, where prevailing winds from the west create areas of heavy snow or rainfall on the western slopes of the Cascade and Sierra Nevada mountain ranges and dry desert and steppe climates in the interior east of the mountains.

Water is essential to all life. Water covers about 71 percent of the Earth's surface, though only a small part of the hydrosphere is suitable or available for use by humans, plants, or animals. The total amount of water on the Earth remains constant as water moves through the **hydrologic cycle,** changing form from vapor to liquid to ice/snow and back again (**Figure 13.6**).

Water plays a critical role in moderating the Earth's climate, reducing temperature extremes, and making the planet habitable. Due to the diversity of Earth's climates, however, the availability of water varies widely across time and space. Ironically, the supply of water is most variable from year to year in the driest locations (**Figure 13.7**). Water, whether in the form of ice or liquid, weathers and erodes the Earth's crust, the lithosphere. As water erodes mountains and plateaus, carves valleys, and transports sediments to floodplains and deltas, it is a major force in the reshaping of the Earth's landforms.

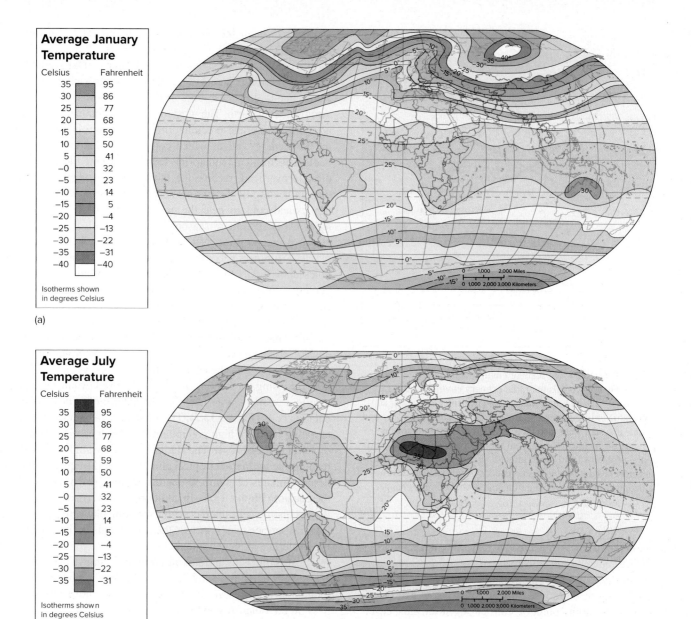

Figure 13.2 Average temperatures in (a) January and (b) July. The rotation of the Earth around the sun, along with the tilt of the Earth's axis, cause more direct solar energy to fall on the Northern Hemisphere in June and on the Southern Hemisphere in January. Note how over water, the temperatures decrease evenly with increased distance from the equator. The effect of land/water contrasts can be seen in the more dramatic temperature differences between January and July at continental locations in the midlatitudes compared to coastal locations.

Plants and animals are adapted to particular climate conditions. Thus, the biosphere is divided into separate groups of biological communities called **biomes,** which are established by the pattern of global climates. In fact, the original classification of global climates was based on observations of plant communities, not temperature and precipitation records, many of which did not exist at that time. In essence, plant communities served as weather stations, recording climate conditions by the ability of different species to survive. We know the Earth's biomes by such descriptive names as *desert, grassland,* or *steppe* or as the *tropical rain forest* and *northern coniferous forest* that were discussed in

Chapter 8. Biomes, in turn, contain smaller, more specialized **ecosystems:** self-contained, self-regulating, and interacting communities adapted to local combinations of climate, topography, soil, and drainage conditions.

The structure of the ecosphere is not eternal and unchanging. On the contrary, alteration is the constant rule of the physical environment and would be so even in the absence of humans and their distorting impacts. Natural climatic change, year-to-year variations in weather conditions, fires, windstorms, floods, diseases, or the unexplained rise and fall of predator and prey populations all call for new environmental configurations and

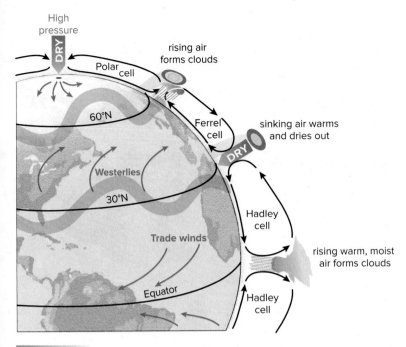

Figure 13.3 Uneven heating of the Earth's surface creates persistent air circulation patterns that move air, moisture, and energy around the Earth. One of those circulation patterns, the Hadley cell, creates wet climates near the tropics and deserts at about 30 degrees north and south.

forever prevent the establishment of a single, constant "balance of nature."

Remember that we began to track cultural geographic patterns from the end of the last continental glaciation, some 11,000 to 12,000 years ago. Our starting point, then, was a time of environmental change when humans were too few in number and primitive in technology to have had any impact on the larger structure of the ecosphere. Their numbers increased and their technologies became vastly more sophisticated and intrusive with the passage of time, but for nearly all of the period of cultural development to modern times, human impact on the world environment was absorbed and accommodated by it with no more than local distress. The rhythm and the regularity of larger global systems proceeded largely unaffected by people.

Impacts on the Atmosphere

Ecosystems have long felt the destructive hand of humans and the cultural landscapes they made. Chapter 2 explored the results of human abuse of the local environment in the Chaco Canyon and Easter Island deforestations. Forest removal, overgrazing, and ill-considered agriculture turned lush hillsides of the Mediterranean Basin into sterile and impoverished landscapes by the end of the Roman Empire. At a global scale, however, human impact was minimal. But slowly, unnoticed at first, human activity began to have a

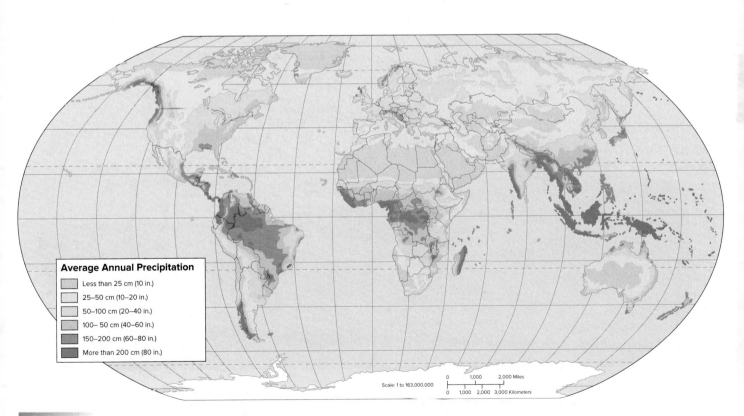

Figure 13.4 Average annual precipitation. Regional contrasts of precipitation clearly demonstrate the truism that natural phenomena are unequally distributed over the surface of the Earth. Global wind patterns create areas of high rainfall in equatorial and tropical areas of Central and South America, Africa, and South and Southeast Asia. A band of deserts is found at about 30 degrees from the equator. Proximity to large water bodies, and position with respect to prevailing winds and mountain ranges, also contribute to the patterns of annual precipitation.

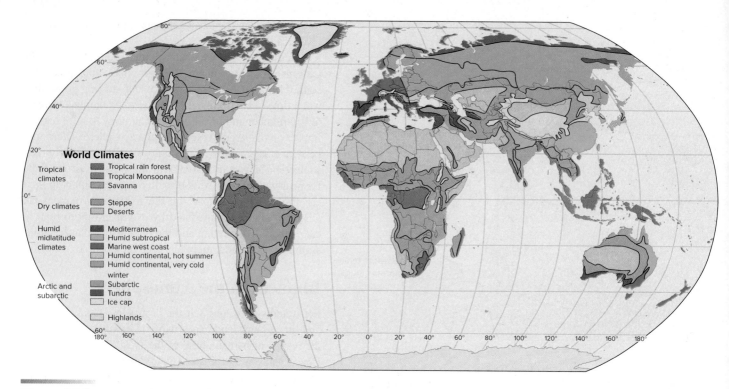

Figure 13.5 Climates of the world. Complex interrelationships of latitude, land and water contrasts, ocean currents, topography, and wind circulation create this generalized global pattern of climates.

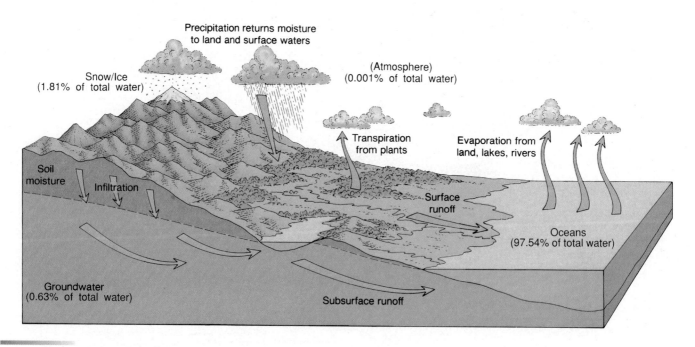

Figure 13.6 The hydrologic cycle circulates water through the atmosphere, lithosphere, biosphere, surface water, and oceans. Water may change form and composition, but under natural environmental circumstances, it is marvelously purified in the recycling process and is again made available with appropriate properties and purity to the ecosystems of the Earth. The sun provides energy for the evaporation of fresh and ocean water. The water is held as vapor until the air becomes supersaturated. Atmospheric moisture is returned to the Earth's surface as solid or liquid precipitation to complete the cycle. Precipitation is not uniformly distributed, and moisture is not necessarily returned to areas in the same quantity as it has evaporated from them. The continents receive more water than they lose; the excess returns to the seas as surface water or groundwater. A global water balance, however, is always maintained.

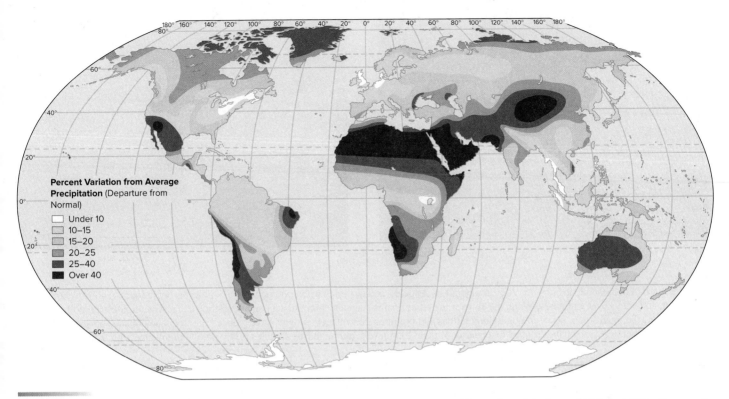

Figure 13.7 Precipitation variability. Note that steppe and desert climate regions have both low total precipitation and high variability. In general, the dryer the climate, the greater the variability. This creates vulnerability to shortages when humans develop agriculture or cities in these regions. Short-run variability in climatic conditions is the rule of nature and occurs independent of any human influence.

global impact, carrying the consequences of human abuse of the ecosphere far beyond the local scene. Air pollution was at first local, in the form of household air pollution and negative health effects from indoor cooking over open fires. However, the Industrial Revolution in the 18th century and continued increase in the use of fossil fuel energy sources has changed the scale of impacts on the atmosphere. At the metropolitan scale, urban air pollution episodes have rendered the air unsafe in many cities, at the continental scale, acid precipitation is carried across international borders, and at the global scale, destruction of the ozone layer and climate change threaten the entire planet's human and environmental systems.

Air Pollution and Acid Precipitation

Every day, thousands of tons of pollutants are discharged into the air by natural events and human actions. Air is polluted when it contains substances in sufficient amounts to have a harmful effect. Truly clean air has never existed, for atmospheric pollution can and does result in nature from ash from volcanic eruptions, marsh gases, smoke from naturally occurring forest fires, and windblown dust. Normally, these pollutants are of low volume, are widely dispersed in the atmosphere, and have no significant long-term effect on human health or ecosystems.

Far more damaging are the substances discharged into the atmosphere by human actions. These pollutants come primarily from burning fossil fuels—coal, oil, and natural gas—in power plants, factories, furnaces, and vehicles, and from fires deliberately set to clear forests and grasslands for agricultural expansion or shifting cultivation clearing and burning. In 1952, smoke from coal-burning factories and home fireplaces combined with fog

to cause a four-day smog episode in London that blackened the skies in daytime and killed an estimated 4,000 people. By the 1950s, air pollution in Los Angeles from car exhausts and other sources stung people's eyes and noses and caused rubber tires to deteriorate rapidly (**Figure 13.8**). Since then, in London, Los Angeles, and other cities in the developed world, environmental regulations and pollution control technologies have dramatically cleared the air (although Los Angeles still does not meet some air quality standards).

Today, air pollution is a global problem; areas far from the polluting source may be adversely affected as atmospheric circulation moves pollutants freely without regard to political boundaries. Just as manufacturing activity has shifted to newly industrializing countries, so too has the world's worst air pollution. For example, current full-color satellite cameras regularly reveal a nearly continuous, 2-mile-thick blanket of soot, organic compounds, dust, ash, and other air debris stretching across much of India, Bangladesh, and Southeast Asia, reaching northward to the industrial heart of China. The pollution shroud in and around India, researchers find, reduces sunlight enough to cut rice yields across much of the country.

Although air quality is generally improving in high-income countries, it has worsened in the developing countries of South, Southeast, and East Asia. The World Health Organization (WHO) estimated that in 2015, air pollution caused 4.2 million premature deaths, mostly in low- and middle-income countries. The WHO reports that of India's 27 cities of more than 1 million people, not one meets the organization's air pollution standards.

In addition to the very serious human health consequences of air pollution, the interaction of pollutants with one another

(a)

(b)

Figure 13.8 Photochemical smog. (*a*) A clear day in Los Angeles, and (*b*) a day when photochemical smog sharply reduces visibility, damages vegetation, and stings eyes and noses. When the air over the city remains stagnant, sunlight acts on automobile exhausts and industrial pollutants to create photochemical smog. Strict air quality regulations and pollution control technologies have reduced the frequency of such episodes. Peak levels of photochemical smog have dropped to a quarter of their 1955 levels, although Los Angeles is still not in compliance with federal air quality standards.

(a) ©Davel5957/Getty Images; (b) ©steinphoto/E+/Getty Images

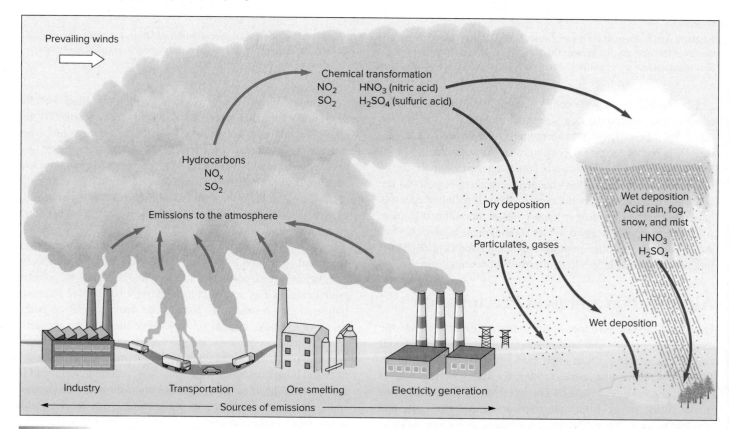

Figure 13.9 Acid precipitation forms as vehicle and industrial emissions are transformed in the atmosphere. Sulfur dioxide and nitrogen oxides produced by the combustion of fossil fuels are transformed into sulfate and nitrate particles; when the particles react with water vapor, they form sulfuric and nitric acids. Often carried long distances on prevailing winds, the acids are deposited on the surface in the form of fog, rain, snow, or particles.

or with natural substances such as water vapor may create secondary pollutants. Among these secondary pollutants is the acid precipitation that occurs when sulfur and nitrogen compounds in emissions from automobiles, coal-fired power plants, and industrial sources produce sulfuric acid and nitric acid in clouds (**Figure 13.9**). Acid precipitation is often carried long distances by prevailing winds to places where it is highly damaging to trees, lakes, and even structures.

Unexpectedly, acid precipitation is a condition in part traceable to actions taken in developed countries to alleviate the smoke and chemicals that poured into the skies from the chimneys of their power plants, mills, and factories. The clean air programs demanded by environmentalists usually incorporated prohibitions against the discharge of atmospheric pollutants damaging to areas near the discharge point. The response was to raise smokestacks to such a height that pollutants were diluted and dispersed far from their origin by higher-elevation winds (**Figure 13.10**).

But when power plants, smelters, and factories were fitted with tall smokestacks to free local areas from pollution, the sulfur dioxide and nitrogen oxides in the smoke were pumped high into the atmosphere instead of being deposited locally. There, they combined with water and other chemicals and turned into sulfuric and nitric acid that was carried to distant areas. They were joined in their impact by other sources of acid gases. Such as motor vehicle exhausts.

Once the pollutants are airborne, winds can carry them thousands of kilometers, depositing them far from their source. In North America, most of the prevailing winds are westerlies (see Figure 13.3), meaning that much of the acid precipitation that falls on the eastern seaboard and eastern Canada originated in the central and upper Midwest. Similarly, airborne pollutants from the United Kingdom, France, and Germany cause acidification problems in Scandinavia (**Figure 13.11**).

When acids from all sources are washed out of the air by rain, snow, or fog, the result is **acid precipitation.** Acidity levels are described by the *pH factor*, the measure of acidity/alkalinity on a scale of 0 to 14. The pH of normal rainfall is 5.6, slightly acidic, but acid rainfalls with a pH of 2.4—approximately the acidity of vinegar and lemon juice—have been recorded.

Figure 13.10 Dilution is not the solution to pollution. Before concern about acid rain became widespread, the U.S. Clean Air Act of 1970 set standards for ground-level air quality that could be met most easily by building tall smokestacks high enough to discharge pollutants into the upper atmosphere. Stacks 300 meters (1,000 feet) and higher became common at utility plants and factories, far exceeding the earlier norm of 60–90 meters (200–300 feet). What helped cleanse one area of pollution greatly increased damage elsewhere.

©Gary Whitton/Shutterstock

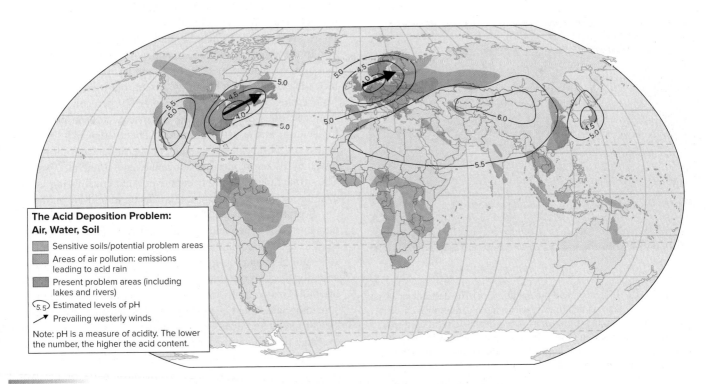

Figure 13.11 Acid precipitation: areas of origin and impacts. Prevailing winds can deposit acid precipitation far from its area of origin and across international boundaries. Prevailing westerly winds carry acid precipitation east from the Ohio and Tennessee River valleys and from the industrial districts of Western Europe toward Scandinavia where it harms vegetation, lakes, and buildings. It is estimated that 90 percent of the sulfuric acid and 80 percent of the nitric acid falling on Norway is from outside the country. Compare this map with the centers of industrial production in Figure 9.20.

Source: Adapted from Student Atlas of World Geography, 4th ed. John Allen, Map 55, p. 70. McGraw-Hill/Dushkin, 2005.

Primarily occurring in industrialized nations, acid precipitation has become a serious problem in many parts of Europe, North America, and East Asia. It expresses itself in several forms, though the most visible are its corrosive effects on marble and limestone sculptures and buildings and on metals such as iron and bronze and in the destruction of forests (**Figure 13.12**). Trees at higher elevations are particularly susceptible, with widespread forest loss clearly apparent on the hillsides and mountain tops of New England, Scandinavia, and Eastern Europe.

(a)

(b)

Figure 13.12 Effects of acid precipitation. *(a)* The destructive effect of acid rain is evident on this limestone carving on an English church. *(b)* The Germans call it *Waldsterben*—forest death—a term used to summarize the destruction of trees by a combination of acid precipitation, ozone, and heavy metals leached by acidic waters. It first strikes at higher elevations where natural stresses are greatest and acidic clouds most prevalent, but it slowly moves downslope until entire forests are gone. Here, at Tatra National Park in Slovakia, *Waldsterben* is thought to result from pollution traveling eastward from industrial areas in Poland, Germany, and the Czech Republic. The forests will recover only in geologic time due to the extremely acidic soils.

(a) ©RMAX/Getty Images; (b) ©Branko Ostojic/123RF

Damage to lakes, fish, and soils is less immediately evident, but more widespread and equally serious. Acid precipitation has been linked to the disappearance of fish in thousands of streams and lakes in New England, Canada, and Scandinavia, and to a decline in fish populations elsewhere. It leaches toxic constituents (such as aluminum) from the soil and kills soil microorganisms that break down organic matter and recycle nutrients through the ecosystem. Acid precipitation has decreased in Europe and North America due to international agreements brokered by the United Nations, strict air emission standards, and deindustrialization—the closure of heavy industrial plants such as steel mills that were once major sources of emissions. The changing geography of industrial production means that the geographic patterns of acid precipitation are shifting toward Asia.

The Trouble with Ozone

The potential destruction of the Earth's ozone layer offers another example of the global nature of air pollution. However, it is also a success story of international cooperation and using science and technology to improve the environment. **Ozone** is a reactive molecule consisting of three oxygen atoms rather than the two of normal oxygen. Ozone can lead to confusion because high in the atmosphere, it offers essential protection against the sun's rays, but when it is near the surface, it is one of the main components in photochemical smog and highly damaging to plants and animals. Sunlight produces it from standard oxygen, and a continuous but thin layer of ozone accumulates at upper levels in the atmosphere. There, it is essential to all life forms because it blocks the cancer-causing ultraviolet (UV) light that damages DNA, the molecule of heredity and cell control.

In the summer of 1986, scientists for the first time verified that a "hole" had formed in the ozone layer over Antarctica. In fact, the ozone was not entirely absent, but it had been reduced from earlier recorded levels by some 40 percent. As a result, Antarctic life—particularly the microscopic ocean plants (phytoplankton) at the base of the food chain—that had lived more or less in UV darkness was suddenly getting a trillionfold (1 followed by 12 zeros) increase above the natural rate of UV receipt. The ozone hole typically occurs over Antarctica during late August through early October and breaks up in mid-November.

Scientists attribute the ozone decline to pollution from human-made chemicals, particularly *chlorofluorocarbons (CFCs)* used as coolants, cleansing agents, propellants for aerosols, and in insulating foams. In a chain reaction of oxygen destruction, each of the chlorine atoms released can destroy upward of 10,000 ozone molecules. Ozone reduction is a continuing and spreading atmospheric problem. A similar ozone hole about the size of Greenland opens in the Arctic, too, and the ozone shield over the midlatitudes has dropped significantly since 1978.

Why should the hole in the ozone layer have appeared first so prominently over Antarctica? In most parts of the world, horizontal winds tend to keep chemicals in the air well mixed. But circulation patterns are such that the freezing whirlpool of air over the south polar continent in winter is not penetrated by air currents from warmer regions of the Earth. In the absence of sunlight and atmospheric mixing, the CFCs work to destroy the ozone. During the Southern Hemisphere summer, sunlight works to replenish it. In either hemisphere, ozone depletion has identical adverse effects. Among other things, greater exposure to UV radiation increases the incidence of skin cancer and, by suppressing bodily defense mechanisms, increases risk from a variety of infectious diseases. Many crop plants are sensitive to UV radiation, and the very existence of the microscopic plankton at the base of the marine food chain is threatened by it.

Production and use of CFCs has been phased out under the Montreal Protocol, a 1987 international agreement made effective in 1992. The developing countries were given a grace period before they began their phaseout schedule for ending use of CFCs. Industries involved in the production or use of CFCs initially resisted the agreement but have since responded by finding substitute chemicals. The Montreal Protocol is an encouraging example of the ability of international agreements to address a global environmental problem successfully. Because of those restrictions, ozone depletion has been slowed or even stopped.

Recent oscillations in the extent of the ozone hole suggest its peak may have been reached by 2003 to 2006 and mending may be complete by the year 2070. The good news has been tempered by recent discovery of rising levels of other ozone-depleting industrial chemicals that are not regulated by the Montreal Protocol.

Global Climate Change

Scientists agree that humans have significantly altered the chemical composition of the atmosphere since the advent of the Industrial Revolution in the 1700s. Human activities have increased the concentrations of three greenhouse gases—carbon dioxide, methane, and nitrous oxide—intensifying the natural **greenhouse effect,** leading to **global climate change.** The greenhouse effect is caused by a group of atmospheric gases that partially capture the heat energy radiated from the Earth back toward space. Without the greenhouse effect, the Earth would be substantially colder and its temperatures would fluctuate wildly. What is of concern today is that human activities have significantly increased the concentrations of greenhouse gases, intensifying the greenhouse effect and causing anthropogenic (human-caused) climate change (**Figure 13.13**). Humankind's

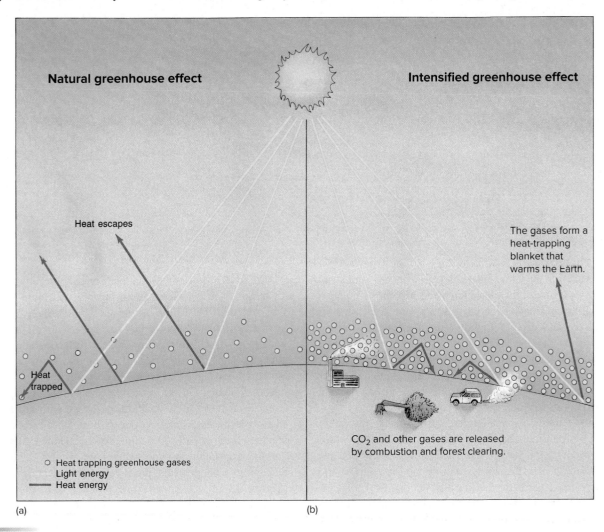

Figure 13.13 Anthropogenic climate change occurs when human activities intensify the natural greenhouse effect. When the level of carbon dioxide (CO_2) in the air is low, as in (a), incoming solar radiation strikes the Earth's surface, heating it up, and the Earth radiates the energy back into space as heat. Some of the heat energy is captured by naturally occurring greenhouse gases, thereby heating the atmosphere. The intensified greenhouse effect depicted in (b) is the result of the higher levels of CO_2 in the atmosphere due to burning of fossil fuels and forest clearing.

Table 13.1

Carbon Dioxide Emissions from Energy Consumption

| | Rank, 2015 | Total Emissions (Millions of Metric Tons/Year) | | | Per Capita Emissions (Metric Tons/Year/Person) |
		1980	2015	Change (1980–2015)	2015
China	1	1,486	8,866	497%	6.5
United States	2	4,680	5,269	13%	16.4
India	3	263	1,894	621%	1.4
Russia	4	N/A	1,687	N/A	11.7
Japan	5	939	1,126	20%	8.9
Germany	6	1,056	743	−30%	9.2
Iran	7	118	654	455%	8.3
Korea, South	8	138	644	369%	12.7
Saudi Arabia	9	177	606	242%	19.2
Canada	10	430	600	40%	16.8
Brazil	11	185	541	193%	2.6
Indonesia	12	85	502	492%	2.0
Mexico	13	239	453	90%	3.6
United Kingdom	14	597	430	−28%	6.6
South Africa	15	225	406	80%	7.4
United Arab Emirates	24	30	237	682%	24.7
Sweden	63	81	47	−43%	4.8
Ghana	96	2	14	515%	0.5
Costa Rica	119	2	7	210%	1.5
WORLD	N/A	18,430	32,722	78%	4.5

Source: *calculated from U.S. Energy Information Administration,* International Energy Statistics, *2018, and Population Reference Bureau* 2015 World Population Datasheet.

massive assault on the atmosphere began with the Industrial Revolution. First, coal and then increasing amounts of petroleum and natural gas have been burned to power industry, heat and cool cities, and drive vehicles. Their burning has turned fuels into carbon dioxide and water vapor. At the same time, the world's forest lands have been cleared for timber and agriculture. With more carbon dioxide in the atmosphere and fewer trees to capture the carbon and produce oxygen, carbon dioxide levels have risen steadily.

The role of trees in managing the carbon cycle is simple: Probably more than half the carbon dioxide put into the atmosphere by burning fossil fuels is absorbed by the Earth's oceans, plants, and soil. The rest of the carbon dioxide remains in the atmosphere where it traps the Earth's heat energy. In theory, atmospheric carbon dioxide could be reduced by expanding plant carbon reservoirs, or *sinks,* on land. Under actual circumstances of expanded combustion of fuels and reduction of forest cover, atmospheric carbon dioxide levels have increased from about 280 parts per million (ppm) at the start of the Industrial Revolution to 403 ppm in 2016.

Yearly carbon emissions from energy consumption totaled 18 billion metric tons in 1980, but reached more than 30 billion tons in 2015. China passed the United States as the single largest emitter in 2006 (**Table 13.1**). Per capita emissions vary widely across world regions, mostly reflecting the vast differences in standards of living. For example, an average resident of the United States emits as much carbon dioxide as 33 residents of Ghana. On the other hand, Japan and many European countries achieve high standards of living with much lower per capita carbon emissions. Sweden's per capita carbon emissions are less than one-third of those of the United States, largely due to government and individual commitments to lifestyle and technology changes that promote environmental sustainability (**Figure 13.14**). Of course, one problem with the statistics in Table 13.1 is that they do not account for the carbon emissions associated with exported or imported products.

Carbon dioxide gets most of the media coverage, but other greenhouse gas concentrations are also increasing (**Figure 13.15**). Methane levels have increased largely due to agricultural expansion to feed a growing world population. Methane is formed by

Figure 13.14 Technologies for reducing greenhouse gas emissions are part of the solution as suggested by the IPAT equation. Many European countries maintain high standards of living with per capita carbon emissions half or one-third that of U.S. residents. Denmark is a world leader in wind energy, generating more than 40 percent of its electricity with renewable wind power, mostly in offshore turbine installations. The Bedding-ton Zero Energy Development (BedZed) is one of many examples of an eco-community. At BedZed in South London, passive solar heating, solar panels to generate electricity, a car-sharing club, and rooftop gardens allow residents to reduce significantly or even eliminate their emissions of greenhouse gases.

©Mark Bjelland

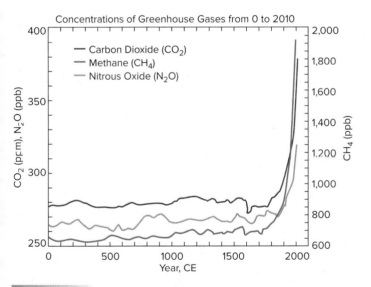

Figure 13.15 Trends in greenhouse gases are upward, and this trend is due to human activities.

Source: Adapted from Intergovernmental Panel on Climate Change, Climate Change 2007. The Physical Science Basis, Figure 6-4, 2007 and Climate Change 2014, Synthesis Report, Figure SPM.1.

decomposition processes and is emitted from the intestinal tracts of livestock and from flooded rice paddies. Nitrous oxide emissions are a byproduct of increased fertilizer use, again a consequence of agricultural expansion and intensification.

Warming of the climate system during the 20th and early 21st centuries was unequivocal. The global average temperature increased 0.6 to 0.7°C from 1951 to 2010 (**Figure 13.16a**). The pattern of increasing temperatures has meant that every decade since the 1980s has been warmer than the previous one (**Figure 13.16b**). The global mean temperature evidence is supported by shrinking glaciers and ice caps, shrinking sea ice (**Figure 13.16c**), rising sea levels (**Figure 13.16d**), rising ocean temperatures, and satellite and weather balloon measurements of temperatures above the Earth's surface.

The most complete assessment of global climate change is done by the United Nations Intergovernmental Panel on Climate Change (IPCC). The panel and reviewers, a group of almost 4,000 scientists from around the world that is affiliated with the United Nations and the World Meteorological Organization, was established in 1988 to assess the science of climate change, determine the impact of any changes on the environment and society, and formulate strategies to respond. In 2014, the IPCC issued results of predictions assuming no major efforts were made to to reduce emissions. They predicted a global temperature increase of 1.4° to 4.8°C (2.5° to 8.6°F) over the 21st century. The effects of such an increase on world climates would be profound.

Although past climates have fluctuated apart from human influence, the IPPC's 2014 report concluded that it was "extremely likely" that human emissions of greenhouse gases rather than natural variations were responsible for the majority of observed warming. They noted that greenhouse gas concentrations are higher today than at any time in the last 800,000 years. Warming that cannot be explained by natural causes has been observed in global oceans and on every continent except Antarctica (**Figure 13.17**). In addition, global rainfall patterns have shifted and extreme rainfall events have become more common.

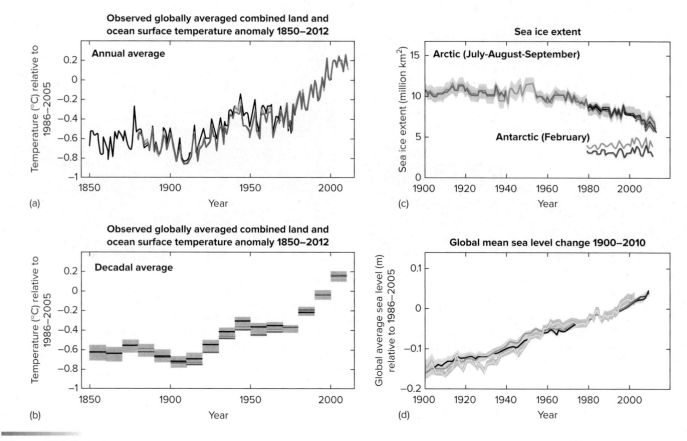

Figure 13.16 Indicators of a changing global climate are all consistent with warming and point to the influence of anthropogenic greenhouse gases.

Source: Intergovernmental Panel on Climate Change, Climate Change 2014 Synthesis Report, *Figure 1.1.*

Climatologists agree on certain of its general consequences in addition to—and the result of—increasing temperatures:

- Arctic summer sea ice is likely to disappear in the second half of the century.
- Sea levels will rise by 28 to 43 cm (11 to 17 in.) by 2100, with an additional 10 to 20 cm (4 to 8 in.) possible if recent accelerated melting of polar ice continues.
- Islands and coastal areas—including densely settled river deltas such as in Bangladesh—will be inundated, affecting the livelihoods and existence of millions.
- Rivers fed by snow or glacier melt will see dramatically altered flows, affecting irrigation.
- There will be spreading droughts in southern Europe, the Middle East, sub-Saharan Africa, the American Southwest, and Mexico. This will impact agriculture and also lead to increased forest fires.
- Many world regions facing the greatest risk or certainty of adverse environmental change are among the world's poorest; damage and misery will not be evenly shared.
- Many parts of the world will see an increase in the number of heat waves and an increase in the intensity of tropical storms.
- If climate change proceeds as the IPPC projects, there will be mass extinctions of perhaps one-fourth of the world's species within 100 years.

The onset of these and other human-induced environmental changes is clearly evident. Melting sea ice itself would have only a modest effect on sea levels; however, the already significant melting of the Greenland ice sheet and accelerating retreat of mountain glaciers throughout the world is adding to ocean volumes. Already, sea levels that rose 19 centimeters (6–9 inches) over the period 1901–2010, led to serious coastal erosion and inundation. A potential 1-meter (3-foot) rise would be enough to cover the Maldives and other low-lying island countries. The homes of between 50 and 100 million people would be inundated, a fifth of Egypt's arable land in the Nile Delta would be flooded, and the impact on the people of the Bangladesh chars would be catastrophic.

Two major interrelated impacts of global climate change will be on water resources and agriculture. Warmer temperatures will increase evaporation, precipitation, and the demand for water, intensifying the movement of water through the hydrologic cycle. In all drought-prone areas of the world, droughts have become more intense and enduring since the 1970s. Earlier IPCC assessments warned that much of the continental interiors of middle latitudes would receive less precipitation than they do now and suffer at least periodic drought, if not absolute aridity. More frequent droughts are likely in the U.S. Corn and Wheat Belts, drastically reducing agricultural productivity and altering world patterns of food supply and trade. Higher temperatures and less snowpack would translate into significantly reduced flows of

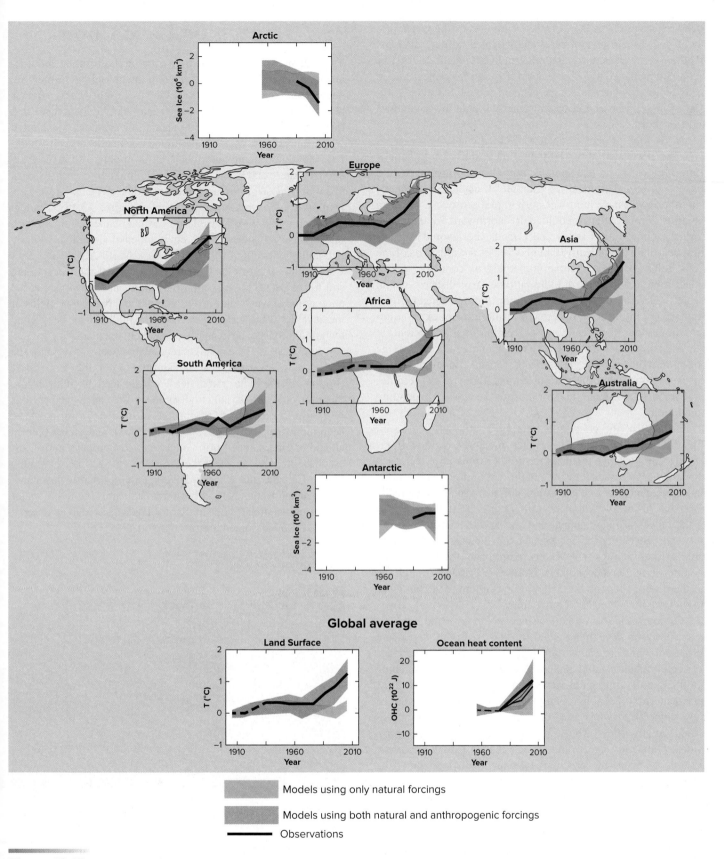

Figure 13.17 Changes in sea ice, global, continental, and ocean temperatures are consistent with predictions for a warming planet. Climate models using only natural causes cannot account for the observed sea ice loss in the Arctic and temperature increases experienced on every continent over the past century. Only by incorporating human influences can the models agree with the observed warming and loss of sea ice. The thickness of the blue and pink bands shows the range of results in different computer simulations using different models.

Source: Intergovernmental Panel on Climate Change, Climate Change 2014 Synthesis Report, *Figure 1.10.*

such western rivers as the Colorado, cutting back the water supply of major southwestern cities and irrigated farming districts. In other world areas, torrential rains and consequent destructive floods are predicted to increase in severity and frequency. And in many regions, winters no longer get cold enough to kill off a variety of insect pests and diseases formerly kept at lower latitudes and elevations.

Not all the projected impacts of climate change are negative. The IPPC also observed that climate shifts could benefit some regions. It projected more rainfall and longer growing seasons in high latitudes, for example; Canada, Scandinavia, and Siberia will have improved agricultural prospects. In North America, crop patterns could shift northward, making the northern Great Lakes states and Canada the favored agricultural heartland climatically, though without the rich soils of the Corn Belt. On average, some climatologists conclude, established middle and upper-middle latitude farm districts would be net beneficiaries of global warming through longer growing seasons and faster crop growth resulting from extra atmospheric carbon. Skeptics, however, remind us that a scientific rule of thumb is that a 1°C (1.8°F) rise in temperature above the optimum reduces grain yields by 10 percent. Long-term studies clearly document the ratio of temperature rise to yield decline for rice, the staple food for most of the world's expanding population.

Because of the effects of decreasing snow and ice cover, higher latitudes will warm faster than equatorial regions. In recent decades, average temperatures in the Arctic have increased almost twice as fast as they have in the rest of the world. Among the consequences of that northward shift of warmth is an already observed 80- to 100-kilometer (50- to 60-mile) poleward shift of the ranges of many animal and bird species. Shifts in structure and distribution of ecosystems and biomes will be inevitable. Arctic seaways will become more open, finally realizing the long-sought and economical Northwest Passage through the northern seas, although the economy and culture of the Inuit will be irreversibly altered at the same time.

Of course, global climate change would have the most severe impact on developing countries that are highly dependent on natural, unmanaged environments for their economic support. Agriculture, hunting and gathering, forestry, and coastal fishing are likely to be more vulnerable than many secondary, tertiary, and quaternary economic activities. Tropical states would be most vulnerable as increased heat and higher evaporation rates would greatly stress wheat, maize, rice, and soybean crops. Global climate change would at the very least disrupt existing patterns of economy, productivity, and population-supporting potential. Certainly, the patterns of climates and biomes developed since the last glaciation would be drastically altered. At the worst, severe and pervasive changes could result in a total restructuring of the present cultural landscapes and human-environmental relationships. Nothing, from population distributions to the relative strength of countries, would ever be quite the same again. Such dire predictions were the background for major international conferences and treaty proposals on global climate change (see the feature "From Kyoto to Paris: Global Climate Change Treaties").

13.2 Impacts on Land Cover

The human population growth, agricultural advances, industrialization, and urbanization detailed in the preceding chapters required a massive transformation of the face of the Earth. Humans have always managed to leave their mark on the landscapes that they occupy. The search for minerals, for example, has altered whole landscapes, beginning with the pockmarks and pits marking Neolithic diggings into chalk cliffs to obtain flints or early Bronze Age excavations for tin and culminating with modern open-pit and strip-mining operations that create massive new landforms of depressions and rubble (**Figure 13.18**). Ancient irrigation systems still visible on the landscape document both the engineering skills and the environmental alterations of early hydraulic civilizations in the Near East, North Africa, and elsewhere. The raised fields built by the Mayas of the Yucatán are still traceable 1,000 years after they were abandoned, and aerial photography reveals the sites of villages and patterns of fields of medieval England. Large sections of the midlatitudes best suited to agriculture have been nearly completely transformed.

Status as a developed core country seems to entail the partial or nearly complete transformation of the land cover (**Figure 13.19**). The places least transformed by humans are those too cold, too dry, too high, or in the least developed regions of the world economy. Vast portions of the Earth's surface have been modified, whole ecosystems destroyed, and global biomes altered or eliminated. North American and European native forests have largely vanished; the grasslands of the interior United States, Canada, and Ukraine have been converted to farmland. Marshes and wetlands have been drained, dams built, and major water impoundments created. Steppe lands have become deserts; deserts have blossomed under irrigation.

Figure 13.18 This open-pit coal mine in Scotland dramatically altered the landscape. On flat or rolling terrain, strip mining leaves a landscape of parallel ridges and trenches, the result of stripping away—overburden—the unwanted surface material. Besides altering the topography, open-pit mining interrupts surface and subsurface drainage patterns, destroys vegetation, and places sterile and frequently highly acidic subsoil and rock on top of the new ground surface.

©Monty Rakusen/Getty Images

From Kyoto to Paris: Global Climate Change Treaties

Our understanding of global climate change depends upon the cooperative work of scientists from around the world. The IPCC, a group of scientists who review and assess the latest scientific, technical, and socio-economic data on climate change and impacts, is a model of international cooperation on scientific issues of climate change. The IPCC doesn't do the actual research; rather, it reviews thousands of published scientific papers and reports. Working groups within the IPCC take on different tasks such as reviewing the scientific data, assessing vulnerability to climate change, or developing strategies for combatting climate change. The IPCC issued its first assessment report in 1990 and in 2014 released its fifth assessment report.

In 1994, the United Nations established a Framework Convention on Climate Change (UNFCCC) that set up arrangements for sharing data on climate, developing national plans for controlling greenhouse gas emissions, and for adapting to climate change. Three years later in Kyoto, Japan, 191 countries signed and ratified a protocol for combatting global climate change. The Kyoto Protocol called for stabilization of greenhouse gases at levels that would not harm the climate system. The agreement assigned different levels of responsibility to countries based on their level of economic development. Arguing that the developed countries had contributed the largest share to the historic increase in greenhouse gases and were the major emitters, the Kyoto Protocol required more of them. The developed countries committed to reducing their greenhouse gas emissions to below 1990 levels by the year 2012. Developing countries were allowed to increase their emissions to meet social and economic needs. The United States signed the protocol but never ratified the treaty. Meanwhile, countries in the European Union made significant progress in honoring their Kyoto commitments.

World leaders met again in Copenhagen (2009), Cancún, Mexico (2010), Durban, South Africa (2011), and Paris (2015). The major complicating factor was the highly uneven geographic distributions of both responsibility and vulnerability. As indicated in Table 13.1, the carbon dioxide emissions of a typical American equal those of 30 sub-Saharan Africans. On the other hand, many of the places most vulnerable to climate change, such as Bangladesh and the Sahel of Africa, are in the developing world.

Meanwhile, some high-latitude regions, mostly in developed countries, will benefit from a longer growing season in a warmer climate.

The 2015 Paris meeting led to a new international climate change agreement to replace the Kyoto Protocol. The centerpiece of the Paris Agreement was a commitment to limit global warming to well below 2°C (3.6°F) above pre-industrial levels. At the request of vulnerable low-lying island states, language was added about working to keep temperature increases below 1.5°C above pre-industrial levels, if feasible. The developed countries also agreed to contribute $100 billion to help countries transition to renewable energy sources and to adapt to the effects of climate change. Each country agreed to submit their own targets for emission reductions called "Nationally Determined Contributions" and begin implementing those reductions in 2020. The Paris Agreement went into effect in 2016 when countries accounting for at least 55 percent of global greenhouse gas emissions had ratified the agreement. On June 1, 2017, President Trump withdrew the United States from the agreement, leaving it the only country to reject the climate accord.

Tropical Deforestation

Forests, we saw in Chapter 8, still cover some 30 percent of the Earth's land surface, though the forest biomes have suffered mightily as human pressures on them have increased. Forest clearing accompanied the development of agriculture and spread of people throughout Europe, Central Asia, the Middle East, and India. European colonization had much the same impact on the temperate forests of eastern North America and Australasia. In most midlatitude developed countries, although original forest cover is largely gone, replanting and reversion of cropland to timber has tended to replenish woodlands at about their rate of cutting.

Now it is the tropical rain forest biome that is feeling the pressure of growing populations, the need for more agricultural land, and expanded demand for fuel and commercial wood. These disappearing forests—covering no more than 6 percent of the planet's land surface—extend across parts of Asia, Africa, and Latin America, and are the world's most diverse and least understood biome.

Tropical forest removal raises three principal global concerns and a host of local ones. First, on a worldwide basis, all forests play a major role in maintaining the oxygen and carbon balance of the Earth. This is particularly true of tropical forests because of their total area and volume. Humans and their industries consume oxygen; vegetation replenishes it through photosynthesis and the release of oxygen back into the atmosphere as a by-product. At the same time, plants extract carbon dioxide from the atmosphere, regulating the levels of this important greenhouse gas. Each year, each hectare (2.5 acres) of Amazon rain forest absorbs one ton of carbon dioxide. When the tropical rain forest is cleared, not only is its role as a carbon sink lost but the act of destruction itself through decomposition or burning releases as carbon dioxide the vast quantities of carbon the forest had stored.

A second global concern is also climate related. Forest destruction changes surface and air temperatures, moisture content, and reflectivity. It is calculated that cutting the forests of South America on a wide scale could raise regional temperatures from 3°C to 5°C (5.5° to 9°F), which in turn would extend the dry season and greatly disrupt not only regional but global climates.

In some ways, the most serious long-term global consequence of the eradication of tropical rain forests will be the loss of a major part of the biological diversity of the planet. Of the estimated 5–10 million plant and animal species believed to exist on Earth, a minimum of 40 percent to 50 percent—and possibly 70 percent or more—are native to the tropical rain forest biome. Many of the plants have become important world staple food crops: rice, millet, cassava, yam, taro, banana, coconut, pineapple, and sugarcane, to name but a few well-known ones. Unknown additional potential food species remain as yet unexploited. Reports from Indonesia suggest that in that country's forests alone, some 4,000 plant species have proved useful to native peoples as foodstuffs of one sort or another, though less than one-tenth have come into wide use. The rain forests are, in addition, the world's main storehouse of drug-yielding plants and insects, including thousands with proven or prospective anticancer properties and many widely used as sources of antibiotics, antivirals, analgesics, tranquilizers, diuretics, and laxatives, among a host of other items. The loss of the zoological and botanical storehouse that the rain forests represent would deprive humans of untold potential benefits that might never be realized.

On a more local basis, tropical forests play for their inhabitants and neighbors the same role taken by forests everywhere. They protect watersheds and regulate water flow. After forest cutting, unregulated flow accentuates the problems of high and low water variations, increases the severity of valley flooding, and makes more serious and prolonged the impact of low water flow on irrigation agriculture, navigation, and urban and rural water supply. Accelerated **soil erosion**—the process of removal of soil particles from the ecosystem, usually by wind or running water—quickly removes the always thin, infertile tropical forest soils from deforested areas. Lands cleared for agriculture almost immediately become unsuitable for that use partially because of soil loss. The surface material removed is transported and deposited downstream, changing valley contours, extending the area subject to flooding, and filling irrigation and drainage channels; or it may be deposited in the reservoirs behind the increasing number of major dams on rivers within the tropical rain forests or rising there (see the feature "Dam Trouble in the Developing World").

Desertification

With no intent to destroy or alter the environment, humans are also negatively affecting the arid and semiarid regions of the world. The process is called **desertification,** the expansion or intensification of areas of degraded or destroyed soil and vegetation cover. While the Earth Summit of 1992 defined desertification broadly as "land degradation in arid, semiarid, and dry subhumid areas, resulting from climatic variations and human activities," the process is often blamed on increasing human pressures exerted through overgrazing, deforestation for fuel wood, clearing of original vegetation for cultivation, and burning.

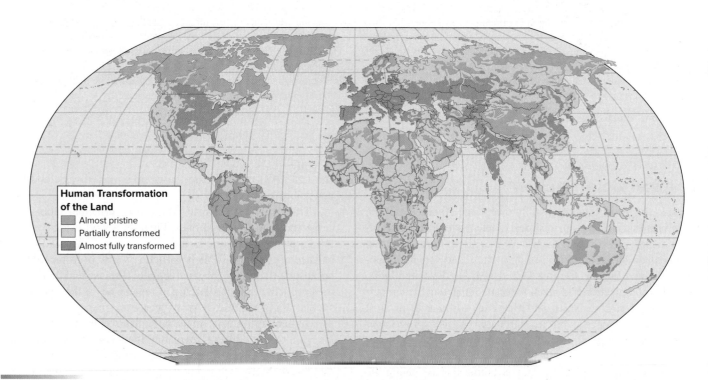

Figure 13.19 Human transformation of the land. Humans have altered much of the Earth's surface in some way. The "almost pristine" areas, covered with original vegetation, tend to be too high, dry, cold, or otherwise unsuitable for human habitation in large numbers. They generally have very low population densities. "Partially transformed" describes areas of secondary vegetation, grown after removal of the original cover. Most are used for agriculture or livestock grazing. "Almost fully transformed" areas are those of permanent and intensive agriculture and urban settlement.

Dam Trouble in the Developing World

By the 1970s, the construction of large dams in the developed countries such as the United States had slowed to a trickle, as the best sites had already been dammed and environmental groups protested the loss of wild rivers. Thus, the focus of dam building shifted to the developing countries. The developing countries have a sizable percentage of the world's undeveloped power potential. The lure of that power and its promise for economic development and national prestige have proved nearly irresistible. China is now the world's leader in large dams and its immense Three Gorges Dam on the Yangtze River is the world's largest hydropower project. China's engineering companies and banks are also actively involved in building large dams in many other developing countries, primarily in Africa and Southeast Asia. The dams (and their reservoirs) often carry a heavy ecological price, and the clearing and development of the areas that they are meant to serve often ensure a shortened life of the dam projects themselves.

The creation of Lake Brokopondo in Suriname in 1964 marked the first large reservoir in a rain forest locale. Without being cleared of their potentially valuable timber, 1,480 square kilometers (570 square miles) of dense tropical forest disappeared underwater. As the trees decomposed, producing hydrogen sulfide, an intolerable stench polluted the atmosphere for scores of miles downwind. For more than two years, employees at the dam wore gas masks at work. Decomposition of vegetation produced acids that corroded the dam's cooling system, leading to costly continuing repairs and upkeep. Identical problems have occurred at the Tucuruí Dam and Reservoir in Brazil, started in 1984 and covering 2,850 square kilometers (1,100 square miles) of uncleared rain forest.

Water hyacinth spreads rapidly in tropical impoundments, its growth hastened by the rich nutrients released by tree decomposition. Within a year of the reservoir's completion, a 130-square-kilometer (50-square-mile) blanket of the weed was afloat on Lake Brokopondo, and after another year, almost half the reservoir was covered. Another 440 square kilometers (170 square miles) were

Figure 13A Hoover Dam on the Colorado River, completed in 1935, created a model for the construction of large, hydroelectric dams.

Source: Natural Resources Conservation Service/U.S. Department of Agriculture (USDA)

claimed by a floating fern, *Ceratopteris.* Identical problems plague most rain forest hydropower projects.

The expense, the disruption of the lives of valley residents whose homes are to be flooded, and the environmental damage of dam projects in the rain forest all may be in vain. Deforestation of river banks and clearing of vegetation for permanent agriculture usually results in accelerated erosion, rapid

sedimentation of reservoirs, and drastic reduction of electrical generating capacity. The Ambuklao Reservoir in the Philippines, built with an expected payback period of 60 years, now appears certain to silt up in half that time. The Anchicaya Reservoir in Colombia lost 25 percent of its storage capacity only 2 years after it was completed and was almost totally filled with silt within 10 years. The Peligre Dam in Haiti was completed

in 1956, with a life expectancy of at least 50 years; siltation reduced its usefulness by some 15 years. El Cajón Dam in Honduras, Arenal in Costa Rica, Chixoy in Guatemala, and many others—all built to last decades or even centuries—have, because of premature siltation, failed to repay their costs or fulfill their promise. The price of deforestation in wet tropics is high indeed.

Certainly much of past desertification has been induced by nature rather than by humans. Over the past 10,000 years, for example, several prolonged and severe droughts far more damaging than the *Dust Bowl* period of the 1930s converted vast stretches of the Great Plains from Texas and New Mexico to Nebraska and South Dakota into seas of windblown sand dunes like those of the Sahara. Such conditions were seen most recently in the 18th and 19th centuries, before the region was heavily settled, but after many explorers and travelers noted—as did one in 1796 in present-day Nebraska—"a great desert of drifting sand, without trees, soil, rock, water, or animals of any kind." Today, those same areas in the Great Plains are covered only thinly by vegetation and could revert to shifting desert—as they almost did in the 1930s—with a prolonged drought of the type that might accompany global climate change.

Whether natural or anthropogenic, every year desertification makes 12 million hectares (46,000 square miles) useless for cultivation. Regardless of its causes, it begins in the same fashion: the disruption or removal of the native cover of grasses and shrubs through farming or overgrazing. If the disruption is severe enough, the original vegetation cannot reestablish itself and the exposed soil is made susceptible to erosion during the brief, heavy rains that dominate precipitation patterns in semi-arid regions. Water runs off the land surface instead of seeping in, carrying soil particles with it and leaving behind an *erosion pavement.* When the water is lost through surface flow rather than seepage downward, the water table is lowered. Eventually, even deep-rooted bushes are unable to reach groundwater, and all natural vegetation is lost. The process is accentuated when too many grazing animals pack the dirt down with their hooves, blocking the passage of air and water through the soil. When both plant cover and soil moisture are lost, desertification has occurred.

Worldwide, desertification affects about 1 billion people in 110 countries and about 1.2 billion hectares of land—about the size of China and India combined. According to the United Nations—which declared 2006 the "International Year of Deserts and Desertification" to address their problems and

solutions—between one-quarter and one-third of the planet's land surface now qualifies as degraded semidesert. Africa is most at risk; the United Nations has estimated that 40 percent of that continent's nondesert land is in danger of human-induced desertification. But nearly a fifth of Latin America's lands and a third of Asia's are similarly endangered. China's Environmental Protection Agency reports, for example, that the country lost 94,000 square kilometers (36,000 square miles)—an area the size of Indiana—to desert from the 1950s to early in the 21st century and each year has an additional 3,900 square kilometers (1,500 square miles) buried by sand.

In countries where desertification is particularly extensive and severe (Algeria, Ethiopia, Iraq, Jordan, Lebanon, Mali, and Niger) per capita food production has declined. The resulting threat of starvation spurs populations of the affected areas to increase their farming and livestock pressures on the denuded land, further contributing to desertification. It has been suggested that Mali may be the first country in the world rendered uninhabitable by environmental destruction. Many of its more than 11 million inhabitants begin their day by shoveling their doorways clear of the night's accumulation of sand (**Figure 13.20a**). The United Nations has identified desertification as a major barrier to poverty elimination in arid regions and has established programs to fight desertification (**Figure 13.20b**).

Soil Erosion

Desertification is but one example of land deterioration. Over much of the Earth's surface, the thin layer of topsoil upon which life depends is only a few inches deep, usually less than 30 centimeters (1 foot). Below it, the lithosphere is nearly as lifeless as the surface of the moon. **Soil** is a complex mixture of rock particles, inorganic mineral matter, organic material, living organisms, air, and water. Under natural conditions, soil is constantly being formed by the physical and chemical decomposition of rock material and by the decay of organic matter. It is simultaneously being eroded, for soil erosion is as natural a process as soil formation and occurs even when land is totally covered by forests

(a)

(b)

Figure 13.20 Desertification is the advance of the world's deserts into adjacent lands. *(a)* In this scene, windblown dust engulfs a scrub forest in a drought-stricken area of Mali, near Timbuktu. This region is part of the Sahel of Africa where desertification is accelerated by climate fluctuations and human pressures on the land. Here, the margins of the Sahara ebb and flow. *(b)* This windbreak and garden in Mali are attempts to combat desertification.

(a) ©Lissa Harrison; (b) ©Jose Azel/Getty Images

or grass. Under most natural conditions, however, the rate of soil formation equals or exceeds the rate of soil erosion, so soil depth and fertility tend to increase with time.

When land is cleared and planted to crops or when the vegetative cover is broken by overgrazing, deforestation, or other disturbances, the process of erosion inevitably accelerates. When its rate exceeds that of soil formation, the life-sustaining veneer of topsoil becomes thinner and eventually disappears, leaving behind only sterile subsoil or barren rock. At that point the renewable soil resource has been converted through human impact into a nonrenewable and dissipated asset. Carried to the extreme of bare rock hillsides or wind-denuded plains, erosion spells the total end of agricultural use of the land. Throughout history, such extreme human-induced destruction has occurred and been observed with dismay.

Any massive destruction of the soil resource could spell the end of the civilization it had supported. For the most part, however, farmers—even those in difficult climatic and topographic circumstances—devised ingenious ways to preserve and even improve the soil resource on which their lives and livelihoods depended. Particularly when farming was carried on outside of fertile, level valley lands, farmers' practices were routinely based on some combination of crop rotation, fallowing, and terracing.

Rotation involves the planting of two or more crops simultaneously or successively on the same area to preserve fertility or to provide a plant cover to protect the soil. **Fallowing** leaves a field idle (uncropped) for one year or more to achieve one of two outcomes. In semiarid areas, the purpose is to accumulate soil moisture from one year to apply to the next year's crop; in tropical wet regions, as we saw in Chapter 8, the purpose is to renew soil fertility of the swidden plot. **Terracing** replaces steep slopes with a series of narrow-layered, level fields, providing cropland where little or none existed previously. In addition, because water moving rapidly down-slope has great erosive power, breaking the speed of flow by terracing reduces the amount of soil lost. Field trials in Nigeria indicate that cultivation on a 1 percent slope (a drop of 1 foot in elevation over 100 feet of horizontal distance) results in soil loss at or below the rate of soil formation; farming there on a 15 percent slope would totally strip a field of its soil cover in only 10 years.

Pressures on farmlands have increased with population growth and the intensification of agriculture. Farming has been forced higher up on steeper slopes, more forest land has been converted to cultivation, grazing and crops have been pushed farther and more intensively into semiarid areas, and existing fields have had to be worked more intensively and less carefully. Many traditional agricultural systems and areas that were ecologically stable and secure as recently as 1950, when world population stood at 2.5 billion and subsistence agriculture was the rule, are disintegrating under the pressures of 7.5 billion people and an integrated global economy.

The evidence of that deterioration is found in all parts of the world (**Figure 13.21**). Soil deterioration expresses itself in two ways: through decreasing yields of cultivated fields themselves and in increased stream sediment loads and downstream deposition of silt. In Guatemala, for example, some 40 percent of the productive capacity of the land has been lost through erosion, and several areas of the country have been abandoned because agriculture has become economically impracticable; the figure is 50 percent in El Salvador. In Turkey, a reported 75 percent of the land is affected, and more than half is severely eroded. Haiti has no high-value soil left at all. A full one-quarter of India's total land area has been significantly eroded. Between 1960 and 2000, China lost more than 15 percent of its total arable land to erosion, desertification, or conversion to nonagricultural use. Its Huang River is the most sediment-laden of any waterway on Earth; in its middle course it is about 50 percent silt by weight, just under the point of liquid mud. Sediment washed into waterways results in reduced reservoir capacity, fish kills, and dredging costs (see the feature "Dam Trouble in the Developing World").

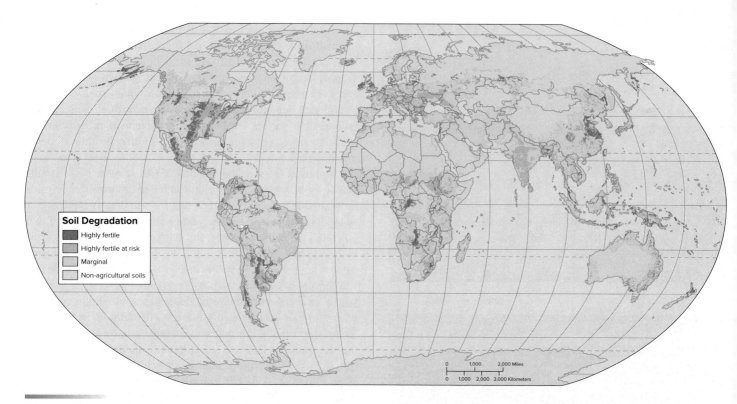

Figure 13.21 Soil fertility and vulnerability to degradation. Between 1945 and 2000, nearly 2 billion hectares (almost 5 billion acres) of the world's 8.7 billion hectares (21.5 billion acres) of cropland, pastures, and forests used in agriculture—an area as large as Russia and India combined— were added to the existing total of degraded soils. Globally, about 18 percent of forest area, 21 percent of pastures, and 37 percent of cropland have undergone moderate to severe degradation. The causes for soil degradation, in order of importance, are water erosion, wind erosion, chemical deterioration due to salinization or nutrient loss, and physical degradation due to compaction or waterlogging.

Map: John L. Allen, Student Atlas of World Geography, 7th edition, McGraw-Hill, p. 124. Data: World Resources Institute and International Soil Reference and Information Center.

Agricultural soil depletion through erosion—and through salt accumulation and desertification—has been called "the quiet crisis." It continues inexorably and unfolds gradually, without the abrupt attention attracted by an earthquake or volcanic explosion. Unfortunately, silent or not, productive soil loss is a crisis of growing importance and immediacy, not just in the countries of its occurrence but—because of international markets and relief programs—throughout the world. Conservation measures, however, can make a difference. In the United States, soil erosion decreased by almost half between 1982 and 2007. In sub-Saharan Africa, natural soil regeneration and agro-forestry projects have succeeded in reducing erosion and restoring soils to productive uses.

13.3 Impacts on Water Resources

Water is essential to all life on Earth. Our bodies are about 60 percent water and about 70 percent of the Earth's surface is covered by water. The supply of water is essentially constant, but highly uneven in its distribution. Careful management of water resources, often for irrigation, has been essential to many early human civilizations such as those of Egypt, Mesopotamia, the Indus Valley, and Mesoamerica (see Figure 2.15). Ancient civilizations that failed to properly manage water resources such as the Anasazi of the American Southwest were subject to collapse

(see the feature "Chaco Canyon Desolation" in Chapter 2). Today, managing water resources is critical because rising human populations and increased agricultural, urban, and industrial development have stressed water resources in many regions.

Water Availability

The problem is not with the global amount of water, but with its distribution, its availability, and its quality. The vast majority of water on Earth (96.5 percent) is saltwater found in oceans, seas, and bays, and most freshwater is in ice caps and glaciers. Only about 1 percent of all water is available as liquid fresh water, and most of that is beneath the surface in groundwater **aquifers** that may be difficult and expensive to pump to the surface. Even so, enough rain and snow fall on the continents each year to cover the Earth's total land area with 83 centimeters (33 inches) of water. Observations about global supplies of fresh water ignore the ever-present geographic reality: things are not uniformly distributed over the surface of the Earth. Populations are rising in many regions where water supplies are limited. For example, the Middle East and North Africa are home to 6.3 percent of the world's population but contain only 1.4 percent of the world's renewable fresh water. Idaho, Nevada, and Utah were the three fastest-growing U.S. states in recent years and are all in arid/semiarid climates. Compounding matters, the demand for water by natural processes of evaporation and transpiration varies spatially as well. In hot, dry subtropical deserts annual evaporation may equal 250 centimeters (100 inches) of water per year or more.

For thousands of years, human cultures have responded to the uneven distribution of water in both time and space by constructing small dams and diversion canals for irrigation. During the height of the Depression, the United States built the Hoover Dam (1931–1936) and Grand Coulee Dam (1933–1942); they were bold engineering feats that raised dams to unprecedented heights and impounded enormous volumes of water. These dams generate large quantities of low-cost, renewable, hydroelectric power for cities and industry, and their reservoirs provide a reliable supply of water for cities and agriculture. Perhaps more significantly, these dams established a model for economic development of arid regions based on large water engineering projects. That model has diffused globally, promoted by international aid agencies and seen as prestigious symbols of development by developing countries. There are many economic, social, and environmental costs of large dams, but they are still being built in developing countries (see the feature "Dam Trouble in the Developing World" earlier in this chapter). Today, the flow in more than half the world's major rivers has been altered significantly by dams.

Within the hydrologic cycle, surface runoff takes place within basins or watersheds that are determined by topography; ridgelines divide the waters flowing toward different streams and rivers. Thus, river basins are the primary areal unit to use when studying water. We can then calculate and compare the quantity of renewable water available per person within different river basins. Areas of high water stress are defined as those with less than 1,000 cubic meters of water per person per year. Of course, not all of this water is needed for human consumption. Rather, it is needed for producing food, sanitation, industry, and other essential uses. It is projected that with current patterns of population growth and climate change, by 2050 half the world's population will live in water-stressed areas. (**Figure 13.22**).

In North America, two large river basins with potential water stress are the Colorado and the Rio Grande, both in the Southwest and both reduced to a trickle at times by diversions for agriculture and cities. Complicating management of the limited water supplies of the Colorado and the Rio Grande is the fact that they are international rivers shared by the United States and Mexico. International treaties govern the obligations of each country with respect to quantity and quality of the river's waters. **Transboundary river basins**—basins straddling two or more countries—cover an estimated 40 percent of the Earth's land area and contain more than 60 percent of the world's population.

Water conflict is a distinct possibility when a substance so essential to life is shared by more than one country. Notable transboundary rivers with potential conflicts include the Indus, Jordan, Nile, and Tigris-Euphrates. The Indus has its source in India but provides irrigation water essential to Pakistan's food supply. Despite a series of wars between India and Pakistan, the two countries continue to observe a negotiated treaty regarding the waters of the Indus. The Jordan River basin takes in parts of Israel, Jordan, Lebanon, the Palestinian Territories, and Syria and has major issues related to scarcity and international security. In fact, the amount of renewable water available per person is below recommended minimums in Israel, Jordan, and Palestine. The Nile River drains 10 countries in Africa and is subject to colonial-era treaties that some believe favor Egypt over the upstream countries such as Ethiopia. The Tigris-Euphrates was among the birthplaces of civilization and today is shared by Turkey, Syria, and Iraq, each with their own water development objectives. Yet, despite the very real possibility of water wars, international cooperation has been more common than hostilities.

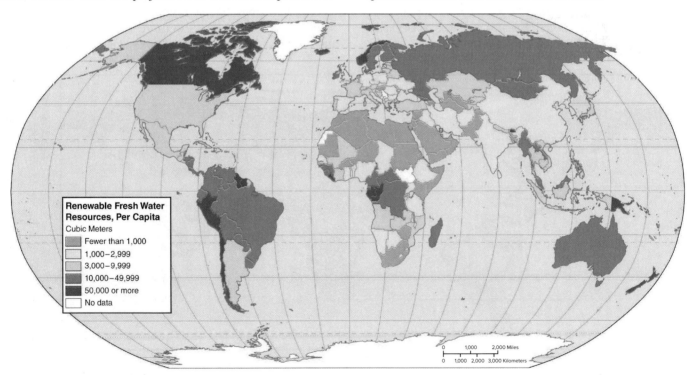

Figure 13.22 Freshwater availability, 2007. Highly stressed areas are those with less than 1,000 cubic meters of water per person per year. In North America, the Rio Grande basin is predicted to be highly stressed and the Colorado River basin stressed.

Source: Sutton C. Student Atlas of World Geography, 8th edition, 2014, based on World Bank, World Development Indicators, *2012.*

The climate system and hydrosphere are intimately connected. Climate change is likely to alter the hydrologic cycle, increasing both the demand for water and annual precipitation in many areas. Climate models forecast altered patterns of water availability with increased dryness in many areas already subject to water scarcity (**Figure 13.23**). Of particular concern is the American Southwest and Indus Basin, where growing populations will need to adapt to declining water supplies. In both regions, declining snowpack in the mountains will lead to decreased river flow in the dry season.

Water supplies and energy supplies are intimately connected. If energy supplies were unlimited, so too would be water supplies. The ocean offers almost unlimited water supplies but making that water drinkable requires expensive, energy-intensive desalination. The world's largest desalination plants are in the Middle East, where water is scarce and energy supplies are abundant. Wastewater reclamation, also quite expensive, is the treatment of sewage so that it can be reused for irrigation or as a drinking water source. For cities in desert regions, their own sewage is often the last remaining untapped water resource. Australia, Israel, Singapore, Florida, and Southern California are leaders in the use of reclaimed water. Use of reclaimed sewage for drinking water was used in the space program and is now growing in acceptance for city supplies. Water reclamation for drinking water uses ultrafiltration technologies and reinjection into aquifers to address health concerns.

Water Use and Abuse

Water supplies and food supplies are intimately connected. Irrigation uses about three-fourths of all water used by humans and a higher percentage in the least developed countries. Irrigation agriculture produces some 40 percent of the world's total harvest and 60 percent of its grain from about 17 percent of its cropland. The productivity gains of the Green Revolution depended on an expansion of irrigation, as will future increases in the food supply. In dry climates, rivers and lakes have shrunk or even disappeared due to irrigation demands. The Aral Sea in Kazakhstan and Uzbekistan was once the world's fourth-largest lake by area before irrigation diversions caused it to shrink by 90 percent (**Figure 13.24**). Groundwater can be used unsustainably if it is pumped faster than it is replenished by natural infiltration. In the United States, a major concern is the dropping groundwater levels in the Ogallala Aquifer, which provides irrigation water for agriculture in eight Great Plains states.

When humans introduce wastes into the biosphere in kinds and amounts that the natural system cannot neutralize or recycle, the result is **environmental pollution.** In the case of water, pollution exists when water composition has been so modified by the presence of one or more substances that either it cannot be used for a specific purpose or it is less suitable for that use than it was in its natural state. In both developed and developing countries, human pressures on freshwater supplies are now serious and pervasive concerns (see the feature "A World of Water Woes").

Water pollution can come in a variety of forms. Human wastes often contain infectious agents that cause waterborne diseases such as cholera, dysentery, and typhoid fever. Waterborne diseases kill an estimated 1.5 million persons each year, largely due to lack of sanitation, sewage treatment, and/or access to safe drinking water supplies. An estimated 2.4 billion people around the world lack basic sanitation such as a latrine or toilet and 1.8 billion use drinking water sources containing fecal contamination. Thus, one of the United Nations' Millennium Development Goals (MDGs) was to cut in half the proportion of the population without access to safe drinking water and basic sanitation between 1990 and 2015. The United Nations' Sustainable Development Goals call for ensuring safe drinking water and basic sanitation for all people by 2030. The world regions that lag behind are sub-Saharan Africa and southern Asia, although all world regions have made substantial progress (**Figure 13.25**). The expansion of cities places

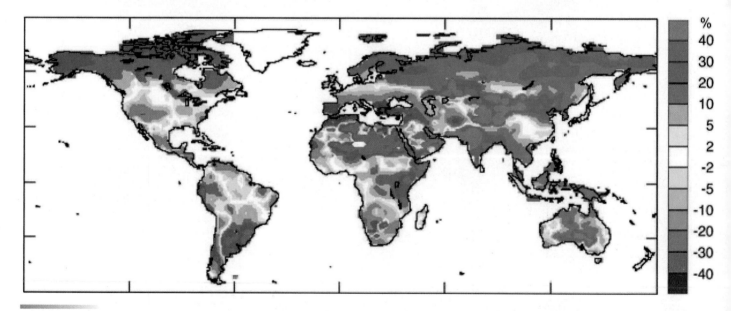

Figure 13.23 Predicted change in annual runoff by the mid-21st century compared to 1900–1970. This map from the IPCC report is based on 12 global climate model simulations using middle-range assumptions about greenhouse gas emissions. A substantial reduction in water availability is predicted for a number of already dry regions such as the American Southwest, Mexico, southern Africa, southern Europe, northern Africa, and the Middle East.

Source: Intergovernmental Panel on Climate Change, 2007. Climate Change 2007: Working Group II: Impacts, Adaptation, and Vulnerability, *Figure 3.4.*

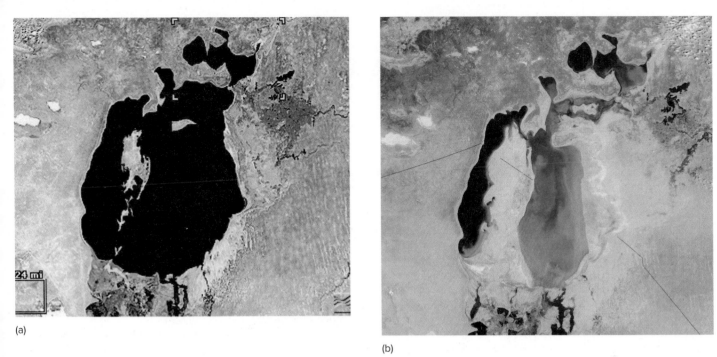

(a)

(b)

Figure 13.24 The Aral Sea has dramatically declined in size and depth. *(a)* This map from 1975 shows the original extent of the sea when it was the fourth-largest inland sea in the world. *(b)* In this satellite image from 2005, the Aral Sea has shrunk due to diversions of water for irrigated agriculture. Water levels have declined by 18 meters (59 feet), causing the lake to split into two parts. Fishing, port, and resort towns are now 80 kilometers (49.6 miles) or farther from the water. The remaining flow into the sea is contaminated by agricultural runoff, yielding a toxic, salty brine. Dust storms whip up particulates and toxic chemicals from the dry lake bed. Restoration work is being done on the Little Aral Sea in the north in Kazakhstan.

Source: EROS Data Center USGS

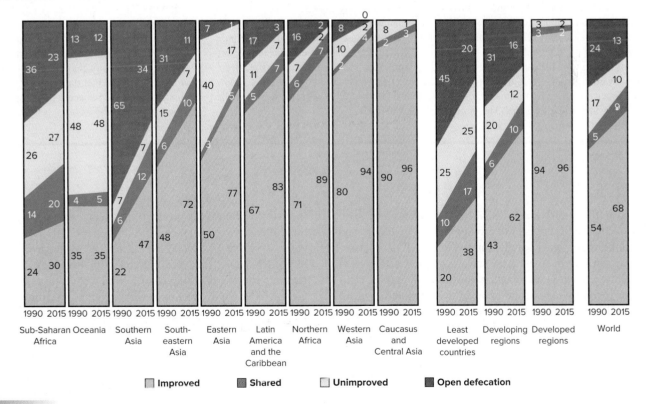

Figure 13.25 The availability of sanitation in each world region improved between 1990 and 2015. While improved facilities (flush toilets and sewage disposal systems) are the norm in developed countries, only 62 percent of people in developing regions had access to this level of sanitation in 2015. The United Nations Millennium Development Goals called for cutting in half the proportion without improved sanitation. Substantial progress on sanitation was made between 1990 and 2015, promising major benefits for human health.

A World of Water Woes

Water covers almost three-quarters of the surface of the globe, yet *scarcity* is the word increasingly used to describe water-related concerns in both the developed and developing world. Globally, fresh water is abundant. Each year, an average of more than 7,000 cubic meters (some 250,000 cubic feet) per person enters rivers and underground reserves. But rainfall does not always occur when or where it is needed. Already, 80 countries with 40 percent of the world's population have serious water shortages that threaten to cripple agriculture and industry; 22 of them have renewable water resources of less than 1,000 cubic meters (35,000 cubic feet) per person—a level generally understood to mean that water scarcity is a severe constraint on the economy and public health. Another 18 countries have less than 2,000 cubic meters per capita on average, a dangerously low figure in years of rainfall shortage. Most of the water-short countries are in the Middle East, North Africa, and sub-Saharan Africa, the regions where populations (and consumption demands) are growing fastest. By 2025, two-thirds of the world's people are likely to be living in areas of acute water stress.

In several major crop-producing regions, water use exceeds sustainable levels, threatening future food supplies. The largest underground water reserve in the United States, stretching from west Texas northward into South Dakota, is drying up, partially depleted by more than 150,000 wells pumping water for irrigation, city supply, and industry. In parts of Texas, Oklahoma, and Kansas, the underground water table has dropped by more than 30 meters (100 feet). In some areas, the wells no longer yield enough to permit irrigation, and farmed land is decreasing; in others, water levels have fallen so far that it is uneconomical to pump it to the surface for any use.

In many agricultural districts of northern China, west and south India, and Mexico, water scarcity limits agriculture even though national supplies are adequate. In Uzbekistan and adjacent sections of Central Asia and Kazakhstan, virtually the entire flow of the area's two primary rivers—the Amu Darya and the Syr Darya—is used for often wasteful irrigation, with little left to maintain the Aral Sea or supply growing urban populations. In Poland, the draining of bogs that formerly stored rainfall, combined with unimaginable pollution of streams and groundwater, has created a water shortage as great as that of any Middle Eastern desert country. And salinity now seriously affects productivity—or prohibits farming completely—on nearly 10 percent of the world's irrigated lands.

Water scarcity is often a regionwide concern. More than 200 river systems draining more than half the Earth's land surface are shared by two or more countries. Egypt draws on the Nile for 86 percent of its domestic consumption, but virtually all that water originates in eight upstream countries. Turkey, Iraq, and Syria have frequently been in dispute over the management of the Tigris and Euphrates rivers, and the downstream states fear the effect on them of Turkish impoundments and diversions. Mexico is angered at the U.S. depletion of the Colorado River before it reaches the international border.

Many coastal communities face saltwater intrusions into their drinking water supplies as they draw down their underlying freshwater aquifers, while both coastal and inland cities dependent on groundwater may be seriously depleting their underground supplies. In China, 110 mostly large cities face acute water shortages; for at least 50 of them, the problem is groundwater levels dropping on average 1 to 2 meters (3 to 6 feet) each year. In Mexico City, groundwater is pumped at rates 40 percent faster than natural recharge; the city has responded to those withdrawals by sinking 30 feet during the 20th century. Millions of citizens of major cities throughout the world have had their water rationed as underground and surface supplies are used beyond recharge or storage capacity.

additional burdens of sewage discharges on rivers, bays, and estuaries. However, it is easier to provide sewage treatment and safe drinking water when populations are clustered in urban areas. Sanitation improvements and sewage treatment have been major components of modernization and the demographic transition, but currently lag in rural areas of the developing world.

Agriculture is the leading cause of poor water quality in the rivers of most countries. Large confined animal feeding operations and fertilizer additions to fields have contributed to rising agricultural productivity. However, both practices lead to excessive nutrients in runoff that create coastal **dead zones** (**Figure 13.26**). These "dead" zones have little or no oxygen because they are choked with decaying algae that grew on the nutrient rich runoff from agricultural lands. The largest dead zones are found where rivers draining rich agricultural regions discharge to coastal bays or estuaries, such as the mouth of the Mississippi River in the Gulf of Mexico (**Figure 13.27**).

Figure 13.26 This slum in Mumbai, India, sits atop a drainage channel and lacks access to clean water and sanitation. Waterborne diseases, mostly stemming from water supplies contaminated with human wastes, are estimated to cause 1.5 million deaths a year.

©McGraw-Hill Education/Barry Barker, Photographer

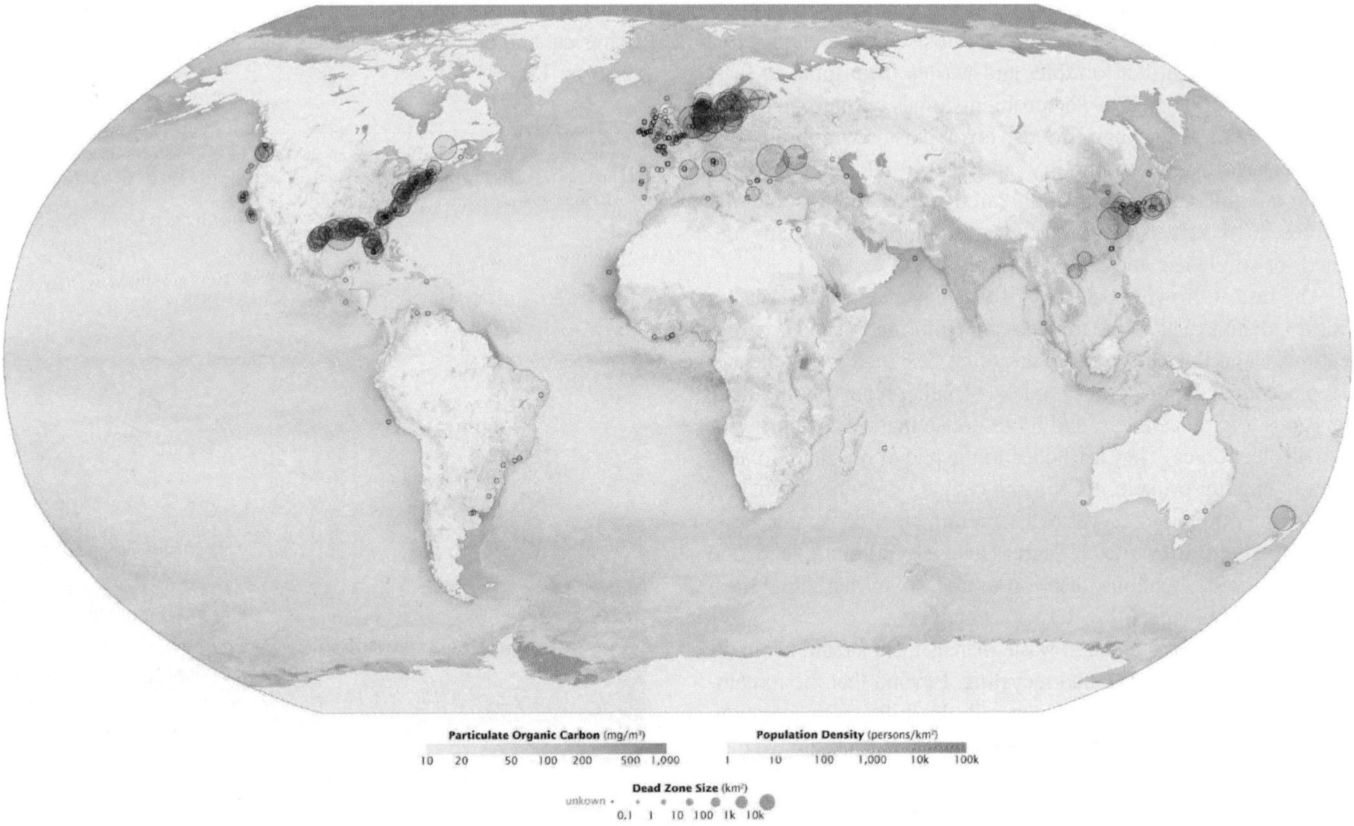

Particulate Organic Carbon (mg/m³)

10   20   50   100   200   500   1,000

Population Density (persons/km²)

1   10   100   1,000   10k   100k

Dead Zone Size (km²)

unkown

0.1   1   10   100   1k   10k

Figure 13.27 Coastal dead zones are places so low in dissolved oxygen that most marine life cannot survive. Nutrients in sewage and agricultural runoff are the most common culprit. Note the prevalence of dead zones where populations are highest in the developed countries.

13.4 Wastes

Humans have always managed to leave their mark on the landscapes that they occupy. Among the most enduring of landscape evidences of human occupancy is the garbage produced and discarded by every society. Prehistoric dwelling sites are located and analyzed by their *middens,* the refuse piles containing the kitchen wastes, broken tools, and other debris of human settlement. We have learned much about Roman and medieval European urban life by examination of the refuse mounds that grew as man-made hills in their vicinities. In the Near East, whole cities gradually rose on the mounds of debris accumulating under them (see Figure 11.9).

Modern cultures differ from their predecessors by the volume and character of their wastes, not by their habits of discard. Generally, the greater the society's population and standard of living, the greater the quantity of its garbage. Developed countries are increasingly discovering that their material wealth and technological advancements are submerging them in a volume and variety of wastes—solid and liquid, harmless and toxic—that threaten both their environments and their established ways of life.

Solid Wastes

Americans produce rubbish, garbage, and other municipal waste at a rate of about 2 kilograms (4.5 pounds) per person per day. As populations grow, incomes rise, and consumption patterns change, the volume of disposable materials continues to expand. Relatively little residue is created in subsistence societies that move food from garden to table, and wastes from table to farm animals or compost heaps. The problem comes with urban dwellers who purchase packaged foods, favor plastic wrappings and containers for every commodity, and seek (and can afford) an ever-broadening array of manufactured goods, both consumer durables such as refrigerators and automobiles and many designed for single use and quick disposal.

The fastest growing category of waste is electronic waste (or e-waste), thanks to the rapid innovation in information and communications technologies. Products are often designed to be cheaper to replace than repair. New models of cell phones, computers, televisions, digital cameras, and iPods create massive quantities of obsolete electronics. They are added to the pile of obsolete fax machines, computer printers, VCRs, and cathode ray tube televisions and computer monitors. E-waste differs from other consumer waste in that it contains toxic metals such as arsenic, cadmium, lead, and mercury. In most locations, disposal of e-waste in municipal trash is forbidden, although still widely practiced.

The first priorities for solid waste management should be, in order, waste reduction, reuse, and recycling. Beyond that the remaining options for solid wastes are generally landfills or incineration. Landfills are an improvement over the unregulated, open dumps they replaced and receive the bulk of solid wastes in most countries. The solid wastes are compacted inside a lined cell and covered with soil at the end of each day. Still, groundwater contamination due to water infiltration through the waste is a major technical challenge to address. Beyond technical details, an increasing problem is where to locate a landfill because most communities will protest, expressing **Not in My Backyard (NIMBY)** sentiments.

Over the years, of course, many filled dumps have posed problems for the cities that gave rise to them. New York City, for example, for years placed all of its daily 14,000 tons of residential waste into the world's largest dump, Fresh Kills on Staten Island. Opened in 1947 as a three-year "temporary" 500-acre facility, it became a malodorous 1,214 hectares (3,000 acres) of decomposing garbage rising 15 stories above the former ground level. Generating 140,000 cubic meters (5 million cubic feet) of methane gas annually and illegally exuding contaminated water, Fresh Kills—finally closed in 2001 at a cost of more than $1 billion—symbolized the rising tide of waste engulfing cities and endangering the environment.

Incineration recovers some of the energy in waste by burning it to produce steam or electricity. Incinerators produce air pollution, including highly toxic metals and organic compounds such as dioxin,[1] so pollution control equipment is required. As with landfills, proposing a location for a solid waste incinerator often leads to community resistance.

For coastal communities around the world the ocean has long been the preferred sink for not only municipal garbage, but for (frequently untreated) sewage, industrial waste, and all the detritus of an advanced urban society. Along the Atlantic coast of North America from Massachusetts to Chesapeake Bay, reports of dead dolphins, raw sewage, tar balls, used syringes, vials of contaminated blood and hospital waste, diapers, plastic products in unimagined amounts and varieties, and other foul refuse has kept swimmers from the beach, closed coastal shellfisheries, and elicited health warnings against entering the water (**Figure 13.28**).

---

[1]Any of several types of hydrocarbon compounds that are extremely toxic, persistent in the environment, and biologically magnified in the food chain. Dioxin is often formed during incineration of waste matter.

Figure 13.28 Warning signs and beaches littered with sewage, garbage, and medical debris are among the increasingly common and distressing evidences of ocean dumping of wastes.

©*ROGER A. CLARK/Science Source*

Environmental Justice

In Houston, a city of some 2 million people, about 25 percent of the population is African American. Yet, when researchers examined the placement of garbage facilities in the city, they found that 11 of 13 solid waste disposal facilities owned by the city were in mostly black areas and all five of the city's garbage incinerators were in black and Hispanic neighborhoods. Thus, when the city proposed establishing a new dump in a primarily black neighborhood in 1979, near houses and a high school, local residents protested and brought the waste management company to court, charging it with racial discrimination in the selection of the landfill site. The court decided in favor of the company, and the landfill was built.

In 1982, a few states away, in Warren County, North Carolina, the rural, mostly African American residents were shocked to learn that the state was proposing their county as the site of a hazardous waste landfill for disposing of polychlorinated biphenyls (PCBs). Their protests resulted in more than 500 arrests, and the effort to block the landfill failed. The Warren County activists were the first to use the term *environmental racism*.

Environmental racism refers to any policy or practice that differentially affects or harms individuals, groups, or communities because of their race or color. The harm may be intentional or unintentional. **Environmental justice** is the fair treatment and meaningful involvement of all people regardless of race, color, national origin, or income with respect to the development, implementation, and enforcement of environmental laws, regulations, and policies. No group of people should bear a disproportionate share of negative environmental consequences. In many cases, environmental racism and injustice result from long-established and unexamined structural inequality: those who have a reduced political voice compared to the dominant groups in society live in the worst environments because authorities and companies have faced little opposition in placing hazardous facilities, such as landfills, in their vicinity.

The problem is not confined to the United States. In most countries, poorer people and minorities tend to live in areas that are disproportionately exposed to polluting industries, waste disposal facilities, toxic soils, and polluted air and water, in. In Kagiso township, southwest of Johannesburg, South Africa, poor African residents live near a gold mine. In the past, when the gold was extracted from the ore, the waste product was pumped into a waste pile a distance from the residential area. However, the pile was eventually expanded to the extent that it now lies less than 27.4 meters (30 yards)—or less than half the length of a typical city block—from some of the houses. The drying waste produces dust that is high in alpha quartz particles, or silica, which cause silicosis. The inhalation of silica causes lung disease characterized by shortness of breath, fever, and cyanosis, and it also leaves its victims susceptible to tuberculosis. The condition is irreversible. The township is dominated by *informal* housing, which has few city services, is overcrowded, and has poor sanitation; it includes areas that were set aside for black Africans during apartheid.

The United States has taken some steps to rectify the wrongs of environmental injustice. Following the Warren County, North Carolina, protests, President Bill Clinton signed an executive order that made environmental justice a national priority and directed all federal agencies to develop policies to reduce environmental inequity, in 1992, the EPA established an Office of Environmental Equity (now called the Office of Environmental Justice). The agency, however, has had an uneven record since its founding. Organizations outside the government continue to call attention to disturbing conditions often using a geographic information systems (GIS) to identify inequalities.

Consider Louisiana's Chemical Corridor, or Cancer Alley, a 130-kilometer (80-mile) stretch of the Mississippi River that is home to a chemical plant every half mile and a predominantly poor, African American population. While community groups protested a proposed new polyvinyl chloride (PVC) plant, the state government gave the owners tax exemptions. The EPA delayed the new plant's permit until the state addressed the citizens' environmental justice concerns. In the end, the owners bowed to community opposition and built the new plant 48 kilometers (30 miles) upstream, which may represent a Pyrrhic victory because the plant will still contaminate the air and water nearby.

Environmental injustice has its origins both in neglect and overt discrimination.

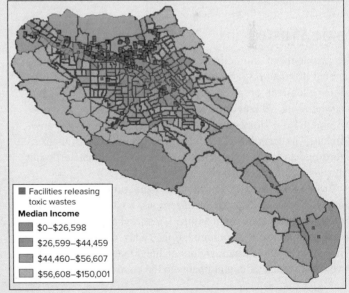

Figure 13B Poverty and Environmental Risk. The distribution of household income and facilities that release toxic materials in Santa Clara County, California. Black lines represent census tracts. (Source: Data from M.R. Meusar and A. Szosz "Environmental Inequality in Silicon Valley." www.mapcnjzin .com/EI/index.hmi.)

There are many instances of intentional placement of environmentally hazardous plants and waste sites in areas already home to minorities and the poor. Government attempts to address the issue of environmental injustice have had an uneven track record. A body of case law regarding environmental justice in the United States has not developed yet, and some scholars contend that, because the issue falls between civil rights law and environmental law, development will continue only slowly.

Questions to Consider

1. Environmentally hazardous institutions—landfills, chemical plants, factories—have to be placed somewhere. How should communities and governments make decisions among different sites? Should the activism of local populations affect government decisions?

2. How would you or your college or university react if a company wanted to site a new landfill for toxic wastes nearby?

3. Where are landfills, polluting industries, and other environmental hazards located in your hometown or where you attend school? Are low-income residents and minorities exposed to greater-than-average amounts of pollution and thus bearing greater health risks than the rest of the population?

Adapted from: *Getis and Getis*, Introduction to Geography, *12th ed.*, McGraw-Hill, 2008, 326–327.

Whether the solution to solid waste disposal is sought by land, by fire, or by sea, humanity's rising tide of refuse threatens to overwhelm the environments that must deal with it. The problems that continue to surface remind us that the Earth's environmental systems are interrelated, so that all too often, our solutions simply move the problem from the land to the sea or air or groundwater. The problem is present, growing, and increasingly costly to manage. Solutions are still to be found, a constant reminder for the future of the threatening impact of the environments of culture upon those of nature.

Toxic Wastes

The problems of municipal and household solid-waste management are daunting; those of treatment and disposal of hazardous chemical or radioactive wastes from industry seem overwhelming. **Hazardous wastes** are substances that pose an immediate or long-term risk to human health or the environment. The EPA has classified more than 400 substances as hazardous, and currently about 10 percent of industrial waste materials are so categorized.

Prior to the introduction of strict regulations governing the transportation and disposal of hazardous wastes in the 1970s, hazardous wastes were often dumped or buried in pits or lagoons, leading to groundwater contamination. Today, they are disposed of in highly regulated incinerators or lined landfills designed to prevent the release of contaminants to the environment.

Radioactive Wastes

Every facility that either uses or produces radioactive materials generates at least *low-level waste,* material whose radioactivity will decay to safe levels in 100 years or less. Nuclear power plants, industries that manufacture radiopharmaceuticals, smoke alarms, and other consumer goods, and research establishments, universities, and hospitals also produce low-level radioactive waste materials.

High-level waste can remain radioactive for 10,000 years and more; plutonium stays dangerously radioactive for 240,000 years. It consists primarily of spent fuel assemblies of nuclear power reactors—termed *civilian waste*—and such *military waste* as the by-products of nuclear weapons manufacture. The volume of civilian waste accumulates rapidly because approximately one-third of a reactor's rods need to be disposed of every year.

Spent fuel is a misleading term: the assemblies are removed from commercial reactors not because their radiation is spent, but because they have become too radioactive for further use. The assemblies will remain radioactively "hot" for thousands of years. By 2018, 90,000 metric tons of accumulated nuclear waste (spent fuel) was in storage in the containment pools or above ground in dry casks at 80 commercial nuclear power reactors, awaiting a more permanent solution (**Figure 13.29**). Stored spent fuel rods were among the most serious concerns and sources of radiation releases when the Fukushima nuclear power plant in Japan was hit by an earthquake and tsunami in 2011. Spent fuel is not contained as well as the fuel rods inside reactors, so they must be constantly covered in cool water to prevent overheating. When the earthquake and tsunami knocked out power to the water pumping systems, the spent fuel ponds overheated, contributing to radiation releases.

The U.S. Department of Energy spent two decades developing plans for a centralized nuclear waste depository inside Yucca Mountain in the desert 90 miles from Las Vegas, Nevada. The plan was to encase the radioactive waste in extremely strong glass inside steel canisters beneath 300 meters (1,000 feet) of volcanic rock. Local opponents to the plan pointed out the potential for accidents in shipment or earthquakes in the area and succeeded in having the project scrapped. No alternative, permanent disposal method has been devised. Much low-level radioactive waste has been placed in tanks and buried in the ground at 13 sites operated by the U.S. Department of Energy and three sites run by private firms. Millions of cubic feet of high-level military

(a)

(b)

Figure 13.29 Spent nuclear fuel from nuclear power plants continues to generate heat and emit radiation long after it is removed from the reactor. *(a)* Initially after it is removed, spent fuel is stored in water-filled pools equipped with continuously circulating water baths to keep it from overheating. Spent fuel remains in the cooling water a year or longer before it can be transferred to *(b)* aboveground dry casks stored at the nuclear reactor site. The lack of a permanent disposal solution in the United States means spent fuel continues to accumulate at nuclear power plants.

(a) ©Steve Allen/DigitalVision/Getty Images; (b) Source: Office of Civilian Waste Management. Department of Energy

waste are temporarily stored in underground tanks at four sites: Hanford, Washington; Savannah River, South Carolina; Idaho Falls, Idaho; and West Valley, New York. Several of these storage areas have experienced leaks, with seepage of waste into the surrounding soil and groundwater. Nuclear power plants continue to store spent fuel on site.

Exporting Wastes

The problem with throwing away wastes is that on Earth, there is no true "away." The Earth's atmosphere, hydrosphere, lithosphere, and biosphere are all interrelated, and every place is somebody's backyard. Almost everywhere that governments or industries have proposed to build landfills, hazardous waste incinerators, or nuclear waste repositories, communities and NIMBY movements have arisen in opposition. As a consequence, unwanted facilities such as these tend to be located in places where there is the least community resistance and least political power, often low-income, minority communities (see the feature "Geography and Citizenship: Environmental Justice").

In a leaked 1991 memorandum, World Bank chief economist Larry Summers shared his opinion that it made perfect economic sense to encourage the movement of dirty industries to poor countries. He argued that residents of less-developed countries were less likely to be concerned about the environment, and if they suffered negative health effects from pollution, the economic cost would be lower because their wages were so much lower. He wrote, "I think the economic logic behind dumping a load of toxic waste in the lowest-wage country is impeccable, and we should face up to that."[2]

Regulations, community resistance, and steeply rising costs of disposal of hazardous wastes in the developed countries encouraged producers of those unwanted commodities to seek alternate areas for their disposal. Transboundary shipments of dangerous wastes became an increasingly attractive option for producers, continuing the globalization of the world economy. In total, such cross-border movement amounted to tens of thousands of shipments annually by the early 1990s, with destinations including debt-ridden Eastern European countries and impoverished developing ones outside of Europe that were willing to trade a hole in the ground for hard currency. It was a trade, however, that increasingly aroused the ire and resistance of destination countries and, ultimately, elicited international agreements among both generating and receiving countries to cease the practice.

The Organization of African Unity (OAU) adopted a 1988 resolution condemning the dumping of all foreign wastes on that continent. Under the sponsorship of the United Nations, 117 countries adopted a treaty—the Basel Convention on the Control of Transboundary Movements of Hazardous Wastes—aimed at regulating the international trade in wastes. That regulation was to be achieved by requiring exporters to obtain consent from receiving countries before shipping waste and by requiring both exporter and importer countries to ensure that the waste would be disposed of in an environmentally sound manner. The Basel Convention came into force in 1992, but the United States has not yet ratified it.

The European Union allows its members to export hazardous wastes only to developed countries that are assumed to have satisfactory treatment capabilities. However, the lack of global

[2]Quoted from David Harvey, *Justice, Nature, and the Geography of Difference*, 366, 2006.

ratification of the Basel Convention means that export of hazardous wastes to developing countries continues.

The line between reusable products and wastes is often fuzzy and depends on local standards of living and costs of labor. Thus, obsolete manufactured goods and recyclable materials are often shipped from high-income countries to low-income developing countries for the labor-intensive and potentially dangerous tasks of sorting, disassembly, and recycling. An estimated 80 percent of e-waste collected in the United States for recycling is exported to areas such as China, India, Pakistan, Nigeria, and Mexico. Reports warn of environmental pollution and accidental toxic exposures due to the unregulated recycling of e-waste in low-income countries.

13.5 Future Prospects and Perspectives

Not surprisingly, the realities of the human impacts upon the environment that we have looked at in this chapter bring us directly back to ideas first presented in Chapter 2, at the start of our examination of culture and the development of human geographic patterns on the surface of the Earth. Humans, in their increasing numbers, growing technical sophistication, rising standards of living, and expanding global reach, have transformed the Earth's landscapes since the end of the last glaciation. Humans have adopted a domineering role in the human-environment relationship, all too often forgetting that they depend upon the environment for their very existence.

That dominance is reflected in the growing divergence of human societies as they distance themselves from common hunting-gathering origins. In creating their differing cultural solutions to common concerns of sustenance and growth, societies altered the environments they occupied. Diverse systems of exploitation of the environment were developed in and diffused from distinctive culture hearths. They were modified by the ever-expanding numbers of people occupying Earth areas of differing carrying capacities and available resources. Spatial interaction among regions did not halt the creation of distinctive regional subsystems of culture or assure common methods of utilization of Earth resources. Sharp contrasts in levels of economic development and well-being emerged and persisted even as cultural convergence through shared technology began to unite societies throughout the world.

Each culture placed its imprint on the environment it occupied. In many cases—Chaco Canyon and Easter Island were our earlier examples—that imprint was ultimately destructive of the resources and local environments upon which the cultures developed and depended. To satisfy their felt needs, humans have learned to manipulate their environment. The greater those needs and the larger the populations with both needs and technical skills to satisfy them, the greater is the manipulation of the natural landscape.

Paralleling the global reach of the world economy, the human impact on the environment has shifted scales from the local or regional to the continental and global scales. Increasingly, the most pressing environmental issues are those that cross international boundaries, such as depletion of the ozone layer, acid precipitation, global climate change, managing transboundary river basins, and transboundary shipment of wastes. This final chapter, detailing a few of the damaging pressures placed upon the environment by today's economies and cultures, is not meant as a litany of despair. Rather, it is a reminder of the potentially destructive ecological dominance of humans alongside examples of humans learning to work together to reduce their impacts. The world's diverse religions, belief systems, and cultures can each offer resources to guide human behavior in ways that are more respectful of the Earth. The scientific and technological advances that lie behind many of our environmental problems can also be used to monitor and restore the environment.

Against the background of our now fuller understanding of human geographic patterns and interactions, this chapter is another reminder of the often repeated truism that everything is connected to everything else; we can never do just *one* thing. The ecological crises described in this chapter show how intimately our created environment is joined to the physical landscape we all share. There is growing awareness of those connections, of the adverse human impacts upon the natural world, and of the unity of all cultural and physical landscapes. Climate change, air and water pollution, soil loss and desertification, toxic wastes, and a host of other environmental problems are all matters of contemporary public debate and international compacts, and treaties. Acceptance of the interconnectedness and indivisibility of cultural and natural environments—the human creation and the physical endowment—is now more the rule than the exception.

SUMMARY

Cultural landscapes may buffer but cannot isolate societies from the physical environments that they occupy. All human activities, from the simplest forms of agriculture to modern industry, have an impact upon the biosphere. Cumulatively, in both developed and developing countries, that impact is now evident in the form of serious and threatening environmental deterioration. As humans have increased in number and affluence, and expanded their use of fossil fuel energy sources, their impact on the Earth has shifted from local to regional, continental, and global scales.

The Earth's environmental systems of atmosphere, hydrosphere, lithosphere, and biosphere are intricately interlinked. Tropical deforestation alters the hydrologic cycle and impacts the climate system. Most threateningly, climate change will have significant effects on all the other environmental systems, altering the availability of water for agriculture and cities and changing the frequency and intensity of floods and droughts. Climate change will lead to rising sea levels and changing geographical limits for the Earth's biomes, allowing some species to expand their range and threatening others with extinction.

The atmosphere unites us all, and its global problems of rising greenhouse gases, ozone depletion, and particulate pollution endanger us all. Desertification, soil erosion, and tropical deforestation may appear to be local or regional problems, but they have worldwide implications of both environmental degradation and reduced population-supporting capacity. Freshwater supplies are deteriorating in quality and decreasing in sufficiency through contamination and competition. Finally, the inevitable end product of human use of the Earth—the garbage and hazardous wastes of civilization—are beginning to overwhelm both sites and technologies of disposal.

We do not end our study of human geography on a note of despondency, however. We end with the conviction that the deeper knowledge that we now have of the spatial patterns and structures of human cultural, economic, and political activities will aid in our understanding of the myriad ways in which human societies are bound to the physical landscapes that they occupy—and which they have so substantially modified.

KEY WORDS

acid precipitation
aquifer
atmosphere
biome
biosphere
dead zones
desertification
ecosphere
ecosystem
environment

environmental justice
environmental pollution
fallowing
global climate change
greenhouse effect
hazardous waste
hydrologic cycle
hydrosphere
IPAT equation
lithosphere

Not In My Backyard (NIMBY)
ozone
rotation
soil
soil erosion
sustainable development
terracing
transboundary river basins

FOR REVIEW

1. What are the components in the IPAT equation? How does the study of human geography inform our understanding of environmental problems?

2. Were there any evidences of human impact upon the natural environment prior to the Industrial Revolution? If so, can you provide examples? If not, can you explain why not?

3. What lines of reasoning and evidence suggest that human activity is altering global climates? What kind of alteration has occurred or is expected to occur? Which places on Earth are most vulnerable to climate change?

4. What is *desertification?* What types of areas are particularly susceptible to desertification? What kinds of land uses are associated with it? How easily can its effects be overcome or reversed?

5. What agricultural techniques have been traditionally employed to reduce or halt soil erosion? Because these are known techniques that have been practiced throughout the world, why is there a current problem of soil erosion anywhere?

6. What effects has the increasing use of fossil fuels over the past 250 years had on the environment? What is *acid precipitation,* and where is it a problem? What factors affect the type and degree of air pollution found at a place?

7. Describe the chief sources of water pollution of which you are aware. How has the supply of fresh water been affected by pollution and human use? When water is used, is it forever lost to the environment? If so, where does it go? If not, why should there be water shortages now in regions of formerly ample supply?

8. What is a transboundary river basin? Which transboundary river basins are most likely to become sites of international water conflicts?

9. Because waste disposal is a technical problem, why should there be any concern with waste disposal in modern advanced economies? Why is there a growing international trade in wastes and recyclable products?

10. What are some examples of success stories of humans working to reduce environmental impacts?

11. Suggest ways in which your study of human geography has increased your understanding of the relationship between culture and nature.

KEY CONCEPTS REVIEW

1. **What are contributing causes and resulting concerns of urban air pollution, acid precipitation, ozone depletion, and global climate change?** Section 13.1
Environmental impact depends on the size of the population, standards of living, and the technologies a culture uses. Combustion of various fuels is the primary source of air pollution problems. Historic urban smog episodes were due to home heating with coal (London) and vehicle exhausts (Los Angeles). Pollution control regulations and technologies have helped clean the air, although many places have a long ways to go. Building taller smokestacks at factories and coal-fired power plants changed the scale of air pollution from local to regional or even continental. Acid precipitation is carried by prevailing winds across international boundaries, corroding stone and metals, destroying forests, and acidifying some lakes and soils. Upper-air ozone depletion, with serious effect on plant and animal life, is due to use of particular chemicals which are now subject to international bans. Rising concentrations of greenhouse gases due to fossil fuel combustion and deforestation are responsible for rising global temperatures. Global climate change is of great concern because the climate, sea levels, water availability, and the biosphere are intimately interrelated.

2. **What human actions have contributed to tropical deforestation, desertification, and soil erosion? What are the consequences?** Section 13.2
Current rapid destruction of tropical forests reflects human intentions to expand farming and grazing areas and harvest tropical wood. Their depletion endangers or destroys the world's richest, most diversified plant and animal biome and adversely affects local, regional, and world patterns of temperature and rainfall. Their loss also diminishes a vital *carbon sink* needed to absorb excess carbon dioxide. Desertification—the expansion of areas of destroyed soil and plant cover in dry climates—results from both natural climatic fluctuations and human pressures from plowing, woody plant removal, or livestock overgrazing. Those same human actions and pressures can accelerate the normal erosional loss of soil beyond natural soil regeneration potential. Such loss reduces total and per capita area of food production, diminishing the human carrying capacity of the land.

3. **How are emerging water resources problems related to human numbers and impacts?** Section 13.3
The hydrologic cycle assures water will be continuously regenerated for further use. But growing demand for irrigation, industrial use, and individual and urban consumption means increasing lack of balance between natural water supplies and consumption demands. Pollution of those supplies by human actions further reduces water availability and utility. A significant portion of the world's land area lies within transboundary river basins. Proper management of these vital water resources requires international cooperation.

4. **How are societies addressing the problems of solid and toxic waste disposal?** Section 13.4–13.5
Increasingly, all societies are becoming more dependent on advanced manufacturing and packaging of industrial, commercial, and consumer products. The easy recycling of waste materials found in subsistence cultures is no longer possible, and humans are presented with increasing needs for sites and facilities to safely dispose of solid wastes. Sanitary landfills and incineration are employed to handle nontoxic wastes. The former demands scarce and expensive land near cities or costly export to distant locations; the latter is often opposed because of unsafe emissions and ash residue. Disposal of toxic and hazardous wastes including radioactive wastes, products of modern societies and technologies, poses problems that have yet to be solved satisfactorily and safely.

MAP PROJECTIONS

A map projection is simply a system for displaying the curved surface of the Earth as a flat image, such as on a sheet of paper or computer screen. The definition is easy; the process is more difficult. No matter how one tries to "flatten" the Earth, it can never be done in such a fashion as to show all Earth details in their correct relative sizes, shapes, distances, or directions. Something is always wrong, and the cartographer's—the mapmaker's—task is to select and preserve those Earth relationships important for the purpose at hand and to minimize or accept those distortions that are inevitable but less important.

Round Globe to Flat Map

The best way to model the Earth's surface accurately, of course, would be to show it on a globe. But globes are not as convenient to use as flat maps and do not allow one to see the entire surface of the Earth all at once. Nor can they show very much detail. Even a very large globe of, say, 1 meter (nearly 3 feet) in diameter, compresses the physical or cultural information of some 130,000 square kilometers (about 50,000 sq mi) of Earth surface into a space 2.5 centimeters (1 in.) on a side.

Geographers make two different demands on the maps they use to represent reality. One requirement is to show at one glance generalized relationships and spatial content of the entire world; the many world maps used in this and other geography textbooks and in atlases have that purpose. The other need is to show the detailed content of only portions of the Earth's surface—cities, regions, countries, hemispheres—without reference to areas outside the zone of interest. Although the needs and problems of both kinds of maps differ, each starts with the same requirement: to transform a curved surface into a flat one.

If we look at the globe directly, only the front—the side facing us—is visible; the back is hidden (**Figure A.1**). To make a world map, we must decide on a way to flatten the globe's curved surface on the hemisphere we can see. Then we have to cut the globe map down the middle of its hidden hemisphere and place the two back quarters on their respective sides of the already visible front half. In simple terms, we have to "peel" the map from the globe and flatten it in the same way we might try to peel an orange and flatten the skin. Inevitably, the peeling and flattening process will produce a resulting map that either shows tears

Figure A.1 An orthographic projection gives us a visually realistic view of the globe; its distortion toward the edges suggests the normal perspective appearance of a sphere viewed from a distance. Only a single hemisphere—one half of the globe—can be seen at a time, and only the central portion of that hemisphere avoids serious distortion of shape.

or breaks in the surface (**Figure A.2a**) or is subject to uneven stretching or shrinking to make it lie flat (**Figure A.2b**).

Projections—Geometrical and Mathematical

Of course, mapmakers do not physically engage in cutting, peeling, flattening, or stretching operations. Their task, rather, is to construct or *project* on a flat surface the network of parallels and meridians (the graticule) of the globe grid (see Section 1.2). The idea of projections is perhaps most easily visualized by thinking of a transparent globe with an imagined light source located

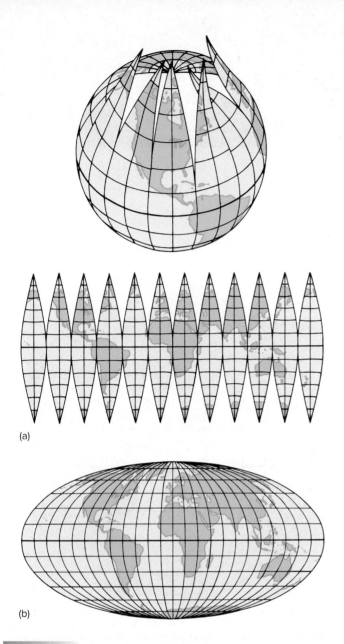

(a)

(b)

Figure A.2 (*a*) A careful "peeling" of the map from the globe yields a set of tapered "gores" that, although individually not showing much stretching or shrinking, do not collectively result in a very useful or understandable world map. (*b*) It is usually considered desirable to avoid or reduce the number of interruptions by depicting the entire global surface as a single flat circular, oval, or rectangular shape. That continuity of area, however, can be achieved only at the cost of considerable alteration of true shapes, distances, directions, or areas. Although the homolographic (Mollweide) projection shows areas correctly, it distorts shapes.

Source: American Congress Surveying and Mapping, "Choosing a World Map," Special Publication No. 2 of the American Cartographic Association, Bethesda, Md. Published in 1988 by CaGIS.

inside. Lines of latitude and longitude (or of coastlines or any other features) drawn on that globe will cast shadows on a nearby surface. A tracing of that shadow globe grid would represent a **geometrical map projection.**

In geometrical (or perspective) projections, the graticule is in theory visually transferred from the globe to a geometrical figure, such as a plane, cylinder, or cone, which, in turn, can be cut and then spread out flat (or *developed*) without any stretching or tearing. The surfaces of cylinders, cones, and planes are said to be **developable surfaces**—cylinders and cones can be cut and laid flat without distortion and planes are flat at the outset (**Figure A.3**). In actuality, geometrical projections are constructed not by tracing shadows but by the application of geometry and the use of lines, circles, arcs, and angles drawn on paper.

The location of the theoretical light source in relation to the globe can cause significant variation in the projection of the graticule on the developable surface. An **orthographic projection** results from placement of the light source at an infinite distance. A **gnomonic projection** is produced when the light source is at the center of the Earth. When the light is placed at the *antipode*—the point exactly opposite the point of tangency (point of contact between globe and map)—a **stereographic projection** is produced (**Figure A.4**).

Although a few useful and common projections are based on these simple geometric methods, most map designs can only be derived mathematically from equations involving angles and trigonometry developed for specific projections. The objective and need for **mathematical projections** is to preserve and emphasize specific Earth relationships that cannot be recorded by the perspective globe and shadow approach. The graticule of each mathematical projection is orderly and "accurate" in the sense of displaying the correct relative locations of lines of latitude and longitude. Each projection scheme, however, presents a different arrangement of the globe grid to minimize or eliminate some of the distortions inherent in projecting from a curved to a flat surface. Every projection represents a compromise or deviation from reality to achieve a selected purpose, but in the process of adjustment or compromise, each inevitably contains specific, accepted distortions.

Globe Properties and Map Distortions

The true properties of the global grid are detailed in Section 1.2. Not all of those grid realities can ever be preserved in any single projection; projections invariably distort some or all of them. The result is that all flat maps, whether geometrically or mathematically derived, also distort in different ways and to different degrees some or all of the four main properties of actual Earth surface relationships: area, shape, distance, and direction. And as our discussion above suggested, we can also see that all projections distort the Earth surface by showing a gap or break in what is actually a continuously connected surface without interruptions.

Area

Cartographers use **equal-area (equivalent) projections** when it is important for the map to show the *areas* of regions in correct or constant proportion to Earth reality—as it is when the map is intended to show the actual areal extent of

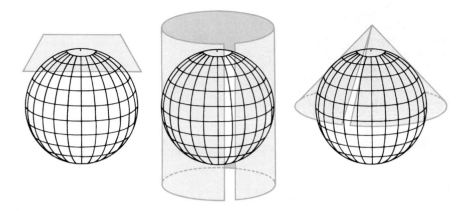

Figure A.3 The theory of *geometrical projections*. The three common geometric forms used in projections are the plane, the cylinder, and the cone.

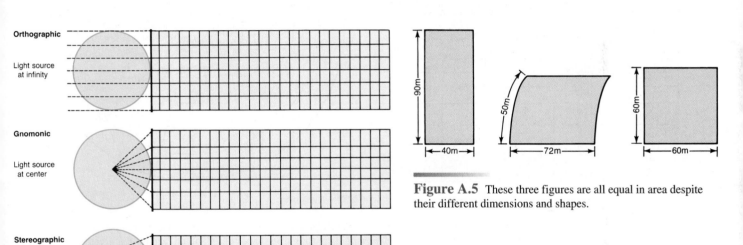

Figure A.4 The effect of light source location on planar surface projections. Note the variations in spacing of the lines of latitude that occur when the light source is moved.

Figure A.5 These three figures are all equal in area despite their different dimensions and shapes.

a phenomenon on the Earth's surface. If we wish to compare the amount of land in agriculture in two different parts of the world, for example, it would be very misleading visually to use a map that represented the same amount of surface area at two different scales.[1] To retain the needed size comparability,

our chosen projection must assure that a unit area drawn anywhere on it will always represent the same number of square kilometers (or similar units) on the Earth's surface. To achieve this *equivalence,* any scale change that the projection imposes in one direction must be offset by compensating changes in the opposite direction. As a result, the shape of the portrayed area is inevitably distorted. A square on the Earth, for example, may become a rectangle on the map, but that rectangle has the correct area (**Figure A.5**). *A map that shows correct areal relationships always distorts the shapes of regions,* as **Figure A.6a** demonstrates.

---

[1]Cartographic scale is the relationship between the size of a feature or length of a line on the map and that same feature or line on the Earth's surface. It may be indicated on a map as a ratio—for example, 1:1,000,000—that tells us the relationship between a unit of measure on the map and that same unit on the Earth's surface. In our example, 1 centimeter of map distance equals 1 million centimeters (or 10 kilometers) of actual Earth distance. See Figure 1.18.

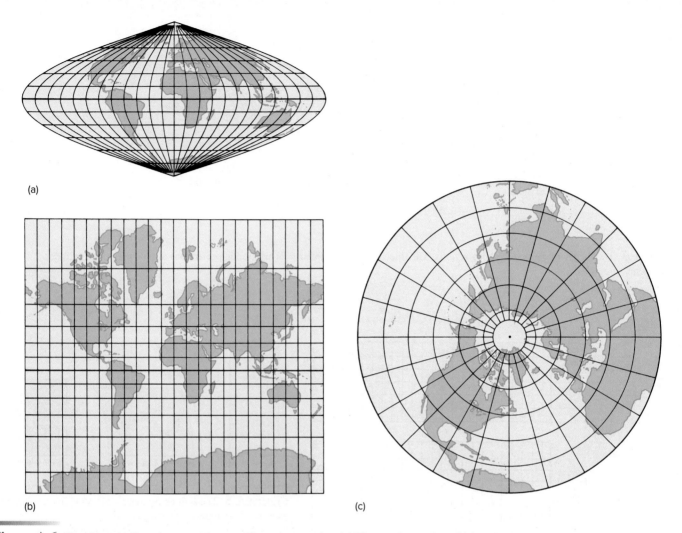

(a)

(b)

(c)

Figure A.6 Sample projections demonstrating specific map properties. (*a*) The equal-area sinusoidal projection retains everywhere the property of *equivalence*. (*b*) The mathematically derived Mercator projection is *conformal*, displaying true shapes of individual features but greatly exaggerating sizes and distorting shapes away from the equator. (*c*) A portion of an azimuthal *equidistant* projection, polar-case. Distances from the center (North Pole) to any other point are true; extension of the grid to the Southern Hemisphere would show the South Pole infinitely stretched to form the circumference of the map.

Shape

Although no projection can reproduce correct shapes for large areas, some do accurately portray the shapes of small areas. These true-shape projections are called **conformal,** and the importance of *conformality* is that regions and features "look right" and have the correct directional relationships. They achieve these properties for small areas by assuring that lines of latitude and longitude cross each other at right angles and that the scale is the same in all directions at any given location. Both these conditions exist on the globe but can be retained for only relatively small areas on maps. Because that is so, the shapes of large regions—continents, for example—are always different from their true Earth shapes even on conformal maps. Except for maps for very small areas, *a map cannot be both equivalent and conformal;* these two properties are mutually exclusive, as **Figure A.6b** suggests. Even on maps of very small areas, these two properties hold only approximately.

Distance

Distance relationships are nearly always distorted on a map, but some projections do maintain true distances in one direction or along certain selected lines. True distance relationships simply mean that the length of a straight line between two points on the map correctly represents the **great circle** distance between those points on the Earth. (An arc of a great circle is the shortest distance between two points on the Earth's curved surface; the equator is a great circle and all meridians of longitude are half great circles.) Projections with this property can be designed, but even on such **equidistant** maps true distance in all directions is shown only from one or two central points. Distances between all other locations are incorrect and, quite likely, greatly distorted as **Figure A.6c** clearly shows.

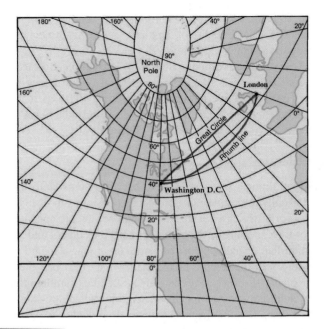

Figure A.7 A gnomonic projection centered on Washington, D.C. In this geometrical projection the light source is at the center of the globe (see Figure A.4), and the capital city marks the "standard point" where the projection plane is in contact with the globe. The rapid outward increase in graticule spacing makes it a projection impractical for more than a portion of a hemisphere. A unique property of the gnomonic projection is that it is the only projection on which all great circles appear as straight lines.

Direction

As is true of distances, directions between all points on a map cannot be shown without distortion. On **azimuthal projections,** however, true directions are shown from one central point to all other points. (An *azimuth* is the angle formed at the beginning point of a straight line, in relation to a meridian.) Directions or azimuths from points other than the central point to other points are not accurate. The azimuthal property of a projection is not exclusive—that is, an azimuthal projection may also be equivalent, conformal, or equidistant. The azimuthal equal-distance (*equidistant*) map shown as Figure A.6c is, as well, a true-direction map from the same North Pole origin. Another more specialized example is the gnomonic projection, displayed as **Figure A.7**.

Classes of Projections

Although there are many hundreds of different projections (and an infinite number of different projections could be created), the great majority of them can be grouped into four primary classes or families based on their origin. Each family has its own distinctive outline, set of similar properties, and pattern of distortions.

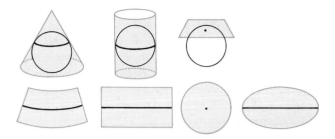

Figure A.8 Shape consistencies within families of projections. When the surface of a cone, cylinder, or plane is made *tangent*—that is, comes into contact with the globe—at either a point or along a circle and then *developed,* a characteristic family outline results. The tangent lines and point are indicated. A fourth common shape, the oval, may reflect a design in which the long dimension is a great circle comparable to the tangent line of the cylinder.

Source: American Congress Surveying and Mapping, "Choosing a World Map," Special Publication No. 2 of the American Cartographic Association, Bethesda, Md. Published in 1988 by CaGIS.

Three of them are easily seen as derived from the geometric or perspective projection of the globe grid onto the developable surfaces of cylinders, cones, and planes. The fourth class is mathematically derived; its members have a variety of attributes but share a general oval design (**Figure A.8**).

Cylindrical Projections

Cylindrical projections are developed geometrically or mathematically from a cylinder wrapped around the globe. Usually, the cylinder is tangent at the equator, which thus becomes the **standard line**—that is, transferred from the globe without distortion. The result is a globe grid network with meridians and parallels intersecting at right angles. There is no scale distortion along the standard line of tangency, but distortion increases with increasing distance away from it. The result is a rectangular world map with acceptable low-latitude representation, but with enormous areal exaggeration toward the poles.

The mathematically derived **Mercator projection** invented in 1569 is a special familiar but commonly misused cylindrical projection (see Figure A.6b). Its sole original purpose was to serve as a navigational chart of the world with the special advantage of showing true compass headings, or **rhumb lines,** as straight lines on the map. That is, any straight line drawn on a Mercator projection will follow a constant compass direction. For instance, one drawn from the lower left to the upper right would follow a northeasterly direction on the Earth surface; a ship could follow this line by constantly heading in the direction northeast (as indicated, for instance, by a magnetic compass). Its frequent use in wall or book maps gives grossly exaggerated impressions of the size of land areas away from the tropics.

The **equirectangular projection** is a cylindrical projection that converts the globe's meridians to equally spaced vertical lines and its parallels to equally spaced horizontal lines.

The result is a rectangular map that resembles the Mercator projection, but is neither equal-area nor conformal. Interestingly, map services on the Web and global satellite images generally use an equirectangular projection despite the massive distortion near the poles. The equirectangular projection's popularity stems from its mathematical simplicity. One version of the equirectangular projection (called Plate Carrée) simply sets the y-values equal to the latitude and the x-values equal to the longitude (**Figure A.9a**).

Equal-area alternatives to the conformal Mercator map are available, and a number of "compromise" cylindrical projections that are neither equal area nor conformal (for example, the Miller projection, **Figure A.9b**) are frequently used for world maps. The Robinson projection (**Figure A.9f**), used for many of the world maps in this textbook, is such a compromise. Designed to show the whole world in a visually satisfactory manner, it does not show true distances or directions and is neither equal-area nor conformal. Instead, it permits some exaggeration of size in the high latitudes in order to improve the shapes of landmasses. Size and shape are most accurate in the temperate and tropical zones, where most people live.

Conic Projections

Of the three developable geometric surfaces, the cone is the closest in form to one-half of a globe. **Conic projections,** therefore, are often employed to depict one hemisphere or smaller parts of the Earth. In the *simple conic* projection, the cone is placed tangent to the globe along a single standard parallel, with the apex of the cone located above the pole. The cone can also be made to intersect the globe along two or more lines, with a *polyconic* projection resulting; the increased number of standard lines reduces the distortion, which increases away from the standard parallel. The projection of the grid on the cone yields evenly spaced straight-line meridians radiating from the pole and parallels that are arcs of circles. Although conic projections can be adjusted to minimize distortions and become either equivalent or conformal, by their nature they can never show the whole globe. In fact, they are most useful for and generally restricted to maps of midlatitude regions of greater east-west than north-south extent. The Albers equal-area projection often used for U.S. maps is a familiar example (**Figure A.9c**).

Planar (Azimuthal) Projections

Planar (azimuthal) projections are constructed by placing a plane tangent to the globe at a single point. Although the plane may touch the globe anywhere the cartographer wishes, the polar case with the plane centered on either the North or the South Pole is easiest to visualize (see Figure A.6c). This equidistant projection is useful because it can be centered anywhere, facilitating the correct measurement of distances from that point to all others. When the plane is tangent at places other than the poles, the meridians and the parallels become curiously curved (**Figure A.9d**).

Planar maps are commonly used in atlases because they are particularly well suited for showing the arrangement of polar land-masses. Depending on the particular projection used, true

shape, equal area, or some compromise between them can be depicted. The special quality of the planar gnomonic projection has already been shown in Figure A.7.

The **perspective projection** is a planar projection used in Google Earth and available in a variety of GIS software packages. The perspective projection depicts the Earth as it would appear when viewed from a finite distance above the Earth's surface such as from a space vehicle (**Figure A.9e**).

Oval or Elliptical Projections

Oval or elliptical projections have been mathematically developed usually as compromise projections designed to display the entire world in a fashion that is visually acceptable and suggestive of the curvature of the globe. In most, the equator and a central meridian (usually the prime meridian) are the standard lines. They cross in the middle of the map, which thus becomes the point of no distortion. Parallels are, as a rule, parallel straight lines; meridians, except for the standard meridian, are shown as curved lines. In some instances the oval projection is a modification of a projection based on a different original shape. Some of the world maps in this textbook (for example, Figures 8.12 and 13.19) are an oval adjustment of the circular (but not azimuthal) Van der Grinten projection, a compromise projection that achieves acceptable degrees of equivalence and conformality in lower and middle latitudes but becomes increasingly and unacceptably distorted in polar regions.

Other Projections and Manipulations

The geometric projections, we have seen, can all be thought of as developed from the projection of the globe grid onto a cylinder, cone, or plane. Many projections, however, cannot be classified in terms of simple geometric shapes. They are derived from mathematical formulas and usually have been developed to display the world or a portion thereof in a fashion that is visually acceptable or in any shape that is desired: ovals are most common, but hearts, trapezoids, stars, and other—sometimes bizarre—forms have been devised for special purposes. One often-seen projection is the equal-area Goode's homolosine, an *interrupted* projection that is actually a product of fitting together the least distorted portions of two different projections (the sinusoidal projection and the Mollweide, or homolographic, projection) and centering the split map along multiple standard meridians to minimize distortion of either (as desired) land or ocean surfaces (**Figure A.10**).

The homolosine map clearly shows how projections may be manipulated or adjusted to achieve desired objectives. Because most projections are based on a mathematically consistent rendering of the actual globe grid, possibilities for such manipulation are nearly unlimited. R. Buckminster Fuller, an architect and designer of the geodesic dome, produced the Fuller dymaxion projection (**Figure A.11**). It consists of 20 equilateral triangles, which can be hinged along different boundaries to show interesting Earth relationships. The projection minimizes distortion of the sizes and shapes of the world's landmasses.

Map properties to be retained, size and shape of areas to be displayed, and overall map design to be achieved may influence the cartographer's choices in reproducing the globe grid on the

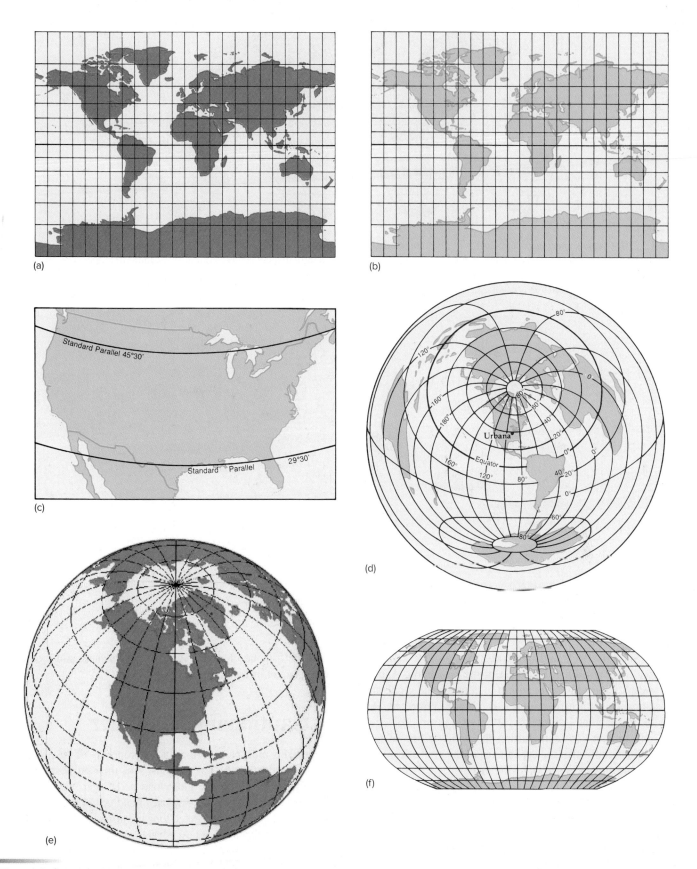

Figure A.9 Some sample members of the principal projection families. (*a*) The equirectangular projection is a simple cylindrical projection. The version shown here was created by setting the x-value equal to the longitude and the y-value equal to the latitude. (*b*) The Miller cylindrical projection is also mathematically derived. (*c*) The Albers equal-area conic projection, used for many official U.S. maps, has two standard parallels: 29 1/2° and 45 1/2°. (*d*) A planar, or azimuthal, equidistant projection centered on Urbana, Illinois. (*e*) The perspective projection is a planar projection that simulates the view from space. (*f*) The Robinson projection of the oval family; neither conformal nor equivalent, it was designed as a visually satisfactory world map.

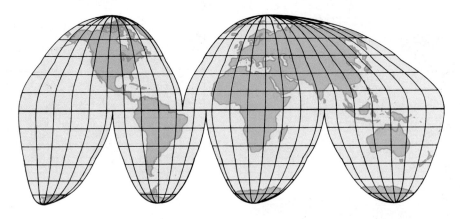

Figure A.10 Goode's interrupted homolosine grafts an upper latitude homolographic (Mollweide) onto a sinusoidal projection at about 40° North and South. To improve shapes, each continent is placed on the middle of a lobe approximately centered on its own central meridian. The projection can also interrupt continents to display the ocean areas intact.

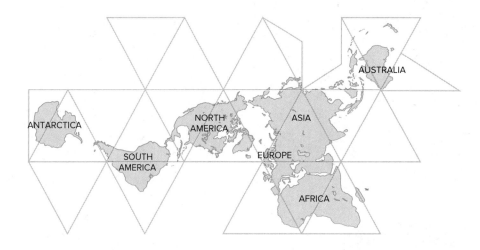

Figure A.11 The Fuller dymaxion projection. The equilateral triangles can be folded into a solid approximating the globe.

flat map. Special effects and properties may also be achieved geometrically by adjusting the aspect of the projection. **Aspect** simply means the positional relationship between the globe and the developable surface on which it is visually projected. Although the fundamental distortion pattern of a given projection system will remain constant, shifting the point or line of tangency will materially alter the appearance of the graticule and of the geographical features shown on the map.

Although an infinite number of aspects are possible for any of the geometric projections, three classes of aspects are most common. Named according to the relation of the axis of the globe to the cylinder, cone, or plane projection surface, the three classes are usually called *equatorial, polar,* and *oblique.* In the equatorial, the axis of the globe parallels the orientation of the plane, cylinder, or cone; a parallel, usually the central equator, is the line of tangency. In the polar aspect, the axis of the globe is perpendicular to the orientation of the developable surface. In the oblique aspect, the axis of the globe makes an oblique angle

with the orientation of the developable surface, and a complex arrangement of the graticule results.

A Cautionary Reminder

Mapmakers must be conscious of the properties of the projections they use, selecting the one that best suits their purposes. It is not possible to transform the globe into a flat map without distortion. But cartographers have devised hundreds of possible mathematical and geometrical projections in various modifications and aspects to display to their best advantage the variety of Earth features and relationships they wish to emphasize. Some projections are highly specialized and properly restricted to a single limited purpose; others achieve a more general acceptability and utility.

If the map shows only a small area, the choice of a projection is less critical because distortion due to projection decreases as map scale increases (i.e., as the Earth surface area shown decreases)—virtually any can be used. The choice becomes more important when the area to be shown extends over a considerable longitude and latitude; then the selection of a projection clearly depends on the purpose of the map. As we have seen, Mercator or gnomonic projections are useful for navigation. Unfortunately, the Mercator projection grossly exaggerates the area of high latitude features, giving a misleading impression when used for books or wall maps. If numerical data are being mapped, the relative sizes of the areal units (countries, states, counties, and so forth) should be correct, and equivalence is the sought-after map property. Conformality and equal distance may be required in other instances.

While selection of an appropriate projection is the task of the cartographer, understanding the consequences of that choice and recognizing and allowing for the distortions inevitable in all flat maps are the responsibility of the map reader. When skillfully designed maps are read by knowledgeable users, they clearly and accurately convey important spatial information and Earth relationships.

KEY WORDS

aspect
azimuthal projection
conformal projection
conic projection
cylindrical projection
developable surface
equal-area (equivalent) projection

equidistant
equirectangular projection
geometrical projection
gnomonic projection
great circle
mathematical projection
Mercator projection

orthographic projection
perspective projection
planar (azimuthal) projection
rhumb line
standard line
stereographic projection

Appendix B

2017 WORLD POPULATION DATA

	Population mid-2017 (millions)	Births per 1,000 Population	Deaths per 1,000 Population	Net Migration Rate per 1,000	Projected Population (millions), Mid–2050	Projected Population Change, 2017–2050, %	Infant Mortality Rate[a]	Total Fertility Rate[b]	Percent of Population Ages <15	Percent of Population Ages 65+	GNI PPP per Capita ($US), 2016	Percent Urban	Population per Square Kilometer of Arable Land (thousands)	Life Expectancy at Birth (years), Males	Life Expectancy at Birth (years), Females	Secondary School Enrollment Ratio[c], Males (2009–2016)	Secondary School Enrollment Ratio[c], Females (2009–2016)
WORLD	**7,536**	**20**	**8**	**0**	**9,846**	**31%**	**32**	**2.5**	**26**	**9**	**16,101**	**54**	**532**	**70**	**74**	**77**	**76**
MORE DEVELOPED	1,263	11	10	3	1,325	5%	5	1.6	16	18	41,421	78	239	76	82	107	107
LESS DEVELOPED	6,273	21	7	0	8,520	36%	35	2.6	28	7	10,822	49	707	69	72	74	73
LESS DEVELOPED (Excl. China)	4,878	24	7	0	7,169	47%	39	2.9	31	5	9,353	47	624	67	71	71	70
LEAST DEVELOPED	1,001	33	8	–1	1,952	95%	52	4.3	40	4	2,566	32	572	63	66	46	41
AFRICA	**1,250**	**35**	**9**	**–1**	**2,574**	**106%**	**51**	**4.6**	**41**	**3**	**4,833**	**41**	**534**	**61**	**64**	**52**	**48**
SUB-SAHARAN AFRICA	**1,021**	**37**	**10**	**0**	**2,193**	**115%**	**56**	**5.0**	**43**	**3**	**3,592**	**39**	**534**	**58**	**62**	**48**	**43**
NORTHERN AFRICA	**230**	**28**	**6**	**–1**	**381**	**66%**	**24**	**3.3**	**31**	**5**	**10,046**	**52**	**538**	**71**	**74**	**77**	**75**
Algeria	42.2	26	4	0	64.8	54%	21	3.1	29	6	14,720	71	565	75	78	98	102
Egypt	93.4	30	7	0	163.5	75%	16	3.3	31	4	11,110	43	3,498	71	73	86	86
Libya	6.4	20	5	–7	8.1	27%	23	2.3	29	4	11,210	79	372	69	75	—	—
Morocco	35.1	19	5	–2	40.2	15%	25	2.4	25	6	7,700	60	432	74	77	74	64
Sudan	40.6	34	8	–2	88.1	117%	46	4.7	41	4	4,290	35	205	63	66	44	41
Tunisia	11.5	20	6	–1	15.3	33%	15	2.4	24	8	11,150	68	397	75	78	90	94
Western Sahara[d]	0.6	20	5	6	0.9	50%	31	2.4	29	3	—	81	15,000	68	71	—	—
WESTERN AFRICA	**371**	**39**	**11**	**1**	**809**	**118%**	**64**	**5.3**	**44**	**3**	**4,095**	**46**	**437**	**55**	**57**	**53**	**47**
Benin	11.2	37	9	–1	23.9	113%	67	5.0	43	3	2,170	44	415	59	62	67	47
Burkina Faso	19.6	41	9	–1	48.4	147%	65	5.7	49	3	1,680	31	327	59	61	35	32
Cape Verde	0.5	20	6	–3	0.6	20%	21	2.3	31	5	6,220	66	909	71	75	88	98
Côte d'Ivoire	24.4	37	13	0	50.1	105%	64	5.0	43	3	3,610	55	841	52	55	51	37

	Population mid-2017 (millions)	Births per 1,000 Population	Deaths per 1,000 Population	Net Migration Rate per 1,000	Projected Population (millions), Mid-2050	Projected Population Change, 2017–2050, %	Infant Mortality Rate[a]	Total Fertility Rate[b]	Percent of Population Ages <15	Percent of Population Ages 65+	GNI PPP per Capita ($US), 2016	Percent Urban	Population per Square Kilometer of Arable Land (thousands)	Life Expectancy at Birth (years), Males	Life Expectancy at Birth (years), Females	Secondary School Enrollment Ratio[c], Males (2009–2016)	Secondary School Enrollment Ratio[c], Females (2009–2016)
Gambia	2.1	40	8	−1	5.1	143%	47	5.5	46	2	1,640	60	477	60	62	59	56
Ghana	28.8	32	8	0	51.2	78%	41	4.0	39	3	4,150	55	613	61	63	63	61
Guinea	11.5	36	10	−3	24.3	111%	59	4.9	43	3	1,200	38	371	59	60	47	31
Guinea–Bissau	1.9	37	11	−1	3.6	89%	75	4.7	42	3	1,580	50	633	55	59	—	—
Liberia	4.7	35	8	−1	9.8	109%	54	4.7	42	3	700	50	940	61	63	42	33
Mali	18.9	43	11	−3	44.8	137%	56	6.0	48	3	2,040	41	295	57	58	46	37
Mauritania	4.4	35	8	2	9	105%	75	4.6	40	3	3,760	60	978	62	65	32	29
Niger	20.6	48	10	0	65.6	218%	61	7.3	50	3	970	19	130	59	61	24	17
Nigeria	190.9	39	13	0	410.6	115%	69	5.5	44	3	5,740	49	562	52	54	58	53
Senegal	15.8	37	6	−1	33.8	114%	39	4.9	43	3	2,480	44	494	65	69	50	49
Sierra Leone	7.6	37	13	−1	13	71%	92	4.9	43	3	1,320	40	480	51	52	46	40
Togo	7.8	35	9	0	15.3	96%	52	4.5	42	3	1,370	40	294	59	61	—	—
EASTERN AFRICA	**422**	**36**	**8**	**−1**	**886**	**110%**	**47**	**4.7**	**43**	**3**	**2,154**	**27**	**636**	**62**	**65**	**38**	**36**
Burundi	10.4	42	11	−1	23.5	126%	47	5.5	45	3	770	12	867	55	59	44	41
Comoros	0.8	33	8	−3	1.5	88%	55	4.3	40	3	1,520	28	1,231	62	65	58	62
Djibouti	1.0	23	8	1	1.3	30%	53	2.9	32	4	—	77	50,000	61	64	53	43
Eritrea	5.9	31	7	−15	8.9	51%	46	4.2	41	4	1,500	23	855	62	67	33	28
Ethiopia	105.0	33	7	0	190.9	82%	48	4.6	42	3	1,730	20	695	63	67	36	34
Kenya	49.7	32	6	0	95.5	92%	37	3.9	41	3	3,130	26	857	64	69	63	57
Madagascar	25.5	33	7	0	48.1	89%	33	4.2	41	3	1,440	36	729	64	67	39	38
Malawi	18.6	35	8	0	37.4	101%	42	4.4	44	3	1,140	16	490	60	65	46	41
Mauritius	1.3	10	8	−1	1.1	−15%	13.7	1.4	20	9	20,980	41	1,733	71	78	94	98
Mayotte	0.2	39	2	0	0.5	150%	4	5.0	44	3	—	47	2,247	75	77	—	—
Mozambique	29.7	39	10	0	67.8	128%	65	5.3	45	3	1,190	33	526	56	60	34	31
Reunion	0.9	17	5	−6	1.1	22%	7	2.5	24	11	—	95	2,528	77	84	—	—
Rwanda	12.3	33	6	−1	24.3	98%	32	4.2	40	3	1,870	30	1,070	65	69	35	38
Seychelles	0.09	17	8	−3	0.1	11%	13.4	2.3	22	8	28,390	54	112,500	68	78	79	84
Somalia	14.7	44	12	−3	35.9	144%	74	6.4	47	3	—	40	1,336	54	58		
South Sudan	12.6	37	11	5	27.9	121%	72	5.1	42	3	1,700	19	—	55	57	12	7
Tanzania	57.5	40	7	−1	152.2	165%	43	5.2	45	3	2,740	32	426	63	67	34	31
Uganda	42.8	40	9	−1	95.6	123%	43	5.4	48	3	1,820	20	620	62	64	24	22
Zambia	16.4	39	8	0	39.3	140%	50	5.2	45	3	3,790	40	432	59	64	—	—
Zimbabwe	16.6	36	9	−2	33.2	100%	50	4.0	41	3	1,920	33	415	59	62	48	47

	Population mid-2017 (millions)	Births per 1,000 Population	Deaths per 1,000 Population	Net Migration Rate per 1,000	Projected Population (millions), Mid–2050	Projected Population Change, 2017–2050, %	Infant Mortality Rate[a]	Total Fertility Rate[b]	Percent of Population Ages <15	Percent of Population Ages 65+	GNI PPP per Capita ($US), 2016	Percent Urban	Population per Square Kilometer of Arable Land (thousands)	Life Expectancy at Birth (years), Males	Life Expectancy at Birth (years), Females	Secondary School Enrollment Ratio[c], Males (2009–2016)	Secondary School Enrollment Ratio[c], Females (2009–2016)
MIDDLE AFRICA	**163**	**42**	**10**	**–1**	**410**	**152%**	**62**	**5.9**	**46**	**3**	**2,688**	**44**	**630**	**57**	**60**	**48**	**32**
Angola	28.6	45	9	0	79.6	178%	44	6.2	47	2	6,220	45	584	58	64	35	23
Cameroon	25.0	36	10	–4	51.9	108%	54	4.8	43	3	3,250	55	403	57	59	63	54
Central African Republic	4.7	36	14	–12	8.9	89%	87	4.9	44	4	700	40	261	50	53	23	12
Chad	14.9	46	13	1	36.8	147%	72	6.4	48	2	1,950	23	304	51	54	31	14
Congo	5.0	36	10	–6	10.2	104%	42	4.7	42	3	5,380	66	909	58	60	58	51
Congo, Dem. Rep.	81.5	44	10	0	215.9	165%	69	6.3	46	3	730	43	1,148	58	61	54	33
Equatorial Guinea	1.3	35	10	15	2.9	123%	65	4.8	37	3	17,020	40	1,083	56	59	—	—
Gabon	2.0	30	8	5	3.5	75%	38	3.9	36	5	16,720	87	615	64	67	—	—
Sao Tome and Principe	0.2	33	7	–5	0.3	50%	38	4.4	42	4	3,240	67	2,299	64	69	81	92
SOUTHERN AFRICA	**65**	**22**	**9**	**2**	**88**	**35%**	**35**	**2.5**	**30**	**5**	**12,467**	**62**	**459**	**61**	**66**	**86**	**108**
Botswana	2.3	24	7	1	3.4	48%	31	2.8	33	5	16,380	58	576	63	69	—	—
Lesotho	2.2	29	13	–2	3.2	45%	59	3.3	36	4	3,390	28	808	51	56	46	62
Namibia	2.5	28	8	0	4.2	68%	38	3.4	38	4	10,550	48	313	62	65	—	—
South Africa	56.5	21	9	2	75.2	33%	33	2.4	30	5	12,860	65	452	61	67	88	112
Swaziland	1.4	29	10	–1	2.1	50%	50	3.3	38	3	7,980	21	800	54	60	66	66
AMERICAS	**1,005**	**15**	**7**	**1**	**1,227**	**22%**	**14**	**2.0**	**23**	**10**	**30,130**	**80**	**267**	**74**	**80**	**94**	**97**
NORTHERN AMERICA	**362**	**12**	**8**	**4**	**444**	**23%**	**6**	**1.8**	**19**	**15**	**56,554**	**81**	**181**	**77**	**81**	**98**	**99**
Canada	36.7	11	8	9	47.1	28%	4.3	1.6	16	17	43,420	82	80	79	84	110	110
United States	325.4	12	8	3	396.8	22%	5.8	1.8	19	15	58,030	81	211	76	81	97	98
LATIN AMERICA AND THE CARIBBEAN	**643**	**17**	**6**	**–1**	**783**	**22%**	**17**	**2.1**	**26**	**8**	**15,001**	**80**	**366**	**73**	**79**	**92**	**97**
CENTRAL AMERICA	**177**	**20**	**5**	**–1**	**232**	**31%**	**19**	**2.3**	**29**	**6**	**15,315**	**74**	**631**	**74**	**79**	**84**	**88**
Belize	0.4	24	6	4	0.6	50%	9	2.6	32	4	8,000	45	513	71	77	80	82
Costa Rica	4.9	14	5	1	6.1	24%	7.9	1.7	23	8	15,750	78	2,111	78	83	121	126
El Salvador	6.4	20	7	–7	8	25%	17	2.3	28	8	8,220	66	853	69	78	79	80
Guatemala	16.9	24	5	–1	27	60%	25	2.9	40	5	7,750	52	1,810	69	76	68	63
Honduras	8.9	22	5	0	12.7	43%	26	2.5	33	4	4,410	55	873	71	76	65	77
Mexico	129.2	20	5	0	164.3	27%	18	2.2	27	6	17,740	80	562	75	79	88	93
Nicaragua	6.2	20	5	–4	7.9	27%	18	2.2	30	5	5,390	59	412	72	78	70	79
Panama	4.1	19	5	2	5.8	41%	13	2.4	27	8	20,990	67	728	75	81	73	78

	Population mid-2017 (millions)	Births per 1,000 Population	Deaths per 1,000 Population	Net Migration Rate per 1,000	Projected Population (millions), Mid-2050	Projected Population Change, 2017–2050, %	Infant Mortality Rate[a]	Total Fertility Rate[b]	Percent of Population Ages <15	Percent of Population Ages 65+	GNI PPP per Capita ($US), 2016	Percent Urban	Population per Square Kilometer of Arable Land (thousands)	Life Expectancy at Birth (years), Males	Life Expectancy at Birth (years), Females	Secondary School Enrollment Ratio[c], Males (2009–2016)	Secondary School Enrollment Ratio[c], Females (2009–2016)
CARIBBEAN	43	17	8	–2	47	9%	29	2.2	25	10	—	71	821	71	76	83	89
Antigua and Barbuda	0.1	15	6	0	0.1	0%	8	1.9	25	7	21,840	23	2,500	74	79	102	104
Bahamas	0.4	13	6	4	0.4	0%	9	1.7	21	8	22,090	83	5,000	72	78	90	95
Barbados	0.3	11	9	1	0.3	0%	9	1.7	19	15	16,070	31	2,727	73	78	108	111
Cuba	11.3	11	9	–5	9.8	–13%	4.3	1.7	17	15	—	77	366	76	81	98	103
Curaçao	0.2	11	9	7	0.2	0%	11.4	1.7	18	16	—	89	—	75	81	86	91
Dominica	0.07	12	8	1	0.08	14%	20	1.8	22	10	10,610	70	1,167	73	78	101	100
Dominican Republic	10.7	21	6	–3	13.2	23%	31	2.5	30	7	14,480	80	1,338	71	77	74	82
Grenada	0.1	17	8	–2	0.1	0%	15	2.1	26	7	13,440	36	3,333	74	79	99	99
Guadeloupe	0.4	12	7	–5	0.4	0%	7.3	2.1	20	17	—	98	1,794	77	85	—	—
Haiti	10.6	23	8	2	14.5	37%	48	2.9	33	4	1,790	60	991	61	66	—	—
Jamaica	2.9	17	7	–6	2.7	–7%	14	2.0	23	9	8,500	55	2,417	73	78	79	85
Martinique	0.4	11	8	–5	0.3	–25%	6	2.0	18	19	—	89	3,960	79	85	—	—
Puerto Rico	3.4	9	8	–5	2.7	–21%	7.0	1.3	16	19	24,020	99	5,574	76	83	79	84
St. Kitts-Nevis	0.05	14	9	3	0.06	20%	17	1.8	21	8	25,940	32	1,000	73	78	88	93
Saint Lucia	0.2	12	6	4	0.2	0%	18	1.5	21	11	11,370	19	6,667	75	83	85	85
St. Vincent and the Grenadines	0.1	16	9	–6	0.1	0%	18	2.1	25	7	11,530	51	2,000	70	75	108	105
Trinidad and Tobago	1.4	14	8	–2	1.3	–7%	13	1.7	19	10	30,810	8	5,600	69	75	—	—
SOUTH AMERICA	423	16	6	0	504	19%	15	1.9	25	8	15,192	83	297	72	79	96	101
Argentina	44.3	17	8	0	54.1	22%	10.1	2.3	25	12	19,480	91	113	74	80	103	110
Bolivia	11.1	24	7	–1	16.5	49%	39	2.9	32	6	7,090	69	248	66	71	87	86
Brazil	207.9	13	6	0	231.1	11%	14	1.6	23	8	14,810	86	260	72	79	97	102
Chile	18.4	14	6	1	21.1	15%	7.3	1.8	20	11	23,270	83	1,427	77	82	100	101
Colombia	49.3	18	6	–1	61.5	25%	14	2.0	26	8	13,910	76	2,943	73	79	95	102
Ecuador	16.8	20	5	0	23.2	38%	20	2.5	30	7	11,070	64	1,651	73	79	105	109
French Guiana	0.3	25	3	4	0.5	67%	9	3.4	34	5	—	85	2,273	77	83	—	—
Guyana	0.8	21	8	–7	0.8	0%	32	2.5	30	5	7,860	29	191	64	69	90	89
Paraguay	6.8	21	6	–3	8.9	31%	28	2.5	30	6	9,060	60	142	71	75	74	79
Peru	31.8	20	6	–1	41.2	30%	17	2.4	28	7	12,480	79	766	72	77	96	96
Suriname	0.6	18	7	–2	0.6	0%	16	2.4	27	7	13,720	66	923	68	75	72	91
Uruguay	3.5	14	9	–1	3.7	6%	11.9	2.0	21	14	21,090	95	145	74	81	90	100
Venezuela	31.4	19	5	–1	40.5	29%	12.5	2.4	28	7	17,700	88	1,163	73	79	86	93

	Population mid-2017 (millions)	Births per 1,000 Population	Deaths per 1,000 Population	Net Migration Rate per 1,000	Projected Population (millions), Mid-2050	Projected Population Change, 2017–2050, %	Infant Mortality Rate[a]	Total Fertility Rate[b]	Percent of Population Ages <15	Percent of Population Ages 65+	GNI PPP per Capita ($US), 2016	Percent Urban	Population per Square Kilometer of Arable Land (thousands)	Life Expectancy at Birth (years), Males	Life Expectancy at Birth (years), Females	Secondary School Enrollment Ratio[c], Males (2009–2016)	Secondary School Enrollment Ratio[c], Females (2009–2016)
ASIA	**4,494**	**18**	**7**	**0**	**5,245**	**17%**	**28**	**2.2**	**24**	**8**	**12,833**	**49**	**933**	**71**	**74**	**80**	**80**
ASIA (Excl. China)	**3,099**	**20**	**7**	**0**	**3,894**	**26%**	**34**	**2.4**	**28**	**7**	**11,454**	**45**	**825**	**69**	**73**	**76**	**75**
WESTERN ASIA	**269**	**21**	**5**	**3**	**390**	**45%**	**22**	**2.8**	**29**	**5**	**27,583**	**71**	**691**	**72**	**76**	**88**	**81**
Armenia	3.0	14	9	–8	2.4	–20%	9	1.6	20	11	9,000	64	670	72	78	88	89
Azerbaijan	9.9	17	6	0	11.7	18%	11	2.0	23	6	16,130	53	514	73	78	—	—
Bahrain	1.5	14	2	19	2.1	40%	6	1.9	21	2	44,690	100	93,750	76	78	102	102
Cyprus	1.2	12	6	7	1.4	17%	2	1.4	17	13	31,420	67	1,504	80	84	100	99
Georgia	3.9	15	14	–2	3.4	–13%	9	1.7	19	14	9,450	57	853	68	77	104	104
Iraq	39.2	32	4	2	76.5	95%	38	4.1	40	3	17,240	70	779	67	72	—	—
Israel	8.3	21	5	1	13.8	66%	3.1	3.1	28	11	37,400	91	2,762	80	84	102	103
Jordan	9.7	26	4	12	12.7	31%	16	3.3	36	4	8,980	84	4,084	73	76	80	85
Kuwait	4.1	15	2	22	5.6	37%	8	2.0	21	2	83,420	98	43,158	74	76	88	103
Lebanon	6.2	14	5	15	5.6	–10%	8	1.7	25	7	13,860	88	4,697	76	79	61	61
Oman	4.7	21	3	36	7.3	55%	9	2.9	22	3	41,320	75	12,368	75	79	101	108
Palestinian Territory	4.9	31	4	–2	8.7	78%	18	4.0	40	3	3,290	75	7,656	71	75	79	87
Qatar	2.7	11	1	36	3.8	41%	7	2.0	14	1	124,740	99	20,611	77	80	82	104
Saudi Arabia	32.6	20	4	7	44.6	37%	12	2.6	25	3	55,760	83	931	73	76	123	94
Syria	18.3	22	6	–21	34	86%	17	2.9	37	4	—	58	393	64	77	50	51
Turkey	80.9	17	5	4	94.8	17%	10	2.1	24	8	23,990	74	391	75	81	104	101
United Arab Emirates	9.4	10	2	9	13.2	40%	6	1.8	14	1	72,850	86	25,067	76	79	—	—
Yemen	28.3	32	7	–1	48.3	71%	45	4.1	41	3	2,490	35	2,268	63	66	57	40
CENTRAL ASIA	**71**	**24**	**6**	**–1**	**104**	**46%**	**26**	**2.8**	**29**	**5**	**10,916**	**41**	**189**	**69**	**75**	**98**	**96**
Kazakhstan	18.0	23	7	1	25	39%	9	3.0	25	7	22,910	55	61	68	77	111	113
Kyrgyzstan	6.2	27	6	–1	9.3	50%	18	3.2	32	4	3,410	34	484	67	75	91	93
Tajikistan	8.8	29	5	–2	14.4	64%	36	3.4	35	3	3,500	27	1,206	68	74	92	83
Turkmenistan	5.8	27	7	–1	8.8	52%	45	3.2	30	4	16,060	50	299	64	71	87	84
Uzbekistan	32.4	23	5	–1	46.5	44%	29	2.5	28	4	6,640	37	736	71	76	97	95
SOUTH ASIA	**1,885**	**22**	**6**	**–1**	**2,406**	**28%**	**40**	**2.4**	**29**	**5**	**6,054**	**35**	**855**	**67**	**70**	**70**	**70**
Afghanistan	35.5	35	7	1	68.9	94%	60	5.3	45	2	1,900	24	457	62	65	71	40
Bangladesh	164.7	19	5	–3	201.9	23%	38	2.3	29	5	3,790	35	2,148	71	74	60	67
Bhutan	0.8	19	6	1	1	25%	27	2.1	27	5	8,070	39	798	70	70	81	87
India	1,352.6	21	7	0	1,675.6	24%	37	2.3	29	6	6,490	33	865	67	70	74	74
Iran	80.6	20	5	–3	92.9	15%	5	1.8	24	5	17,370	73	549	75	77	89	89
Maldives	0.4	20	3	8	0.6	50%	8	2.2	23	4	11,970	46	10,256	76	78	—	—

	Population mid-2017 (millions)	Births per 1,000 Population	Deaths per 1,000 Population	Net Migration Rate per 1,000	Projected Population (millions), Mid-2050	Projected Population Change, 2017–2050, %	Infant Mortality Rate[a]	Total Fertility Rate[b]	Percent of Population Ages <15	Percent of Population Ages 65+	GNI PPP per Capita ($US), 2016	Percent Urban	Population per Square Kilometer of Arable Land (thousands)	Life Expectancy at Birth (years), Males	Life Expectancy at Birth (years), Females	Secondary School Enrollment Ratio[c], Males (2009–2016)	Secondary School Enrollment Ratio[c], Females (2009–2016)
Nepal	29.4	20	6	–2	33.3	13%	32	2.3	31	5	2,520	20	1,391	69	71	67	72
Pakistan	199.3	29	7	–1	310.5	56%	67	3.6	35	4	5,580	39	655	65	67	49	39
Sri Lanka	21.4	16	6	–4	21.3	0%	7	2.1	25	9	11,970	18	1,646	72	78	97	102
SOUTHEAST ASIA	**644**	**18**	**7**	**0**	**789**	**23%**	**23**	**2.3**	**27**	**6**	**11,376**	**48**	**926**	**68**	**73**	**85**	**87**
Brunei	0.4	16	4	1	0.5	25%	6	1.9	24	4	83,250	77	8,000	75	79	96	96
Cambodia	15.9	24	6	–2	21.8	37%	25	2.6	32	4	3,510	21	418	66	71	—	—
Indonesia	264.0	19	7	–1	321.6	22%	23	2.4	28	5	11,220	54	1,123	67	71	86	86
Laos	7.0	24	7	–4	9.3	33%	43	2.8	34	4	5,920	40	459	65	68	64	59
Malaysia	31.6	17	5	4	41.7	32%	7	2.0	25	6	26,900	75	3,312	73	77	75	81
Myanmar	53.4	18	8	–1	62.4	17%	52	2.3	28	5	5,070	35	495	64	69	51	52
Philippines	105.0	23	7	–1	151.4	44%	21	2.8	32	5	9,400	45	1,878	66	73	84	93
Singapore	5.7	9	5	8	6.5	14%	2.4	1.2	15	12	85,050	100	1,017,857	81	85	—	—
Thailand	66.1	11	8	0	62.6	–5%	10	1.5	18	11	16,070	49	393	72	79	133	125
Timor-Leste	1.3	38	10	–8	2.4	85%	39	5.6	44	3	4,340	33	839	67	70	74	80
Vietnam	93.7	16	7	0	108.2	15%	15	2.1	24	8	6,050	33	1,462	71	76	—	—
EAST ASIA	**1,625**	**12**	**7**	**0**	**1,557**	**–4%**	**9**	**1.8**	**16**	**12**	**18,561**	**62**	**1,414**	**76**	**79**	**94**	**96**
China	1,386.8	13	7	0	1,342.5	–3%	10	1.8	17	11	15,500	57	1,312	75	78	93	96
China, Hong Kong SAR[e]	7.4	8	6	7	8.2	11%	1.5	1.2	11	16	60,530	100	238,710	81	87	103	99
China, Macao SAR[e]	0.6	11	3	11	0.8	33%	2	1.1	12	10	98,450	100	—	80	86	97	96
Japan	126.7	8	10	1	101.9	–20%	1.9	1.5	12	28	42,870	94	3,000	81	87	102	102
Korea, North	25.5	14	9	0	26.8	5%	16	1.9	21	10	—	61	1,085	68	75	93	94
Korea, South	51.4	8	6	1	49.2	–4%	2.7	1.2	13	14	35,790	83	3,482	79	85	99	98
Mongolia	3.2	27	6	–1	4.7	47%	15	3.0	29	4	11,290	67	564	66	75	91	92
Taiwan	23.6	9	7	0	22.7	–4%	4.1	1.2	14	13	—	77	3,969	77	83	—	—
EUROPE	**745**	**11**	**11**	**2**	**736**	**–1%**	**4**	**1.6**	**16**	**18**	**33,677**	**74**	**269**	**75**	**81**	**111**	**111**
EUROPEAN UNION	**511**	**10**	**10**	**3**	**515**	**1%**	**4**	**1.6**	**15**	**19**	**39,480**	**75**	**472**	**78**	**83**	**115**	**115**
NORTHERN EUROPE	**104**	**12**	**9**	**5**	**121**	**16%**	**4**	**1.8**	**18**	**18**	**44,333**	**81**	**529**	**79**	**83**	**126**	**132**
Channel Islands	0.2	10	8	8	0.2	0%	8.0	1.5	16	17	—	32	4,796	80	85	—	—
Denmark	5.8	11	9	6	6.3	9%	3.1	1.8	17	19	51,040	88	239	79	83	128	133
Estonia	1.3	11	12	1	1.1	–15%	2.5	1.6	16	19	28,920	68	201	73	82	116	115
Finland	5.5	10	10	3	5.9	7%	1.9	1.6	16	21	43,400	84	247	78	84	143	156
Iceland	0.3	12	7	12	0.4	33%	1.7	1.8	20	14	52,490	94	248	81	84	116	121
Ireland	4.8	14	7	1	5.9	23%	3.3	1.9	21	13	56,870	64	454	79	84	126	129

	Population mid-2017 (millions)	Births per 1,000 Population	Deaths per 1,000 Population	Net Migration Rate per 1,000	Projected Population (millions), Mid-2050	Projected Population Change, 2017–2050, %	Infant Mortality Rate[a]	Total Fertility Rate[b]	Percent of Population Ages <15	Percent of Population Ages 65+	GNI PPP per Capita ($US), 2016	Percent Urban	Population per Square Kilometer of Arable Land (thousands)	Life Expectancy at Birth (years), Males	Life Expectancy at Birth (years), Females	Secondary School Enrollment Ratio[c], Males (2009–2016)	Secondary School Enrollment Ratio[c], Females (2009–2016)
Latvia	1.9	11	14	−5	1.5	−21%	4.1	1.7	15	20	26,090	68	157	70	79	120	119
Lithuania	2.8	11	14	−10	2.4	−14%	4.5	1.7	15	19	28,840	67	119	69	80	110	106
Norway	5.3	11	8	5	6.7	26%	2.2	1.7	18	17	62,510	81	657	81	84	115	111
Sweden	10.1	12	9	12	12.4	23%	2.5	1.9	18	20	50,000	86	390	81	84	132	150
United Kingdom	66.2	12	9	5	77.7	17%	3.9	1.8	18	18	42,100	83	1,062	79	82	125	130
WESTERN EUROPE	**195**	**10**	**10**	**4**	**207**	**6%**	**3**	**1.7**	**15**	**20**	**47,708**	**80**	**574**	**79**	**84**	**113**	**112**
Austria	8.8	10	9	7	9.8	11%	3.1	1.5	14	19	49,990	66	651	79	84	102	98
Belgium	11.3	11	10	4	12.7	12%	3.4	1.7	17	19	46,010	98	1,383	79	83	156	178
France	65.0	12	9	−3	72.3	11%	3.5	1.9	18	19	42,380	80	355	79	85	110	111
Germany	83.1	9	11	9	83.2	0%	3.3	1.5	13	21	49,530	76	700	78	83	106	100
Liechtenstein	0.04	9	7	5	0.04	0%	3.3	1.3	15	17	—	14	1,724	81	83	131	102
Luxembourg	0.6	10	7	16	0.7	17%	3.8	1.4	16	14	75,750	90	958	81	85	101	103
Monaco	0.04	7	7	10	0.04	0%	—	1.5	13	26	—	100	—	—	—	—	—
Netherlands	17.1	10	9	5	18.1	6%	3.5	1.7	16	18	50,320	91	1,636	80	84	135	136
Switzerland	8.5	11	8	9	9.9	16%	3.4	1.5	15	18	63,660	85	2,124	81	85	103	99
EASTERN EUROPE	**293**	**11**	**13**	**1**	**267**	**−9%**	**6**	**1.6**	**16**	**15**	**21,068**	**70**	**150**	**68**	**78**	**104**	**102**
Belarus	9.5	12	13	1	9.1	−4%	3.2	1.7	17	15	17,210	78	168	69	79	108	106
Bulgaria	7.1	9	15	−1	5.8	−18%	6.5	1.5	14	21	19,020	73	204	71	78	101	97
Czech Republic	10.6	11	10	2	10.4	−2%	2.8	1.6	15	18	32,710	73	337	76	81	105	106
Hungary	9.8	10	13	0	9.5	−3%	3.9	1.5	14	18	25,640	72	223	72	79	105	105
Moldova	3.6	11	11	0	2.9	−19%	9	1.3	16	11	5,670	43	198	68	76	86	86
Poland	38.4	10	10	0	32.6	−15%	4.0	1.4	15	16	26,770	60	351	74	82	110	106
Romania	19.6	10	13	−2	13.9	−29%	7.3	1.2	16	17	22,950	55	223	72	79	93	92
Russia	146.8	13	13	2	144.8	−1%	6.0	1.7	17	14	22,540	74	119	66	77	106	103
Slovakia	5.4	11	10	1	5	−7%	5.1	1.4	15	14	29,910	54	387	73	80	92	93
Ukraine	42.3	9	14	0	33.5	−21%	7.4	1.5	15	16	8,190	70	130	66	76	100	98
SOUTHERN EUROPE	**153**	**9**	**10**	**1**	**141**	**−8%**	**4**	**1.4**	**14**	**20**	**32,803**	**70**	**538**	**79**	**84**	**112**	**110**
Albania	2.9	11	7	−3	2.6	−10%	8.7	1.6	18	13	11,880	56	471	77	80	99	93
Andorra	0.08	9	4	−4	0.07	−13%	3.4	1.2	14	14	—	85	2,857	—	—	—	—
Bosnia-Herzegovina	3.5	9	11	−1	2.7	−23%	5	1.2	14	16	12,140	40	346	74	79	—	—
Croatia	4.1	9	12	−5	3.4	−17%	4.1	1.4	15	19	22,880	59	504	74	80	96	101
Greece	10.7	8	11	−4	9.1	−15%	4.0	1.3	14	21	26,900	78	412	78	84	109	103
Italy	60.5	8	10	2	57.5	−5%	3.0	1.3	14	22	38,230	69	899	81	85	104	102
Kosovo[f]	1.8	13	5	2	1.8	0%	12	1.8	24	8	10,200	38	—	74	79	—	—

	Population mid-2017 (millions)	Births per 1,000 Population	Deaths per 1,000 Population	Net Migration Rate per 1,000	Projected Population (millions), Mid–2050	Projected Population Change, 2017–2050, %	Infant Mortality Rate[a]	Total Fertility Rate[b]	Percent of Population Ages <15	Percent of Population Ages 65+	GNI PPP per Capita ($US), 2016	Percent Urban	Population per Square Kilometer of Arable Land (thousands)	Life Expectancy at Birth (years), Males	Life Expectancy at Birth (years), Females	Secondary School Enrollment Ratio[c], Males (2009–2016)	Secondary School Enrollment Ratio[c], Females (2009–2016)
Macedonia[g]	2.1	11	10	1	1.9	–10%	12	1.5	17	13	14,480	57	507	73	77	80	78
Malta	0.4	10	8	10	0.4	0%	5.7	1.5	14	19	35,720	96	4,459	80	84	92	98
Montenegro	0.6	12	10	2	0.7	17%	3.5	1.6	18	14	17,090	64	6,873	74	79	90	90
Portugal	10.3	8	11	–1	9.2	–11%	3.2	1.4	14	21	29,990	64	906	78	83	121	117
San Marino	0.03	8	8	5	0.03	0%	2.3	1.4	15	19	—	94	3,000	85	89	93	96
Serbia	7.0	9	14	0	5.3	–24%	5.4	1.5	14	19	13,680	60	269	73	78	96	97
Slovenia	2.1	10	10	0	1.9	–10%	1.8	1.6	15	17	32,360	50	1,140	78	84	111	111
Spain	46.6	9	9	2	44.4	–5%	2.6	1.3	15	19	36,340	80	380	80	86	130	130
OCEANIA	**42**	**16**	**7**	**4**	**63**	**50%**	**20**	**2.3**	**23**	**12**	**33,668**	**69**	**87**	**75**	**79**	**103**	**98**
Australia	24.5	13	7	8	37.1	51%	3.2	1.8	19	15	45,970	90	52	80	85	141	134
Federated States of Micronesia	0.1	22	6	–13	0.1	0%	32	3.0	34	4	4,330	23	5,000	68	70	—	—
Fiji	0.9	18	9	–6	1.1	22%	14	3.1	29	6	9,140	51	546	67	73	84	93
French Polynesia	0.3	14	5	–5	0.3	0%	7.5	1.8	24	7	—	56	12,000	74	78	—	—
Guam	0.2	21	6	–5	0.2	0%	10.6	2.9	25	8	—	94	20,000	76	82	—	—
Kiribati	0.1	29	7	–4	0.2	100%	44	3.8	35	4	3,240	44	5,000	63	70	—	—
Marshall Islands	0.06	27	4	–17	0.08	33%	22	4.1	41	3	5,280	74	3,000	71	73	73	80
Nauru	0.01	34	8	–8	0.01	0%	18	3.9	37	13	17,520	100	—	63	71	82	83
New Caledonia	0.3	15	5	3	0.4	33%	12	2.3	24	9	—	71	4,808	74	80	—	—
New Zealand	4.8	13	7	3	5.3	10%	3.6	1.9	20	15	37,860	86	814	80	83	113	120
Palau	0.02	13	9	0	0.02	0%	12	2.2	21	7	14,740	88	2,000	70	77	96	95
Papua New Guinea	8.3	28	7	0	14	69%	47	3.7	37	4	2,700	13	2,767	63	68	46	35
Samoa	0.2	26	5	–13	0.2	0%	17	4.2	37	5	6,200	18	2,500	72	78	81	90
Solomon Islands	0.7	29	5	–4	1.1	57%	26	3.9	40	3	2,150	20	3,500	69	72	50	47
Tonga	0.1	24	6	–11	0.1	0%	20	3.6	36	6	5,760	24	556	70	76	86	94
Tuvalu	0.01	25	9	–3	0.01	0%	10	3.6	33	5	5,920	61	—	67	72	76	97
Vanuatu	0.3	27	5	0	0.6	100%	22	3.4	36	4	3,050	26	1,500	70	74	53	56

(—) Indicates data unavailable or inapplicable.

a. Infant deaths per 1,000 live births.

b. Average number of children born to a woman during her lifetime.

c. The ratio of number of students enrolled in secondary school to the population ages 12–17. Values may exceed 100 when the secondary school population exceeds the size of the relevant age group.

d. The status of Western Sahara is disputed by Morocco.

e. Special Administrative Region.

f. Kosovo declared independence from Serbia on Feb. 17, 2008. Serbia has not recognized Kosovo's independence.

g. The former Yugoslav Republic.

Table modified from the **2017 World Population Data Sheet** prepared by the Population Reference Bureau.

CANADA, MEXICO, AND UNITED STATES REFERENCE MAP

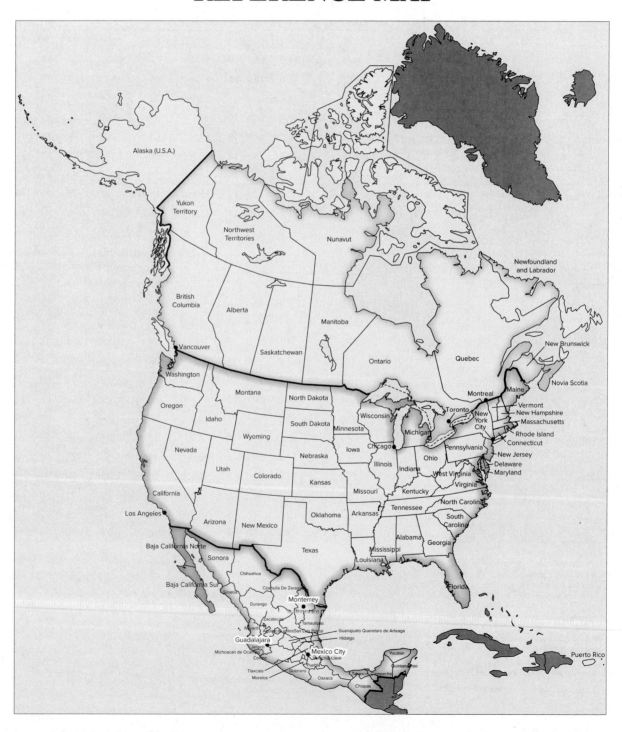

Glossary

Terms in italics identify related glossary items.

A

absolute direction Direction with respect to global location references, such as cardinal directions or macroscopic features.

absolute distance The physical separation between two places measured on a standard unit of length (e.g., miles or kilometers).

absolute location The position of a feature or place expressed in spatial coordinates of a grid system designed for locational purposes. In geography, the most common reference system is the *globe grid* or *graticule* of *parallels of latitude* north or south of the *equator* and of *meridians of longitude* east or west of a *prime meridian.* Absolute globe locations are cited in degrees, which are often subdivided for greater precision into minutes and seconds, or tenths and hundredths of degrees, of latitude and longitude north or south and east or west of the equatorial and prime meridian base lines.

accessibility The relative ease with which a destination may be reached from other locations; the relative opportunity for *spatial interaction.* May be measured in spatial, social, or economic terms.

acculturation Cultural modification or change that results when one *culture* group or individual adopts traits of a dominant or *host society;* cultural development or change through "borrowing."

acid precipitation *Precipitation* that is unusually acidic; created when oxides of sulfur and nitrogen change chemically as they dissolve in water vapor in the *atmosphere* and return to Earth as acidic rain, snow, or fog.

activity space The area within which people move freely on their typical rounds of regular activity (not unusual exceptions).

administrative region Geographic *region* created by law, treaty, or regulation; includes political regions such as countries and states, and internal regions such as school and voting districts

agglomeration The spatial grouping of people or activities for mutual benefit; in *economic geography,* the concentration of productive enterprises for collective or cooperative use of *infrastructure* and sharing of labor resources and market access.

agglomeration economy The savings to an individual enterprise derived from locational association with a cluster of other similar economic activities, such as other factories or retail stores. Example of an external economy.

agricultural density The number of rural residents per unit of agriculturally productive land; a variant of *physiological density* that excludes urban population.

agriculture The practice of farming, including the cultivation of the soil and the rearing of livestock.

amalgamation theory In *ethnic geography,* the concept that multiethnic societies become a merger of the *culture traits* of their member groups.

animism A belief that natural objects may be the abode of dead people, spirits, or gods who occasionally give the objects the appearance of life.

antecedent boundary A *boundary* line established before the area in question is well populated.

aquaculture Farming of cultivated fish and shellfish under controlled conditions, in contrast to the harvesting of wild fish and shellfish.

aquifer A porous, water-bearing layer of rock, sand, or gravel below ground level.

arithmetic density See *crude density.*

artifact A material manifestation of *culture,* including tools, housing, systems of land use, clothing, and the like. An element in the material/*technological subsystem* of culture.

artificial boundary See *geometric boundary.*

aspect In *map projections,* the positional relationship between the globe and the *developable surface* on which it is visually projected: the polar aspect is tangent at the pole; the equatorial aspect is tangent at the equator; the oblique aspect is tangent anywhere else.

assimilation A process by which a minority population reduces or loses completely its identifying cultural characteristics and blends into the *host society.* One component is *spatial assimilation.*

atmosphere The air or mixture of gases surrounding the Earth.

attitude Belief and feeling about places, people, or events.

autonomous nationalism Movement by a dissident minority intent on achieving partial or total independence of territory it occupies from the *state* within which it lies.

awareness space Locations or places about which an individual has knowledge even without visiting all of them; includes *activity space* and additional areas newly encountered or about which one acquires information.

azimuthal projection See *planar projection.*

B

barrier A *geographic feature* that impedes spatial interaction, either by blocking it totally, slowing it down, or redirecting it. Barriers may be physical, socio-cultural, or psychological.

basic sector Those products or services of an *urban* economy that are exported outside the city itself, earning income for the community.

behavior Coordinated and goal-directed action by people or institutions.

behavioral approach A way of doing *human geography* that focuses on a disaggregate (individual) level of analysis, appreciating the role of *cognition* and emotion in determining human actions.

bilingualism Describing a person's or society's use of two *languages.*

biome A major ecological community, including plants and animals, occupying an extensive Earth area.

biosphere (See *ecosphere*)

birth rate See *crude birth rate.*

Boserup thesis The view that population growth independently forces a conversion from extensive to intensive *subsistence* agriculture.

brain drain The loss of a developing country's most educated citizens as they emigrate in search of better educational and career opportunities in developed countries.

break-of-bulk point A location where goods are transferred from one type of carrier to another (e.g., from barge to railroad).

brownfield site A former industrial or commercial site that is under-used, vacant, or abandoned where there is the potential for environmental contamination.

Buddhism A *universalizing religion,* primarily of eastern and central Asia, based on teachings of Siddhartha Gautama, the Buddha, that suffering is inherent in all life but can be relieved by mental and moral self-purification.

C

carrying capacity The maximum population numbers that an area can support on a continuing basis without experiencing unacceptable deterioration; for humans, the numbers supportable by an area's known and used resources—including agricultural and energy resources.

cartogram A *thematic map* that shows the values of a quantitative variable for each region by shrinking or expanding the sizes of the regions to correspond with the variable's value.

cartography The art and science of maps and mapping.

caste One of the hereditary social classes in *Hinduism* that determines one's occupation and position in society.

central business district (CBD) The nucleus or "downtown" of a city, where retail stores, offices, and cultural activities are concentrated, mass transit systems converge, and land values and building densities are high.

central city That part of the *metropolitan area* contained within the boundaries of the main city around which suburbs have developed.

central place An *urban* or other settlement node whose primary function is to provide goods and services to the consuming population of its *hinterland, complementary region,* or trade area.

central place theory A deductive theory formulated by Walter *Christaller* (1893–1969) to explain the size and distribution of settlements through reference to competitive supply of goods and services to dispersed rural populations.

centrifugal force 1: In *urban geography,* economic and social forces pushing households and businesses outward from central and inner-city locations. 2: In *political geography,* forces of disruption and dissolution threatening the unity of a *state.*

centripetal force 1: In *urban geography,* economic and social forces attracting establishments or activities to central and inner-city locations. 2: In *political geography,* forces tending to bind together the citizens of a state, thus promoting its unity.

chain migration The process by which *migration* movements from a common home area to a specific destination are sustained by links of friendship or kinship between first movers and later followers.

channelized migration The tendency for *migration* to flow between areas that are socially and economically allied by past migration patterns, by economic and trade connections, or by some other affinity.

charter group In plural societies, the early arriving ethnic group that created the *first effective settlement* and established the recognized cultural norms to which other, later groups are expected to conform.

choropleth map A *thematic map* that shows the values of a quantitative variable for each region by shading or coloring the regions to correspond with the variable's value.

Christaller, Walter Walter Christaller (1893–1969), German geographer credited with developing *central place theory* (1933).

Christianity A *monotheistic, universalizing religion* based on the teachings of Jesus Christ and of the Bible as sacred scripture.

circular and cumulative causation A process through which tendencies for economic growth are self-reinforcing; an expression of the *multiplier effect,* it tends to favor major cities and *core* regions over less-advantaged *peripheral* regions.

city A multifunctional nucleated settlement with a *central business district* and both residential and nonresidential land uses.

cluster migration A pattern of movement and settlement resulting from the collective action of a distinctive social or *ethnic group.*

cognition Knowledge and beliefs about something, and the thinking and memory processes that create and modify them; it varies somewhat across individual people and cultural groups; *perception* is sometimes used broadly as a synonym for cognition.

cognitive map See *mental map.*

cohort A population group unified by a specific common characteristic, such as age, and subsequently treated as a statistical unit for data analysis.

colony In *ethnic geography,* an urban ethnic area serving as point of entry and temporary *acculturation* zone for a specific immigrant group.

commercial economy (*syn:* **market economy**) A system of production of goods and services for exchange in competitive markets where price and availability are determined by supply and demand forces.

commodity chain The set of activities involved in the production of a single good or service. A commodity chain encompasses the relationships between buyers and suppliers and the flows of materials, finance, and knowledge.

compact state A *state* whose territory is nearly circular.

comparative advantage The principle that an area produces the items for which it has the greatest ratio of advantage or the least ratio of disadvantage in comparison to other areas, assuming free trade exists.

complementarity The actual or potential relationship of two places or regions that each produce different goods or services for which the other has an effective demand, resulting in an exchange between the locales.

concentration In *spatial distributions,* the clustering of a phenomenon around a central location.

concentric zone model A model describing urban land uses as a series of circular belts or rings around a core *central business district,* each ring housing a distinct type of land use. It also describes a common urban residence pattern corresponding to different family life stages (family status).

conformal projection A *map projection* that retains correct shapes of small areas; lines of *latitude* and *longitude* cross at right angles and *scale* (1) is the same in all directions at any point on the map.

Confucianism A Chinese *value system* and *ethnic religion* emphasizing ethics, social morality, tradition, and ancestor worship.

conic projection A *map projection* employing a cone placed over the globe as the presumed *developable surface.*

connectivity The directness of routes linking pairs of places; an indication of the degree of internal connection in a transport *network.* More generally, all of the tangible and intangible means of connection and communication between places.

consequent boundary (*syn*: ethnographic boundary) A *boundary* line that coincides with some cultural divide, such as religion or language.

consumer service A portion of the service (tertiary) sector of the economy involved in providing services to individuals and households. Examples include hair salons and retailers.

contagious diffusion A pattern of *diffusion* in which cultural innovations spread to closer places before they spread to further places (it reflects distance decay). Contagious expansion diffusion typically results from direct contact between actual and potential adopters of the *innovation*, in a manner analogous to the spread of contagious diseases.

conurbation A continuous, extended *urban* area formed by the growing together of several formerly separate, expanding cities. Called Consolidated Metropolitan Area by the U.S. Census.

core area 1: In *economic geography*, a "core region," the national or world districts of concentrated economic power, wealth, innovation, and advanced technology. 2: In *political geography*, the heartland or nucleus of a *state*, containing its most developed area, greatest wealth, densest populations, and clearest national identity.

core-periphery model A model of the spatial structure of an economic or political system in which underdeveloped or declining peripheral areas are defined with respect to their dependence on a dominating developed *core region*.

counter migration (*syn*: return migration) The return of migrants to the regions from which they earlier emigrated.

creole A *language* developed from a *pidgin* to become the native tongue of a society.

critical distance The distance beyond which cost, effort, and/or means play a determining role in the willingness of people to travel.

crude birth rate (CBR) The ratio of the number of live births during one year to the total population, usually at the midpoint of the same year, expressed as the number of births per year per 1,000 population.

crude death rate (CDR) (*syn*: **mortality rate**) The ratio of the number of deaths during one year to the total population, usually at the midpoint of the same year, expressed as the number of deaths per year per 1,000 population.

crude density (*syn*: arithmetic density) The number of people per unit area of land.

cultural autonomy The view that cultures can develop any particular set of cultural traits independent of their environmental circumstances, a view opposite of *environmental determinism*.

cultural convergence The tendency for *cultures* to become more alike as they increasingly share *technology* and organizational structures in a modern world united by improved developments to transportation and communication.

cultural divergence The likelihood or tendency for *cultures* to become increasingly dissimilar with the passage of time.

cultural ecology The study of the interactions between societies and the natural *environments* they occupy.

cultural integration The interconnectedness of all aspects of a *culture*: no part can be altered without creating an impact on other components of the culture.

cultural landscape The *natural landscape* as modified by human activities and bearing the imprint of a *culture* group or society; the *built environment*.

cultural system A generalization suggesting shared, identifying traits uniting two or more *culture complexes*.

culture 1: A society's collective beliefs, symbols, values, forms of behavior, and social organizations, together with its tools, structures, and artifacts created according to the group's conditions of life; transmitted as a heritage to succeeding generations and undergoing adoptions, modifications, and changes in the process. 2: A collective term for a group displaying uniform cultural characteristics.

culture complex A related set of *culture traits* descriptive of one aspect of a society's behavior or activity. Culture complexes may be as basic as those associated with food preparation, serving, and consumption, or as involved as those associated with religious beliefs or business practices.

culture hearth A nuclear area within which an advanced and distinctive set of *culture traits*, ideas, and *technologies* develops and from which there is *diffusion* of those characteristics and the *cultural landscape* features they imply.

culture realm A collective of *culture regions* sharing related culture systems; a major world area having sufficient distinctiveness to be perceived as set apart from other realms in terms of cultural characteristics and complexes.

culture rebound The readoption by later generations of *culture traits* and identities associated with immigrant forebears or ancestral homelands.

culture region A *thematic* or *functional region* within which common cultural characteristics prevail. It may be based on single *culture traits*, on *culture complexes*, or on political, social, or economic integration.

culture trait A single distinguishing feature of regular occurrence within a *culture*, such as the use of chopsticks or the observance of a particular caste system. A single element of learned behavior.

custom The body of traditional practices, usages, and conventions that regulate social life; an expression of *culture*.

cylindrical projection A *map projection* employing a cylinder wrapped around the globe as the presumed *developable surface*.

D

dead zone Coastal zone, often near the mouth of major rivers, where waters are very low in oxygen and subject to major die-offs of marine life. Coastal dead zones are generally caused by *environmental pollution* from fertilizers and manure in agricultural runoff, as well as urban sewage.

death rate See *crude death rate*.

deforestation The clearing of land through total removal of forest cover.

deglomeration The process of deconcentration; the location of industrial or other activities away from established *agglomerations* in response to growing costs of congestion, competition, and regulation.

deindustrialization The cumulative and sustained decline of manufacturing activities in a regional or national economy, involving the loss of both firms and jobs.

demographic equation A mathematical expression that summarizes the contribution of different demographic processes to the population change of a given area during a specified time period. $P_2 = P_1 + B_{1-2} - D_{1-2} + IM_{1-2} - OM_{1-2}$, where P_2 is population at time 2; P_1 is population at beginning date; B_{1-2} is the number of births between times 1 and 2; D_{1-2} is the number of deaths during that period; IM_{1-2} is the number of in-migrants and OM_{1-2} the number of out-migrants between times 1 and 2.

demographic momentum See *population momentum*.

demographic transition A model of the effect of economic development on population growth, originally expressed in terms of four stages but now including a final fifth stage.

demography The scientific study of population, with particular emphasis upon quantitative aspects of births and deaths.

density The quantity of some feature (people, buildings, animals, traffic, etc.) per unit area or size.

dependency ratio The number of dependents, old or young, that each 100 persons in the economically productive years must on average support.

dependency theory A theory that attempts to explain patterns and processes of economic development by extending the *core-periphery model* to the international scene, arguing that the development of the advanced core nations has depended upon the underdevelopment of the peripheral nations. Dependency theory argues that developing countries were made poor by their interactions with the advanced countries, starting with colonialism and the slave trade and continuing in new forms to the present.

desertification Extension of desert-like landscapes as a result of overgrazing, destruction of the forests, or other human-induced changes, usually in semiarid regions.

develop Representing the Earth's curved surface as a flat map, as when creating a *map projection.*

developable surface *Map projection* surface (including the plane, cone, or cylinder) that can be made flat without further distorting the Earth surface it depicts, beyond the distortion arising from projecting onto that surface as it is wrapped around the earth and then unwrapping it to split the surface.

development The process of growth, expansion, or realization of potential; bringing regional resources into full productive use.

devolution The transfer of certain powers from the *state* central government to separate political subdivisions within the state's territory.

dialect A *language* variant marked by vocabulary, grammar, or pronunciation differences from other variants of the same common language. When those variations are spatial or regional, they are called *geographic dialects;* when they are indicative of socioeconomic or educational levels, they are called *social dialects.*

diffusion The spread or movement of a phenomenon over space and through time. The dispersion of a *culture trait,* such as a new idea or practice, from an origin area (e.g., *language,* plant *domestication,* new industrial *technology*).

dispersion In *spatial distributions,* a statement of the amount of spread of

a phenomenon over area or around a central location. Dispersion in this sense represents a continuum from clustered, concentrated, or agglomerated (at one end) to dispersed or scattered (at the other).

distance decay The declining intensity of any *spatial interaction* with increasing distance from its point of origin.

domestication The successful transformation of plant or animal species from a wild state to a condition of dependency on human management, usually with distinct physical change from wild forebears.

dot map A *thematic map* that shows the occurrence of one or more instance of a feature with a dot at a particular location.

doubling time The time period required for a population to double in size.

E

economic base The manufacturing and service activities performed by the *basic sector* of a city's labor force; functions of a city performed to satisfy demands external to the city itself and, in that performance, earning income to support the urban population.

economic geography The branch of *systematic geography* concerned with how people support themselves, with the spatial patterns of production, distribution, and consumption of goods and services, and with the areal variation of economic activities over the surface of the Earth.

ecosphere The region of air, water, solid Earth, and living organisms where life is found. The ecosphere includes the *atmosphere,* the *hydrosphere* of surface and subsurface waters, the *lithosphere* of the upper reaches of the Earth's crust, the *biosphere* of living organisms.

ecosystem A population of organisms existing together in a small, relatively homogeneous area (pond, forest, small island), together with the energy, air, water, soil, and chemicals upon which it depends.

ecumene That part of the Earth's surface physically suitable for permanent human settlement; the permanently inhabited areas of the Earth.

edge city Distinct sizable nodal concentration of retail and office space of lower than central city densities and situated on the outer fringes of older metropolitan areas; usually localized near major highway intersections.

Ehrlich, Paul R. Paul Ehrlich (1932–), population biologist and author of the influential *neo-Malthusian* book, *The Population Bomb (1968).*

electoral geography The study of the geographical elements of the organization and results of elections, including the design of electoral districts.

elongated state A *state* whose territory is long and narrow.

enclave A small bit of foreign territory lying within a *state* but not under its jurisdiction.

environment Surroundings; the totality of things that in any way may affect a person or other living organism, including both biophysical and sociocultural conditions.

environmental determinism The view that the physical *environment,* particularly *climate,* controls human action, molds human behavior, and conditions cultural development.

environmental justice The notion that all people, regardless of race, ethnicity, or income, should live in a safe environment and be equally protected from environmental hazards and pollution.

environmental pollution See *pollution.*

equal-area (equivalent) projection A *map projection* with the property that a unit area drawn anywhere on the map always represents the same area on the Earth's surface.

equator An imaginary east-west line (*parallel of latitude*) that encircles the globe halfway between the North and South Poles.

equidistant The property of *map projections* showing true distances in all directions only from the center of the projection; all other distances are incorrect.

equirectangular projection A cylindrical projection that converts the globe's meridians to equally spaced vertical lines and its parallels to equally spaced horizontal lines.

erosion See *soil erosion.*

ethnic cleansing The attempt by a dominant ethnic or national group to violently eliminate a less powerful ethnic or national group from a particular geographic area, to achieve racial or cultural homogeneity and an expanded settlement area for the perpetrating group.

ethnic enclave A small area occupied by a distinctive minority *culture.*

ethnic geography The study of spatial distributions and interactions of *ethnic groups* and of the cultural characteristics on which they are based.

ethnic group People who share common identity as members of a distinctive group, based on common national origin or heritage, *religion, language,* ideology, or *race.*

ethnic island A small rural area settled by a single, distinctive *ethnic group* that placed its imprint on the landscape.

ethnic province A large territory, urban and rural, dominated by or closely associated with a single *ethnic group.*

ethnic religion A *religion* identified with a particular *ethnic group* and largely exclusive to it. Such a religion does not seek converts (does not proselytize).

ethnicity Ethnic quality; affiliation with a group whose racial, cultural, religious, or linguistic characteristics or national origins distinguish it from a larger population within which it is found.

ethnoburb A politically independent suburban community with a significant, though not exclusive, concentration of a single ethnic group.

ethnocentrism Evaluating other ethnic groups based exclusively on the perspective of one's own *ethnic group.*

European Union (EU) An economic association established in 1957 by a number of Western European countries to promote free trade among members; often called the Common Market. It currently has almost 30 member countries.

exclave A portion of a *state* that is separated from the main territory and surrounded by another country, inside of which it is an enclave.

exclusive economic zone (EEZ) As established in the *United Nations Convention on the Law of the Sea,* a zone of exploitation extending 200 nautical miles (370 km) seaward from a coastal state that has exclusive mineral and fishing rights over it.

expansion diffusion The spread of ideas, behaviors, or other culture traits from one place to another through direct or indirect contact and exchange of information; the diffusion increases the number of people or cultural groups practicing the trait while leaving the trait intact or intensified in its area of origin.

extensive commercial agriculture A crop or livestock system in a commercial economy characterized by low inputs of labor and capital, and low value, per unit area of land. Range herding and consumption grain are important examples.

extensive subsistence agriculture A crop or livestock system in a subsistence economy characterized by low inputs of labor and capital, and low value, per unit area of land. Nomadic herding and swidden agriculture are important examples.

external economy benefits that firms enjoy due to factors outside the firm, including the benefits of *agglomeration economies.*

extractive industry *Primary activity* involving the mining and quarrying of *non-renewable* metallic and nonmetallic mineral resources.

F

fallowing The practice of allowing plowed or cultivated land to remain (rest) uncropped or only partially cropped for one or more growing seasons.

federal state A *state* made up of more or less equal provinces each with relatively large *regional autonomy* and government responsibility; the "states" of the United States are, in actuality, such provinces.

field *Geographic feature* thought of as a continuously varying surface that completely covers the space of the landscape it occupies without overlapping other fields of the same type. Examples include average precipitation or landform elevation.

first effective settlement The influence that the characteristics of an early dominant settlement group exert on the later *social* and *cultural geography* of an area.

First Law of Geography (*syn:* distance decay) "Everything is related to everything else, but near things are more related than distant things"; attributed to the American geographer Waldo Tobler (1930–).

folk culture The body of institutions, customs, dress, *artifacts,* collective wisdoms, and traditions of a homogeneous, isolated, largely self-sufficient, and relatively static social group. Unlike *popular culture,* folk culture varies greatly over space but not much over time.

food security Refers to the situation wherein every person has access to safe and nutritious food of sufficient quantity for an active and healthy lifestyle.

footloose firm A descriptive term applied to manufacturing activities for which the cost of transporting material or product is not important in determining location of production; an industry or firm showing neither *market* nor *material orientation.*

forced migration When people are compelled by someone or some event to relocate their residence.

Fordism The manufacturing economy and system derived from assembly-line mass production and the mass consumption of standardized goods. Named after Henry Ford, who innovated many of its production techniques.

foreign direct investment (FDI) the purchase or construction of factories and other fixed assets by *transnational corporations.*

fragmented state A *state* whose territory contains isolated parts, separated and discontinuous.

freight rate The charge levied by a transporter for the loading, moving, and unloading of goods; includes *line-haul costs* and *terminal costs.*

friction of distance A measure of the retarding or restricting effect of distance on *spatial interaction.* Generally, the greater the distance, the greater the "friction" and the less the interaction, or the greater the cost of achieving the interaction.

functional dispute (*syn:* boundary dispute) In *political geography,* a disagreement between neighboring *states* over policies to be applied to their common border; often induced by differing customs regulations, movement of nomadic groups, or illegal immigration or emigration.

functional region (*syn:* nodal region) Geographic *region* emerging from patterns of interaction over space and time that connect places.

G

galactic city model See *peripheral model.*

gated community A restricted access subdivision or neighborhood, often surrounded by a barrier, with entry permitted only for residents and their guests; usually totally planned in land use and design, with "residents only" limitations on public streets and parks.

gathering industry *Primary activity* involving the *subsistence* or *commercial* harvesting of *renewable* natural resources of land or water. Primitive gathering involves local collection of food and other materials of nature, both plant and animal; commercial gathering usually implies forestry and fishing industries.

gender In the cultural sense, a reference to socially created—not biologically based—distinctions between femininity and masculinity.

genetically modified (GM) crops Food or fiber crops whose genetic material has been altered through biotechnology in ways that do not occur naturally. Genetic modification takes place through transferring individual genes between organisms or between species.

gentrification The movement into the inner portions of American cities of middle- and upper-income people who replace low-income populations, rehabilitate the structures they occupied, and change the social character of neighborhoods.

geodemographic analysis A form of marketing analysis that uses GIS to infer demographic and lifestyle characteristics of potential consumers according to their residential location.

geographic dialect (*syn:* regional dialect) See *dialect.*

geographic feature Natural or cultural entity on the landscapes of the Earth's surface, e.g., mountain, river, forest, cornfield, city, country.

geographic information system (GIS) Integrated computer software and hardware for storing, processing, analyzing, and displaying data specifically referenced to locations on the surface of the Earth.

geometric boundary (syn: artificial boundary) A *boundary* line based on a coordinate system such as the latitude-longitude *graticule*, rather than physiographic features such as mountains or rivers.

geometrical projection (*syn:* perspective projection; visual projection) A *map projection* theoretically created by tracing the *graticule* shadow projected on a *developable surface* from a light source placed relative to a transparent globe.

geopolitics Study of how spatial relations among *regions* influence their current and past political activities and relations.

gerrymandering To redraw electoral voting district boundaries in such a way as to give particular candidates or classes of candidates an electoral advantage beyond the share of the electorate that supports them. Different varieties of gerrymandering may be seen as motivated by nondemocratic objectives or attempts to implement social justice.

ghetto A forced or voluntarily segregated residential area housing a racial, ethnic, or religious minority.

GIS See *geographic information system.*

global climate change Change in the Earth's climate system, whether natural or caused by humans.

globalization A reference to the increasing interconnection of all parts of the world as the full range of social, cultural, political, and economic processes becomes international in scale and effect. One result of *space-time compression.*

globe A spherical physical model of the Earth.

glocalization The adaptation of globalized products to local tastes and contexts.

GDI See *gross domestic income.*

GNI See *gross national income.*

gnomonic projection A *geometrical projection* produced with the light source at the center of the Earth.

graduated circle map A *proportional area symbol* map using circles.

graphic scale A visual expression of cartographic *scale* in a map legend, consisting of a graduated line showing how much distance on the map represents a particular distance on the Earth surface.

graticule (*syn:* globe grid) The network of *meridians of longitude* and *parallels of latitude* that make up a coordinate reference system on the globe.

gravity model A mathematical prediction of the interaction between two places as a function of their size (or other measure of attractiveness to interaction) and some measure of the distance separating them.

great circle Line formed by the intersection with the Earth's surface of a plane passing through the center of the Earth; definition of a straight line on the surface of a sphere, an arc of a great circle is the shortest earth-surface distance between two points on the surface.

Green Revolution A series of agricultural technology transfers during the 1940s–70s from the United States to subtropical areas of undeveloped and developing countries, accomplished through the introduction of very high-yielding hybrid grain crops, particularly wheat, maize, and rice; increased irrigation infrastructure; synthetic fertilizers and pesticides; and new forms of agricultural management. It has been estimated to have saved the lives of over a billion people from starvation but has been heavily criticized as well.

greenhouse effect Heating of the Earth's surface as shortwave solar energy passes through the *atmosphere,* which is transparent to it but opaque to reradiated long-wave terrestrial energy; also, increasing the opacity of the atmosphere through addition of increased amounts of carbon dioxide and other gases that trap heat.

gross domestic income (GDI) The total value of goods and services produced per year within a country by domestically- or foreign-owned interests; formerly called "gross domestic product."

gross national income (GNI) The total value of goods and services produced per year at home or abroad by domestically-owned interests within a country; formerly called "gross national product."

H

hazardous waste Discarded solid, liquid, or gaseous material that poses a substantial threat to human health or to the *environment* when improperly disposed of or stored.

heritage landscape Landscape strongly associated with the history of a particular cultural group.

hierarchical diffusion A pattern of *diffusion* in which cultural *innovations* spread by "jumping" between places of more importance (such as larger cities) before they spread to places of less importance; reverse hierarchical diffusion from less important to more important places occurs less often.

hierarchy of central places The step-like series of *urban* units in classes differentiated by both size and function.

Hinduism An ancient and now dominant *value system* and *religion* of India, closely identified with Indian *culture* but without central creed, single doctrine, or religious organization. Dharma (customary duty and divine law) and *caste* are uniting elements.

hinterland The rural market area or region served by an *urban* center.

host society The established and dominant society within which immigrant groups seek accommodation.

human geography One of the two major divisions (the other is *physical geography*) of *systematic geography;* the spatial analysis of human populations, their *cultures,* their activities and behaviors, and their interrelationships with the physical landscapes they occupy.

hunter-gatherer/hunting-gathering An economic and social system based primarily or exclusively on the hunting of wild animals and the gathering of food, fiber, and other materials from uncultivated plants, insects, eggs, and so on.

hydrologic cycle The natural system by which water is continuously circulated through Earth systems by evaporation, condensation, and *precipitation.*

hydrosphere All water at or near the Earth's surface that is not chemically bound in rocks, including the oceans, surface waters, *groundwater,* and water held in the *atmosphere.* It also includes frozen water, although this is sometimes distinguished as the *cryosphere.*

I

ideological subsystem The complex of ideas, beliefs, knowledge, and means of their communication that characterize a *culture,* along with the *technological* and *sociological subsystems.*

independent invention (*syn:* parallel invention) *Innovations* developed in two or more unconnected locations by individuals or groups acting independently. See also *multilinear evolution.*

Industrial Revolution The term applied to the rapid economic and social changes in manufacturing that followed the introduction of mechanized production to the textile industry of England in the last quarter of the 18th century, and subsequently to other economic activities and places around the globe since then.

infant mortality rate A refinement of the *death rate* to specify the ratio of deaths of infants age 1 year or less per 1,000 live births.

informal economy That part of a national economy that involves productive labor not subject to formal systems of control or payment, such as taxation; economic activity or individual enterprise operating without official recognition or measured by official statistics. It ranges from "under the table" employment to buying and selling on the "black market."

infrastructure The basic structure of services, installations, and facilities needed to support industrial, agricultural, and other economic activity, including transportation and communication, along with water, power, and other utilities.

innovation Introduction of new *culture traits,* whether ideas, practices, or material objects.

insolation The solar radiation received at the Earth's surface (from *int*ercepted *sol*ar radi*ation*).

intensive commercial agriculture A crop or livestock system in a commercial economy characterized by high inputs of labor and/or capital, and high value, per unit area of land. *Truck farming* is an important example.

intensive subsistence agriculture A crop or livestock system in a subsistence economy characterized by high inputs of labor and/or capital, and high value, per unit area of land. Consumption rice is an important example.

intervening opportunity The concept that closer opportunities will materially reduce the attractiveness of interaction with more distant—even slightly better—alternatives; a closer alternative source of supply between a demand point and the original source of supply.

IPAT equation An equation relating the environmental impact of a society to the key factors of population, affluence, and technology.

irredentism The policy of a *state* wishing to incorporate within itself territory inhabited by people who have ethnic or linguistic links with the people of the state but that lies within a neighboring state.

Islam A *monotheistic, universalizing* religion that includes belief in Allah as the sole deity and in Mohammed as his prophet completing the work of earlier prophets of *Judaism* and *Christianity.*

isogloss A mapped boundary line marking the limits of a particular linguistic feature.

isoline A map line connecting points of equal value on some variable, such as elevation, precipitation, or travel time.

isotropic plain See *uniform plain.*

J

J-curve A curve shaped like the letter J, depicting exponential or geometric population growth (e.g., 1, 2, 4, 8, 16 …).

Judaism A *monotheistic, ethnic religion* first developed among the Hebrew people of the ancient Near East; its determining conditions include descent from Israel (Jacob), the Torah (law and scripture), and tradition.

L

landlocked Describing a *state* that lacks a sea coast.

language The system of words, their pronunciation, and methods of combination used and mutually understood by a community of individuals.

language family A group of *languages* thought to have descended from a single, common ancestral tongue.

Latin American city model A description of land uses in Latin American cities. The model combines wedge-shaped sectors and concentric rings emanating from a central business district. The wealthy live along a well-served commercial spine and the poorest residents live in peripheral squatter settlements.

latitude Angular distance of a location north or south of the *equator,* measured in degrees, minutes, and seconds. Grid lines marking latitude are called *parallels.* The equator is 0°; the North Pole is 90°N; the South Pole is 90°S.

law of retail gravitation Any *gravity model* of shopping behavior, such as *Reilly's Breaking-Point Law* or the *potential model* when applied to multiple stores or shopping centers.

least-cost theory (*syn:* Weberian analysis) The view that the optimum location of a manufacturing establishment is at the place where the costs of transporting raw materials and finished products, as well as labor, and the advantages of *agglomeration* or *deglomeration* are most favorable.

line-haul cost (*syn:* over-the-road cost) A cost involved in the actual physical movement of goods (or passengers); cost of haulage (including equipment and route costs), excluding *terminal costs.*

lingua franca Any of various auxiliary *languages* used as common tongues among people of an area where several languages are spoken; literally, "Frankish language."

linguistic geography (*syn:* dialect geography; dialectology) The study of the spatial distribution of languages, including the study of language groups and families, dialects, creoles and pidgins, and so on.

link A transportation or communication connection or route within a *network.*

lithosphere The Earth's solid crust and mantle.

locational interdependence The circumstance under which the locational decision of a particular firm is influenced by the locations chosen by competitors. For example, retail businesses often choose their location in order to attract customers away from competing retail businesses.

longitude Angular distance of a location east or west of a designated *prime meridian,* measured in degrees, minutes, and seconds. Grid lines marking longitude are called *meridians.* Distances are measured from 0° at the *prime meridian* to 180° both east and west, with 180°E and W being the same line. For much of its extent, the 180° meridian also serves as the *International Date Line.*

long-lot system A survey system that divides land into long, narrow strips extending back from a river or road, found in areas settled by French colonists.

M

Malthus, Thomas Robert Thomas R. Malthus (1766–1843). English economist, demographer, and cleric who suggested that unless self-control, war, or natural disaster checks population, it will inevitably increase faster than will the food supplies needed to sustain it. This view is known as Malthusianism. See also *neo-Malthusianism.*

map projection A systematic method of transferring the *globe grid* system from the Earth's curved surface to the flat surface of a map. Projection automatically incurs error, but an attempt is usually made to preserve one or more (though never all) of the characteristics of the spherical surface: equal area, correct distance, true direction, proper shape.

map scale (*syn:* cartographic scale) See *scale.*

maquiladora (also maquila) Foreign-owned manufacturing plant located in Mexico for the low cost assembly of clothing, electronics, automobiles, and other export products.

market economy An economic system in which most goods and services are privately produced and distributed for monetary exchange; an economy characterized by free market exchange (freely varying prices and production systems) with no or only minimum state intervention.

market equilibrium The theoretically stable point of intersection of demand and supply curves of a given commodity; at equilibrium the market is cleared of the commodity.

market orientation The tendency of an economic activity to locate close to its market; a reflection of large and variable costs of transporting finished products.

Marx, Karl Karl Marx (1818–1883), German philosopher, economist, and social theorist who decried the inequalities produced by capitalism and advocated revolutionary change in order to bring about a communist system.

material culture The tangible, physical items produced and used by members of a specific *culture* group and reflective of their traditions, lifestyles, and technologies.

material orientation The tendency of an economic activity to locate near or at its source of raw material; this is experienced when the costs of transporting materials are highly variable spatially and/or represent a significant share of total costs.

mathematical projection The systematic rendering of the *globe grid* on a *developable surface* to achieve *graticule* characteristics not obtainable by visual means of *geometrical projection.*

maximum sustainable yield The maximum rate at which a *renewable resource* can be exploited without impairing its ability to be renewed or replenished.

Mediterranean agriculture An agricultural system based upon the mild, moist winters; hot, sunny summers; and rough terrain such as that found in the Mediterranean basin. It involves cereals as winter crops, summer tree and vine crops (olives, figs, dates, citrus and other tree fruits, and grapes), and grazing animals (sheep and goats).

mental map (*syn:* cognitive map) By analogy to a cartographic map, the set of mental representations people hold in their mind that expresses their beliefs and knowledge about the layout of the environment at different scales, whether neighborhoods,

cities, regions, countries, or the entire world. The representations are subjective and influenced by personal feelings, and may be quite incomplete and distorted as compared to the actual layouts.

mentifact A central element of a *culture* expressing its values and beliefs, including *language, religion, folklore,* artistic traditions, and the like. An element in the *ideological subsystem* of culture.

Mercator projection A true *conformal cylindrical projection* first published in 1569; because straight lines drawn on this projection are lines of constant compass direction (*rhumb lines*), it is especially useful for navigation.

meridian A north-south line of *longitude* indicating distance east or west of the *prime meridian* running through Greenwich, England.

metropolitan area In the United States, a large functionally integrated settlement area comprising one or more whole county units and usually containing several *urbanized areas;* discontinuously built up, it operates as a coherent economic whole.

migration The permanent (or relatively permanent) relocation of an individual or group to a new place of residence.

migration field The area from which a given city or region draws the majority of its in-migrants.

mobility General term for all types of human movement through space and time, including *temporary travel* and *migration.*

model An idealized representation, abstraction, or simulation of reality. It is designed to simplify real-world complexity and eliminate extraneous phenomena in order to isolate for detailed study causal factors and interrelationships of *spatial systems.*

modernization theory An influential theory that attempts to explain patterns and processes of economic development. It sees societies arranged on a continuum between traditional and modern with modernization occurring as a society adopts advanced technologies and market mechanisms and experiences industrialization, urbanization, and rising prosperity.

monolingual A society or country that uses only one *language* for all purposes of communication.

monotheism The belief that there is but a single God.

mortality rate (*syn:* **death rate**) See *crude death rate.*

movement bias Any aggregate control on or regularity of movement of people, commodities, or communication. Included are *distance bias, direction bias,* and *network bias.*

multilinear evolution A concept of independent but parallel cultural development advanced by the anthropologist Julian Steward (1902–1972) to explain cultural similarities among widely separated peoples existing in similar environments but who could not have benefited from shared experiences, borrowed ideas, or diffused technologies. See *independent invention.*

multilingualism The common use of two or more *languages* by an individual or in a society or country.

multiple-nuclei model The postulate that large cities develop by peripheral spread not from one *central business district* but from several nodes of growth, each of specialized use. The separately expanding use districts eventually coalesce at their margins. It also describes a common urban residence pattern corresponding to different ethnic groups (ethnic status).

multiplier effect The direct, indirect, and induced consequences of change in an activity. 1: In industrial *agglomerations,* the cumulative processes by which a given change (such as a new plant opening) sets in motion a sequence of further industrial employment and *infrastructure* growth. 2: In *urban geography,* the expected addition of *nonbasic* workers and dependents to a city's total employment and population that accompanies new *basic sector* employment.

N

nation A culturally distinctive group who self-identity as a separate group, may or may not occupy a territorial homeland, and are bound together by a sense of unity arising from shared *ethnicity,* beliefs, and *customs.*

nationalism A sense of unity binding the people of a *state* together; devotion to the interests of a particular country, an identification with the state and an acceptance of its goals.

nation-state A *state* whose territory is identical to that occupied by a single particular *ethnic group* or *nation.*

natural boundary (*syn:* physical boundary) A *boundary* line based on recognizable physiographic features, such as mountains or rivers.

natural hazard A process or event in the natural environment that has consequences harmful to humans.

natural landscape The physical *environment* unaffected by human activities. The duration and near totality of human occupation of the Earth's surface assures that little or no "natural landscape" strictly defined remains intact. Opposed to *cultural landscape.*

natural resource A feature or material in the environment that a population perceives to be necessary and useful to its maintenance and well-being.

natural selection the mechanism of biological evolution; changes to the gene pool of a population of living organisms over time according to variation and survival of the fittest in particular environmental conditions.

neocolonialism A disparaging reference to economic and political policies by which major developed countries are seen to retain or extend influence over the economies of less developed countries and peoples. A continuing expression of dependency theory.

neoliberal globalization Approach introduced during the 1980s that revived faith in the market mechanism and the private sector as a means of promoting regional development.

neolocalism A social movement advocating a return to local products, locally owned businesses, and locally controlled institutions in reaction against mass popular culture and globalization.

neo-Malthusianism The advocacy of population control programs to preserve and improve general national prosperity and well-being, and avoid the catastrophic consequences of overpopulation.

net migration The difference between in-migration and out-migration of an area.

network The areal pattern of sets of places (*nodes*) and the routes (*links*) connecting them, along which movement or communication can take place.

network bias The view that the pattern of *links* in a *network* will affect the likelihood of flows between specific *nodes*.

network city One of two or more nearby cities, potentially or actually complementary in function, that cooperate by developing transportation links and communications infrastructure joining them.

new international division of labor (NIDL) A spatial rearrangement of production in which developing countries capture more of the world's manufacturing activity while developed countries shift to services.

New Urbanism A planning movement that promotes walkability and mixed-use buildings with offices and residences on the upper floors.

node An origin, destination, or intersection place in a communication or transportation *network*.

nomadic herding Migratory but controlled movement of livestock solely dependent on natural forage. A type of *extensive subsistence agriculture*.

nonbasic sector Those economic activities of an urban area that supply the resident population with goods and services, have no "export" implication, and do not bring in wealth from outside the city.

nonecumene (*syn:* anecumene). That portion of the Earth's surface that is uninhabited by people or only temporarily or intermittently inhabited. See also *ecumene*.

nonmaterial culture The oral traditions, songs, and stories of a *culture* group along with its beliefs and customary behaviors.

nonrenewable resource A *natural resource* that is not replenished or replaced by natural processes or is used at a rate that exceeds its replacement rate.

Not-In-My-Backyard (NIMBY) Protests from local residents that often appear when companies or public agencies try to select a location for a waste treatment facility or other potentially polluting facility.

O

object *Geographic feature* thought of as a discrete bounded entity separated from other entities by space that can be conceived of as empty. Examples include mountain peaks or roads.

official language A governmentally designated *language* of instruction, of government, of the courts, and other official public and private communication.

offshoring The relocation of business processes and services to a lower-cost foreign location; the offshore *outsourcing* of, particularly, white-collar technical, professional, and clerical services.

orthographic projection A *geometrical projection* that results from placing the light source at infinity.

outsourcing 1: Producing abroad parts or products for domestic use or sale; 2: Subcontracting production or services rather than performing those activities "in house."

overpopulation A judgment that the resources of an area are insufficient to sustain adequately its present population numbers.

ozone A gas molecule consisting of three atoms of oxygen (O_3) formed when diatomic oxygen (O_2) is exposed to *ultraviolet radiation*. In the upper *atmosphere* it forms a normally continuous, thin layer that blocks ultraviolet light; in the lower atmosphere it constitutes a damaging component of *photochemical smog*.

P

palimpsest Originally a reused leather document with visible traces of previous writing. In cultural geography, the landscape may be viewed as a palimpsest containing evidence of previous cultures and previous land uses.

parallel invention See *independent invention*.

parallel An east-west line of *latitude* indicating distance north or south of the equator.

partial displacement migration *Migrations* wherein migrants move to a new residence nearby, with a new activity spaces that overlap some with their former home ranges.

pattern The design or spatial arrangement of phenomena on the Earth surface.

peak land value intersection The most accessible and costly parcel of land in the *central business district* and, therefore, in the entire *urbanized area*.

perceptual region (*syn:* cognitive region) Geographic *region* created informally to reflect the subjective beliefs and feelings of individuals or cultural groups (in the latter case, they are also known as *vernacular regions*).

perforated state A *state* whose territory is interrupted ("perforated") by a separate, independent state totally contained within its borders.

peripheral model A depiction of the contemporary metropolitan area emphasizing patterns of suburban location and functions.

periphery/peripheral In urban geography, the outer regions or boundaries of an area. In economic geography, peripheral areas tend to be underdeveloped and have lower levels of productivity than core areas. See also *core-periphery model*.

personal communication field An area defined by the distribution of an individual's short-range informal communications. The size and shape of the field are defined by work, recreation, school, and other regular contacts and are affected by age, sex, employment, and other personal characteristics.

personal space An invisible, usually irregular area around a person into which he or she does not willingly admit others. The sense (and extent) of personal space is a situational and cultural variable.

physical boundary See *natural boundary*.

physical geography One of two major divisions (the other is *human geography*) of *systematic geography;* the study of the structures, processes, distributions, and change through time of the natural biophysical phenomena of the Earth's surface that are significant to human life.

physiological density The number of persons per unit area of arable (cultivable) land.

pidgin An auxiliary *language* derived, with reduced vocabulary and simplified structure, from combinations of other languages. Not a native tongue, it is a cultural syncretism used for limited communication among people with different languages, such as in situations of trade.

place A particular geographic location with its unique biophysical, cultural, and social characteristics.

place perception Beliefs and attitudes people have about particular places, regions, or landscapes.

place stereotype A simplified belief or set of beliefs about a place that often reflect actual characteristics of the place somewhat inaccurately.

place utility 1: In human movement and *migration* studies, a measure of an individual's perceived satisfaction or approval of a place in its social, economic, or environmental attributes. 2: In *economic geography,* the value imparted to goods or services by *tertiary* activities that provide things needed in specific markets.

placelessness The loss of locally distinctive characteristics and identity and replacement by standardized landscapes.

planar projection (*syn:* azimuthal projection) A *map projection* employing a plane as the presumed *developable surface.*

planned economy A system of production of goods and services, usually consumed or distributed by a governmental agency, in quantities, at prices, and in locations determined, at least in part, by governmental programs.

plantation agriculture A large commercial agricultural holding, frequently foreign owned, devoted to the production of a single export crop.

political geography A branch of *human geography* concerned with the spatial analysis of political phenomena.

polytheism Belief in or worship of more than one god.

popular culture The constantly changing mix of material and nonmaterial *culture traits* available through mass production and the mass media to an urbanized, heterogeneous, nontraditional society. Unlike *folk culture,* popular culture varies quickly over time but not much over space.

popular region See *vernacular region.*

population density A measurement of the numbers of persons per unit area of land within predetermined limits, usually political or census boundaries. See also *physiological density.*

population geography A division of *human geography* concerned with spatial variations in distribution, composition, growth, and movements of population.

population momentum (*syn:* demographic momentum) The tendency for population growth to continue despite rapid changes to fertility rates (such as due to stringent family planning programs) because of a relatively high concentration of people in the childbearing years.

population projection A prediction of a population's future size, age, and sex composition based on the application of stated assumptions to current data.

population pyramid A bar graph in pyramid form showing the age and sex composition of a population, usually a national one.

positional dispute (*syn:* boundary dispute) In *political geography,* disagreement about the actual location of a *boundary.*

possibilism The philosophical viewpoint that the physical *environment* offers human beings a set of opportunities from which (within limits) people may choose according to their cultural needs and technological awareness. The emphasis is on a freedom of choice and action not allowed under *environmental determinism,* while still recognizing the influence of the environment on culture.

postindustrial A stage of economic development in which service activities become relatively more important than goods production; professional and technical employment supersedes employment in agriculture and manufacturing; and level of living is defined by the quality of services and amenities rather than by the quantity of goods available.

potential model A measurement of the total interaction opportunities available under *gravity model* assumptions to a center in a multicenter system.

primary activity A part of the economy involved in making *natural resources* available for use or further processing; included are mining, *agriculture,* forestry, and fishing and hunting.

primate city When a country's leading city is disproportionately larger and functionally more complex than any other city; a city dominating an urban hierarchy composed of a base of small towns and an absence of intermediate-sized cities. Contrary to the *rank-size rule,* primate cities are considerably larger than twice the size of the second largest city in a country or region.

prime meridian An imaginary line passing through the Royal Observatory at Greenwich, England (now a suburb of London), serving by international agreement as the 0 degree line of *longitude.*

producer service A service sector activity performed for other businesses such as accounting, advertising, engineering consulting, and public relations.

projection See *map projection.*

proportional area symbol A *thematic map* symbol that varies the size of two-dimensional shapes to show quantities of some variable (such as a *graduated circle*) circleamplee)

proportional area symbols Symbols of different size on maps to indicate the magnitude of a variable of interest in different places; the larger the symbol, the greater the magnitude of the variable.

prorupt state A *state* of basically *compact* form but with one or more narrow extensions of territory.

protolanguage An assumed, reconstructed, or recorded ancestral *language.*

Public Land Survey System (PLSS) A *rectangular survey* system adopted in the Land Ordinance of 1785, used to divide land over much of the United States from Ohio to the West Coast. The PLSS creates a checkerboard township and range pattern, dividing land into square townships six miles on a side. Townships are further subdivided into 36 sections of land with each section one mile on a side.

pull factor Characteristic of a locale that acts as an attractive force, drawing migrants from other regions.

purchasing power parity (PPP) A measurement of a country's wealth that takes account of what money actually buys in the country, relative to the cost of living.

push factor Unfavorable characteristic of a locale that contributes to the dissatisfaction of its residents and impel their emigration.

Q

quaternary activity A specialized subset of service activities involving research, information, and administration.

R

rank-size rule An observed regularity in the city-size distribution of some countries. In a rank-size hierarchy, the population of any given town will be inversely proportional to its rank in the hierarchy; that is, the *n*th-ranked city will be 1/*n* the size of the largest city.

raster approach A data model for digital geographic information in which the landscape is broken into small cells, each of which contains a numerical value reflecting the degree of presence of some feature; computational expression of the *field* conception of *geographic features*.

rate The frequency of an event's occurrence during a specified time period.

rate of natural increase *Birth rate* minus the *death rate*, suggesting the annual rate of population growth without considering *net migration*.

reapportionment The process and outcome of a reallocation of electoral seats to defined territories, such as congressional seats to states of the United States.

rectangular survey A survey system that superimposes a rectangular grid upon the land rather than using natural features to describe and divide land.

redistricting The drawing of new electoral district boundary lines in response to changing patterns of population or changing legal requirements.

reference map A general-purpose map that attempts to show *geographic features* such as roads and landforms accurately and in detail.

refugee *Forced* or *reluctant* migrant, usually at the international scale, fleeing difficult or dangerous environmental, military, economic, or political conditions.

region (syn: geographic region) Any Earth area with distinctive and unifying physical or cultural characteristics that set it off and make it substantially different from surrounding areas. Regions and their boundaries are devices of areal generalization, intellectual concepts rather than just visible landscape entities.

regional concept The view that physical and cultural phenomena on the surface of the Earth are rationally arranged by complex, diverse, but comprehensible interrelated spatial processes.

regional geography The study of the natural and cultural characteristics of geographic *regions;* the study of areal differentiation.

regionalism In *political geography,* group—frequently ethnic group—identification with a particular region of a *state* rather than with the state as a whole.

Reilly's Breaking-Point Law A *law of retail gravitation* proposed by William J. Reilly that finds the breaking point or boundary line of the market area *functional regions* around two cities' trade areas. It predicts that consumers will make shopping trips to the city within the market area in which they live.

relative direction Direction with respect to personal or cultural ideas rather than objective systems such as cardinal directions or landmarks.

relative distance A transformation of *absolute distance* into such relative measures as time or monetary costs. Such measures yield different explanations of human spatial behavior than do linear distances alone. Distances between places are constant by absolute terms, but relative distances may vary with improvements in transportation or communication technology or with different perceptions of space.

relative location The position of a place or activity in relation to other places or activities (see *situation*). Relative location implies spatial relationships and usually suggests the relative advantages or disadvantages of a location with respect to all competing locations.

relic boundary A former *boundary* line that is still discernible and marked by some *cultural landscape* feature, such as a fence.

religion A personal or institutionalized system of worship and of faith in the sacred and divine.

relocation diffusion The transport of ideas, behaviors, or articles from one place to another through the *migration* of those possessing the feature transported.

reluctant relocation When people relocate their residence (migrate) somewhat involuntarily.

remittance Money sent by international migrants back to family members in their home country.

remote sensing Any of several techniques of obtaining images of an area or object without having the sensor in direct physical contact with it, as by aerial photography or satellite sensors.

renewable resource A *natural resource* that is potentially inexhaustible either because it is constantly (as solar radiation) or periodically (as *biomass*) replenished as long as its use does not exceed its *maximum sustainable yield*.

replacement level (syn: replacement fertility rate) The number of children per woman that will supply just enough births to replace parents and compensate for early deaths, keeping the population size of an area constant, with no allowance for *migration* effects; depending on the rate of survival into the reproductive years, calculated at between 2.1 and 2.5 children per woman.

representative fraction The *scale* of a map expressed as a ratio of a unit of distance on the map to distance measured in the same unit on the ground, e.g., 1:250,000 means that 1 inch on the map represents 250 thousand inches (almost 4 miles) on the Earth surface.

residential dissimilarity index A measure of how segregated a group is from other groups.

resource See *natural resource.*

resource dispute In *political geography,* disagreement over the control or use of shared resources, such as boundary rivers or jointly claimed fishing grounds.

return migration See *counter migration.*

rhumb line A directional line that crosses each successive *meridian* at a constant angle; a rhumb line accurately shows a constant compass direction. All straight lines are rhumb lines on the *Mercator projection* map.

rotation The agricultural practice of planting two or more crops simultaneously or successively on the same area to preserve fertility or to provide a plant cover to protect the soil.

S

sacred place Location with special significance to a religious group, and often attracting pilgrimages or worship rituals. Sacred places are often natural features or religious structures directly connected with a deity or associated with significant events in the history of a particular religion.

satisficing location A less-than-ideal best location, but one providing an acceptable level of utility or satisfaction.

scale 1: In cartography, the ratio between the size of area on a map and the actual size of that same area on the Earth's surface. 2: In more general terms, scale refers to the size of the area studied or over which some phenomenon exists, from local to global.

S-curve The horizontal bending, or leveling, of an exponential or J-*curve* of population growth, reflecting the *carrying capacity.*

secondary activity Part of the economy involved in the processing of raw materials derived from *primary activities* and in altering or combining materials to produce commodities of enhanced utility and value; included are manufacturing, construction, and power generation.

sector model A description of urban land uses as wedge-shaped sectors radiating outward from the *central business district* along transportation corridors. The radial access routes attract particular uses to certain sectors, with high-status residential uses occupying the most desirable wedges. Thus, it also describes a common urban residence pattern corresponding to different levels of socioeconomic status (social status).

secularism A rejection of or indifference to *religion* and religious practice.

segregation A measure of the degree to which members of a minority group are not uniformly distributed among the total population.

semi-periphery Newly industrializing countries, such as South Korea and Brazil, that occupy an intermediate position between core countries, such as the United States, and peripheral countries, such as Liberia.

semi-periphery Regions occupying an intermediate position between the core and the periphery within the world system.

separatism Desired *regional autonomy* expressed by a culturally distinctive group within a larger, politically dominant *culture.*

service activity See *tertiary activities.*

sequential occupation The use and modification of the cultural landscape by successive cultural groups, reflecting differing cultural values, technologies, and social relations.

sex ratio The ratio of the number of one sex to that of the other in a population; typically the number of males relative to the number of females

shamanism A set of beliefs and practices in some *tribal religions* based on belief in a hidden world of gods, ancestral spirits, and demons responsive only to a shaman or interceding priest.

shifting cultivation (*syn:* slash-and-burn agriculture; swidden agriculture) Crop production on tropical forest clearings kept in cultivation until their quickly declining fertility is lost. Cleared plots are then abandoned and new sites are prepared. A type of *extensive subsistence agriculture.*

Shinto The *polytheistic, ethnic religion* of Japan that includes reverence of deities of natural forces and veneration of the emperor as descendent of the sun-goddess.

Simon, Julian Julian Simon (1932–1998), American economist who rejected *neo-Malthusian* arguments, instead arguing that population growth generates economic growth and innovation.

site A concept of *absolute location,* describes a place by reference to characteristics at the location of the place itself, such as local landforms, climate, ethnicity of residents, and other physical or cultural characteristics.

situation A concept of *relative location,* describes a place by reference to characteristics that derive from the place's location relative to other places or the larger regional or *spatial system* of which it is a part. Situation implies spatial interconnection and interdependence.

social distance A measure of the perceived degree of social separation between individuals, *ethnic groups,* neighborhoods, or other groupings; the voluntary or enforced *segregation* of two or more distinct social groups for most activities.

sociofact A rule, custom, or institution that links individuals and groups as part of a *culture,* including family structure and political, educational, and religious institutions. An element in the *sociological subsystem* of culture.

sociological subsystem The totality of expected and accepted patterns of interpersonal relations and social rituals that characterize a *culture,* along with the *ideological* and *technological subsystems.*

soil The complex mixture of loose material including minerals, organic and inorganic compounds, living organisms, air, and water found at the Earth's surface and capable of supporting plant life.

soil erosion The wearing away and removal of rock and soil particles from exposed surfaces by agents such as moving water, wind, or ice.

space As used by geographers, it does not refer to outer space but to areal extent on the Earth's surface, in and around which all humans exist and their activity occurs.

space-time compression/convergence Expressions of the extent to which improvements in transportation and communication have reduced the *friction of distance* and permitted, for example, the very rapid *diffusion* of ideas across space. *Globalization* depends in part on space-time compression.

space-time path A diagram of the line through space and time which describes where we are at any given time, how long we spend there, and how fast we move between locations; they are usually described at the scale of single days but may be monthly, yearly, or lifetime paths. Space-time paths must fit within space-time prisms.

space-time prism A diagram of the volume of space and the length of time within which our activities are confined by constraints of our bodily needs (eating, resting), our daily responsibilities, and the means of mobility at our command.

spatial assimilation A part of the assimilation process in which a minority population leaves its segregated enclaves and becomes widely distributed through a territory.

spatial association When the spatial arrangements of two distributions of features correspond or covary with each other in some way.

spatial diffusion See *diffusion.*

spatial distribution The arrangement of things on the Earth's surface; the descriptive elements of spatial distribution are *density, dispersion,* and *pattern.*

spatial interaction The movement (e.g., of people, goods, information) between different places; an indication of contact and interdependence between different geographic locations or areas.

spatial margin of profitability The set of points delimiting the area within which an economic activity can be profitably carried out.

spatial search The process by which individuals evaluate the alternative locations to which they might move.

spatial system The arrangement and integrated operation of phenomena produced by or responding to spatial processes on the Earth's surface.

spatially fixed cost An input cost in manufacturing that remains constant wherever production is located; locational decisions are not influenced much by such costs.

spatially variable cost An input cost in manufacturing that changes significantly from place to place in its amount and its relative share of total costs; locational decisions are influenced considerably by such costs.

speech community A group of people having common characteristic patterns of vocabulary, word arrangement, and pronunciation.

spread effect (*syn:* trickle-down effect) The diffusion outward of the benefits of economic growth and prosperity from the power center or *core area* to poorer districts and people.

standard language A *language* substantially uniform with respect to spelling, grammar, pronunciation, and vocabulary and representing the approved community norm of the tongue.

standard line Line of contact between a projection surface and the globe; locations covered by the line are transformed from the sphere to the plane surface without distortion.

state (*syn:* country) An independent political unit occupying a defined, permanently populated territory and internationally recognized as having full sovereign control over its internal and foreign affairs.

statistical map A *thematic map* that shows counts of a mapped item per unit area or location.

step migration A *migration* in which an eventual long-distance relocation is undertaken in stages as, for example, from farm to village to small town to city. See also *hierarchical migration.*

stereographic projection A *geometrical projection* that results from placing the light source at the *antipode.*

subnationalism The feeling that one owes primary allegiance to a traditional group or nation rather than to the state.

subsequent boundary A *boundary* line that is established after the area in question has been settled and that reflects the cultural characteristics of the bounded area.

subsistence agriculture Any of several agricultural economies in which crops are grown or livestock are raised nearly exclusively for local or family consumption.

subsistence economy An economic system of relatively simple technology in which people produce most or all of the goods to satisfy their own and their family's needs; little or no exchange occurs outside of the family or social group.

substitution principle In industry, the tendency to substitute one factor of production for another in order to achieve optimum plant location.

suburb A functionally specialized segment of a large *urban* complex located outside the boundaries of the *central city;* usually, a relatively homogeneous residential community, separately incorporated and administered.

superimposed boundary A *boundary* line placed over and ignoring an existing cultural pattern.

supranationalism Term applied to associations created by three or more states for their mutual benefit and achievement of shared objectives.

sustainable development A concept popularized by the World Commission on Environment and Development's Bruntland Report (1987) calling for development that meets the needs of the present without endangering the ability of future generations to meet their needs; the concept of sustainable development seeks to balance the desire for economic growth with the recognition of environmental limits to growth.

syncretism The development of a new form of *culture trait* by the fusion of two or more distinct parental traits.

systematic geography An approach to geographic study that selects a particular aspect of the physical or cultural *environment* for detailed study of its areal differentiation and interrelationships. Branches of systematic geography are labeled according to the topic studied (e.g., recreational geography) or the related science with which the branch is associated (e.g., *economic geography*).

T

Taoism (*syn:* Daoism) A Chinese *value system* and *ethnic religion* emphasizing conformity to Tao (Way), the creative reality ordering the universe.

tapering principle The diminution or tapering of costs of transportation with increasing distance from the point of origin of the shipment because of the averaging of *fixed costs* over a greater number of miles of travel.

technological subsystem The complex of material objects together with the techniques of their use by means of which people carry out their productive activities and that characterize a *culture,* along with the *ideological* and *sociological subsystems.*

technology The integrated system of knowledge, skills, tools, and methods developed within or used by a *culture* to successfully carry out purposeful and productive tasks.

technology gap The contrast between the *technology* available in developed *core regions* and that present in *peripheral areas* of *underdevelopment.*

technology transfer The *diffusion* to or acquisition by one *culture* or *region* of the *technology* possessed by another, usually more developed, society.

temporary travel Short-term *mobility,* such as journeys to stores, workplaces, school, entertainment locales, or vacation destinations, in which people intend to return home at the end of the day or soon thereafter.

terminal cost (*syn:* fixed cost of transportation) A cost incurred, and charged, for loading and unloading freight at origin and destination points and for the paperwork involved; cost charged each shipment for terminal facility use and unrelated to distance of movement or *line-haul cost.*

terracing The practice of planting crops on steep slopes that have been converted into a series of horizontal step-like level plots (terraces).

territorial dispute (*syn:* boundary dispute; functional dispute) In *political geography,* disagreement between *states* over the control of surface area, including the locations of boundaries.

territoriality An individual or group attempt to identify and establish control over a defined territory considered partially or wholly an exclusive domain; the behavior associated with the defense of the home territory.

terrorism Systematic open and covert action employing the inducement of fear and terror as a means of political coercion.

tertiary activity (*syn:* service sector) A part of the economy that fulfills the exchange function, that provides market availability of commodities, and that brings together consumers and providers of services; included are wholesale and retail trade, associated transportation and government services, and personal and professional services of all kinds.

thematic map (*syn:* statistical map) A specific-purpose map that shows the distribution of one or a few themes or variables, such as unemployment rates by county.

thematic region (sometimes *formal region*) Geographic *region* based on the pattern of one or more objectively measurable themes or properties, such as soil types or linguistic dialects.

themed landscape Built environments designed to suggest or simulate a distinctive place, created to attract visitors by allowing them to fantasize about being in different times, places, and events. Disneyland and Las Vegas are two of the most famous examples, but more modest examples can be found in many shopping malls, chain restaurants, housing subdivisions, and the like.

Third World Originally (1950s) designating countries uncommitted to either the "First World" Western capitalist bloc or the "Second World" Eastern communist bloc; subsequently, a term applied to countries considered in a state of *underdevelopment* in economic and social terms.

threshold In *economic geography* and *central place theory,* the minimum market area or population needed to support the supply of a product or service.

time geography The study of temporal and spatial properties of human activity, particularly temporary travel.

tipping point The degree of neighborhood racial or ethnic mixing that induces the former majority group to move out rapidly.

toponym A place or geographic feature name.

toponymy The place names of a region or, especially, the study of place names.

total displacement migration *Migrations* wherein migrants move far enough so their new activity spaces do not overlap at all with their former home ranges.

total fertility rate (TFR) The average number of children that would be born to each woman during her childbearing years if she bore children at the current year's rate for women of that age.

town A nucleated settlement that contains a *central business district* but that is small and less functionally complex than a *city*.

tragedy of the commons The observation that in the absence of collective control over the use of a resource available to all, it is to the advantage of all users to maximize their separate shares even though their collective pressures may diminish total yield or destroy the resource altogether.

transboundary river basin Drains land from two or more countries, thus requiring international cooperation over the management of water resources.

transferability Acceptable costs of a spatial exchange; the cost of moving a commodity relative to the ability of the commodity to bear that cost.

transnational corporation (TNC) A large business organization operating in at least two separate national economies; a form of *multinational corporation*.

transnationalism The practice of immigrants maintaining close social or economic ties to their country of origin through frequent travel or communications.

tribal religion (*syn:* traditional religion) An *ethnic religion* specific to a small, localized, preindustrial culture group.

trickle-down effect See *spread effect*.

truck farm (*syn:* horticultural farm; market gardening) A farm that intensively produces fruits and vegetables for market rather than for processing or canning.

U

ubiquitous industry A *market-oriented* industry whose establishments are distributed in direct proportion to the distribution of population.

underdevelopment A level of economic and social achievement below what could be reached—given the natural and human resources of an area—were necessary capital and technology available.

uneven spatial development The uneven spatial pattern observed in standards of living and levels of economic development.

uniform (istotropic) plain A hypothetical portion of the Earth's surface assumed to be an unbounded, uniformly flat plain with uniform and unvarying distribution of population, purchasing power, transport costs, accessibility, and the like. Commonly assumed as a simplification in geographic models of spatial interaction and economic activity.

unitary state A *state* in which the central government dictates the degree of local or *regional autonomy* and the nature of local governmental units; a country with few cultural conflicts and with a strong sense of national identity.

United Nations Convention on the Law of the Sea (UNCLOS) A code of maritime law approved by the United Nations in 1982 that authorizes, among other provisions, territorial waters extending 12 nautical miles (22 km) from shore and 200-nautical-mile-wide (370-km-wide) *exclusive economic zones*.

universalizing religion A *religion* that claims global truth and applicability, regardless of ethnicity or culture group, and seeks the conversion of all humankind via proselytizing.

urban geography The geographical study of cities; the branch of *human geography* concerned with the spatial aspects of (1) the locations, functional structures, size hierarchies, and intercity relationships of national or regional systems of cities, and (2) the *site,* evolution, *economic base,* internal land use, and social geographic patterns of individual cities.

urban hierarchy A ranking of cities based on their size and functional complexity.

urban influence zone An area outside of a *city* that is nevertheless affected by the city.

urbanization Transformation of a population from rural to *urban* status; the process of city formation and expansion.

urbanized area A continuously built-up *urban* landscape defined by building and population densities with no reference to the political boundaries of the city; it may contain a *central city* and many contiguous towns, *suburbs,* and unincorporated areas.

usable reserves Mineral deposits that have been identified and can be recovered at current prices and with current technology.

V

vector approach A data model for digital geographic information in which meaningful discrete objects are represented as points connected with lines; computational expression of the *object* conception of *geographic features*.

vernacular 1: The nonstandard indigenous *language* or *dialect* of a locality. 2: Of or related to indigenous arts and architecture, such as a *vernacular house.* 3: Of or related to the perceptions and understandings of the general population, such as a *vernacular region*.

vernacular house An indigenous style of building constructed of native materials to traditional plan, without formal drawings.

vernacular region A *perceptual region* defined informally by inhabitants of a cultural group, usually with a popularly given or accepted nickname.

voluntary migration When people relocate their residence by free choice, without being forced or compelled.

von Thünen model Model developed by Johann Heinrich von Thünen (1783–1850), German economist and landowner, to explain the forces that control the prices of agricultural commodities and how those variable prices affect spatial patterns of agricultural land utilization.

W

Weberian analysis See *least-cost theory*.

world city One of a small number of interconnected, internationally dominant centers (e.g., New York, London, Tokyo) that together control the global systems of finance and commerce.

X

xenophobia A hatred or fear of that which is foreign, often directed at immigrant minority groups.

Z

zero population growth (ZPG) A term suggesting a population in equilibrium, fully stable in numbers with births (plus immigration) equaling deaths (plus emigration).

Index

497